Chiral Separations and Stereochemical Elucidation

Chiral Separations and Stereochemical Elucidation

Fundamentals, Methods, and Applications

Edited by

Quezia Bezerra Cass
Universidade Federal de São Carlos
São Carlos, SP, Brazil

Maria Elizabeth Tiritan
Universidade do Porto
Porto, Portugal

João Marcos Batista Junior
Universidade Federal de São Paulo
São José dos Campos, SP, Brazil

Juliana Cristina Barreiro
Universidade de São Paulo
São Carlos, SP, Brazil

WILEY

Published by John Wiley & Sons, Inc., Hoboken, New Jersey.
Published simultaneously in Canada.

For general information on our other products and services or for technical support, please contact our Customer Care Department within the United States at (800) 762-2974, outside the United States at (317) 572-3993 or fax (317) 572-4002.

Wiley also publishes its books in a variety of electronic formats. Some content that appears in print may not be available in electronic formats. For more information about Wiley products, visit our web site at www.wiley.com.

Library of Congress Cataloging-in-Publication Data applied for:
ISBN: 9781119802259 (HB); ePDF: 9781119802273; epub: 9781119802266

Cover Design: Wiley
Cover Image: © kcarper/500px/Getty Images

Set in 9.5/12.5pt STIXTwoText by Straive, Pondicherry, India

Contents

List of Contributors

Daniel W. Armstrong
Department of Chemistry and Biochemistry
University of Texas
Arlington, TX, USA

Saba Aslani
Department of Chemistry and Biochemistry
University of Texas
Arlington, TX, USA

Juliana Cristina Barreiro
Instituto de Química de São Carlos
Universidade de São Paulo
São Carlos, SP, Brazil

João Marcos Batista Junior
Instituto de Ciência e Tecnologia
Universidade Federal de São Paulo
São José dos Campos, SP, Brazil

Alain Berthod
Institute of Analytical Sciences
University of Lyon 1
CNRS, Villeurbanne, France

Quezia Bezerra Cass
Departamento de Química
Universidade Federal de São Carlos
São Carlos, SP, Brazil

Robert A. Burrow
Departamento de Química
Universidade Federal de Santa Maria
RS, Brazil

Charles Clark
Leiden Academic Centre for Drug Research, Metabolomics and Analytics Centre, Leiden University
Leiden, the Netherlands

Fernando Martins dos Santos Junior
Instituto de Química
Universidade Federal Fluminense
Niterói, RJ, Brazil

Carla Fernandes
Laboratório de Química Orgânica e Farmacêutica
Departamento de Ciências Químicas
Faculdade de Farmácia da Universidade do Porto
Porto, Portugal
and
Centro Interdisciplinar de Investigação Marinha e Ambiental (CIIMAR)
Matosinhos, Portugal

Pilar Franco
Chiral Technologies Europe
Illkirch Cedex, France

Jun Haginaka
Institute for Biosciences
Mukogawa Women's University
Koshien Kyuban-cho
Nishinomiya, Japan

Isabelle Kohler
Division of BioAnalytical Chemistry
Amsterdam Institute of Molecular Life Sciences (AIMMS)
Vrije Universiteit Amsterdam
Amsterdam, the Netherlands
and
Center for Analytical Sciences Amsterdam
Amsterdam, the Netherlands

Mari-Luiza Konjaria
Department of Pharmaceutical/Medicinal Chemistry
Friedrich Schiller University Jena
Jena, Germany

Sulaiman Krait
Department of Pharmaceutical/Medicinal Chemistry
Friedrich Schiller University Jena
Jena, Germany

Michael Lämmerhofer
Institute of Pharmaceutical Sciences
University of Tübingen
Tübingen, Germany

Xiaolu Li
Group of Structural Chemistry and Computational Biophysics
Leibniz-Forschungsinsitut für Molekulare Pharmakologie
Berlin, Germany

Wolfgang Lindner
Institute of Analytical Chemistry
University of Vienna
Vienna, Austria

Emmanuelle Lipka
Laboratoire de Chimie Analytique – UFR3S-Faculté
de Pharmacie de Lille, Inserm U1167, Université
de Lille, Lille Cedex, France

Larry Miller
Amgen Research
One Amgen Center Drive
Thousand Oaks, CA, USA

Ademir F. Morel
Departamento de Química
Universidade Federal de Santa Maria
RS, Brazil

Takafumi Onishi
Daicel Corporation
CPI Company
Myoko-shi, Niigata, Japan

Madalena Pinto
Laboratório de Química Orgânica e Farmacêutica
Departamento de Ciências Químicas
Faculdade de Farmácia da Universidade do Porto
Porto, Portugal
and
Centro Interdisciplinar de Investigação Marinha e Ambiental (CIIMAR)
Matosinhos, Portugal

Gerhard K. E. Scriba
Department of Pharmaceutical/
Medicinal Chemistry
Friedrich Schiller University Jena
Jena, Germany

Govert W. Somsen
Division of BioAnalytical
Chemistry
Amsterdam Institute of Molecular
Life Sciences (AIMMS)
Vrije Universiteit Amsterdam
Amsterdam, the Netherlands
and
Center for Analytical Sciences
Amsterdam
Amsterdam, the Netherlands

Han Sun
Group of Structural Chemistry and
Computational Biophysics
Leibniz-Forschungsinsitut für
Molekulare Pharmakologie
Berlin, Germany
and
Institute of Chemistry
Technical University of Berlin
Berlin, Germany

Maria Elizabeth Tiritan
Laboratório de Química Orgânica e
Farmacêutica
Departamento de Ciências Químicas
Faculdade de Farmácia da
Universidade do Porto
Porto, Portugal
and
Centro Interdisciplinar de
Investigação Marinha e Ambiental
(CIIMAR)
Matosinhos, Portugal

Oliver Trapp
Department of Chemistry
Ludwig-Maximilians-
University Munich
Munich, Germany

Weston Umstead
Chiral Technologies Inc.
West Chester, PA, USA

Xiaoliang Yang
State Key Laboratory of
Coordination Chemistry and
Jiangsu Key Laboratory of
Advanced Organic Materials
School of Chemistry and Chemical
Engineering, Nanjing University
Nanjing, Jiangsu, China

Preface

The enchantment of reading about the first developments on chiral selectors for enantioseparation persists up to now. The papers from Hesse and Hagel [1], Pirkle [2, 3], Allenmark [4, 5], Okamoto [6], Armstrong [7] back in the 1970s–1980s, to cite a few, marked the vertiginous development of chiral stationary phases (CSPs).

Pirkle's paper [8] about ionically bonded CSPs is a milestone paper. It pointed to the need of stablishing a relationship between absolute configuration (AC) and elution order providing a rationale for chiral recognition. This publication came alongside with the 3,5-dinitrobenzoyl phenylglycine CSP, the first commercial chiral column for liquid chromatography (LC) [9]. Ariens' crusade [10] to raise awareness about the implication of stereochemistry in pharmacokinetics studies and, specially, in drug development programs and in clinical practice gave the tone for the development of appropriate chiral drug analysis procedures. The chiral analysis of drugs in biological fluids was, then, frenetically pursued in the 1980s and resulted in the FDA's regulations on pharmaceutical development of single enantiomers and racemates [11]. The approval of a new chiral drug by regulatory agencies requires a complete dossier of the pharmacology and pharmacokinetic profiles of the single enantiomers and their mixtures. Methods to monitor the biological effects of a chiral drug are required since pre-clinical level with a few milligrams of racemates and single enantiomers moving to kilogram quantities at clinical trials. At each stage, it is important to know which is the most appropriate procedure either to the analysis or production of enantiomers. Such knowledge is thus considered a very important asset [12].

The advent of direct enantioseparation by chromatography and related techniques has just completed half a century, and it is nowadays used as routine in research and industry laboratories around the world. This book covers the main features of chiral separation by LC, gas chromatography

(GC), capillary electrophoresis (CE), and supercritical fluid chromatography (SFC). Chiral separation at semi- and preparative scales are also included.

With the five fundamental chapters (1–5) covering LC, GC, CE, SFC, and preparative chromatography, the book encompasses the main advances of each technique and discusses the application of chiral separation in different areas of science, such as enantioselective synthesis, chiral drug designing and development, bio, forensic, and environmental markers, chiral materials, quality control in pharmaceuticals, food, and fragrances. Chiral separations in metabolomics and lipidomics to disclose the enantiomeric signature in various biological processes is another topic that has been positively impacted by the advance in chiral selector technologies alongside with the appropriate analytical platforms. Application examples can be found throughout the book. The role of 2D-LC in chiral separation method development and applications is discussed in Chapter 1 as well as in other chapters of the book.

The quality of the separation in CE, GC, LC, and SFC is dictated by the chiral selector, and with so many interconnected interactions playing a role in chiral discrimination, it is expected that there will never be either a universal chiral selector or CSP.

The advances in chiral selectors and their use for all these separation technologies are thoroughly discussed from Chapters 6 to 11, including new CSP for LC and/or SFC with high-throughput and ultra-high-efficiency capabilities. These chapters portray the most used chiral selectors including their designing, mechanism, and applications for a variety of fields with practical examples. The impact on the chiral recognition mechanism caused by the elution mode in LC is also reviewed as well as the best chiral selectors either for SFC or for preparative separation.

Following chiral separation, the next challenge is the determination of molecular stereochemistry. Methods for stereochemical elucidation are widely used across biochemistry, chemistry, biology, and physics, but there is a clear need for streamlining their application for characterization of small chiral organic molecules.

The 3D structure characterization of a given chiral molecule is of prime importance, and, although not trivial, it is routinely required in many research areas, especially in drug discovery programs. For chiral molecules, prior to setting up several biological experiments, it is indispensable to unambiguously determine not only their stereoisomeric composition but also their AC [13, 14].

This approach ensures that a correct structure-activity relationship is established. The AC assignment of a given molecule is usually carried out by different methods in which X-ray crystallography is considered the gold standard approach. Nuclear magnetic resonance (NMR) is routinely used for establishing relative configurations, while the AC can be determined either by diastereomeric derivatizations or by a combination of anisotropic NMR and chiroptical spectroscopy. Chiroptical methods, mainly associated with quantum-mechanical calculations, have proved to be an excellent choice to assign AC especially for compounds in which a well-defined single crystal is not available.

In this regard, Chapters 12–14 cover these three main stereochemical elucidation methods. Information on their theoretical backgrounds, advantages, and limitations, as well as examples of application are provided. With that, we do hope many research projects dealing with small organic molecules will benefit from a streamlined approach to both enantiomeric separation and AC determination.

By
Quezia Bezerra Cass, Maria Elizabeth Tiritan,
João Marcos Batista Junior, and Juliana Cristina Barreiro,
São Carlos, August 2022

References

1 Hesse, G. and Hagel, R. (1973). Eine vollständige Recemattennung durch eluitons-chromagographie an cellulose-tri-acetat. *Chromatographia* 6: 277–280. https://doi.org/10.1007/BF02282825.

2 Pirkle, W.H. and Sikkenga, D.L. (1976). Resolution of optical isomers by liquid chromatography. *J. Chromatogr. A* 123: 400–404. https://doi.org/10.1016/S0021-9673(00)82210-4.

3 Pirkle, W.H., House, D.W., and Finn, J.M. (1980). Broad spectrum resolution of optical isomers using chiral high-performance liquid chromatographic bonded phases. *J. Chromatogr. A* 192: 143–158. https://doi.org/10.1016/S0021-9673(00)81849-X.

4 Allenmark, S. and Bomgren, B. (1982). Direct liquid chromatographic separation of enantiomers on immobilized protein stationary phases. *J. Chromatogr. A* 252: 297–300. https://doi.org/10.1016/S0021-9673(01)88421-1.

5 Allenmark, S., Bomgren, B., and Borén, H. (1983). Direct liquid chromatographic separation of enantiomers on immobilized protein stationary phases. *J. Chromatogr. A* 264: 63–68. https://doi.org/10.1016/S0021-9673(01)95006-X.

6 Okamoto, Y., Kawashima, M., and Hatada, K. (1984). Chromatographic resolution. 7. Useful chiral packing materials for high-performance liquid

chromatographic resolution of enantiomers: phenylcarbamates of polysaccharides coated on silica gel. *J. Am. Chem. Soc.* 106: 5357–5359. https://doi.org/10.1021/ja00330a057.
7 Armstrong, D.W. and DeMond, W. (1984). Cyclodextrin bonded phases for the liquid chromatographic separation of optical, geometrical, and structural isomers. *J. Chromatogr. Sci.* 22: 411–415. https://doi.org/10.1093/chromsci/22.9.411.
8 Pirkle, W.H., Finn, J.M., Schreiner, J.L., and Hamper, B.C. (1981). A widely useful chiral stationary phase for the high-performance liquid chromatography separation of enantiomers. *J. Am. Chem. Soc.* 103: 3964–3966. https://doi.org/10.1021/ja00403a076.
9 Pirkle, W.H., Myung, H.H., and Bank, B. (1984). A rational approach to the design of highly-effective chiral stationary phases. *J. Chromatogr. A* 316: 585–604. https://doi.org/10.1016/S0021-9673(00)96185-5.
10 Ariëns, E.J. (1984). Stereochemistry, a basis for sophisticated nonsense in pharmacokinetics and clinical pharmacology. *Eur. J. Clin. Pharmacol.* 26: 663–668. https://doi.org/10.1007/BF00541922.
11 (1992). FDA'S policy statement for the development of new stereoisomeric drugs. *Chirality* 4: 338–340. https://doi.org/10.1002/chir.530040513.
12 Tarafder, A. and Miller, L. (2021). Chiral chromatography method screening strategies: past, present and future. *J. Chromatogr. A* 1638: 461878. https://doi.org/10.1016/j.chroma.2021.461878.
13 Bogaerts, J., Aerts, R., Vermeyen, T. et al. (2021). Tackling stereochemistry in drug molecules with vibrational optical activity. *Pharmaceuticals* 14: 877. https://doi.org/10.3390/ph14090877.
14 Batista, A.N.L., dos Santos, F.M., Batista, J., and Cass, Q. (2018). Enantiomeric mixtures in natural product chemistry: separation and absolute configuration assignment. *Molecules 23*, 492: https://doi.org/10.3390/molecules23020492.

Part I

Fundamentals of Chiral Separation

1

Chiral Separation by LC

Juliana Cristina Barreiro[1] *and Quezia Bezerra Cass*[2]

[1]*Instituto de Química de São Carlos, Universidade de São Paulo, São Carlos, SP, Brazil*
[2]*Departamento de Química, Universidade Federal de São Carlos, São Carlos, SP, Brazil*

1.1 Introduction

Liquid chromatography (LC) plays a central role in enantioseparation as it is used for analytical and preparative purposes.

Indirect chiral separation is still important in diverse application fields, mainly in metabolomic and lipidomics non-target analysis as well as in forensics [1–3]. For that, the formed diastereomeric mixtures are separated by achiral columns, mostly under reverse elution mode. Impurities of derivatization reagents that can lead to inaccuracies is one of the main drawbacks of the indirect approach. In the field of non-target metabolomics, several approaches have been described for derivatization of —OH/—NH_2 moiety-containing metabolites. In this regard, the use of diacetyl-tartaric anhydride (DATAN) has demonstrated its utility for identifying the enantiomers of hydroxycarboxylic acids (HAs) and amino acids (AAs) through the reversal in the elution order of the diastereomers formed using either (*RR*)-DATAN or (*SS*)-DATAN. Since the order of elution does not change for achiral metabolites, this approach has been used also to differentiate achiral metabolites from the chiral ones [4] (Scheme 1.1).

Direct chiral separation in LC can be achieved either by addition of a chiral selector to the mobile phase or by using chiral stationary phases (CSPs); herein we will discuss only the latter mode.

Chiral Separations and Stereochemical Elucidation: Fundamentals, Methods, and Applications, First Edition. Edited by Quezia Bezerra Cass, Maria Elizabeth Tiritan, João Marcos Batista Junior, and Juliana Cristina Barreiro.

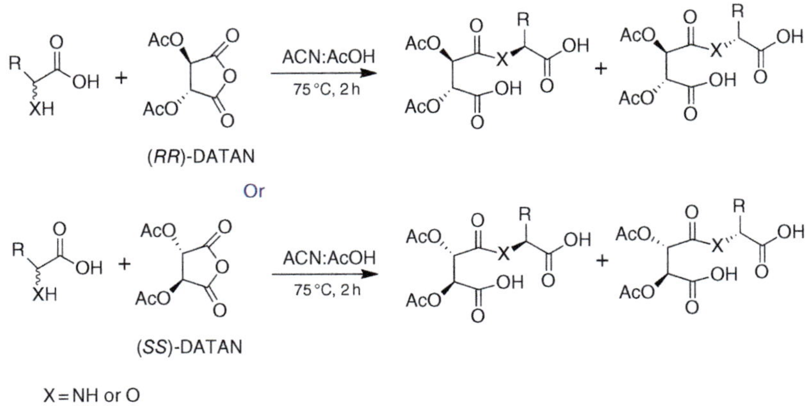

Scheme 1.1 Derivatization reaction of HAs and AAs either with (*RR*)-DATAN or with (*SS*)-DATAN for producing diastereomers. *Source:* Oliveira et al. [2]/with permission of MDPI/Public Domain CC BY 4.0.

Back in the 1980s when the chiral columns started to be commercialized, it was believed that the enantioresolution depended only on the CSP and, thus, the mobile phases and organic modifiers were chosen based only on solubility parameters of the chiral solutes.

Moreover, most of the chiral selectors had elution mode restrictions, and thus, it was usual to change the chiral selector without exploring the mobile phase. Nowadays, the interdependent role of chiral selectors and mobile phases is acknowledged.

The illustrated pattern in Figure 1.1 should be considered for starting a direct chiral separation.

Nevertheless, in method development one should have in mind the intended use. The reason is that the application imposes restrictions. For instance, for measuring enantiomeric ratios of synthetic products one can easily use the normal elution mode, but if the application is for determining enantiomeric fraction in environmental matrices, which implies mass spectrometer detection, the normal elution mode is impaired, and the chromatographic conditions should be developed under polar organic or reversed phase elution modes [1]. Thus, for selecting the separation conditions one should bare this in mind.

The influence of the elution mode and organic modifier can be perceived in the enantiomeric separation of the proton pump inhibitors: omeprazole, lansoprazole, and pantoprazole in four different polysaccharide-based chiral selectors [5]. Taking as an example, the separation values (enantioselectivity [α] and resolution [Rs]) obtained for omeprazole at the CSPs

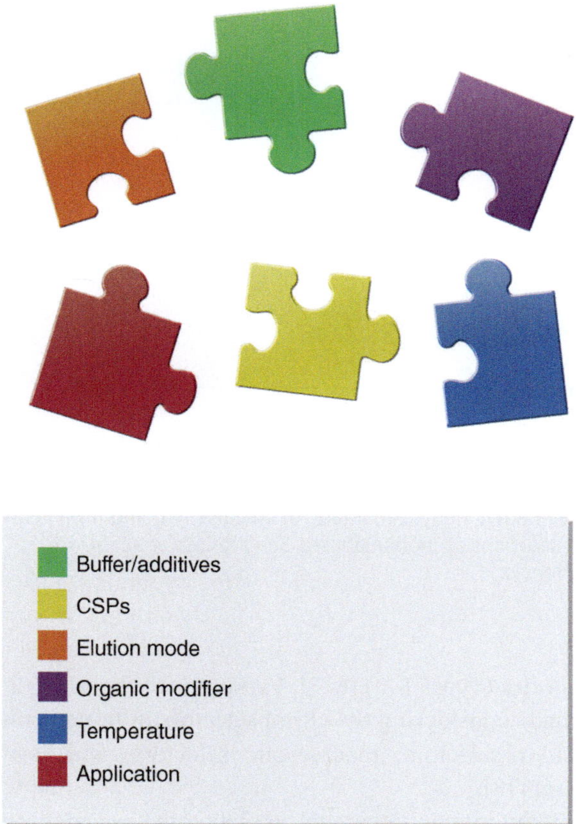

Figure 1.1 Illustration of a pattern that should be considered for starting a direct chiral separation.

tris-(3,5-dimethylphenylcarbamate) and tris-[(*S*)-1-phenylethylcarbamate] of amylose were $\alpha = 1.57$ and Rs = 3.15 and $\alpha = 1.48$ and Rs = 1.71, respectively, using Hex:EtOH (70/30, v/v) as mobile phase, while at tris-(3,5-dimethylphenylcarbamate) of cellulose no separation was obtained in any of the elution conditions examined. The graphic at Figure 1.2 [5] illustrates the influence of the elution mode in the separation of omeprazole in going from polar organic to reverse phase at the CSP of tris-(3,5-dimethylphenylcarbamate) of amylose.

At the end of the 1980s, the high number of commercial chiral columns available demanded a classification of the chiral selectors. The classification was then based on their recognition interactions or, in other words, in the formation of solute–CSP complexes. The CSPs were

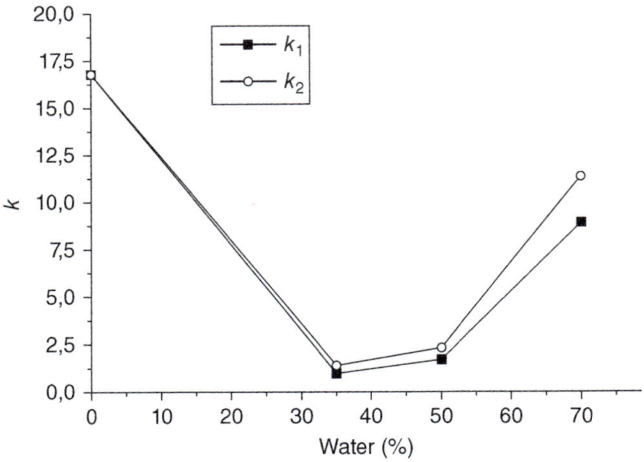

Figure 1.2 Graphic showing the retention factors (k_1 and k_2) of omeprazole enantiomers at tris-(3,5-dimethylphenylcarbamate) of amylose in going from neat MeCN to aqueous MeCN solutions as mobile phases. *Source:* Cass et al. [5]/with permission of Taylor & Francis.

classified in five categories (Types I–V) [6, 7]. Variations of these classifications have been made considering the chiral selectors in three main groups: macromolecular selectors, macrocyclic selectors, and low-molecular mass selectors [8].

Although the three-point interaction model, as elaborated by Dalgliesh for paper chromatography separation of AAs [9], didactically explains the formation of the transient diastereomeric complex between solute and CSP, the interactions accountable for the chiral discrimination still demand clarification [8]. Moreover, a CSP encompasses several heterogeneous non-selective and stereoselective active sites that contribute to the resolution of an enantiomeric mixture by the CSP at a given mobile phase and temperature [10].

It is important to stress that, due to the complexity of chiral discrimination mechanisms, there is no preset-up condition for achieving resolution for a given application. For method development, understanding the most important interactions should somehow help in a planned workflow.

The most important types of CSPs and their properties will be discussed in detail throughout the chapters of this book.

1.2 Workflow for LC Chiral Method Development

For method development, automated column and eluent scouting LC systems have been preferred for sequential column and mobile phase evaluation. The number of scouting columns and mobile phases that can be used varies according to the manufacturer. Small columns with small silica particles as the support are preferred for these systems. The software packages to assist in method setup and data analysis are also of great help in these scouting systems (Figure 1.3).

It is acknowledged that the CSPs used in an initial screening should have complementary chiral discrimination, and the selected elution mode and organic modifier should be in accordance with the intended use [11].

As there are many columns of the same chiral selector type, but with different enantioselectivities, ideally the number of examined columns should be high. It is, however, impossible by economic reasons to cover the vast range of CSPs (above 150 commercially available) – thus, select CSPs of the same selector type and of different interaction categories. To minimize trial time, select CSPs with the highest degree of successful rate for the intended

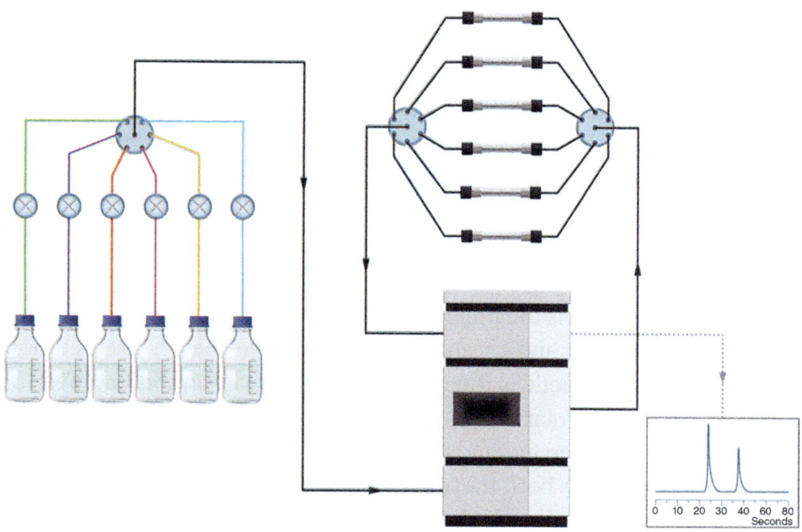

Figure 1.3 Scouting LC system for sequential column and mobile phase evaluation.

class of chiral analytes. For preparative separation, the load capacity should also be considered (see Chapter 5).

In exploring multimodal elution, either sequentially in a single column or using multiple columns in a scouting system, one should be aware of solvent miscibility. An excellent choice for changing elution mode is 100% ethanol. Give time for column equilibration between elution modes and modifiers.

The use of either acidic or basic additives and buffers does not affect the enantioselectivity for some CSPs, and it should be used only if the enantiomeric mixtures demand it, either because they are ionizable or to facilitate detection. For instance, the enantiomeric resolutions of both ketamine and norketamine are affected by the used pH (pH 5.0, 7.8, and 9.0) of the 10 $mmol^{-1}$ ammonium bicarbonate/acetonitrile (54/46, v/v) mobile phase in a CHIRALPAK AS-3R column (3 μm particles, 4.6 mm I.D. × 100 mm) [12].

On the other hand, CSPs of the following types: macrocyclic antibiotics, protein, cinchona alkaloids, and cyclodextrin are all affected by the use of buffer or additives, and the chiral discrimination changes even to neutral analytes. Volatile buffers are required for detectors such as MS (mass spectrometry), CAD (charged-aerosol detection), ELSD (evaporative light-scattering detection). Make sure to wash off the buffer or additives before changing elution mode.

In selecting elution mode and mobile phases for measuring enantiomeric ratio consider that the limit of quantification (LOQ), concentration, matrix effect, and detection mode limit elution conditions and can drastically affect the enantioselectivity and/or resolution. On the other hand, solubility is an important parameter when the intended use is preparative separation (see Chapter 5).

The main interactions involving the elution mode are listed in Table 1.1 [13].

The normal elution is the preferred mode to measure enantiomeric ratio of synthesis products and in quality control, while the reversed elution mode or polar ionic mode is the most used in bioanalysis and for environmental matrices [1]. Polar organic and normal elution modes are preferred for preparative scale separation [14].

Temperature affects both thermodynamic and kinetic aspects of chiral discrimination. Inversion of the elution order can be obtained even in the allowed CSPs temperature range, and the effects on retention factor and enantioselectivity should be tuned in a case-to-case basis and, for that, it is important to determine the temperature of isoelution [15].

Table 1.1 Main interactions in different elution modes.*

Elution mode	Interactions	Strength
Normal	Hydrogen bonding	Strong
	π–π	Strong
	Dipole–dipole	Strong
	Dispersive (steric)	Weak
Polar organic	Hydrogen bonding	Strong
	π–π	Moderate to weak
	Dipole–dipole	Moderate
	Dispersive (steric)	Weak
Polar ionic	Ionic	Strong
	Hydrogen bonding	Moderate
	π–π	Moderate to weak
	Dipole–dipole	Moderate
	Dispersive (steric)	Weak
Reversed	Hydrophobic	Strong
	Ionic	Strong
	π–π	Strong
	Hydrogen bonding	Weak
	Dipole–dipole	Weak
	Dispersive (steric)	Weak

Source: Adapted with permission from reference [13].

Please notice that reversal of elution order can be obtained by changing CSP, and in the same CSP by influence of temperature and/or of the mobile phase used [15, 16], however, be aware not to interpret reversal of optical signal rotation with reversal of elution order [17, 18].

1.3 New Column Technologies

Chromatographic particle technology has played an important role in speeding up chiral analysis time. Major manufacturers have produced CSPs (such as Pirkle type, polysaccharides, macrocycles antibiotics, and cinchona-based selectors) using either fully porous (FPP) or core shell (superficially porous – SPP) silica particles (varying from sub-2 to sub-3 μm particles size). These chromatographic supports have allowed gain in column efficiency and throughput [19].

Regarding particle morphology, an important feature to be observed when using either FFP or SPP is the adsorption–desorption kinetics dependence on the surface density of chiral selectors. Another important parameter to observe is the effect of frictional heating, which is generated by the stream of the mobile phase against the packed bed of the column through which it percolates under a significant pressure gradient. While it is expected that FPP has higher amount of bonded chiral selector due to its specific surface area ($m^2\ g^{-1}$), it is accepted, as it happens in achiral columns, that the SPP offers better heat dissipations and more homogeneous packing [19].

FPP or SPP sub-2 to sub-3 μm silica presents pros and cons when used for preparing CSPs. It has been reported that the packing of SPP CSP is not as smooth as it is for achiral alkyl columns. On the other hand, sub-2 particles tend to aggregate and interfere in the synthesis/preparation of the stationary phases. Either way, deeper investigation on mass transfer phenomena is necessary and the column efficiency limitation will probably come from the chiral selector bonding density, which varies from different chiral selectors and the particle diameter [20]. Anyhow, chiral columns using fully porous 5 μm silica as the support had plate number in the order of 40,000 m^{-1} and now we can find commercial columns with plate number in the order of 300,000 m^{-1} that can operate at high flow rates [21, 22].

As with achiral separation, the system extra volume is a technical constraint that impacts ultrafast chiral separations. Detector sampling rates and response time affect peak shape, peak width, and baseline noise, and they are important features in ultrafast chiral separation. The effect on chromatographic signal caused by detector sampling rate and response times is illustrated in Figure 1.4 [23]. The cycle time of the autosamplers (usually 20–60s) and the injection disturbance in the flow rate are also constrictive system limitations. To overcome speed limitations, analytical tools such as MISER (multiple injections in a single experimental run) have been explored using an autosampler with a dual needle principle to achieve injection cycle time as low as 10.5 seconds for analysis of thalidomide enantiomers [24]. In fast separation one should pay attention also to the internal diameter of the tubing used in the LC system [23]. Overall, these impacts can be noticed much easily with supercritical fluid chromatography (SFC), which in general has worse technical system performance [19, 25].

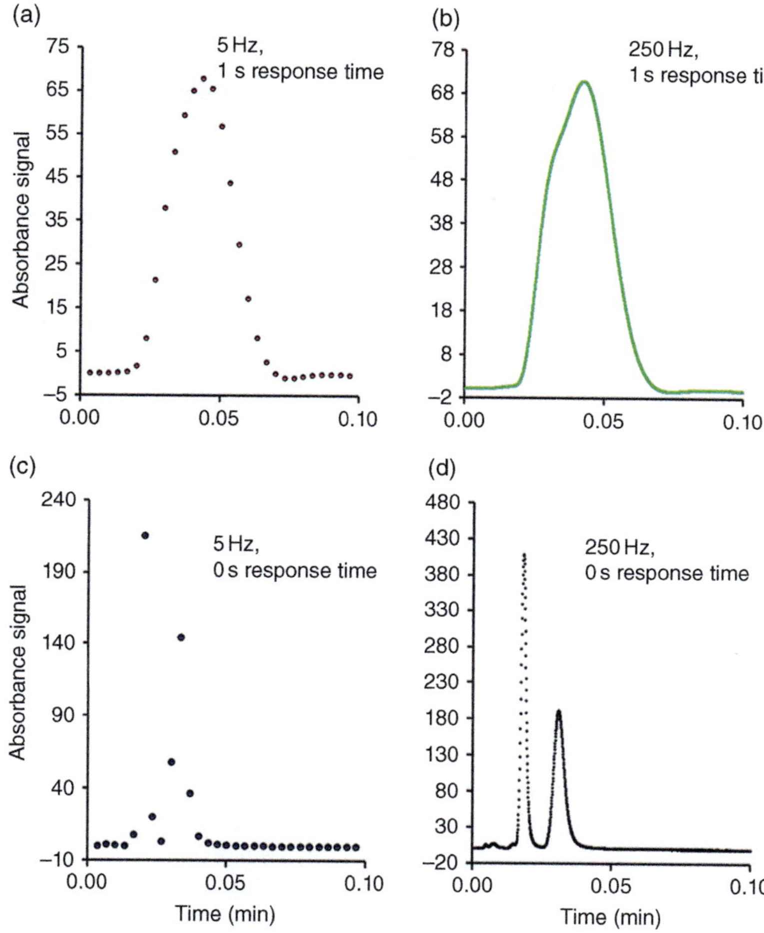

Figure 1.4 The effect on chromatographic signal caused by detector sampling rate and response times on the ultrafast separation of oxazepam enantiomers on a 2.7 μm 1 cm SPP teicoplanin column. The sampling rate and response times were: (a) 5 Hz, 1 s; (b) 250 Hz, 1 s; (c) 5 Hz, 0 s and (d) 250 Hz, 0 s. *Source:* Wahab et al. [23] with permission from ACS.

Despite the restriction caused by the system's or column's state-of-the-art, chiral analysis in sub-minute and/or sub-seconds time is now a reality. Ultrafast chiral chromatography offers a plethora of advantages, since not only enantiomeric separation of several pharmaceutical-related drugs and intermediates have been obtained in seconds but also ultra-rapid analysis can be developed for monitoring asymmetric synthesis and for other applications. They have been most useful in 2D-LC systems [1, 2, 26].

1.4 Selected Examples of Fast Separation

In direct enantioselective separation by LC, one of the most difficult goal is to find the appropriated chromatographic conditions to suit a multicomponent enantiomeric mixture, as for instance in environmental or metabolomic analysis, in which many times the mixtures are composed of stereoisomers and achiral analytes [1]. The main difficulty is related to enantioselectivity; usually, a condition that will separate a chiral analyte will not separate an analog [27]. Another inconvenience is sometimes due to coelution with isobaric analytes as it happens in the separation of AAs [2]. The following selected examples show the developments and applications of chiral analysis for complex mixtures.

The cannabinoid 1-(pentyl-1*H*-indol-3-yl)-1-naphthalenyl-methanone (JWH-018) and the fluorinated analog [1-(5-fluoropentyl)-1*H*-indol-3-yl]-1-naphthalenyl-methanone (AM2201) represent cannabinoids found in synthetic marijuana commercialized products. The metabolism of JWH-018 and AM2201 by the cytochrome-P450 is responsible for their three primary alkyl side chain metabolites: (ω)-monohydroxyl, (ω)-carboxyl, and (ω-1)-monohydroxyl (Figure 1.5) [28]. Thus, evaluation of the clinical consequences of the enantiomers of chiral (ω-1)-monohydroxyl metabolites is important.

For that, a targeted human metabolomic chiral LC–MS/MS method was developed for studying metabolism and toxicity of these two synthetic cannabinoids [28]. A PHENOMENEX LUX® Cellulose-3 (3 µm, 2 mm I.D. × 150 mm) was used as analytical column, at 40 °C with a 20 mmol ammonium bicarbonate (A) and acetonitrile (B) gradient elution as follows: 40–95% B in 10 minutes with additional 2 minutes at 95% before returning to initial conditions over 3 minutes and equilibration of 1 minute. The total analysis run time was of 16 minutes with baseline separation for all human

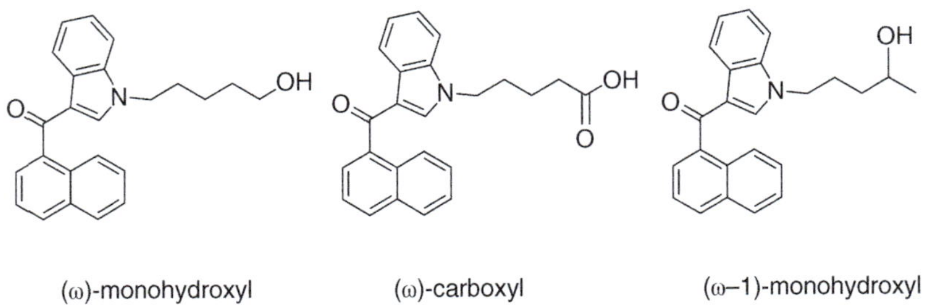

Figure 1.5 Structures of alkyl side chain metabolites from JWH-018 and AM2201 – (ω)-monohydroxyl, (ω)-carboxyl, and (ω-1)-monohydroxyl.

primary metabolites found in urine of JWH-018 and AM2201, including the (*R*) and (*S*)-enantiomers of (ω-1)-monohydroxyl metabolites. The results of a pilot clinical investigation showed that the (*S*)-enantiomer of JWH-018-(ω-1)-monohydroxyl metabolite was predominantly excreted (>87%) in human urine as the glucuronic acid conjugate, while the excretion of the (*S*)-enantiomer of AM2201-(ω-1)-metabolite was low (<5%), providing conditions for further pharmacokinetic properties of these synthetic cannabinoids [29].

The use of sub-2 μm immobilized polysaccharide CSP under normal and reverse phase conditions has been explored for the analysis of a mixture of chiral and achiral analytical standards of cannabinoids. Under normal elution conditions, CHIRALPAK IB-U and IH-U® provided two sub-15 minutes analysis while under reverse phase elution, CHIRALPAK ID-U, ID-U+IC-U, and IG-U® provided similarly fast analyses with baseline resolution of all 10 cannabinoids, including the enantiomeric resolution of the cannabichromene racemate (see also Chapter 6) [30].

The simultaneous quantification of 19 enantiomeric pairs of proteinogenic AAs and achiral glycine in serum, from both hepatocellular carcinoma (HCC) patients and healthy individuals, was carried out using CROWNAK CR-I (+)® column (5 μm, 3.0 mm I.D. × 150 mm) in a total run of 13 minutes with gradient elution using ethanol, water, and trifluoroacetic acid (TFA) (50:50:0.4, v/v/v) (A) and acetonitrile and TFA (100:0.4, v/v) (B) as mobile phase. For the quantification, single reaction monitoring (SRM) transitions were used and, due to the chromatographic coelution, L-glutamine and D-lysine were not differentiated. From the list of monitored AAs, D,L-proline was the only one that was not resolved in the used CSP. Comparing the samples from healthy volunteers and HCC patients, D-glutamate and D-glutamine have emerged as the most downregulated AAs and, thus, as potential HCC serum tumor markers [31].

A liquid chromatography–tandem mass spectroscopy (LC–MS/MS) enantioselective analysis was carried out for a series of oxylipin (chiral polyunsaturated fatty acids) using a CHIRALPAK IA-U® (1.6 μm, 3.0 mm I.D. × 100 mm) under reverse phase gradient elution with acetonitrile selected as organic modifier, covering hydrocarbon chain lengths between C6 and C14. The method showed to be useful for comprehensive profiling of regio- and stereoisomeric oxylipins in biological and clinical samples with short analysis run [32].

The use of multicolumn screening platforms integrated with the state-of-the-art of CSPs have been finally catching up in areas such as enantioselective synthesis biocatalysis, peptide synthesis, and pharmaceutical process.

To demonstrate the versatility of an integrated workflow method development [33], 14 columns and 10 eluent lines scouting LC system were used to sequentially screen 18 pharmaceutical racemates (6 hours per racemate). For that, 14 orthogonal chiral columns from different manufacturers and chiral selector types were used at multimodal elution conditions either in gradient (utilized for Whelk-O1 and polysaccharide CSPs) or in isocratic modes (for all AZYP CSP). At least one baseline enantiomeric resolution was obtained for all racemates, and most of them were resolved in more than one of the screened conditions.

To provide high-throughput enantiomeric analysis [33], the sizes of the selected chiral columns were reduced from 100 to 50 mm length, and with straightforward optimization, high speed separations (32 to 145 seconds) were obtained for 14 out of 18 enantiomeric mixtures. The resolution of the four remaining racemates was obtained in analysis runs of three to eight minutes with columns of 100 mm length. The isocratic elution mode was used and with that no column equilibration between runs were needed, an advantage to fast enantiomeric analysis.

1.5 Chiral 2D-LC

The advances gained in LC system performance and in CSP technologies have furnished the appropriate ground for higher efficiency with high-throughput enantioseparations as we have discussed in previous topics. These advances impacted chiral separations carried out not only in one dimension but also by 2D-LC. The recent developments were initially focused on AA analysis or pharmaceutical products, but new systems setup has provided conditions to a variety of applications, and we will discuss these developments in this topic.

To talk about 2D-LC, it is important initially to differentiate target analysis by LC–LC (heart-cut), multiple heart-cutting (mLC–LC), comprehensive (LC × LC), and selective comprehensive (sLC × LC) 2D separations. For a better understanding of the basic concepts, please refer to these reviews [34–36]. Herein, we are going to discuss only online 2D-LC analysis.

1.5.1 LC–LC and mLC–LC

The chromatographic space in LC–LC separation is the target analytes; thus, the concept of peak capacity is not much used. In this regard, applications with chiral column in the second dimension date to the 1990s [37–40].

In these applications, protein or macromolecules depletion has been carried out in the first dimension (^{1}D), usually by a restrict access media column, while the chiral separation takes placed in the second dimension (^{2}D). The columns are coupled via a switching valve, and the target analytes are transferred using the back- or forward-flush system configuration mode [41]. The use of achiral columns of diverse types, such as bare silica, aminopropyl silica, or alkyl silicas, and working in different elution modes have also been used in the first dimension for analyte retention and/or online sample cleanup [42, 43].

In LC–LC, the transfer of the entire peak volume from the ^{1}D to the ^{2}D allows quantification and, although it does not require fast columns in either dimension, their use increases analysis turnover. The use of MS detection in the ^{2}D has been explored for simultaneous enantiomer quantification in mixtures of related analytes [44]. While the number of chiral LC–LC publications is not large, we can find applications for determination of enantiomeric ratio and for the analysis of a variety of biological and environmental samples [1, 42, 43]

Recently, an LC–LC platform based on the use of a single achiral column in the ^{1}D and six selected chiral columns for the ^{2}D allowed automated chiral screening and provided resolution for a mixture of five racemates with a broad polarity range. The utility of the developed platform for measuring chiral, achiral, or diastereoisomeric impurities of chiral active pharmaceutical ingredients (APIs) have also been demonstrated with coeluted impurities quantified down to 0.05% level [45].

With the progress of the LC systems and CSP technologies, LC–LC have evolved to mLC–LC. In this latter mode, multiple peak single fractions are transferred one at a time from the ^{1}D to the ^{2}D, allowing more targets analytes to be separate in a single analytical run. Hence, mLC–LC is an extension of LC–LC with a wider capacity of analysis (Figure 1.6).

The mLC–LC instrumental configurations are designed to collect a fraction from the ^{1}D, while simultaneously separating another fraction in the ^{2}D. The cuts can be either time or peak based. A fixed sampling based on the loop volume and flow rate of the ^{1}D are used for time-based cuts, while either the selected threshold or slope for the ^{1}D detector is used for peak-based cuts.

Several configurations have been devised for working with mLC–LC, as the one used for the target analysis of eight chiral and five achiral AAs in samples from the Yamato 791191 meteorite. In this case, one multiloop deck for collecting the peaks from the achiral column in the ^{1}D were connected through high pressure valves to two narrow-bore chiral columns [46].

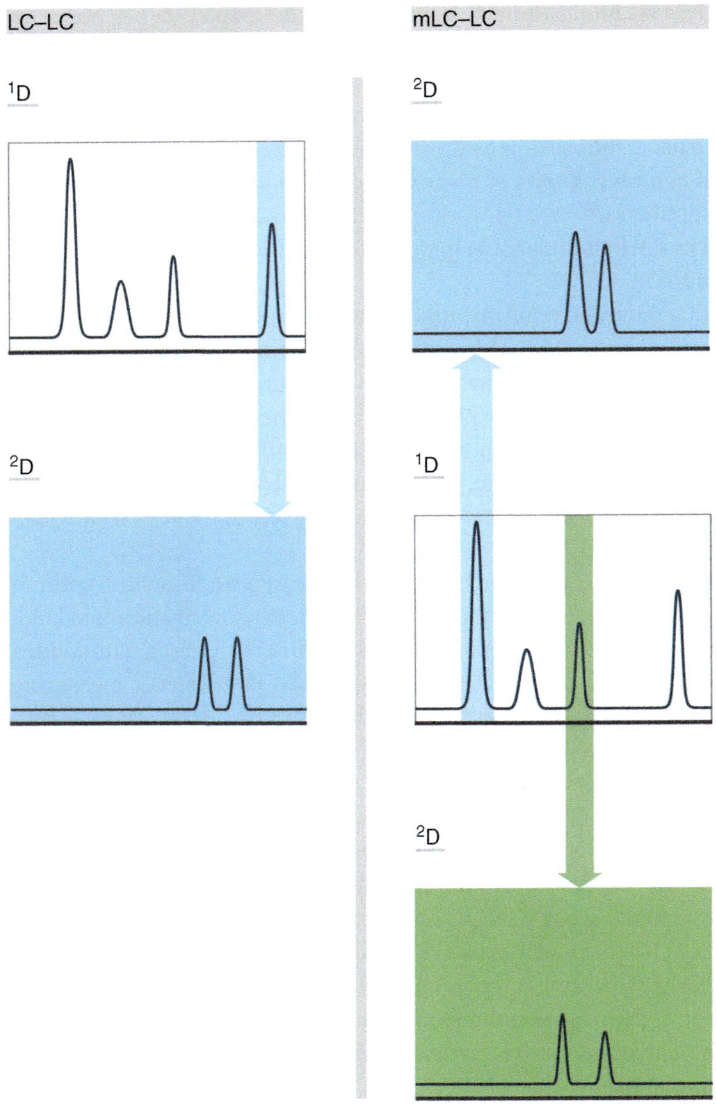

Figure 1.6 Illustration of two different modes of 2D-LC separation for target analysis – heart-cut (LC–LC) and multiple heart-cutting (mLC–LC).

An mLC–LC enantioselective separation of D-asparagine, D-serine, D-alanine, and D-proline enantiomers as 4-fluoro-7-nitro-2,1,3-benzoxadiazole (NBD-F) derivatives have been carried out for analysis of clinical samples. Taking advantage of adding an extra dimension, the LC system was configured as a 3D–LC containing a reversed-phase column, an anion-exchange column, and an enantioselective column. The target NDB-amino acids (NBD-AA) were separated from other proteinogenic AA by hydrophobicity in the ^{1}D with a C18 narrow-bore column and the fractions collected in a multiloop deck before transferring to the anion-exchange ^{2}D column for further cleanup. The fractions that were collected by their difference in the anionic strength were then transferred to a Pirkle-type Sumichiral KSAACSP-001S column in ^{3}D. The method provided conditions for the diagnosis of chronic kidney disease by enantioselective determination of D-AA in clinical samples [47].

In this past decade, commercial 2D-LC system can be acquired in different configurations. To follow the excretion of propranolol and their hydroxy metabolites, mLC–LC was conducted using Agilent 1290 Infinity 2D-LC system. For the analysis, a Poroshell 120 CHIRAL-T® (2.7 μm, 4.6 mm I.D. × 100 mm) column was used in the ^{2}D, while urine samples were injected onto a Poroshell 120 PHENYL-HEXYL® (2.7 μm, 2.1 mm I.D. × 100 mm) column. The sampling was based on a 40 μl loop deck [48].

A multicolumn 2D-LC system has been designed to connect multiple columns in both dimensions (modified version of Agilent 1290 Infinity 2D-LC). In this configuration, a six-column selection valve combines four columns in the ^{1}D with four columns in the ^{2}D. The innovation with this design is that the screening and selection of columns and mobile phases in both dimensions can speed up method development for multicomponent mixtures since any given peak in the ^{1}D can be transferred to multiple columns in ^{2}D [49].

Commercial 2D-LC systems using multiloop decks have broaden the possibilities of applications for chiral separations [26, 50].

1.5.2 LC × LC and sLC × LC

An LC × LC separation is suitable for non-targeted and comprehensive analysis of a sample; it allows broader information in a single 2D chromatogram. For that, all fractions from the ^{1}D are transferred into the ^{2}D for further analysis (Figure 1.7). To avoid undersampling and remixing fractions, 4 fractions per 8σ peak width must be collected in the ^{1}D and transferred into the ^{2}D. Moreover, the ^{2}D column needs to provide very fast separations, and it is expected that SPP will outperform FPP CSPs. As for heart-cut, the switching valve configuration (such as of 8-, 10-, or two 6-port valves) plays an important role in the interface between the columns of both dimensions but, in this case, the modulation interface plays a much greater role, since it is responsible for transferring the fractions with lower undersampling effect [3, 35, 36].

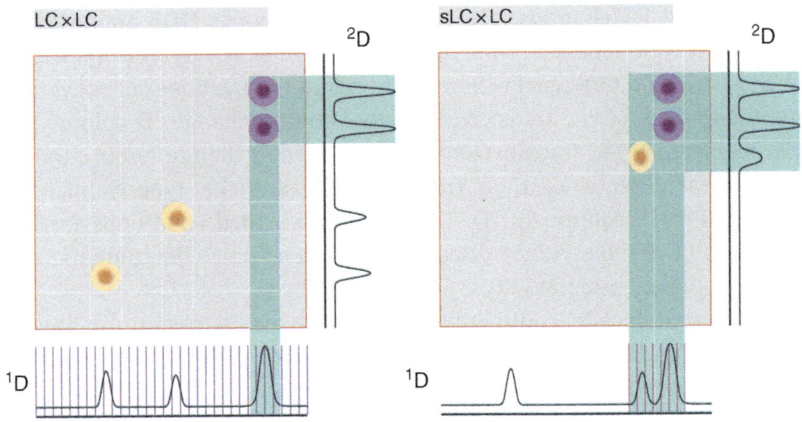

Figure 1.7 Illustration of two different modes of 2D-LC separation for non-targeted and comprehensive analysis – comprehensive (LC × LC) and selective comprehensive (sLC × LC).

The interface can be based on either loops or trap columns. The loop interface is the most used either for heart-cut or for comprehensive 2D-LC; however, the trap columns interface is preferable when the mobile phase of the ^{1}D is immiscible with the mobile phase of the ^{2}D or when there are great differences in solvent strength. The trap columns interface should enhance sensitivity and diminish injection problems; nevertheless, the selected trap columns need to show good retention/desorption efficiency and do not affect the plate capacity in the ^{2}D [40].

Since orthogonality can be achieved in coupling a CSP to an achiral column, reversed-phase elution mode has been preferred for 2D-LC, since alkyl columns are an obvious choice in LC×LC for either dimension. Software for data analysis is indispensable in LC×LC, and it is still the Achilles' heel.

In a very elegant LC×LC approach, two CSPs (quinine and quinidine carbamates) with similar chemoselectivity but orthogonal stereochemistry have been used for enantioselective AA analysis of peptide and protein hydrolysates. With that, the achiral analytes elute along a diagonal line, while the AA enantiomers appeared in the same retention but in opposite position of the diagonal line in the 2D plot. Therewith, stereochemical information for unknown peptides in complex samples has been obtained [51]. For further discussion about this example, see Chapter 11 and Figure 11.12.

For enantioselective separation of AAs in honey samples, derivatization by Sanger's reaction with 1-fluoro-2,4-dinitrobenzene to form dinitro-phenyl amino acids (DNP-AAs) has been first carried out to enhance sensitivity and, thus, detection at 365 nm. In an unusual approach and to take advantage of the much slower enantioselective separation, a quinine-bonded SPP CSP was used in the ^{1}D, while a C18 column was used in the ^{2}D. This LC × LC setup resolved most enantiomers of 20 DNP-AAs in this natural complex sample [52].

sLC × LC is the notation given to LC × LC when only a target region of the ^{1}D chromatogram is comprehensively sampled to the ^{2}D. It differs from mLC–LC in the sense that the achieved resolution for the selected area of the ^{1}D chromatogram does not diminish by sampling it to the ^{2}D (Figure 1.7). It is important to say that the LC system used for sLC × LC is essentially the same as that used for mLC–LC.

Stoll and Carr [35] call attention to the fact that the main advantage of sLC × LC is that the separation obtained in the ^{1}D is not lost in ^{2}D regardless of how narrow the width of the peak is, with no risk of peak remixing. Moreover, the sLC × LC allows quantification without the need of transferring a fraction larger than the peak volume to avoid losing analytes from the ^{1}D, as usually is the case with LC–LC.

Taking advantage of both mLC–LC and sLC–LC modes, the enantiomeric resolution of the 20 proteinogenic AAs (and 5 others) were obtained in a commercial 2D-LC system. The DNP-AA was separated in the ^{1}D using an achiral C18 column (1.8 μm, 2.1 mm I.D. × 100 mm) with gradient elution, whereas the enantiomers were resolved in the ^{2}D using a *tert*-butylcarbamoylquinine-based SPP CSP (2.7 μm, 3.0 mm I.D. × 50 mm) in isocratic elution. The ^{1}D peaks were sampled as 40 μl fractions to the ^{2}D by means of a two-position four-port dual valve connected to a double loop deck, each with six 40 μl loops in a total run time of 130 minutes. The absolute configuration of the AAs in the peptide antibiotic drugs gramicidin and bacitracin have been determined by the use of this 2D-LC method [50] (see also Chapter 11).

Despite the advances in CSP technology, the number of reported enantioselective comprehensive methods is so far very low. With the capability to solve non-targeted analysis, an increase in publications in this topic is expected soon.

1.6 Future and Perspectives

The advances in CPS technologies and in LC multicolumn systems have allowed fast and ultrafast analysis runs that can provide a diversity of platforms for fast method development and for profiling enantiomeric ratios in a variety of sample matrices and applications, such as reaction monitoring, enantiomerization and/or racemization studies, metabolic profiling, multi-residues analysis, and targeting biomarkers, to cite a few. Nevertheless, it is still noticeable the need for more innovative sample preparation protocols for a valuable step forward in enantioselective analysis.

2D-LC is expected to impact online reaction monitoring especially on the field of flow chemistry in which a continuous flow microreactor can be coupled to an LC system for product reaction purification in the ^{1}D, followed by a chiral column in the ^{2}D for analysis of the conversion ratio and stereoselectivity.

References

1 Barreiro, J.C., Tiritan, M.E., and Cass, Q.B. (2021). Challenges and innovations in chiral drugs in an environmental and bioanalysis perspective. *TrAC Trends Anal. Chem.* 142: 116326. https://doi.org/10.1016/j.trac.2021.116326.

2 Oliveira, R.V., Simionato, A.V.C., and Cass, Q.B. (2021). Enantioselectivity effects in clinical metabolomics and lipidomics. *Molecules* 26: 5231. https://doi.org/10.3390/molecules26175231.

3 Calderón, C. and Lämmerhofer, M. (2021). Enantioselective metabolomics by liquid chromatography-mass spectrometry. *J. Pharm. Biomed. Anal.* 207: 114430. https://doi.org/10.1016/j.jpba.2021.114430.

4 Pandey, R., Collins, M., Lu, X. et al. (2021). Novel strategy for untargeted chiral metabolomics using liquid chromatography-high resolution tandem mass spectrometry. *Anal. Chem.* 93: 5805–5814. https://doi.org/10.1021/acs.analchem.0c05325.

5 Cass, Q.B., Degani, A.L.G., and Cassiano, N.M. (2003). Effects on enantioselectivity by the use of polysaccharide-based columns by multimodal elution. *J. Liq. Chromatogr. Relat. Technol.* 26: 2083–2101. https://doi.org/10.1081/JLC-120022395.

6 Wainer, I.W. (1987). Proposal for the classification of high-performance liquid chromatographic chiral stationary phases: how to choose the right column. *TrAC Trends Anal. Chem.* 6: 125–134. https://doi.org/10.1016/0165-9936(87)87055-3.

7 Lough, W.J. (2014). Classification of LC chiral stationary phases: Wainer Types I–V revisited. *J. Chromatogr. B* 968: 1–7. https://doi.org/10.1016/j.jchromb.2014.04.044.

8 Lämmerhofer, M. (2010). Chiral recognition by enantioselective liquid chromatography: mechanisms and modern chiral stationary phases. *J. Chromatogr. A* 1217: 814–856. https://doi.org/10.1016/j.chroma.2009.10.022.

9 Dalgliesh, C.E. (1952). 756. The optical resolution of aromatic amino-acids on paper chromatograms. *J. Chem. Soc.* 3940: https://doi.org/10.1039/jr9520003940.

10 Asnin, L. (2012). Adsorption models in chiral chromatography. *J. Chromatogr. A* 1269: 3–25. https://doi.org/10.1016/j.chroma.2012.08.096.

11 Batista, A.N.L., dos Santos, F.M., Batista, J., and Cass, Q. (2018). Enantiomeric mixtures in natural product chemistry: separation and absolute configuration assignment. *Molecules 23*: 492. https://doi.org/10.3390/molecules23020492.

12 Toki, H., Ichikawa, T., Mizuno-Yasuhira, A., and Yamaguchi, J. (2018). A rapid and sensitive chiral LC–MS/MS method for the determination of ketamine and norketamine in mouse plasma, brain and cerebrospinal fluid applicable to the stereoselective pharmacokinetic study of ketamine. *J. Pharm. Biomed. Anal.* 148: 288–297. https://doi.org/10.1016/j.jpba.2017.09.033.

13 Lang, J.C. and Armstrong, D.W. (2017). Chiral surfaces: the many faces of chiral recognition. *Curr. Opin. Colloid Interface Sci.* 32: 94–107. https://doi.org/10.1016/j.cocis.2017.10.004.

14 Tarafder, A. and Miller, L. (2021). Chiral chromatography method screening strategies: past, present and future. *J. Chromatogr. A* 1638: 461878. https://doi.org/10.1016/j.chroma.2021.461878.

15 Panella, C., Ferretti, R., Casulli, A., and Cirilli, R. (2019). Temperature and eluent composition effects on enantiomer separation of carvedilol by high-performance liquid chromatography on immobilized amylose-based chiral stationary phases. *J. Pharm. Anal.* 9: 324–331. https://doi.org/10.1016/j.jpha.2019.04.002.

16 Gegenava, M., Chankvetadze, L., Farkas, T., and Chankvetadze, B. (2014). Enantioseparation of selected chiral sulfoxides in high-performance liquid chromatography with polysaccharide-based chiral selectors in polar organic mobile phases with emphasis on enantiomer elution order. *J. Sep. Sci.* 37: 1083–1088. https://doi.org/10.1002/jssc.201301318.

17 Barreiro, J.C., Paixão, M.W., Lourenço, T.C. et al. (2016). A high-resolution magic angle spinning NMR study of the enantiodiscrimination of 3,4-meth ylenedioxymethamphetamine (MDMA) by an immobilized polysaccharide-based chiral phase. *PLoS One* 11: e0162892. https://doi.org/10.1371/journal.pone.0162892.

18 Roussel, C., Vanthuyne, N., Serradeil-Albalat, M., and Vallejos, J.C. (2003). True or apparent reversal of elution order during chiral high-performance liquid chromatography monitored by a polarimetric detector under different mobile phase conditions. *J. Chromatogr. A* 995: 79–85. https://doi.org/10.1016/S0021-9673(03)00533-8.

19 Catani, M., Ismail, O.H., Gasparrini, F. et al. (2017). Recent advancements and future directions of superficially porous chiral stationary phases for ultrafast high-performance enantioseparations. *Analyst* 142: 555–566. https://doi.org/10.1039/C6AN02530G.

20 Catani, M., Felletti, S., Ismail, O.H. et al. (2018). New frontiers and cutting edge applications in ultra high performance liquid chromatography through latest generation superficially porous particles with particular emphasis to the field of chiral separations. *Anal. Bioanal. Chem.* 410: 2457–2465. https://doi.org/10.1007/s00216-017-0842-4.

21 Ismail, O., Felletti, S., Luca, C. et al. (2018). The way to ultrafast, high-throughput enantioseparations of bioactive compounds in liquid and supercritical fluid chromatography. *Molecules* 23: 2709. https://doi.org/10.3390/molecules23102709.

22 Roy, D., Weatherly, C.A., Wahab, M.F., and Armstrong, D.W. (2018). Frontiers in ultrafast chiral chromatography. *LCGC N. Am.* 36: 510–519.

23 Wahab, M.F., Roy, D., and Armstrong, D.W. (2021). The theory and practice of ultrafast liquid chromatography: a tutorial. *Anal. Chim. Acta* 1151: 238170. https://doi.org/10.1016/j.aca.2020.12.045.

24 Zawatzky, K., Barhate, C.L., Regalado, E.L. et al. (2017). Overcoming "speed limits" in high throughput chromatographic analysis. *J. Chromatogr. A* 1499: 211–216. https://doi.org/10.1016/j.chroma.2017.04.002.

25 Kaplitz, A.S., Kresge, G.A., Selover, B. et al. (2020). High-throughput and ultrafast liquid chromatography. *Anal. Chem.* 92: 67–84. https://doi.org/10.1021/acs.analchem.9b04713.

26 Barhate, C.L., Regalado, E.L., Contrella, N.D. et al. (2017). Ultrafast chiral chromatography as the second dimension in two-dimensional liquid chromatography experiments. *Anal. Chem.* 89: 3545–3553. https://doi.org/10.1021/acs.analchem.6b04834.

27 Tanács, D., Berkecz, R., Misicka, A. et al. (2021). Enantioseparation of ß2-amino acids by liquid chromatography using core-shell chiral stationary phases based on teicoplanin and teicoplanin aglycone. *J. Chromatogr. A* 1653: 462383. https://doi.org/10.1016/j.chroma.2021.462383.

28 Chimalakonda, K.C., Seely, K.A., Bratton, S.M. et al. (2012). Cytochrome P450-mediated oxidative metabolism of abused synthetic cannabinoids found in K2/spice: identification of novel cannabinoid receptor ligands. *Drug Metab. Dispos.* 40: 2174–2184. https://doi.org/10.1124/dmd.112.047530.

29 Patton, A.L., Seely, K.A., Chimalakonda, K.C. et al. (2013). Targeted metabolomic approach for assessing human synthetic cannabinoid exposure and pharmacology. *Anal. Chem.* 85: 9390–9399. https://doi.org/10.1021/ac4024704.

30 Onishi, T. and Umstead, W.J. (2021). The separation of cannabinoids on sub-2 μm immobilized polysaccharide chiral stationary phases. *Pharmaceuticals* 14: 1250. https://doi.org/10.3390/ph14121250.

31 Han, M., Xie, M., Han, J. et al. (2018). Development and validation of a rapid, selective, and sensitive LC–MS/MS method for simultaneous determination of D- and L-amino acids in human serum: application to the study of hepatocellular carcinoma. *Anal. Bioanal. Chem.* 410: 2517–2531. https://doi.org/10.1007/s00216-018-0883-3.

32 Cebo, M., Fu, X., Gawaz, M. et al. (2020). Enantioselective ultra-high performance liquid chromatography-tandem mass spectrometry method based on sub-2μm particle polysaccharide column for chiral separation of oxylipins and its application for the analysis of autoxidized fatty acids and platelet r. *J. Chromatogr. A* 1624: 461206. https://doi.org/10.1016/j.chroma.2020.461206.

33 Losacco, G.L., Wang, H., Haidar Ahmad, I.A. et al. (2022). Enantioselective UHPLC screening combined with in silico modeling for streamlined development of ultrafast enantiopurity assays. *Anal. Chem.* 94: 1804–1812. https://doi.org/10.1021/acs.analchem.1c04585.

34 Cassiano, N., Barreiro, J., Oliveira, R., and Cass, Q. (2012). Direct bioanalytical sample injection with 2D LC-MS. *Bioanalysis* 4: 2737–2756. https://doi.org/10.4155/bio.09.39.

35 Stoll, D.R. and Carr, P.W. (2017). Two-dimensional liquid chromatography: a state of the art tutorial. *Anal. Chem.* 89: 519–531. https://doi.org/10.1021/acs.analchem.6b03506.

36 Pirok, B.W.J., Stoll, D.R., and Schoenmakers, P.J. (2019). Recent developments in two-dimensional liquid chromatography: fundamental improvements for practical applications. *Anal. Chem.* 91: 240–263. https://doi.org/10.1021/acs.analchem.8b04841.

37 Cassiano, N.M., Cass, Q.B., Degani, A.L.G., and Wainer, I.W. (2002). Determination of the plasma levels of metyrapone and its enantiomeric metyrapol metabolites by direct plasma injection and multidimensional achiral-chiral chromatography. *Chirality* 14: 731–735. https://doi.org/10.1002/chir.10133.

38 Cassiano, N.M., Lima, V.V., Oliveira, R.V. et al. (2006). Development of restricted-access media supports and their application to the direct analysis of biological fluid samples via high-performance liquid chromatography. *Anal. Bioanal. Chem.* 384: 1462–1469. https://doi.org/10.1007/s00216-005-0253-9.

39 Schlauch, M., Fulde, K., and Frahm, A.W. (2002). Enantioselective determination of (R)- and (S)-sotalol in human plasma by on-line coupling of a restricted-access material precolumn to a cellobiohydrolase I-based chiral stationary phase. *J. Chromatogr. B Anal. Technol. Biomed. Life Sci.* 775: 197–207. https://doi.org/10.1016/S1570-0232(02)00287-8.

40 Iguiniz, M. and Heinisch, S. (2017). Two-dimensional liquid chromatography in pharmaceutical analysis. Instrumental aspects, trends and applications. *J. Pharm. Biomed. Anal.* 145: 482–503. https://doi.org/10.1016/j.jpba.2017.07.009.

41 Cassiano, N.M., Barreiro, J.C., Moraes, M.C. et al. (2009). Restricted-access media supports for direct high-throughput analysis of biological fluid samples: review of recent applications. *Bioanalysis* 1: 577–594. https://doi.org/10.4155/bio.09.39.

42 Mechelke, J., Rust, D., Jaeger, A., and Hollender, J. (2020). Enantiomeric fractionation during biotransformation of chiral pharmaceuticals in recirculating water-sediment test flumes. *Environ. Sci. Technol.* 54: 7291–7301. https://doi.org/10.1021/acs.est.0c00767.

43 Peñín-Ibáñez, M., Santos-Delgado, M.J., and Polo-Diez, L.M. (2018). Determination of profens in pharmaceuticals by heart-cut achiral-chiral two-dimensional HPLC. *Sep. Sci. Plus* 1: 168–176. https://doi.org/10.1002/SSCP.201700038.

44 Alvim, J., Lopes, B.R., and Cass, B. (2016). Simultaneous enantioselective quantification of fluoxetine and norfluoxetine in human milk by direct sample injection using 2-dimensional liquid chromatography-tandem mass spectrometry. *J. Chromatogr. A* 1451: 120–126. https://doi.org/10.1016/j.chroma.2016.05.022.

45 Hegade, R.S., Chen, K., Boon, J.-P. et al. (2020). Development of an achiral-chiral 2-dimensional heart-cutting platform for enhanced pharmaceutical impurity analysis. *J. Chromatogr. A* 1628: 461425. https://doi.org/10.1016/j.chroma.2020.461425.

46 Hamase, K., Nakauchi, Y., Miyoshi, Y. et al. (2014). Enantioselective determination of extraterrestrial amino acids using a two-dimensional chiral high-performance liquid chromatographic system. *Chromatography* 35: 103–110. https://doi.org/10.15583/jpchrom.2014.014.

47 Furusho, A., Koga, R., Akita, T. et al. (2019). Three-dimensional high-performance liquid chromatographic determination of Asn, Ser, Ala, and Pro enantiomers in the plasma of patients with chronic kidney disease. *Anal. Chem.* 91: 11569–11575. https://doi.org/10.1021/acs.analchem.9b01615.

48 Harps, L.C., Schipperges, S., Bredendiek, F. et al. (2020). Two dimensional chromatography mass spectrometry: quantitation of chiral shifts in

metabolism of propranolol in bioanalysis. *J. Chromatogr. A* 1617: 460828. https://doi.org/10.1016/j.chroma.2019.460828.

49 Wang, H., Herderschee, H.R., Bennett, R. et al. (2020). Introducing online multicolumn two-dimensional liquid chromatography screening for facile selection of stationary and mobile phase conditions in both dimensions. *J. Chromatogr. A* 1622: 460895. https://doi.org/10.1016/j.chroma.2020.460895.

50 Woiwode, U., Neubauer, S., Lindner, W. et al. (2018). Enantioselective multiple heartcut two-dimensional ultra-high-performance liquid chromatography method with a Coreshell chiral stationary phase in the second dimension for analysis of all proteinogenic amino acids in a single run. *J. Chromatogr. A* 1562: 69–77. https://doi.org/10.1016/j.chroma.2018.05.062.

51 Woiwode, U., Reischl, R.J., Buckenmaier, S. et al. (2018). Imaging peptide and protein chirality via amino acid analysis by chiral × chiral two-dimensional correlation liquid chromatography. *Anal. Chem.* 90: 7963–7971. https://doi.org/10.1021/acs.analchem.8b00676.

52 Acquaviva, A., Siano, G., Quintas, P. et al. (2020). Chiral x achiral multidimensional liquid chromatography. Application to the enantioseparation of dintitrophenyl amino acids in honey samples and their fingerprint classification. *J. Chromatogr. A* 1614: 460729. https://doi.org/10.1016/j.chroma.2019.460729.

2

Chiral Separation by GC

Oliver Trapp

Department of Chemistry, Ludwig-Maximilians-University Munich, Munich, Germany

2.1 Introduction

The separation of stereoisomers and in particular enantiomers is highly important to analyze the enantiomeric excess of enantiomerically pure compounds [1], to investigate the stereochemical integrity [2–6], and in general to isolate enantiomerically pure compounds. The key to the separation of enantiomers is the successful development of chiral stationary phases (CSPs) employing chiral selectors that interact with the chiral analytes by forming diastereomeric interactions, which leads to a separation of the analyte enantiomers. In the last decades, a large number of highly versatile and broadly applicable selectors and CSPs [7–9] were developed for the separation of enantiomers by gas chromatography (GC) [10–16], high performance liquid chromatography (HPLC) [17–35], supercritical fluid chromatography (SFC) [36–39] and capillary electrochromatography (CEC) [40–47]. Chiral separations and CSPs can be characterized by their enantioselectivities for the separation of enantiomers. Modern CPSs are often applicable to a broad spectrum of different chiral compound classes with typically high separation factors α. This is important for a rapid analysis of enantiomeric ratios (er) or enantiomeric excesses (ee). Another important aspect is that absolute configurations can be assigned with known reference compounds. In the recent years, preparative separation of

Chiral Separations and Stereochemical Elucidation: Fundamentals, Methods, and Applications, First Edition. Edited by Quezia Bezerra Cass, Maria Elizabeth Tiritan, João Marcos Batista Junior, and Juliana Cristina Barreiro.

enantiomers became more important [48, 49]. Compared to asymmetric synthesis or catalysis, which is often time consuming and expensive, the racemate is prepared and separated by chiral chromatography.

It has to be pointed out that the achievements made in asymmetric synthesis [50] and catalysis [51–55]; asymmetric organocatalysis [56–64]; high-throughput screening [65–67] of chiral compounds; the assignment of relative and absolute configurations [68–70]; and achievements made with chiral compounds in chemical biology, medicine, and pharmaceutical sciences rely on the development of highly versatile CPSs. Furthermore, these developments enabled the investigation of asymmetric autocatalysis [71–73], self-amplification of chirality [74–78], or the homochirality in the context of the emergence of life [79–81]. Another example is the chiral analysis of compounds in meteorites [82–86]. CPSs are also used to determine physical organic properties [87, 88] such as supramolecular interactions [89–92]. Furthermore, the interconversion barrier of stereochemically labile compounds can be determined by (enantioselective) chiral dynamic chromatography, where the interconversion becomes visible by a distinct plateau formation in the chromatographic separation of the enantiomers [2–34, 56, 93–99]. This is of particular interest for drug molecules.

This chapter is a guide for the use of important CSPs for chiral GC, their development [100–104], and their applications in the analysis of chiral volatile compounds. An important property of all CSPs for GC applications is that they form a liquid film at the inner wall of the separation capillary. This can be sometimes very challenging for high melting or crystalline selectors, e.g. native cyclodextrins (CDs) or zwitterionic amino acids. Therefore, two strategies are applied: (i) the selector is modified and dissolved in a polymeric matrix, e.g. polysiloxane, and (ii) the selector is modified with long chain alkyl groups to reduce the polarity. It is important to point out that it is sufficient if the CSP is liquid or a supercooled melt at the analysis temperature.

This principle of chemical modification of selectors was first applied by Emanuel Gil-Av and co-workers leading to the first separation of enantiomers by GC in 1966 [105]. They used the amino acid derivative *N*-trifluoroacetyl-L-isoleucine lauryl ester **1** to avoid zwitterion formation, as CSP (Figure 2.1). The chiral separation of chiral proteinogenic α-amino acids derivatized as the *N*-trifluoroacetyl alkyl esters of alanine, valine, and leucine on a glass capillary column (100 m × 0.25 mm inner diameter [i.d.]) was achieved.

Besides the principle of three-point interactions [106] between the chiral analyte and the chiral selector, the recognition principle is based on complementary supramolecular hydrogen bonding. This type of complementary

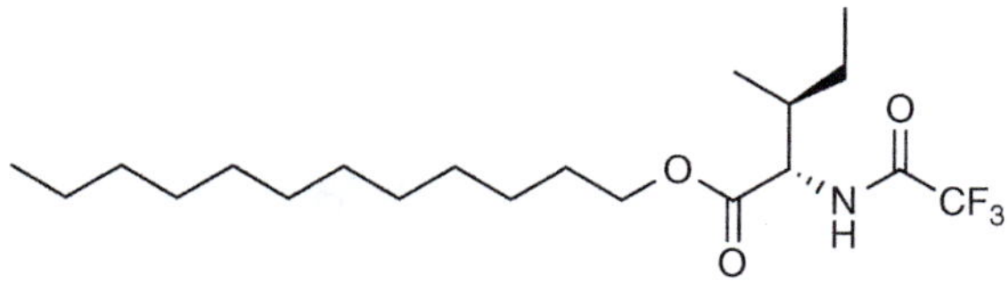

Figure 2.1 Enantioseparation of proteinogenic α-amino acids, derivatized as *N*-trifluoroacetyl alkyl esters, by the CSP *N*-trifluoroacetyl-L-isoleucine lauryl ester developed by Gil-Av and coworkers. *Source:* Gil-Av et al. [105]/Elsevier.

supramolecular hydrogen bonding is also found in nature in peptide–peptide (β-sheet in the secondary peptide structure) interactions.

The demand for a CSP to separate amino acids and the development of *N*-trifluoroacetyl-L-isoleucine lauryl ester as CSP by Gil-Av originated from the Apollo program [107] to explore potential organic compounds on the moon and in moon rocks. In general, the analysis of enantiomerically enriched amino acids in extraterrestrial samples is of great interest. In context of the Apollo program, no amino acids were found in lunar samples from the Sea of Tranquility.

The same design principle of the CSP *N*-trifluoroacetyl-L-isoleucine lauryl ester was transferred and further improved by permanent bonding of the L-valine diamide selector to a polysiloxane (Chirasil-Val) by Bayer and co-workers. This CSP is turned out to be very efficient in the chiral analysis of amino acids [108–110]. So far, the recognition principle relied on hydrogen bonding. Schurig et al. overcame this limit by developing CSPs for GC based on metal complexes [111–113], which can even interact with double bonds or other functional groups. In particular, chiral β-diketonate complexes were employed as selectors, which are also broadly used as chiral shift reagents in NMR spectroscopy [114–116].

The introduction of modified CDs as selectors for CSPs in GC broadened the spectrum of separable chiral compounds substantially because of the high flexibility of supramolecular selector–analyte complexes covering many compound classes. An important aspect is that even non-functionalized chiral molecules, e.g. chiral alkanes, can be separated.

2.2 Chiral Recognition in Gas Chromatography

The separation of enantiomers relies on "chiral recognition" of the analyte by a selector. The selector interacts with the analyte enantiomers and forms adducts based on the nature of the interaction, which can be hydrogen bonding, complexation, or supramolecular interaction. These interactions are typically weak so that the analyte is also released by the selector. Because the analyte and the selector are chiral, they form transient diastereomeric complexes.

GC separations are typically characterized by high separation efficiencies with 1000–5000 theoretical plates per meter of GC capillary column. This is of great advantage because excellent separations can be achieved even for only small energetic differences for these transient diastereomeric complexes. Because the equilibria of these transient complexes are adjusted in each theoretical separation plate of a separation column, very high efficiencies can be achieved.

The three-point interaction model is very descriptive to explain chiral recognition between the selector and the selectand (cf. Figure 2.2). The interactions can be by hydrogen bonds, charge-transfer interactions, coordination, and steric hindrance.

The two transient diastereomeric adducts that form differ in their complexation/interaction energies and therefore in their equilibrium constants. As already explained, these equilibria are adjusted in every theoretical plate of a GC column, which leads then to a difference in the retention times for the individual enantiomers.

In the following the different recognition mechanisms are discussed, which is of importance to understand the applicability of certain selectors and should be considered for future developments of CSPs.

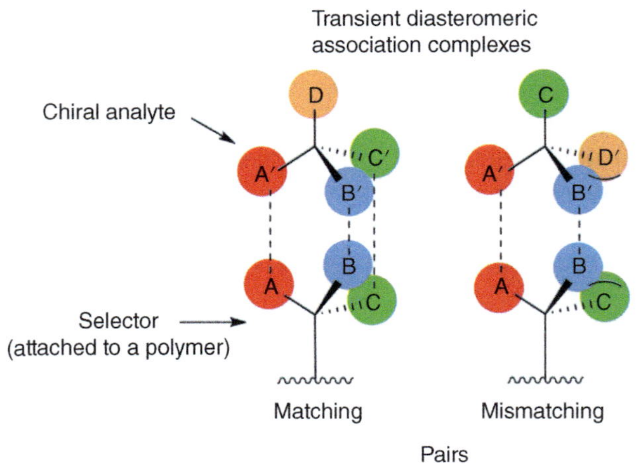

Figure 2.2 Representation of the three-point interaction model for the transient diastereomeric complexes formed between the chiral selector and the enantiomers.

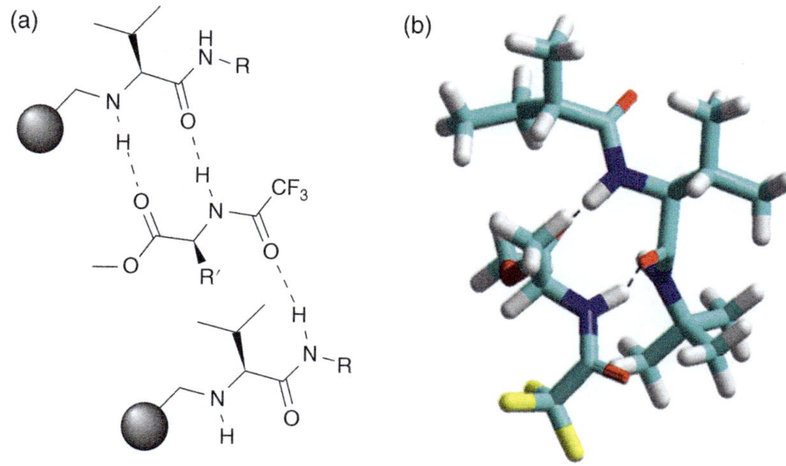

Figure 2.3 Chiral recognition by hydrogen bonding. (a) Interaction with a diamide selector and an amino acid derivative forming a 1:1 or 2:1 complex, and (b) DFT B3LYP/6-31G* optimized structure of a supramolecular hydrogen bonding complex.

2.2.1 Chiral Recognition by Hydrogen Bonding

Hydrogen bonding is maybe the most important binding motif in nature. It is not surprising that hydrogen bonding is highly efficient for interactions between the selector and the selectand/analyte molecule as we have already seen in diamide-type CSPs like the *N*-trifluoroacetyl-L-isoleucine lauryl ester [105] or Chirasil-Val [108–110] (Figure 2.3), where the carbonyl group serves as hydrogen bonding acceptor and the amine group as hydrogen bonding donor.

2.2.2 Chiral Recognition Using Chiral Metal Complexes

Interaction between unsaturated organic compounds such as the strong π-acid ethylene to platinum in the *Zeise*-salt [117] are very well documented. Chiral β-diketonate complexes with various metal ions are well suited and can be prepared from natural mixtures of chiral terpenes [118, 119]. Broadly used are phases with nickel (II) and europium (III) complexes. Figure 2.4 shows the interaction of (*S*,*S*)-2,3-dimethyloxirane [120] with the selector nickel(II)-bis [(1*R*)-3-(heptafluorobutyryl)-camphorate] [121].

2.2.3 Chiral Recognition by Host–Guest Interactions

Chiral recognition by host–guest interactions (Figure 2.5) is a highly versatile approach since a supramolecular chiral host, such as CDs [122], can

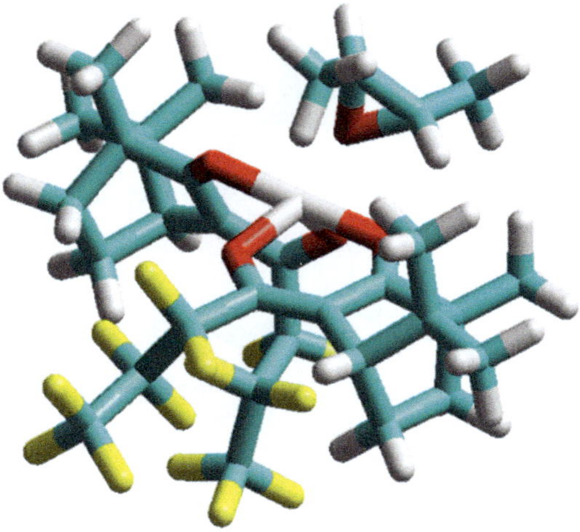

Figure 2.4 Transient diastereomeric complex of (*S*,*S*)-2,3-dimethyloxirane and nickel(II)-bis [(1*R*)-3-(heptafluorobutyryl)-camphorate]. DFT B3LYP/6-31G* optimized structure.

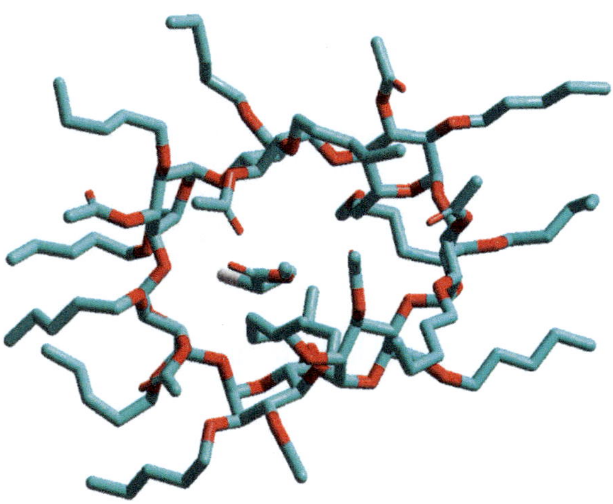

Figure 2.5 Supramolecular interaction between methyl (*S*)-2-chloropropionate and heptakis(3-*O*-acetyl-2,6-di-*O*-pentyl)-b-CD. DFT B3LYP/6-31G* optimized structure. The atoms of hydrogen have been hidden for simplicity.

interact through multiple non-covalent interactions. Such host–guest interactions can be mixed with H-bonding by means of either hydrogen-bond acceptor or hydrogen-bond donor groups. Successful examples include modified resorcinarenes functionalized with *N*-acetyl-L-valine-*tert*-butylamide moieties [123]. Resorcinarenes, which are structurally similar to calixarenes, consist of cyclic methylene-bridged resorcin tetramers. Recent developments include metal organic frameworks with chiral recognition sites [124]. Such supramolecular interactions are in particular important for separating enantiomers lacking functional groups like in chiral hydrocarbons. The recognition in the presence of CDs is ascribed to weak van der Waals forces between the supramolecular structure and the analyte. Such interactions were studied by spectroscopic techniques and supported by quantum chemical calculations [125–127].

2.3 Preparation of Fused-Silica Capillaries for GC with CSPs

Nowadays many chiral GC columns are commercially available; nevertheless, several CSPs that are not commercially available can be prepared by coating onto bare fused-silica capillaries. Herein, practical guides for proper treatment of capillaries and for preservation of high separation efficiency over their lifetime will be presented.

Capillary columns made of fused silica with lengths of 12.5–50 m and inner diameter typically of 0.25 mm must first be prepared for coating with the desired stationary phases. In the first step, the capillaries prepared on capillary cages are dehydrated at 260 °C under a weak carrier gas stream for 48 hours and then coated directly with the corresponding CSP by using the static method [128] at room temperature. In the case of very nonpolar CSP, e.g. the Lipodex phases, developed originally for pyrex glass capillaries, a pretreatment is required. It can be done by silanization of the capillary surface or by coating with hydridomethyldimethylpolysiloxane containing up to 10% hydrido groups and coating thickness within 5–10 nm.

Stationary phases in the form of polymers, pure selectors, or as selectors in a polymer [129] are subsequently dissolved in a volatile solvent, e.g. anhydrous diethyl ether. The resulting film thickness is determined by the concentration (w/w) of the solution and can be by Eq. (2.1):

$$\text{Coating solution } (\%) = \frac{\text{film thickness } (\mu\text{m})}{2.5 \times \text{column ID (mm)}} \tag{2.1}$$

By using a coating apparatus, the capillary is first filled with this solution and the end of the capillary is then sealed, for example with silicone. Under vacuum, the solvent is subsequently removed from the filling side, and the progress of the coating can be followed visually. Due to the excellent heat transfer properties of fused-silica capillaries, the evaporation procedures can be facilitated with the use of a water bath.

A temperature program from 50 to 180 °C, with a heating rate of 1 K min^{-1}, is used for conditioning of the coated capillaries and the final temperature is kept for 24 hours.

In the case of Chirasil phases, the thermal immobilization is performed by heating the conditioned capillaries at 220 °C for 48 hours under a very slow carrier gas stream of about 50–80 bubbles min^{-1} in ether. Immobilized phases can be purged, in contrast to dissolved phases, which is very useful if one desires to regenerate immobilized phases from time to time.

Quality control must be carried out periodically to document any changes in retention or selectivity. For this purpose, test mixtures, e.g. the *Schurig* test [130] (α-pinene, *trans*- and *cis*-pinane, *rac*- and *meso*-2,3-butanediol, γ-valerolactone, 1-phenylethylamine, 1-phenylethanol, and 2-ethylhexanoic acid), are commonly used.

2.4 Application of CSPs in Chiral Gas Chromatography

2.4.1 CSPs with Diamide Selectors

2.4.1.1 Chirasil-Val

Chirasil-Val (Figure 2.6) is a very versatile CSP with excellent chiral recognition properties to separate amino acids derivatized as *N*(*O*,*S*)-pentafluoropropionyl/isopropyl esters. The selector is a diamide derived from D- or L-valine. Diamides tend to form insoluble solids because of the favored formation of self-association complexes and the high polarity of the two diamide bonds. To use a valine–diamide selector as a CSP in GC, it is necessary that the diamide selector is in a liquid, supercooled state. Gil-Av and co-workers modified their amino acid selector with lauryl ester groups [105] to prevent agglomeration and precipitation. Furthermore, the

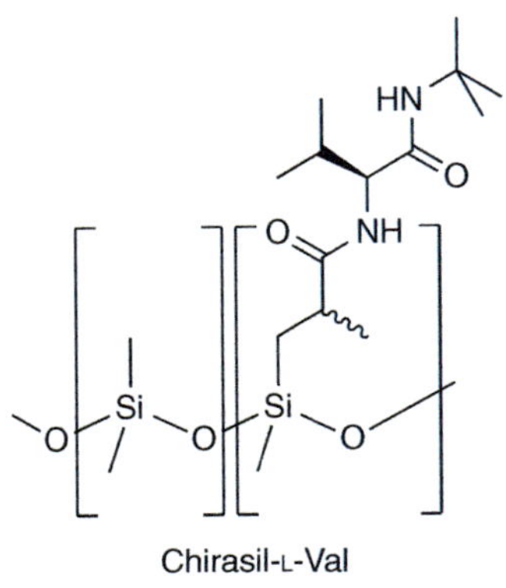

Figure 2.6 The CSP Chirasil-Val.

phase becomes nonpolar and improves the coating properties. Bayer and co-workers [108–110] utilized polysiloxanes as inert polymer backbone and attached the selector to a functionalized copolymer that circumvents formation of solid agglomerates, which is a problem if the selector is only dissolved in a polymer. The advantage of immobilization is that these phases are more stable and can be used in GC–MS measurements because column bleeding of impurities or fragments is reduced.

Chirasil-Val can be used between 70 and 240 °C. A broad range of chiral compounds can be separated such as amino acids (Figure 2.7), hydroxy acids, alcohols, amines, and biphenyl derivatives. L-Amino acids elute after the D-isomer because of the matching L-amino acid/L-amino acid interactions (Table 2.1).

König and co-workers developed similar CSPs for the separation of amines, amino acids, carbohydrates, and hydroxy carboxy acids. They used cyanoethyl polysiloxane (XE-60) [131, 132] and cyanopropyl polysiloxane (OV-225) as polymeric backbones and modified these with diamide selectors. They replaced the *tert*-butylamide by (*S*)- or (*R*)-1-phenylethylamide in the XE-60-Val phase (Figure 2.8).

The CSP XE-60-L-valine-(*S*)-1-phenylethylamide **1** can be used for the separation of carbohydrate enantiomers, either as trifluoroacetates or as trifluoroacetate/methylglycosides [131]. The OV-225-based CSP **2** has an inverse recognition pattern using benzyloxycarbonyl-valine or -L-leucine as selector (Figure 2.8) [133].

2.4.2 CSPs with CD Selectors

Cyclodextrin-based CSPs dominate chiral GC separations because of their very broad applicability and excellent separation factors α. Cyclodextrins form selective inclusion complexes with enantiomers and therefore they were considered as selectors in GC early on [134, 135]. Sybilska et al. and Smolková-Keulemansová et al. made pioneering contributions to this field by their investigation of interactions of packed CD columns with chiral analytes such as α- and β-pinene [136, 137] and achiral analytes [138–140]. The breakthrough in the field was due to liquid derivatized CDs and CDs dissolved in polysiloxane polymer became available. The problem of CDs and modified CDs is that they typically crystallize or even supercooled stationary phases form crystal seeds, which destroys the ability to interact with analytes in the gas phase. Therefore, it is necessary to keep the CDs dissolved in a polymer by immobilizing them to a polymer or by introducing them to long chain alkyl groups, which lowers the polarity and tendency of CDs to crystalize. These developments led to high separation efficiencies

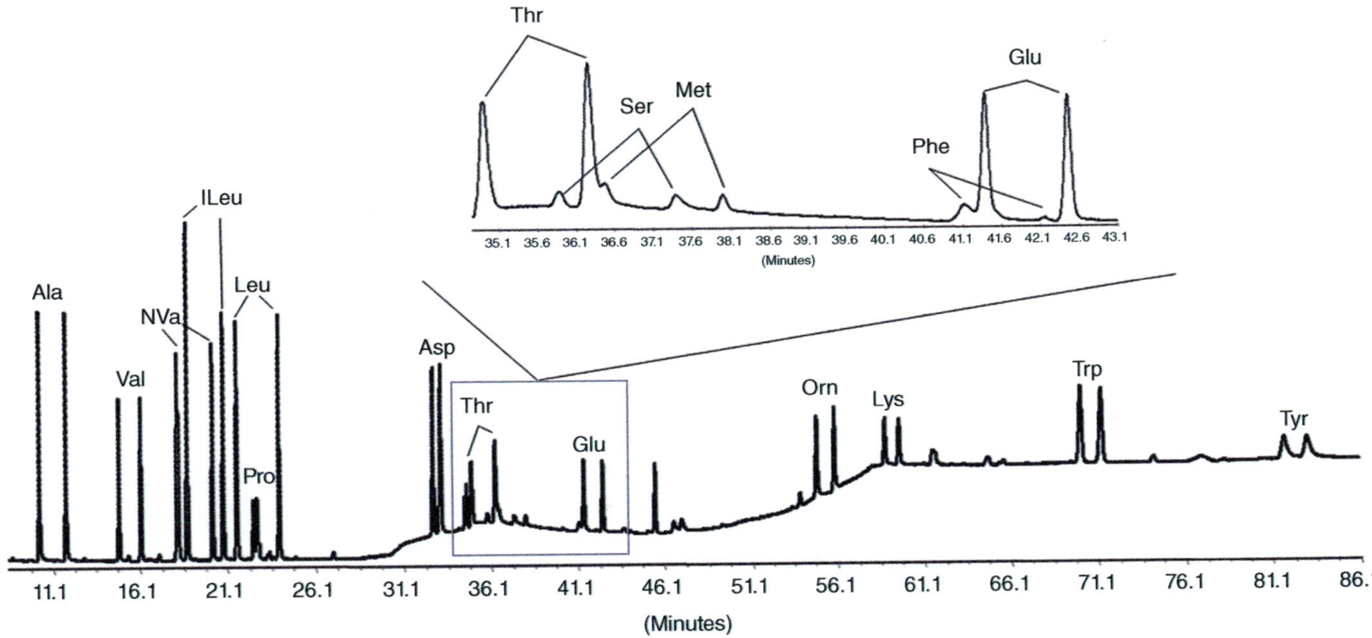

Figure 2.7 Amino acid separation using the CSP Chirasil-Val.

Table 2.1 Enantioseparations of canonical amino acids, as *N*(*O*,*S*)-pentafluoropropionyl/isopropyl esters derivatives.

Analyte	Separation temperature T (°C)	Separation factor α
Alanine, A	100	1.129
Valine, V	100	1.117
Isoleucine, I	100	1.158
Leucine, L	100	1.222
Threonine, T	100	1.069
Serine, S	120	1.052
Proline, P	100	1.005
Cysteine, C	120	1.060
Methionine, M	140	1.079
Phenylalanine, F	140	1.086
Ornithine, Orn	170	1.081
Lysine, K	170	1.065
Aspartic acid, D	120	1.031
Glutamic acid, E	140	1.108
Tryptophane, W	170	1.078
Tyrosine, Y	170	1.024

The enantiomer with D configuration is the first eluted (20 m Schott glass capillary column, 0.30 mm i.d., coated with L-Chirasil-Val).
Source: Adapted from Frank et al. [108].

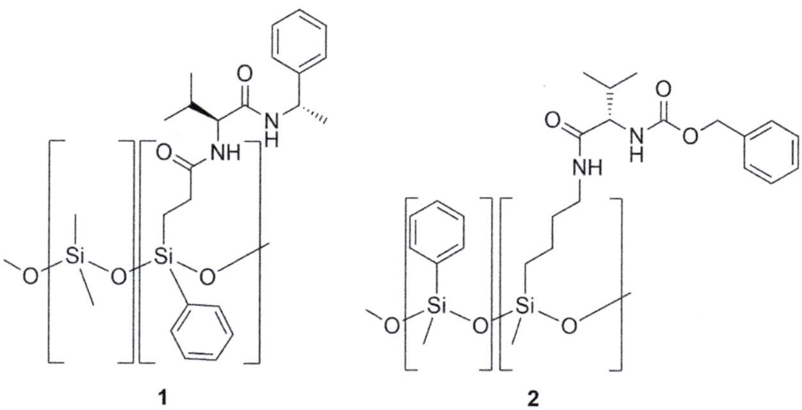

Figure 2.8 The CSPs XE-60-L-valine-(*S*)-1-phenylethylamide **1** and OV-225-benzyloxycarbonyl-L-valine **2**.

using coated capillary columns. The benefit of CDs is that they have a high number of chiral centers, which results in increased enantioselectivity by supramolecular interactions. CDs are thermally stable and can be operated at temperatures between 40 and 220 °C.

Cyclodextrins used as chiral selectors in GC are typically composed of 6 (α-CD), 7 (β-CD), or 8 (α-CD) D-glucopyranoside units in the ring, which are 1,4-linked. Depending on the ring size of the CDs and the substitution pattern of modified CDs, different analytes can be recognized. Selectively modified CDs were prepared and exhibit high enantioselectivities and can be broadly used for the separation of enantiomers by GC [141–144]. In the following the most important selectors and stationary phases are discussed.

2.4.2.1 Heptakis(2,3,6-tri-*O*-Methyl)-β-Cyclodextrin (Permethyl-β-Cyclodextrin)

Permethylated β-CD **3** is a widely used selector because of the straightforward preparation and the excellent enantioselectivity for a broad range of analytes. This selector and stationary phase are also commercially available, e.g. Astec CHIRALDEX B-PM, Supelco β-DEX 110, Supelco β-DEX 120 (Figure 2.9). This selector was initially directly and neatly coated on glass capillary columns and heated to higher temperatures. Upon cooling, a "supercooled liquid state" of the CSP was obtained [145].

The handling of such CSPs is not practical because the supercooled liquid state leads to spontaneous crystallization and loss of the enantioselectivity. To overcome this, Schurig et al. dissolved permethylated β-CD **3** in the semipolar polysiloxanes OV-1701. Dissolving the selector in a polysiloxane phase resulted in reproducible enantioselectivities and chromatographic efficiencies [146].

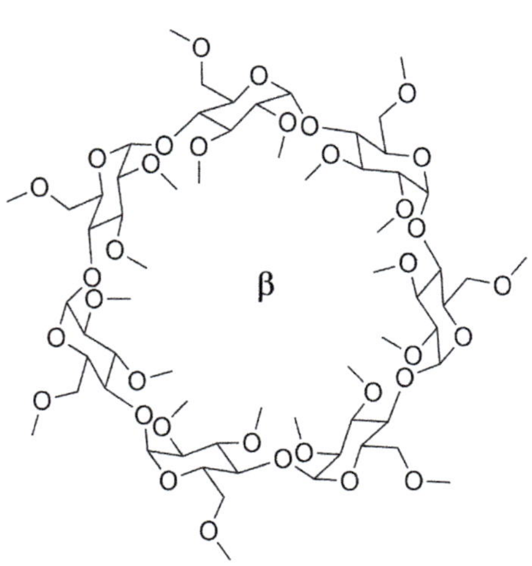

3

Figure 2.9 Chiral selector heptakis(2,3,6-tri-*O*-methyl)-β-CD (permethyl-β-CD) **3**.

Diluting selectors in polymers is still broadly used, and most commercially available chiral GC columns are based on this concept [147]. Permethylated β-CD **3** is the first choice for developing a chiral GC method because of the wide range of separable enantiomers and chiral compound classes.

2.4.2.2 Heptakis(2,3,6-tri-*O*-Methyl)-β-Cyclodextrin Immobilized to Hydrido Dimethyl Polysiloxane (Chirasil-β-Dex)

Similar to the case of Chirasil-Val, where the immobilization prevents self-association, precipitation, and crystallization of the chiral selector, permethylated β-CD **3** was attached via a polymethylene linker to polydimethylsiloxane. This immobilized version of permethylated β-CD **3** is known as Chirasil-β-Dex **4** [148–150]. An advantage of Chirasil-β-Dex is that it can be thermally immobilized on the inner capillary wall of fused silica capillaries. This allows to purge columns with solvents for removing non-volatile analytes or impurities and it opened the possibility to use these columns also in liquid separations. Early versions of Chirasil-Dex employed trimethylene [148], pentamethylene [149], and octamethylene [149, 150] spacers juxtaposed between the permethylated CD and the polysiloxane backbone (Figure 2.10). It was initially assumed that the trimethylene

Figure 2.10 Chirasil-β-Dex **4**: Immobilized heptakis(2,3,6-tri-*O*-methyl)-β-CD via an octenyl spacer to O-2 in the β-cyclodextrin.

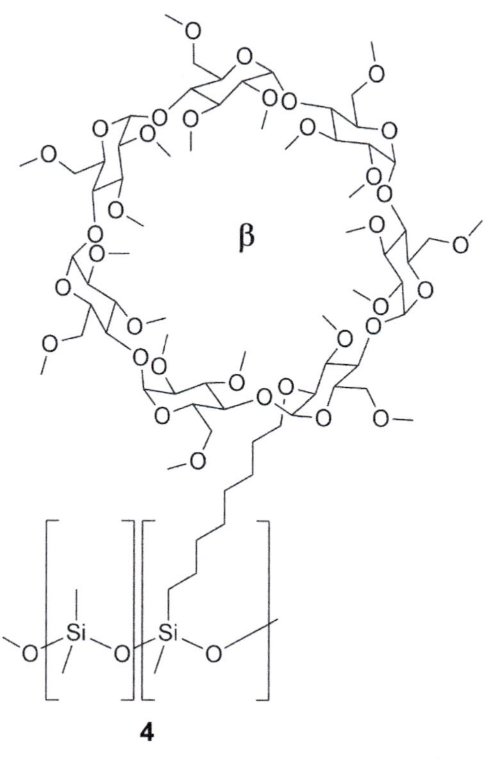

4

spacer was attached at the primary O-6 position of the cycloxdextrin ring. Systematic investigations indicated that the mono-oct-7-enyl ether linker was ideal [149, 151]. Comprehensive synthetic and analytical investigations of the permethylated mono-*O*-pent-1-enyl-β-CD obtained in dimethyl sulfoxide (DMSO) in the presence of sodium hydroxide demonstrated the ether linker to be predominantly formed (96%) at the O-2 position [152, 153].

It is interesting to mention that regioselective O-2-Chirasil-Dex columns show the highest separation factors *α*. Originally formulated as O-6, the non-regioselective Chirasil-Dex columns exist as a mixture of both O-2- and O-6-Chirasil-Dex with a O-2-Chirasil-Dex dominance and operate at optimum enantioselectivities in various applications of different chromatographic techniques [149]. The differences between O-2, O-3, and O-6-Chirasil-Dex regioisomers are not significant enough, which makes the tedious synthetic pathway to each selected regioisomers necessary.

Chirasil-β-Dex together with permethylated β-CD **3** are recommended as starting points for chiral separation method development due to their versatility. A very broad range of chiral analytes can be separated with excellent separation factors *α* and resolution *R* (Table 2.2). Chirasil-β-Dex is also highly versatile to separate compounds with stereogenic nitrogen such as *Tröger*'s base [154] or stereochemically labile aziridines and diaziridines [155–158]. Even stereochemically flexible compounds like 3,4-di-*tert*-butyl-1,3,4-oxadiazolidine [157] can be separated using the CSP Chirasil-β-Dex with an excellent separation factor *α* of 1.17 (Figure 2.11). Chiral GC offers not only the separation of chiral compounds but also the determination of thermodynamic and even kinetic data. In this case, the inversion barrier or in general the enantiomerization barrier can be determined by dynamic GC (DGC) because the interconversion of the enantiomers in the course of the separation leads to a pronounced plateau formation, which can be used to determine the rate constant of the interconversion process. The rate constants of enantiomerization can be directly accessed using the unified equation of chromatography [4]. The advantage is that modern GC instrumentation allows highly precise temperature control and thus a precise determination of kinetic parameters. By temperature-dependent measurements, the Gibbs free enthalpy $\Delta H^{\ddagger}$ and entropy $\Delta S^{\ddagger}$ can be determined. In this particular example the enantiomerization barrier $\Delta G^{\ddagger}$ and activation parameters using stopped-flow multidimensional GC (sfMDGC), the barrier was determined in the presence of the stationary phases Chirasil-β-Dex (90.0–160.0 °C) and Chirasil-Ni(II) (120.0–150.5 °C) using DGC and computer simulation. With the activation parameters in the inert gas phase, the catalytic effect of CSP of Chirasil-β-Dex and Chirasil-Ni(II) could be

Table 2.2 Enantioseparations on Chirasil-β-Dex **4** (25 m, 0.25 mm i.d., 250 nm film thickness).

Compound class	Analyte	Separation factor α	Optimum temperature T (°C)
Amino acid	*N*-TFA-Ala-*iso*-propyl ester	1.184	80
Amino acid	*N*-TFA-Ser-*iso*-propyl ester	1.184	90
Amino acid	*N*-TFA-Cys-*iso*-propyl ester	1.105	100
Amino acid	*N*-TFA-Phe-*iso*-propyl ester	1.022	120
Alcohol	1-(2-Methylphenyl) ethanol	1.343	110
Alcohol	1-(2-Bromophenyl) ethanol	1.440	130
Alcohol	1-(3-Methylphenyl) ethanol	1.124	100
Alcohol	1-Phenylethanol	1.140	100
Hydroxy carboxylic acid	Mandelic acid propyl ester	1.100	130
Ketone	3,5-Dimethyl-2-cyclohexen-1-on	1.110	90
Alkene	2,4-Dimethylheptene	1.034	40
Alkane	3-Methylheptane	1.034	40
Heterocycle	1,2-Di-*tert*-butyldiaziridine	1.440	80
Heterocycle	1-Isopropyl-2-*n*-propyldiaziridine	1.180	100
Heterocycle	1,2-Diisopropyldiaziridine	1.240	100
Heterocycle	3,4-Di-*tert*-butyl-1,3,4-oxadiazolidine	1.170	85

quantified. The highest barrier for a 5-membered heterocycle with stereogenic nitrogen was determined to be 131.8 kJ mol^{-1} at 56.5 °C.

Axially chiral allenes are an interesting class of compounds. The enantiomeric separation of some aliphatically substituted allenes on CD-containing stationary phases has been described in the literature [159], where no interconversion can be observed. Allenes that have a conjugated double bond to the cumulative one can – compared to aliphatically substituted allenes – interconvert much easier, since the π-electrons are delocalized [160]. The enantiomerization barrier of methyl 1,3-allendicarboxylate was determined at 120 °C by DGC in the presence of Chirasil-β-Dex ($\Delta G^{\ddagger}$(120 °C) = 114.5 kJ mol^{-1}) (cf. Figure 2.12), Chirasil-Nickel(II) ($\Delta G^{\ddagger}$(120 °C) = 117.5 kJ mol^{-1}), and in the achiral environment by sfMDGC ($\Delta G^{\ddagger}$(120 °C) = 128 kJ mol^{-1}).

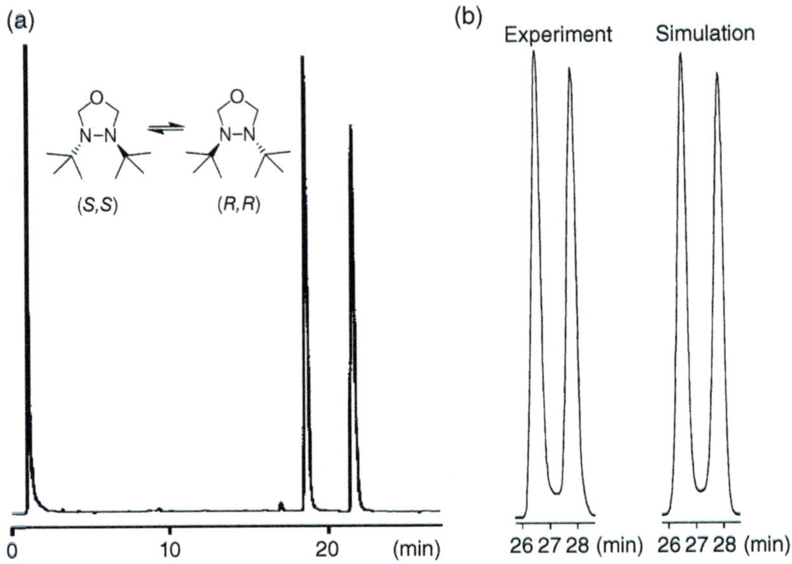

Figure 2.11 (a) Enantiomeric separation of 3,4-di-*tert*-butyl-1,3,4-oxadiazolidine at 80 °C ($\alpha = 1.17$). 20 m Chirasil-β-Dex (0.25 mm i.d., film thickness 0.4 μm). Carrier gas: H_2. (b) Determination of the enantiomerization barrier of 3,4-di-*tert*-butyl-1,3,4-oxadiazolidine by DGC. 25 m Chirasil-β-Dex (0.25 mm i.d., film thickness 0.5 μm). Carrier gas: N_2. Separation at 130 °C, $p = 0.2$ bar (left), simulation of the profile using the theoretical plate model (right).

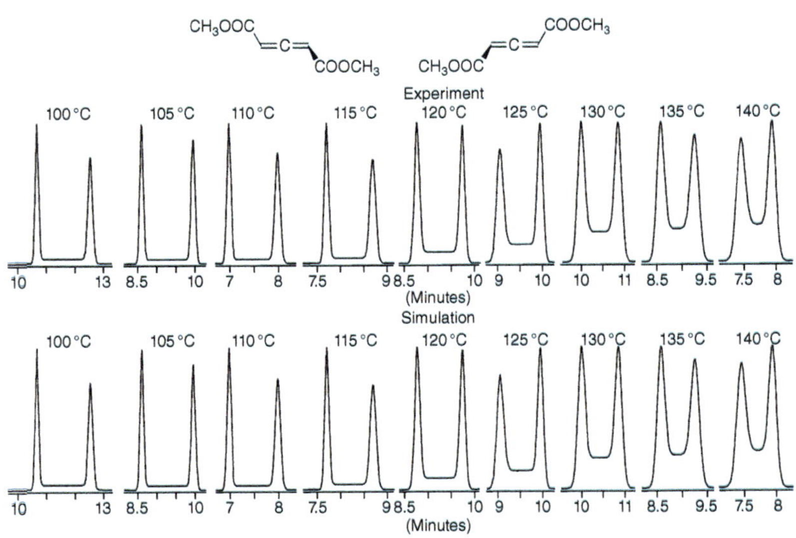

Figure 2.12 DGC of 1,3-allendicarboxylic acid dimethyl ester at various temperatures. 12.5 m Chirasil-β-Dex (0.25 mm i.d., film thickness 0.4 μm). Carrier gas: He. Top: experimental chromatograms. Bottom: simulated elution profiles (ChromWin, SM+).

Spiroketals are a class of compounds widely found in nature. Examples include the reveromycins, which show potential antiproliferative activity against certain human tumor cells; ionophore antibiotics such as calcimycin (A-23187) from *Streptomyces chartreusensis* and routiennocin from *Streptomyces routiennii*; cell growth inhibitors such as the cephalostatins of the South African sea worm *Cephalodiscus gilchristi*; steroids in solanaceous plants (solanaceae) such as diosgenin, yamogenin, hecogenin, and willagenin; the toxic metabolites calyculins A, B, and C of the marine sponge *Discodermis calyx*; and volatile pheromones for communication between insects [161, 162]. Francke et al. isolated [163] chalcogran ((2RS, 5RS)-2-ethyl-1,6-dioxaspiro[4.4]nonane) (cf. Figure 2.13), which is the major component of the aggregation pheromone of the bark beetle *Pityogenes chalcographus* (copper engraver), a pest of Norway spruce. Chalcogran can be separated in all four stereoisomers by using Chirasil-β-Dex [164]. With increasing temperature between 70 and 120 °C, chalcogran shows typical interconversion profiles between the epimeric pairs (2*R*,5*S*)/(2*R*,5*R*) and (2*S*,5*S*)/(2*S*,5*R*) (cf. Figure 2.13). Since there is no superposition of the individual interconversion profiles, it can be clearly shown that epimerization occurs exclusively at the spiro atom (C5).

2.4.2.3 Heptakis(2,6-di-*O*-Methyl-3-*O*-Pentyl)-β-Cyclodextrin

As already pointed out, there is also the strategy to modify CDs with long chain alkyl groups to lower the melting point. This was applied in the development of the CSP heptakis(2,6-di-*O*-methyl-3-*O*-pentyl)-β-CD **5**, which is also commercially available as MEGA-DEX DMP-Beta (Figure 2.14). This CSP was developed by König [165] targeting the separation of chiral compounds in essential oils and by Bicchi [166].

Bicchi et al. reported better reproducibility and higher performance consistency of the CSPs based on heptakis(2,6-di-*O*-methyl-3-*O*-pentyl)-β-CD **5** dissolved in OV-1701 polysiloxane compared to their permethylated β-CD analogues [159]. It should be noted that heptakis(2,6-di-*O*-methyl-3-*O*-pentyl)-β-CD **5** was designed for the coating of glass capillaries. For the use with modern fused-silica capillaries, the use of a polymer like OC-1701 or OV-1701OH is recommended to achieve reproducible coatings. Heptakis(2,6-di-*O*-methyl-3-*O*-pentyl)-β-CD **5** is an excellent choice for the separation of chiral carboxylic acids, hydroxy carboxylic acids, epoxides, terpenes, and terpene alcohols while exhibiting more moderate separation factors for compound classes such as amino acids.

Selected separation parameters of chiral compounds separated on heptakis(2,6-di-*O*-methyl-3-*O*-pentyl)-β-CD by König [159] are summarized in Table 2.3.

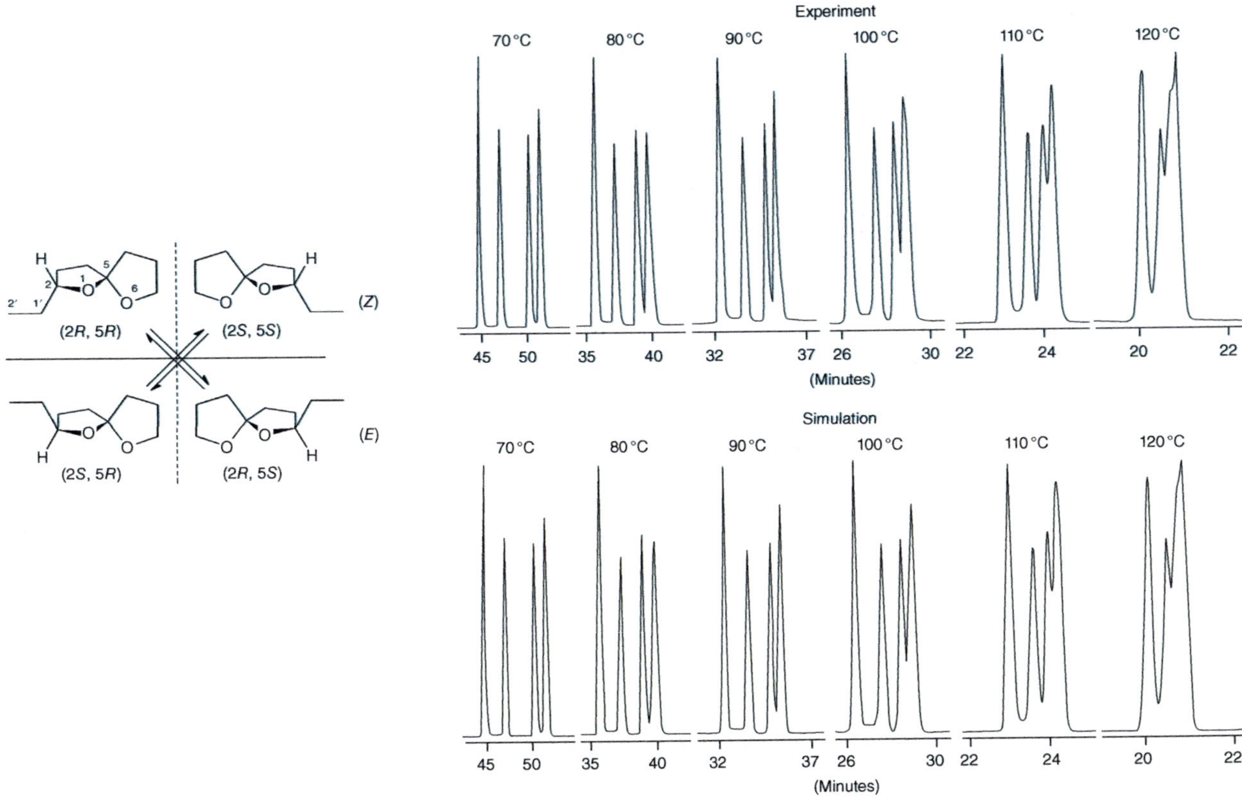

Figure 2.13 DGC of chalcogran at different temperatures. 50 m Chirasil-Dex (0.25 mm i.d., film thickness 0.3 μm). Carrier gas: H_2. Top: experimental chromatograms. Bottom: simulated elution profiles (ChromWin, SM+).

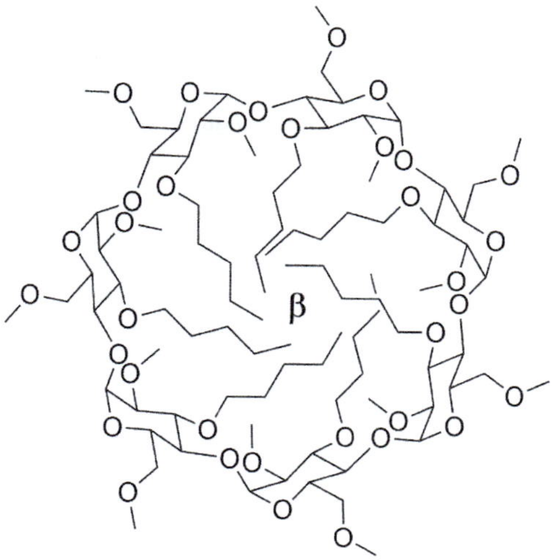

5

Figure 2.14 Chiral selector heptakis(2,6-di-*O*-methyl-3-*O*-pentyl)-β-CD **5**.

Table 2.3 Enantiomeric separation of a series of racemates: The amino acids were derivatized as *N*-trifluoroacetyl amino acid methyl esters.

Compound class	Analyte	Separation factor α	Optimum temperature T (°C)
Amino acid	Valine, V	1.037	80
Amino acid	Proline, P	1.075	100
Amino acid	Methionine, M	1.023	100
Amino acid	Phenylalanine, F	1.031	110
Amino acid	Aspartic acid, D	1.041	105
Amino acid	Glutamic acid, E	1.050	110
Heterocycle	5-Phenylhydantoin[a]	1.287	175
Terpene	α-Pinene[b]	1.057	70
Terpene	β-Pinene	1.021	70
Terpene	Limonene	1.070	70
Terpene	Camphene[b]	1.087	70
terpene	Sabinene	1.119	70
Terpene	α-Thujene	1.059	70
Terpene	α-Copaene	1.094	100

(Continued)

Table 2.3 (Continued)

Compound class	Analyte	Separation factor α	Optimum temperature T (°C)
Cycloalkene	1-*tert*-butyl-1,2 cyclooctadiene	1.411	80
Hydroxy carboxylic acid	2-Hydroxyoctanoic acid (OMe)[c]	1.602	80
Hydroxy carboxylic acid	2-Hydroxydodecanoic acid (OMe)[c]	1.153	125
Alcohol	1-Phenyl-3-butanol[c]	1.104	95
Alcohol	1-Phenylethanol (TFA)	1.111	90
Alcohol	1-Octen-3-ol	1.049	98
Alcohol	2-Octanol	1.016	90
drug	Ibuprofen (OMe)	1.042	135
Hydroxy carboxylic acid	Malic acid (OMe)	1.149	100
Hydroxy carboxylic acid	Mandelic acid (OMe)[c]	1.120	105
Terpene	Myrtenol[d]	1.078	105
Terpene	Menthol[d]	1.092	100
Terpene	Linalool	1.082	95
Terpene	Isoborneol	1.030	95
Oxirane	Phenyloxirane	1.127	105
Oxirane	(*E*)-1,2-Diphenyloxirane[c]	1.108	135
Terpene	Citronellal[d]	1.027	75
Terpene	Menthone[d]	1.098	90
Terpene	Carvone[d]	1.068	100

Trifluoroacetylated alcohols are marked in brackets as TFA and methylated carboxylic acids as OMe. 25 m fused-silica capillary column, 0.25 mm i.d., coated with heptakis(2,6-di-*O*-methyl-3-*O*-pentyl)-β-CD **5**.
[a] Column length: 4.5 m.
[b] Column length: 50 m.
[c] Column length: 8 m.
[d] Column length: 30 m.
Source: Adapted from Bertsch et al. [160].

It is interesting to mention that CSPs with a variation in the substitution pattern can show very different selectivities and are useful to separate enantiomers. An example is the related CSP octakis(6-*O*-methyl-2,3-di-*O*-pentyl)-γ-CD, which can be used to separate the enantiomers of the organophosphorus pesticide fonofos [167], which contains stereogenic phosphorous. There is considerable interest in the separation of the enantiomers of agrochemicals because often the toxicity and the metabolism differs for the enantiomers, which requires a stereospecific analytical technique to study such effects.

2.4.2.4 Hexakis-(2,3,6-tri-*O*-Pentyl)-α-Cyclodextrin

Hexakis-(2,3,6-tri-*O*-pentyl)-α-CD **6** (Figure 2.15) is a CD-based CSP for GC capable of separating various chiral compounds such as carbohydrates, polyols, diols, hydroxy acid esters, (epoxy-) alcohols, glycerol derivatives, spiroketals, ketones, alkylhalogenides, and even carbohydrates, derivatized as trifluoroacetylates. This CSP is also commercially available as e.g. Lipodex A [168, 169]. It should be mentioned that the polarity of hexakis(2,3,6-tri-*O*-pentyl)-α-CD **6** is comparable to that of the methyl-phenyl polysiloxane OV-17.

Excellent separation factors *α* can be even achieved for carbohydrates and amino acids as shown in Table 2.4. It must be pointed out that the analysis of carbohydrate is often complicated by the formation of α/β anomers and formation of furanosides and pyranosides in the same mixture. For the amino acids, the D-amino acids are eluted prior to the L-amino acids.

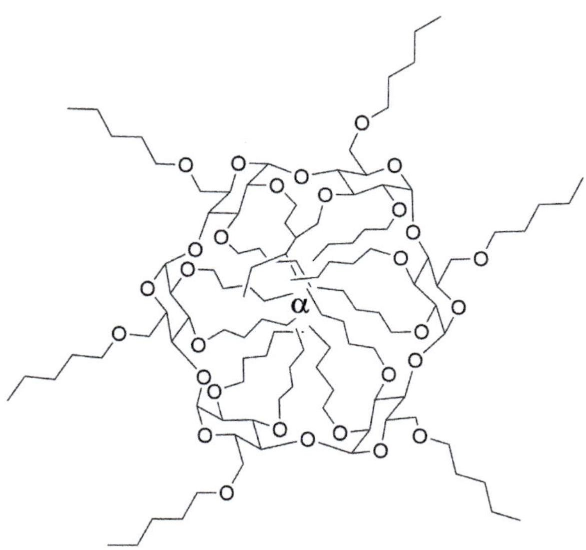

6

Figure 2.15 Chiral selector hexakis-(2,3,6-tri-*O*-pentyl)-α-CD **6**.

Table 2.4 Enantioseparation of trifluoroacetylated carbohydrates and their α/β anomers, polyols, and amino acid derivatives on a 20 m Pyrex glass capillary column, 0.25 mm i.d., coated with hexakis-(2,3,6-tri-*O*-pentyl)-α-CD **6**.

Analyte	Separation factor α	First eluted	Optimum temperature T (°C)
α-Glucose	1.119	L	100
β-Glucose	1.140	L	100
α-Galactose	1.070	L	100
β-Galactose	1.080	L	130
α-Allose	1.171	D	120
β-Allose	1.064	D	120
α-Mannose	1.000	—	100
β-Mannose	1.110	L	100
α-Gulose	1.000	—	120
β-Gulose	1.043	D	120
α-Talose	1.099	D	110
β-Fucose (furanoside)	1.039	L	100
α-Methylgalactoside	1.091	D	100
α-Methylglucoside	1.035	L	100
α-Methylmannoside	1.051	L	100
α-Methylidoside	1.040	D	90
α-Methylriboside	1.075	D	110
Sorbitol	1.042	D	100
Mannitol	1.019	D	90
Arabinitol	1.175	D	100
1,5-Anhydrofucitol	1.035	D	80
1,5-Anhydrolyxitol	1.064	D	80
1,5-Anhydroarabinitol	1.074	D	80
Alanine, A	1.036	D	80
Phenylalanine, F	1.028	D	130
Threonine, T	1.074	D	95

Source: Adapted from Francke et al. [163].

2.4.2.5 Heptakis(2,3,6-tri-*O*-Pentyl)-β-Cyclodextrin

Very similar to hexakis-(2,3,6-tri-*O*-pentyl)-α-CD also β-CD has been modified with pentyl groups to form the selector heptakis(2,3,6-tri-*O*-pentyl)-β-CD **7** (Figure 2.16), which is also commercially available as e.g. Lipodex C [170].

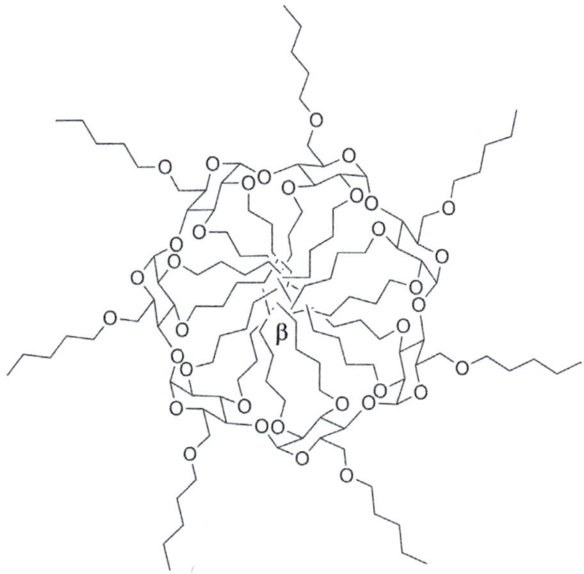

7

Figure 2.16 Chiral selector heptakis-(2,3,6-tri-*O*-pentyl)-β-CD **7**.

This perpentylated β-CD shows in comparison to the α-CD derivative some remarkable enantioselectivities in the separation of chiral compounds and has a good thermal stability of up to 200 °C. A number of inseparable alcohols can be separated on this phase, as well as cyanohydrins and carbohydrates. One example is the alcohol pheromone grandisol **8** (Figure 2.17). Perpentylated β-CD is in particular very versatile to separate enantiomers of nonpolar olefins and alkyl halides.

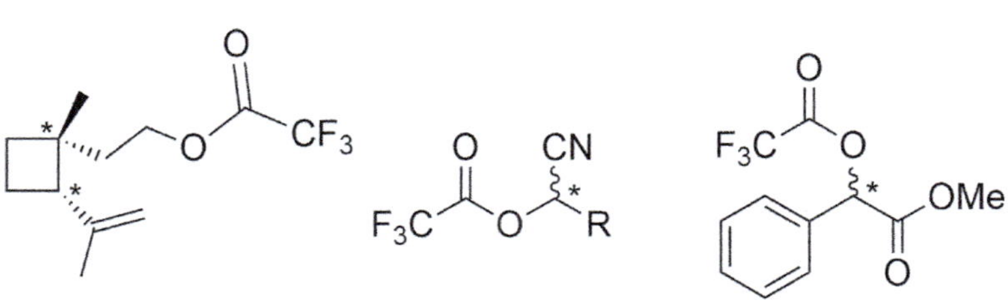

8 **9** **10**

Figure 2.17 Trifluoroacetylated derivatives of grandisol **8**, cyanohydrins **9**, and mandelic acid methyl ester **10**.

Even though there are many options to chromatographically separate trifluoroacetylated carbohydrates using CD phases, the very best separation of 6-deoxy sugars is achieved on perpentylated β-CD **7**. Another very interesting compound class are cyanohydrins **9** (Figure 2.17) because they represent important precursors in the synthesis of sugars (Kiliani-Fischer reaction), amino, and hydroxyl acids as their enantiomeric purity is of great importance [171]. Furthermore, homologues of α-hydroxy acid methyl esters can be separated with the CSP heptakis(2,3,6-tri-*O*-pentyl)-β-CD **7**. The advantage of CD phases is that they typically cover whole families of related compounds. For example, in 3- or 4-position, hydroxy- and methoxy-substituted derivatives of mandelic acid **10** (Figure 2.17) can be separated with the CSP heptakis(2,3,6-tri-*O*-pentyl)-β-CD **7**.

Sybilska et al. investigated enantioselective interaction of unmodified α-CD with olefins and saturated hydrocarbons like β-pinene and its hydrogenation products [172]. König et al. reported that perpentylated β-CD **7** allows to separate many other chiral olefins and dienes with high efficiency and selectivity. It must be mentioned that structural changes can completely change the chiral olefins recognition. 3-Methyl-1-hexene **11** (Figure 2.18) and 3-methyl-cyclohexene **12** (Figure 2.18) can be separated very well, while for terpenes such as limonene **13**, no separation could be achieved (Figure 2.18).

In general, the size of the macrocycle in pentylated α- and β-CD affects the interaction with the analyte.

Excellent enantioselectivities of heptakis(2,3,6-tri-*O*-pentyl)-β-CD **7** are observed for chiral chloro- and bromoalkanes. The enantiomers of 2-chlorobutane **14** up to the homologous compound 2-chlorooctane **15** are completely separated (Figure 2.18), however, with decreasing α-values. The

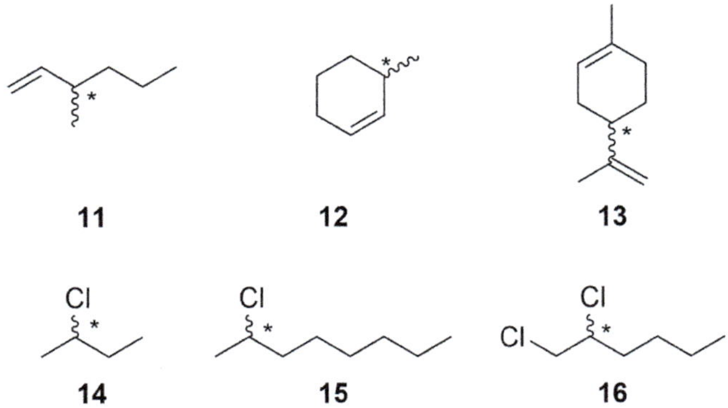

Figure 2.18 Chemical structures of 3-methyl-1-hexene **11**, 3-methyl-cyclohexene **12**, limonene **13**, 2-chlorobutane **14**, 2-chlorooctane **15**, and 1,2-dibromohexane **16**.

disubstituted 1,2-dibromohexane **16** can also be separated, but not the higher homologues 1,2-dibromoheptane and 1,2-dibromoctane [166].

2.4.2.6 Hexakis-(3-*O*-Acetyl-2,6-di-*O*-Pentyl)-α-Cyclodextrin

Hexakis-(3-*O*-acetyl-2,6-di-*O*-pentyl)-α-CD **17** (Figure 2.19a) is a more polar CD derivative compared to the completely alkylated derivatives [173, 174], which can be used in a temperature range between room temperature and 200 °C. This CD derivative is very useful for the separation of chiral diols, lactones, cyclic carbonates, amino alcohols, aldol compounds (as O-TFA derivative), and asymmetric glycerol derivatives. This CSP is also commercially available as e.g. Lipodex B.

The separation of lactones is summarized in Table 2.5 and shows good separation factors α for hexakis-(3-*O*-acetyl-2,6-di-*O*-pentyl)-α-CD **17**.

It is important to note that α- and β-CDs with the same substitution pattern 3-*O*-acetyl and 2,6-di-*O*-pentyl, i.e. hexakis-(3-*O*-acetyl-2,6-di-*O*-pentyl)-α-CD and heptakis(3-*O*-acetyl-2,6-di-*O*-pentyl)-β-CD, respectively, show remarkable differences in their enantioselective properties.

2.4.2.7 Heptakis(3-*O*-Acetyl-2,6-di-*O*-Pentyl)-β-Cyclodextrin

In the series of 3-*O*-acetyl-2,6-di-*O*-pentylated CDs, the CSP heptakis(3-*O*-acetyl-2,6-di-*O*-pentyl)-β-CD **18** (Figure 2.19b) is a commonly used CSP in capillary GC because of the broad applicability in chiral separations. This CSP is also commercially available as e.g. Lipodex D [175]. High enantioselectivity is observed for trifluoro-acetylated α- and β-chiral amines, amino alcohols, α- and β-amino acid esters, and cyclic *trans*-diols.

König's studies have shown that many functional groups of CD molecules offer many possibilities of further modifications. Properties such as proper solubility in organic solvents, low melting points, and high temperature stability were achieved by introducing alkyl substituents. Since the hydroxyl groups in positions 2, 3, and 6 of the hydroglucose residues have different reactivities, it is generally possible to introduce various types of substituents into the molecule.

For example, after pentylation of the hydroxyl groups in positions 2 and 6, the remaining and less reactive hydroxyl groups in position 3 can be acetylated. Thus, opportunities for weak hydrogen bonding or dipole–dipole interactions can be introduced without altering significantly the hydrophobic character of the perpentylated CD. As a result of the above approach, in the case of the 3-acetyl-2,6-di-pentyl derivative of β-CD, a strong increase in enantioselectivity is observed for substrates with the ability to form hydrogen bonds. This type of modification gives rise to exceptionally large separation factors for amines and amino compounds with additional functional

(a)

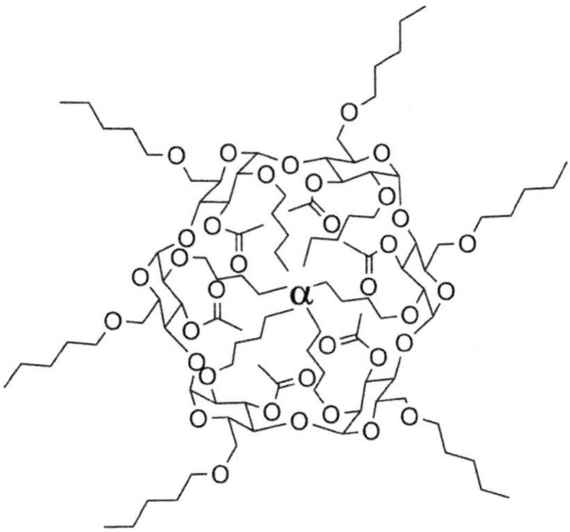

17

(b)

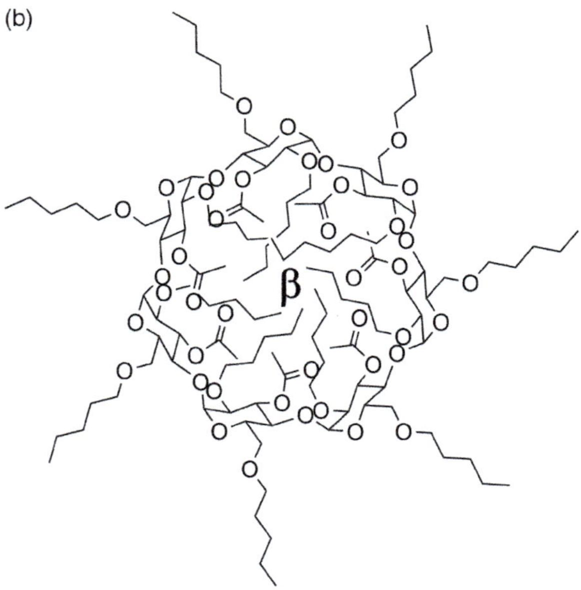

18

Figure 2.19 Chiral selector (a) hexakis-(3-*O*-acetyl-2,6-di-*O*-pentyl)-α-CD and (b) heptakis(3-*O*-acetyl-2,6-di-*O*-pentyl)-β-CD.

Table 2.5 Enantioseparation of target chiral lactones.

Analyte	Separation factor α	Separation temperature T (°C)
2-Methylbutyrolactone	1.106	120
2-Ethylbutyrolactone	1.063	120
2-*n*-Propylbutyrolactone	1.031	120
2-*n*-Pentylbutyrolactone	1.036	120
2-Methylvalerolactone	1.026	120
3-Ethylvalerolactone	1.035	140
(*E*)-4-*n*-butyl-3-methylbutyrolactone	1.060	140
(*Z*)-4-*n*-butyl-3-methylbutyrolactone	1.021	140
(*E*)-3,4-dimethylbutyrolactone	1.121	140
(*Z*)-3,4-dimethylbutyrolactone	1.094	140
4-Methylbutyrolactone	1.108	150
4-Ethylbutyrolactone	1.104	150
4-*n*-Propylbutyrolactone	1.048	150
4-*n*-Butylbutyrolactone	1.048	150
4-*n*-Pentylbutyrolactone	1.066	150
4-*n*-Hexylbutyrolactone	1.035	170
4-*n*-Octylbutyrolactone	1.040	170
4-*n*-Decylbutyrolactone	1.044	170

38 m Pyrex glass capillary column, 0.25 mm i.d., coated with hexakis-(3-*O*-acetyl-2,6-di-*O*-pentyl)-α-CD) **17**.
Source: Adapted from König [159].

groups. In Figure 2.20, six representative examples of such enantiomer separations are listed.

The separation of the amphetamine derivative **22**, pholedrine **20**, or the aminomethyl-norbornenes **23** and **24** with the center of chirality in β-position to the functional group can be achieved on heptakis(3-*O*-acetyl-2,6-di-*O*-pentyl)-β-CD **18**.

Besides these amino compounds, high enantioselectivity is obtained for diols including propane-1,2-diol or butane-1,3-diol.

2.4.2.8 Octakis(3-*O*-Butyryl-2,6-di-*O*-Pentyl)-γ-Cyclodextrin

Octakis(3-*O*-butyryl-2,6-di-*O*-pentyl)-γ-CD **25**, commercially available as e.g. Lipodex E and Astec CHIRALDEX G-BP [176], is one of the most

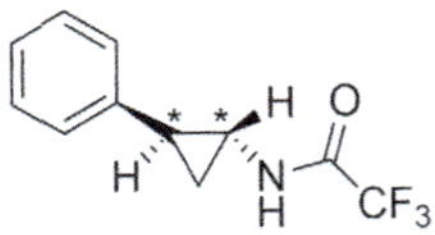

19

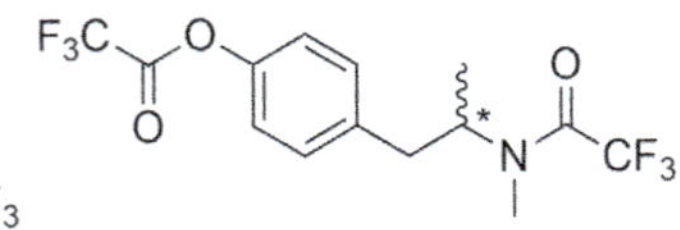

20

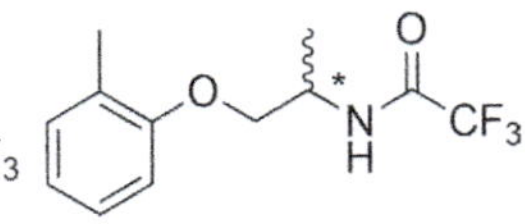

21

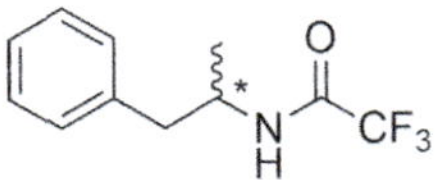

22

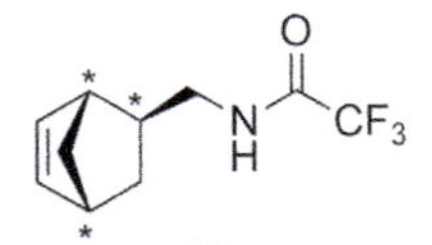

23

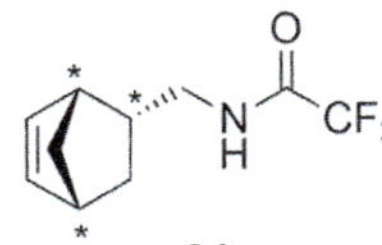

24

Figure 2.20 Chemical structures of tranylcypromine **19**, pholedrin **20**, mexiletine **21** and amphetamine **22**, exo-**23** and endo-5-aminomethyl-norbornene **24**.

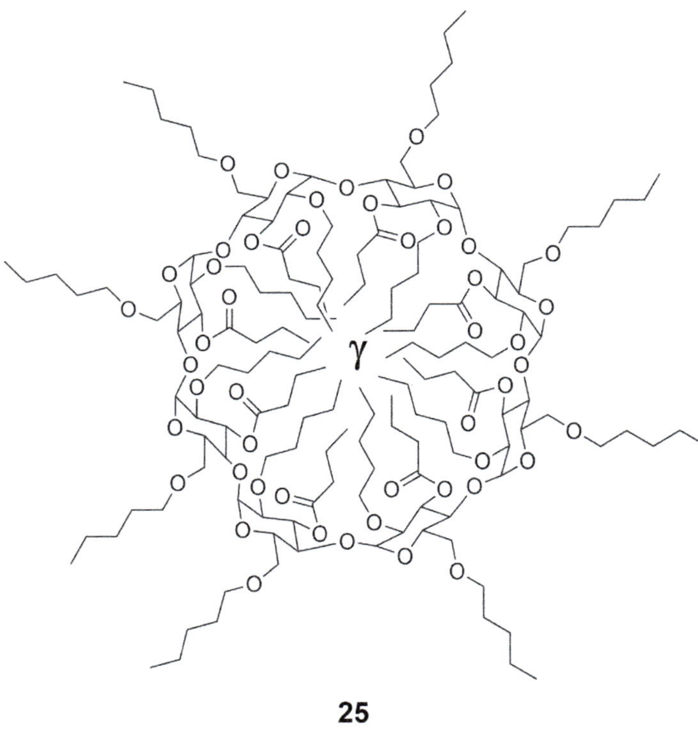

25

Figure 2.21 Chiral selector octakis(3-*O*-butyryl-2,6-di-*O*-pentyl-γ-CD) **25**.

versatile CSPs [169] (Figure 2.21). This CSP can separate a variety of chiral compounds, i.e. most of the canonical amino acids and some unusual amino acids, amines, alcohols, diols, triols, ethers, α- and β-hydroxy carboxylic acids, ketones, bicyclic and tricyclic acetals, alkyl halogen compounds, lactones, and chiral cyclopropanes. Amino acids are typically analyzed as *N*-trifluoroacetyl amino acid methyl esters and in most cases the D-α-amino acids are eluted prior to the L-α-amino acids (reversed elution order for proline) with excellent separation factors α. Permethylated-β-CD or Chirasil-β-Dex, and octakis(3-*O*-butyryl-2,6-di-*O*-pentyl)-γ-CD are the most promising CSPs to start developing a GC separation method for chiral compounds.

Octakis(3-*O*-butyryl-2,6-di-*O*-pentyl)-γ-CD **25** was originally developed for use as CSP in pyrex glass columns. Its use as CSP in fused-silica columns requires dilution of the selector (5–30% w/w) in the polysiloxane PS 255 (copolymer of dimethylsiloxane and 1–3% methylvinylsiloxane).

Table 2.6 summarizes the separation of *N*-trifluoroacetyl amino acid methyl esters on octakis(3-*O*-butyryl-2,6-di-*O*-pentyl)-γ-CD reinforcing its complementarity to the Chirasil-Val CSP.

Table 2.6 Enantioseparation of amino acids, as *N*-trifluoroacetyl amino acid methyl esters derivatives, hydroxy carboxylic acids (as *O*-trifluoroacetyl/methyl esters), and alcohols as *O*-trifluoroacetyl derivatives.

Compound class	Analyte	Separation factor α	Separation temperature T (°C)
Amino acid	Alanine, A	1.100	130
Amino acid	Valine, V	1.148	130
Amino acid	Isoleucine, I	1.139	130
Amino acid	Leucine, L	1.126	130
Amino acid	Threonine, T	1.069	130
Amino acid	Serine, S	1.153	130
Amino acid	Proline, P	1.173	130
Amino acid	Cysteine, C	1.042	160
Amino acid	Methionine, M	1.053	160
Amino acid	Phenylalanine, F	1.038	170
Amino acid	Ornithine, Orn	1.112	180
Amino acid	Lysine, K	1.184	180
Amino acid	Aspartic acid, D	1.163	160
Amino acid	Glutamic acid, E	1.065	160
Amino acid	Tryptophane, W	1.014	190
Amino acid	Tyrosine, Y	1.024	175
Amino acid	α-Aminobutyric acid	1.107	150

(Continued)

Table 2.6 (Continued)

Compound class	Analyte	Separation factor α	Separation temperature T (°C)
Hydroxy carboxylic acid	Lactic acid	1.320	70
Hydroxy carboxylic acid	2-Hydroxybutyric acid	1.065	100
Hydroxy carboxylic acid	3-Hydroxybutyric acid	1.926	80
Hydroxy carboxylic acid	2-Hydroxyisovaleric acid	1.024	100
Hydroxy carboxylic acid	2-Hydroxycaproic acid	1.125	100
Hydroxy carboxylic acid	2-Hydroxyoctanoic acid	1.121	100
Hydroxy carboxylic acid	2-Hydroxydodecanoic acid	1.021	150
Hydroxy carboxylic acid	Tartaric acid	1.081	110
Hydroxy carboxylic acid	Malic acid	1.148	130
Hydroxy carboxylic acid	Mandelic acid	1.090	140
Alcohol	2-Methyl-1-penten-3-ol	1047	55
Alcohol	1-Octen-3-ol	1.053	75
Alcohol	2-Octanol	1.021	60
Terpene	Myrtenol	1.025	90
Terpene	Borneol	1.031	90
Terpene	β-Citronellol	1.014	90

50 m Pyrex glass capillary column, 0.25 mm i.d., coated with octakis(3-*O*-butyryl-2,6-di-*O*-pentyl)-γ-CD) **25**.

An outstanding separation factor (α of 10.6 at 26 °C) was obtained using octakis(3-*O*-butyryl-2,6-di-*O*-pentyl)-γ-CD dissolved in PS 255 for the degradation product of the anesthetic sevoflurane (2-(fluoromethoxy)-1,1,1,3,3,3-hexafluoropropane) compound B (2-(fluoromethoxy)-3-methoxy-1,1,1,3,3-pentafluoropropane) **26** (Figure 2.22), which forms under elimination and CH_3OH addition [170–172].

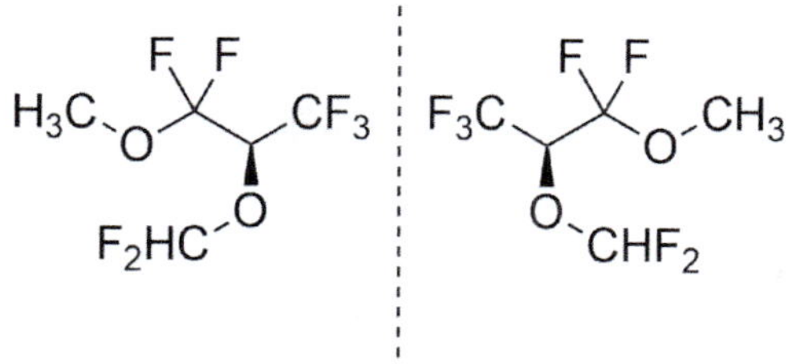

(*S*)-26 **(*R*)-26**

Figure 2.22 Enantiomers for the compound B (2-(fluoromethoxy)-3-methoxy-1,1,1,3,3-pentafluoropropane **26**) degradation product of sevoflurane.

2.4.2.9 Hexakis/Heptakis/Octakis(2,6-di-*O*-Alkyl-3-*O*-Trifluoroacetyl)-α/β/γ-Cyclodextrins

A further modification of the acetylated CDs are 2,6-dialkyl-3-trifluoroacetyl functionalized CDs, which were introduced by Schurig et al. as the CSP heptakis(2,6-di-*O*-methyl-3-*O*-trifluoroacetyl)-β-CD **27** (Figure 2.23) [177, 178] and the functionalized α- and γ-CDs by Bicchi et al. [166, 179, 180]. Armstrong et al. prepared 2,6-di-*O*-pentyl-3-*O*-trifluoroacetyl functionalized α-, β-, and γ-CDs, which are commercially available as e.g. Astec CHIRALDEX A-TA, Astec CHIRALDEX B-TA, and Astec CHIRALDEX C-TA [181].

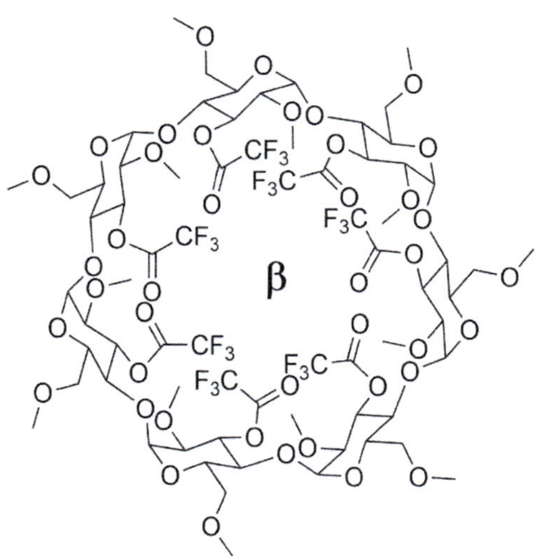

27

Figure 2.23 Chiral selector heptakis(2,6-di-*O*-methyl-3-*O*-trifluoroacetyl)-β-CD **27**.

The dilution of the 2,6-di-*O*-methyl-3-*O*-trifluoroacetyl-CDs in polysiloxane OV-1701 and their successful application as CSPs to the separation of chiral lactones, dioxolanes, spiroketals, and terpenes, often showing complementary selectivity to their permethylated analogues was achieved by Bicchi and Schurig [175–178]. However, the column performance for α- and β-CDs functionalized with 2,6-di-*O*-methyl-3-*O*-trifluoroacetyl decreased faster when compared to their permethylated counterparts [177, 178]. The susceptibility to hydrolysis of trifluoroacetylated columns is an important point of concern that may lead to reduced column lifetimes, especially under higher separation temperatures [177, 178]. A reactivation is possible by on-column reaction with trifluoroacetic anhydride.

Armstrong and co-workers applied the 2,6-di-*O*-pentyl-3-*O*-trifluoroacety l-α/β/γ-CD selectors in the separation of more than 150 racemic compounds. Selected results of the screening are summarized in Table 2.7 [178].

The CDs functionalized with 2,6-di-*O*-pentyl-3-trifluoroacetyl present high separation ability for many compound classes, such as amino alcohols, amines, alcohols, diols, carboxylic acid esters, epoxides, and lactones. Especially the γ-CD phase proved to be the most versatile CSP separating over 120 chiral compounds.

2.4.2.10 Heptakis(2,3-di-*O*-Acetyl-6-*O*-*tert*-Butyldimethylsilyl)-β-Cyclodextrin (DIAC-6-TBDMS-β-CD)

Per-*O*-alkylated and dialkylated/monoacylated derivatives of CDs are broadly used CSPs for the separation of chiral flavor compounds [141]. The use of 6-*tert*-butyldimethylsilylated and 2,3-diacylated β-CD (Figure 2.24) [182], which are commercially available as e.g. Supelco beta-DEX 225, MEGA-DEX DAC-Beta, or 2,3-dialkylated β-CD derivatives [183] was reported by Mosandl et al. for the simultaneous chiral separation of a wide range of chiral flavor and fragrance compounds with different functionalities [184].

Chiral lactones and 2-alkylalkanols (e.g. 2-methylbutanol and 1-octen-3-ol) can be separated using DIAC-6-TBDMS-β-CD **28** dissolved in the polysiloxane PS086 (50%/50%). This CD derivative was also applied to the chiral GC separation of rose oxides, such as citronellol, linalool, and carvone.

2.4.2.11 Heptakis(2,3-di-*O*-Methyl-6-*O*-*tert*-Butyldimethylsilyl)-β-Cyclodextrin (DIME-6-TBDMS-β-CD)

In addition to heptakis(2,3-di-*O*-acetyl-6-*O*-*tert*-butyldimethylsilyl)-β-CD **28**, heptakis(2,3-di-*O*-methyl-6-*O*-*tert*-butyldimethyl-silyl)-β-CD **29** (Figure 2.25), which is commercially available as e.g. CycloSil-B, Astec CHIRALDEX B-DM, MEGA-DEX DMT-Beta, and Supelco beta-DEX 325, is another excellent

Table 2.7 Enantioseparation of several racemates.

Analyte	Separation factor α	CD	Separation temperature T (°C)
2-Butanol	1.22	β	40
	1.16	γ	40
2-Octanol	1.06	β	70
	1.15	γ	40
Trans-1,2-cyclohexane-diol	1.58	γ	70
2-Amino-1-propanol	1.07	α	100
	1.16	β	110
	1.99	γ	100
Leucinol	1.06	α	120
	1.14	β	110
2-Aminobutane	1.04	γ	80
Lactic acid	1.47	γ	50
Tartaric acid	1.04	β	90
Mandelic acid	1.04	β	110
2-Bromo-1-chloro-propane	1.06	α	40
	1.12	β	40
	1.05	γ	30
Phenyloxirane	1.01	γ	80
	1.57	γ	80
(*E*)-1,2-diphenyloxirane	1.02	γ	140
β-Butyrolactone	1.14	α	110
	1.62	β	70
	1.20	γ	80
Isoborneol	1.05	γ	70
Carvone	1.04	α	90
	1.09	b	110
	1.01	g	100

Trifluoroacetic anhydride was used for derivatization of the amino and hydroxyl moieties and the carboxylic acids were derivatized as methyl esters. 10 m fused-silica capillary column, 0.25 mm i.d., coated with hexakis/heptakis/octakis(2,6-di-*O*-pentyl-3-*O*-trifluoroacetyl)-α/β/γ-CD.
Source: Adapted from Schurig et al. [178].

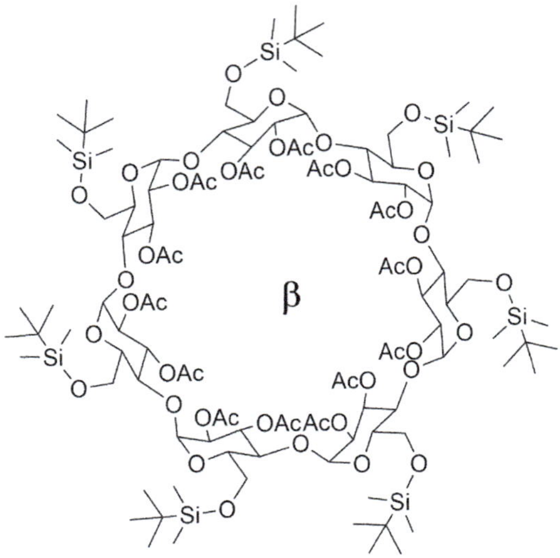

28

Figure 2.24 Chiral selector heptakis(2,3-di-*O*-acetyl-6-*O*-*tert*-butyldimethylsilyl)-β-CD (DIAC-6-TBDMS-β-CD).

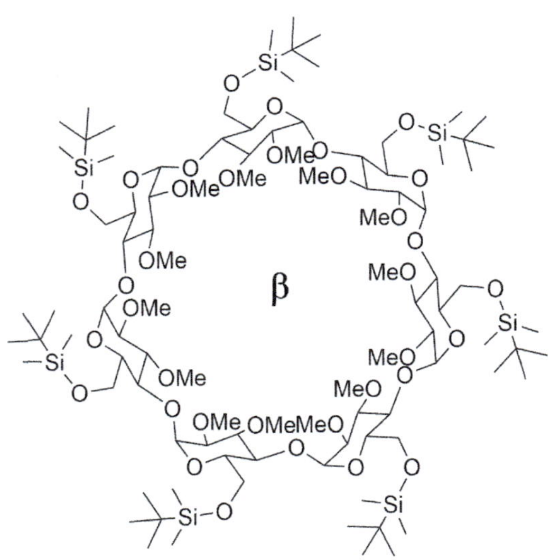

29

Figure 2.25 Chiral selector DIME-6-TBDMS-β-CD (heptakis(2,3-di-*O*-methyl-6-*O*-*tert*-butyldimethylsilyl)-β-CD) **29**.

CSP [166]. The selectivities are comparable; however, the methylated derivative is better soluble in polymers.

Compared to permethyl-β-CD **3**, the separations of various enantiomers (e.g. methyl jasmonate, tetramezine, filbertone, borneol, β-pinene, 3-mercaptohexanol [185]) were improved using DIME-6-TBDMS-β-CD **29** dissolved in the polysiloxane PS086 (50%/50%) [186]. Additionally, *cis*- and *trans*-nerolidol were separated at 80 °C. Furthermore, important flavors such as alkyl-branched acids and their corresponding esters could be separated using DIME-6-TBDMS-β-CD **29** [187]. In this context, another versatile derivative is heptakis(6-*O*-*tert*-butyldimethylsilyl-2,3-di-*O*-ethyl)-β-CD [183, 188], which can be also dissolved in PS086 (50%/50%) and used as CSP in GC separations. The separation, for example, of chiral diaziridine derivatives with small alkyl substituents at the stereogenic nitrogen atoms [189, 190] is possible with this CSP.

An interesting application of DIME-6-TBDMS-β-CD **29** is the separation of the ansa-compounds paracyclophanes, in particular [11]paracyclophanes, because they can interconvert. These compounds are planar chiral and interconvert by rotation of the alkyl loop [191, 192]. This is an excellent example showing that only minute amounts of material are needed to determine kinetic activation parameters of the interconversion. For the separation of the enantiomers, the CSPs DIME-6-TBDMS-β-CD **29** and heptakis(2,6-di-*O*-methyl-3-*O*-pentyl)-β-cyclodextrin were employed and DGC experiments performed (cf. Figure 2.26).

The animal drug tetramezine (1,2-bis(3,3-dimethyldiaziridin-1-yl)ethane) shows antidepressant and dopamine-positive activities, which inhibits monoamineoxidase (MAO). The bioavailability [193] for dogs from intestine-insoluble tablets, its pharmacokinetics [194], and metabolic pathway [195] have been investigated. Information about the stereochemical integrity of chiral drugs is of great importance because one of the stereoisomers can exhibit toxic effects [196] and a low interconversion barrier makes stereoselective synthesis or separation impossible [197].

A separation of the three stereoisomers was achieved by chiral GC using the CSP DIME-6-TBDMS-β-CD **29** (50% w/w) dissolved in the polydimethylsiloxane PS086 (50% w/w) [185]. An excellent separation was observed on separation capillaries as short as only 5 m (cf. Figure 2.27).

As already mentioned, DIME-6-TBDMS-β-CD **29** dissolved in polydimethylsiloxane PS086 (50% w/w) was found to be the CSP of choice for the investigation of the small and volatile diaziridines by chiral GC [189]. This phase is superior in respect of separation of enantiomers even at elevated temperatures, which are sometimes mandatory to determine reaction rate constants by temperature-dependent measurements. Even a baseline

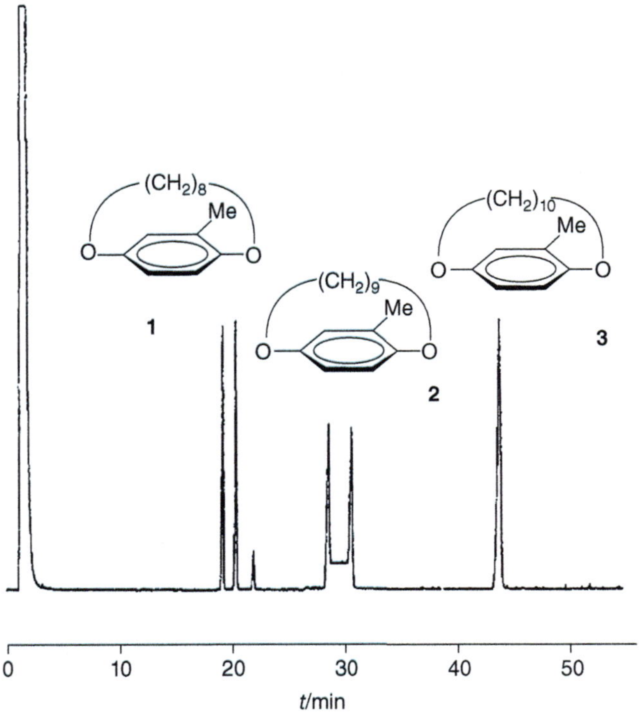

Figure 2.26 The configurational stability depending on the bridge length. CSP heptakis(2,6-di-*O*-methyl-3-*O*-pentyl)-β-cyclodextrin, 20% (w/w) in polysiloxane OV 1701, 25-m fused silica capillary, 150 °C isothermal, 0.5 bar H_2.

separation of the enantiomers of highly volatile 1,2-dimethyldiaziridine (bp: 49 °C) can be achieved (cf. Figure 2.28).

2.4.3 Cyclofructans

Cyclofructans (CFs) are cyclic carbohydrates that can consist of 6 (α-CF, cycloinulohexaose, Figure 2.29) **30**, 7 (β-CF, Cycloinuloheptaose), or 8 (γ-CF, Cycloinulooctaose) fructose units connected by β-(2,1)-glycosidic bonds. Due to their linkage, CFs form a crown ether core, which can be made accessible for substrate binding by derivatization of the 3-hydroxy groups. Armstrong et al. pioneered the use of derivatized CFs as CSPs in GC employing permethylated α-CF, permethylated β-CF, and 4,6-di-*O*-pentyl α-CF [198].

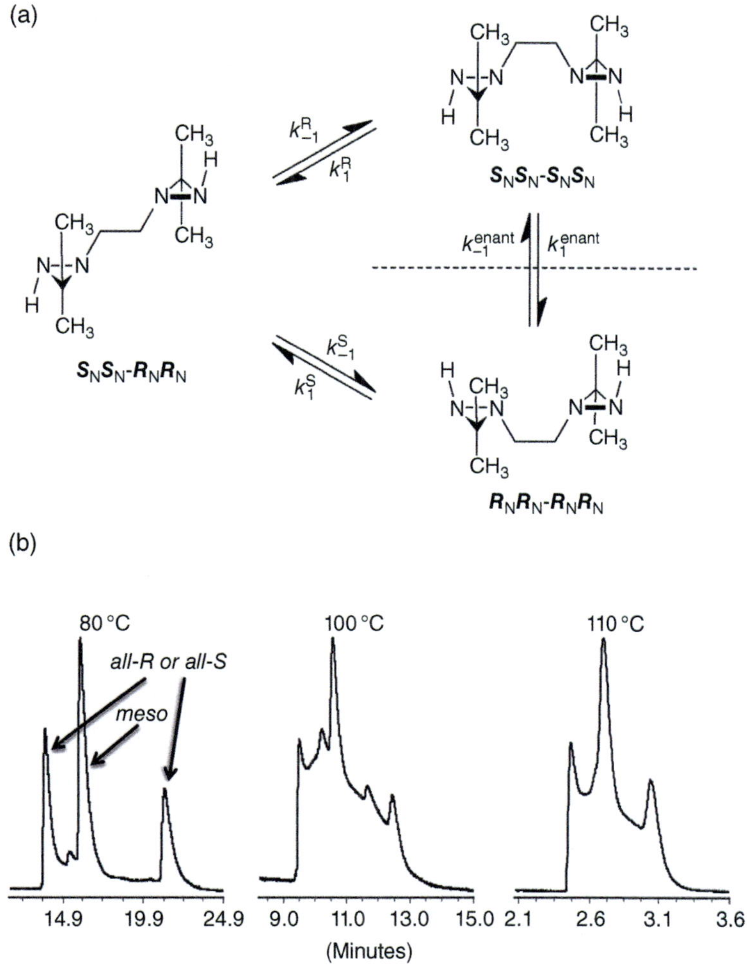

Figure 2.27 (a) Stereoisomers of tetramezine (1,2-bis(3,3-dimethyldiaziridin-1-yl) ethane) and interconversion by epimerization of single diaziridines moieties or simultaneous double interconversion (enantiomerization). (b) Selected experimental interconversion profiles of the stereoselective DGC experiments of tetramezine between 80 and 110 °C. Gas chromatographic conditions: 5-m fused-silica capillary coated with DIME-6-TBDMS-β-CD **29** 50% (w/w) PS086 (0.25-mm i.d., film thickness 0.25 μm). Carrier gas: 80 kPa H_2.

The permethylated CF selectors can be used to separate chiral alcohols, esters, β-lactams, and derivatized amino acids. A selection of compounds that were separated on permethylated α-CF is given in Table 2.8.

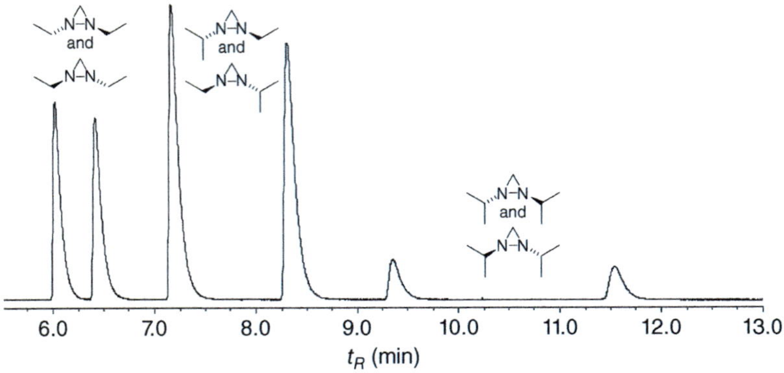

Figure 2.28 Chromatogram of 1,2-diethyl-, 1-ethyl-2-isopropyl- and 1,2-diisopropyldiaziridine. Separation conditions: 25 m × 250 μm i.d. fused-silica capillary coated with 250 nm heptakis(2,3-*di*-*O*-ethyl-6-*O*-*tert*-butyldimethylsilyl)-β-cyclodextrin dissolved in polydimethylsiloxane PS086 (50% w/w), 80 °C, 40 kPa He.

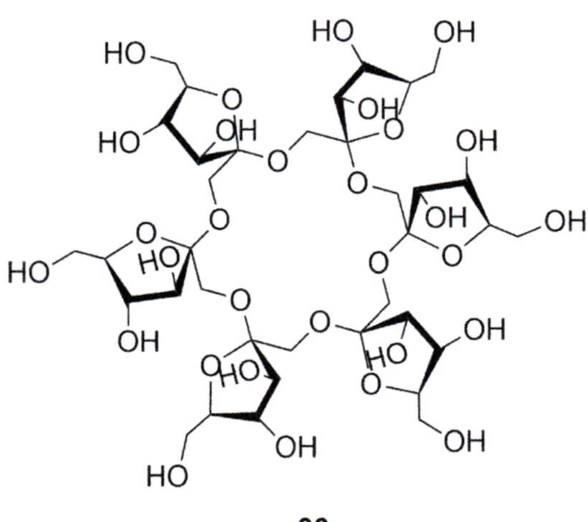

30

Figure 2.29 α-CF **30** (cycloinulohexaose).

Even though many amino acids were separated on permethylated α-CF, permethylated β-CF led to improved separation of β-lactams. On 4,6-di-*O*-pentyl α-CF, on the other hand, the separation factors were diminished for most compounds, indicating a negative influence of the free 3-hydroxy groups for chiral recognition. Compared to their CD analogues,

Table 2.8 Enantioseparations of amino acids, alcohols, esters, and β-lactams.

Analyte	Separation factor α	Separation temperature T (°C)
Alanine, A	1.03	50
Valine, V	1.03	60
Isoleucine, I	1.13	50
Leucine, L	1.05	60
Methionine, M	1.02	100
Phenylalanine, F	1.01	100
Aspartic acid, D	1.02	70
(±)-α-(Trifluoromethyl)-benzylalcohol	1.05	50
Tartaric acid	1.03	65
cis-7-Azabicyclo[4.2.0]-oct-3-en-8-one	1.05	100

The amino groups have been trifluoroacetylated, while the carboxyl groups methylated. 10 m fused-silica capillary column, coated with permethylated α-CF dissolved in polysiloxane OV-1701 (15%).
Source: Adapted from Maas et al. [187].

separation factors are lower on CFs. This can be attributed to a lack of inclusion interaction in CF selectors. Armstrong et al. also prepared the 3-trifluoroacetyl derivative and the 3-propionyl derivative of 4,6-di-*O*-pentyl α-CF [188]. The separation of derivatized amino acids, amino alcohols, amines, alcohols, tartrates, and lactones were investigated on both CSPs generally showing similar separation factors.

The absence of an inclusion complex for CFs was further supported by thermodynamic analysis [188].

2.4.4 CSPs with Metal Complexes

Chiral metal chelates are a privileged class of highly versatile chiral selectors successfully applied in enantioselective complexation GC. Metal complexes, such as metal 3-(trifluoroacetyl)-(1*R*)-camphorates [189, 190], were utilized as CSPs because of their extraordinarily high enantioselectivities in separating chiral compounds. Later this concept was extended to Mn(II) [191], Co(II) [191], and Ni(II) β-diketonate complexes,

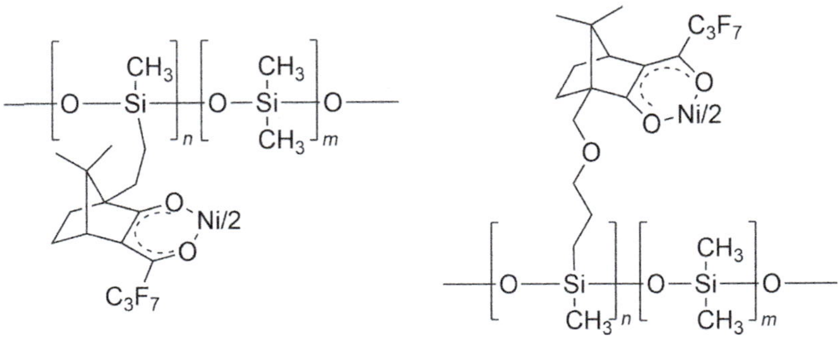

31 32

Figure 2.30 Chirasil-Ni **31** and Chirasil-Ni-OC$_3$ **32**.

e.g. Chirasil-Ni **31** or Chirasil-Ni-OC$_3$ **32** [192] (cf. Figure 2.30). In general, such diketonate ligands are easily accessible and can be used to improve chemical properties of the chiral complexes, e.g. decreasing the high volatility, which typically limits the applicable temperature range in enantioselective complexation GC. Higher temperatures lead unfortunately to leaching of the chiral selector, which decreases the separation efficiency and limits the overall lifetime. Thermal stability can be improved and leaching decreased by linking the chiral metal-containing selector to hydridomethyldimethylpolysiloxane by Pt-catalyzed hydrosilylation [150]. The resulting stationary phases present improved thermal stabilities and decreased column bleeding, which is important for MS detection.

High separation factors α are obtained for a broad range of compounds using these CSPs, in particular for chiral oxirane derivatives like (*R*,*S*)-methyloxirane ($\alpha = 1.27$) (Table 2.9).

Room temperature ionic liquids (ILs), which are low melting salts (<100 °C), were used as polymers to stabilize chiral selectors or as starting material to chemically modify and introduce chiral selectors. ILs are characterized by high viscosity, good wettability, and high thermal stability, and they exhibit often dual nature, i.e. they are both polar and nonpolar. As already mentioned, ILs can be used in chiral separations as achiral ionic liquids to solve chiral selectors or the ionic liquid itself is chiral and used as CSP [129, 199]. Chiral IL stationary phases in GC were investigated for the enantiomeric separation of alcohols, diols, sulfoxides, epoxides, and acetylated amines [200, 201]. Because of the exceptional polarity and low volatility properties, and as demonstrated by the broad applicability in chiral GC separations, ILs are promising (chiral) CSPs.

Table 2.9 Enantioseparation on the CSP Chirasil-Ni-OC_3.

Analyte	Separation factor α	Separation temperature T (°C)
	1.27	40
	1.15	90
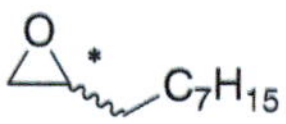	1.05	110
	1.06	100
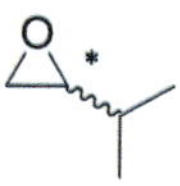	1.12	45
	1.09	40
	1.13	45
	1.10	120
	1.04	90
	1.05	100
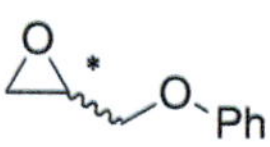	1.06	100
	1.20	45
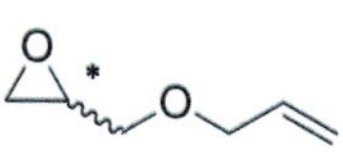	1.06	80
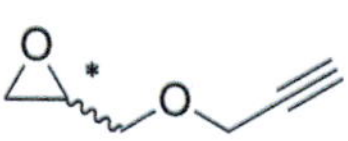	1.04	80
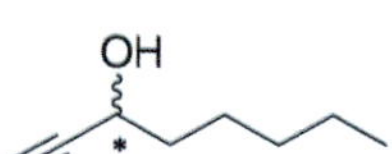	1.11	120

(*Continued*)

Table 2.9 (Continued)

Analyte	Separation factor α	Separation temperature T (°C)
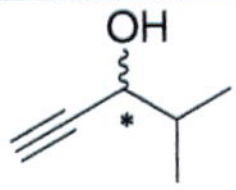	1.18	110
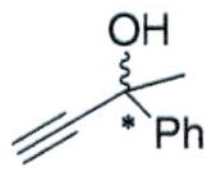	1.12	100
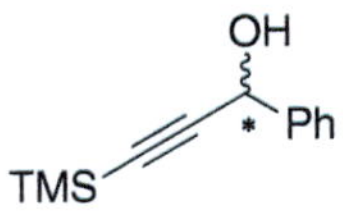	1.66	143
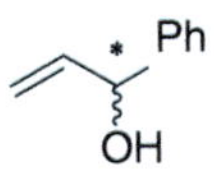	1.07	110
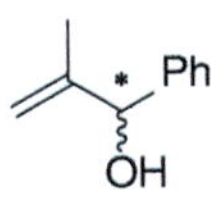	1.08	90
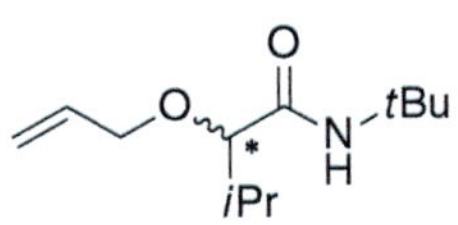	1.02	140
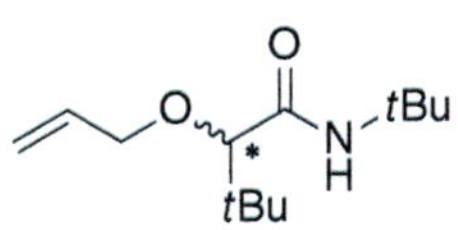	1.02	120
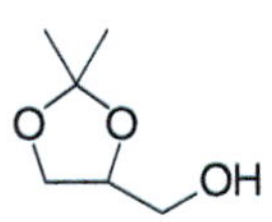	1.20	130
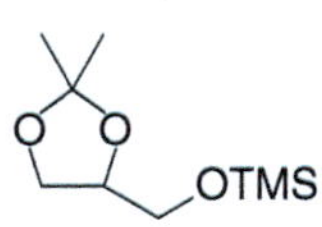	1.04	110
(2*R*,5*R*)-,(2*S*,5*S*)-chalcogran	1.39	100
(2*R*,5*S*)-,(2*S*,5*R*)-chalcogran	1.35	100
(±)-Camphor	1.06	80
(±)-Menthol	1.12	140
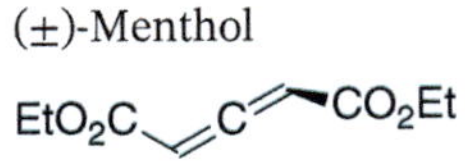	1.31	120

2.5 Conclusion

Chiral GC is a highly important technique for the separation of volatile chiral compounds. In this chapter, the most important CSPs were presented and discussed, ranging from diamide selectors, CDs to chiral metal complexes. A large number of applications were presented, demonstrating that chiral GC is an indispensable tool for the characterization and quantification of enantiomeric ratios in asymmetric synthesis, catalysis, medicinal chemistry, chemical biology, and drug research.

References

1 Maier, N.M., Franco, P., and Lindner, W. (2001). Separation of enantiomers: needs, challenges, perspectives. *J. Chromatogr. A* 906: 3–33.

2 Trapp, O., Schoetz, G., and Schurig, V. (2001). Determination of enantiomerization barriers by dynamic and stopped flow chromatographic methods. *Chirality* 13: 403–414.

3 Wolf, C. (2005). Stereolabile chiral compounds: analysis by dynamic chromatography and stopped-flow methods. *Chem. Soc. Rev.* 34: 595–608.

4 Trapp, O. (2006). Unified equation for access to rate constants of first-order reactions in dynamic and on-column reaction chromatography. *Anal. Chem.* 78: 189–198.

5 D'Acquarica, I., Gasparrini, F., Pierini, M. et al. (2006). Dynamic HPLC on chiral stationary phases: a powerful tool for the investigation of stereomutation processes. *J. Sep. Sci.* 29: 1508–1516.

6 Wolf, C. (2008). *Dynamic Stereochemistry of Chiral Compounds – Principles and Applications*. Cambridge: RSC Publishing.

7 Koppenhöfer, B., Piras, P., and Roussel, C. (1994). Chirbase, a molecular database for the separation of enantiomers by chromatography. *J. Chromatogr. A* 666: 557–563.

8 Piras, P., Roussel, C., and Pierrot-Sanders, J. (2001). Reviewing mobile phases used on Chiralcel OD through an application of data mining tools to CHIRBASE database. *J. Chromatogr. A* 906: 443–458.

9 Sheridan, R., Schafer, W., Piras, P. et al. (2016). Toward structure-based predictive tools for the selection of chiralstationary phases for the chromatographic separation of enantiomers. *J. Chromatogr. A* 1467: 206–213.

10 Betzenbichler, G., Huber, L., Kräh, S. et al. (2022). Chiral stationary phases and applications in gas chromatography. *Chirality* 34: 732–759.

11 Berthod, A., Li, W.Y., and Armstrong, D.W. (1990). Chiral recognition of racemic sugars by polar and nonpolar cyclodextrin-derivative gas chromatography. *Carbohydr. Res.* 201: 175–184.

12 Schurig, V. (1984). Gas chromatographic separation of enantiomers on optically active metal-complex-free stationary phases. New analytical methods (24). *Angew. Chem. Int. Ed. Eng.* 23: 747–765.

13 Schurig, V. (2001). Separation of enantiomers by gas chromatography. *J. Chromatogr. A* 906: 275–299.

14 Schurig, V. (2005). Contributions to the theory and practice of the chromatographic separation of enantiomers. *Chirality* 17: S205–S226.

15 Schurig, V. (2002). Chiral separations using gas chromatography. *Trends Anal. Chem.* 21: 647–661.

16 Cagliero, C., Sgorbini, B., Cordero, C. et al. (2021). Separation of stereoisomers by gas chromatography. In: *Handbooks in Separation Science: Gas Chromatography* (ed. C.F. Poole). Amsterdam: Elsevier.

17 Pirkle, W.H. and Sikkenga, D.L. (1976). Resolution of optical isomers by liquid chromatography. *J. Chromatogr.* 123: 400–404.

18 Pirkle, W.H., House, D.W., and Finn, J.M. (1980). Broad spectrum resolution of optical isomers using chiral high-performance liquid chromatographic bonded phases. *J. Chromatogr.* 192: 143–158.

19 Pirkle, W.H. and Welch, C.J. (1984). Chromatographic separation of the enantiomers of acylated amines on chiral stationary phases. *J. Organomet. Chem.* 49: 138–140.

20 Pirkle, W.H., Welch, C.J., and Lamm, B. (1992). Design, synthesis and evaluation of an improved enantioselective naproxen selector. *J. Organomet. Chem.* 57: 3854–3560.

21 Okamoto, Y. and Kaida, Y. (1994). Resolution by high-performance liquid chromatography using polysaccharide carbamates and benzoates as chiral stationary phases. *J. Chromatogr. A* 666: 403–419.

22 Uray, G. and Lindner, W. (1990). (S,S)-Diphenyl-ethanediamine (DPEDA) derivatives as chiral selectors. Part I: undecenoyl bound dinitrobenzoyl-DPEDA as a broadly applicable chiral stationary phase. *Chromatographia* 30: 323–327.

23 Schleimer, M., Pirkle, W.H., and Schurig, V. (1994). Enantiomer separation by high-performance liquid chromatography on polysiloxane-based chiral stationary phases. *J. Chromatogr. A* 679: 23–34.

24 Armstrong, D.W., Liu, Y., and Ekborgott, H. (1995). A covalently bonded teicoplanin chiral stationary phase for HPLC enantioseparations. *Chirality* 7: 474–497.

25 Yashima, E. and Okamoto, Y. (1995). Chiral discrimination on polysaccharides derivatives. *Bull. Chem. Soc. Jpn.* 65: 3289–3307.

26 Lämmerhofer, M. and Lindner, W. (1996). Quinine and quinidine derivatives as chiral selectors I. Brush type chiral stationary phases for high-performance liquid chromatography based on cinchonan carbamates and their application as chiral anion exchangers. *J. Chromatogr. A* 741: 33–48.

27 Berthod, A., Liu, Y., Bagwill, C., and Armstrong, D.W. (1996). Facile liquid chromatographic enantioresolution of native amino acids and peptides using a teicoplanin chiral stationary phase. *J. Chromatogr. A* 731: 123–137.

28 Okamoto, Y. and Yashima, E. (1998). Polysaccharide derivatives for chromatographic separation of enantiomers. *Angew. Chem. Int. Ed.* 37: 1020–1043.

29 Kontrec, D., Vinkovic, V., and Sunjic, V. (1999). New chiral stationary phases based on (R)-1-Naphthylethylamine bound to 2,4,5,6-tetrachloro-1,3-dicyanobenzene. *Chirality* 11: 722–730.

30 Chankvetadze, B., Yamamoto, C., and Okamoto, Y. (2001). Enantioseparation of selected chiral sulfoxides using polysaccharide-type chiral stationary phases and polar organic, polar aqueous-organic and normal-phase eluents. *J. Chromatogr. A* 922: 127–137.

31 Gasparrini, F., Misiti, D., and Villani, C. (2001). High-performance liquid chromatography chiral stationary phases based on low-molecular-mass selectors. *J. Chromatogr. A* 906: 35–50.

32 Yamamoto, C., Hayashi, T., and Okamoto, Y. (2003). High-performance liquid chromatographic enantioseparation using chitin carbamate derivatives as chiral stationary phases. *J. Chromatogr. A* 1021: 83–91.

33 Ikai, T. and Okamoto, Y. (2009). Structure control of polysaccharide derivatives for efficient separation of enantiomers by chromatography. *Chem. Rev.* 109: 6077–6101.

34 Franco, P., Klaus, P.M., Minguillon, C., and Lindner, W. (2001). Evaluation of the contribution to enantioselectivity of quinine and quinidine scaffolds in chemically and physically mixed chiral selectors. *Chirality* 13: 177–186.

35 Huil, F., Ekborg-Ott, K.H., and Armstrong, D.W. (2001). High-performance liquid chromatographic and capillary electrophoretic enantioseparation of plant growth regulators and related indole compounds using macrocyclic antibiotics as chiral selectors. *J. Chromatogr. A* 906: 91–103.

36 Schleimer, M. and Schurig, V. (1992). Enantiomer separation by capillary supercritical fluid chromatography. In: *Analysis with Supercritical Fluids: Extraction and Chromatography* (ed. B. Wenclawiak), 135–150. Heidelberg: Springer Verlag.

37 Terfloth, G.J., Pirkle, W.H., Lynam, K.G., and Nicolas, E.C. (1995). Broadly applicable polysiloxane-based chiral stationary phase for high performance liquid chromatography and supercritical fluid chromatography. *J. Chromatogr. A* 705: 185–194.

38 Pirkle, W.H., Brice, J.L., and Terfloth, G.J. (1996). Liquid and subcritical CO2 separations of enantiomers on a broadly applicable polysiloxane chiral stationary phase. *J. Chromatogr. A* 753: 109–119.

39 Anton, K., Eppinger, J., Frederiksen, L. et al. (1994). Chiral separations by packed-column super- and subcritical fluid chromatography. *J. Chromatogr. A* 666: 395–401.

40 Mayer, S. and Schurig, V. (1992). Enantiomer separation by electrochromatography on capillaries coated with Chirasil-Dex. *J. High Resolut. Chromatogr.* 15: 129–131.

41 Wistuba, D., Czesla, H., Roeder, M., and Schurig, V. (1998). Enantiomer separation by pressure supported electrochromatography using capillaries packed with a permethyl-b-cyclodextrin stationary phase. *J. Chromatogr. A* 815: 183–188.

42 Wistuba, D. and Schurig, V. (2000). Enantiomer separation of chiral pharmaceuticals by capillary electrophochromatography. *J. Chromatogr. A* 875: 255–276.

43 Chankvetadze, B., Kartozia, I., Breitkreutz, J. et al. (2001). Effect of organic solvent, electrolyte salt and a loading of cellulose tris(3,5-dichlorophenylcarbamate) on silica gel on enantioseparation characteristics in capillary electrochromatography. *Electrophoresis* 22: 3327–3334.

44 Chankvetadze, L., Kartozia, I., Yamamoto, C. et al. (2002). Enantioseparations in capillary liquid chromatography and capillary electrochromatography using amylose tris(3,5-dimethylphenylcarbamate) in combination with aqueous organic mobile phase. *J. Sep. Sci.* 25: 653–660.

45 Mayer, S., Briand, X., and Francotte, E. (2000). Separation of enantiomers by packed capillary electrochromatography on a cellulose-based stationary phase. *J. Chromatogr. A* 875: 331–339.

46 Wistuba, D. and Schurig, V. (2000). Recent progress in enantiomer separation by capillary electrochromatography. *Electrophoresis* 21: 4136–4158.

47 Chankvetadze, B., Ikai, T., Yamamoto, C., and Okamoto, Y. (2004). High-performance liquid chromatographic enantioseparations on monolithic silica columns containing a covalently attached 3,5-dimethylphenylcarbamate derivative of cellulose. *J. Chromatogr. A* 1042: 55–60.

48 Francotte, E. (1994). Contribution of preperative chromatographic resolution to the investigation of chiral phenomena. *J. Chromatogr. A* 666: 565–601.

49 Francotte, E.R. (2001). Enantioselective chromatography as a powerful alternative for the preparation of drug enantiomers. *J. Chromatogr. A* 906: 379–397.

50 Armstrong, D.W., Lee, J.T., and Chang, L.W. (1998). Enantiomeric impurities in chiral catalysts, auxiliaries and synthons used in enantioselective synthesis. *Tetrahedron Asymmetry* 9: 2043–2064.

51 Noyori, R. (1990). Chiral metal complexes as discriminating molecular catalysts. *Science* 248: 1194–1199.

52 Noyori, R. (2013). Facts are the enemy of truth-reflections on serendipitous discovery and unforeseen developments in asymmetric catalysis. *Angew. Chem. Int. Ed.* 52: 79–92.

53 Sharpless, K.B. (2002). Searching for new reactivity (Nobel lecture). *Angew. Chem. Int. Ed.* 41: 2024–2032.

54 Bäckvall, J.-E., Nordberg, R.E., Bjorkman, E.E., and Moberg, C. (1980). Stereochemistry of nucleophilic attack on [p]-allylpalladium complexes. Evidence for cis-migration of acetate from palladium to carbon. *J. Chem. Soc. Chem. Commun.* 943–944.

55 Reetz, M.T. (2008). Combinatorial transition-metal catalysis: mixing monodentate ligands to control enantio-, diastereo-, and regioselectivity. *Angew. Chem. Int. Ed.* 47: 2556–2588.

56 List, B., Lerner, R.A., and Barbas, C.F. (2000). Proline-catalyzed direct asymmetric aldol reactions. *J. Am. Chem. Soc.* 122: 2395–2396.

57 Ahrendt, K.A., Borths, C.J., and MacMillan, D.W.C. (2000). New strategies for organic catalysis: the first highly enantioselective organocatalytic Diels–Alder reaction. *J. Am. Chem. Soc.* 122: 4243–4244.

58 Jen, W.S., Wiener, J.J.M., and MacMillan, D.W.C. (2000). New strategies for organic catalysis: the first enantioselective organocatalytic 1,3-dipolar cycloaddition. *J. Am. Chem. Soc.* 122: 9874–9875.

59 List, B. (2000). The direct catalytic asymmetric three-component Mannich reaction. *J. Am. Chem. Soc.* 122: 9336–9337.

60 List, B. (2002). Proline-catalyzed asymmetric reactions. *Tetrahedron* 58: 5573–5590.

61 List, B. (2004). Enamine catalysis is a powerful strategy for the catalytic generation and use of carbanion equivalents. *Acc. Chem. Res.* 37: 548–557.

62 Mukherjee, S., Yang, J.W., Hoffmann, S., and List, B. (2007). Asymmetric enamine catalysis. *Chem. Rev.* 107: 5471–5569.

63 Beeson, T.D., Mastracchio, A., Hong, J.-B. et al. (2007). Enantioselective organocatalysis using SOMO activation. *Science* 316: 582–585.

64 MacMillan, D.W.C. (2008). The advent and development of organocatalysis. *Nature* 455: 304–308.

65 Reetz, M.T. (2002). New methods for the high-throughput screening of enantioselective catalysts and biocatalysts. *Angew. Chem. Int. Ed.* 41: 1335–1338.

66 Wolf, C. and Hawes, P.A. (2002). A high-throughput screening protocol for fast evaluation of enantioselective catalysts. *J. Organomet. Chem.* 67: 2727–2729.

67 Trapp, O. (2008). Gas chromatographic high-throughput screening techniques in catalysis. *J. Chromatogr. A* 1184: 160–190.

68 Herwig, P., Zawatzky, K., Grieser, M. et al. (2013). Imaging the absolute configuration of a chiral epoxide in the gas phase. *Science* 342: 1084–1086.

69 Zawatzky, K., Herwig, P., Grieser, M. et al. (2014). Coulomb explosion imaged cryptochiral (R,R)-2,3-dideuterooxirane: unambiguous access to the absolute configuration of (+)-glyceraldehyde. *Chem. Eur. J.* 20: 5555–5558.

70 Trapp, O. and Zawatzky, K. (2016). Synthesis of cryptochiral (R,R)-2,3-dideuterooxirane as stereochemical reference compound and chemical correlation with D-(+)-glyceraldehyde. *Isr. J. Chem.* 56: 1082–1090.

71 Soai, K., Shibata, T., Morioka, H., and Choji, K. (1995). Asymmetric autoctalysis and amplification of enantiomeric execss of a chiral molecule. *Nature* 378: 767–768.

72 Trapp, O., Lamour, S., Maier, F. et al. (2020). In situ mass spectrometric and kinetic investigations of Soai's asymmetric autocatalysis. *Chem. Eur. J.* 26: 15871–15880.

73 Trapp, O. (2020). Efficient amplification in Soai's asymmetric autocatalysis by a transient stereodynamic catalyst. *Front. Chem.* 8: 1173.

74 Maier, F. and Trapp, O. (2014). Selector-induced dynamic deracemization of a selectand-modified tropos BIPHEPO-ligand: application in the organocatalyzed asymmetric double-aldol-reaction. *Angew. Chem. Int. Ed.* 53: 8756–8760.

75 Storch, G. and Trapp, O. (2015). Temperature controlled bidirectional Enantioselectivity in a dynamic catalyst for asymmetric hydrogenation. *Angew. Chem. Int. Ed.* 54: 3580–3586.

76 Storch, G. and Trapp, O. (2017). By-design enantioselective self-amplification based on non-covalent product-catalyst interactions. *Nat. Chem.* 9: 179–187.

77 Scholtes, J.F. and Trapp, O. (2019). Inducing enantioselectivity in a dynamic catalyst by supramolecular interlocking. *Angew. Chem. Int. Ed.* 58: 6306–6310.

78 Scholtes, J.F. and Trapp, O. (2021). Asymmetric induction and amplification in stereodynamic catalytic systems by noncovalent interactions. *Synlett* 32: 971–980.

79 Kawasaki, T., Suzuki, K., Hakoda, Y., and Soai, K. (2008). Achiral nucleobase cytosine acts as an origin of Homochirality of biomolecules in conjunction with asymmetric autocatalysis. *Angew. Chem. Int. Ed.* 47: 496–499.

80 Blackmond, D.G. (2020). Autocatalytic models for the origin of biological homochirality. *Chem. Rev.* 120: 4831–4847.

81 Buhse, T., Cruz, J.-M., Noble-Terán, M.E. et al. (2021). Spontaneous deracemizations. *Chem. Rev.* 121: 2147–2229.

82 Meierhenrich, U.J., Thiemann, W.H.-P., Goesmann, F. et al. (2001). Enantiomer separation of hydrocarbons in preparation for ROSETTA's chirality-experiment. *Chirality* 13: 454–457.

83 Bocková, J., Jones, N.C., Meierhenrich, U.J. et al. (2021). Chiroptical activity of hydroxycarboxylic acids with implications for the origin of biological homochirality. *Commun. Chem.* 4: 86.

84 Caro, G.M.M., Meierhenrich, U.J., Schutte, W.A. et al. (2002). Amino acids from ultraviolet irradiation of interstellar ice analogues. *Nature* 416: 403.

85 Myrgorodska, I., Meinert, C., Martins, Z. et al. (2015). Molecular chirality in meteorites and interstellar ices, and the chirality experiment on board the ESA cometary Rosetta Mission. *Angew. Chem. Int. Ed.* 54: 1402–1412.

86 Meinert, C., Myrgorodska, I., Marcellus, P.D. et al. (2016). Ribose and related sugars from ultraviolet irradiation of interstellar ice analogs. *Science* 352: 208–212.

87 Rundlett, K.L. and Armstrong, D.W. (2001). Methods for the determination of binding constants by capillary electrophoresis. *Anal. Chem.* 22: 1419–1427.

88 Trapp, O. and Schurig, V. (2010). Nonlinear effects in enantioselective chromatography. Prediction of unusual elution profiles of enantiomers on an achiral stationary phase doped with small amounts of a chiral selector. *Tetrahedron Asymmetry* 21: 1334–1340.

89 Pirkle, W.H. and Murray, P.G. (1993). Chiral stationary phase design use of intercalative effects to enhance enantioselectivity. *J. Chromatogr.* 641: 11–19.

90 Pirkle, W.H., Murray, P.G., Rausch, D.J., and McKenna, S.T. (1996). Intermolecular 1H-1H two-dimensional nuclear overhauser enhancements in the characterization of a rationally designed chiral recognition system. *J. Organomet. Chem.* 61: 4769–4774.

91 Storch, G., Haas, M., and Trapp, O. (2017). Attracting enantiomers: chiral analytes that are simultaneously shift reagents allow rapid screening of enantiomeric ratios by NMR spectroscopy. *Chem. Eur. J.* 23: 5414–5418.

92 Scholtes, J.F. and Trapp, O. (2019). Design and synthesis of a stereodynamic catalyst with reversal of selectivity by enantioselective self-inhibition. *Chirality* 31: 1028–1042.

93 Schurig, V., Bürkle, W., Zlatkis, A., and Poole, C.F. (1979). Quantitative resolution of pyramidal nitrogen invertomers by complexation chromatography. *Naturwissenschaften* 66: 423–424.

94 Bürkle, W., Karfunkel, H., and Schurig, V. (1984). Dynamic phenomena during enantiomer resolution by complexation gas chromatography. *J. Chromatogr.* 288: 1–14.

95 Gasparrini, F., Misiti, D., Pierini, M., and Villani, C. (1997). Enantiomerization barriers by dynamic HPLC. Stationary phase effects. *Tetrahedron Asymmetry* 8: 2069–2073.

96 Maier, F. and Trapp, O. (2013). The stereodynamics of 5,5′-disubstituted BIPHEPs. *Chirality* 25: 126–132.

97 Trapp, O. (2013). Interconversion of stereochemically labile enantiomers (enantiomerization). *Top. Curr. Chem.* 341: 231–270.

98 Maier, F. and Trapp, O. (2012). Stationary phase and solvent effects on the stereodynamics of tropos BIPHEP ligands revealed by a novel HPLC technique. *Angew. Chem. Int. Ed.* 51: 2985–2988.

99 Ciogli, A., Bicker, W., and Lindner, W. (2010). Determination of enantiomerization barriers of hypericin and pseudohypericin by dynamic high-performance liquid chromatography on immobilized polysaccharide-type chiral stationary phases and off-column racemization experiments. *Chirality* 22: 463–471.

100 Pirkle, W.H., Murray, P.G., and Burke, J.A. (1993). Use of homologous series of analytes as mechanistic probes to investigate the origins of enantioselectivity two chiral stationary phases. *J. Chromatogr.* 641: 21–29.

101 Pirkle, W.H. and Welch, C.J. (1994). Use of simultaneous face to face and face to edge pi-pi interactions to facilitate chiral recognition. *Tetrahedron Asymmetry* 5: 777–780.

102 Welch, C.J. (1994). Evolution of chiral stationary phase design in the Pirkle laboratories. *J. Chromatogr. A* 666: 3–26.

103 Yamamoto, C., Yashima, E., and Okamoto, Y. (2002). Structural analysis of amylose tris(3,5-dimethylphenylcarbamate) by NMR relevant to its chiral recognition mechanism in HPLC. *J. Am. Chem. Soc.* 124: 12583–12589.

104 Kellner, H.-K., Blasch, A., Chmiel, H. et al. (1997). Enantioseparation of N-protected a-amino acid derivatives by liquid-liquid extraction technique employing stereoselective ion-pair formation with a carbamoylated quinine derivative. *Chirality* 9: 268–273.

105 Gil-Av, E., Feibush, B., and Charles-Sigler, R. (1966). Separation of enantiomers by gas liquid chromatography with an optically active stationary phase. *Tetrahedron Lett.* 7: 1009–1015.

106 Davankov, V.A. (1997). The nature of chiral recognition: is it a three-point interaction? *Chirality* 9: 99–102.

107 Oro, J., Updegrove, S., Gilbert, J. et al. (1970). Organic elements and compounds in surface samples from the sea of tranquility. *Science* 167: 765–767.

108 Frank, H., Nicholson, G.J., and Bayer, E. (1977). Rapid gas chromatographic separation of amino acid enantiomers with a novel chiral stationary phase. *J. Chromatogr. Sci.* 15: 174–176.

109 Frank, H., Nicholson, G.J., and Bayer, E. (1978). Gas chromatographic-mass spectrometric analysis of optically active metabolites and drugs an a novel chiral stationary phase. *J. Chromatogr.* 146: 197–206.

110 Frank, H., Nicholson, G.J., and Bayer, E. (1978). Chiral polysiloxanes for resolution of optical antipodes. *Angew. Chem. Int. Ed. Eng.* 17: 363–365.

111 Schurig, V. and Gil-Av, E. (1976/77). Chromatographic resolution of chiral olefins. Specific rotation of 3-methylcyclopentene and related compounds. *Isr. J. Chem.* 15: 96–98.

112 Schurig, V. (1977). Resolution of a chiral olefin by complexation chromatography on an optically active rhodium(I) complex. *Angew. Chem. Int. Ed.* 28: 110.

113 Schurig, V. and Bürkle, W. (1978). Quantitative Enantiomerentrennung von trans-2,3-Epoxybutan durch Komplexierungschromatographie an einem optisch aktiven Nickel(II)-Komplex. *Angew. Chem. Int. Ed. Eng.* 17: 132–133.

114 Whitesides, G.M. and Lewis, D.W. (1970). Tris[3(tert-butylhydroxy methylene)-D-camphorato]europium(III). A reagent for determining enantiomeric purity. *J. Am. Chem. Soc.* 92: 6979–6980.

115 Schurig, V. (1972). Chiral shift reagents for NMR-spectroscopy. A simple and improved access to lanthanide-tris-chelates of D-3-TFA-camphor. *Tetrahedron Lett.* 13: 3297–3300.

116 Schurig, V. (1972). Chiral d8 metal ion coordination compounds. The preparation of D-3-trifluoroacetylcamphorato complexes of rhodium, palladium, and nickel. *Inorg. Chem.* 11: 736–738.

117 Zeise, W.C. (1831). Von der Wirkung zwischen Platinchlorid und Alkohol, und von den dabei entstehenden neuen Substanzen. *Ann. Phys.* 97: 497–541.

118 Schurig, V. (1976). The preparation of Mn(II), Co(II) and Ni(II)-bis-3-trifluoroacetyl-camphorates. *J. Coord. Chem.* 6: 63–64.

119 Schurig, V. (1977). Selektivität und Stereochemie der Olefin-Metall-pi-Komplexierung. *Chemiker-Zeitung* 101: 173–183.

120 Schurig, V., Koppenhoefer, B., and Bürkle, W. (1980). Preparation and determination of configurationally pure trans -(2S,3S)-2,3-epoxybutane. *J. Organomet. Chem.* 45: 538–541.

121 Schurig, V. and Bürkle, W. (1982). Extending the scope of enantiomer resolution by complexation gas chromatography. *J. Am. Chem. Soc.* 104: 7573–7580.

122 Freudenberg, K., Plankenhorn, E., and Knauber, H. (1945). Über Schardingers Dextrine aus Stärke. *Liebigs Ann.* 558: 1–10.

123 Levkin, P.A., Ruderisch, A., and Schurig, V. (2006). Combining the enantioselectivity of a cyclodextrin and a diamide selector in a mixed binary gas-chromatographic chiral stationary phase. *Chirality* 18: 49–63.

124 Valente, C., Choi, E., Belowich, M.E. et al. (2010). Metal-organic frameworks with designed chiral recognition sites. *Chem. Commun.* 46: 4911–4913.

125 Köhler, J.E.H., Hohla, M., Richters, M., and König, W.A. (1992). Cyclodextrin derivatives as chiral selectors – investigation of the interaction with (R,S)-methyl-2-chloropropionate by enantioselective gas chromatography, NMR spectroscopy, and molecular dynamics simulation. *Angew. Chem. Int. Ed. Eng.* 31: 319–320.

126 Köhler, J.E.H., Hohla, M., Richters, M., and König, W.A. (1994). A molecular-dynamics simulation of the complex formation between methyl R/S-2-chloropropionate and heptakis(3-O-acetyl-2,6-di-O-pentyl)-ß-cyclodextrin. *Chem. Ber.* 127: 119–126.

127 Kobor, F., Angermund, K., and Schomburg, G. (1993). Molecular modelling experiments on chiral recognition in GC with specially derivatized cyclodextrins as selectors. *J. High Resolut. Chromatogr.* 16: 299–311.

128 Grob, K. (1986). *Making and Manipulating Capillary Columns for Gas Chromatography*. Heidelberg: Huethig.

129 Berthod, A., He, L., and Armstrong, D.W. (2001). Ionic liquids as stationary phase solvents for methylated cyclodextrins in gas chromatography. *Chromatographia* 53: 63–68.

130 Mayer, S., Schmalzing, D., Jung, M., and Schleimer, M. (1992). A chiral test mixture for permethylated b-cyclodextrin-polysiloxane gas-liquid gas chromatographic phases: the Schurig test mixture. *LCGC Int.* 10: 58.

131 König, W.A., Benecke, I., and Sievers, S. (1981). New results in the gas chromatographic separation of enantiomers of hydroxy acids and carbohydrates. *J. Chromatogr. A* 217: 71–79.

132 Abe, I., Kuramoto, S., and Musha, S. (1983). Chiral phases derived from XE-60 for glass capillary gas chromatography of amino acid enantiomers. *J. Chromatogr. A* 258: 35–42.

133 König, W.A. and Benecke, I. (1981). Gas chromatographic separation of enantiomers of amines and amino alcohols on chiral stationary phases. *J. Chromatogr.* 209: 91–95.

134 Smolková-Keulemansová, E. and Krýsl, S. (1980). Inclusion compounds in chromatography. *J. Chromatogr. A* 184: 347–361.

135 Smolková-Keulemansová, E. (1982). Cyclodextrins as stationary phases in chromatography. *J. Chromatogr. A* 251: 17–34.

136 Kościelski, T., Sybilska, D., and Jurczak, J. (1983). Separation of α- and β-pinene into enantiomers in gas-liquid chromatography systems via α-cyclodextrin inclusion complexes. *J. Chromatogr. A* 280: 131–134.

137 Kościelski, T., Sybilska, D., Belniak, S., and Jurczak, J. (1984). Gas-liquid chromatography system with α-cyclodextrin as an analytical tool for the studies of stereoselective hydrogenation of α-pinene. *Chromatographia* 19: 292.

138 Smolková, E., Králová, H., Krýsl, S., and Feltl, L. (1982). Study of the properties of cyclodextrins as stationary phases in gas chromatograpy. *J. Chromatogr. A* 241: 3–8.

139 Mráz, J., Feltl, L., and Smolková-Keulemansová, E. (1984). Cyclodextrins and methylated cyclodextrins as stationary phases in gas-solid chromatography. *J. Chromatogr. A* 286: 17–22.

140 Smolková-Keulemansová, A., Neumannová, E., and Feltl, L. (1986). Study of the stereospecific properties of cyclodextrins as gas-solid chromatographic stationary phases. *J. Chromatogr. A* 365: 279–288.

141 Schurig, V. and Nowotny, H.-P. (1990). Gas chromatographic separation of enantiomers on cyclodextrin derivatives. *Angew. Chem. Int. Ed. Eng.* 29: 939–957.

142 König, W.A. (1992). *Enantioselective Gas Chromatography with Modified Cyclodextrins*. Heidelberg: Hüthig.

143 Schreier, P., Bernreuther, A., and Huffer, M. (1995). *Analysis of Chiral Organic Molecules, Methodology and Applications, Chapter 3.5.2.3*. Berlin and New York: Walter de Gruyter.

144 Tang, W., Ng, S., and Sun, D. (2013). *Modified Cyclodextrins for Chiral Separation*. Heidelberg: Springer.

145 Juvancz, Z., Alexander, G., and Szeitli, J. (1987). Permethylated b-cyclodextrin as stationary phase in capillary gas chromatography. *J. High Resolut. Chromatogr.* 10: 105–107.

146 Schurig, V. and Nowotny, H.-P. (1988). Separation of enantiomers on diluted permethylated β-cyclodextrin by high-resolution gas chromatography. *J. Chromatogr. A* 441: 155–163.

147 Schreier, P. (1993). cf. Table on p. 55. In: *Progress in Flavour Precursor Studies* (ed. P. Schreier and P. Winterhalter). Carol Stream, IL: Allured Publishing Corporation.

148 Fischer, P., Aichholz, R., Bölz, U. et al. (1990). Permethyl-β-cyclodextrin, chemically bonded to polysiloxane: a chiral stationary phase with wider application range for enantiomer separation by capillary gas chromatography. *Angew. Chem. Int. Ed. Eng.* 29: 427–429.

149 Schurig, V., Schmalzing, D., Mühleck, U. et al. (1990). Gas chromatographic enantiomer separation on polysiloxane-anchored permethyl-beta-cyclodextrin (Chirasil-Dex). *J. High Resolut. Chromatogr.* 13: 713–717.

150 Schurig, V., Schmalzing, D., and Schleimer, M. (1991). Enantiomer separation on immobilized Chirasil-metal and Chirasil-Dex by gas chromatography and supercritical fluid chromatography. *Angew. Chem. Int. Ed.* 30: 987–989.

151 Jung, M. and Schurig, V. (1993). Enantiomeric separation by GC on Chirasil-Dex: systematic study of cyclodextrin concentration, polarity, immobilization, and column stability. *J. Microcolumn Sep.* 5: 11–22.

152 Ciucanu, I. and König, W.A. (1994). Immobilization of peralkylated β-cyclodextrin on silica gel for high-performance liquid chromatography. *J. Chromatogr. A* 685: 166–171.

153 Cousin, H., Trapp, O., Peulon-Agasse, V. et al. (2003). Synthesis, NMR characterisation and polysiloxane-based immobilization of the three regioisomeric monooctenylpermethyl-b-cyclodextrins – application in enantioselective GC. *Eur. J. Org. Chem.* 3273–3287.

154 Trapp, O. and Schurig, V. (2000). Stereointegrity of Tröger's base: gas-chromatographic determination of the enantiomerization barrier. *J. Am. Chem. Soc.* 122: 1424–1430.

155 Trapp, O., Schurig, V., and Kostyanovsky, R.G. (2004). The control of the nitrogen inversion in alkylsubstituted diaziridines. *Chem. Eur. J.* 10: 951–957.

156 Trapp, O., Sahraoui, L., Hofstadt, W., and Könen, W. (2010). The stereodynamics of 1,2-dipropyldiaziridines. *Chirality* 22: 284–291.

157 Kostyanovsky, R.G., Kadorkina, G.K., Kostyanovsky, V.R. et al. (2000). Pronounced steric hindrance for nitrogen inversion in 1,3,4-oxadiazolidines. *Angew. Chem. Int. Ed.* 39: 2938–2940.

158 Kostyanovsky, R.G., Schurig, V., Trapp, O. et al. (2002). Chiral 1-alkoxyaziridines: resolution, nitrogen inversion, structure, and diastereomeric transformations. *Mendeleev Commun.* 12: 137–140.

159 König, W.A. (1993). Collection of enantiomer separation factors obtained by capillary gas chromatography on chiral stationary phases. *J. High Resolut. Chromatogr.* 16: 338–352.

160 Bertsch, K., Rahman, M.A., and Jochims, J.C. (1979). Enantiomere Pentatetraene und deren thermische Racemisierungsbarrieren. *Chem. Ber.* 112: 567–576.

161 Francke, W., Reith, W., Bergström, G., and Tengö, J. (1980). Spiroketals in the mandibular glands of Andrena bees. *Naturwissenschaften* 67: 149–150.

162 Baker, R., Herbert, R., Howse, P.E. et al. (1980). Identification and synthesis of the major sex pheromones of the olive fly. *Chem. Commun.* 52–53.

163 Francke, W., Heemann, V., Gerken, B. et al. (1977). 2-Ethyl-1,6-dioxaspiro[4.]nonane, principal aggregation pheromone of pityogenes chalcographus (L.). *Naturwissenschaften* 64: 590–591.

164 Trapp, O. and Schurig, V. (2001). Determination of interconversion barriers by dynamic gas chromatography: epimerization of chalcogran. *Chem. Eur. J.* 7: 1495–1502.

165 König, W.A., Gehrcke, B., Icheln, D. et al. (1992). New, selectively substituted cyclodextrins as stationary phases for the analysis of chiral constituents of essential oils. *J. High Resolut. Chromatogr.* 15: 367–372.

166 Bicchi, C., Artuffo, G., D'Amato, A. et al. (1992). Cyclodextrin derivatives in the GC separation of racemic mixtures of volatile compounds: part IV. *J. High Resolut. Chromatogr.* 655–658.

167 König, W.A., Icheln, D., Runge, T. et al. (1991). Gas chromatographic enantiomer separation of agrochemicals using modified cyclodextrins. *J. High Resolut. Chromatogr.* 14: 530–536.

168 Wenz, G., Mischnick, P., Krebber, R., and König, W.A. (1990). Preparation and characterization of per-O-pentylated cyclodextrins. *J. High Resolut. Chromatogr.* 13: 724–727.

169 König, W.A., Mischnick-Lübbecke, P., Brassat, B. et al. (1988). Improved gas chromatographic separation of enantiomeric carbohydrate derivatives using a new chiral stationary phase. *Carbohydr. Res.* 183: 11–17.

170 König, W.A., Lutz, S., Hagen, M. et al. (1989). Cyclodextrins as chiral stationary phases in capillary gas chromatography. Part IV: heptakis(2,3,6-tri-O-pentyl)-β-cyclodextrin. *J. High Resolut. Chromatogr.* 12: 35–39.

171 Effenberger, F., Ziegler, T., and Förster, S. (1987). Enzyme-catalyzed cyanohydrin synthesis in organic solvents. *Angew. Chem. Int. Ed. Eng.* 26: 458–460.

172 Kośćielski, T., Sybilska, D., Belniak, S., and Jurczak, J. (1986). Application of a gas-liquid chromatography system with α-cyclodextrin for monitoring the stereochemical course of β-pinene hydrogenation. *Chromatographia* 21: 413–416.

173 König, W.A., Lutz, S., Colberg, C. et al. (1988). Cyclodextrins as chiral stationary phases in capillary gas chromatography. Part III: hexakis(3-O-acetyl-2,6-di-O-pentyl)-α-cyclodextrin. *J. High Resolut. Chromatogr.* 11: 621–625.

174 Ciucanu, I. and Kerek, F. (1984). A simple and rapid method for the permethylation of carbohydrates. *Carbohydr. Res.* 131: 209–217.

175 König, W.A., Lutz, S., Wenz, G., and Bey, E.V.D. (1988). Cyclodextrins as chiral stationary phases in capillary gas chromatography. Part II: heptakis(3-O-acetyl-2,6-di-O-pentyl)-ß-cyclodextrin. *J. High Resolut. Chromatogr.* 11: 506–509.

176 König, W.A., Krebber, R., and Mischnick, P. (1989). Cyclodextrins as chiral stationary phases in capillary gas chromatography – part V: octakis(3-O-butyryl-2,6-di-O-pentyl)-y-cyclodextrin. *J. High Resolut. Chromatogr.* 12: 732–738.

177 Nowotny, H.-P., Schmalzing, D., Wistuba, D., and Schurig, V. (1989). Extending the scope of enantiomer separation on diluted methylated ß-cyclodextrin derivatives by high-resolution gas chromatography. *J. High Resolut. Chromatogr.* 12: 383–393.

178 Schurig, V., Jung, M., Schmalzing, D. et al. (1990). CGC enantiomer separation on diluted cyclodextrin derivatives coated on fused silica columns. *J. High Resolut. Chromatogr.* 13: 470–474.

179 Bicchi, C., Artuffo, G., D'Amato, A., and Nano, G.M. (1991). Permethylated cyclodextrins in the GC separation of racemic mixtures of volatiles. *J. High Resolut. Chromatogr.* 301–305.

180 Bicchi, C., Artuffo, G., D'Amato, A. et al. (1992). Cyclodextrin derivatives in the GC separation of racemic mixtures of volatile compounds, part V heptakis 2,6-dimethyl-3-Pentyl-~-cyclodextrins. *J. High Resolut. Chromatogr.* 15: 710–714.

181 Li, W.-Y., Jin, H.L., and Armstrong, D.W. (1990). 2,6-Di-O-pentyl-3-O-trifluoroacetyl cyclodextrin liquid stationary phases for capillary gas chromatographic separation of enantiomers. *J. Chromatogr. A* 509: 303–324.

182 Dietrich, A., Maas, B., Karl, V. et al. (1992). Stereoisomeric flavor compounds. Part LV. Stereodifferentiation of some chiral volatiles on heptakis(2,3-di-O-acetyl-6-O-tert-butyldimethylsilyl)-ß-cyclodextrin. *J. High Resolut. Chromatogr.* 15: 176–179.

183 Dietrich, A., Maas, B., Messer, W. et al. (1992). Stereoisomeric flavor compounds. Part LVIII. The use of heptakis(2,3-di-O-methyl-6-O-tert-butyldimethylsilyl)-ß-cyclodextrin as a chiral stationary phase in flavor analysis. *J. High Resolut. Chromatogr.* 15: 590–593.

184 Cagliero, C., Sgorbini, B., Cordero, C. et al. (2016). Enantioselective gas chromatography with derivatized cyclodextrins in the flavour and fragrance field. *Isr. J. Chem.* 56: 925–939.

185 Kamuf, M. and Trapp, O. (2011). Stereodynamics of tetramezine. *Chirality* 23: 113–117.

186 Blum, W. and Aichholz, R. (1990). Gas chromatographic enantiomer separation on tert-butyldimethylsilylated β-cyclodextrin diluted in PS-086. A simple method to prepare enantioselective glass capillary columns. *J. High Resolut. Chromatogr.* 13: 515–518.

187 Maas, B., Dietrich, A., Karl, V. et al. (1993). Tert-Butyldimethylsilyl-substituted cyclodextrin derivatives as versatile chiral stationary phases in capillary GC. *J. Microcolumn Sep.* 5: 421–427.

188 Bicchi, C., D'Amato, A., and Rubiolo, P. (1999). Cyclodextrin derivatives as chiral selectors for direct gas chromatographic separation of enantiomers in the essential oil, aroma and flavour fields. *J. Chromatogr. A* 843: 99–121.

189 Kamuf, M. and Trapp, O. (2013). Stereodynamics of small 1,2-dialkyldiaziridines. *Chirality* 25: 224–229.

190 Zawatzky, K., Kamuf, M., and Trapp, O. (2015). Chiral 1,2-dialkenyl diaziridines: synthesis, enantioselective separation and nitrogen inversion barriers. *Chirality* 27: 156–162.

191 Hochmuth, D.H. and König, W.A. (1999). Synthesis, resolution and determination of energy barriers to rotation of atropisomeric, planar-chiral [n]paracyclophanes by dynamic enantioselective gas chromatography and computer simulation. *Tetrahedron Asymmetry* 10: 1089–1097.

192 Hochmuth, D.H. and König, W.A. (1996). Determination of the rotational energy barrier of planar-chiral Cyclophanes using dynamic enantioselective gas chromatography and computer simulation. *Liebigs Ann.* 947–951.

193 Prokopov, A.A., Kotlova, L.I., and Berlyand, A.S. (2006). Experimental evaluation of bioavailability of tetramezine from tablets. *Pharm. Chem. J.* 40: 463–464.

194 Prokopov, A.A., Kotlova, L.I., and Berlyand, A.S. (2005). Experimental pharmacokinetics of tetramezine in rats. *Pharm. Chem. J.* 39: 345–349.

195 Prokopov, A.A., Kotlova, L.I., and Berlyand, A.S. (2007). Tetramezine metabolism in rats. *Pharm. Chem. J.* 41: 7–9.

196 Ito, T., Ando, H., Suzuki, T. et al. (2010). Identification of a primary target of thalidomide teratogenicity. *Science* 327: 1345–1350.

197 Schoetz, G., Trapp, O., and Schurig, V. (2000). Dynamic micellar electrokinetic chromatography – determination of the enantiomerization barrier of oxazepam, temazepam and lorazepam. *Anal. Chem.* 72: 2758–2764.

198 Zhang, Y., Breitbach, Z.S., Wang, C., and Armstrong, D.W. (2010). The use of cyclofructans as novel chiral selectors for gas chromatography. *Analyst* 135: 1076–1083.

199 Huang, K., Zhang, X., and Armstrong, D.W. (2010). Ionic cyclodextrins in ionic liquid matrices as chiral stationary phases for gas chromatography. *J. Chromatogr. A* 1217: 5261–5273.

200 Ding, J., Welton, T., and Armstrong, D.W. (2004). Chiral ionic liquids as stationary phases in gas chromatography. *Anal. Chem.* 76: 6819–6822.

201 Sun, X., Xu, J., Zhao, X. et al. (2013). Study of chiral ionic liquid as stationary phases for GC. *Chromatographia* 76: 1013–1019.

3

Chiral Separation by Supercritical Fluid Chromatography

Emmanuelle Lipka

Laboratoire de Chimie Analytique – UFR3S-Faculté de Pharmacie de Lille, Inserm U1167, Université de Lille, Lille Cedex, France

3.1 Introduction

"Perhaps looking-glass milk is not good to drink..." said Alice to her black kitten in the Lewis Carroll's novel, *Through the Looking-Glass and What Alice Found There* (Figure 3.1).

This odd idea gives a great image to merely explain chirality features, and foreshadows, in 1871, what can be the consequence of this phenomenon. Chiral molecules cannot be superimposed on their mirror image, known as enantiomers or optical isomers, and different enantiomers may have very different effects. Indeed, while the origin of chirality in life is still not elucidated, its profound impact on most biological mechanisms is now well established. This is mainly due to the *thalidomide* story providing one of the most dramatic examples of the influence of chirality. Indeed, thalidomide (also named Contergan) was sold as a racemate and in the late 1950s led to severe malformations in children. It was later found that this was due to the teratogenic nature of the (*S*)-enantiomer [1]. However, things maybe not as simple as it seems. Another point of view is that the enantioselectivity exists in the teratogenicity of thalidomide (i.e. one enantiomer is teratogenic, whereas the other is safe), which results from the fast hydrolysis to chirally stable teratogenic metabolites, rather than thalidomide enantiomers themselves [2, 3].

Chiral Separations and Stereochemical Elucidation: Fundamentals, Methods, and Applications, First Edition. Edited by Quezia Bezerra Cass, Maria Elizabeth Tiritan, João Marcos Batista Junior, and Juliana Cristina Barreiro.

Figure 3.1 Alice stepping through the looking-glass. *Source:* Through the looking-glass and what Alice found there, illustration by John Tenniel, Public domain, via Wikimedia Commons.

The importance of chirality is involved in many different fields such as therapeutics, organic synthesis, ecology, and food chemistry [4]. The global market for chiral chemicals was valued at US$39.79 billion in 2015 and is projected to expand to US$96.89 billion by 2023. This market comprises flavors and fragrances; agrochemicals such as herbicides, pesticides (30% of registered pesticides are chiral compounds), plant growth regulators, and fungicides; and lastly pharmaceuticals that accounted for 72.5% of the global market in 2015. Due to the impact of chirality on both the physical and biological properties of molecules, the separation of enantiomers (and stereoisomers in broader terms) has been extensively studied for two centuries now and remains a hot topic regarding the expansion of the number of chiral chemicals. Two strategies can be envisaged to obtain the enantiomers of a compound: the first consists of designing an enantioselective synthesis of the desired enantiomer. Two independent syntheses must be developed if both enantiomers are wanted, and typically the enantiomeric excess must be determined for each enantiomer obtained. The second approach relies on the synthesis of a racemate that must be subsequently separated into the corresponding enantiomers. Therefore, the resolution can be implemented through different methods: crystallization, enzymatic bioconversion, and chromatographic separation. Within analytical chromatography, gas chromatography (GC), high-performance liquid chromatography (HPLC), and supercritical fluid chromatography (SFC) can be used [5]. Upscaling the chromatography to a preparative scale is possible for both HPLC and SFC. Thanks to recent upgrades in instrumentation [6] SFC has completely caught up with HPLC, both in analytical and in preparative scales, taking advantage of its intrinsic properties and green features [7]. Nowadays chiral SFC has a broad range of application areas. The main one is pharmaceuticals where drug candidates are largely concerned by enantioseparation, natural products comprising chiral fragrances, agrochemicals, forensics, and cosmetics.

3.2 Characteristics and Properties of Supercritical Fluids

Supercritical fluids have properties intermediate between those of gases and liquids. The fluid can be adjusted for compounds to be sufficiently soluble to be eluted, while at the same time the viscosity and diffusion coefficients can be high enough to bring about relatively rapid mass transport. Table 3.1 shows typical values for the density and viscosity of a gas, supercritical fluid, and liquid, taking carbon dioxide as an example. Density is more than half that of the liquid, giving rise to reasonable solubility. In contrast, however, the viscosity of a supercritical fluid is much closer to that of a gas than that of a liquid. Thus pressure drop through a packed column is less than for the equivalent liquid process. Diffusion coefficients, also shown in Table 3.1, for naphthalene in carbon dioxide (CO_2) are higher in a supercritical fluid than in a liquid.

This has the advantage of faster transport. Supercritical fluid is the main constituent of the mobile phase used in SFC. So the consequences of these particular properties (i.e. low viscosity, high diffusivity of the mobile phase,

Table 3.1 Density, ρ, and viscosity, η, of carbon dioxide and the molecular diffusion coefficient, D, for naphthalene in carbon dioxide, D, under gas, supercritical, and liquid conditions.

	ρ (kg m^{-3})	η (μPa s)	D (m^2 s^{-1})
Gas, 313 K, 1 bar	2	16	5.1×10^{-6}
Supercritical, 313 K, 100 bar	632	17	1.4×10^{-8}
Liquid, 300 K, 500 bar	1029	133	8.7×10^{-9}

and low pressure drop) permit high flowrate (then short run time) with similar peak efficiency. Among fluids, such as fluorocarbons or ammonia, CO_2 is the most used one because its critical point (P_{crit} = 73 bar, T_{crit} = 31.1 °C) is easily accessible in the laboratory and compatible with the stability of the compounds separated. In addition, it is miscible with various organic solvents. In order to get to the supercritical fluid state, CO_2 has to be pumped and then heated. CO_2 is stored in B50 cylinders. The SFC system can be fed with either liquid CO_2 or gas CO_2, but in this latter case CO_2 must be cooled below 0 °C with a pressure higher than 55 bar before the SFC pump. However, although increasing with the pressure, the polarity of CO_2 is low, considered as similar to the eluting strength of pentane, which is the reason why SFC implementing pure CO_2 was used for nonpolar compounds. The addition of an organic solvent called co-solvent strongly increases the compound solubility for both nonpolar and polar ones. This co-solvent is usually added to the mobile phase up to 50% and can be an alcohol, acetonitrile (ACN), or even dichloromethane (DMC) or tetrahydrofuran (THF) most of the time. It is worth noting that the addition of a liquid modifier moves conditions to the subcritical area. Obviously enriching the mobile phase with CO_2 will reduce the volume of the organic solvent used. This reduction in solvent volumes allows higher product concentration and reduced time and cost of purification. It is noteworthy that CO_2 is a nontoxic solvent, nonflammable, and a renewable resource and can be considered as eco-friendly compared to the organic solvents used in chiral HPLC. This feature fully answers the new guidelines leading to environmentally friendly chromatography. Thus, SFC is starting to carve out a niche. Like in HPLC, the main parameter to obtain a chiral separation is the stationary phase, then the type and the percentage of co-solvent in the CO_2-based mobile phase. Flow rate and temperature are operating parameters that can be optimized. The main difference with the HPLC is that, in SFC, a back-pressure is applied after the column to ensure a constant pressure drop along the column so this parameter can also be optimized. The first application of a supercritical fluid as mobile phase for chromatography was reported by Klesper et al. [8] in 1962. The survival of SFC in the 1980s and 1990s is mainly due to its recognized advantages for enantioseparations, where it often provided improved resolution at higher speed than HPLC. Indeed, chiral separation is one field where the benefits of SFC have always been recognized, since the first SFC separation of enantiomers by Mourier et al. [9]. Nowadays, apart from enantioseparations in agrochemicals, forensics, natural products, and cosmetics, SFC has found applications in different domains such as food sciences, fossil fuels, polymers, and bioanalysis.

3.3 Development of a Chiral SFC Method

Among the different possibilities developed for enantioseparation, chromatographic resolution of a racemate is the most widely used method. Two approaches can be pursued: the first, “indirect” one, forms true diastereoisomers through derivatization of the compound using an enantiomerically pure chiral derivatization agent (CDA), which can then be subsequently separated in an achiral environment. The second, “direct” one, forms transient diastereoisomers between the solute and either a chiral stationary phase (CSP) or a chiral additive to the mobile phase (CAMP). While the CAMP method is classically employed in capillary electrophoresis (CE) using most of the time cyclodextrins, this direct approach is not very popular in HPLC nor in SFC, mainly because it is only suitable for analytical-scale and not preparative-scale separations. Direct methodologies based on the use of CSPs are most conveniently employed [10].

3.3.1 Chiral Stationary Phases

Many different types of CSPs are nowadays commercially available. Those are the following: polysaccharide, cyclodextrins, Pirkle or brush-type, cinchona-alkaloid, cyclic oligosaccharide, macrocyclic glycopeptide, synthetic polymeric and molecularly imprinted polymer, and immobilized chiral protein-based and crown-ether stationary phases. It is not the intention here to deeply discuss their characteristics or enantioseparation mechanism, as this it will be addressed in other chapters of this book. However, one must be aware that CSPs used in HPLC are used in SFC as well, except a few (i.e. immobilized chiral protein-based, cyclodextrins, and crown-ether stationary phases). Unlike reversed-phase chromatography having its C18 column, no universal stationary phase is able to resolve all enantiomeric pairs. Thus a diversity of CSPs is needed. With regard to this specificity, the new analytical SFC devices purchased by Sepiatec, Waters, Agilent, Shimadzu, Berger, and Pic solution to cite a few offer the possibility of working with several columns placed in parallel in the oven, facilitating the screening and saving time. Among the existing CSPs, the polysaccharide based are the most widely used because of their broad range of applications and high loading capacity, which is an advantage for preparative scale (see Chapter 5). The tris-(3,5-dimethylphenylcarbamate) of cellulose and amylose are extremely popular, together with the cellulose tris-(4-methylbenzoate) and the amylose tris-[(*S*)-α-methylbenzylcarbamate]; these columns form the so-called “golden four” and cover more than 80% of enantioseparations. In 2005, Maftouh et al. proposed a very useful screening strategy with those four

CSPs and achieved the enantioseparation of the 3,5-difluoro mandelic acid [11]. An example of the enantioseparation of a proprietary compound bearing two stereogenic centers (four stereoisomers) is shown in Figure 3.2 for four different CSPs, corresponding to the "golden four."

New halogenated polysaccharides have reached the market, for instance amylose tris-(5-chloro-2-methylphenylcarbamate) or cellulose tris-(3-chloro-4-methylphenylcarbamate). Some of them have proven their complementary selectivity to non-chlorinated CSPs [12]. This is illustrated in Figure 3.3, on the amylose tris-(5-chloro-2-methylphenylcarbamate) CSP, for a proprietary compound presenting four stereoisomers.

The limited resistance of these coated phases was a major drawback. This was overcome by an approach aiming to immobilize polysaccharide derivatives, allowing the expansion of the choice of co-solvents. It is worth to note that coated and immobilized CSPs having the same chiral selector can exhibit different enantioselectivities toward a racemate; moreover, the

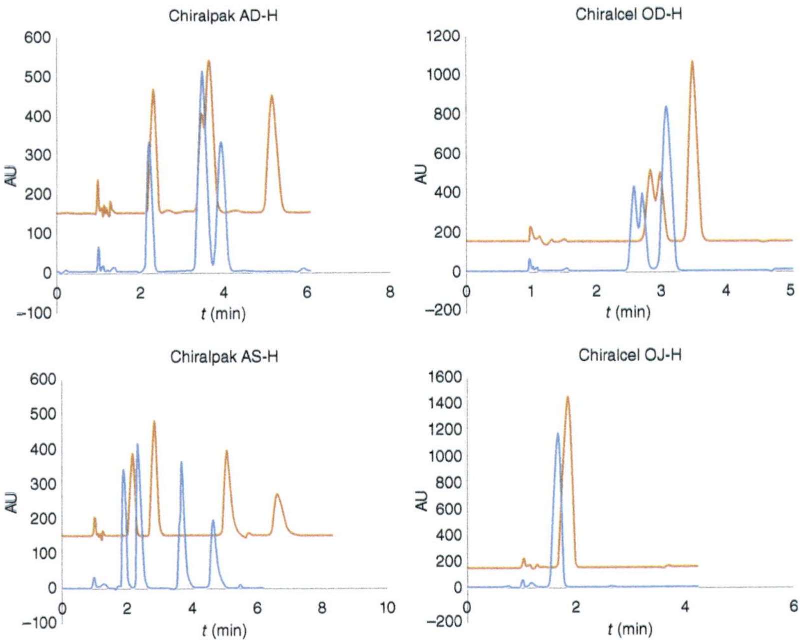

Figure 3.2 Proprietary compound (1 mM) on the tris-(3,5-dimethylphenylcarbamate) of amylose CSP (Chiralpak AD-H), on the tris-((*S*)-α-methylbenzylcarbamate) of amylose CSP (Chiralpak AS-H), on the tris-(3,5-dimethylphenylcarbamate) of cellulose CSP (Chiralcel OD-H), and on the tris-(4-methylbenzoate) of cellulose CSP (Chiralcel OJ-H), 20% MeOH (blue line) or EtOH (orange line), flow rate of 3 ml·min^{-1}, 120 bar, 40 °C, λ = 210 nm.

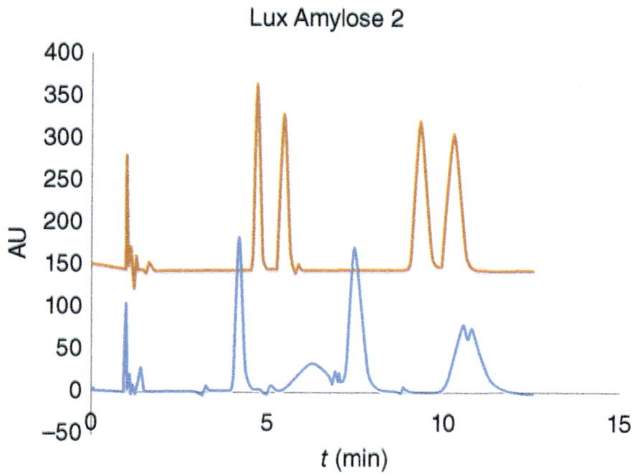

Figure 3.3 Proprietary compound (1 mM) on the tris-(5-chloro-2-methylphenylcarbamate) of amylose (Lux amylose-2) CSP, 20% MeOH (blue line) or EtOH (orange line), flow rate of 3 ml·min^{-1}, 120 bar, 40 °C, λ = 210 nm.

immobilized versions of the coated polysaccharide CSPs do not seem to display higher separation capability over the coated ones. This is illustrated in Figure 3.4, on the tris-(3,5-dimethylphenylcarbamate) of cellulose and amylose CSP immobilized versions, for a proprietary compound presenting four stereoisomers.

The last way of development of these polysaccharides relates to the particle size that was, until very recently, 5 μm fully porous particle (FPP). Coated polysaccharide phases based on 2.5 μm FPP were marketed by Waters and immobilized polysaccharide phases based on 1.6 μm FPP were marketed by Daicel.

3.3.2 Mobile Phases

In the following section we will study the effect of the type and the percentage of the co-solvent chosen in order to improve the resolution. Through the Purnell equation, one can see that retention, selectivity, and, most of all, the number of theorical plates influence resolution:

$$\mathrm{Rs} = \frac{1}{4}\sqrt{N_2} * \frac{k_2}{1+k_2} * \frac{\alpha - 1}{\alpha} \tag{3.1}$$

where the retention factor (k) is a mean of measuring the retention of an analyte on the chromatographic column for the second eluted enantiomer; k_2 was calculated by $k_2 = (t_{r2} - t_0)/t_0$, where t_0 was measured using

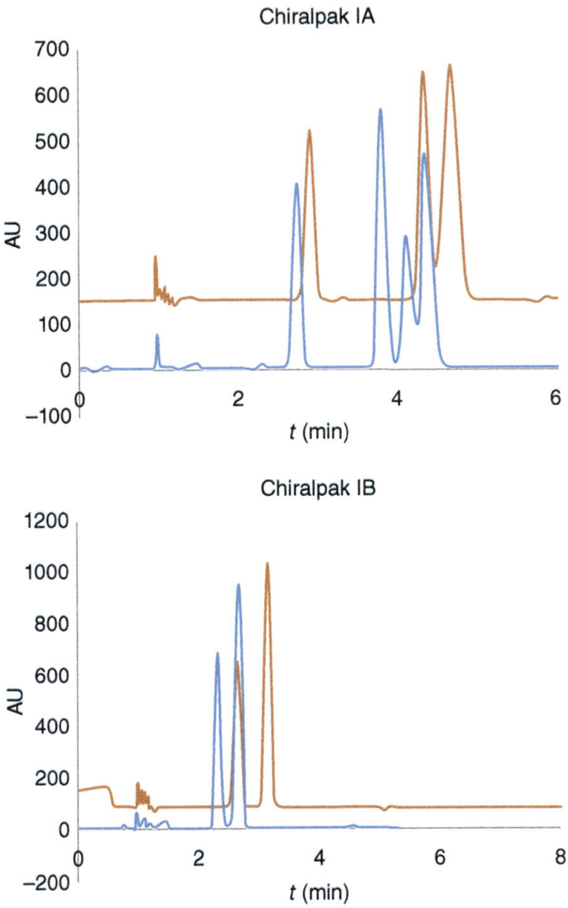

Figure 3.4 Proprietary compound (1 mM) on immobilized tris-(3,5-dimethylphenylcarbamate) of amylose CSP (Chiralpak IA) and on immobilized tris-(3,5-dimethylphenylcarbamate) of cellulose CSP (Chiralpak IB), 20% MeOH (blue line) or EtOH (orange line), flow rate of 3 ml·min^{-1}, 120 bar, 40 °C, λ = 210 nm.

1,3,5-tri-*tert*-butylbenzene (TTBB). N, the number of theoretical plates (or efficiency) was calculated for the second eluted peak (N_2), using this equation: $N = 16\ (t_{r2}/\omega_2)^2$. The selectivity α was equal to k_2/k_1 and ω was the width at baseline. Generally speaking, regarding the mobile phase, it is necessary to add a co-solvent to elute polar compounds. This co-solvent is most of the time a short-chain alcohol, such as methanol (MeOH), ethanol (EtOH), or isopropanol (2-PrOH); ACN can also be used, in proportions that may vary between 1 and 99% in isocratic or gradient elution mode, even if the range 5–50% is more usual, particularly in chiral separations. The immobilized CSPs also represent an opportunity to explore different modifiers as organic solvents, such as DCM, THF, or methyl *tert*-butyl ether (MTBE), which are all miscible with supercritical CO_2, in a quest for novel enantioselectivities. For highly hydrophilic analytes (peptides, sugars, or nucleobases), the addition of water to the mobile phase is possible to achieve reasonable retention. However, the proportion of water is limited due to miscibility limits and increase in detection noise. It must be highlighted that because of the high pressures applied to maintain the fluid in a sub- or supercritical state, the eluting strength of mixtures of organic solvents with CO_2 is comparable to that of the pure liquid solvent until significant levels (40–60%) of CO_2 are used. In SFC, the co-solvent has two main effects: first, the organic solvent changes the polarity of the mobile phase, modifying its solubilizing properties; second, it changes the density of the mobile phase, particularly when pressure and temperature conditions are such that fluid compressibility is high (close to the critical point and when the fluid is more gas-like) [13]. However, this effect is of little impact under the operating conditions used in applications of chiral SFC, where the fluid is denser, thus more liquid-like.

3.3.2.1 Mobile Phase: Type of Co-solvent Used

As aforementioned, co-solvent is most of the time MeOH, EtOH, 2-PrOH, or ACN. Retention has been seen to increase when modifier polarity decreases upon moving from MeOH to EtOH and 2-PrOH, making it possible to consider chiral SFC as a normal-phase chromatographic mode; however, some studies have shown totally different retention variations. Thus, the effect of the nature of the modifier on retention seems to be both compound- and CSP-dependent. Compared to MeOH, ACN as a co-solvent often results in an increase in retention. Concerning the selectivity, no trend has been drawn as a function of the type of co-solvent. In theory, band broadening partly arises from slow diffusion between mobile and stationary phases, and thus, efficiency should increase when the mobile phase viscosity decreases (as diffusion rates increase in less viscous solvents). So, efficiency would be expected to increase from higher long-chain alcohols to shorter ones (MeOH). However, this trend is not systematically observed either.

In conclusion, the co-solvent type strongly impacts the retention and the enantioselectivity but varies in an unpredictable way. This is illustrated in Figures 3.2–3.4, with EtOH and MeOH toward the same compound on various polysaccharide CSPs. For this reason, several mobile phase compositions are usually tested in systematic screening strategies.

3.3.2.2 Mobile Phase: Percentage of Co-solvent Used

Once a co-solvent is chosen, the effects of varying its proportions in CO_2 are generally investigated to improve resolution. In SFC, by increasing the percentage of the co-solvent (the most polar constituent of the mobile phase), the retention decreases up to a certain point where retention might increase again. Relatively large changes in retention are seen under small percentages (for instance going from 5 to 10%) of the co-solvent, whereas the relative variations are smaller at higher percentages of the modifier. The selectivity usually remains constant or slightly decreases under larger modifier proportions. Concerning the efficiency, two successive trends have been observed: the efficiency increases with the first few percent of the co-solvent because of mobile phase absorption on the CSP and then decreases when solvent portion further increases because the diffusion rates of the solutes in the mobile phase decrease with the addition of a viscous solvent. Nevertheless, variations in selectivity and efficiency are generally less significant than the variation in retention.

Thus, the effect of changing the solvent proportion essentially results in changes in retention.

3.3.2.3 Mobile Phase: Use of Additives

A third mobile phase component is often added, typically at a small level of about 0.1–0.5% and named "additive." It can be used for different purposes: either for detection issues, to reduce excessive retention, or for the purpose of achieving good resolution, as it is often observed that introduction of small amounts of an acid or a base in the co-solvent contributes to improve peak shapes. The effect of the nature of the additive and of its concentration in the co-solvent on the enantioselectivity is a controversial topic. Usually, basic additives, such as triethylamine (TEA), diethylamine (DEA), or isopropylamine (IPA), are employed for the separations of basic compounds; acidic additives, such as trifluoroacetic acid (TFA) or formic acid (FA), are used for acidic compounds, but a basic additive for an acidic compound or vice versa can also generate good resolution. For a comprehensive study on this topic, the reader is referred to reference [14].

3.4 Operating Parameters

The following parameters are rarely optimized in order to obtain a separation. Most of the time, their effects are studied in order to reduce analysis time or to further improve resolution once the enantioseparation has already been established.

3.4.1 Effect of the Flow Rate

Nowadays, flow rates in analytical SFC are between 2 and 6 ml·min^{-1}. It is noteworthy that the low viscosity of the fluids permits using elevated flow rates while pressure drop still remains under the limits of pumping systems found on conventional devices. Usually, flow rate is optimized to shorten analysis time. Increasing the flow rate decreases the retention times. Indeed, as the fluid is compressible, changing its linear velocity in the column affects its density and, thus, affects its eluting strength. Considering efficiency, it remains unchanged in the column due to the high diffusivity of the solutes, which is the main advantage of the technique. Thus, considering the Purnell equation (Eq. 3.1), in theory, resolution decreases with increasing flow rates. Notwithstanding, several studies show that increasing the flow rate led to a marked decrease in retention but only a slight decrease in resolution. This is well illustrated in Figure 3.5, for the chiral separation of *trans*-stilbene oxide under flow rates ranging from 2 to 6 ml·min^{-1}.

3.4.2 Effect of the Outlet Pressure (Back-pressure)

The main difference in comparison with HPLC remains in the fact that in SFC a back-pressure also called "outlet pressure" is applied after the column to ensure a constant pressure drop along the stationary phase. This

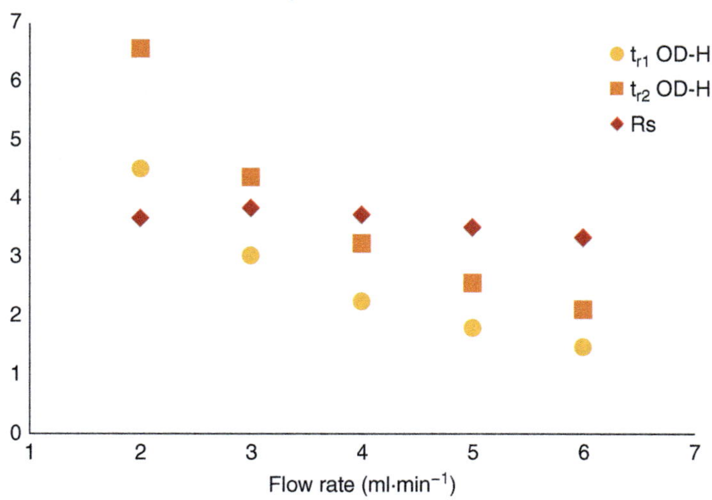

Figure 3.5 Plots of retention time of the first enantiomer (t_{r1}), second enantiomer (t_{r2}), and resolution (Rs) on tris-(3,5-dimethylphenylcarbamate) of cellulose CSP (Chiralcel OD-H) for trans-stilbene oxyde, against the flow rate; 150 bar back-pressure; 40 °C; λ = 210 nm.

specificity was underlined by one of the SFC gurus, Larry Taylor, who used to explain that SFC is just a chromatographic technique with a new dimension of flexibility: the outlet pressure. So, this parameter could also be optimized even if its impact is not so strong. The stated back-pressure values usually range from 80 to 200–250 bar. It is widely recognized that with using high portions of co-solvent, low pressure values should be avoided, because of the risk of occurrence of a two-phase system, producing noisy baseline in the detector during the enantioseparation. When optimizing outlet pressure, two cases can be encountered.

3.4.2.1 Effect of Pressure When the Mobile Phase is a Gas-Like Fluid

Similar to the flow rate, an elevation of the pressure causes an increase in the density of the fluid. Thus, the elution strength increases leading to a decrease in retention times while separation factors decrease to a lesser extent or remain constant. Back-pressure also influences the efficiency of the column. Indeed, diffusivity is reduced under high pressure conditions, and thus, column efficiency may be altered. The range of variation depends on mobile phase composition. When the fluid is more gas-like, i.e. when pure CO_2 or small percentages of modifier (below 5%) are employed as mobile phase, the effect of outlet pressure on resolution is significant, to such an extent that pressure gradients can be implemented similar to mobile phase gradients. This effect is illustrated in Figure 3.6 for the chiral separation of *trans*-stilbene oxide, under back-pressures ranging from 80 to 200 bar, with 5% of EtOH as co-solvent.

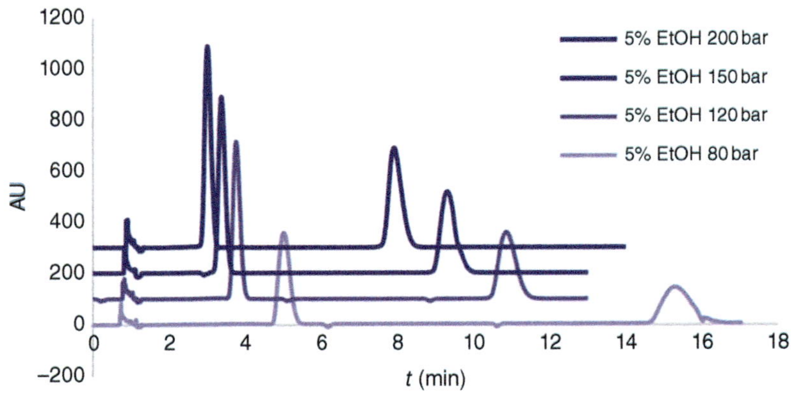

Figure 3.6 Overlayed chromatograms for trans-stilbene oxyde, with a carbon dioxide mobile phase containing 5% of EtOH at different bar back-pressure ranging from 80 to 200 bar on the tris-(3,5-dimethylphenylcarbamate) of amylose CSP (Chiralpak AD-H); 40 °C; 4 ml·min^{-1}; λ = 210 nm.

3.4.2.2 Effect of Pressure When the Mobile Phase is a Liquid-Like Fluid

When the mobile phase is more liquid-like, containing above 10% of co-solvent, the pressure effect is less important. This effect is illustrated in Figure 3.7 for the chiral separation of *trans*-stilbene oxide, under back-pressures ranging from 80 to 200 bar, but with 40% of EtOH as co-solvent.

Globally, the influence on selectivity and column efficiency is generally less significant than the effect on retention, thus resolution is less affected by a change in pressure.

3.4.3 Effect of Temperature

Temperature is an important parameter to optimize as it deeply influences enantioseparation, in spite of the fact that temperature has the most complex effect as it may act in two opposite ways on the retention. Indeed, at constant pressure drop, an increase in the temperature reduces the density of CO_2 (thus elution strength), leading to an increased retention. Simultaneously, an increase in temperature enhances the coefficient of diffusion of solutes and then decreases their retention.

The relationship between the temperature and the retention factor can be expressed by the Van't Hoff equation:

$$\ln(k) = -\left(\frac{\Delta H^\circ}{RT}\right) + \left(\frac{\Delta S^\circ}{R}\right) + \ln(\beta) \tag{3.2}$$

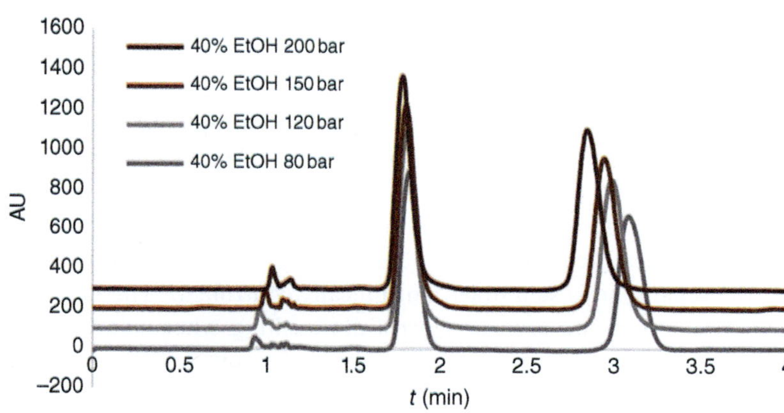

Figure 3.7 Overlayed chromatograms for trans-stilbene oxyde, with a carbon dioxide mobile phase containing 40% of EtOH at different bar back-pressure ranging from 80 to 200 bar on the tris-(3,5-dimethylphenylcarbamate) of amylose CSP (Chiralpak AD-H); 40 °C; 4 ml·min^{-1}; λ = 210 nm.

where T is the absolute temperature, R is the ideal gas constant, β is the phase ratio, and ΔH° and ΔS° are the standard molar enthalpy and standard molar entropy of the solute transfer (from mobile to stationary phase), respectively. Therefore, in theory, $\ln(k)$ plotted against $1/T$ should be linear. As for outlet pressure two main trends of the Neperian logarithm of the retention factor versus reversed temperature are reported as a function of the mobile phase state.

3.4.3.1 Effect of Temperature When the Mobile Phase is a Gas-Like Fluid

The $\ln(k)$ versus $1/T$ plot is decreasing, meaning that retention increases with an increased temperature. This trend is classically observed in SFC under low proportions of co-solvent, as the fluid density decreases, and thus the solvating power, when temperature rises, resulting in a reduction of the elution strength of the mobile phase.

3.4.3.2 Effect of Temperature When the Mobile Phase is a Liquid-Like Fluid

The $\ln(k)$ versus $1/T$ plot is crescent indicating that retention decreases with an increase in the temperature. This HPLC-close trend is explained by the high proportion of co-solvent used and high back-pressure employed causing an increase in the mobile phase density that results in a more liquid-like fluid. Therefore, rising the temperature leads to an improved solute solubility in the mobile phase, thus decreasing retention times.

The variation of the enantioselectivity as a function of the temperature can be expressed by the following equation, deduced from Eq. (3.2):

$$\ln(\alpha) = -\left(\frac{\Delta\Delta H^\circ}{RT}\right) + \left(\frac{\Delta\Delta S^\circ}{R}\right) \tag{3.3}$$

where $\Delta\Delta H^\circ$ and $\Delta\Delta S^\circ$ are the enthalpy and entropy change of the solute-stationary phase interactions, respectively. Briefly, if the slope of the $\ln(\alpha)$ versus $1/T$ plot is positive, this indicates that a decrease in temperature leads to a separation factor increase, then the separation is said to be "enthalpically controlled." On the contrary, when the slope is negative, the separation is said to be "entropically controlled." The fact that separation factors are seen as always diminishing when temperature increases is not totally true and the variation of enantioselectivity is not predictable.

Most of the time in SFC (i.e. with low percentage of co-solvent), high temperature leads to an increase in column efficiency, through increased mobile phase diffusivity. This is even more true when the mobile phase contains high percentages of co-solvent.

Besides, temperature may affect the rigidity of the stationary phase backbone and analytes, thus changing the analyte entrance into the chiral cavities. In addition, the temperature effect is also related to the back-pressure. Employing pressures well above the critical pressure can be a way to overcome this problem.

In most current works, temperature is maintained at 30–40 °C, while early works were often conducted at lower temperatures close to room temperature (20–25 °C), and one must keep in mind that polysaccharide CSPs should not be operated above 50 °C. Thus, as most of the commercially available ovens are unable to cool the temperature below room temperature, the available range of temperature variation is not so broad, being between 15–20 and 50 °C in practice.

3.5 Detection

In earlier works, pure CO_2 mobile phase was essentially used, providing the advantage of possibly using the flame ionization detector (FID) as a universal detector. Nevertheless, this is not the case anymore. The most popular detectors used in chiral SFC are the same as those in chiral HPLC: UV–visible or diode-array detection (DAD). The CO_2-based mobile phase is also well adapted to implement evaporative light scattering detector (ELSD) or charged-aerosol detection (CAD) because of the ease of vaporization of the mobile phase after its nebulization. Even in the field of enantioseparation, optical rotation detectors (ORDs) or circular dichroism detectors (CDDs) are not commonly encountered. A short word should be said about the recent use of mass spectrometry with atmospheric pressure chemical ionization (APCI) or electrospray ionization (ESI) sources. This development is due to improved hyphenation of modern devices and may be useful when very low concentrations are targeted in environmental applications or in bioanalysis for instance.

3.6 Scale-Up to Preparative Separation

Whatever the quantity of racemates to be separated, the method is first developed on analytical system before upscaling on the preparative system. Usually an analytical screening is performed with a set of columns and various co-solvents in order to get the best separation. Once the best couple CSP/co-solvent is selected, the percentage of modifier is then chosen, based on the retention time. The strategy to determine the percentage of modifier is empirical and based on experience. Once the conditions are determined, various strategies can be used to scale up. Commonly the choice depends on the quantity of racemates to be separated. For instance, in drug discovery during the lead optimization stage less than 100 mg of racemic mixture is submitted for separation. Usually a good separation obtained on the analytical unit is transferred on the preparative one using the following process with no issue. The total amount of compound is dissolved in the co-solvent, with a concentration mainly based on the solubility of the compounds in the co-solvent; a few hundred milliliters are injected (depending on the column size), and depending on the separation, additional injections are done with larger volumes until touching the band resolution, meaning that the two peaks are just baseline resolved. This process is called volume overloading as opposed to concentration overloading, in which the volume of injection is fixed and the concentration is varied. Regarding the scaling described above and used mainly for small quantities, it is applied either with the same particle size used to pack both analytical and preparative column or with a different size. Whatever the column size used to develop the method, when the overloading is determined for the analytical column, by using the same stationary phase, the same column length, and the same injection concentration used afterwards on the preparative columns, the same peak shape and separation are obtained. In this case, the flow rate is tuned based on the ratio between the volume of the preparative column and the volume of the analytical column. For example, for an analytical column with dimension (250×4.6 mm I.D.) and a flow rate of 3 $ml{\cdot}min^{-1}$, to get the same peak shape and the same retention times, the flow rate for a preparative column with dimensions (250×21 mm I.D.) is 62.5 $ml{\cdot}min^{-1}$ generating the same pressure drop.

3.7 Conclusion

For sure, chiral liquid chromatography and chiral SFC are complementary methods as enantiomeric pairs can frequently be resolved with one and not with the other and vice versa. While reading these lines, it seems that method development for enantioseparation in SFC is more complex and unpredictable and requires more optimization steps, which can appear as time consuming. However, this drawback is overcome by short equilibration times and possible high flow rates reducing analysis time. The higher flow rates lead also to lower energy consumption and higher productivity compared with HPLC. SFC is relatively non-polluting as CO_2 is nontoxic, nonflammable, and a renewable resource, and only small amounts of hazardous solvents are used. This low solvent consumption is also an advantage from an economic point of view, and this is particularly true when the preparative scale is considered. The last point that deserves to be mentioned is the possibility to serially couple columns in SFC. Obviously it is also possible in HPLC, but the number of columns is quite limited because of the pressure drop. This is not the case in SFC as the high diffusivity and low viscosity of the CO_2 mobile phase ensure reasonable pressure drop while high efficiency is achieved by an extended length of columns. In addition, the possibility to connect both achiral and chiral columns in series, or even columns with different polarities, creates unique selectivity. Thus, improved efficiency and selectivity together offer the promise of high-resolution separations.

References

1 Blaser, H.U. (2013). Chirality and its implications for the pharmaceutical industry. *Rendiconti Lincei. Scienze Fisiche e Naturali* 24: 213–216.

2 Knoche, B. and Blaschke, G. (1994). Investigations on the in vitro racemization of thalidomide by high-performance liquid chromatography. *Journal of Chromatography A* 666: 235–241.

3 Tian, C., Xiu, P., Meng, Y. et al. (2012). Enantiomerization mechanism of thalidomide and the role of water and hydroxide ions. *Chemistry - A European Journal* 18: 14305–14313.

4 Tiritan, M.E., Fernandes, C., Maia, A.S. et al. (2018). Enantiomeric ratios: why so many notations? *Journal of Chromatography A* 1569: 1–7.

5 Tarafder, A. and Miller, L. (2021). Chiral chromatography method screening strategies: past, present and future. *Journal of Chromatography A* 1638: 461878.

6 West, C. (2019). Recent trends in chiral supercritical fluid chromatography. *Trends in Analytical Chemistry* 120 (2019): 115648.

7 Taylor, L.T. (2009). Supercritical fluid chromatography for the 21st century. *Journal of Supercritical Fluids* 47: 566–573.

8 Klesper, E., Corwin, D.A., and Turner, D. (1962). High pressure gas chromatography above critical temperatures. *Journal of Organic Chemistry* 27: 700–706.

9 Mourier, P.A., Eliot, E., Caude, M.H. et al. (1985). Supercritical and subcriticalfluid chromatography on a chiral stationary phase for the resolution of phosphine oxide enantiomers. *Analytical Chemistry* 129: 2819–2823.

10 Calcaterra, A. and D'Acquarica, I. (2018). The market of chiral drugs: chiral switches *versus de novo* enantiomerically pure compounds. *Journal of Pharmaceutical and Biomedical Analysis* 147: 323–340.

11 Maftouh, M., Granier-Loyaux, C., Chavana, E. et al. (2005). Screening approach for chiral separation of pharmaceuticals. Part III. Supercritical fluid chromatography for analysis and purification in drug discovery. *Journal of Chromatography A* 1088: 67–81.

12 Khater, S., Zhang, Y., and West, C. (2014). Insights into chiral recognition mechanism in supercritical fluid chromatography IV. Chlorinated polysaccharide stationary phases. *Journal of Chromatography A* 1363: 294–310.

13 West, C. (2014). Enantioselective separations with supercritical fluids. *Current Analytical Chemistry* 10: 99–120.

14 Speybrouck, D., Doublet, C., Cardinael, P. et al. (2017). The effect of high concentration additive on chiral separations in supercritical fluid chromatography. *Journal of Chromatography A* 1510: 89–99.

4

Chiral Separation by Capillary Electrophoresis and Capillary Electrophoresis–Mass Spectrometry

Fundamentals, Recent Developments, and Applications

Charles Clark[1], *Govert W. Somsen*[2,3], *and Isabelle Kohler*[2,3]

[1] *Leiden Academic Centre for Drug Research, Metabolomics and Analytics Centre, Leiden University, Leiden, the Netherlands*
[2] *Division of BioAnalytical Chemistry, Amsterdam Institute of Molecular Life Sciences (AIMMS), Vrije Universiteit Amsterdam, Amsterdam, the Netherlands*
[3] *Center for Analytical Sciences Amsterdam, Amsterdam, the Netherlands*

4.1 Introduction

Molecular chirality is a key parameter in understanding biological function. Molecules are chiral when their structures have non-superimposable mirror images. The enantiomers of one compound may have substantially different biological activities, chemical reactivities, and/or metabolic rates or pathways. For example, many drugs are chiral. The two enantiomers often present (vastly) different pharmacological effects and/or toxicity, which can have very serious consequences: one enantiomer can be more active, inactive, or toxic. There is also a growing attention for the role of molecular chirality in biomedical, clinical, forensic, food, and environmental fields. Longitudinal changes in enantiomer ratios of chiral compounds can be indicative of certain disorders, the time elapsed since an individual's death (i.e. postmortem interval), or microbiological contamination.

The separation of enantiomers and assessment of enantiomeric purity have thus become essential in various fields. However, separating enantiomers – i.e. compounds with the same molecular mass, size, functional groups, and highly similar structures – with sufficient resolution can be quite challenging. Moreover, enantiomers present in biological or environmental samples are often (minor) components of complex samples.

Chiral Separations and Stereochemical Elucidation: Fundamentals, Methods, and Applications, First Edition. Edited by Quezia Bezerra Cass, Maria Elizabeth Tiritan, João Marcos Batista Junior, and Juliana Cristina Barreiro.

This means that besides achieving enantioresolution, chiral molecules also have to be separated from sample matrix components and often preconcentrated in order to allow their unambiguous determination. Overall, this entails highly selective and sensitive analytical approaches.

The separation of enantiomers can only be achieved in a chiral environment or using a pretreatment that introduces enantioselectivity. Nowadays, enantiomer separations are predominantly achieved by liquid chromatography (LC), gas chromatography (GC), supercritical fluid chromatography (SFC), and capillary electrophoresis (CE) techniques. In the direct approach, separation media employing chiral stationary phases (for LC, GC, and SFC) or dissolved chiral selectors (predominantly for CE, occasionally for LC) are used to achieve enantioresolution. The enantiomers are separated based on their different affinities to the chiral phase or selector. In the indirect approach, enantiomers are derivatized with an enantiomerically pure chiral reagent prior to their separation leading to the formation of their respective diastereomers, which subsequently can be separated with conventional (non-chiral) methods.

Since the first enantiomeric separations by CE were demonstrated in 1985 [1], it has developed into a powerful technique for chiral analysis. CE is a microscale technique that employs open capillary separation tubes of typically 50–75 μm internal diameter. CE is a voltage-driven separation technique particularly well suited for the analysis of ionic or ionogenic species, although separation of neutral compounds is also possible after addition of charged micelles or selectors to the separation medium. CE provides high efficiencies (i.e. narrow peaks), which are also beneficial in enantioselective separations using chiral selectors (CSs). In CE, the CSs are simply dissolved in the separation buffer (referred to as background electrolyte [BGE]). In contrast to LC, GC, and SFC, there are no concerns with costly chiral columns with limited lifespan, because the commonly used fused-silica capillaries are inexpensive and disposable. Moreover, as the required BGE volumes are small (i.e. in the μl range), consumption of (pricey) CSs is very low, and different selectors can be screened easily and rapidly. Finally, injection volumes in CE are minute (i.e. in the nl range), which is advantageous when sample volumes are limited and/or repeated analysis of the same sample is needed.

This chapter discusses the principles, state-of-the-art approaches, as well as recent developments and applications of chiral CE. First, the fundamentals of CE and how these can be employed to achieve chiral separation are described. After giving a brief historical overview of chiral CE and its different separation modes, common and new CSs, as well as the utilization of microfluidics in enantioselective CE, are discussed. Selected applications (e.g. pharmaceutical, food, and environmental fields) are provided to underline the scope and benefits of using CE in chiral analysis. In the last part, special attention is paid to the coupling of chiral CE to mass spectrometry (MS), outlining its rationale, challenges, and strategies.

4.2 Principles of Chiral CE

4.2.1 Electrophoretic Mobility

In CE, separation is achieved by differential migration of charged solutes in a BGE-filled capillary with length L across which a high voltage V is applied. The resulting electric field E – which equals V/L – in the capillary will cause an ion to move with a constant velocity v through the solvent. The electrophoretic mobility μ_e of an ionic species is then defined as:

$$\mu_e = \frac{v}{E} = \frac{vL}{V} \tag{4.1}$$

If we assume the moving ion to be a sphere with hydrodynamic radius r and consider that the electrostatic force on the ion by the electric field is fully counterbalanced by the frictional or viscous force on the ion by the solvent, it follows that:

$$\mu_e = \frac{q}{6\pi\eta r} \tag{4.2}$$

where q is the charge of the ion and η the solvent viscosity. Hence, although molecular analyte ions rarely are perfectly spherical, we can generally state that ion electrophoretic mobility is a function of the molecular charge-to-size ratio.

When using fused-silica capillaries in combination with aqueous BGEs (as is most common in CE), the capillary inner wall usually carries negative charge, mostly due to the (partial) deprotonation of the surface silanol groups. In this context, there is a local accumulation of cations from BGE to neutralize these fixed negative charges. Upon application of voltage across the capillary, part of this cationic layer will migrate toward the cathode and cause the whole solution inside the capillary to move forward at a constant velocity. This bulk flow is called the electroosmotic flow (EOF). All species (both neutral and charged) in the capillary move along with the EOF, and hence for an analyte ion we observe an apparent electrophoretic mobility μ_a, which is the resultant of its μ_e and the EOF mobility (μ_{EOF}) according to:

$$\mu_a = \mu_e + \mu_{EOF} \tag{4.3}$$

Note that for systems with a cathodic EOF (as is usual for bare fused-silica capillaries), the μ_e of anions has a negative sign. As the surface silanols are weakly acidic exhibiting a range of pK_as, the magnitude of the EOF depends on the pH of the BGE. The EOF is minimal at pH 2 and lower (i.e. slight or no silanol deprotonation) and increases with raising pH until a plateau at pH 8–9 and higher is reached (i.e. full silanol deprotonation). The μ_e of an analyte ion can be experimentally assessed by measuring the migration times of the analyte (t_m) and that of a neutral compound or EOF marker (t_{EOF}) and applying the following relation:

$$\mu_e = \frac{L_d L_t}{V}\left(\frac{1}{t_m} - \frac{1}{t_{EOF}}\right) \tag{4.4}$$

where L_d is the distance from the capillary inlet to the point of detection and L_t the total capillary length.

4.2.2 CE Separation Efficiency

In comparison with chromatographic techniques, CE can provide superior separation efficiencies. Indeed, relatively small peak widths are obtained, leading to high plate numbers (N):

$$N = \left(\frac{t_m}{\sigma_t}\right)^2 \tag{4.5}$$

where σ_t is the standard deviation of a peak in time units. As there is no stationary phase in CE, typical chromatographic sources of band broadening, such as Eddy diffusion (due to multiple flow paths in a packed column; A term of the van Deemter equation) and resistance to mass transfer (related to analyte exchange between mobile and stationary phase; C term of the van Deemter equation), are absent. Moreover, in contrast to pressure-driven chromatographic techniques in which flows are laminar with a parabolic profile, in CE the EOF exhibits a flat, plug-like profile. Hence, the EOF does not contribute to band broadening. Consequently, in CE, the main potential factors causing band broadening are Joule heating and longitudinal diffusion.

In CE, the voltage across the capillary induces an electric current. This current through the BGE inherently produces resistive heat (so-called Joule heating) and, therefore, an increase in temperature in the capillary. This heat can only be dissipated through the capillary wall, triggering a radial temperature gradient in the capillary. The accompanying radial viscosity gradient causes band broadening as the electrophoretic mobility of the analyte ion becomes dependent on its radial position. The rise of temperature also increases molecular diffusion that adds to band broadening as well (see next). Fortunately, even when applying high voltages (10–30 kV), the adverse effects of Joule heating generally become negligible when (i) capillaries with internal diameters below 100 μm are used to minimize the current, (ii) effective heat dissipation is achieved by applying active capillary thermostating (as provided by commercial CE instruments), and (iii) the BGE conductivity is kept below 1 Ω^{-1} m^{-1} (which is true for many typically used buffers up to 50 mmol L^{-1} concentration).

Molecular diffusion is an inevitable physical phenomenon and, therefore, axial or longitudinal diffusion always adds to band broadening in CE. However, molecular diffusion coefficients in liquid phase are quite small, so that their contribution on the separation efficiency is relatively modest. When we suppose longitudinal diffusion to be the predominant source of band broadening in CE and consider the point of detection to be close to the capillary outlet (i.e. $L_d \approx L_t$), the plate number N obtained for an analyte ion can be reliably estimated from:

$$N = \frac{\mu_a V}{2D} \tag{4.6}$$

where D is the diffusion constant of the analyte. Clearly, high voltages promote high plate numbers (narrow peaks) and, therefore, voltages of up to 30,000 V are very commonly applied in CE. This implies that for low-molecular-weight compounds (as most chiral compounds of interest are), which have D values of about 10^{-9} m^2 s^{-1}, typical achievable plate numbers range between 100,000 and 500,000, which is significantly higher than those attained with modern LC techniques.

4.2.3 Chiral Resolution in CE

In CE, under the assumption that band broadening is primarily caused by longitudinal diffusion, the resolution R_S between two adjacently migrating compounds (1 and 2) with electrophoretic mobilities $\mu_{e,1}$ and $\mu_{e,2}$, respectively, is given by:

$$R_S = \frac{1}{4}\sqrt{\frac{V}{2D}}\left(\frac{\Delta\mu_e}{\sqrt{\mu_{e,avg} + \mu_{EOF}}}\right) \tag{4.7}$$

where $\Delta\mu_e$ is $(\mu_{e,1} - \mu_{e,2})$ and $\mu_{e,avg}$ is the average of $\mu_{e,1}$ and $\mu_{e,2}$. Obviously, CE separation requires compounds to have different electrophoretic mobilities ($\Delta\mu_e \neq 0$). Considering Eq. (4.6), it is also clear that high plate numbers favor resolution. Notably, optimal resolution is obtained when $\mu_{e,avg}$ and μ_{EOF} have an opposite sign, i.e. when analyte ions migrate against the EOF.

Two enantiomers of an ionic chiral compound have identical ion mobilities and, therefore, cannot be separated by CE using a conventional BGE. However, chiral separation can be achieved quite straightforwardly by adding a CS, such as cyclodextrins (CDs), to the BGE. CS can be added to both aqueous and organic BGEs (the latter being referred to as non-aqueous capillary electrophoresis, NACE) provided that the selected CS is soluble in the chosen BGE. Dynamic complexation between the analyte ions (A) and the CS during analysis (A + CS ↔ A – CS) causes a shift in the effective mobility of the analyte ions due to a change in the total hydrodynamic volume upon temporary association.

The effective mobility (μ_e^{eff}) of the ion is the weighted average of the mobility of the free ion (μ_e^{free}) and the mobility of the complexed ion ($\mu_e^{complex}$) according to:

$$\mu_e^{eff} = \left(\frac{1}{1+[CS]K}\right)\mu_e^{free} + \left(\frac{[CS]K}{1+[CS]K}\right)\mu_e^{complex} \tag{4.8}$$

where $[CS]$ is the CS concentration in the BGE and K the complexation or binding constant of the ion and the CS. Differences in K (i.e. affinity) of the respective enantiomers of an analyte ion for the CS result in different degrees of association, and, thus, in different μ_e^{eff} values for the enantiomers. Assuming that $\mu_e^{complex}$ is the same for the two enantiomers (1 and 2) of an ionic compound, Wren and Stove derived the following principal equation for enantiomer separation by CE [2]:

$$\Delta\mu_e = \mu_{e,1}^{eff} - \mu_{e,2}^{eff} = \frac{[CS]\left(\mu_e^{free} - \mu_e^{complex}\right)\left(K_2 - K_1\right)}{1+[CS]\left(K_1+K_2\right)+K_1K_2[CS]^2} \tag{4.9}$$

where K_1 and K_2 are the CS-binding constants of enantiomer 1 and 2, respectively. Equation (4.9) shows that enantioseparation (i.e. larger $\Delta\mu_e$) is favored by larger differences between μ_e^{free} and $\mu_e^{complex}$, as well as between K_1 and K_2, and that no enantiomer separation (i.e. $\Delta\mu_e = 0$) is achieved when $\mu_e^{free} = \mu_e^{complex}$ or when $K_1 = K_2$. Moreover, for the enantiomers, $\Delta\mu_e = 0$ when $[CS] = 0$ (i.e. no CS present) or when the $[CS]$ is very high. In the latter case, both enantiomer ions are predominantly associated with the

CS. For a certain combination of chiral analyte and CS, the optimum $[CS]$ for which $\Delta\mu_e$ is maximal is found by:

$$[CS]_{opt} = \frac{1}{\sqrt{K_1 K_2}} \tag{4.10}$$

Equations (4.9) and (6.7) essentially encapsulate the power of CE for chiral separation: adding a suitable CS at the adequate concentration to the BGE induces mobility differences among the two enantiomers, while high plate numbers facilitate the enantioresolution, even when the achieved $\Delta\mu_e$ is relatively modest.

Chankvetadze et al. pertinently pointed out that differences in the mobilities of the diastereomeric complexes of two enantiomers with the CS ($\mu_{e,1}^{complex}$ and $\mu_{e,2}^{complex}$) may also add a factor to enantioselectivity in CE [3, 4]. Strikingly, enantiomer separation by CE in principle is possible even when the affinities of enantiomers to the CS are the same ($K_1 = K_2 = K$):

$$\Delta\mu_e = \frac{K[CS]\left(\mu_{e,2}^{complex} - \mu_{e,1}^{complex}\right)}{1 + K[CS]} \tag{4.11}$$

Moreover, by adjusting the separation conditions in chiral CE, it is also possible to change the enantiomer migration order without altering the fundamental chiral recognition (i.e. using the same CS) [5]. These remarkable aspects arise from the migration principles of CE, where enantiomers, CS, and enantiomer–CS complexes are mobile and migrate in one liquid phase under an electric field. Notably, separation of enantiomers with equal CS affinity, as well as alteration of the enantiomer order without changing the character of the CS, is practically impossible with chromatographic techniques using chiral stationary phases.

4.2.4 Chiral Micellar Electrokinetic Chromatography and Capillary Electrochromatography

Micellar electrokinetic chromatography (MEKC) and capillary electrochromatography (CEC) are often considered CE modes, as they employ narrow capillaries, high voltages, and the same instrumentation but have chromatographic separation principles.

MEKC allows separation of neutral compounds using a charged surfactant added to the aqueous BGE above its critical micellar concentration.

The resulting micelles act as a pseudo-stationary phase, which moves through the capillary when the voltage is applied at a velocity determined by the sum of the micellar and the electroosmotic mobility. The BGE moves with the speed of the EOF only. Neutral analytes can dynamically distribute (according to their partition coefficient) between the micellar phase and the aqueous phase. Differences in distribution will lead to overall different apparent mobilities and, thus, separation of the neutral compounds. Notably, charged analytes can also be analyzed by MEKC, in which case the micellar phase adds extra selectivity to the separation which is no longer based on ion electrophoretic mobility only (as in CE).

Chiral separation by MEKC can be achieved in several ways. First, a chiral surfactant (e.g. a bile acid) can be employed, which forms charged micelles and acts as CS at the same time. Analyte interactions will include both micelle partitioning as well as CS-complex formation, potentially providing chiral resolution next to chemical separation. Second, a more common approach is to add both a CS, such as CDs, and a micelle-forming surfactant to the BGE, creating a mixed-mode separation mechanism that offers both chiral and chemical selectivity. A third way of using MEKC is by derivatization of analyte enantiomers with an enantiomerically pure reagent followed by the separation of the formed diastereomers by MEKC employing common non-chiral surfactants. Diastereomers often exhibit equal or very similar electrophoretic mobility and, thus, typically cannot be separated by CE only, requiring some extra chromatographic selectivity.

In CEC, the capillary is packed with stationary phase particles, creating a chromatographic column. The high voltage is used to drive the BGE through the column by electroosmosis. Hence, in contrast with conventional LC, no pressure is needed, and potentially very small stationary phase particles can be used. Enantiomer separations by CEC require the use of capillaries packed with a chiral stationary phase, which acts as CS by itself or holds chemically bound CS ligands. The selectivity of chiral separation is fully determined by chiral recognition, i.e. by the difference between the affinities of two enantiomers with the immobile CS. The migration of ionic analytes through the column will be the resultant of both chromatographic retention and electrophoretic mobility. In combination with the flat flow profile provided by the EOF, this means that relatively favorable plate numbers can be achieved with CEC (higher than for LC, but lower than for CE). Both neutral and charged compounds can be analyzed by CEC.

4.3 Short History of Chiral CE Modes

4.3.1 Chiral CE

After CE using fused-silica capillaries was developed and introduced into practice in the early 1980s, about four years elapsed before Gassman et al. reported the first chiral CE separations [1]. Their work showed that the DL-enantiomers of dansylated amino acids (e.g. tyrosine, phenylalanine, asparagine, and glutamic acid) could be resolved within 10 minutes by employing a bare fused-silica capillary with a simple BGE composed of 10 mM ammonium acetate (pH 7–8) containing Cu(II)-L-histidine as the CS. The authors proposed applications of their method in geological dating and clinical diagnosis, but the uses of chiral CE would rapidly surpass these ideas. CDs were introduced as convenient CSs for enantioselective CE in the late 1980s. For example, Fanali used derivatized β-CD for the separation of ephedrine enantiomers and some of its analogues [6]. Soon, chiral CE was seen as an attractive alternative to chiral GC, thin-layer chromatography, and LC, specifically in the field of pharmaceutical analysis, which later resulted in several CE methods entering pharmacopeias for the purpose of drug quality control. From the early 1990s, many new CSs for chiral CE were successfully introduced, such as charged CDs, macrocyclic antibiotics (e.g. vancomycin), (cyclic) peptides, polysaccharides, proteins, and crown ethers. Although development and study of novel CSs for chiral CE has never stopped (see Section 4.4), for now CDs in all their variety and availability remain by far the most popular type of CSs to achieve efficient enantiomer separations by CE. Today, routine applications of chiral CE can be found in pharmaceutical, bioanalytical, clinical, food, beverage, environmental, and forensic fields.

4.3.2 Chiral MEKC and Chiral CEC

After introducing MEKC as a CE mode for separation of neutral compounds in 1984, Terabe's group was also the first to publish on chiral MEKC. They showed in 1989 that dansylated amino acids could be chemically and enantiomerically resolved using a bile salt as micelle-forming species, which served both as CS and pseudo-stationary phase [7]. Since then, several types of natural and synthetic chiral surfactants have been introduced and used for chiral MEKC. Next to bile salts, amino acid-derived, glycosidic, and polymeric surfactants have been employed to achieve chiral MEKC. Application of numerous mixtures of two chiral surfactants or a chiral and non-chiral surfactant has been reported over the past 20 years, making use of the synergistic effects of two selectors. The use of ligand-exchange complexes as CS together with micellar systems gave rise to the distinct mode of ligand-exchange MEKC. Still, the most common strategy to perform chiral MEKC is by using an achiral surfactant (frequently sodium dodecyl sulfate [SDS]) in combination with a CD as CS. In the indirect mode of chiral MEKC, chiral analytes are derivatized with an enantiomerically pure chiral reagent in order to produce the corresponding stable diastereomers, which are then separated under achiral MEKC conditions. For example, the reagent (+ or −)-1-(9-fluorenyl)ethyl chloroformate (FLEC) has shown to be very useful for this purpose.

Chiral CEC was first reported in 1992, when Mayer et al. demonstrated enantiomer separation using a glass capillary whose wall was coated with a chiral stationary phase (Chirasil-Dex) [8]. The enantiomers of 1-phenylethanol were separated with a resolution of 1.05 within 16 minutes. This is an example of open-tubular CEC. In other more frequent chiral CEC modes, the capillary is packed with a chiral stationary phase as also used for chiral LC. Practical performance issues related to the frits used to keep the stationary phase particles in the capillary were circumvented by the introduction of chiral monolithic columns for CEC. CEC is mainly applicable for separating chiral drugs and pesticides, specifically with regard to quality control testing during production. Chiral CEC research is currently ongoing and new dedicated chiral phases are proposed regularly. However, so far, chiral CEC never turned into a routine method as the required columns are only scarcely commercially available.

4.4 State of the Art and Recent Developments

4.4.1 Common Chiral Selectors

Cyclodextrins were among the first compounds used as CSs, and they remain one of the most popular choices for enantiomer separations. In their native form of α-, β-, or γ-CD (Figure 4.1a, 4.1b, 4.1c), enantioselectivity is achieved by inclusion of the analyte into the hydrophobic core of the CD, where noncovalent interactions with the chiral glucopyranose interior cause different host–guest complexation affinities for chiral analytes. The driving factor in forming these CD–analyte complexes is the polarity difference between the electrolyte solution, the analyte, and the CD itself. Though the exterior of the CD molecule is polar, and thus water-soluble, the interior cavity is nonpolar and will form CD–analyte complexes via an inclusion

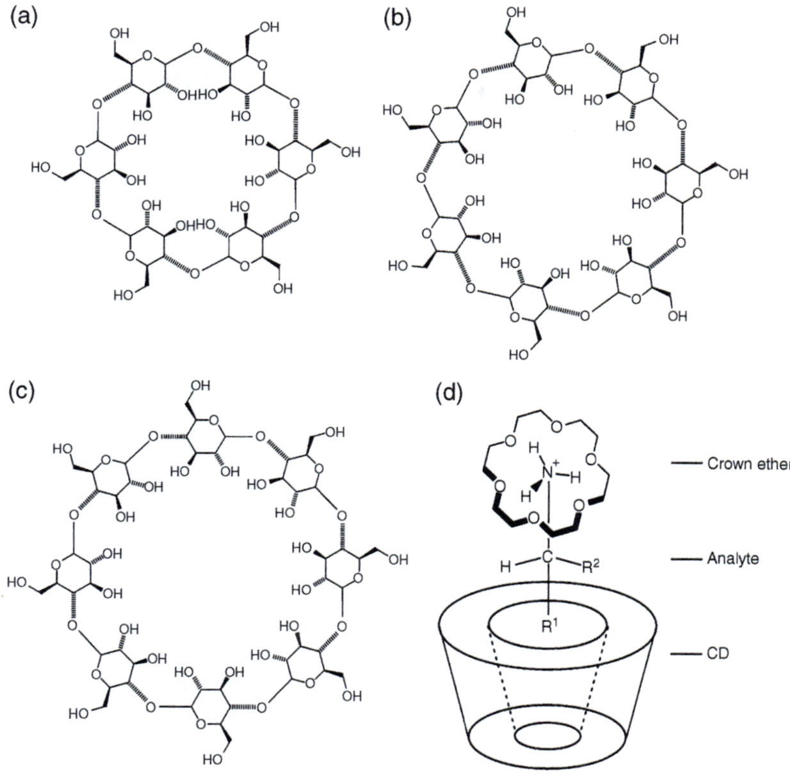

Figure 4.1 Molecular structure of cyclodextrins (CDs). (a) α-CD; (b) β-CD; (c) γ-CD; (d) supramolecular interaction between a crown ether (18-crown-6), a cationic analyte with a primary amine, and a CD. *Source:* Figure D is adapted from Koide and Ueno [9] (open access article under the terms of the Creative Commons Attribution License).

mechanism that replaces water from the core of the CS. The strength of these complexes depends in part on how well the analyte fits inside the CD cavity. When the size of the CD compound closely matches the analyte of interest, then the orientation of the analyte also impacts complex stability. This is precisely what causes enantioselectivity among native CDs: the relative fit and complexation strength of chiral, nonpolar analytes within the CS molecules. Simply by varying the number of glucose units composing the CD, the cavity shape and size will be altered, generating new chiral recognition sites and expanding the range of enantioselective capabilities.

While native CDs are still used in chiral separations today, state-of-the-art techniques will often use derivatized CDs with special functionality. This allows for custom separations tailored to provide extra resolution for specific analytes. Derivatized CDs will often possess regions of different charge states or specific functional groups within the molecule. Chiral interactions with functional groups on the ring of the CD cavity provide additional enantioselectivity by increasing the strength of certain CD–analyte complexes based on their orientation. Derivatized CDs are nowadays widely used but may potentially lead to poor reproducibility due to batch-to-batch differences. A derivatized CD is in fact composed of a complex mixture of CD isomer species, which can vary in both the degree and position(s) of substitution. Manufacturers usually declare the degree of substitution (i.e. the average number of derivatized OH groups) but cannot guarantee the same substitution pattern for every lot. To circumvent this, single isomer CD derivatives have been synthetized, which are typically mono-substituted CDs (with only one substituent per CD molecule in a defined position) and persubstituted CDs (with all OH groups substituted in the same position). Such well-characterized, single-component charged derivatives of CDs have certainly contributed to improving the reproducibility of chiral CE and its use in industrial settings [4, 10].

Another common approach is to use multiple CDs for chiral recognition, referred to as "dual-cyclodextrin systems." These separations will combine multiple classes of CDs, often pairing neutral CDs with anionic or cationic CD derivatives. This strategy enhances chiral separations for some enantiomers and has been commonly applied in pharmaceutical analysis. Regardless of which specific molecule is employed, CDs remain the most popular CSs due to their low cost and the low quantities required, high solubility in water, low UV absorbance, large degree of customization, and compatibility with a wide range of chiral analytes.

One type of CS that is especially prevalent for MEKC separations are the bile salts. Even though bile salts were used for the first reported chiral MEKC separation 1989 [7], significant advances over the past three decades have kept them relevant CSs today. Bile salts are naturally occurring compounds with a steroid-like structure and will differ based on their side chains and stereochemistry of hydroxyl groups (Figure 4.2) [11].

Bile salts contain a hydrophilic surface composed of several hydroxyl groups, as well as a hydrophobic surface with multiple methyl groups. Both regions provide different binding possibilities, and the ionic head group of the carbon side chain provides a charged region to enhance electroosmotic mobility. The amphipathic nature of bile salts means that they will form micelles in an aqueous solution. However, these micelles are not the spheres formed by SDS or other compounds with a polar head and long nonpolar tail; rather, the hydrophilic and hydrophobic surfaces are on opposite faces

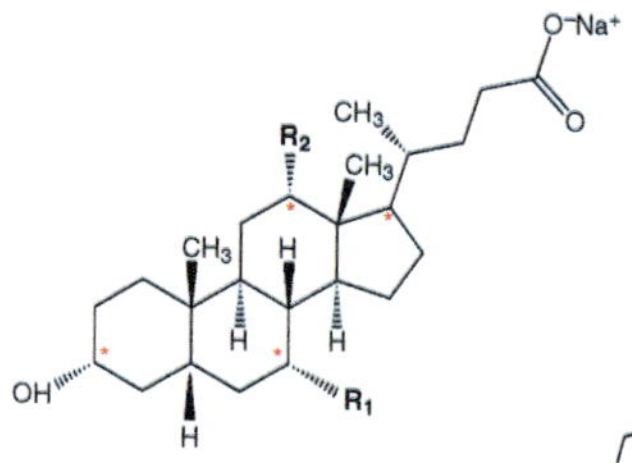

Bile salt	R_1	R_2
Sodium cholate	OH	OH
Sodium deoxycholate	H	OH
Sodium chenodeoxycholate	OH	H

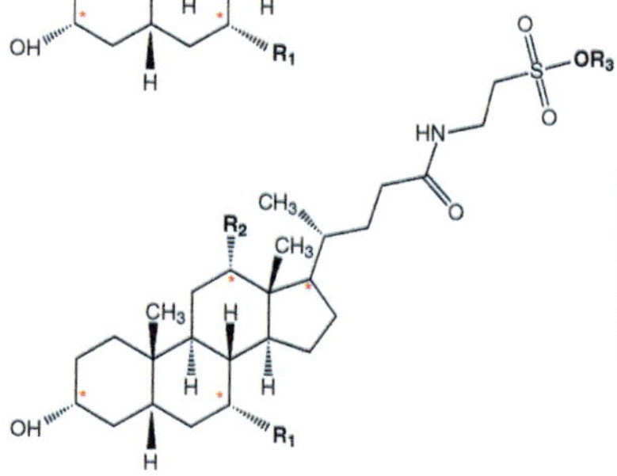

Bile salt	R_1	R_2	R_3
Sodium taurocholate	OH	OH	Na^+
Sodium taurodeoxycholate	OH	H	Na^+
Taurochenodeoxycholic acid	H	OH	H

Figure 4.2 Structures of bile salts commonly used in chiral MEKC separations. The composition of the carbon side chain, character of the R groups, and presence of between one and three hydroxyl groups (occurring at the indicated positions) differentiate the compounds. Figure generated in ChemDraw Professional 16.0.

of the bile salt and cause formation of disc-shaped or helical micelles [12]. For this reason, bile salts are employed as CSs almost exclusively in MEKC separations. Sodium cholate is one of the more common bile salts used as a CS in MEKC, along with sodium taurocholate and sodium taurochenodeoxycholate. These bile salts will often be used in combination with CDs forming another type of dual-selector system. Bile salts are a chiral surfactant, which shows the advantage of adding a secondary, enantioselective process to the CD–analyte complexation already occurring. Dual-selector systems like these can be combined with advanced preconcentration techniques such as sweeping, stacking, or partial filling in order to further enhance separation and detection limits [13].

Another fascinating class of CS compounds are antibiotics. These naturally occurring molecules with antibacterial activity are appealing for enantioselectivity due to their large number of chiral centers, variation in size, and the range of chemical functionalities present throughout different classes of antibiotics. A notable advantage over CDs is that antibiotics are more soluble in organic solutions, meaning that both aqueous and nonaqueous separations are possible. Antibiotics are much larger than CDs, and thus their chiral selectivity is not due strictly to cavity inclusion at the center of the molecule. Rather, antibiotics will possess many different regions where hydrogen bonding, dipole interactions, steric hindrance,

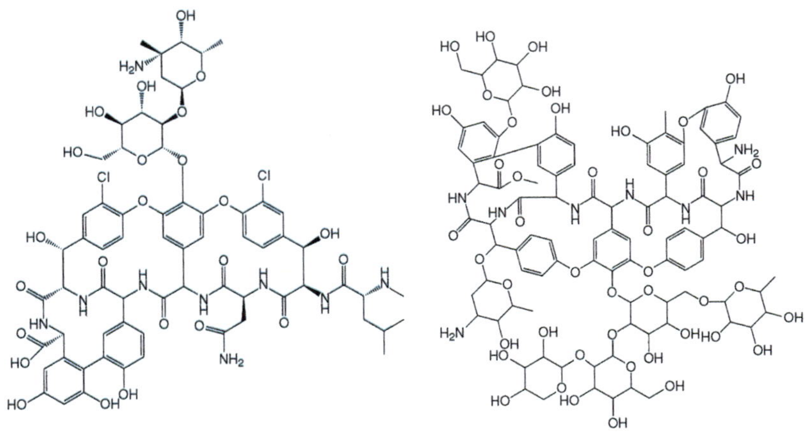

Vancomycin **Ristocetin A**

Figure 4.3 Structures of vancomycin and ristocetin A – two glycopeptide antibiotics that are commonly used as chiral selectors. These macrocyclic antibiotics contain several regions where host–guest complexation occurs during separation. Figure generated in ChemDraw Professional 16.0.

ionic interactions, as well as hydrophobic interactions can all impact the host–guest complex formation (Figure 4.3). This wide range of possibilities makes antibiotics incredibly versatile in their uses. Certain classes have been identified as favorable for different analytes. For example, vancomycin, a macrocyclic antibiotic that is one of the most commonly used CSs, is known to be excellent for separating analytes with carboxylic acid groups. Macrocyclic antibiotics as a whole, however, are typically poor CSs for neutral analytes [14]. One downside to implementing antibiotics as CSs is their tendency to self-aggregate, reducing the availability for complexing with the analytes of interest. This can be minimized by tuning the antibiotic concentration, as well as adding a small proportion of organic solvent to the separation [15].

Crown ethers have also been used as CSs for chiral CE, typically when the goal is to resolve enantiomers that contain primary amines, such as amino acids. These interconnected polyether molecules form the host–guest complex via hydrogen bonding and ion–dipole interactions and are often used in combination with other CSs. A dual-selector system of crown ethers and CDs functions by complexing analytes between the CD core and crown ether, enhancing chiral recognition (Figure 4.1d). Compared to a single CD selector system, dual selectors of crown ethers with a derivatized CD often produce better enantioresolution for amino acids, although with an increased migration time [16].

4.4.2 Ionic Liquids as Chiral Selectors

An ionic liquid (IL) is defined as an organic salt that is liquid below 100 °C. ILs show a large variety in their composition, physiochemical properties, and classification, leading to their use in synthesis, catalysis, separation science, drug delivery, and many other applications [17]. Some of the important properties of ILs include high conductivity, solubility in both organic and inorganic solvents, thermal stability, and low volatility. When implemented in chiral CE, ILs can serve many different roles. The most common use of ILs is in combination with another CS to form a synergistic separation, although chiral ILs are also used as sole CSs in MEKC and in ligand-exchange systems [18]. In fact, the first reported use of ILs in a chiral CE separation was not as a CS. Rather, achiral ILs were included in MEKC in order to improve separation of hydrophobic compounds, while maintaining stable currents during the separation due to the high conductivity of the IL [19]. Soon after, Rizvi et al. showed the first enantiomeric separation using chiral ILs as the pseudo-stationary phase in CE for separating two acidic chiral analytes [20]. For a comprehensive overview of the use of ILs in chiral CE, reviews from Greño [21] and Zhang [22] may be consulted.

ILs are applied to a variety of chiral separations because ILs can often improve enantioresolution in a twofold manner. First, their unique properties as modifiers impact the conductivity and solubility of the BGE. Second, chiral ILs participate in the host–guest complexation with analytes and when combined with another CS such as β-CD, the improvement in resolution can be significant.

4.4.3 Nanoparticles as Chiral Selector Carriers

A recent topic of research in chiral analysis is the incorporation of nanomaterials for enantiomer separations. In principle, nanoparticles can be implemented in CE or CEC separations as a chiral additive or as a pseudo-stationary phase, possessing different affinities for enantiomers in solution and thus generating chiral separation. There are several reasons underlining the growing interest in nanoparticles for this type of analysis, including their high degree of customizability. Nanoparticles can be easily fabricated from materials such as gold, silver, carbon, or metal oxide, and their size and shape can be adjusted for different desired properties. The high surface area-to-volume ratio of nanoparticles may increase enantioselectivity [23], and there is near limitless flexibility in choosing which chiral ligands to coat the particles with. Furthermore, nanoparticles possess high mechanical strength and are chemically stable, making them compatible with many applications. In most applications, nanoparticles are combined with molecularly imprinted polymers (MIPs) to generate so-called MIP-NPs. MIPs are well-established for use as chemical sensors [24], in solid-phase extraction (SPE) [25], and for various biological applications [26], and have been applied with nanoparticles for chiral separations. Specific to CE, MIP-NPs were employed by We et al. on a microfluidic CEC chip [27]. They fabricated MIPs with magnetic nanoparticles as a chiral stationary phase for their microchip, and used that to demonstrate a separation of *R*- and *S*-mandelic acid. Their separation had a resolution factor of 1.82, exceeding that achieved via high-performance liquid chromatography (HPLC) with a substituted β-CD CS (1.29) or with IL additives (1.62) [28, 29]. Nanoparticles can also be used in conjunction with chiral CE, but as a sample preparation step rather than as the CS molecule. Wu et al. performed SPE using magnetic nanoparticles for a highly efficient extraction of three β-blocker compounds [30]. They were then separated via chiral CE with modified β-CD as the CS phase.

The application of chiral nanoparticles is not just limited to separations. Enantiorecognition is also valuable, as sensors can be designed that indicate excess of a certain enantiomer, for instance during pharmaceutical production. Nanoparticle-based enantiomer recognition is appealing as it can be observed via color change, thus leading to optical sensors for chiral recognition [31]. Asymmetric synthesis is another intriguing application of chiral nanomaterials, as it allows for generation of 100% enantiopure compounds. Although it can be costly, using nanoparticles for asymmetric catalysis is quite desirable for synthesizing medicines and pesticides [32].

4.4.4 Microfluidic Chiral CE

Microfluidic platforms are popular throughout analytical chemistry, as they offer the well-known advantages of reduced sample and reagent volumes, low-cost analyses, rapid separations, and the potential for point-of-care applications [33]. These benefits can also translate to chiral CE. Beyond the potential for more efficient and lower cost separations, performing chiral microchip electrophoresis (MCE) can also open the door for integrated devices, combining sample preparation workflows with separation and detection.

The first report of chiral MCE was from Hutt et al. in 1999, in an attempt to develop a separation method that could detect life in extraterrestrial environments [34]. This fascinating premise illustrates perfectly the different applications made possible by a miniaturized, portable system that can deliver rapid and sensitive results. Hutt tested their device on a meteorite sample and successfully separated the enantiomers of four amino acids in less than four minutes using an SDS/CD electrolyte solution [34]. As with conventional CE, CDs were the most popular choice of CS when developing chiral MCE, and they remain so today. Other projects have applied chiral MCE to separation of pharmaceutical compounds [35], or coupled it to MS for human neuron analysis [36], both also employing CD-based systems.

There are certain challenges unique to microfluidic systems, which can be a disadvantage for chiral MCE compared to benchtop CE separations. One major goal for MCE is to reduce the footprint of the device, thus creating a smaller and possibly portable platform. However, this also requires shrinking the power supplies, supporting hardware, and detection system in addition to the microchip itself. These factors can often prevent a microfluidic platform from being truly point-of-care. In addition, one intrinsic property of microchips is that they are composed of very small fluidic channels. This is advantageous for many reasons, but in terms of different detection methods, it can pose a challenge. When performing optical detection in MCE, the small channel dimensions (typically tens of micrometers) result in a reduced pathlength and, thus, lower sensitivity. This can be alleviated through creative microchip design [37] or by simply coupling the chiral MCE separation to a detection mode that is not reliant on optical pathlength. Finally, the durability of MCE platforms can be problematic depending on the materials used and desired application. While glass is a preferred substrate for many electrodriven microchips, it is not as rugged as many polymer alternatives. If the chiral MCE device is intended for use in the field or frequent transport, then a more durable material such as poly(methyl methacrylate), cyclic olefin copolymer, or polydimethylsiloxane is preferable. Each of these materials comes with its own considerations regarding surface charge, biocompatibility, and fabrication requirements, which must be tailored to each application.

4.5 Applications of Chiral CE

4.5.1 Pharmaceutical Analysis

Quality control is a crucial aspect of any commercial production process, and the same is true for the development of new pharmaceutical entities. Racemates as well as enantiomerically pure drugs produced in the pharmaceutical industry are subject to rigorous quality guidelines, brought on in no small part due to the notorious thalidomide disaster of the late 1950s and early 1960s [38]. Thalidomide, which was prescribed to treat morning sickness in pregnant women, contains a chiral carbon center and exists as *R*- and *S*-enantiomers. After it was observed that children were born with limb malformations in some cases when the mother was prescribed thalidomide, it was discovered that the *S*-enantiomer of the drug has teratogenic effects, disturbing fetal development during pregnancy [39]. Tens of thousands of babies were maimed or killed from side effects of the drug, and this tragedy has become the classic example of the importance of enantiomeric purity for many pharmaceuticals. Today, any drug that is marketed as enantiomerically pure undergoes thorough enantioselective quality control to ensure that potentially hazardous enantiomer impurities are not present in the final drug. Furthermore, any drug marketed as a racemate must first demonstrate that both enantiomers are nontoxic, or that one is inactive in the final product. Even then, a validated chiral separation method needs to be in place for quality control.

Chiral CE for pharmaceutical analysis typically employs CDs as CS, although the detection method vary. Depending on the compound being analyzed, methods can employ UV absorbance detection or laser-induced fluorescence (LIF) detection in conjunction with analyte derivatization to measure enantiomeric ratios. The choice of CS and detection method will of course vary with the target analyte. For example, when analyzing the enantiomeric purity of tramadol, Sarkany et al. used carboxymethyl-β-CD in a chiral CE-UV system [40]. This analysis is necessary not to produce an enantiomerically pure compound for distribution but rather to ensure that the correct racemic ratio is achieved. Tramadol enantiomers possess different binding and activity characteristics, which complement each other for improved effectiveness as an opioid analgesic. This reported chiral CE method is a rapid, low-cost, and straightforward way to characterize enantiomeric ratios in tramadol formulations. Similar approaches have been used for quantitation and/or enantioseparation of blood pressure medication [41, 42] and Parkinson's disease medication [43], by employing chiral CE with modified CDs in combination with UV detection. For a more complete list of various modes of chiral CE for pharmaceutical applications, the recent review from Bernardo-Bermejo et al. can be consulted [44].

4.5.2 Food Analysis

Chiral analysis is useful in investigation of various food products, providing information on the quality of food, degree of adulteration or food processing, and effects of storage under different conditions [45]. Like many applications of chiral separations, food science relies heavily on HPLC, but chiral CE is also employed in some cases. One interesting application of chiral CE is for assessing the geographical origin of green teas. Modified β-CDs as the achiral selector in MEKC could resolve enantiomers of catechin, a natural phenol with four diastereomers. The different enantiomer contents could then be used to identify the geographical origin of various commercial tea products [46]. Additional information beyond the growing region of various teas can also be determined via chiral separations. Fiori et al. recently measured (−)-catechin and D-theanine using MEKC-UV with a BGE containing SDS and a modified-CD CS that allowed for the enantioseparation of seven compounds in green teas (Figure 4.4) [47]. A relatively high content of D-theanine (about 18%, m/m with respect to L-theanine) was found in the samples. The authors hypothesized that the presence of (−)-catechin is a useful marker of thermal degradation, whereas D-theanine could be the result of microbial or enzymatic processes, not necessarily related to thermal treatment.

Chiral electromigration techniques have further been applied to study storage effects on pomegranate juice and adulterations in apple juice by measuring D-amino acids using MEKC-LIF after derivatization with fluorescein isothiocyanate [48]. Detection of D-amino acids is particularly useful for identifying illegally altered food products, as chiral separations can identify the non-naturally occurring enantiomers. Certain food industries have their own regulations for legal additives. For example, winemakers are not allowed to artificially acidify their products with tartaric acid. Chiral ligand-exchange CE can be employed to measure D-tartaric acid in wine samples since it successfully separates the compound from the similar (but legal) acidifying agent D-malic acid [49].

In some cases, the moderate sensitivity of UV/vis absorbance detection is not sufficient for detecting trace amounts of enantiomers in food products. In this case, chiral CE can be coupled with MS detection to enhance sensitivity. Lee et al. used a chiral CE-MS method with a crown ether CS to measure D,L-amino acids in vinegars, which allowed them to study the degree of fermentation of different products [50]. More information on the coupling of chiral CE to MS will be discussed in Section 4.6.

4.5.3 Environmental Analysis

Chiral separations are valuable to environmental analysis to identify synthetic or non-naturally occurring compounds in various samples. Chiral CE is typically applied for the analysis of water samples, ranging from lakes and rivers to wastewater and groundwater samples. UV and LIF are desirable as detection methods for their relative simplicity; however, due to lower sensitivity of UV and the often trace amounts of analytes present, preconcentration techniques are commonly necessary.

One recent study showed the development of chiral CE methods for the separation of neutral and acidic amino acids from lake water samples [51]. By first derivatizing samples with the fluorescent label 5-carboxyfluorescein succinimidyl ester, the water samples were measured with either MEKC-LIF (neutral analytes) or CE-LIF (acidic analytes) separations, in both cases using a CD. Figure 4.5 shows the chromatograms observed for the chiral

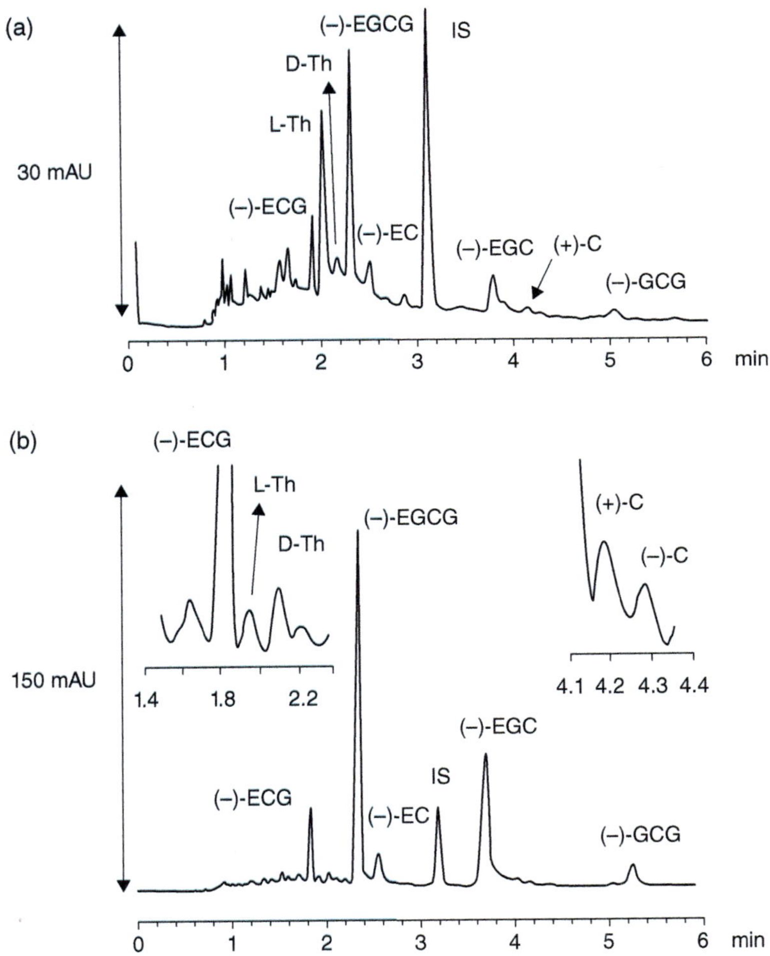

Figure 4.4 CD-MEKC electropherograms of (a) Oolong Formosa tea sample (60% fermentation) and (b) Oolong tea sample (15% fermentation). In the inset, the peak of (−)-ECG is out of the response scale; (c) Bi Luo Chun tea sample (green tea as such). U* = unknown. Experimental conditions: analyte derivatization using OPA/NAC; BGE, pH 2.5 phosphate-borate buffer (25 mM) containing 65 mM SDS and 28 mM DM-β-CD. *Source:* Reproduced from Fiori et al. [47]/with permission of Elsevier.

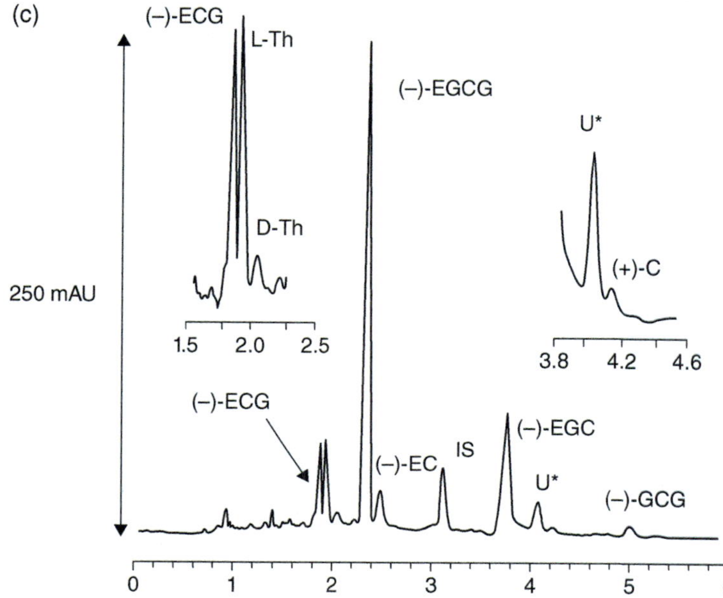

Figure 4.4 (Continued)

separation of neutral (Figure 4.5a) and acidic (Figure 4.5b) amino acids, as well as the Mono Lake water sample compared with a blank (Figure 4.5c). The authors proposed that this method could be compatible with extraterrestrial samples in future work.

Other work has examined environmental samples for the presence of β-blockers [52], herbicides [53], and other drugs that may contaminate water sources [54]. Chiral CE is often difficult to apply specifically for pesticide measurements in environmental samples due to the low solubility of pesticides in aqueous BGE. However, this can be overcome by modifying the composition of the BGE with organic solvents, such as methanol, which resulted in the baseline separation of three finamiphos enantiomers, but at the cost of a lengthy (47 minutes) run time [55].

4.5.4 Bioanalysis

Performing chiral analysis of biological samples is typically useful for studying drugs or other pharmaceutical compounds in living organisms, as the chirality of certain molecules can provide information on metabolism and chemical pathways within the body. The most common sample matrices for chiral bioanalysis are plasma, urine, and cerebrospinal fluid (CSF) [56].

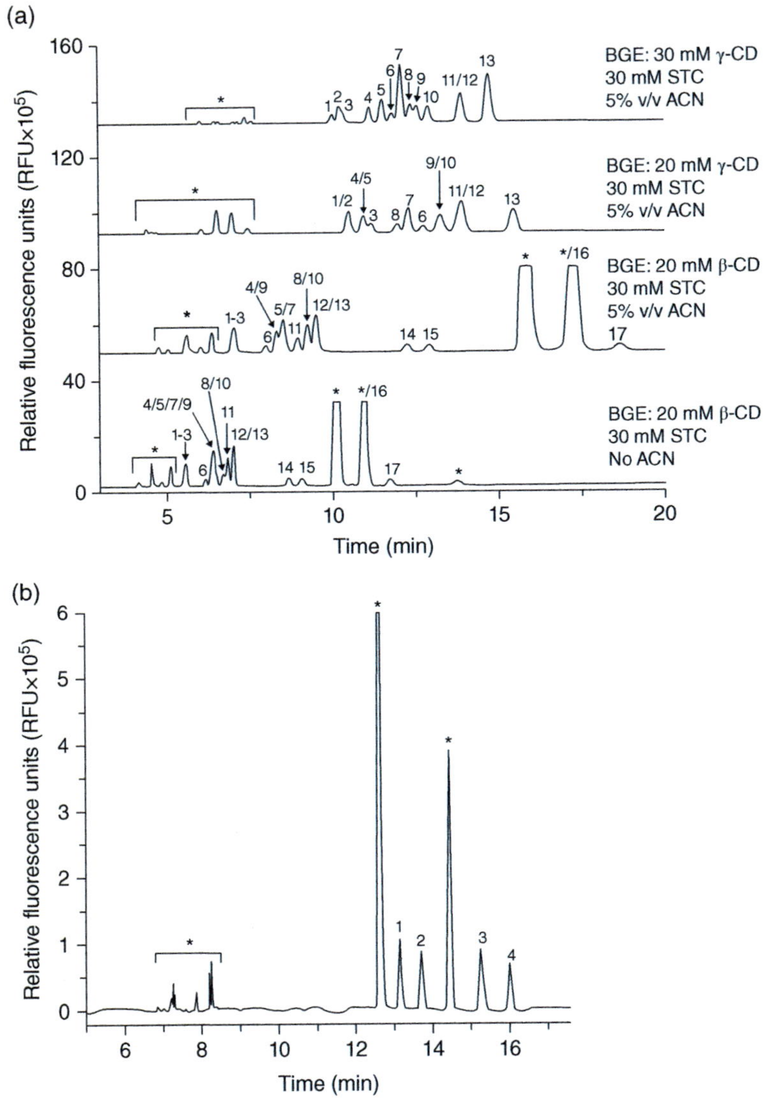

Figure 4.5 Chiral CE of amino acids. (a) Chiral separation of neutral amino acids using a BGE of 80 mM sodium tetraborate pH 9.2 and different CDs, namely, β-CD and γ-CD, combined with sodium taurocholate (STC). Peak identities: (1) D-His, (2) D-Leu, (3) D-Val, (4) L-His, (5) L-Leu, (6) D-Ser, (7) GABA, (8) L-Val, (9) D-Ala, (10) L-Ser, (11) D-Ala, (12) L-Ala, (13) Gly, (14) D-Glu, (15) D-Asp, (16) L-Glu, (17) L-Asp. (b) Chiral separation of acidic amino acids using a BGE of 80 mM sodium tetraborate with 30 mM γ-CD. Peak identities: (1) D-Glu, (2) D-Asp, (3) L-Glu, (4) L-Asp. (c) Chiral CE of Mono Lake sample compared to a blank. Separation conditions optimized for neutral amino acids: BGE, 80 mM sodium tetraborate pH 9.2, 30 mM γ-CD, 30 mM STC, and 5% v/v ACN. Peak identities: (4) L-His, (5) L-Leu, (7) GABA, (8) L-Val, (9) D-Ala, (10) L-Ser, (12) L-Ala, (13) Gly. *Source:* Reproduced from Creamer et al. [51]/with permission of ACS.

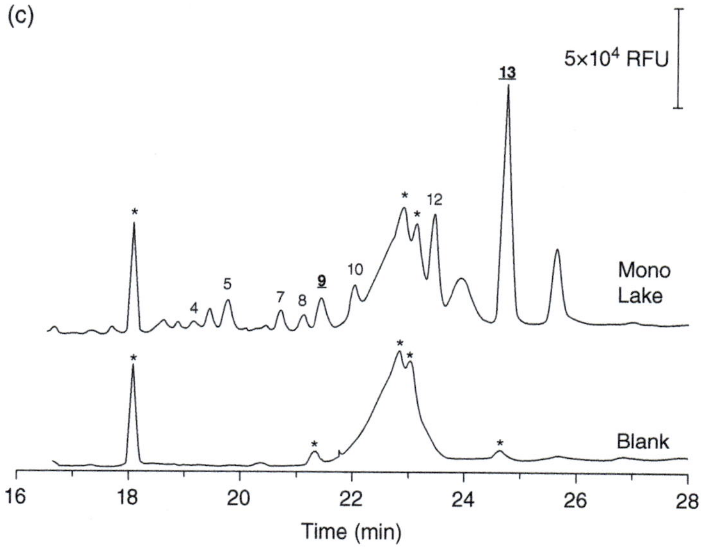

Figure 4.5 (Continued)

Like other applications, the specific type of chiral separation and detection method will vary based on concentration and physicochemical properties of the compounds of interest. Regardless of the biological matrix chosen, a sample preparation step is usually required prior to the separation, using SPE, liquid–liquid extraction (LLE), or microextraction approaches, such as liquid-phase microextraction (LPME). These sample preparation steps can often be automated and have been characterized in depth for many different biological applications [57].

Studying the metabolism of ketamine is one important application of chiral CE, as this anesthetic drug is approved for use in both humans and veterinary medicine. Its enantiomers possess different activities; however, (*S*)-(+)-ketamine has greater potency and fewer adverse side effects. Ketamine has a primary active metabolite, norketamine, which is known to impact multiple biological pathways including glycolysis and amino acid metabolism [58]. To better understand ketamine metabolism in dogs, Theurillat et al. [59] developed a chiral CE separation using sulfated CDs for enantioselective analysis of ketamine and its three major metabolites, reaching a limit of quantification (LOQ) of 10 μg ml^{-1} with samples of 50 μl. In applications where higher sensitivity is required, such as for the analysis of amino acids in plasma and CSF, chiral CE-MS can be employed [60, 61]. Additional ketamine studies are of interest, as its potential for abuse is high and chiral analysis is capable of determining if the drug source was a single enantiomer or a racemate. As the *R*- and *S*-enantiomers of ketamine possess different pharmacodynamic behavior, and different mixtures of both are available, accurate chiral analysis can indicate the type of drug consumed and potentially its origin. By performing LLE on hair collected from drug users and analyzing it via chiral CE with modified CDs, Porpiglia et al. [62] were capable of separating enantiomers of ketamine and norketamine with a limit of detection (LOD) of 0.8 ng mg^{-1} (Figure 4.6).

4.5.5 Forensic Analysis

The fields of forensic sciences also rely on chiral separations, as they can help distinguish between recreational and medically prescribed drugs, which will have different degrees of enantiopurity. For example, dexamphetamine, a stimulant that is medically prescribed to treat attention-deficit disorders, is an enantiomer of amphetamine (along with levoamphetamine). Analysis of the enantiomer ratios in amphetamine samples can indicate if the drug is in fact a medical treatment, and for this reason there are chiral CE protocols using CDs for this type of analysis [63]. In some cases, the so-called designer drugs can be created that technically avoid controlled substance regulations by virtue of substituting or modifying chemical groups of controlled substances with other moieties [64]. These compounds can possess drastically different pharmacological activities and potency compared to the original molecule. In the case of designer drugs of the classes cathinones and phenethylamines, the majority of these drugs are chiral. For example, a chiral CE-UV method was developed using modified β-CD to resolve cathinone and phenethylamine enantiomers (Figure 4.7), being successfully applied to the enantioselective analysis of five contraband samples [64]. Recent work by Hägele et al. [65] used chiral CE to analyze over 50 novel psychoactive substances (NPS) that extend beyond just amphetamine and cathinone-like molecules, such as benzofurans and phenidines. By employing native β-CD as well as three derivatized CDs, they resolved 50 out of 51 NPS, and showed that all were racemates.

4.6 Chiral CE-MS: Strategies and Challenges

In many applications, such as bio- and environmental analyses, MS detection is frequently required as UV/vis absorbance detection typically does not provide sufficient sensitivity and selectivity. Compared to conventional CE-MS, the combination of chiral CE with MS comes with additional challenges due to the presence of the non-volatile CS in the BGE. Possible

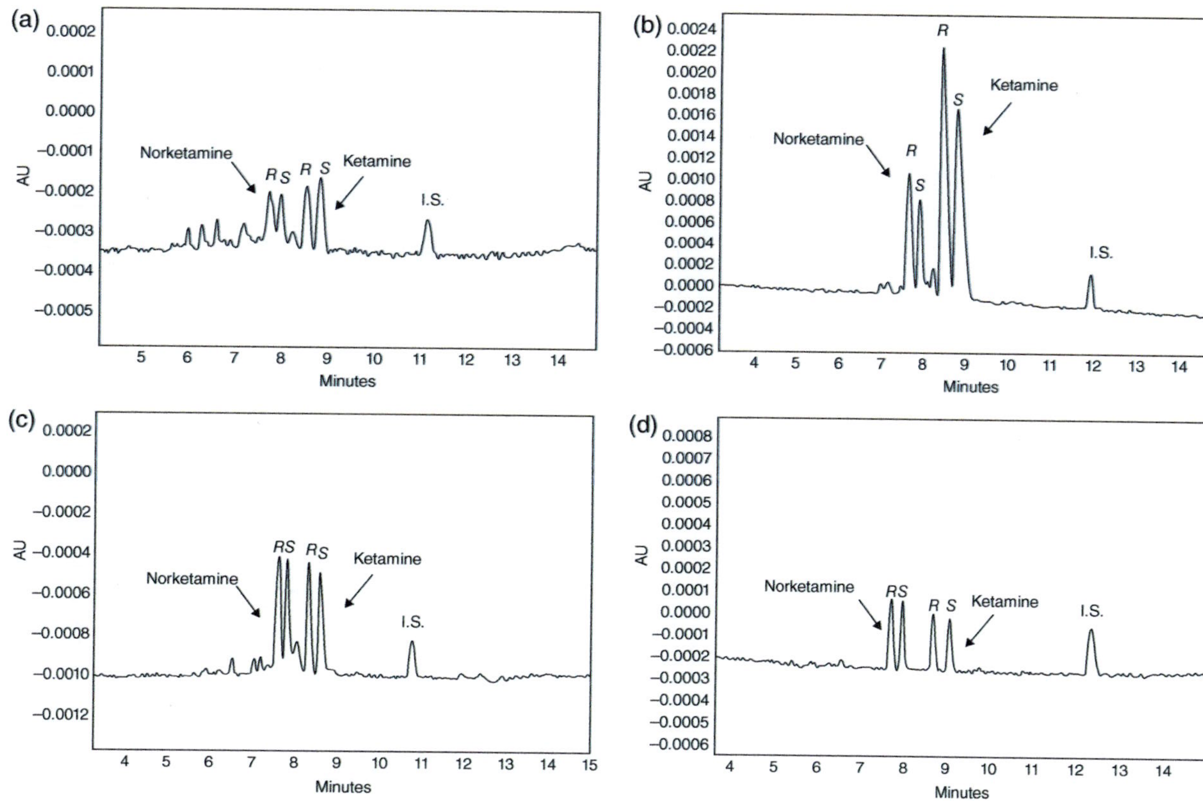

Figure 4.6 Chiral CE analysis of ketamine and norketamine in three different hair samples (a–c) and a standard solution (d). Field amplified sample stacking was used to increase the sensitivity. The BGE was 15 mM Tris phosphate (pH 2.5) containing 0.1% of HS-γ-CDs (w/v). *Source:* Reproduced from Porpiglia et al. [62]/with permission of Elsevier.

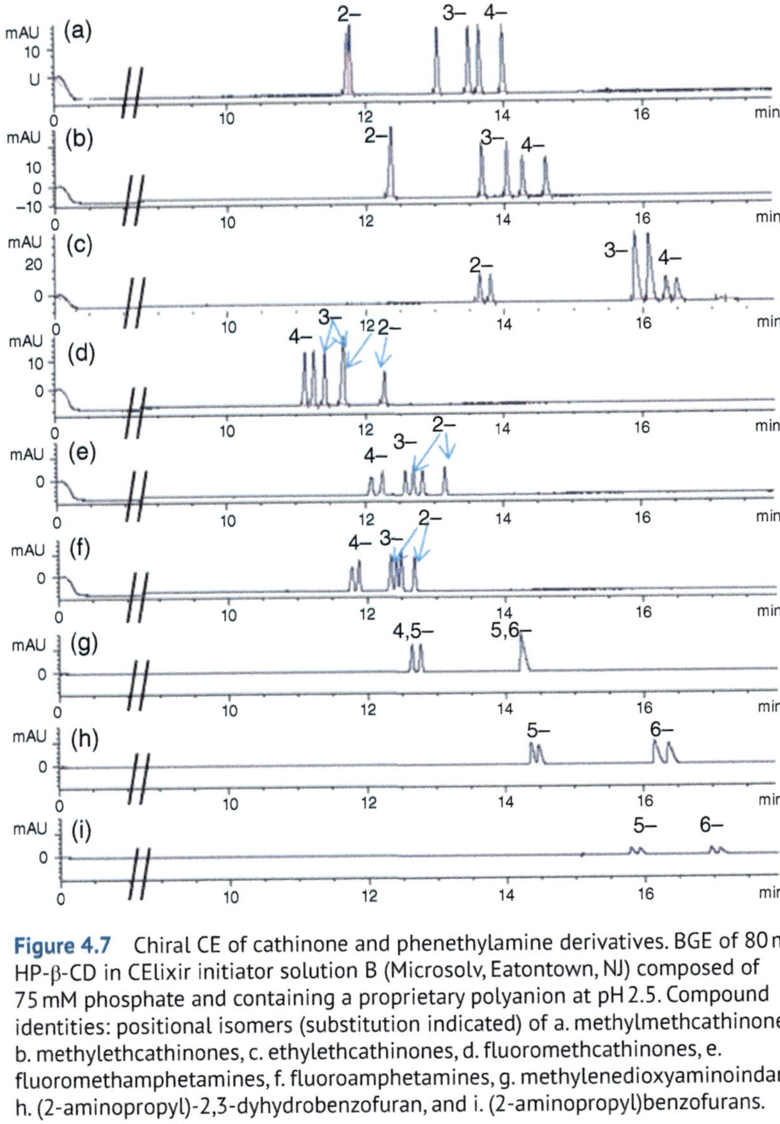

Figure 4.7 Chiral CE of cathinone and phenethylamine derivatives. BGE of 80 mM HP-β-CD in CElixir initiator solution B (Microsolv, Eatontown, NJ) composed of 75 mM phosphate and containing a proprietary polyanion at pH 2.5. Compound identities: positional isomers (substitution indicated) of a. methylmethcathinones, b. methylethcathinones, c. ethylethcathinones, d. fluoromethcathinones, e. fluoromethamphetamines, f. fluoroamphetamines, g. methylenedioxyaminoindanes, h. (2-aminopropyl)-2,3-dyhydrobenzofuran, and i. (2-aminopropyl)benzofurans.

strategies that can be implemented to avoid the contamination of the ionization source with CSs include partial-filling and counter-migration techniques, which will be explained hereafter [66]. Compared with chiral CE, chiral CEC does not suffer from the same challenges due to the immobilization of the CS on the capillary surface or stationary phase. However, due to the complexity of chiral CEC-MS (e.g. clogging issues, fragility of some of the columns, stability of the coupling), comparatively few studies using chiral CEC-MS have been reported so far [67].

Similar to conventional CE-MS, the classical non-volatile buffers (e.g. phosphate and borate buffers) should be replaced by volatile BGEs, i.e. ammonium-based (formate and acetate) in chiral CE-MS. These changes in the composition of the BGE may also impact the separation and resolution of enantiomers.

4.6.1 Hyphenation Approaches

4.6.1.1 Sheath-Liquid and Sheathless CE-MS Interfacing

Hyphenating CE with MS, usually using electrospray ionization (ESI), requires a dedicated interface, which ensures the closure of the electrical circuit while providing a suitable flow rate. Commercially available CE-MS instrumentation employs either sheath liquid or sheathless interfacing. With the former, during the ESI process a sheath liquid is added to the CE effluent, providing both electrical contact and the adequate flow rate (in the $\mu l\ min^{-1}$ range) to ensure a stable electrospray. The sheath liquid is typically composed of a hydro-organic mixture containing a small proportion of a volatile acid or base, which improves the ionization of the target analytes [68]. This type of interface is more stable than the sheathless interface, enabling the analysis of a larger number of samples but may lead to a lower sensitivity due to the dilution of the CE effluent.

With sheathless interfacing, as highlighted in its name, no extra liquid is added to the ESI source, but the capillary is modified with a porous or metal-coated tip enabling electrical contact and used in combination with a nanospray source. This interface typically results in higher sensitivities, as there is no dilution involved and nanoESI is employed, although it suffers from a lower stability and is more costly due to the use of modified capillaries [68].

Both interface principles can be used in chiral CE-MS. Nevertheless, without preventive measures, the CS will enter the mass spectrometer (due to the EOF or via suction effect) and contaminate the ionization source, potentially leading to ion suppression, loss of sensitivity, and/or inaccurate quantitation – and this independently of the interface used. One strategy to minimize contamination is to suppress the EOF or use lower concentrations of the CS. For instance, successful chiral CE-MS of amino acids has been reported using native β-CD with a moderate and acceptable ionization suppression [60]. However, such approaches are often not sufficient to fully overcome this issue. Other strategies include the partial-filling and counter-migration techniques, which have been both commonly used for chiral CE-MS applications [67].

4.6.1.2 Partial-Filling Techniques

In the partial-filling technique, the CS is prevented from entering the mass spectrometer (Figure 4.8). In order to achieve this, a plug of BGE (10–80% of capillary volume) containing the CS is introduced in the first part of the capillary, instead of being added to the running BGE, prior to the injection of the sample. In this configuration, the analytes interact first with the CS but will leave the CS-containing plug prior to migrating toward the detector, after which the analysis is stopped before the CS reaches the outlet of the capillary. This is possible when the apparent mobility of the analytes is higher than the EOF mobility. The partial-filling technique is easily integrated into either sheath–liquid or sheathless interfacing schemes and

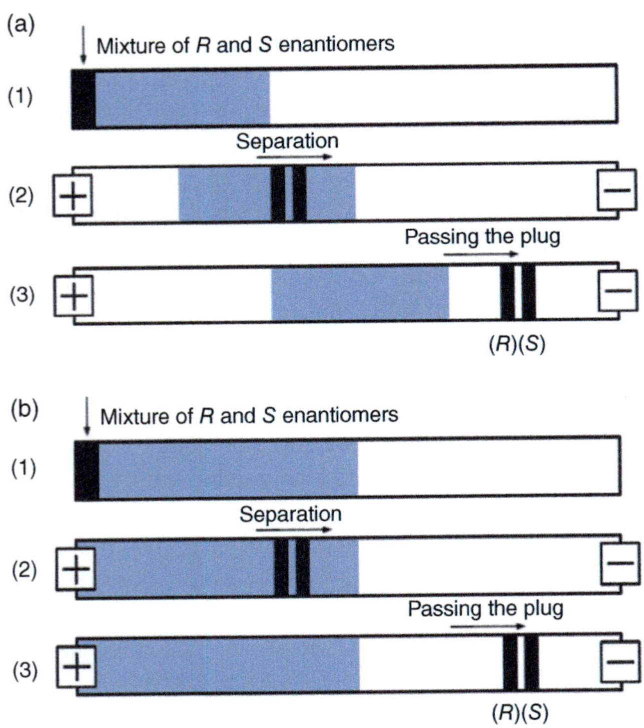

Figure 4.8 Illustration of the partial-filling technique applied to the separation of two positively charged enantiomers using a neutral CS. (a) Separation under EOF with a CS plug (in blue) that has a lower apparent mobility than the chiral analyte. The racemic mixture enters the selector plug (1) where the enantiomers are separated (2). Once the selector plug is passed, the two enantiomers move with equal mobility toward the detector (3). (b) Separation when the chiral selector plug remains stationary, e.g. when there is no EOF. *Source:* Reproduced from de Koster et al. [67]/John Wiley & Sons/Public Domain CC By 4.0.

shows the advantage of a reduced consumption of the CS, enabling the use of more expensive or exotic options. Compared with conventional chiral CE and the counter-migration technique (see next), the partial-filling approach can lead to lower enantiomer resolutions due to the shorter chiral separation length and possible peak broadening arising from differences in ionic strength and viscosity in the zones present in the capillary [67]. Moreover, careful optimization of the separation conditions is needed for each chiral analyte and type of sample.

Tanaka et al. [69] reported in 2000 the successful separation of several enantiomers with a primary amine group using partial filling with a chiral crown ether derivative ((1)-(18-crown-6)-2,3,11,12-tetracarboxylic acid; 18C6H4). The length of the CS plug accounted for ca. 70% of the total capillary volume. More recently, the same approach was used by Lee et al. for the enantioseparation of amino acids in vinegar [50]. In another recent study, Sánchez-López et al. [61] used partial filling for the chiral separation of neurotransmitters and metabolites involved in the phenylalanine-tyrosine pathway in rat plasma using CE-MS/MS and a dual CD system (i.e. methyl-β-CD and 2-hydroxypropyl-β-CD).

4.6.1.3 Counter-Migration Techniques

With the counter-migration technique, the electrophoretic mobility of the CS is directed toward the capillary inlet so that it will not enter the ion source of the mass spectrometer (Figure 4.9) [70]. Charged CSs (e.g. sulfated β-cyclodextrins) are therefore required in combination with oppositely charged analytes to achieve mutually reverse electrophoretic mobilities [71]. Interactions between the CS and the analyte ion should occur, but these should not be too strong as the analytes may then have an apparent mobility toward the capillary inlet and will not be detected.

The counter-migration technique has for instance been used by Cui et al. [72] for the enantioseparation of amphetamine-like stimulants and ephedrine (positively charged) in seized drug material using a sulfated--γ-CD (negatively charged) as CS in a CE-TOF/MS system. Another interesting application comes from Švidrnoch et al., who used vancomycin (migrating toward the cathode) as CS for the separation of 2-hydroxyglutaric acid enantiomers (negatively charged) in urine samples for the diagnosis of D-2-hydroxyglutaric aciduria in children [73]. Moini et al. [71] demonstrated the use of both chiral CE-MS approaches, i.e. partial-filling technique and counter-migration technique for the enantioseparation of synthetic cathinones and their optical isomers using sheathless CE-MS. Under low-flow conditions, there was no significant suppression due to the presence of HS-γ-CD in the electrospray ESI process. Both approaches

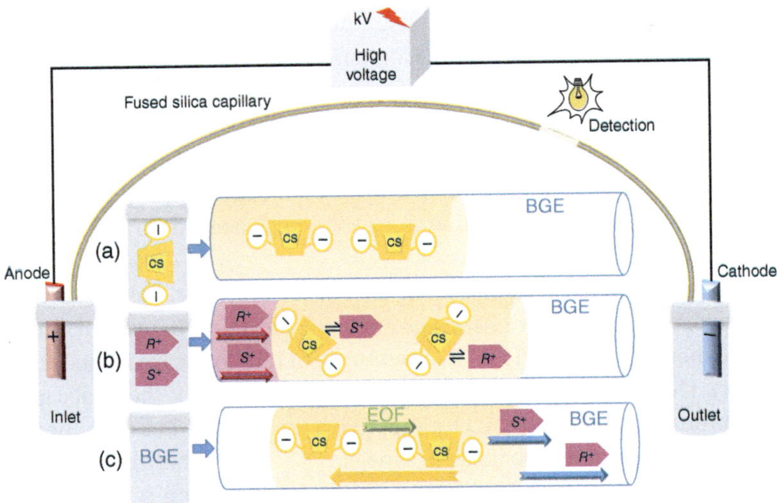

Figure 4.9 Illustration of the counter-migration technique applied to the separation of two cationic enantiomers (R^+ and S^+) using a negatively charged CS in the BGE. (a) Introduction of CS into the capillary. (b) Injection of the two enantiomers. (c) CE analysis. Pink, apparent mobility of the analytes; green, EOF mobility; yellow, apparent mobility of the CS. The CS moves toward the anode at the capillary inlet, while the enantiomeric pair moves toward the detector. *Source:* Deeb et al. [70]/MDPI/Public Domain CC BY 4.0.

led to satisfactory results, but an enhanced chiral resolution was observed for the counter-migration technique.

4.6.2 Chiral MEKC-MS

Similar to what is observed for CSs, the use of non-volatile surfactants in MEKC-MS can lead to severe ion suppression and possible contamination of the ionization source. The stability of the electrospray is also hampered due to the high surface activity of surfactant molecules. Two strategies are usually considered for coupling chiral MEKC with MS, namely, the partial-filling approach and the use of specific micelles, notably volatile surfactants and polymeric micelles. Enantioselective polymeric micelles (also referred to as molecular micelles) share similar chiral recognition as conventional surfactant-based micelles. However, due to their covalent nature, they do not dissociate into monomers in the ESI source. Moreover, they result in less ion suppression due to a low surface activity and also because no surfactant monomers are present in the BGE. Polymeric micelles also show the advantage of being compatible with organic solvents and a broad pH range, besides having a highly customizable structure. Most of the applications reported so far for MEKC-MS using polymeric micelles involve the use of polymers derived from undecylenic acid, with carbamate and amide linkages functionalized with amino acids [66, 74].

As an example of the use of volatile surfactants in chiral MEKC-MS, Somsen and coworkers employed ammonium perfluorooctanoate (APFO) for the chiral analysis of amino acids in CSF [75]. (−)-1-(9-fluorenyl) ethyl chloroformate (FLEC) was used as a chiral amino acid derivatizing agent and APFO as volatile pseudo-stationary phase to separate the formed diastereomers. APFO presented significantly less ion suppression than equal concentrations of ammonium acetate. Figure 4.10 shows the electropherograms obtained with chiral CE-MS analysis of CSF spiked with DL-amino acids. With the optimized method, 14 chiral proteinogenic amino acids were resolved (Rs between 1.2 and 8.6) and quantified. In another study, amino acids were in-capillary derivatized with FLEC using a fully automated approach [76]. A baseline resolution was obtained for eight pairs of amino acids, while partial separation was achieved for six other pairs.

Shamsi and coworkers reported a number of studies using polymeric micelles for the chiral analysis of different analytes. A relevant study reports the use of poly-sodium *N*-undecenoyl-L,L-leucylalaninate (poly-L,L-SULA) for the separation of the enantiomers of venlafaxine and its metabolite *O*-desmethylvenlafaxine using MEKC-MS [77]. The developed method was successfully applied to quantify plasma concentrations after oral administration of venlafaxine to human patients.

4.6.3 Chiral CEC-MS

Compared with other chiral CE-MS approaches, chiral CEC-MS shows the benefit of having immobilized CSs, which circumvents the risks for ionization suppression and source contamination. Most of the applications of chiral CEC-MS involve the use of a sheath–liquid interface [78–81], even though some studies report that sheathless interfacing can also be used [82]. The application of chiral CEC-MS would certainly benefit from a dedicated, commercially available interface. Among the possible chiral stationary phases, enantioselective monoliths have been used successfully in some applications, showing good separation stability and repeatability of enantiomer migration times. As an example, Gu et al. [83] developed a CEC-MS method based on a glycidyl methacrylate-bonded β-CD monolith for the enantioseparation of basic, neutral, and acidic compounds. This monolith showed high chemical and mechanical stability, homogenous microflow, and no bleeding of material was observed at the CEC-MS interface.

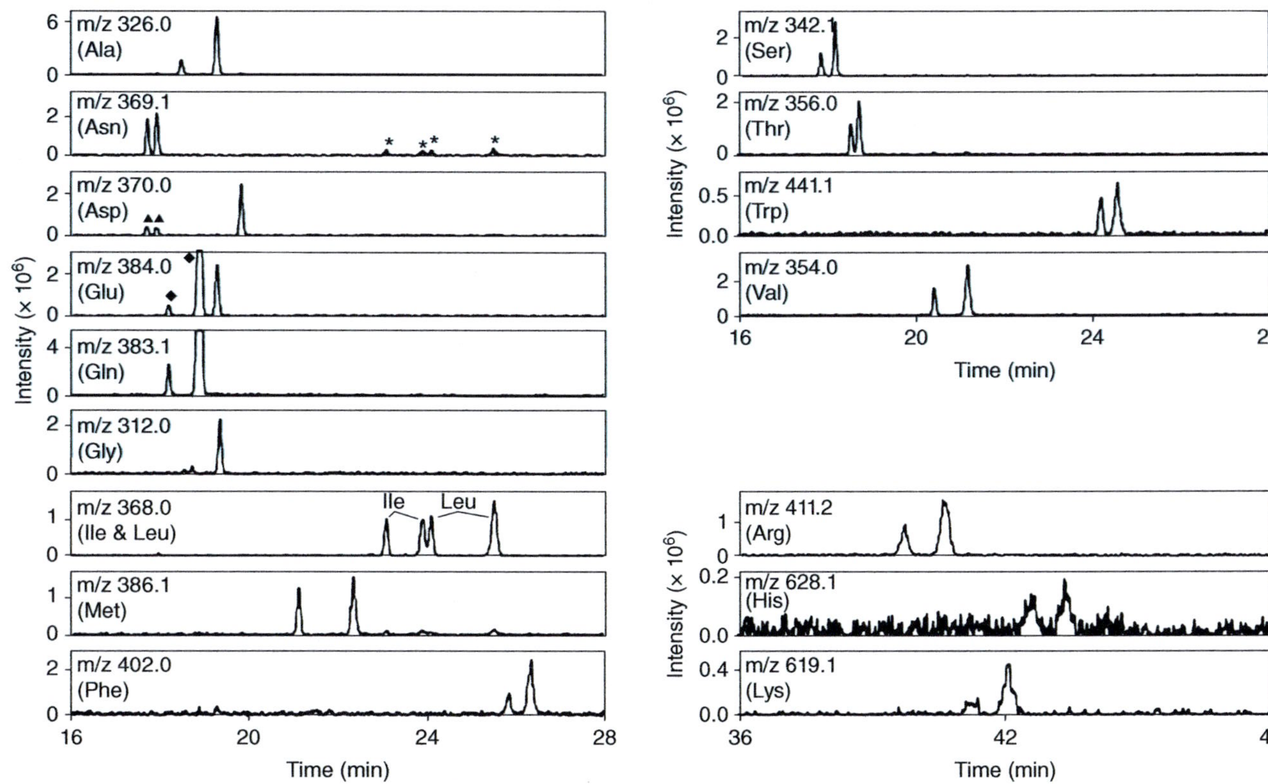

Figure 4.10 Extracted ion electropherograms obtained with the chiral CE-MS analysis of CSF spiked with DL-amino acids. Experimental conditions: BGE, 150 mM APFO (pH 9.5); spiked concentration per amino acid enantiomer, 3.2 mg ml^{-1} in CSF; sheath liquid, isopropanol-water-formic acid (90 : 10 : 0.1, v/v/v); sheath liquid flow rate, 3 µl min^{-1}, nebulizer gas pressure, 2 psi. Asterisks indicate signals of the mono C13 isotope peaks of D- and L-isoleucine and D- and L-leucine. Triangles indicate the mono C13 isotope peaks of D- and L-asparagine. Diamonds indicate mono C13 isotope peaks of D- and L-glutamine. *Source:* Reproduced from Prior et al. [75]/with permission of Elsevier.

4.7 Conclusions and Perspectives

Chiral separations by CE offer a wide array of approaches for resolving enantiomers, as illustrated by numerous applications in the fields of pharmaceutical, food, clinical forensic, and environmental analysis. Diverse choices in CSs; various separation modes including CE, MEKC, and CEC; and the ability to couple with different detectors allow for a flexible and customizable analysis. Over the past years, numerous innovative approaches, novel additives, and CEC stationary phases have been developed and successfully applied to the analysis of chiral analytes. CDs, antibiotics, bile acids, and crown ethers are commercially available and can be readily used. Novel approaches based on ILs, chiral nanoparticles, and microfluidic platforms appear very promising, but need further developments to be widely accepted in the field of chiral CE. In CEC, progress has been observed with the implementation of packed-column, monolithic-, and OT-CEC. Further improvements are needed in the fabrication procedures of stationary phases and the implementation of standardized procedures, which may foster the use of CEC beyond academic research. Coupling chiral CE with MS detection using the partial filling or counter-migration techniques represents an essential step to meet the needs in many applications where high sensitivity and high selectivity are required, particularly for biomedical, forensic, clinical, agrochemical, and food applications, as well as in metabolomics-based studies. Further technological developments in CE-MS instrumentation will certainly help fostering a broader acceptance of this technique and chiral CE in academic, governmental, and industrial environments. Supported by the remarkable technological developments carried out over the past years in the field of MS (notably ion mobility spectrometry), CE-MS will probably attract a growing interest for chiral CE-based analysis.

References

1 Gassmann, E., Kuo, J.E., and Zare, R.N. (1985). Electrokinetic separation of chiral compounds. *Science (80-)* 230 (4727): 813–814.

2 Wren, S.A.C. and Rowe, R.C. (1992). Theoretical aspects of chiral separation in capillary electrophoresis. I. Initial evaluation of a model. *J. Chromatogr. A* 603 (1–2): 235–241.

3 Chankvetadze, B., Linder, W., and Schriba, G.K.E. (2004). Enantiomer separations in capillary electrophoresis in the case of equal binding constants of the enantiomers with a chiral selector: commentary on the feasibility of the concept. *Anal. Chem.* 76 (14): 4256–4260.

4 Fanali, S. and Chankvetadze, B. (2019). Some thoughts about enantioseparations in capillary electrophoresis. *Electrophoresis* 40 (18–19): 2420–2437.

5 Chankvetadze, B. (2002). Enantiomer migration order in chiral capillary electrophoresis. *Electrophoresis* 23 (22–23): 4022–4035.

6 Fanali, S. (1989). Separation of optical isomers by capillary zone electrophoresis based on host-guest complexation with cyclodextrins. *J. Chromatogr. A* 474 (2): 441–446.

7 Terabe, S., Shibata, M., and Miyashita, Y. (1989). Chiral separation by electronkinetic chromatography while bile salt micelles. *J. Chromatogr. A* 480(C): 403–411.

8 Mayer, S. and Schurig, V. (1992). Enantiomer separation by electrochromatography on capillaries coated with Chirasil-Dex. *J. High Resolut. Chromatogr.* 15: 129–131.

9 Koide, T. and Ueno, K. (2000). Enantiomeric separations of cationic and neutral compounds by capillary electrochromatography with monolithic chiral stationary phases of β-cyclodextrin-bonded negatively charged polyacrylamide gels. *J. Chromatogr. A* 893 (1): 177–187.

10 Fejős, I., Kalydi, E., Malanga, M. et al. (2020). Single isomer cyclodextrins as chiral selectors in capillary electrophoresis. *J. Chromatogr. A* 1627: 461375.

11 Yu, R.B. and Quirino, J.P. (2021). Bile salts in chiral micellar electrokinetic chromatography: 2000–2020. *Molecules* 26 (18): 2000–2020.

12 Ninomiya, R., Matsuoka, K., and Moroi, Y. (2003). Micelle formation of sodium chenodeoxycholate and solubilization into the micelles: comparison with other unconjugated bile salts. *Biochim. Biophys. Acta Mol. Cell Biol. Lipids* 1634 (3): 116–125.

13 Lin, E.P., Lin, K.C., Chang, C.W., and Hsieh, M.M. (2013). On-line sample preconcentration by sweeping and poly(ethylene oxide)-mediated stacking for simultaneous analysis of nine pairs of amino acid enantiomers in capillary electrophoresis. *Talanta* 114: 297–303.

14 Blaschke, G. and Chankvetadze, B. (2000). Enantiomer separation of drugs by capillary electromigration techniques. *J. Chromatogr. A* 875 (1–2): 3–25.

15 Prokhorova, A.F., Shapovalova, E.N., and Shpigun, O.A. (2010). Chiral analysis of pharmaceuticals by capillary electrophoresis using antibiotics as chiral selectors. *J. Pharm. Biomed. Anal.* 53 (5): 1170–1179.

16 Elek, J., Mangelings, D., Iványi, T. et al. (2005). Enantioselective capillary electrophoretic separation of tryptophane- and tyrosine-methylesters in a dual system with a tetra-oxadiaza-crown-ether derivative and a cyclodextrin. *J. Pharm. Biomed. Anal.* 38 (4): 601–608.

17 Lei, Z., Chen, B., Koo, Y.M., and Macfarlane, D.R. (2017). Introduction: ionic liquids. *Chem. Rev.* 117 (10): 6633–6635.

18 Zhang, Q., Xue, S., Li, A., and Ren, S. (2021). Functional materials in chiral capillary electrophoresis. *Coord. Chem. Rev.* 445: 214108.

19 Mwongela, S.M., Numan, A., Gill, N.L. et al. (2003). Separation of achiral and chiral analytes using polymeric surfactants with ionic liquids as modifiers in micellar electrokinetic chromatography. *Anal. Chem.* 75 (22): 6089–6096.

20 Rizvi, S.A.A. and Shamsi, S.A. (2006). Synthesis, characterization, and application of chiral ionic liquids and their polymers in micellar electrokinetic chromatography. *Anal. Chem.* 78 (19): 7061–7069.

21 Greño, M., Marina, M.L., and Castro-Puyana, M. (2018). Enantioseparation by capillary electrophoresis using ionic liquids as chiral selectors. *Crit. Rev. Anal. Chem.* 48 (6): 429–446.

22 Zhang, Q. (2018). Ionic liquids in capillary electrophoresis for enantioseparation. *TrAC Trends Anal. Chem.* 100: 145–154.

23 Sarkaya, K., Gokturk, I., Yilmaz, F., and Denizli, A. (2021). Chiral separations by capillary electrophoresis and related techniques with different chiral selectors: a review [Farklı Kiral Selektörler ile Kapiler Elektroforez ve İlgili Tekniklerle Kiral Ayrımlar: Bir Derleme]. *Hacettepe J. Biol. Chem.* 49 (3): 253–303.

24 Belbruno, J.J. (2019). Molecularly imprinted polymers. *Chem. Rev.* 119 (1): 94–119.

25 Sanagi, M.M., Salleh, S., Ibrahim, W.A.W. et al. (2013). Molecularly imprinted polymer solid-phase extraction for the analysis of organophosphorus pesticides in fruit samples. *J. Food Compos. Anal.* 32 (2): 155–161.

26 El-Schich, Z., Zhang, Y., Feith, M. et al. (2020). Molecularly imprinted polymers in biological applications. *BioTechniques* 69 (6): 407–420.

27 Wu, L.L., Liang, R.P., Chen, J., and Qiu, J.D. (2018). Separation of chiral compounds using magnetic molecularly imprinted polymer nanoparticles as stationary phase by microchip capillary electrochromatography. *Electrophoresis* 39 (2): 356–362.

28 Tong, S., Zhang, H., Shen, M. et al. (2014). Enantioseparation of mandelic acid derivatives by high performance liquid chromatography with substituted β-cyclodextrin as chiral mobile phase additive and evaluation of inclusion complex formation. *J. Chromatogr. B Anal. Technol. Biomed. Life Sci.* 962: 44–51.

29 Zhou, J., Zhao, S., Fu, G., and Zhang, Z. (2014). Isoleucine ionic liquids as additives to separate mandelic acid and their derivative enantiomers by HPLC. *Anal. Methods* 6 (15): 5627–5631.

30 Wu, J., Xiao, X., Li, Z., and Jia, L. (2019). Enantioseparation of chiral B-blockers using polynorepinephrine-coated nanoparticles and chiral capillary electrophoresis. *Anal. Bioanal. Chem.* 411 (10): 2121–2129.

31 Tan, X., Li, Q., Shen, Y. et al. (2010). 3,3 0-Diaryl-BINOL phosphoric acids as enantioselective extractants of benzylic primary amines. *Chirality* 2014 (43): 34–43.

32 Yasukawa, K. (2020). Chiral metal nanoparticles for asymmetric catalysis. *Top. Organomet. Chem.* 279–314.

33 Whitesides, G.M. (2006). The origins and the future of microfluidics. *Nature* 442 (7101): 368–373.

34 Hutt, L.D., Glavin, D.P., Bada, J.L., and Mathies, R.A. (1999). Microfabricated capillary electrophoresis amino acid chirality analyzer for extraterrestrial exploration. *Anal. Chem.* 71 (18): 4000–4006.

35 Zeid, A.M., Nasr, J.J.M., Belal, F. et al. (2021). Microfluidic fast chiral separation of baclofen and phenylalanine enantiomers based on cyclodextrin-electrokinetic chromatography. *Microchem. J.* 160 (PB): 105770.

36 Li, X., Xiao, D., Ou, X.M. et al. (2013). A microchip electrophoresis-mass spectrometric platform for fast separation and identification of enantiomers employing the partial filling technique. *J. Chromatogr. A* 1318: 251–256.

37 Ro, K.W., Lim, K., Shim, B.C., and Hahn, J.H. (2005). Integrated light collimating system for extended optical-path-length absorbance detection in microchip-based capillary electrophoresis. *Anal. Chem.* 77 (16): 5160–5166.

38 Burley, D.M. (1962). Thalidomide and congenital abnormalities. *Lancet* 2: 271–272.

39 Blaschke, G., Kraft, H., Fickentscher, K., and Köhler, F. (1979). Chromatographic separation of racemic thalidomide and teratogenic activity of its enantiomers (author's transl). *Arzneimittelforschung* 29 (10): 1640–1642.

40 Sarkany, A., Hancu, G., Cârje, A. et al. (2019). Chiral separation of tramadol enantiomers by capillary electrophoresis using cyclodextrins as chiral selectors and experimental design method optimization. *Chem. Pap.* 73 (9): 2363–2370.

41 Kanizsová, L., Ansorge, M., Zusková, I., and Dubský, P. (2018). Using single-isomer octa(6-O-sulfo)-γ-cyclodextrin for fast capillary zone electrophoretic enantioseparation of pindolol: determination of complexation constants, software-assisted optimization, and method validation. *J. Chromatogr. A* 1568: 214–221.

42 Hancu, G., Papp, L.A., Tóth, G., and Kelemen, H. (2021). The use of dual cyclodextrin chiral selector systems in the enantioseparation of pharmaceuticals by capillary electrophoresis: an overview. *Molecules* 26 (8): 2261.

43 Szabó, Z.I., Ludmerczki, R., Fiser, B. et al. (2019). Chiral separation of rasagiline using sulfobutylether-β-cyclodextrin: capillary electrophoresis, NMR and molecular modeling study. *Electrophoresis* 40 (15): 1897–1903.

44 Bernardo-Bermejo S, Sánchez-López E, Castro-Puyana M, Marina ML. Chiral capillary electrophoresis *TrAC Trends Anal. Chem.* 2020;124 115807.

45 D'Orazio, G., Fanali, C., Asensio-Ramos, M., and Fanali, S. (2017). Chiral separations in food analysis. *TrAC Trends Anal. Chem.* 96: 151–171.

46 Pasquini, B., Orlandini, S., Goodarzi, M. et al. (2016). Chiral cyclodextrin-modified micellar electrokinetic chromatography and chemometric techniques for green tea samples origin discrimination. *Talanta* 150: 7–13.

47 Fiori, J., Pasquini, B., Caprini, C. et al. (2018). Chiral analysis of theanine and catechin in characterization of green tea by cyclodextrin-modified micellar electrokinetic chromatography and high performance liquid chromatography. *J. Chromatogr. A* 1562: 115–122.

48 Tezcan, F., Uzaşçi, S., Uyar, G. et al. (2013). Determination of amino acids in pomegranate juices and fingerprint for adulteration with apple juices. *Food Chem.* 141 (2): 1187–1191.

49 Kamencev, M., Komarova, N., and Morozova, O. (2016). Enantioseparation of tartaric and malic acids in wines by ligand exchange capillary electrophoresis using uncoated fused silica capillary. *Chromatographia* 79 (13–14): 927–931.

50 Lee, S., Kim, S.J., Bang, E., and Na, Y.C. (2019). Chiral separation of intact amino acids by capillary electrophoresis-mass spectrometry employing a partial filling technique with a crown ether carboxylic acid. *J. Chromatogr. A* 1586: 128–138.

51 Creamer, J.S., Mora, M.F., and Willis, P.A. (2017). Enhanced resolution of chiral amino acids with capillary electrophoresis for biosignature detection in extraterrestrial samples. *Anal. Chem.* 89 (2): 1329–1337.

52 Silva, M., Morante-Zarcero, S., Pérez-Quintanilla, D. et al. (2018). Environmental chiral analysis of β-blockers: evaluation of different n -alkyl-modified SBA-15 mesoporous silicas as sorbents in solid-phase extraction. *Environ. Chem.* 15 (6): 362–371.

53 Valimaña-Traverso, J., Morante-Zarcero, S., Pérez-Quintanilla, D. et al. (2018). Cationic amine-bridged periodic mesoporous organosilica materials for off-line solid-phase extraction of phenoxy acid herbicides from water samples prior to their simultaneous enantiomeric determination by capillary electrophoresis. *J. Chromatogr. A* 1566: 146–157.

54 Valimaña-Traverso, J., Morante-Zarcero, S., Pérez-Quintanilla, D. et al. (2018). Periodic mesoporous organosilica materials as sorbents for solid-phase extraction of drugs prior to simultaneous enantiomeric separation by capillary electrophoresis. *J. Chromatogr. A* 1566: 135–145.

55 Lecoeur-Lorin, M., Delépée, R., and Morin, P. (2009). Simultaneous enantioselective determination of fenamiphos and its two metabolites in soil sample by CE. *Electrophoresis* 30 (16): 2931–2939.

56 Scriba, G.K.E. (2003). Pharmaceutical and biomedical applications of chiral capillary electrophoresis and capillary electrochromatography: an update. *Electrophoresis* 24 (15): 2409–2421.

57 Fumes, B.H., Silva, M.R., Andrade, F.N. et al. (2015). Recent advances and future trends in new materials for sample preparation. *TrAC Trends Anal. Chem.* 71: 9–25.

58 Dinis-Oliveira, R.J. (2017). Metabolism and metabolomics of ketamine: a toxicological approach. *Forensic Sci. Res.* 2 (1): 2–10.

59 Theurillat, R., Sandbaumhüter, F.A., Bettschart-Wolfensberger, R., and Thormann, W. (2016). Microassay for ketamine and metabolites in plasma and serum based on enantioselective capillary electrophoresis with highly sulfated γ-cyclodextrin and electrokinetic analyte injection. *Electrophoresis* 37 (9): 1129–1138.

60 Prior, A., Sánchez-Hernández, L., Sastre-Toraño, J. et al. (2016). Enantioselective analysis of proteinogenic amino acids in cerebrospinal fluid by capillary electrophoresis–mass spectrometry. *Electrophoresis* 37 (17–18): 2410–2419.

61 Sánchez-López, E., Marcos, A., Ambrosio, E. et al. (2016). Enantioseparation of the constituents involved in the phenylalanine-tyrosine metabolic pathway by capillary electrophoresis tandem mass spectrometry. *J. Chromatogr. A* 1467: 372–382.

62 Porpiglia, N., Musile, G., Bortolotti, F. et al. (2016). Chiral separation and determination of ketamine and norketamine in hair by capillary electrophoresis. *Forensic Sci. Int.* 266: 304–310.

63 Kokiashvili, N.G., Wongwan, S., Landgraf, C. et al. (2009). Profiling of levoamphetamine and related substances in dexamphetamine sulfate by capillary electrophoresis. *J. Pharm. Biomed. Anal.* 50 (5): 1050–1053.

64 Li, L. and Lurie, I.S. (2015). Regioisomeric and enantiomeric analyses of 24 designer cathinones and phenethylamines using ultra high performance liquid chromatography and capillary electrophoresis with added cyclodextrins. *Forensic Sci. Int.* 254: 148–157.

65 Hägele, J.S., Hubner, E.M., and Schmid, M.G. (2020). Determination of the chiral status of different novel psychoactive substance classes by capillary electrophoresis and β-cyclodextrin derivatives. *Chirality* 32 (9): 1191–1207.

66 Castro-Puyana, M. and Marina, M.L. (2019). Chiral capillary electrophoresis-mass spectrometry. *Methods Mol. Biol.* 1985: 391–405.

67 de Koster, N., Clark, C.P., and Kohler, I. (2021). Past, present, and future developments in enantioselective analysis using capillary electromigration techniques. *Electrophoresis* 42 (1–2): 38–57.

68 Bonvin, G., Schappler, J., and Rudaz, S. (2012). Capillary electrophoresis-electrospray ionization-mass spectrometry interfaces: fundamental concepts and technical developments. *J. Chromatogr. A* 1267: 17–31.

69 Tanaka, Y., Otsuka, K., and Terabe, S. (2000). Separation of enantiomers by capillary electrophoresis-mass spectrometry employing a partial filling technique with a chiral crown ether. *J. Chromatogr. A* 875 (1–2): 323–330.

70 El, D.S., Silva, C.F., Junior, C.S.N. et al. (2021). Chiral capillary electrokinetic chromatography: principle and applications, detection and identification, design of experiment, and exploration of chiral recognition using molecular modeling. *Molecules* 26 (10): https://doi.org/10.3390/molecules26102841.

71 Moini, M. and Rollman, C. (2015). Compatibility of highly sulfated cyclodextrin with electrospray ionization at low nanoliter/minute flow rates and its application to capillary electrophoresis/electrospray ionization mass spectrometric analysis of cathinone derivatives and their optical i. *Rapid Commun. Mass Spectrom.* 29 (3): 304–310.

72 Cui, X., Liang, C., Gong, F. et al. (2018). Simultaneous chiral analysis of amphetamine-type stimulants and ephedrine by capillary electrophoresis coupled to time-of-flight mass spectrometry. *Chirality* 30 (9): 1079–1087.

73 Švidrnoch, M., Přibylka, A., Bekárek, V. et al. (2016). Enantioseparation of D,L-2-hydroxyglutaric acid by capillary electrophoresis with tandem mass spectrometry – fast and efficient tool for D- and L-2-hydroxyglutaracidurias diagnosis. *J. Chromatogr. A* 1467: 383–390.

74 Merino, F., Rubio, S., and Pérez-Bendito, D. (2005). Supramolecular systems-based extraction-separation techniques coupled to mass spectrometry. *J. Sep. Sci.* 28 (14): 1613–1627.

75 Prior, A., Moldovan, R.C., Crommen, J. et al. (2016). Enantioselective capillary electrophoresis-mass spectrometry of amino acids in cerebrospinal fluid using a chiral derivatizing agent and volatile surfactant. *Anal. Chim. Acta* 940: 150–158.

76 Moldovan, R.C., Bodoki, E., Kacsó, T. et al. (2016). A micellar electrokinetic chromatography–mass spectrometry approach using in-capillary diastereomeric derivatization for fully automatized chiral analysis of amino acids. *J. Chromatogr. A* 1467: 400–408.

77 Liu, Y., Jann, M., Vandenberg, C. et al. (2015). Development of an enantioselective assay for simultaneous separation of venlafaxine and

O-desmethylvenlafaxine by micellar electrokinetic chromatography-tandem mass spectrometry: application to the analysis of drug-drug interaction. *J. Chromatogr. A* 1420: 119–128.

78 Kato, M., Onda, Y., Sekimoto, M. et al. (2009). A capillary electrochromatography-electron spray ionization-mass spectrometry method for simultaneous analysis of charged and neutral constituents of a hepatocarcinoma cell metabolome. *J. Chromatogr. A* 1216 (47): 8277–8282.

79 Kitagawa, F., Inoue, K., Hasegawa, T. et al. (2006). Chiral separation of acidic drug components by open tubular electrochromatography using avidin immobilized capillaries. *J. Chromatogr. A* 1130 (2 SPEC. ISS): 219–226.

80 Zheng, J. and Shamsi, S.A. (2006). Simultaneous enantioseparation and sensitive detection of eight β-blockers using capillary electrochromatography-electrospray ionization-mass spectrometry. *Electrophoresis* 27 (11): 2139–2151.

81 Bragg, W., Norton, D., and Shamsi, S.A. (2008). Optimized separation of β-blockers with multiple chiral centers using capillary electrochromatography-mass spectrometry. *J. Chromatogr. B Anal. Technol. Biomed. Life Sci.* 875 (1): 304–316.

82 Chen, Z., Boggess, B., and Chang, H.C. (2007). Open-tubular capillary electrochromatography-mass spectrometry with sheathless nanoflow electrospray ionization for analysis of amino acids and peptides. *J. Mass Spectrom.* 42 (2): 244–253.

83 Gu, C. and Shamsi, S.A. (2011). Evaluation of a methacrylate bonded cyclodextrins as a monolithic chiral stationary phase for capillary electrochromatography (CEC)-UV and CEC coupled to mass spectrometry. *Electrophoresis* 32 (19): 2727–2737.

5

Chiral Separations at Semi and Preparative Scale

Larry Miller

Amgen Research, One Amgen Center Drive, Thousand Oaks, CA, USA

5.1 Introduction

The generation of individual enantiomers is a requirement for pharmaceutical research and development (R&D). There are two approaches to generating enantiomerically pure materials: (i) asymmetric synthesis of the desired enantiomer and (ii) resolution of a racemate. The development of an asymmetric synthesis can be time-intensive and is not always practical at the scales required (10 mg to 1 kg) during pharmaceutical discovery and early development. In addition, asymmetric synthesis by design generates only one enantiomer. Resolution of a racemate generates both enantiomers, which is essential for initial biological assays. Resolution approaches include recrystallization, enzymatic resolution, indirect chromatographic, and direct chromatographic resolution. The time required for development of a recrystallization method or an enzymatic resolution is ineffective for the generation of small amounts of material. Indirect resolution: derivatization of the racemate to form a pair of diastereomers that are separated on an achiral column is useful for the generation of individual enantiomers [1]. While the chromatographic separation of diastereomers is often easier than the separation of enantiomers, there are limitations to this process. First, high enantiomeric purity of the derivatizing agent is essential. In addition, removal of the derivative after separation should not cause racemization. Finally, the use of indirect separation adds two synthetic steps, one to form the derivative and the other to remove it post purification.

Chiral Separations and Stereochemical Elucidation: Fundamentals, Methods, and Applications, First Edition. Edited by Quezia Bezerra Cass, Maria Elizabeth Tiritan, João Marcos Batista Junior, and Juliana Cristina Barreiro.

The number of unique racemates to be resolved and the scale of separation, as well as time and cost considerations, vary greatly depending on the phase of pharmaceutical R&D. During pharmaceutical discovery thousands of different compounds are synthesized to identify a molecule with the desired biological and physical properties. Often these compounds are only synthesized once, and the quantities required are small – less than 50 mg. At this stage the most important consideration is time. The compound must be synthesized, purified, and tested as quickly as possible. Medicinal chemist's focus is on maximizing the number of compounds prepared and tested, and less so on developing asymmetric syntheses or chemical resolution routes to enantiomerically pure material. In addition, both enantiomers are required at this stage. For this work chromatographic separation of racemates is ideal. A skilled chromatographer can quickly develop a chromatographic method, perform the purification, and return the individual enantiomers rapidly to the chemist, often in less than one day. As a compound moves through discovery and the chemical requirements increase (1–100 g), chromatographic resolution often remains a part of the synthetic route. For all these reasons the use of chromatography using chiral stationary phases (CSP) has become the preferred method for generation of enantiomers in pharmaceutical discovery [2–7].

When a compound moves into development the chemical quantity requirements increase drastically. Quantities of 1–10 kg of final compound are required for Phase I studies. At this stage cost of goods becomes more important and routes that avoid chromatography are considered. During early development, time is still a critical consideration. It is important to get active pharmaceutical ingredient (API) prepared and into the clinic as soon as possible. An approach that numerous companies utilize is to use chromatographic resolution to prepare Phase I supplies, even if a cheaper synthetic route is possible. As many of the compounds entering Phase I will ultimately fail, it may make economic sense to avoid spending process chemistry resources until satisfactory Phase I results are obtained, and Phase II supplies must be produced. Depending on the biological activities as well as where in the synthetic route the separation occurs (early vs. late), the amount of racemate to be separated can vary from 2 to 100 kg. At these scales separation time can be on the order of a week to a month depending on chromatographic productivity. Techniques utilized at this scale include batch chromatography utilizing larger I.D. columns, as well as steady-state recycle (SSR) and simulated moving bed (SMB).

During late stage development and manufacturing, cost is the overriding consideration. At this stage a chromatographic resolution must be cost effective relative to alternative approaches such as asymmetric synthesis, crystallization, or enzymatic resolution. For these quantities (100 kg to metric tons) SMB is the only suitable chromatographic technique. There are numerous marketed products that utilize an SMB step. Additional details will be presented later in this chapter.

5.2 Selection of Operating Conditions

Chromatographic purification of racemates involves four steps: (i) selection of CSP and mobile phase, (ii) choice of separation technique, (iii) preparative separation, and (iv) recovery of product from the mobile phase. By far the most important step is the selection of the chiral phase and mobile phase. The selection of the separation conditions has a direct impact on productivity, the amount of solvent required, and ultimately the cost for the separation. For a highly productive separation, one desires a retention factor (k) of less than 5 for the second eluting enantiomer as well as maximum selectivity. The low k reduces cycle time for the separation, allowing maximum productivity. The maximum selectivity allows larger quantities of racemate to be loaded onto the column. Any purification has three major attributes: throughput (or speed), yield, and purity. These three attributes are the points of the "purification triangle." In practice, only two of the attributes can be optimized. For example, increasing purity is at the expense of throughput or yield. It is impossible to optimize all three attributes at the same time.

There are currently hundreds of CSP available in the market. The main types of CSP are polysaccharides [8–10], Pirkle type [11–13], protein-based [14], cyclodextrin [15, 16], and macrocyclic glycopeptides [17]. Each of these phases has different characteristics that are discussed in other chapters of this book. While all these phases are useful for analytical separations, not all are suitable for preparative separations. For a preparative separation a CSP must be stable, available in larger column sizes or as a bulk packing and have a high loading capacity. The characteristics of the types of CSP are summarized in Table 5.1. Loading capacity is the most critical requirement for a highly productive purification. A CSP that has a high loading capacity will give Gaussian peaks at a higher amount of injected racemate relative to a CSP with low loadability. This will allow increased amounts of pure enantiomers to be produced per unit time. While productivity is dependent on the racemate to be resolved as well as loading capacity, the general trend for loadability is polysaccharide > Pirkle type > macrocyclic glycopeptides > cyclodextrin > protein based. Most preparative separations are performed using polysaccharide-based or Pirkle-type

Table 5.1 Characteristics of chiral stationary phases (CSPs).

	Solvent limitations	Loadability	Range of resolution	Large column sizes/ bulk availability
Polysaccharides	Severe	High	High	Yes
Immobilized polysaccharides	None	Medium	High	Yes
Pirkle type	None	Medium	Low	Yes
Protein based	Reversed phase only	Very low	Medium	No
Cyclodextrin	None	Low	Low	2 cm I.D. and less
Macrocyclic glycopeptides	None	Medium	Medium	2 cm I.D. and less

phases. The main limitation of polysaccharide-based CSP is their limited solvent compatibility. Most polysaccharide-based chiral selectors are adsorbed on silica and can be dissolved and washed off the silica if an incompatible solvent is used. This limitation has been reduced with the introduction of immobilized polysaccharide-based CSP that will be discussed later.

The importance of loading capacity on productivity is shown in Figure 5.1. The racemate was resolved with methanol on two amylose-based CSPs. For CSP1 selectivity is 2.25, and for CSP2 selectivity is 2.68. Due to a larger selectivity, one might predict that CSP2 would generate the highest productivity. When these methods were scaled up to preparative loadings, it was shown that CSP1 had a higher loading capacity than CSP2. The increased loading produced a nearly twofold increase in productivity for CSP1 even though the analytical separation was inferior to CSP2.

Another factor to consider during method development is solubility. As a rule, a mobile phase that affords higher racemate solubility will have a higher preparative productivity. In addition, higher solubility often translates to reduced solvent requirements for the purification.

5.3 Batch HPLC Purification

5.3.1 Analytical Method Development for Preparative Separations

The first step in any preparative separation is the development of an analytical separation. Time spent on analytical method development determines the time required for the preparative resolution as well as whether the

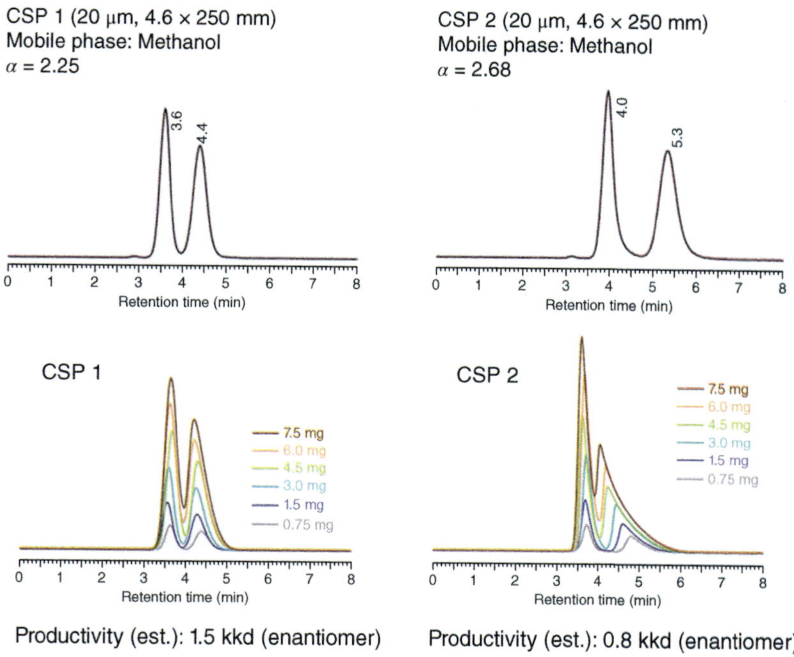

Figure 5.1 Impact of CSP loading capacity on purification productivities. *Source:* Reproduced from Caille et al. [18] with permission of American Chemical Society. kkd = kilograms product/kg CSP/day.

preparative separation will be successful. The amount of time spent on method development is highly dependent on the amount of racemate to be resolved. When performing medicinal chemistry purifications where the amount of racemate is small (often <100 mg) and the quantity of different molecules to be resolved is large, an approach for analytical method development must be utilized that is both time and compound efficient. As separation scale increases, more extensive method development may be warranted. Approaches to method development are discussed in other chapters of this book.

Alcohol-hydrocarbon mobile phases will often provide enantioselectivity. Unfortunately, many pharmaceutical compounds have poor solubility in these mobile phases. As discussed previously, high solubility in the mobile phase is a requirement for highly productive preparative separations. Polar organic mode utilizes methanol, ethanol, or acetonitrile, alone or in combination for the mobile phase. Many racemates have increased solubility in these solvents compared to alcohol-hydrocarbon mixtures. The use of CSP in the polar organic mode has desirable characteristics for preparative separations and have shown to generate highly productive preparative separations [6, 7, 19].

5.3.2 Batch HPLC Examples

Most small-scale separations are performed using batch operation. Batch chromatography involves injection, elution, and collection. Due to the high cost of most CSP, a smaller column is utilized, and multiple injections are used to process the material. Scale up from analytical to batch preparative high-performance liquid chromatography (HPLC) is a straightforward process and has been shown to be linear over a large range [20]. Due to the rapid turnaround times necessary and small quantity of racemate, limited preparative optimization is performed for most small-scale separations. Often a small amount of racemate is applied to the column and the load increased until the elution bands of each enantiomer are touching (touching band). Injections can be performed manually, or an automated injection system can be employed to allow unattended operation. Fraction collection is performed via manual collection or an automated system that allows the user to set UV thresholds to collect each enantiomer into a separate container. Mass spectrometry has been used for direct collection of non-UV-active racemates [21]. Approaches for rapid, efficient liquid chromatographic resolution of racemates are illustrated in the following examples.

5.3.2.1 Batch HPLC Example 1

It is preferable to generate the desired product(s) from a preparative separation via one pass through the column. Unfortunately, that is not always achievable, and a two-step purification process may be required. 66 g of compound 1, a racemic pharmaceutical intermediate, needed to be resolved. Analytical screening produced two methods with good enantioselectivity. Method 1: Chiralpak AD with an ethanol-heptane (40:60, v/v) mobile phase. Method 2: Chiralpak AS with an acetonitrile mobile phase. Both methods had desirable retention for the second eluting enantiomer ($k < 3$) and good selectivity ($\alpha > 2$). The Chiralpak AS/acetonitrile was chosen due to better racemate solubility. The analytical separation of compound 1 is shown in Figure 5.2. The desired enantiomer eluted second and gave a broad peak, especially for a short retention ($k \sim 0.4$). Isolation of the second eluting enantiomer at high purity and yield is more difficult due to tailing of the first peak into the second.

Utilizing a 50 cm × 10 cm I.D. column, studies were performed to determine preparative loading. It was determined that except at low loadings, peak 1 tailed into peak 2, preventing isolation of peak 2 at required purity. A two-step purification approach was required. The first step removed the bulk of the first eluting enantiomer. The fraction enriched in the second eluting enantiomer was evaporated to dryness and repurified.

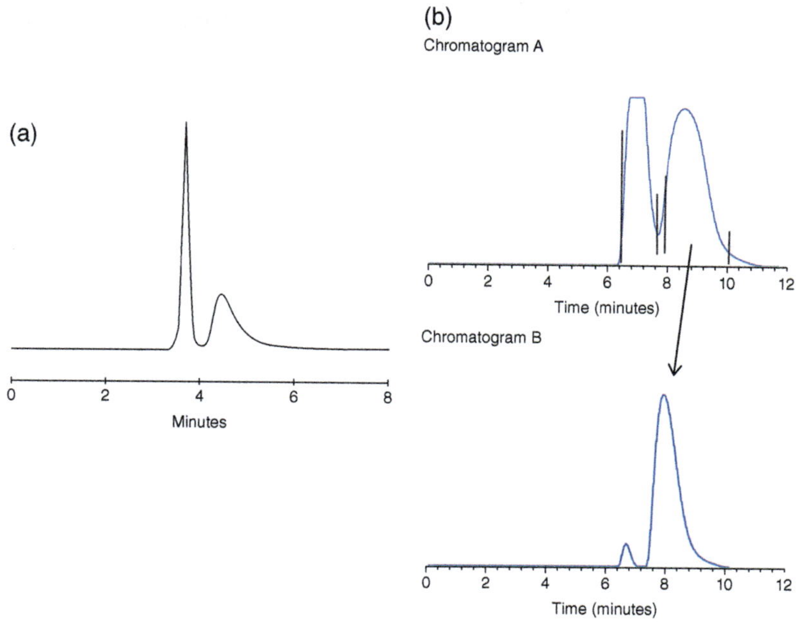

Figure 5.2 (a) Analytical HPLC separation of compound 1. Analysis conducted on Chiralpak AS column 10 μm (25 cm × 2 cm I.D), detection at 254 nm. Mobile phase: acetonitrile, flow rate 1 ml min^{-1}. *Source:* Reproduced from Miller et al. [7] with permission of Elsevier. (b) Preparative HPLC separation of compound 3. The purification was conducted on Chiralpak AS column 20 μm (50 cm × 10 cm I.D.) with a mobile phase of acetonitrile. A flow rate of 500 ml min^{-1} and detection at 254 nm was used for both separations. Chromatogram A: Loading of 7 g of racemate (20 mg ml^{-1} in acetonitrile). Chromatogram B: Loading of 4.5 g of enriched peak 2 from chromatogram A (14 mg ml^{-1} in acetonitrile). *Source:* Reproduced from Miller et al. [7] with permission of Elsevier.

The two-step preparative separation is shown in Figure 5.2. 7 g of racemate (chromatogram A) was injected every five minutes. Fraction collection generated the first enantiomer at e.r. > 99.5 : 0.5 and the second enantiomer at e.r. = 97 : 3. Enriched peak 2 was dried and repurified at a load of 4.5 g injection^{-1} (chromatogram B). The two-step process generated the first enantiomer at e.r. > 99.5 : 0.5 and a yield of 98%. The desired second enantiomer was isolated at e.r. > 99 : 1 and a yield of approximately 93%.

5.3.2.2 Batch HPLC Example 2

As a compound moves into pharmaceutical development, time to generate clinical supplies is often critical. Chemical routes used to prepare discovery supplies (often performed at less than 100 g scale) many times are not suitable at the larger scale required for Phase I supplies. This could be due to starting materials not being available or being cost prohibitive at larger quantities. Alternately, chemistry that was suitable at small scale may not be possible at larger scale.

The racemate (compound 2) to be resolved was an intermediate to an API in early development [22]. The discovery route to the API utilized enantiomerically pure starting material. While this route was sufficient for discovery supplies, the route had issues (low temperature reactions, low yield steps, non-crystalline intermediates) that made scale-up not feasible. Alternate synthetic approaches were investigated with the most promising route involving the synthesis of a racemate followed by chromatographic resolution. This racemate was investigated for large-scale preparative resolution. Analytical method development results are summarized in Table 5.2. The best analytical separation was achieved using Chiralpak AD with methanol. This method was highly desirable from a preparative perspective – short retention (k' peak 2 of 1.31), high selectivity ($\alpha = 3.95$), and high racemate solubility ($> 200\,\mathrm{mg\,ml^{-1}}$). In addition, the mobile phase, methanol, is inexpensive and has a low viscosity, allowing higher flow rates and increased purification throughputs.

As the separation was early in the synthetic process, the separation needed to be performed at multi-kg scale (77 kg racemate) to produce sufficient Phase I supplies. In addition, to meet chemical demands the racemate needed to be resolved in one month. A 15 cm I.D. axial compression column was packed with 3 kg of 20 μm Chiralpak AD to a bed length of 28 cm. Due to the large amount of racemate to be resolved, additional

Table 5.2 Chromatographic results for compound 2.

Column	Mobile phase	k'_2[a]	α	Solubility
Chiralpak AD	20/80 (v/v) Isopropanol/heptane	2.24	1.52	Minimal
Chiralpak AD	Methanol	1.31	3.95	$> 200\,\mathrm{mg\,ml^{-1}}$
Chiralpak AD	Ethanol	0.57	1.79	Not determined
Chiralpak AD	Acetonitrile	6.50	2.78	$10\,\mathrm{mg\,ml^{-1}}$
Chiralpak AD	30/70 (v/v) Acetonitrile/methanol	1.58	4.93	Not determined
Chiralpak AS	Acetonitrile	1.05	2.82	$10\,\mathrm{mg\,ml^{-1}}$
Chiralpak AS	Methanol	0.03	1.00	$> 200\,\mathrm{mg\,ml^{-1}}$
Chiralpak AS	Ethanol	0.27	1.42	Not determined

[a] Retention capacity factor for second eluting enantiomer.

Source: Reproduced from Miller et al. [22] with permission of Elsevier.

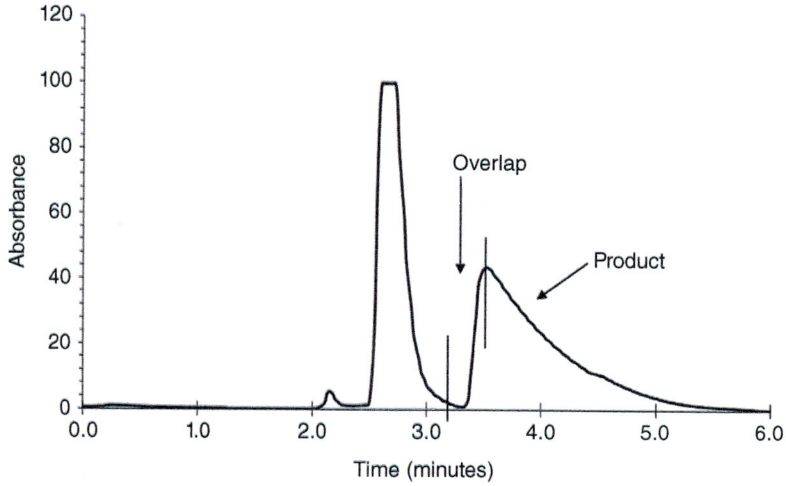

Figure 5.3 Preparative HPLC separation of compound 2. Purification conducted on a Chiralpak AD column (28 cm × 15 cm I.D.) with a mobile phase of methanol. A flow rate of 1615 ml min^{-1}, detection at 240 nm, and a loading of 11 g were used. *Source:* Reproduced from Miller et al. [22] with permission of Elsevier.

preparative method development beyond loading was explored. Load and flow rate were optimized to produce maximum productivity while meeting the required purity (e.r. > 99 : 1) and yield (>90%) for the desired second eluting enantiomer as well as the time (< one month) requirements. The final conditions were 11 g of racemate per injection with a flow rate of 1615 ml min^{-1}. The separation is shown in Figure 5.3. Injections were made every 3.75 minutes, and the separation process runs 24 h day^{-1}, 5 day $week^{-1}$ for nearly 4 weeks. Approximately 7000 injections were performed to generate the desired enantiomer at a purity of e.r. = 99.7 : 0.3 with an isolated yield of 94%.

5.4 Steady-State Recycle Introduction

Steady-state recycle (SSR) is a chromatographic technique used mainly for the separation of binary mixtures, making it ideally suited for chiral separations. A binary chromatographic process is one in which only two fractions or product streams are collected. SSR is a semi-continuous process; pure fractions are collected at the front and the end of the elution profile; impure material is circulated back through the column and fresh feed is injected into the middle of the recirculating profile. A schematic

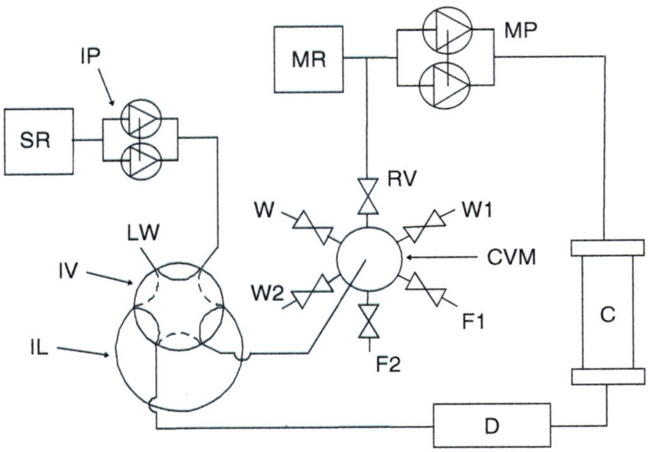

Figure 5.4 Schematic diagram of a closed-loop SSR system. MR, mobile phase reservoir; MP, mobile phase pump; C, column; D, detector; CVM, collection valve manifold; W1, waste 1 fraction collection valve; F1, fraction 1 collection valve; F2, fraction 2 collection valve; W2, waste 2 collection valve; W, waste valve; RV, recycle valve; SR, sample reservoir; IP, injection pump; IV, injection valve; IL, injection loop; LW, injection valve waste port. *Source:* Reproduced from Grill and Miller [23] with permission of Elsevier.

diagram of an SSR system is shown in Figure 5.4. SSR has been shown to achieve productivities and solvent consumption intermediate between batch HPLC and SMB chromatography [24–27]. SSR is a technique designed for the separation of larger quantities of racemate, usually greater than 25–100 g. For continuous, reproducible operation of an SSR separation, steady state must be reached. Steady state is only achieved after 10–20 cycles. At steady state the amount of material collected is equal to the amount of material injected and the profiles are nearly identical from cycle to cycle. Due to the need to reach steady state after each change to the process, the time for method development and optimization is longer for SSR than it is for preparative HPLC or supercritical fluid chromatography (SFC). SSR is basically a closed-loop recycling system to which an automated injection pump/loop has been added. Detailed description regarding the operation of an SSR system can be found in reference [27]. The costs of an SSR system are only marginally higher than the underlying HPLC system but are significantly lower than those of an SMB system of comparable production capacity. SSR fills a niche for the lower to moderate scales where the savings in operational costs do not justify the capital investment of an SMB system.

5.4.1 SSR Example 1

A detailed account of this example can be found in literature [24]. SSR was investigated for the chromatographic resolution of compound 1. The preparative HPLC separation of this racemate is described in Section 5.3.2.1, and the preparative separation is shown in Figure 5.2. Due to extensive tailing of peak 1 into peak 2, a two-step HPLC purification process was required to isolate peak 2 at required purity. Experiments were designed to compare productivities and solvent consumption for preparative HPLC and SSR. The optimized SSR separation is shown in Figure 5.5. Peak 2 is the desired enantiomer and the separation had a purity and yield specification of e.r. > 99 : 1 and > 90%, respectively.

Comparison of the purification results shows a nearly fourfold increase in productivity (435–1662 g racemate/kg CSP/day). The increased productivity of SSR is due to the need for a two-step process to isolate the second eluting enantiomer at required purity using preparative HPLC. This illustrates one of the main advantages of SSR relative to HPLC. In preparative HPLC it is always more difficult to isolate the second eluting enantiomer under

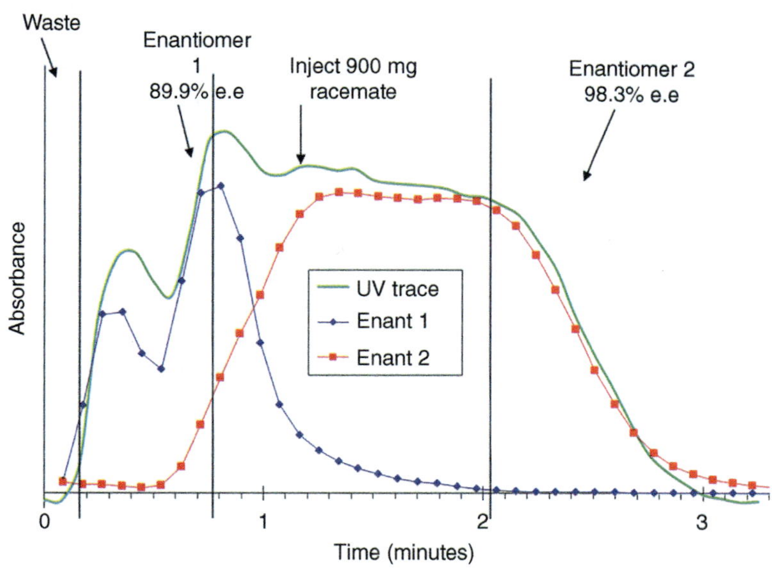

Figure 5.5 SSR separation of compound 3 in which 900 mg of racemate were injected in each cycle. Column dimensions were 5.0 cm I.D. × 20 cm. The UV trace is indicated by the line with no symbols, the ◆ symbol (diamond) indicates the concentration profile of enantiomer 1, and the ■ symbol (square) indicates the concentration profile of enantiomer 2. *Source:* Reproduced from Grill et al. [24] with permission of Elsevier.

overload conditions due to tailing of the first eluting peak into the band of the second eluting enantiomer. This is not the case with SSR; productivities are approximately the same regardless of whether the first or second eluting enantiomer is required at high purity. As expected SSR required less solvent, nearly a two-thirds reduction (0.71–0.26 l solvent g^{-1} racemate).

5.5 Simulated Moving Bed Chromatography – Introduction

As a molecule moves through development, the required quantities of API increase drastically. Previous chromatographic processes utilized to prepare enantiopure material such as batch HPLC, SFC, or SSR may not be sufficient. During preparation of Phase II/III manufacturing supplies cost of goods becomes the main driver. It is rare that a batch HPLC process will have costs low enough for cost-effective production of the hundreds of kilos required for Phase II/III or the tons of enantiomer required for manufacturing. At this scale the use of SMB chromatography is essential. This technique is also known as multicolumn chromatography (MCC). SMB was introduced by Universal Oil Products in the 1960s [28]. SMB is a continuous binary process that simulates the counter current movement of the CSP by shifting of the inlet and outlet ports. A schematic of SMB process can be found in Figure 5.6. A thorough explanation of the SMB process is beyond the scope of this chapter. Interested readers are directed to references [29–32] for further information. SMB has been shown to increase productivity and require less solvent, as well as requiring less operator time to monitor the separation [22]. SMB is a green technology, providing optimal utilization of the CSP and mobile phase. Numerous SMB users have reported excellent CSP lifetimes, often exceeding four years. With the addition of solvent recycling, solvent usage can be extremely low. UCB, a Belgium-based pharmaceutical company, has coupled solvent recycling with one of their large SMB processes [33]. Solvent recovery rates of 99.7% resulted in a solvent usage of approximately 1 l kg^{-1} of product.

SMB has been utilized in the petrochemical and sugar industries for over four decades. SMB separations in these industries are performed on the 100,000 ton $year^{-1}$ scale. Over the past 20 years SMB has moved into the pharmaceutical industry and is now a routine operation for generation of enantiopure materials [22, 34–36]. SMB is currently or has in the past been used for commercial production of Keppra ® [37], Xyzal ® [38, 39], Zoloft ® [40–42], Citalopram ® [43], and Provigil ® [44]. While SMB is an important technology for pharmaceutical manufacturing, it is also useful on smaller scale. SMB is

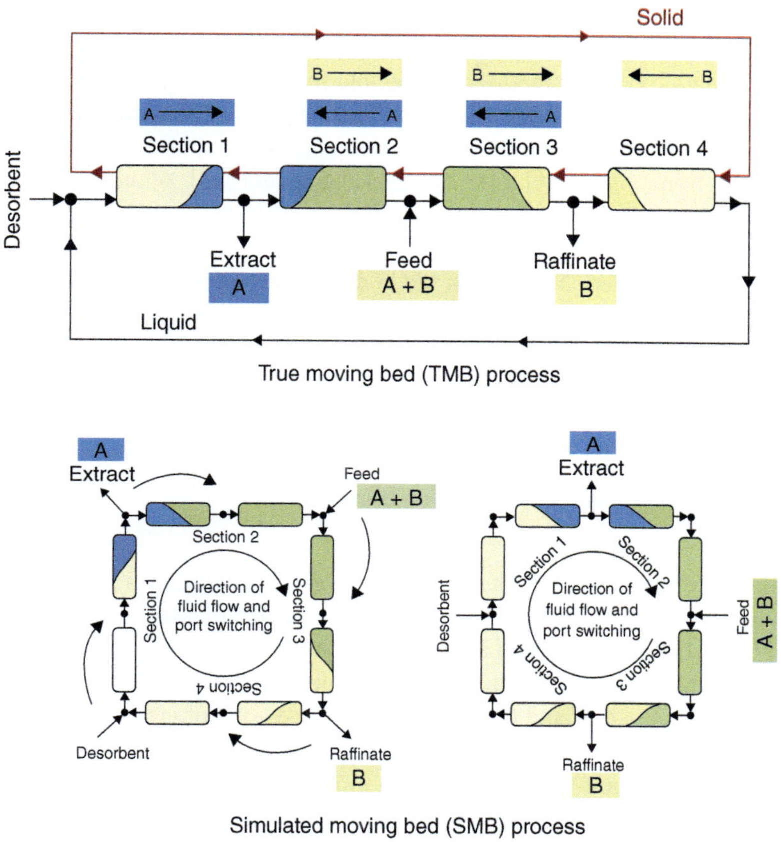

Figure 5.6 Schematic of four-section true moving bed (TMB) and simulated moving bed (SMB) units separating a binary mixture (A and B). The arrows in the TMB scheme indicate the direction of the species fluxes in each section of the unit working under complete separation conditions. The dashed arrows in the SMB scheme represent the port switch. *Source:* Reproduced from Rajendran et al. [29] with permission of Elsevier.

routinely used for the production of API and/or intermediates when separation quantities as low as 5 kg are required. SMB systems utilizing 5 cm I.D. columns that have a capacity of 100 g to 50 kg have been installed in many pharmaceutical companies and are utilized for early development and Phase I production [30].

The majority of reported SMB enantioseparations operate using liquid mobile phases and operate under isocratic conditions. SMB operations utilizing carbon dioxide mobile phases (SF-SMB) have the advantage of being able to modify the elution strength in different zones via modification of pressure and temperature. The concept of SF-SMB was first reported by Mazzotti et al. [45]. The first enantioseparation using SF-SMB was performed on tetralol and reported a threefold increase in productivity when operated in pressure gradient vs. isocratic mode [46]. SMB separation using carbon dioxide based mobile phases have also been reported for ibuprofen [47], bi-naphthol [48], and 1-phenyl-1-propanol [49]. While there are advantages to operating a carbon dioxide based SMB system, there has been little recently reported in the literature and there are no known large-scale SF-SMB operations. This is most likely due to the success of existing LC based SMB operations and the high cost of manufacturing an SMB unit able to withstand the high pressure seen under supercritical conditions. The increased cost and complexity of an SF-SMB system quickly overwhelm the operating cost savings with carbon dioxide based mobile phases.

5.5.1 SMB Examples for R&D and Separation of Compound 2

Soon after completion of the preparative HPLC separation described in Section 5.3.2, an additional 70 kg of racemate needed to be resolved. While the batch HPLC process had an acceptable productivity (1.41 kg racemate/kg CSP/day), a more efficient, cost-effective process needed to be developed for future campaigns and SMB was evaluated. A total of eight 5 cm (I.D.) columns were each packed with 110 g of Chiralpak AD to a bed length of 9.5 cm. One column was used for SMB modeling experiments. Increasing quantities of racemate were injected and retention changes measured to determine adsorption isotherms. Van Deemter and pressure drop measurements were also made. All measurements were performed at 25, 30, and 40 °C. This data was entered into simulation software that predicted starting conditions for the SMB including flow rates, switching period, feed concentration, purity/concentrations of streams, productivity, and solvent consumption. At 40 °C maximum productivity and minimum solvent consumption were predicted. The predicted operating conditions were used as a starting point for the SMB system. Conditions were modified slightly during process optimization to produce material at desired purity and yield. Additional information on operating conditions can be found in literature [22]. Utilizing these conditions, both streams were greater than 99.5% pure. Productivity under these conditions was 2.05 kg racemate/kg CSP/day. Solvent consumption was 0.171 g^{-1} racemate. The SMB was operated under these conditions for approximately 24 days straight.

Four additional SMB campaigns for compound 2 were performed at two chromatography contracting organizations to produce additional

Table 5.3 Comparison of campaign results.

Technique	Amount processed (kg)	Productivity (kkd[a])	Solvent consumption ($l\,g^{-1}$ racemate)
Batch	77	1.41	0.55
SMB (Campaign 1)	70	2.05	0.174
SMB (Campaign 2)	37	2.8	0.127
SMB (Campaign 3)	297	1.65	0.24
SMB (Campaign 4)	289	1.67	0.26
SMB (Campaign 5)	304	1.50	0.286

[a] kilograms racemate/kg CSP/day.
Source: Reproduced from Miller et al. [22] with permission of Elsevier.

enantiopure intermediate for the production of API for Phase II supplies. A summary of all campaigns for this separation is in Table 5.3. SMB Campaigns 3–5 were conducted at the same contractor using the same batch of CSP. During these three campaigns a slow degradation of the performance was observed. This degradation was not observed in the first two SMB campaigns. The performance degradation may be due to an unknown impurity being generated during racemate synthesis or present in the technical grade methanol. The impurity may be strongly absorbed on the CSP, reducing the availability of the chiral sites for the separation. This illustrates the importance of a clean feed solution for successful SMB operation. As SMB is a continuous operation, any need to stop the system results in loss of product as well as a reduction in productivity. In batch HPLC it is possible to incorporate a wash step into your process with limited impact on productivity. A wash step in SMB is undesirable due to the time and material required to wash out the SMB plant as well as restart the process after cleaning of the CSP.

A total of 1070 kg of racemate was resolved in six campaigns. These campaigns used both batch and SMB chromatography and occurred at three unique sites (Pharmacia, Carbogen, and Aerojet). Both SMB and batch processes were fully optimized. Further optimization could be achieved by utilizing alternate purification techniques such as SSR or Varicol [50]. The results from these campaigns are summarized in Table 5.3. Not unexpected, SMB was more productive and used less solvent than batch chromatography. One major advantage not evident from the data is the personnel requirements for equipment operation. Both batch separation and the first SMB campaign were performed in the same laboratory, allowing direct comparison of labor requirements. Operation of the batch system required two people per shift. The SMB required approximately 6 hours of attendance per 24 hour day. Normalizing operator hours utilized to kg racemate resolved, one obtains a value of 11.35 operator hours/kg racemate for the batch separation and 3.33 operator hours/kg racemate for the SMB separation. While these figures are not all inclusive, for example time for column packing and process development is higher for the SMB separation, it is evident that SMB requires fewer personnel relative to batch chromatography. This difference is mainly due to the large differences in solvent consumption between the processes. Batch chromatography requires near constant attendance due to the larger volumes of solvent used per day. At a flow rate of 1615 ml min^{-1}, a 55-gal drum was used approximately every two hours. The SMB unit operated for approximately 14 hours using the same volume of solvent. This allows unattended overnight operation, something not possible with the batch chromatography operation. An additional advantage of the SMB is higher product concentrations, peak 1 was almost five times more concentrated (19 vs. 4.5 mg ml^{-1}) and peak 2 more than twofold more concentrated (4.2 vs. 1.8 mg ml^{-1}). The higher product concentrations translate to reduced distillation times, resulting in additional equipment, electrical, and operator savings for SMB over batch chromatography.

5.5.2 Development of a Manufacturing SMB Process (Compound 1)

As this project moved into development, a decision was necessary on the process to be utilized for generation of Phase I supplies. Development of a synthetic route that did not involve a chromatographic enantioseparation was estimated to require multiple process chemists and take upwards of one year. To generate required quantities of API using chromatographic resolution, 20 kg of compound 1 needed to be resolved. Using a 10 cm I.D. × 50 cm column with the HPLC conditions listed in Section 5.3.2.1, the separation would require ~80 days. Based on the success of this separation using preparative HPLC and the need to decrease time for separation, it was decided to explore SMB for generation of Phase I supplies. The SMB process gave a production rate of 1105 g racemate/kg CSP/day; a purity and recovery of the desired enantiomer of e.r. > 99 : 1 and 98.4%, respectively; and a solvent usage of 0.161 l g^{-1} racemate. While the SMB process had much improved economics relative to the preparative HPLC separation (production rate = 435 g racemate/kg CSP/day, solvent usage = 0.711 l g^{-1} racemate), the question of whether this process was suitable for Phase II/III and manufacturing supplies still needed to be addressed.

Required cost of goods had yet to be calculated for the API, so it was unknown whether the current process was cost effective for manufacturing. Additional work was performed to drive the SMB costs as low as possible. The first step was evaluation of all CSP that were available in bulk supplies. From this extensive method development, a second analytical method was identified that was suitable for SMB scale-up. This method used a 3,5-dimethylbenzoyl tartardiamide CSP (Kromasil 3,5 DMB) with a mobile phase of 70/30 (v/v) toluene/methyl *tert* butyl ether. This method gave lower selectivity (2.3 vs. 3.6) relative to the original method (Chiralpak AS/acetonitrile) but had increased solubility (50 vs. 30 g l^{-1}). The 3,5 DMB method utilized a binary mobile phase. The use of this mobile phase would complicate mobile phase preparation and solvent recycling. The main advantage of these solvents was the elimination of acetonitrile, which when present at low levels in the recovered enantiomer caused side reactions in the next synthetic step. If a method without acetonitrile could be identified, the concentrated SMB product solution could be used directly in the next synthetic step, eliminating the process steps to isolate and thoroughly dry a solid prior to performing the next synthetic step.

Both methods were modeled using simulation software that showed the productivity for the 3,5 DMB process was 50% lower than the productivity for the AS/acetonitrile process and utilized approximately three times the solvent; requiring this method to be eliminated. This difference in productivity was mainly due to the lower saturation capacity for the 3,5 DMB CSP under the operating conditions.

500 kg of racemate was ultimately separated using an SMB system with six 20 cm I.D. columns containing a total of 12.8 kg of 20 μm CSP. A productivity of 4100 g racemate/kg CSP/day and a solvent usage of 0.11 l g^{-1} racemate were obtained. Purity and recovery of the desired enantiomer were e.r. = 99.2 : 0.8 and 93%, respectively. Under the separation conditions, approximately 50 kg of racemate was processed per day. Comparing the original HPLC and the optimized SMB method (Table 5.4), productivity was increased more than ninefold and solvent consumption was reduced 85%.

The final costs of the SMB process are proprietary, but based on the results from the 500 kg separation it was decided to forgo any additional process development work and use this route for manufacturing. The characteristics that made this separation cost effective were (i) high productivity for the separation, (ii) low solvent usage, and (iii) a single solvent mobile phase, making solvent recycling fast and efficient. In addition, the undesired enantiomer was easily racemized and could be reprocessed, increasing the yield to >90 vs. 46% if the undesired enantiomer was not racemized and reprocessed. Process

Table 5.4 Comparison of HPLC and SMB separation of compound 1.

Technique	Production rate (g racemate/kg CSP/day)	Solvent usage (l solvent g^{-1} racemate)	Amount processed (kg)	Enantiomer 2 purity (e.r.)	Enantiomer 2 recovery (%)
HPLC (2 step process)	435	0.71	1.05	> 99 : 1	93.0
SMB (lab scale)	1105	0.16	19	>99 : 1	98.4
SMB (process scale)	4100	0.11	500	99.2 : 0.8	93.0

Source: Reproduced from Grill et al. [24] with permission of Elsevier.

economics are always improved if the undesired enantiomer can be racemized and reprocessed. A thorough discussion of implementing racemization with SMB separations can be found in Reference [30].

5.5.3 Cost for SMB Processes

The cost for an SMB separation will vary depending on numerous parameters. These parameters include:

1) Separation productivity
2) Separation yield
3) Solvent usage
4) Solvent recycling efficiency
5) CSP cost
6) CSP lifetime
7) SMB plant utilization
8) Ability to racemize wrong enantiomer

It is difficult to estimate SMB costs without performing experiments to determine actual values for the aforementioned information. It is possible to generate a range of estimated SMB costs depending on productivity, CSP cost/lifetime, and SMB plant utilization rate. For generation of these estimates the following ranges of values were used:

1) CSP cost
   a) High cost: 26 K US$ kg^{-1}
   b) Low cost: 16 K US$ kg^{-1}

2) CSP lifetime
   a) Long lifetime: 5 years
   b) Short lifetime: 1 year
3) Productivity
   a) High productivity: 5 kg feed/kg CSP/day
   b) Low productivity: 1 kg feed/kg CSP/day
4) SMB plant size
   a) 20 cm I.D. columns
   b) 80 cm I.D. columns
   c) 100 cm I.D. columns

Further assumptions include (i) yield of 47% (94% of available desired enantiomer), (ii) wrong enantiomer not racemized, (iii) volume discount for CSP as purchased quantities increase, (iv) solvent is recycled, and (v) product recovered as solid, no further processing required. This data generates a range of costs based on scale. The estimated costs based on production scale (metric tons annually, MTA) are:

- 1–10 MTA, 350 to 2000 US\$ kg^{-1} product (using 20 cm SMB plant)
- 30–170 MTA, 50 to 250 US\$ kg^{-1} product (using 80 cm SMB plant)
- 50–280 MTA, 40–250 US\$ kg^{-1} product (using 100 cm SMB plant)

As an example, for the production of 280 metric tons of product per year using a 100 cm I.D. SMB system with a productivity of 5 kg feed/kg CSP/day and using a low cost CSP with a five-year lifetime, the estimated cost is 40 US\$ kg^{-1} product. Actual costs are highly dependent on the separation problem. The aforementioned ranges are rough estimates, and it is highly recommended to work with an SMB outsourcing company to obtain accurate separation costs.

5.6 Introduction to Supercritical Fluid Chromatography

The use of SFC has become the predominant technique for chiral preparative resolution, especially in pharmaceutical research [51–62]. Additional background information on SFC can be found in Chapter 3. A major advantage of preparative SFC vs. preparative HPLC is lower organic solvent usage. The lower solvent usage in preparative SFC is achieved by replacing a majority of the mobile phase with CO_2. CO_2 is removed post chromatography by decreasing pressure, leaving only the modifier. The result is higher product concentrations post chromatography, reducing the time and energy required for post purification solvent removal and product isolation. Additionally, SFC is an environmentally conscious technology. CO_2 used in SFC is generally recovered as a byproduct of manufacturing processes, resulting in no net increase in CO_2 [63]. Overall organic solvent volumes for SFC purifications are 2–10 times less than seen with HPLC purifications. The reduction in solvent volumes results in reduced time and cost to isolate the purified material. Due to the many advantages of SFC over HPLC for preparative purifications, in a pharmaceutical discovery purification laboratory most users prefer preparative SFC for preparative enantioseparations. If adequate separation is not seen by SFC, then HPLC is evaluated. As purification scale increases HPLC may be evaluated.

5.6.1 Analytical Method Development for Scale-up to Preparative SFC

The approach for analytical method development of SFC enantioseparations for scale-up to preparative is closely related to the approach used for HPLC-based purifications [5, 64–68]. A desirable separation maximizes selectivity while minimizing retention. The modifiers used during method development are methanol, ethanol, or isopropanol and at times acetonitrile. Other parameters that can be investigated include temperature and pressure, although these have minimal impact and are rarely exploited for preparative separations. An additive may be added to the modifier at low percentage (0.1–0.2%) to improve peak shape of ionizable compounds. For basic racemates a basic additive such as diethylamine (DEA), ammonia, or ammonium hydroxide is added [69, 70]. Under SFC conditions it has been demonstrated that carbon dioxide can interact with aliphatic alcohols such as methanol, forming carbonate species [71]. The apparent pH in carbon dioxide-methanol mobile phases is close to 5 [72]. Due to the presence of the acidic carbonate species in the SFC mobile phase, some acidic racemates may exhibit excellent peak shape under neutral conditions. Some acidic racemates may require the addition of low level of acidic additives such as acetic, formic, or trifluoroacetic acid. Increased enantioselectivity has been reported with both acidic and basic additives [73]. Combined additives are rarely used for preparative separations due to potential formation of non-volatile salts that would need to be removed post purification.

Many of the CSP developed for normal phase preparative HPLC separations can be utilized with SFC. The high flow rates possible in SFC result in reduced time for screening of chiral phases and modifiers. Flow rates of 4–5 $ml\,min^{-1}$ are standard for 4.6 mm I.D. columns. This compares to 1–2 $ml\,min^{-1}$ for HPLC screening with 4.6 mm I.D. columns. Using SFC a typical screening method is often less than five minutes [5, 54, 64, 65, 67]. Most analytical SFC equipment available offers column and modifier switching valves to automate the method development process. To decrease method development time many chromatographers utilize "ballistic" gradients (i.e. 5–70% modifier over 2–3 minutes). Studies in our laboratories have shown little difference between ethanol and methanol as a modifier for SFC enantioseparations. Ethanol is only evaluated if acceptable separation is not obtained with methanol or isopropanol. Initial evaluation is performed under gradient conditions due to the wide range of compounds evaluated in our laboratories. Once the best CSP/modifier combination has been identified, an isocratic method is quickly developed prior to preparative separation.

5.6.2 Preparative SFC Example 1

A 300 mg of a small molecule pharmaceutical racemate (compound 3) was submitted for resolution and screened using a variation of conditions. Enantioseparation was seen with numerous CSP/modifier combinations. The best analytical separation was seen with Chiralpak AD and methanol/DEA modifier. Isocratic conditions (40% methanol w/0.2% DEA/60% CO_2) were quickly developed and scaled to preparative loading. Under analytical conditions retentions of $k' = 1.78$ and 4.45 for peaks 1 and 2 and selectivity of 2.50 were measured. The preparative separation of 30 mg of compound 3 is shown in Figure 5.7 chromatogram A. Using stacked injections, the sample was processed in less than 30 minutes. Both enantiomers were isolated at near quantitative yield with e.r. > 99.5 : 0.5.

5.6.3 Preparative SFC Example 2

A 330 g of a small molecule racemic pharmaceutical intermediate (compound 4) was submitted for resolution. The best analytical separation was observed with Chiralpak AD-H and a mobile phase of 35/65 methanol w/0.2% DEA/CO_2. During the startup of the preparative separation, it was determined that the racemate had poor solubility in methanol and the addition of dichloromethane was necessary to achieve dissolution. Standard practice in many laboratories is to continue to use coated CSP with up to 50% dichloromethane in the sample solution. Above 50% the risk of CSP dissolution is increased, and it is advisable to switch to an immobilized CSP. 75% dichloromethane in methanol was required to achieve a 50 mg ml^{-1} solution of compound 4, and it was decided to switch to an immobilized CSP. The analytical separation was evaluated using Chiralpak IA with a methanol/CO_2 mobile phase. Separation was near identical to that seen with Chiralpak AD-H, and the separation was

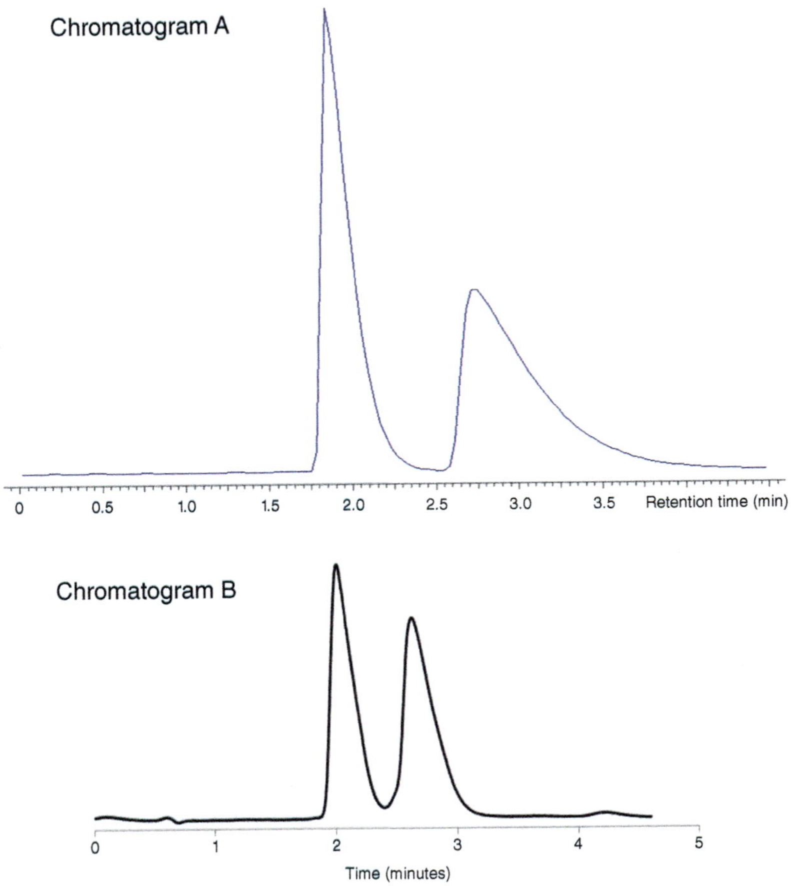

Figure 5.7 (Chromatogram A) Preparative SFC separation of compound 3. Purification conducted on a Chiralpak AD-H column (15 cm × 2 cm I.D.) with a mobile phase of 40/60 methanol w/0.2% DEA/CO_2. A flow rate of $80\,ml\,min^{-1}$, detection at 254 nm, and a loading of 30 mg were used. (Chromatogram B) Preparative SFC separation of compound 4. Purification conducted on a Chiralpak IA column (15 cm × 5 cm I.D.) with a mobile phase of 35/65 methanol w/0.2% DEA/CO_2. A flow rate of $350\,ml\,min^{-1}$, detection at 287 nm, and a loading of 200 mg were used.

scaled to a preparative column. Loading studies were conducted; 200 mg of racemate could be injected with near baseline separation achieved. The preparative separation is shown in Figure 5.7 chromatogram B. Stacked injections allowed injections to occur every 1.5 minutes. Approximately 1600 injections were performed to process the material. Peaks 1 and 2 were isolated at ~95% yield with purities greater than e.r. = 99.5 : 0.5.

5.7 Options for Increasing Purification Productivity

5.7.1 Closed-Loop Recycling

There are currently dozens of CSP that can be utilized for preparative HPLC resolution of racemates. Even with all the mobile phase and stationary phase possibilities, there are still times when the best resolution possible is not suitable for scale-up to preparative loadings. Closed-loop recycling and peak shaving is a technique shown to improve throughput for preparative chromatography [74]. Preparative HPLC systems can be modified for recycling experiments via the addition of a three-way valve installed after the detector that allows mobile phase to flow either to the fraction collection system or to the inlet of the mobile phase pump. In closed-loop recycling an initial overload injection is made. During the first pass through the column, little if any separation is achieved and the eluent is recycled back through the pump and onto the column. In subsequent passes through the column, the leading and trailing edges of the elution band are collected to isolate the first and second eluting enantiomer, respectively. The impure portion of the elution band is recycled back to the column. This process is repeated as necessary until baseline resolution is achieved. Closed-loop recycling and peak shaving allows purifications with higher throughput and lower solvent consumption compared to normal elution processes.

A comparison of recycling and peak shaving technique and normal elution chromatography for the preparative resolution of 2,2,2-trifluoro-1-(9-anthryl) ethanol was evaluated by Dingenen et al. [74]. For the resolution of 1.25 g racemate, recycling had higher yields (96–82% for peak 1, 93–42% for peak 2) and a 25% decrease in solvent usage relative to batch operation. This example shows one of the main advantages of recycling, the ability to isolate the second eluting enantiomer at high purities without sacrificing productivity or yield. During batch preparative chromatography, tailing of the first eluting enantiomer into the second eluting enantiomer is common, resulting in either low purity or a large decrease in yield.

The advantages of closed-loop recycling is demonstrated for the separation of compound 1 (Figure 5.8) [24]. 1200 mg of racemate was injected on a 5 cm × 20 cm column. During the first pass through the column, only the tail of peak 2 was collected. Peak 1 and peak 2 were both collected on subsequent recycles. The resolution of the enantiomers increased with each cycle, and the leading and trailing edges of the profile were shaved as necessary to achieve the desired purity and recovery of peak 2. Compared to the two-step batch process described in Section 5.3.2.1, productivity was 18% higher (514 vs. 435 g racemate/kg CSP/day) for the closed-loop recycling

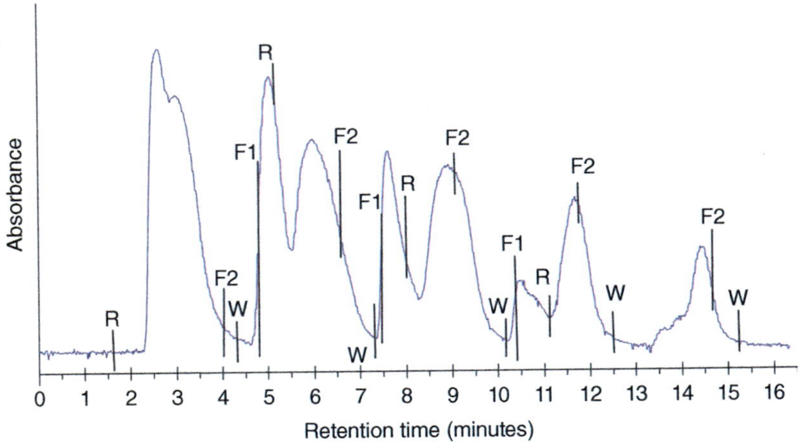

Figure 5.8 Closed-loop recycling with peak shaving in which 1200 mg of compound 1 was injected. Column dimensions were 5.0 cm I.D.× 20 cm. R = recycling, W = waste, F1 = fraction 1, and F2 = fraction 2. *Source:* Reproduced from Grill et al. [24] with permission of Elsevier.

process. Solvent usage was higher (0.94 vs. 0.71 $l g^{-1}$ racemate) for the closed-loop recycling process. Purity (e.r. > 99 : 1) and yield (>90%) were near identical for both processes. Total time for resolution with the closed-loop recycling process was significantly less due to the elimination of the need to dry material prior to step two of the batch process.

5.7.2 Stacked Injections

Another technique for increasing purification throughput is stacked injections. This technique is known by many names, including overlap injections and boxcar injections. With this technique a second injection is made prior to the elution of all peaks from the first injection. An example of this technique is shown in Figure 5.9. Chromatogram A shows a single preparative injection of racemate with a total run time of approximately 110 seconds. Examination of the chromatogram shows no product eluting during the first 50 seconds of the separation with the elution time for the enantiomers being only 60 seconds. To fully optimize this separation, stacked injection technique was utilized, and an injection performed every 60 seconds. The separation of five stacked injections is shown in chromatogram B. Using this technique, the time required for separation was reduced by 45% with a corresponding reduction in solvent requirement of 45%.

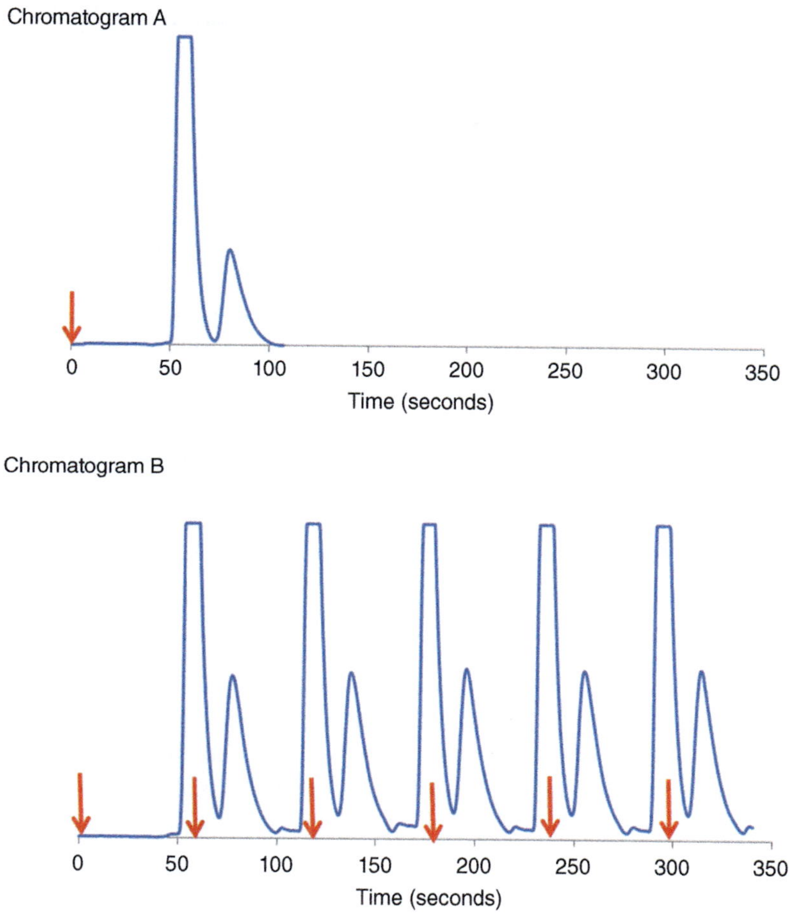

Figure 5.9 Stacked injection example. Chromatogram A: single injection at time zero; chromatogram B: stacked injection, injection made every 60 seconds. Injection indicated by arrow.

5.7.3 Choosing the Best Synthetic Intermediate for Separation

During drug discovery it may be preferable to separate the final compound vs. separation at an earlier intermediate. With this approach the enantiomeric purity is fixed at the last step and the risk of racemization in subsequent chemical steps is eliminated. Also, since the scale of separation is small, even a difficult separation can be completed quickly. As a compound moves into development and the separation scale increases, the choice of which synthetic intermediate to resolve should be considered. Besides maximum selectivity and high racemate solubility, additional points to consider include (i) racemate purity, (ii) mobile phase additives, (iii) mobile phase cost, (iv) mobile phase distillation time, (v) physical state of racemate, (vi) stability of racemate in solution, and (vii) potential for racemization at subsequent steps. It is preferable to perform the separation as early as possible in the synthetic route. This will reduce the scale of subsequent reactions, eliminating waste and reducing processing time. For larger scale separations, it is desirable to have racemate of high chemical purity. Achiral impurities can co-elute with the enantiomers or elute such that collection using UV detection is complicated. Highly retained impurities can result in pollution of the CSP and ultimate loss of separation. Ideally a separation should be performed on a racemate that does not require the use of an acidic or basic additive. The presence of an additive complicates the solvent recycling process. If possible, a single solvent mobile phase should be chosen, making solvent recycling straightforward. The cost of mobile phase and time for distillation become more important as separation scale increases. Another point to consider is the physical state of the racemate. Is it a solid or an oil? At larger synthetic and purification scale an oil is difficult to weigh for dissolution and will make for a complicated isolation post chromatography (inability to crystallize). Upon chromatographic scale-up the racemate and isolated enantiomers are in solution for longer periods of time. Stability studies in the mobile phase should be performed and solvents with the potential to degrade the racemate should be avoided. Finally, subsequent chemical steps should be evaluated for racemization potential. If the potential for racemization is identified, the enantioseparation should take place after this synthetic step or the chemistry should be changed to reduce the chance of racemization.

5.7.3.1 Choosing Synthetic Step for Separation – HPLC/SMB Example

The process of choosing the best intermediate to resolve is illustrated in the following example. The synthetic route had four intermediates where the separation could occur. Intermediate 1 had a primary amine and carboxylic acid functional groups, and for Intermediate 2 the primary amine was protected with a Boc group. Intermediate 3 contained two amine functional groups, both protected (one with Boc; one with CBz); Intermediate 4 had the CBz group removed to generate a primary amine. All four intermediates were evaluated using analytical HPLC. The decision of which intermediate to separate was based on selectivity, solubility, racemate purity, and mobile phase. Intermediates 1 and 4 contain acidic and/or basic functional groups. Enantioseparations of racemates with these functional groups often require acidic or basic additives. These functional groups may exhibit non-specific interactions on the CSP and often have lower solubility compared to neutral racemates. In addition, mobile phases containing additives are more difficult to prepare and to recycle. For these reasons the separation was not considered for these intermediates. The best separation was observed with Intermediate 2. This racemate also had good solubility. Unfortunately, the reaction of Intermediate 1 to Intermediate 2 formed several achiral impurities, leading to Intermediate 2 purity of only 65%. In addition, Intermediate 2 was non-crystalline and chemical purity could not be increased by crystallization. An achiral chromatographic purification was required prior to chiral separation. While Intermediate 3 did not offer the best separation, it was adequate for scale-up. Also, the racemate had good solubility and high purity (~99%). Intermediate 3 was purified using batch HPLC (560 g) and SMB (10 kg) using Chiralcel OD and a mobile phase of 10/90 (v/v) isopropanol/heptane to generate product with e.r. > 99 : 1. Productivity and solvent consumption for the batch HPLC process was 0.67 kg racemate/kg CSP/day and 0.84 $l g^{-1}$ racemate and for the SMB process was 2.5 kg racemate/kg CSP/day and 0.084 $l g^{-1}$ racemate.

5.7.3.2 Choosing Synthetic Step for Separation – SFC Example

A more detailed account of this example can be found in literature [62]. In support of late stage discovery approximately 300 g of enantiomerically pure material needed to be generated. Review of the synthetic scheme showed three racemates that could be separated: two intermediates and the final compound. All three racemates were subject to chiral SFC method development. Intermediate 2 and the final product had low selectivities (< 1.7). High selectivity was achieved for Intermediate 2 using Chiralpak IC CSP and either methanol, ethanol, or isopropanol modifier. The next step was to measure racemate solubility. Solubility measurements in CO_2 or CO_2/modifier mixtures require specialized equipment [75]. For this experiment solubility was measured in pure organic solvents (methanol, ethanol, and isopropanol) only. Solubility of greater than 100 mg ml^{-1} racemate was measured in each of the solvents. As chromatographic separation or selectivity could not help select the best method, loading studies were performed. These results are summarized in Table 5.5 with the preparative chromatograms shown in Figure 5.10. The highest productivity was seen with ethanol/DEA modifier, although the separation with methanol/DEA was nearly as productive. Approximately 1 kg of Intermediate 2 needed to be resolved using a 5 cm I.D. × 25 cm column. Under these conditions separation time with ethanol/DEA would be approximately 22 hours, with methanol/DEA the time would be approximately 27 hours.

Table 5.5 Preparative SFC loading studies for Intermediate 2.

Modifier	Injection amount (mg)	Cycle time (min)	Productivity (kkd)	Solvent usage ($l\,g^{-1}$ racemate)
Methanol	200	1.8	5.16	0.29
Ethanol	400	3.0	6.19	0.30
Isopropanol	600	7.0	3.98	0.47

Conditions: Column, 2×15 cm; flow rate 80 ml min^{-1}; detection at 320 nm; back-pressure regulator at 100 bar; racemate dissolved at 100 mg ml^{-1} in modifier (without diethylamine); all modifiers include 0.2% diethylamine (kkd = kg racemate/kg CSP/day).

Two purification methods with approximately the same productivity, separation time, solubility, and solvent usage were developed for the resolution of Intermediate 2. The final purification method chosen used a methanol/DEA modifier. Methanol was chosen over ethanol based on solvent cost and distillation time. Methanol is less expensive than ethanol and has a lower boiling point, resulting in decreased time to remove post chromatography. For small scale separation where solvent volumes are small, this is often not a factor in the separation decision. For this separation approximately 300 l of solvent was required, resulting in numerous hours or days of additional distillation time depending on the equipment used. Finally, while it did not have an impact on this separation, ethanol has a higher viscosity than methanol, resulting in increased systems pressures. For separations operated near the equipment pressure limit, this could limit maximum flow rates for the separation.

5.7.4 Use of Non-Commercialized CSP

While there are dozens of CSP available for preparative separations, there are times when a suitable separation cannot be developed. The racemate of interest was an intermediate to an API in early development. During the discovery process the racemate was separated at 100 g scale using batch HPLC and SSR. The racemate contained a primary amine and was difficult to resolve. The analytical separation of this racemate is shown in Figure 5.11, chromatogram A. 20 kg of racemate needed to be resolved for Phase I supplies. The SMB productivity for this separation was estimated to be ~300 g/kg CSP/day, translating to 70 days to resolve 20 kg of racemate. Additional intermediates in the synthetic route were evaluated for preparative resolution without success.

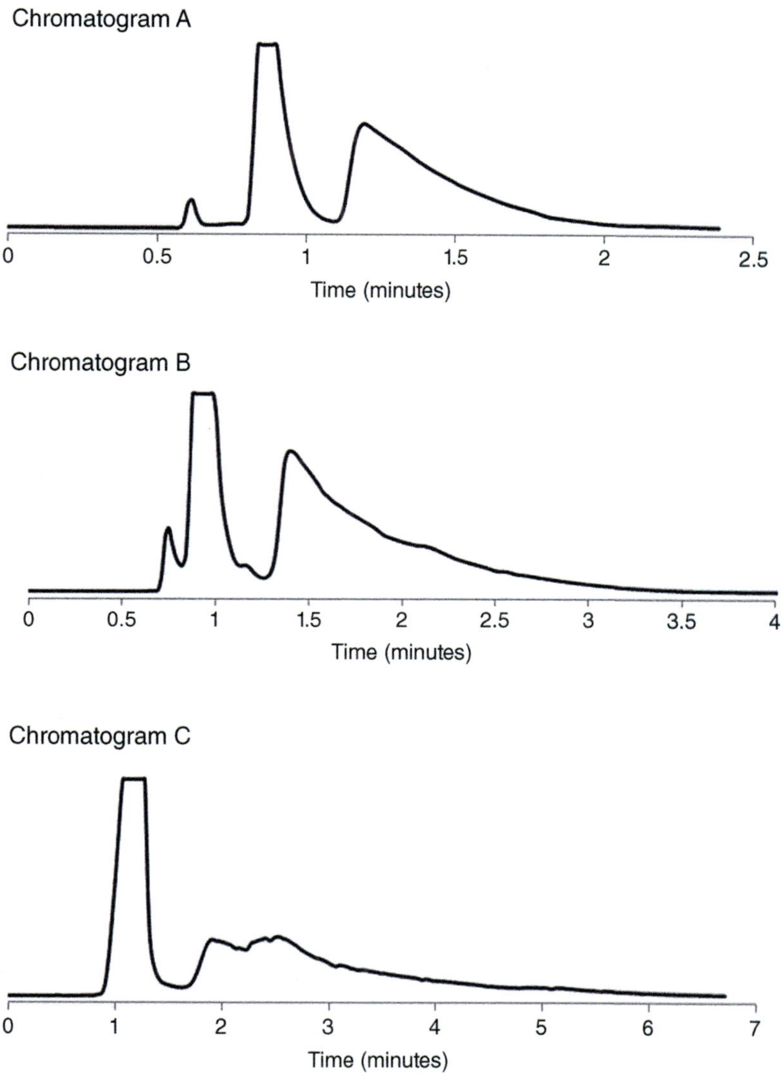

Figure 5.10 Preparative SFC separation of Intermediate 2. Purification conducted on a Chiralpak IC column (15 cm × 2 cm I.D.) with a flow rate of 80 ml $min^{-1}$and detection at 320 was used. Chromatogram A: 40% methanol w/0.2% diethylamine, loading of 200 mg; Chromatogram B: 50% ethanol w/0.2% diethylamine, loading of 400 mg; Chromatogram C: 50% isopropanol w/0.2% diethylamine, loading of 600 mg.

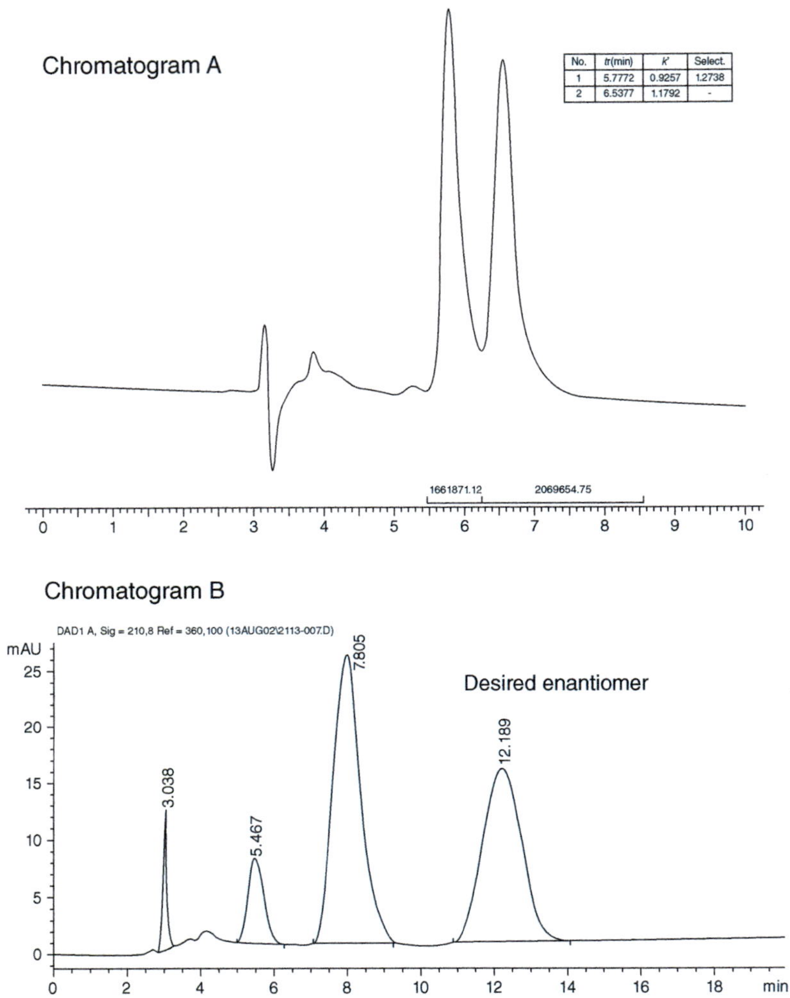

Figure 5.11 (Chromatogram A) Analytical HPLC separation. Analysis conducted on Chiralpak AD column 10 μm (2 cm I.D. × 25 cm), detection at 254 nm. Mobile phase: ethanol w/0.1% diethylamine, flow rate ml min^{-1}. (Chromatogram B) Analytical HPLC separation. Analysis conducted on Chiralpak 122,873 column 10 μm (25 cm × 2 cm I.D.), detection at 254 nm. Mobile phase: methanol, flow rate ml min^{-1}.

Only a small percentage of CSPs that have been synthesized are currently commercially available. One of the CSP suppliers has an inventory of "non-marketed" polysaccharide based CSP that are available for larger scale separations. A sample of the racemate was sent to this supplier for evaluation. The best separation from the non-marketed screen is shown in Figure 5.11, chromatogram B. These conditions were scaled to an SMB separation generating the desired enantiomer at e.r. > 99.5 : 0.5 and > 94% yield with a productivity of 1.23 kg racemate/kg CSP/day and a solvent usage of 0.34 l g^{-1} racemate. Comparison of the Chiralpak AD method to the improved method showed a 78% reduction in separation time (73–16 days) and a 72% reduction in solvent usage (24,000 to <7000 l).

5.7.5 Immobilized CSP for Preparative Resolution

Polysaccharide-based CSP have been utilized for over 30 years for the preparative resolution of racemates. While these phases have proven extremely useful, they have severe solvent restrictions due to the derivatized cellulose or amylose being only adsorbed onto the silica gel. Contact with certain solvents can result in dissolution of the cellulose/amylose and loss of resolution and/or column destruction. Over the past 20 years a series of immobilized cellulose/amylose CSP (Chiralpak IA, IB, IC, ID, IE, and IF) have been introduced [76–79]. These CSPs have no solvent restrictions and have proven useful where solvents other than traditional solvents used for chiral separations (alkanes, alcohols, and acetonitrile) are required for resolution. These CSPs have also proven useful for preparative resolutions where the racemate to be separated is poorly soluble in traditional solvents [78].

5.7.5.1 Processing of Low Solubility Racemate

Development of an analytical chiral method requires identification of a suitable CSP and mobile phase. Under analytical conditions, racemate solubility has minimal impact on separation. Solubility is more critical under preparative conditions, and high racemate solubility is required for a successful purification. 2.5 g of a discovery pharmaceutical small molecule (compound 5) was submitted for resolution. The racemate was impure, containing a 1 : 1 : 2 mixture of product:regioisomer:unknown impurity as well as other lower level impurities. To make the chiral resolution less complicated, the impurity was removed using preparative reverse phase chromatography. The resulting product/regioisomer mixture (822 mg) was subjected to polar organic chiral HPLC method development. Separation of the product and regioisomer enantiomers was observed with Chiralpak AD and a methanol mobile phase. While this method gave adequate analytical separation, it was quickly determined that this approach was not suitable for preparative work due to extremely poor solubility in methanol (< 1 mg ml^{-1}).

Solubility studies showed the material was soluble in dichloromethane, ethyl acetate, methyl *tert*-butyl ether, or tetrahydrofuran (THF). These solvents were used as starting points for chiral method development using

Table 5.6 Chromatographic results for compound 5.

CSP	Mobile phase	Regio. peak 1 t_r	Regio. peak 2 t_r	Product peak 1 t_r	Product peak 2 t_r
Chiralpak IA	50/50 (v/v) DCM/heptane	15.07	16.40	15.64	17.04
Chiralpak IA	30/70 (v/v) THF/heptane	15.93	21.05	10.75	14.12
Chiralpak IA	50/50 (v/v) MTBE/ heptane	4.64	4.82	5.10	5.28
Chiralpak IA	50/50 (v/v) EA/heptane	6.40	6.40	6.29	6.9

DCM = dichloromethane, EA = ethyl acetate, MTBE = methyl tert-butyl ether, THF = tetrahydrofuran.
Source: Reproduced from Miller and Potter [5] with permission of Elsevier.

Chiralpak IA (immobilized version of Chiralpak AD), which is summarized in Table 5.6. The best separation was observed with 30/70 (v/v) THF/heptane. The analytical separation and preparative separation are shown in Figure 5.11. The sample was dissolved in THF at a concentration of 134 mg ml^{-1}. Final purification conditions separated ~34 mg of sample/injection (Figure 5.12, chromatogram B). 822 mg of sample was processed using these conditions to generate 170 mg of enantiomer 1 and 166 mg of enantiomer 2, both with e.r. = 99.4 : 0.6.

5.7.5.2 Preparative Resolution of EMD 53986

High sample solubility is required for a highly productive purification method. When developing a preparative enantioseparation method one must balance sample solubility, enantioselectivity, and resolution. The introduction of immobilized CSP with their wider range of solvent possibilities allows sample solubility to be the guiding criteria during method development. This approach was utilized during the development of a large-scale separation of EMD 53986. The initial purification method utilized Chiralpak AD with ethanol as the mobile phase [80]. While this method had good enantioselectivity ($\alpha = 3.10$), the low solubility of the racemate in ethanol limited the purification productivity.

The preparative separation of EMD 53986 was investigated using Chiralpak IA [77]. Initial experiments were performed to measure racemate solubility. Highest solubility (75–85 g l^{-1}) was measured in THF and 1,4-dioxane, with moderate solubility (35–38 g l^{-1}) measured for chloroform

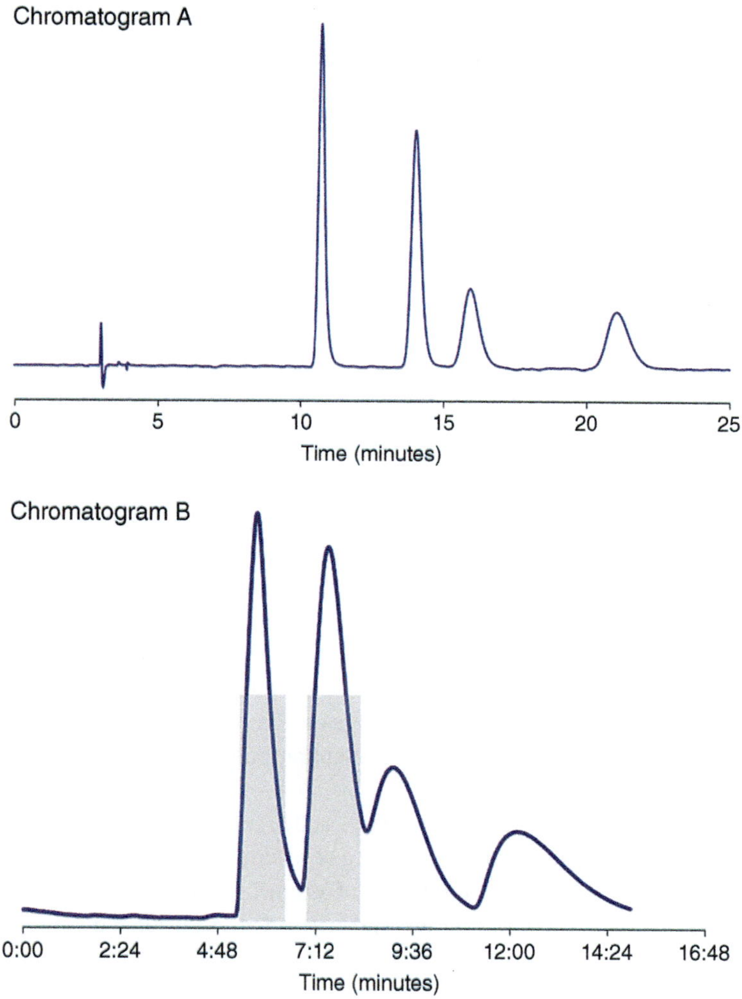

Figure 5.12 Chromatogram A: Analytical HPLC separation of compound 5. Analysis conducted on Chiralpak IA (0.45 cm I.D. × 15 cm), mobile phase of 30/70 (v/v) THF/heptane, and a flow rate of 1 ml min^{-1}. Peaks at 10.7 and 14 minutes are product enantiomers, and peaks at 15.9 and 21 minutes are regioisomer enantiomers. Chromatogram B: preparative HPLC separation of compound 5. The purification was conducted on Chiralpak IA (25 cm × 2 m I.D.) with a mobile phase of 30/70 THF/heptane and a flow rate of 40 ml min^{-1}. 34 mg of material was injected. *Source:* Reproduced from Miller and Potter [5] with permission of Elsevier.

and dichloromethane. Analytical method development was performed utilizing combinations of the aforementioned solvents. The optimum analytical method for preparative scale-up was determined to be Chiralpak IA with a mobile phase of 70/30 (v/v) dichloromethane (CH_2Cl_2)/THF. Loading studies were performed to measure adsorption isotherms and the SMB separation was modeled. A comparison of the original (Chiralpak AD) and improved (Chiralpak IA) method can be found in [77]. The use of a mobile phase offering improved solubility generated an SMB method with 6.5-fold higher productivity. In addition, the use of a lower viscosity mobile phase allowed the SMB unit to be operated at higher flow rates, also improving productivity.

5.8 Choosing a Technique for Preparative Enantioseparation

In a well-equipped purification laboratory, there are often numerous separation techniques available (batch HPLC, SFC, SSR, SMB) to complete a separation. The criteria used for this decision is highly dependent on the equipment available in the individual laboratory. The amount of racemate to be resolved is the most important consideration for choice of separation technique. As the scale of separation increases, it is normal for different separation techniques to be utilized. The separation of a racemic thiourea (Figure 5.13) illustrates this point.

The initial separation of this racemate (~ 800 g) was accomplished using closed-loop recycling preparative HPLC. 2 g of racemate was injected onto a 10 cm I.D. × 40 cm column containing 2 kg of 20 μm Chiralpak AD CSP. After one recycle the separated enantiomers were collected. Additional details on the recycling experiment can be found in the literature [81]. Recycling was performed to improve the productivity over batch HPLC purification. The next batch of thiourea was submitted at 5.9 kg scale. For this separation SSR was utilized. SSR will often offer higher productivities relative to batch HPLC or closed-loop recycling. The separation was also evaluated using preparative SFC. The SFC separation of 350 mg of racemate is shown in Figure 5.13. While SFC offered good productivity for the separation, at the time of this work at Amgen SFC equipment was limited to a maximum 3 cm I.D. column. The use of 10 cm I.D. columns with the SSR equipment resulted in shorter processing times for SSR relative to SFC. As the project moved into development, the quantity of material to be separated increased to multi kilogram and it was appropriate to evaluate SMB for the resolution. At this point a decision on whether SMB may be an option as a manufacturing

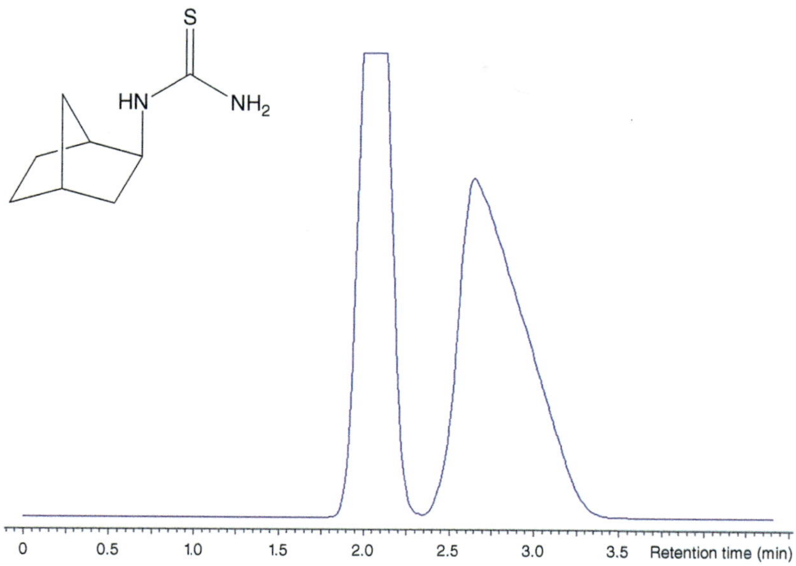

Figure 5.13 Preparative SFC separation of thiourea. Purification conducted on a Chiralpak AD-H column (15 cm × 5 cm I.D.) with a mobile phase of 25/75 methanol w/0.2% diethylamine/CO_2. A flow rate of 350 ml min^{-1}, detection at 225 nm, and a loading of 350 mg were used.

route needed to be made. Details on the SMB work can be found in the literature [18].

A comparison of the four separation techniques is in Table 5.7. While SMB has the highest productivity and lowest solvent consumption, it is not the technique of choice for all scales. SMB is a powerful technique but is not suited for separation of low amounts of racemate or for short separation campaigns. Besides the additional cost of an SMB unit relative to

Table 5.7 Comparison of separation techniques for thiourea resolution.

Technique	Productivity (g racemate/ kg CSP/day)	Solvent Usage (l g^{-1} racemate)	Purity (e.r.)	Yield (%)
Closed-loop recycle	110	2.10	98 : 1	97.5
SSR	358	0.80	98 : 1	95.2
SFC	1600	0.44	9 : 1	99
SMB	1800	0.17	96 : 2	95

HPLC, SSR, or SFC, there are additional considerations. SMB requires the packing and testing of multiple columns. The additional time required preseparation needs to be considered. Also, the training requirements of the chromatographer are much higher for SMB relative to other techniques. Finally, the increased time for pre- and post-run cleaning of an SMB adds time to the separation. Assuming an SMB unit with 5 cm I.D. columns, it is the author's opinion that SMB should not be used for quantities less than 5 kg. For difficult separations (selectivity <1.5) this threshold will decrease.

SSR is a good choice for separation quantities from 50 g to 5 kg. The cost of an SSR system is 25–50% higher than a standard preparative HPLC system. In addition, an SSR system can also be utilized for batch HPLC and closed-loop recycling and utilizes standard preparative HPLC columns. Preparative SFC has many advantages and should be considered the technique of choice for enantioseparations of less than 10 g. Separation of 10 g of racemate using SFC can be easily achieved with 2 cm I.D. columns. Current SFC equipment utilizing 5 cm I.D. columns can be used for separations up to 1 kg. Preparative HPLC has been used for enantioseparations for almost 30 years and is still the major player. While preparative HPLC often offers lower productivity and higher solvent consumption relative to other techniques, for small-scale separations the best method is often not the main goal of the chromatographers. Preparative HPLC has the advantage of being very easy to scale-up from analytical conditions and is already available in many laboratories. In addition, the level of training is by far the lowest of the available preparative techniques.

5.9 Conclusion

Chromatographic resolution of enantiomers is the most efficient approach for generating gram to kilogram quantities of enantiopure materials. The use of preparative SFC has been shown to reduce the time and solvent required for separations and is the method of choice for separations up to 10 g scale. Above this scale separation using SSR and larger column preparative HPLC is often the method of choice. SMB technology has proven to be extremely useful for the rapid separation of racemates at the 5–1000 kg scale required for pharmaceutical development. At the ton scale SMB has been shown to be the method of choice for the manufacturing of numerous pharmaceutical intermediates and APIs.

References

1 Brocks, D.R., Pasutto, F.M., and Jamali, F. (1992). Analytical and semi-preparative high-performance liquid chromatographic separation and assay of hydroxychloroquine enantiomers. *J. Chromatogr.* 581 (1): 83–92.

2 Andersson, S. (2007). Preparative chiral chromatography – a powerful and efficient tool in drug discovery. In: *Chiral Separation Techniques, A Practical Approach*, 3rde (ed. G. Subramanian), 585–600. Wiley-VCH.

3 Cox, G.B. (2008). *Preparative Enantioselective Chromatography*. Wiley.

4 Francotte, E. (2007). Isolation and production of optically pure drugs by enantioselective chromatography. In: *Chirality in Drug Research*, vol. 33, 155.

5 Miller, L. and Potter, M. (2008). Preparative chromatographic resolution of racemates using HPLC and SFC in a pharmaceutical discovery environment. *J. Chromatogr. B Anal. Technol. Biomed. Life Sci.* 875 (1): 230–236.

6 Miller, L., Orihuela, C., Fronek, R. et al. (1999). Chromatographic resolution of the enantiomers of a pharmaceutical intermediate from the milligram to the kilogram scale. *J. Chromatogr. A* 849 (2): 309–317.

7 Miller, L., Orihuela, C., Fronek, R., and Murphy, J. (1999). Preparative chromatographic resolution of enantiomers using polar organic solvents with polysaccharide chiral stationary phases. *J. Chromatogr. A* 865: 211–226.

8 Ali, I. and Aboul-Enein, H.Y. (2008). Role of polysaccharides in chiral separations by liquid chromatography and capillary electrophoresis. In: *Chiral Separation Techniques: A Practical Approach* (ed. G. Subramanian), 29–98. Wiley-VCH Verlag GmbH & Co. KGaA.

9 Yashima, E. (2001). Polysaccharide-based chiral stationary phases for high-performance liquid chromatographic enantioseparation. *J. Chromatogr. A* 906 (1–2): 105–125.

10 Okamoto, Y. and Yashima, E. (1998). Polysaccharide derivatives for chromatographic separation of enantiomers. *Angew. Chem. Int. Ed. Eng.* 37 (8): 1020–1043.

11 Welch, C.J. (1994). Evolution of chiral stationary phase design in the Pirkle laboratories. *J. Chromatogr. A* 666 (1): 3–26.

12 Gasparrini, F., Misiti, D., and Villani, C. (2001). High-performance liquid chromatography chiral stationary phases based on low-molecular-mass selectors. *J. Chromatogr. A* 906 (1): 35–50.

13 Fernandes, C., Phyo, Y.Z., Silva, A.S. et al. (2018). Chiral stationary phases based on small molecules: an update of the last 17 years. *Sep. Purif. Rev.* 47 (2): 89–123.

14 Haginaka, J. (2001). Protein-based chiral stationary phases for high-performance liquid chromatography enantioseparations. *J. Chromatogr. A* 906 (1): 253–273.

15 Cabrera, K. and Ludbda, D. (1992). Chemically-bonded β-cyclodextrin as chiral stationary phase for the separation of enantiomers of pharmaceutical drugs. *GIT Spez. Chromatogr.* 12 (2): 77–79.

16 Menges, R.A. and Armstrong, D.W. (1991). Chiral separations using native and functionalized cyclodextrin-bonded stationary phases in high-pressure liquid chromatography. In: *Chiral Separations by Liquid Chromatography* (ed. S. Ahuju), 67–100. American Chemical Society.

17 Ward, T.J. and Farris, A.B. III (2001). Chiral separations using the macrocyclic antibiotics: a review. *J. Chromatogr. A* 906 (1): 73–89.

18 Caille, S., Boni, J., Cox, G.B. et al. (2010). Comparison of large-scale routes to manufacture chiral exo-2-norbornyl thiourea. *Org. Process Res. Dev.* 14 (1): 133–141.

19 Suteu, C. (2005). Method development for preparative enantioselective chromatography. *Prep. Enantioselect. Chromatogr.* 78–109.

20 Christopher, J., Welch, P.S., Spencer, G. et al. (2008). Microscale HPLC predicts preparative performance at millionfold scale. *Org. Process Res. Dev.* 12: 674–677.

21 Kagan, M., Chlenov, M., Greenfield, A., and Ho, D.M. (2006). Normal-phase chiral liquid chromatography–mass spectrometry of non-UV-active compounds: applications for pharmaceutically relevant racemates. *J. Chromatogr. A* 1120 (1–2): 82–88.

22 Miller, L., Grill, C., Yan, T. et al. (2003). Batch and simulated moving bed chromatographic resolution of a pharmaceutical racemate. *J. Chromatogr. A* 1006 (1–2): 267–280.

23 Grill, C.M. and Miller, L. (1998). Separation of a racemic pharmaceutical intermediate using closed-loop steady state recycling. *J. Chromatogr.* 827: 359–371.

24 Grill, C.M., Miller, L., and Yan, T.Q. (2004). Resolution of a racemic pharmaceutical intermediate. A comparison of preparative HPLC, steady state recycling, and simulated moving bed. *J. Chromatogr. A* 1026 (1–2): 101–108.

25 Kennedy, J.H., Belvo, M.D., Sharp, V.S., and Williams, J.D. (2004). Comparison of separation efficiency of early phase active pharmaceutical intermediates by steady state recycle and batch chromatographic techniques. *J. Chromatogr. A* 1046 (1–2): 55–60.

26 Yan, T.Q. and Orihuela, C. (2007). Rapid and high throughput separation technologies – steady state recycling and supercritical fluid

chromatography for chiral resolution of pharmaceutical intermediates. *J. Chromatogr. A* 1156 (1): 220–227.
27 Grill, C.M. and Miller, L.M. (2005). Steady-state recycling and its use in chiral separations. In: *Preparative Enantioselective Chromatography* (ed. G.B. Cox), 149–175. Blackwell Publishing.
28 Broughton, D.B. (1961). Continuous sorption process employing fixed bed. Google Patents US2985589A.
29 Rajendran, A., Paredes, G., and Mazzotti, M. (2009). Simulated moving bed chromatography for the separation of enantiomers. *J. Chromatogr. A* 1216 (4): 709–738.
30 Abel, S. and Juza, M. (2007). Less common applications of enantioselective HPLC using the SMB technology in the pharmaceutical industry. In: *Chiral Separation Techniques: A Practical Approach*, 3e (ed. G. Subramanian), 203–273. Wiley-VCH Verlag GmbH & Co. KGaA.
31 Gomes, P.S., Minceva, M., Pais, L.S., and Rodrigues, A.E. (2008). Advances in simulated moving bed chromatographic separations. In: *Chiral Separation Techniques: A Practical Approach* (ed. G. Subramanian), 29–98. Wiley-VCH Verlag GmbH & Co. KGaA.
32 Juza, M., Mazzotti, M., and Morbidelli, M. (2000). Simulated moving-bed chromatography and its application to chirotechnology. *Trends Biotechnol.* 18 (3): 108–118.
33 Hamende, M. (2005). Case study in production-scale multicolumn continuous chromatography. *Prep. Enantioselect. Chromatogr.* 253–276.
34 Rouhi, A.M. (2003). Chiral business. *Chem. Eng. News* 81 (18): 45–55.
35 McCoy, M. (2000). SMB emerges as chiral technique. *Chem. Eng. News* 78 (25): 17–19.
36 Nicoud, R.-M. (1999). The separation of optical isomers by simulated moving bed chromatography. *Pharm. Technol. Eur.* 11 (3): 36.
37 Canvat, J.-P., Cavoy, E., Deleers, M. et al., inventors (2000). Process for the preparation of levetiracetam.
38 Cavoy, E., Hamende, M., Zimmerman, V., inventors (2005). Google Patents, assignee. Process for preparing (S) and (R)—2-[4-(4-chlorobenzhydryl) piperazin-1-yl]-ethoxyacetamide.
39 Pflum, D.A., Wilkinson, H.S., Tanoury, G.J. et al. (2001). A large-scale synthesis of enantiomerically pure cetirizine dihydrochloride using preparative chiral HPLC. *Org. Process Res. Dev.* 5 (2): 110–115.
40 Hawkins, J.M. and Watson, T.J. (2004). Asymmetric catalysis in the pharmaceutical industry. *Angew. Chem. Int. Ed. Eng.* 43 (25): 3224–3228.
41 Houlton, S. (2001). SMB chromatography: a fast way forward? *Manuf. Chem.* 72 (11): 23–26.

42 Dapremont, O., Geiser, F., et al., inventors (2002). Process for the production of enantiomerically pure or optically enriched sertraline-tetralone using continuous chromatography.

43 Bech Sommer, M., Nielsen, O., Petersen, H., et al. (2003). Method for the preparation of escitalopram. Inventors.

44 McCoy, M. (2004). Divide and conquer: beyond chiral technologies' growth is drug industry acceptance of chiral chromatography. *Chem. Eng. News* 82 (41): 20.

45 Mazzotti, M., Storti, G., and Morbidelli, M. (1997). Supercritical fluid simulated moving bed chromatography. *J. Chromatogr. A* 786 (2): 309–320.

46 Denet, F., Hauck, W., Nicoud, R.M. et al. (2001). Enantioseparation through supercritical fluid simulated moving bed (SF-SMB). *Chromatography* 40 (21): 4603–4609.

47 Peper, S., Lübbert, M., and Johannsen, M. (2002). Separation of ibuprofen enantiomers by supercritical fluid simulated moving bed chromatography. *Sep. Sci. Technol.* 37 (11): 2545–2566.

48 Johannsen, M., Peper, S., and Depta, A. (2002). Simulated moving bed chromatography with supercritical fluids for the resolution of bi-naphthol enantiomers and phytol isomers. *J. Biochem. Biophys. Methods* 54 (1–3): 85–102.

49 Rajendran, A., Peper, S., Johannsen, M. et al. (2005). Enantioseparation of 1-phenyl-1-propanol by supercritical fluid-simulated moving bed chromatography. *J. Chromatogr. A* 1092 (1): 55–64.

50 Ludemann-Hombourger, O., Nicoud, R., and Bailly, M. (2000). The "VARICOL" process: a new multicolumn continuous chromatographic process. *Sep. Sci. Technol.* 35 (12): 1829–1862.

51 Welch, C.J., Leonard, W.R. Jr., Dasilva, J.O. et al. (2005). Preparative chiral SFC as a green technology for rapid access to enantiopurity in pharmaceutical process research. *LCGC N. Am.* 23 (1): 24.

52 White, C. and Burnett, J. (2005). Integration of supercritical fluid chromatography into drug discovery as a routine support tool. *J. Chromatogr. A* 1074 (1–2): 175–185.

53 Toribio, L., Alonso, C., del Nozal, M.J. et al. (2006). Semipreparative enantiomeric separation of omeprazole by supercritical fluid chromatography. *J. Chromatogr. A* 1137 (1): 30–35.

54 Maftouh, M., Granier-Loyaux, C., Chavana, E. et al. (2005). Screening approach for chiral separation of pharmaceuticals. Part III. Supercritical fluid chromatography for analysis and purification in drug discovery. *J. Chromatogr. A* 1088 (1–2): 67–81.

55 Miller, L. (2012). Preparative enantioseparations using supercritical fluid chromatography. *J. Chromatogr. A* 1250: 250–255.

56 Speybrouck, D. and Lipka, E. (2016). Preparative supercritical fluid chromatography: a powerful tool for chiral separations. *J. Chromatogr. A* 1467: 33–55.

57 Noireau, A., Lemasson, E., Mauge, F. et al. (2019). Purification of drug degradation products supported by analytical and preparative supercritical fluid chromatography. *J. Pharm. Biomed. Anal.* 170: 40–47.

58 Speybrouck, D., Howsam, M., and Lipka, E. (2020). Recent developments in preparative-scale supercritical fluid- and liquid chromatography for chiral separations. *TrAC Trends Anal. Chem.* 133.

59 Yip, S.H., Wu, D.R., Li, P. et al. (2019). Separation of Bruton's tyrosine kinase inhibitor atropisomers by supercritical fluid chromatography. *J. Chromatogr. A* 1586: 106–115.

60 Wu, D.R., Yip, S.H., Li, P. et al. (2016). From analytical methods to large scale chiral supercritical fluid chromatography using chlorinated chiral stationary phases. *J. Chromatogr. A* 1432: 122–131.

61 Ventura, M.C. (2014). Chiral preparative supercritical fluid chromatography. In: *Supercritical Fluid Chromatography: Advances and Applications in Pharmaceutical Analysis* (ed. G.K. Webster), 171. Pan Stanford Publishing.

62 Miller, L.M., Pinkston, J.D., and Taylor, L.T. (2019). *Modern Supercritical Fluid Chromatography: Carbon Dioxide Containing Mobile Phases*. Wiley.

63 Pierantozzi, R. (2001). *Carbon Dioxide Kirk-Othmer Enyclopedia of Chemical Technology*. Wiley.

64 White, C. (2005). Integration of supercritical fluid chromatography into drug discovery as a routine support tool. Part I. fast chiral screening and purification. *J. Chromatogr. A* 1074 (1–2): 163–173.

65 Hamman, C., Wong, M., Hayes, M., and Gibbons, P. (2011). A high throughput approach to purifying chiral molecules using 3μm analytical chiral stationary phases via supercritical fluid chromatography. *J. Chromatogr. A* 1218 (22): 3529–3536.

66 De Klerck, K., Parewyck, G., Mangelings, D., and Vander, H.Y. (2012). Enantioselectivity of polysaccharide-based chiral stationary phases in supercritical fluid chromatography using methanol-containing carbon dioxide mobile phases. *J. Chromatogr. A* 1269: 336–345.

67 Hamman, C., Wong, M., Aliagas, I. et al. (2013). The evaluation of 25 chiral stationary phases and the utilization of sub-2.0μm coated polysaccharide chiral stationary phases via supercritical fluid chromatography. *J. Chromatogr. A* 1305: 310–319.

68 De Klerck, K., Vander Heyden, Y., and Mangelings, D. (2014). Generic chiral method development in supercritical fluid chromatography and ultra-performance supercritical fluid chromatography. *J. Chromatogr. A* 1363: 311–322.

69 Hamman, C., Schmidt, D.E. Jr., Wong, M., and Hayes, M. (2011). The use of ammonium hydroxide as an additive in supercritical fluid chromatography for achiral and chiral separations and purifications of small, basic medicinal molecules. *J. Chromatogr. A* 1218 (43): 7886–7894.

70 Ventura, M., Murphy, B., and Goetzinger, W. (2012). Ammonia as a preferred additive in chiral and achiral applications of supercritical fluid chromatography for small, drug-like molecules. *J. Chromatogr. A* 1220: 147–155.

71 West, K.N., Wheeler, C., McCarney, J.P. et al. (2001). In situ formation of alkylcarbonic acids with CO2. *J. Phys. Chem. A* 105 (16): 3947–3948.

72 West, C., Melin, J., Ansouri, H., and Mengue, M.M. (2017). Unravelling the effects of mobile phase additives in supercritical fluid chromatography. Part I: polarity and acidity of the mobile phase. *J. Chromatogr. A* 1492: 136–143.

73 De Klerck, K., Mangelings, D., Clicq, D. et al. (2012). Combined use of isopropylamine and trifluoroacetic acid in methanol-containing mobile phases for chiral supercritical fluid chromatography. *J. Chromatogr. A* 1234: 72–79.

74 Dingenen, J. and Kinkel, J.N. (1994). Preparative chromatographic resolution of racemates on chiral stationary phases on laboratory and production scales by closed-loop recycling chromatography. *J. Chromatogr. A* 666 (1): 627–650.

75 Johannsen, M. and Brunner, G. (1997). Determination of solubility data with on-line packed column supercritical fluid chromatography. In: *Supercritical Fluid Chromatography with Packed Columns; Techniques and Applications* (ed. K. Anton and C. Berger), 1–17. Marcel Dekker, Inc.

76 Francotte, E. and Huynh, D. (2002). Immobilized halogenophenylcarbamate derivatives of cellulose as novel stationary phases for enantioselective drug analysis. *J. Pharm. Biomed. Anal.* 27 (3–4): 421–429.

77 Zhang, T., Schaeffer, M., and Franco, P. (2005). Optimization of the chiral separation of a Ca-sensitizing drug on an immobilized polysaccharide-based chiral stationary phase: case study with a preparative perspective. *J. Chromatogr. A* 1083 (1–2): 96–101.

78 Zhang, T. and Franco, P. (2007). Analytical and preparative potential of immobilized polysaccharide-derived chiral stationary phases. In: *Chiral Separation Techniques: A Practical Approach* (ed. G. Subramanian), 99–134. Wiley-VCH Verlag GmbH & Co. KGaA.

79 Lee, J., Lee, J.T., Watts, W.L. et al. (2014). On the method development of immobilized polysaccharide chiral stationary phases in supercritical fluid chromatography using an extended range of modifiers. *J. Chromatogr. A* 1374: 238–246.

80 Schulte, M., Ditz, R., Devant, R. et al. (1997). Comparison of the specific productivity of different chiral stationary phases used for simulated moving-bed chromatography. *J. Chromatogr. A* 769 (1): 93–100.

81 Yan, T.Q., Orihuela, C., and Swanson, D. (2008). The application of preparative batch HPLC, supercritical fluid chromatography, steady-state recycling, and simulated moving bed for the resolution of a racemic pharmaceutical intermediate. *Chirality* 20 (2): 139–146.

Part II

Chiral Selectors

6

Polysaccharides

Weston Umstead[1], Takafumi Onishi[2], and Pilar Franco[3]

[1] *Chiral Technologies Inc., West Chester, PA, USA*
[2] *Daicel Corporation, CPI Company, Myoko-shi, Niigata, Japan*
[3] *Chiral Technologies Europe, Illkirch Cedex, France*

6.1 Introduction

Polysaccharide-based chiral stationary phases (CSPs) have been reported in the literature for more than 50 years as of the writing of this chapter. Since the first reports in 1973 by Hesse and Hagel [1], the applications have grown from simple models like the separations of Troger's Base or *trans*-stilbene oxide to more modern and complex examples like large molecule pharmaceuticals, polypeptides, and natural products. Over this stretch of time, the technology behind the production of these materials has also grown. Early reported chiral phases were typically cellulose-based polymers derivatized with simple chiral moieties/functionalities like acetate coated onto a nominally sized macroporous silica support. Now, new polysaccharide-based chiral phases are typically immobilized, providing added robustness and expanded solvent compatibility not previously accessible. The chiral selectors have added complexity too, with the most widely applied phases being functionalized with mono- or disubstituted carbamate or benzoate selectors. This provides for a chromatographic environment where often unpredictable intermolecular interactions can take place, which leads to a broader range of selectivity across a wider variety of chiral compounds, in particular with the expanded access to more organic solvents. Technology around the production of silica gel support materials has also evolved, which has resulted in the reliable production of smaller and smaller particle sizes, thus

Chiral Separations and Stereochemical Elucidation: Fundamentals, Methods, and Applications, First Edition. Edited by Quezia Bezerra Cass, Maria Elizabeth Tiritan, João Marcos Batista Junior, and Juliana Cristina Barreiro.

improving the quality and speed of separations. Particle size offerings now typically cover the range from 1.6 to 20 μm. Collectively, all of these changes have transformed the polysaccharide-based CSP technology to produce more than 100 commercially available phases, which warrants a review from initial conception to modern application.

6.2 The Early Years

The first functionally useful polysaccharide-based CSP was reported in 1973 by Hesse and Hagel and demonstrated the successful resolution of Troger's base [1]. This cellulose-based CSP was microcrystalline triacetylcellulose (MCTA) and showed quite an extensive range of applicability toward separating aliphatic and aromatic enantiomers via liquid chromatography (LC). As an added advantage, this CSP showed high loadability when used for preparative applications and remained a highly utilized material for such purposes for many years [2–4]. Despite these utilities, it also exhibited a number of drawbacks, specifically under the high-pressure conditions of high-performance liquid chromatography (HPLC). Limitations included instability and swelling under certain solvent conditions and low chromatographic efficiency.

From these initial reports on the utility of MCTA and the knowledge gained from using a more rigid support material like macroporous silica, the natural progression of the research was to identify and explore what other support materials, polymers, and selectors could be of value. Professor Yoshio Okamoto in Japan made a breakthrough step demonstrating that by supporting polysaccharide derivatives onto a silica backbone, stable and efficient CSPs could be prepared and used under HPLC conditions [5–7]. Such CSPs overcame the drawbacks of the materials exclusively based on polysaccharides with no support and reached the market in a collaboration with DAICEL. The first polysaccharide-based and silica-supported columns commercially available contained coated cellulose tribenzoate and cellulose triacetate (commercialized as CHIRALCEL OA and CHIRALCEL OB). Both of these materials exhibited a broad range of selectivity and loadability for analytical and preparative applications [2–4].

Beyond triacetate and tribenzoate, the expansion of selectors grew to include derivatives of functionalized tribenzoates and triphenyl carbamates [5–12]. Much like triacetate and tribenzoate, these derivatives could be prepared in a straightforward reaction of cellulose with unfunctionalized or functionalized benzoyl chlorides (for benzoates) or phenyl isocyanates (for phenylcarbamates) – see Figures 6.1 and 6.2 for chemical structures and column names. Once coated onto macroporous silica, they exhibited

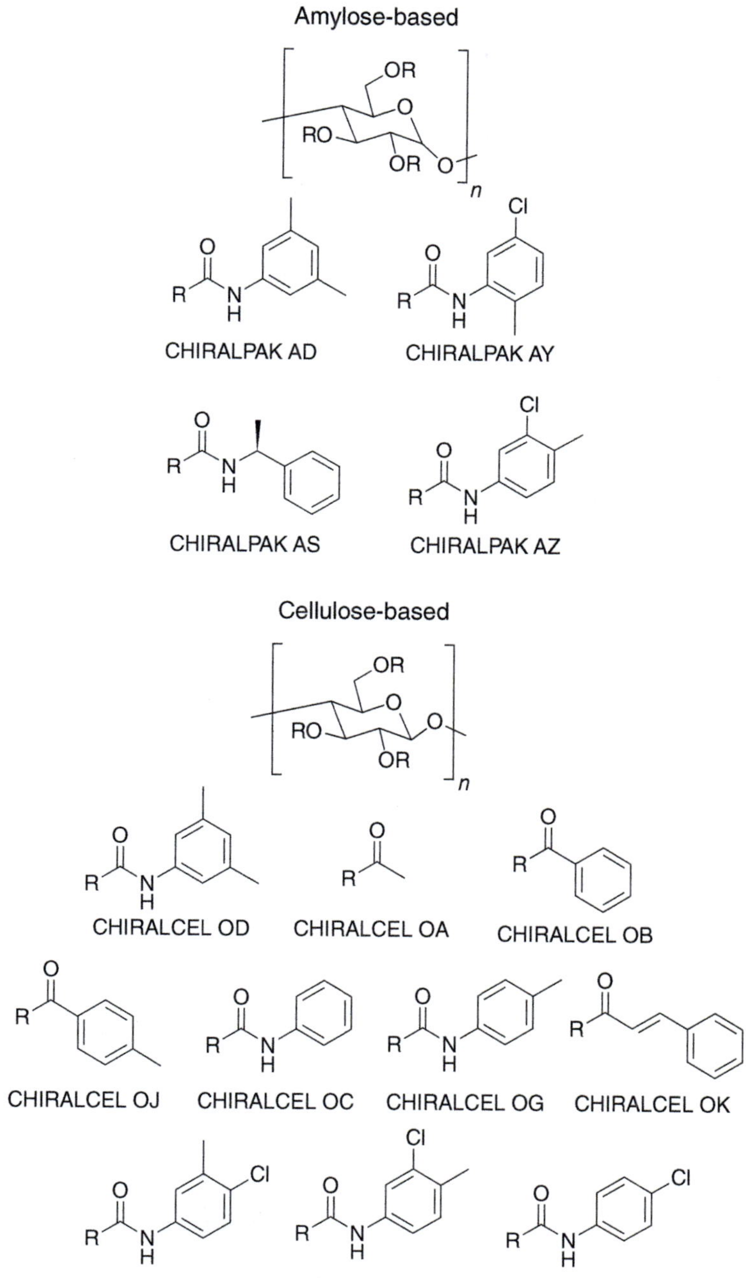

Figure 6.1 Examples of structures of chiral selectors and names of coated polysaccharide-derived CSPs.

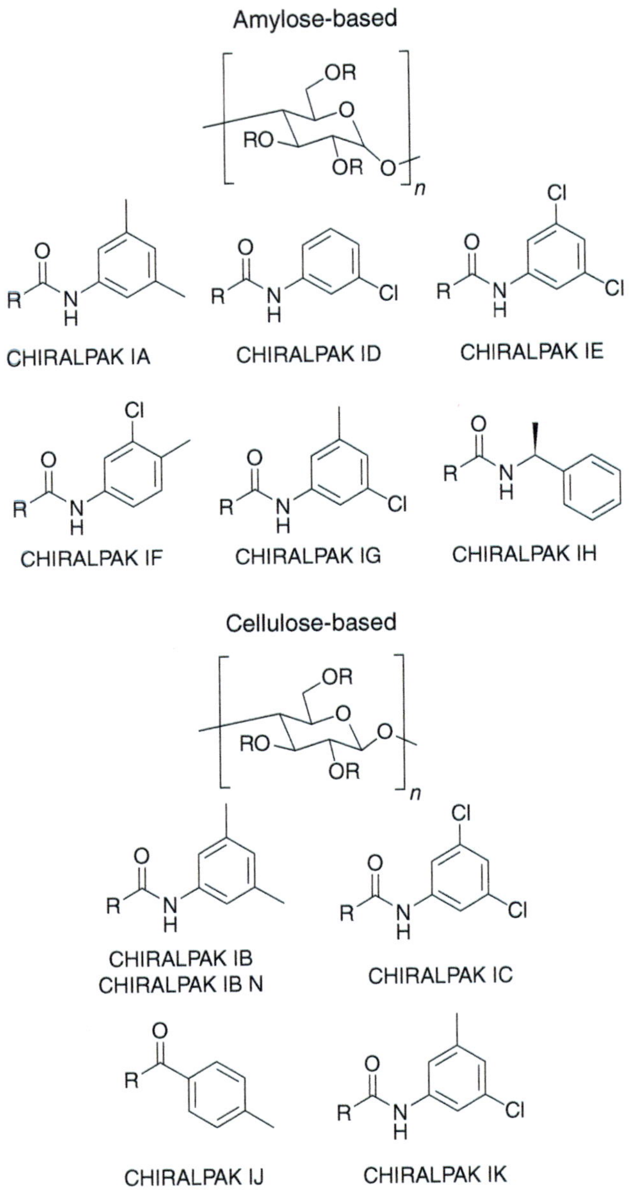

Figure 6.2 Examples of structures of chiral selectors and names of immobilized polysaccharide-derived CSPs.

similar robustness toward high-pressure applications. Although the preparation of the derivatized cellulose polymer was and is a relatively straightforward procedure, the effectiveness, consistency, and uniformity of the performance of the resulting CSPs is highly dependent upon the consistency and uniformity of the coating process. This was a critical aspect to master in order to yield the reproducible chromatographic performance for eventual commercialization.

N.B.: For practical reasons, the authors will present mostly data on the columns commercialized by DAICEL, because of availability and copyright aspects. However, literature citations and coverage of examples are extended to all polysaccharide-derived phases available in the market.

6.3 Polysaccharide Chiral Separation Mechanism

Before further discussion on the innovations of coated and immobilized polysaccharide CSPs, and before introducing the numerous applications of polysaccharide CSPs, it is imperative to briefly discuss the means by which polysaccharide CSPs separate enantiomers. This should provide for better context later as to why there are as many selectors available, with more being commercialized annually and why technology advancements like increased solvent resistance are so important.

Achiral separations on widely available phases like octadecyl (ODS or C_{18}) or silica are generally governed by polarity. That is, the difference in polarity between the achiral analytes and the polarity of the stationary phase (SP). Therefore, with a rationalized model, one can easily predict elution order based simply on the chemical structure (or polarity) of the analyte and the polarity of the SP. As a simple example, for the separation of 1-phenylethanol and toluene on a standard C_{18} column with a 50/50 mixture of methanol/water, one could expect that 1-phenylethanol would elute first as it is more polar than toluene, which will be more strongly attracted to the nonpolar SP. The same exercise cannot be performed for chiral separations, however, as enantiomers are equal in their polarity. Although enantiomers have the same functional groups or elements bonded to their stereogenic centers, they can differ from each other either in the tetrahedral configuration of said elements around the chiral center or by other elements of asymmetry (e.g. plane and/or axis of symmetry where two distinct confirmations are formed, as is the case with atropisomers). Importantly, this is an exploitable difference! An environment that is able to distinguish

Table 6.1 Several intermolecular forces known to occur between analyte and polysaccharide-derived CSPs.

Type of interaction	Strength	Direction	Working distance
Hydrogen bonding	Very strong	Attractive	Long range
Steric hindrance	Weak to very strong	Repulsive	Short range
π–π Interaction	Strong	Attractive	Medium range
Dipole–dipole	Intermediate	Attractive	Short range

Source: Adapted from Berthod [13].

the two enantiomers via this configurational difference and the subtle discriminating interactions that arise from it (see Table 6.1), if contained within an enclosed system (i.e. a column), should therefore be capable of producing an enantioseparation of said chiral compound. This is the role of the chiral selector: a molecule and/or a 3D structure interacting with the compounds to be separated via hydrogen bonding, π–π stacking, dipole forces, inclusion, repulsion, etc. that can exploit the subtle differences between the enantiomers [14].

Polysaccharide-based selectors are unique in their ability to combine all those types of differentiating interactions (Table 6.1) into a macromolecule that is still capable of interaction with the two enantiomers or a racemate. The interactions are highly dependent on the type of polysaccharide backbone (e.g. cellulose or amylose), the functionalization of the chiral selectors (e.g. carbamates, benzoates, and their respective substituents), the combined 3D conformation onto the silica support and furthermore the solvation, swelling, or shrinking of said derivatized polymer in the presence of certain solvents or the addition of certain additives. In such light, the interactions that take place on polysaccharide CSPs are much more unpredictable and systematic screening becomes an essential tool for their effective use.

The first step in the production of polysaccharide-based CSPs is to react cellulose or amylose with small, functionalized isocyanates or benzoyl halides, commonly referred to here-on-out as chiral selectors or just selectors. The native polymers themselves (cellulose and amylose) already differ from each other in the physical linkage between their individual glucose monomers. For cellulose, glucose monomers are connected via β-linkages, whereas for amylose, they are connected via α-linkages. The addition of these small selectors magnifies this difference, resulting in left-handed helical structures with differences in appearance and functionality. The helical structure forms channels or "chiral grooves" that wrap around the outer edges of the macrostructure. It is in these grooves where the chiral selectors orient in varying and unpredictable ways and where the potential for enantiomeric separation takes place. Molecular modeling of these polymers has been done in some cases and shows a clear difference between the physical structure of the helix and the orientation of the chiral selectors within the chiral grooves [10]. Because of this, it is not a stretch to assume that they should function differently in a chiral separation. Furthermore, it is well established that the solvent environment can change the shape and orientation of the chiral selectors within the grooves, further altering the performance between cellulose and amylose-based derivatives. Multiple examples have been described in the literature illustrating the recognition changes of the same molecule with the same selectors but a diversity of mobile phases [15]. Such changes lead to different resolution degrees but also in many cases to different elution orders.

Figure 6.3 demonstrates the changes in solvation status from switching mobile phases from *n*-hexane/ethanol to *n*-hexane/2-propanol (both including diethylamine (DEA) as additive). For the separation of alprenolol on CHIRALPAK ID, the separation is lost.

Sometimes the difference comes from the polysaccharide backbone. Figure 6.4 shows that even with the same derivatization and mobile phase, large variations in a separation can be expected between amylose and cellulose. The separation of sulconazole enantiomers is nearly but not quite baseline on CHIRALPAK IA-3 with *n*-hexane/2-propanol. In contrast, the corresponding cellulose-based CSP (namely, CHIRALPAK IB N-3) leads to a baseline resolution in the same conditions.

As previously mentioned, there are several well-known intermolecular interactions that can take place between a chiral analyte and a chiral selector. Examples of these forces, listed in Table 6.1, are well known to govern the chiral separation mechanism, but how these interactions occur, to what frequency, and the difference in energy between the two enantiomers is what makes predicting a separation especially challenging. It only takes a 0.11 kcal difference in energy to produce a chiral separation [16], meaning that subtle changes in selector orientation and frequency can have dramatic effects on said separation.

The prediction is further complicated by the solvation status of the analyte and chiral selector. Certain mobile phase combinations or even a specific percentage of a mobile phase component/additive can shield or promote a particular intermolecular interaction. This can result in further alterations of the chiral separation, be it a loss of separation, an

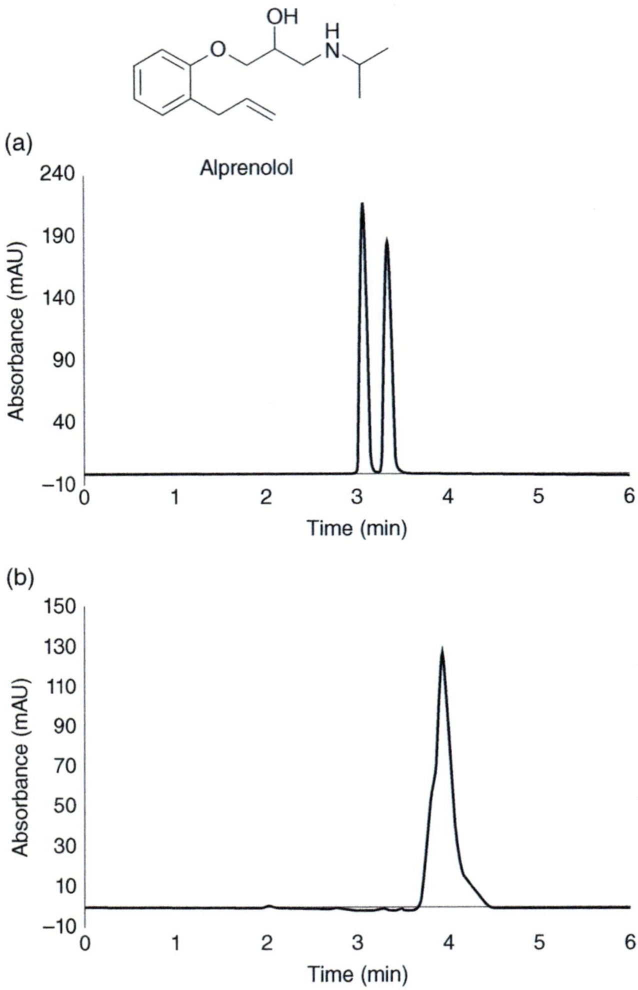

Figure 6.3 Separation of alprenolol on CHIRALPAK ID with (a) *n*-hexane/ethanol/DEA 90/10/0.1 and (b) *n*-hexane/2-propanol 90/10/0.1.

improvement in separation, poor peak shape, or reversal of elution order. It is for this reason that method development screening is an often recommended starting point for finding a new chiral separation. The types of changes that dramatically affect chiral selectivity are changes to the phase (mobile phase or chiral phase), hence more options equate to a higher probability of finding a separation (more on this later as we discuss the different CSPs).

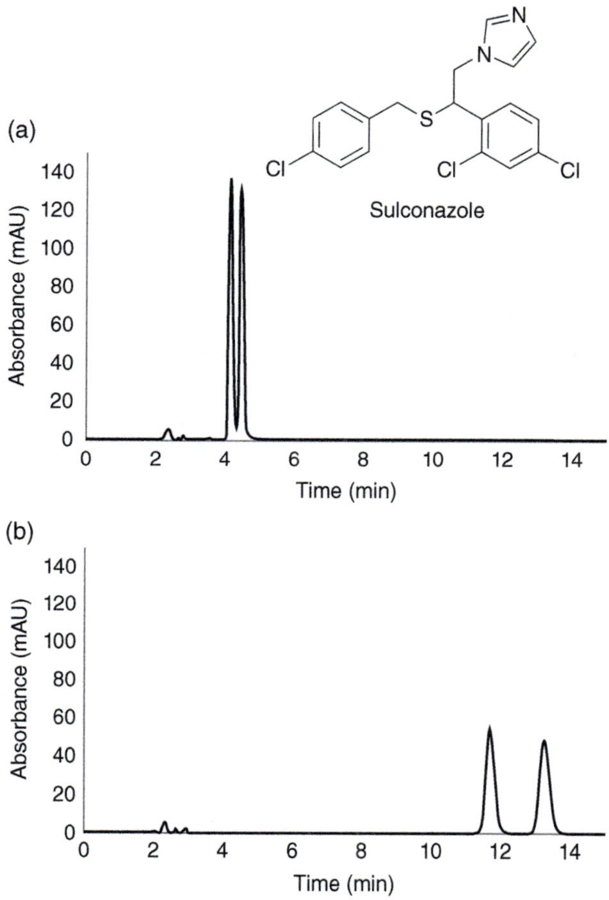

Figure 6.4 Separation of sulconazole in *n*-hexane/2-propanol/DEA 70/30/0.1: (a) CHIRALPAK IA-3; (b) CHIRALPAK IB N-3.

6.4 Coated Chiral Stationary Phases

The earliest reports of polysaccharide-based CSPs were primarily centered around cellulose as a polymer, functionalized with small functional groups, and coated onto the macroporous silica support. Given that the derivatized polymers are soluble in a number of organic solvents like methylene chloride (DCM) or tetrahydrofuran (THF), once solubilized, the mixture can be slurried with macroporous silica and dried to coat the silica surface [6, 17, 18]. In 1984, the first applications of a version of this process were undertaken by Okamoto et al. and Ichida et al. [5–7]. Rather than depositing the derivatized polymer directly onto the bare silica support, the silica was first prepared by depositing a γ-aminopropylsilane matrix onto its surface. This can be accomplished under anhydrous conditions by the condensation reaction of silanol (on the silica surface) with triethoxypropylsilane. The resulting reaction produces ethanol and a new silane bond between silica and propylsilane. The addition of heat and continuous removal of moisture can push this reaction to near full material coverage. Full material coverage in this case does not equate to a complete covering of the surface or reaction of all free silanols on the silica surface necessarily. As the surface is reacted and functionalized, steric hindrance begins to limit further functionalization. The primary amino groups on the surface can be readily charged via an internal proton transfer from the unreacted silanols on the silica surface or from an external proton source. This creates an ammonium cation surface, which is electrostatically charged, and capable of forming intermolecular interactions with the derivatized cellulose polymer. This fixes the polymer to the silica support reversibly.

The chiral selectors for these works were triacetate and tribenzoate, primarily. To demonstrate the utility of these CSPs, the separation of several test chiral compounds, including but not limited to Troger's base, *trans*-stilbene oxide (TSO), and helicene-type compounds were used, resulting in baseline or partial separations in most cases [5]. From a practical standpoint, these early phases were readily compatible with alkane/alcohol mixtures of *n*-hexane/ethanol or *n*-hexane/2-propanol, and in some cases, pure organic solvents like methanol, ethanol, or some mixture of the two or aqueous mixtures of a polar organic solvent like ethanol or 2-propanol.

From these two original selectors came the exploration of cellulose triphenylcarbamate (derivatized and underivatized) and the separation of a series of trialkylsilyl ethers of 4-hydroxy-2-cyclopenternone [19]. This work introduced a number of now commercially available selectors, including CHIRALCEL OC (cellulose *tris*(phenylcarbamate)), CHIRALCEL OF (cellulose *tris*(4-chlorophenylcarbamate)), CHIRALCEL OG (cellulose *tris*(4-methylphenylcarbamate)), and CHIRALCEL OD (cellulose *tris*(3,5-dimethylphenylcarbamate)) (see Figure 6.1 for structures).

The uses of coated cellulose *tris*(4-phenylazophenylcarbamate) as a chiral selector were reported in 1986 [20]. This was a unique exploration into the differences in the separation characteristics of the two forms of the azophenylcarbamate selector. The nature of the molecule is such that the azo bond resulted in both a *cis* and *trans* configuration of the selector. The conversion between the two configurations could be photoinduced with exposure to 370 or 470 nm wavelength light. The performance characteristics of the CSP could then be modulated depending on the percentage of which configuration was present. The observation was that less retention was achieved with a higher percentage of the *cis* form. While retention is typically controlled by increasing the elution strength of the mobile phase (as is the case today with most polysaccharide-derived CSPs), this was a novel solution for a photoinduced option. However, such a selector was never commercialized.

A number of publications dealing with practical applications of the newly developed cellulose selectors for the separation of relevant racemates (e.g. β-blockers) appeared soon after [21–35]. Among them, two selectors seemed to draw major interest due to their recognition capacity; namely, those based on coated cellulose *tris*(3,5-dimethylphenylcarbamate) and cellulose *tris*(3,5-dichlorophenylcarbamate).The CSP bearing 3,5-dimethyl selectors has become one of the most used CSPs to date, commercialized under the name CHIRALCEL OD. This CSP showed high selectivity and good robustness, allowing the separation of preparative quantities of alprenolol, propranolol, and oxyprenolol. Unfortunately, for the 3,5-dichloro analog, while the selectivity was nearly as good as the 3,5-dimethyl derivative, it was observed that the polymer was solubilized under the testing conditions of *n*-hexane/2-propanol, resulting in a deterioration of the column performance. This is important to note as this selector would eventually be revisited and investigated as an immobilized CSP. Once immobilized, it no longer suffered from the same performance losses and was successfully commercialized.

In 1987, the exploration of amylose as a polymer was more widely investigated [36]. The work by Okamoto et al. highlighted the effectiveness of coated amylose *tris*(3,5-dimethylphenylcarbamate) (later commercialized as CHIRALPAK AD) as a chiral selector for the separation of TSO, Troger's base, and a number of other chiral analytes. The use of starch as a support was also explored, which typically contains 20–25% linear amylose and 75–80% branched amylopectin. Although the starch derivative was effective, it showed a lower chiral recognition ability compared to the commercially available amylose.

From these early works, the substitution pattern of the derivatized phenylcarbamates and benzoates seemed to favor the use of methyl or chloro groups in the 3, 4, and/or 5 position on the aromatic rings. This was explored a bit further in a 1987 work on the separation of chiral alcohols [37]. OD-type selectors again showed the greatest selectivity of the derivatives tried. A brief explanation was included as to the potential separation mechanism for these cellulose *tris*(phenylcarbamate) derivatives. As discussed in Section 6.3, intermolecular forces like hydrogen bonding and dipole–dipole interactions were included as possible contributors to some derivatives being more effective than others.

In 1988, the separation of various chiral acids was explored [38]. This work contained the first examples of the separation of non-neutral compounds, which up until this work were not explored. The first observations were no elution under what had become standard normal phase (NP) conditions of *n*-hexane/2-propanol. However, with the addition of an acidic additive, such as acetic or formic acid, the chiral analytes eluted and were well resolved in most cases. To this day, the addition of the proper additive is critical to achieve a quality chiral separation.

From this point forward, having a number of established protocols in place for the chiral resolution of various chemical classes, the limitations of the technology became less about the physical limitations of the CSPs and more about the limitations of the end user, i.e. their imagination. A persistent issue, however, was still the limited solvent compatibility for all coated phases, with some having such heavy restriction that they could only be used with polar organic mobile phases or were simply too unstable in any solvent to be used.

Daicel Chemical Industries was the first to commercialize the technology developed by Professor Okamoto et al. The earliest phases included CHIRALCEL OA (cellulose triacetate), CHIRALCEL OB (cellulose tribenzoate), CHIRALCEL OC (cellulose *tris*(phenylcarbamate)), CHIRALCEL OF (cellulose *tris*(4-chlorophenylcarbamate)), CHIRALCEL OG (cellulose *tris*(4-methylphenylcarbamate)), and CHIRALCEL OK (cellulose tricinnamate). These phases were rather limited in some cases as to what solvents they could be used with. CHIRALCEL OF, for instance, can only be used with *n*-hexane/2-propanol mixtures between 100% *n*-hexane and 50% *n*-hexane.

The more popular coated CSPs undoubtedly have been the so-called golden four: CHIRALPAK AD (amylose *tris*(3,5-dimethylphenylcarbamate)), CHIRALCEL OD (cellulose *tris*(3,5-dimethylphenylcarbamate)), CHIRALPAK AS (amylose *tris*[(*S*)-α-methylbenzylcarbamate]), and CHIRALCEL OJ (cellulose *tris*(4-methylphenylcarbamate)). As highlighted, AD and OD were two of the first selectors published in the early works on polysaccharide CSPs and have remained two of the most widely utilized selectors to date. Additional coated phases like CHIRALPAK AY (amylose *tris*(5-chloro-2-methylphenylcarbamate)), CHIRALPAK AZ (amylose *tris*(3-chloro-4-methylphenylcarbamate)), CHIRALCEL OX (cellulose *tris*(4-chloro-3-methylphenylcarbamate)), and CHIRALCEL OZ (cellulose *tris*(3-chloro-4-methylphenylcarbamate)) have become available over the years providing additional selectivity and complementarity (see Figure 6.1 for structures).

Standard screening protocols have been adopted by laboratories or research groups worldwide, usually with three or four analytical columns in the primary screening. Such configurations with polysaccharide-derived columns, tested on a randomly composed sample pool and with four solvent mixtures, demonstrated being able to afford complete enantiomer resolution for over 90% of the racemates examined and enantiorecognition above 95%. A very substantial number of publications and reviews covering their applications have been published [39–55 and citations therein] with polysaccharide-derived CSPs of different column suppliers. Some of them will be discussed in the next sections.

In addition to amylose and cellulose, other polysaccharide polymers have been utilized but not yet commercialized. As these are not the focus of this chapter, we will not include an in-depth discussion on them. However, they are interesting, valuable, and worth mentioning. Chitosan, xylan, curdlan, dextran, chitin, and inulin have all been explored to some extent in the literature. Okamoto et al. published findings on the first chitosan-based CSP in 1984, which was the same work containing the first reports of cellulose-based CSPs [6]. Many additional applications have since followed [56–59]. The first reports using chitin were made by Cass et al. in 1996, demonstrating the separation ability of two arylcarbamate CSPs [60]. A more thorough review of relevant applications for these two was published recently by Ribeiro et al. [61]. Xylan, curdlan, dextran, and inulin were all reported as capable of performing chiral separations by Okamoto et al. in 1984 and reports continued into the mid-1990s [6, 8, 41]. More recent works have been published on the applications of these polymers as well [62–64].

6.5 Immobilized Chiral Stationary Phases

As covered in the previous sections, while many of the coated CSPs showed utility in a wide range of applications, they all exhibited limitations in solvent compatibility, some to a point where they could not be used entirely. Therefore, the idea of preparing CSPs with the polysaccharide being immobilized onto the silica support was soon envisaged. However, much like the importance of a robust procedure for producing coated-type CSPs, the preparation of efficient and reproducible immobilized CSPs was not a straightforward exercise.

Beyond the potential chemical solutions to physically bind or reticulate the polymers onto silica particles, any method to be used had to find an equilibrium between the immobilization features and the retention of the 3D polymer conformation that allows chiral recognition. Any very substantial modification of such tridimensional structure could and would lead to a dramatic drop of the chiral recognition. Any heterogeneous or non-uniform binding would also have a negative consequence on the chromatographic efficiency due to the creation of an irregular network of binding sites within the column.

With these important requirements in mind, a number of immobilization methods were investigated, many patents were submitted, and numerous publications devoted to this topic appeared since the 1980s [65 and citations therein]. This section will describe some relevant examples, but does not aim to be an exhaustive review of the immobilization topic.

As a brief overview, polysaccharide derivatives have been immobilized via two main approaches:

1) Chemical bonding to the silica surface
2) Reticulation of the polymer around the silica particles

Both approaches ultimately lead to the insolubilization of the polymer and, consequently, to the production of CSPs with extended solvent versatility.

In 1987, Okamoto et al. described the first process by which the solubility issues of the polymer could be solved, that being an immobilization process [66]. The process was initially described to fix the polysaccharide polymer to the aminopropylsilane surface with a covalent linkage, which in theory would prevent the redissolving of the polymer from the silica support. The original method for immobilization utilized a diisocyanate linker that would react with the free amino group from the aminopropylsilane matrix. As shown in Figure 6.5, the amino functionalized silica was first coated with 6-*O*-tritylcellulose.

6-*O*-tritylcellulose was used in this procedure rather than unprotected cellulose, as the former was soluble under the standard coating procedures

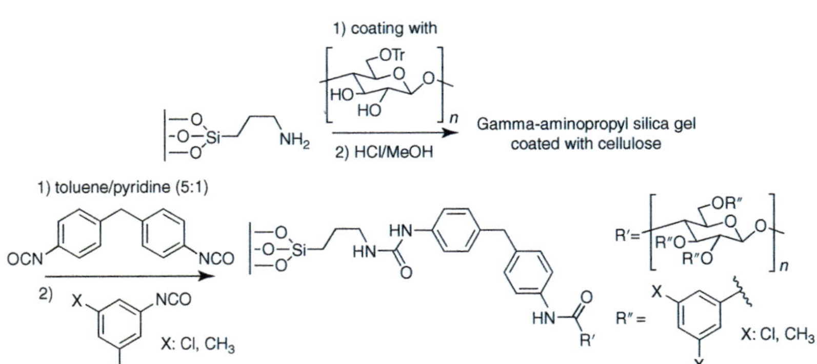

Figure 6.5 Synthesis of first immobilized CSPs. *Source:* Adapted from Okamoto et al. [66].

already established in previous works. Unprotected cellulose was restored by an acidic de-tritylation step that deprotected the free 6′-hydroxyl group. The cellulose-coated silica was then reacted with a less than 1 : 1 ratio of the diisocyanate linker to partially react with some of the free hydroxyl groups. The remaining free hydroxyl groups were then reacted with functionalized chiral selector isocyanates to produce the immobilized CSP. As a part of this work, three different linker types were used, including 4,4′-methylene-diphenyl diisocyanate, 1,6-hexamethylene diisocyanate, and 1,4-phenylene diisocyanate. Additionally, the length and the quantity of the linker (for 4,4′-methylene-diphenyl diisocyanate) were evaluated.

The selectors were limited to immobilized cellulose *tris*(3,5-dichloro) and *tris*(3,5-dimethylphenylcarbamate) and were tested against their coated counterparts under NP conditions of *n*-hexane/2-propanol. One might remember from the discussion on coated CSPs that while the coated cellulose 3,5-dichloro CSP was effective as separating agent for a range of different compounds, the polymer was solubilized under *n*-hexane/2-propanol conditions and the performance of the columns deteriorated. With the immobilized version, this was no longer the case and the column held up well under the testing conditions.

For most of the compounds tested, the selectivity of the immobilized CSP was lower than or slightly lower than the coated CSP. It was posited at that time that this was likely due to the additional restriction in polymer movement coming from the immobilization process. Any sort of hindrance from chemical modification has the potential to limit the polymer from flexing to achieve a better fit or interaction with the analyte. While the structure of the linker seemed to have no effect on the loss of selectivity, the amount of the linker did, whereby more linker resulted in a greater loss of selectivity.

Interesting was the result that came from a thermal treatment of the cellulose 3,5-dimethylphenyl carbamate immobilized CSP and the separation of one of the test compounds. Unexpectedly, the selectivity increased quite dramatically after the thermal treatment. It was supposed that the conformation of cellulose is locked to some degree after the deposition and that further functionalization and immobilization do not allow for an environment where the polymer can "re-adjust" to its preferred conformation. For coated CSPs they still have some freedom to conform as they are not as restricted by the diisocyanate linker. The energy added to the material by heating it allowed for the reconfiguration of the polymeric structure to a more energetically favorable (and effective) conformation for chiral recognition.

A further modification of this procedure was published in 1994 using only the 3,5-dimethylphenylcarbamate selector [67]. Whereas the original procedure allowed for an uncontrolled derivatization and immobilization at the 2, 3, or 6-positions on the polysaccharide, these altered procedures were regioselective in the sense that the polysaccharide was limited as to where it could react. One route used a trityl blocking group at the 6-position, limiting the immobilization to only the 2 and/or 3 positions (Figure 6.6) or derivatized the 2 and 3 positions first, leaving only the 6-position free for immobilization (Figure 6.7).

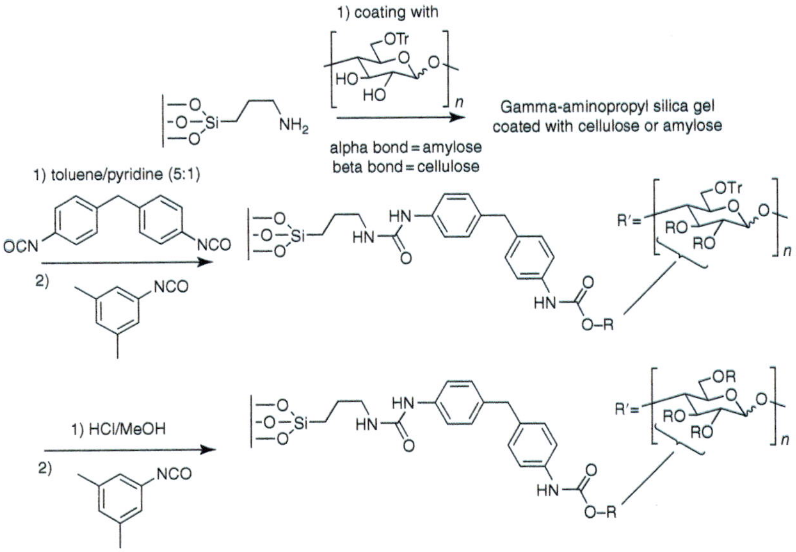

Figure 6.6 Regioselective immobilization of 3,5-dimethylphenylcarbamate of cellulose and amylose by positions 2- and 3- of the glycosidic rings.

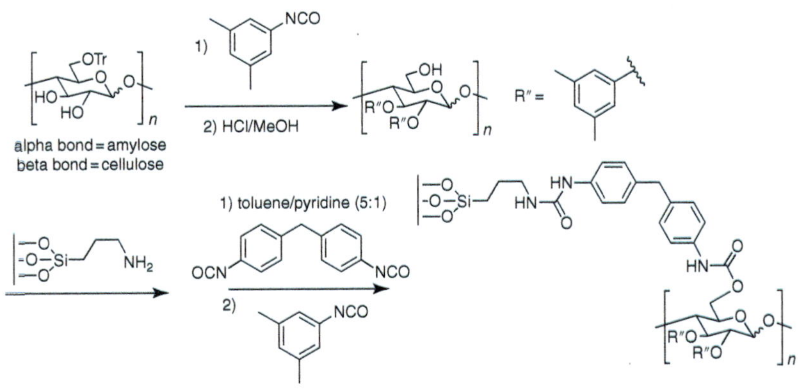

Figure 6.7 Regioselective immobilization of 3,5-dimethylphenylcarbamate of cellulose and amylose by position 6 of the glycosidic rings.

Not surprisingly both routes still resulted in a loss of selectivity compared to the coated *tris*(3,5-dimethylphenylcarbamate) CSPs. As before, the performance of the immobilized CSPs was directly correlated to the amount of linker used (3% linker had a greater performance than 10% linker). Under standard NP test conditions, there was no obvious difference between the two regioselective immobilization procedures and the performance of the cellulose derivatives. There was, however, a better performance for the amylose *tris*(3,5-dimethylphenylcarbamate) CSP noted when the immobilization was done via the 6-hydroxy position. Enantioselective modulation (i.e. an improvement of the separation) was achieved in several cases when an extended solvent range of organic solvents was added to the mobile phase setup, namely chloroform.

As latter rationalized, the reaction of the diisocyanate linker with the coated CSP has a higher probability of having the diisocyanate react with the unprotected hydroxyl groups of the polymer than the surface of the macroporous silica [65]. The reaction of the diisocyanate with the surface requires an unabated diffusion through the entirety of the macroporous structure. Given the surface and the pores are partially blocked by the polymer, it seems unlikely that only the reaction with the surface is taking place. In all likelihood, at least a portion of the diisocyanate is resulting in polymer cross-linking. This cross-linking is therefore responsible for the newly formed insolubility of the polymer in solvents like THF and methylene chloride. This also helps to explain why the size of the linker was not a contributing factor to the performance of the immobilized CSPs.

In 1996, a specific procedure for the immobilization of amylose was published that focused on the reduction of the terminal monomer of each polymer, using the enzymatic polymerization of α-D-glucose-1-phosphate [68]. There were two methods used for immobilization, both starting as follows: the desired amylose chain length and molecular weight were synthesized by enzymatic polymerization, followed by an oxidative lactone ring opening to produce a terminal potassium carboxylate salt. This common intermediate was then handled in two different methods (Method I and Method II). For Method I, the potassium carboxylate intermediate was ring closed or lactonized using acidic amberlite, resulting in an oxidized terminal glucose monomer, which was allowed to react with 3-aminotriethoxysilane. This reacted aminosilane intermediate was introduced into the glucose polymerization reaction by addition of potato phosphorylase, followed by immobilization to the silica support material, and derivatization with the chiral selectors. Method II was essentially a reversal of Method I, whereby the potassium carboxylate intermediate was first introduced into the glucose polymerization reaction with potato phosphorylase, followed by the acidic lactonization, reaction with γ-aminopropylsilane silica, and the derivatization with chiral selectors.

The major advantage of this route for amylose CSP synthesis was the prevention of structural alterations to amylose, which were noted in the previously documented procedures. Between the two methods, it was noted that Method II yielded a CSP that was both more robust and exhibited a high enantioselectivity compared to that of Method I. The authors also noted that the degree of polymerization (DP) was important, whereby the longer the polymer the closer its performance was to that of the coated analog. A DP of 120 was found to be an equivalent point. This comparable performance was likely arising from the increased flexibility that the polymer had to accommodate the chiral analytes.

There have been several reports and patents in the area of photochemical immobilization as well [69–72]. In these reports, there have been three different approaches: (i) photochemical insolubilization of the polysaccharide bearing photopolymerizable functional groups, (ii) photochemical insolubilization of the polysaccharide not bearing photopolymerizable functional groups, and (iii) thermal treatment of macroporous γ-aminopropylsilane silica coated with cellulose *tris*(carbamate) in the presence of the radical initiator AIBN (azobisisobutyronitrile). In these reports, no mechanistic explanation is provided by the authors; however, the immobilization of the polysaccharide is observed. It is not unlikely to surmise a similar cross-linking mechanism as responsible.

Additional immobilization procedures have been published following these original works exploring different linkers, different covalent bonding procedures, and different supports. While they all serve the same purpose (increased robustness), each different material presents its own unique set of separation characteristics [65, 73–86].

As was briefly discussed in the section on separation mechanism, the interactions that can take place on the column are varied and unpredictable. It is not entirely uncommon to find a separation under NP conditions with, for example, *n*-hexane/ethanol, and for that same separation to be lost when switching to 2-propanol (Figure 6.3). The expanded access to new organic solvents resulting from the immobilization process is also helpful in promoting an environment where improved chiral separations can be found (see example in Figure 6.8 adapted from Ref. [87] for the separation of bupivacaine).

The first immobilized-type CSPs were commercialized in 2004. CHIRALPAK IA and IB, which are the immobilized versions of the popular coated CHIRALPAK AD and CHIRALCEL OD, respectively, reached the market first and were progressively adopted while the series of immobilized

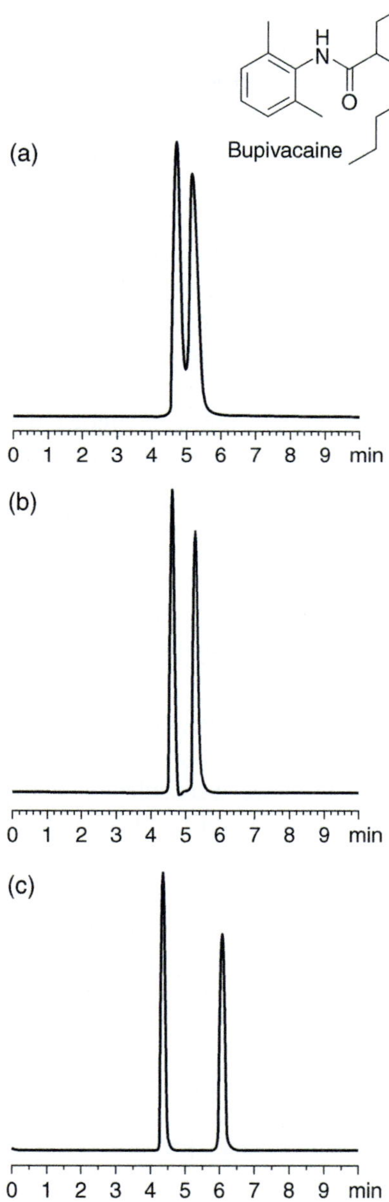

Figure 6.8 Separation of bupivacaine enantiomers with (a) CHIRALPAK AD-H in acetonitrile/DEA 100/0.1, (b) CHIRALPAK IA in acetonitrile/DEA 100/0.1, and (c) CHIRALPAK IA in *n*-hexane/dichloromethane/DEA 50/50/0.1.

CSPs available was expanded in the search of complementary recognitions. The immobilized CSP series reaches as we write this chapter CHIRALPAK IK and covers a broad diversity of functional substitution of cellulose and amylose derivatives (see Figure 6.2 for details). The list includes today the immobilized versions of the classical "golden four" coated CSPs, with the addition of CHIRALPAK IH (immobilized AS-type) and CHIRALPAK IJ (immobilized OJ-type). Prior to the commercialization of IH and IJ, several other immobilized phases were introduced and containing chiral selectors not previously available in the coated series, namely, chlorinated selectors showing a high solubility in the usual NP chromatographic solvents. Some of those phases, such as CHIRALPAK IC (immobilized cellulose *tris*(3,5-dichlorophenylcarbamate)) and CHIRALPAK IG (immobilized amylose *tris*(3-chloro-5-methylphenyl carbamate)), are now among the most useful and widely applied selectors. These are phases that offer expanded selectivity and improve the likelihood of finding a chiral separation and have been investigated and applied to diverse applications involving enantioseparations [15, 52–55, 87–138].

6.6 Applications of Polysaccharide-Derived CSPs

Polysaccharide-derived CSPs are now considered the first choice to setup screening platforms for enantiomer separations by LC and supercritical fluid chromatography (SFC). As previously discussed, a considerable number of publications over three decades have been reported and their applications reviewed. It would be beyond the scope of this chapter to attempt to cover them all. Nevertheless, we will try to give a general overview of the techniques, where they are used, and the industries taking advantage of their features, historically but also with the context of recent developments.

The reasons behind the wide acceptance of polysaccharide CSPs can be distilled down to two main points:

1) a limited number of columns can lead to a relatively high success rate for the separation of many different racemates;
2) their ability to perform well under both analytical and preparative applications in different chromatographic modes.

Finding both features in the same selector type is rather unusual, as can be seen for example when talking about protein-derived CSPs. Some of them (e.g. CHIRALPAK AGP) show very high and broad recognition ability on a wide range of different chiral analytes, but they are limited to analytical LC separations. This is due to the fact that they can be easily overloaded with large injection volumes or high concentration injections, thus hindering their applications under preparative conditions. And under SFC conditions, the protein may lose/modify their recognition ability.

These limitations are overcome by polysaccharide-derived columns, where we will often find them as validated tools for analytical applications, utilizing NP, polar organic, or water-compatible solvents in LC, in addition to compatibility in SFC mode. They are the selectors of choice for preparative and industrial applications, as they can handle large injection volumes/high concentration injections without being as easily overloaded (see Chapter 5).

For historical reasons (see thalidomide as an example), the pharmaceutical industry has been the pioneer in implementing their use, somehow encouraged by the obligation to control and isolate pure enantiomeric active pharmaceutical ingredients (APIs). It is still one of the major application areas, although evolving in the type of applications or target molecules, as the therapeutic arsenal evolves also in their structures. Moreover, pharma applications have been linked to clinical monitoring in drug metabolism, also including investigations of metabolic pathways with endogenic molecules, which very often present stereogenic centers and isomeric transformations in the body fluids. The studies of metabolic pathways, including their chiral implications, are attracting a growing interest as early diagnostic tools for metabolic diseases.

Other very relevant fields of application are for the analysis of agrochemicals and food analysis, including flavor and fragrances.

Currently, a number of polysaccharide-derived CSPs are commercially available, either as coated or as immobilized phases, with particle sizes ranging from 20 μm for preparative separations, over 5 μm for conventional method development, and to 3 μm and sub-2 μm for ultrafast analysis of enantiomers. We can schematically explain that their use in analytical/preparative applications in LC and SFC is often set as shown in Figure 6.9.

Smaller particle sizes are generally preferred for analytical applications, both for LC and SFC. Because of the more efficient packing and naturally higher theoretical plate count (or smaller theoretical plate height), these smaller particles produce sharper peaks and provide an improvement in resolution for high sensitivity applications, for instance when low limits of detection or quantification (LOD or LOQ) are required.

Larger particle sizes are more suited for preparative applications. While 5 μm particles can be and are often used for analytical applications, they also find much utility in semi and preparative applications, along with 10 μm, particularly for SFC preparative applications. For LC preparative use, either traditional single column processing or multicolumn processing

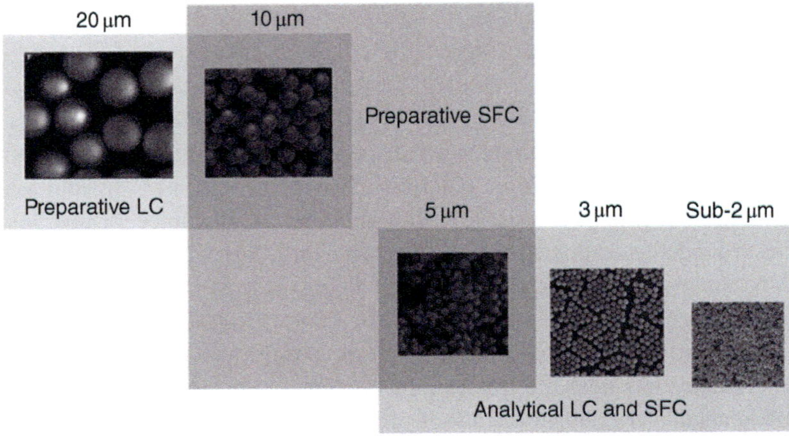

Figure 6.9 Most popular particle sizes of polysaccharide-derived CSPs and their application scale.

(i.e. simulated moving bed (SMB) for instance), 10 or 20 μm CSPs are the most commonly applied.

Presented hereafter are sections on analytical and preparative applications, focusing on a selection of several classes of compounds that have heavily benefited from the features of polysaccharide-based CSPs. This is by no means a comprehensive review of all applications in the literature.

6.6.1 Analytical Applications

Given the increasing number of chiral compounds in drug development and other domains, efficient and automated LC and SFC screening using columns packed with various CSPs has been a challenging subject and discussed by several authors [52, 53, 55, 139–142]. The strategies to achieve enantiomer separations are quite variable from one laboratory to another depending on the type of substances in screening, the equipment available, and the R&D background. However, the motivation for the development of efficient screening approaches remains common. We all search for the easiest, the fastest, the most successful, and the most reliable way to analyze a series of molecules with diverse structures and chemical properties in a reasonable time frame.

Until the advent of mass-compatible SFC systems with limited background noise, most analytical applications in the field were made in LC. The situation is changing with the growing installation of many SFC systems, but LC still remains the most broadened technique in a global dimension.

Most screening setups are based on a short selection of columns (usually 4–8) with a limited choice of mobile phase composed by organic solvents (see Chapter 1). This approach is the first choice at the early development of a chiral method. However, such screening may be unsuitable for certain molecules (e.g. insoluble molecules in organic media) but also for certain applications where the sample will be in a water solution (e.g. pharma formulations, biological fluids, etc.). The use of reversed-phase (RP) screenings will become then the first choice. Examples of all those scenarios will be presented in the current section.

6.6.1.1 Pharmaceuticals

One area of massive adoption is in the pharmaceutical industry, where the biological effects of enantiomers were first felt. There have been countless reports on separations of β-blockers, proton pump inhibitors (PPIs), nonsteroidal anti-inflammatory drugs (NSAIDs), steroids, and antihistamines under LC conditions (NP and reversed phase (RP)) as well as SFC.

6.6.1.1.1 Nonsteroidal Anti-inflammatory Drugs (NSAIDs)

In 2011, Zhang et al. reported the separation of six "profens" including carprofen, fenoprofen, flurbiprofen, ibuprofen, indoprofen, and ketoprofen under SFC conditions [110]. In 2014, Lee et al. reported the separation of many different compound classes, including fenoprofen and suprofen, using extended range of SFC co-solvents like dichloromethane (DCM), THF, and ethyl acetate (EtOAc) [117]. Using common SFC co-solvents like methanol and ethanol, on immobilized-type CSPs (CHIRALPAK IA, IB, IC, and ID), separations for all compounds tested were found. Earlier, Wilson separated similar compounds on CHIRALPAK AD under SFC conditions with methanol only as a co-solvent [143], followed by Johannsen exploring the separation of ibuprofen only with 2-propanol as a co-solvent [144]. Wenda et al. reported the separation of flurbiprofen, also on CHIRALPAK AD and explored the effects on the separation under standard and overloaded conditions [145]. Gyllenhaal et al. separated ibuprofen, ketoprofen, and naproxen with several alternative alcohols (2-butanol and 2-pentanol) on CHIRALPAK AD-H, which resulted in several reversals of elution order [146]. Much earlier in 1989, Okamoto et al. separated ibuprofen, ketoprofen, flurbiprofen, and tiaprofenic acid on coated cellulose and amylose *tris*(3,5-dimethylphenylcarbamate) CSPs under standard NP conditions of *n*-hexane/alcohol [147].

He et al. separated the enantiomers of ketorolac on CHIRALCEL OJ under polar organic conditions, with a focus on both analytical and preparative separation conditions [148]. Ing-Lorenzini et al. resolved the enantiomers of ketorolac on CHIRALPAK AD-RH with acidic water and acetonitrile [149]. Caccamese reported the separation of etodolac on CHIRALCEL OD in 1993 with the NP conditions [150].

6.6.1.1.2 Beta-Blockers

Much like NSAIDs, there have been a copious number of reports related to beta-blockers. Pandya et al. reported the separation of atenolol, metoprolol, and propranolol on immobilized amylose *tris*(3-chloro-5-methylphenylcarbamate) with a 25% modifier of a (1 : 1, v/v) mixture of 2-propanol and methanol [151]. Garzotti repeated the separation of a series of pharmaceutical compounds (including the beta-blockers alprenolol, atenolol, and pindolol) on coated-type CSPs CHIRALPAK AD, CHIRALCEL OD, and CHIRALCEL OJ with ethanol as a modifier [152]. Maftouh et al. reported separations of acebutolol, alprenolol, metoprolol, nadolol, oxprenolol, pindolol, and propranolol on the "golden four" coated columns as a part of an optimized screening study for NP separations [153].

6.6.1.1.3 Proton Pump Inhibitors (PPIs)

There are several PPIs commonly prescribed or available over-the-counter – omeprazole and esomeprazole being the two most common options. Chennuru et al. developed LC and SFC separations on a series of PPIs using CHIRALPAK IA, IB, and IC. They found IC to be a particularly useful phase for these compounds (see Figure 6.10) [154].

Rahman et al. reported the method validation for the separation of omeprazole under NP conditions with CHIRALCEL OD-H [155]. Nozal et al. investigated the separation of omeprazole and several related compounds under SFC conditions using CHIRALPAK AD-H [156]. Kim et al. reported the validated separation of rabeprazole with CHIRALPAK IC under NP conditions [157]. Wang et al. implemented the RP separation of lansoprazole in dog plasma samples using CHIRALCEL OZ-RH [158]. Similarly, Sun et al. described the RP separation of lansoprazole from human plasma samples with CHIRALPAK IC [159]. Cirilli et al. reported the improved separation of lansoprazole on CHIRALPAK IA with a mixture of methyl *tert*-butyl ether and EtOAc, highlighting the importance of expanded solvent compatibility that comes along with immobilization [126].

6.6.1.1.4 Antihistamines

Eom et al. reported the subcritical fluid separation of cetirizine on CHIRALPAK IE, in a tandem mass spectrometry method for the quantification in human plasma samples [160]. Kannappan et al. investigated the separation of cetirizine and the enantiopurity assessment of levocetirizine under RP conditions with CHIRALPAK AS-3R [161]. Ali et al. reported the separations of pheniramine and cetirizine from human plasma samples on

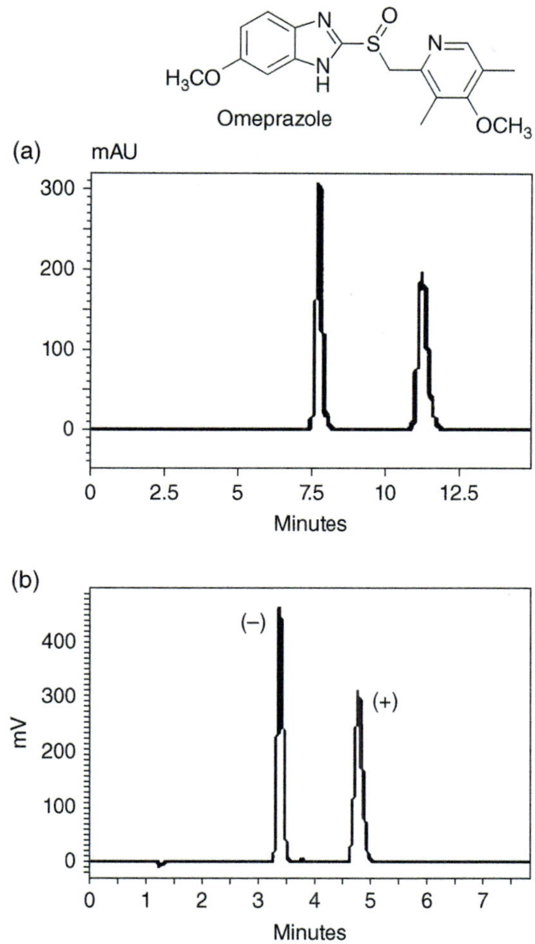

Figure 6.10 Separation of omeprazole enantiomers on CHIRALPAK IC with (a) CHIRALPAK IC in normal phase mode with hexane/ethanol/DEA 50/50/0.1 and (b) CHIRALPAK IC in SFC mode with CO_2/EtOH/DEA 60/40/0.1. *Source:* Reproduced from Chennuru et al. [154].

CHIRALPAK AD under NP conditions [162]. Toribio et al. described a similar separation on CHIRALPAK AD, with a focus on method optimization and the effects that certain chromatographic parameters like temperature can have to that end [163]. Liu et al. reported the separation of cetirizine under RP conditions with CHIRALCEL OD-R [164], which was followed up recently by Liu et al. on the corresponding 5 μm equivalent column and the addition of five other antihistamines, including orphenadrine, meclizine, terfenadine, dioxopromethazine, and carbinoxamine [165].

6.6.1.1.5 Selective Serotonin Reuptake Inhibitors (SSRIs)

Rosetti et al. described the separation of sertraline and its related impurities on CHIRALPAK IG-3 under RP conditions [166]. Jiang et al. reported the NP separation of sertraline on CHIRALPAK AD-H [167]. Olsen et al. published the NP separation on CHIRALCEL OD-H and compared it to the separation on an ovomucoid column, demonstrating better performance from the polysaccharide phase because of expanded solvent compatibility [168]. Umstead recently reported the separation of fluoxetine under NP conditions with CHIRALPAK IK (see Figure 6.11) [169].

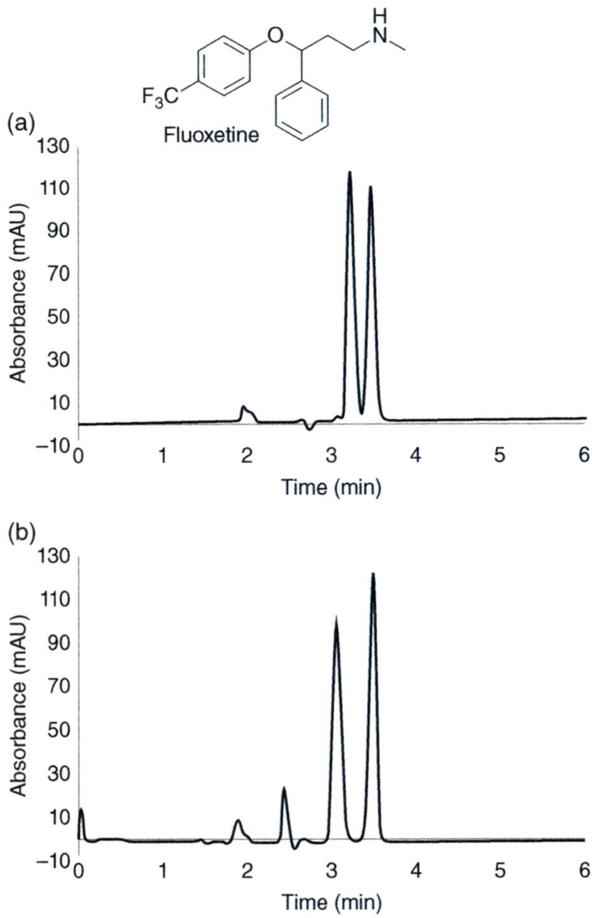

Figure 6.11 Separation of fluoxetine enantiomers on CHIRALPAK IK with (a) in hexane/ethanol/DEA 95/5/0.1 and (b) in hexane/isopropanol/DEA 90/10/0.1.

Geryk et al. described the NP separation of citalopram and its precursor, citadiol, on CHIRALCEL OD-H [170]. Rochat et al. used CHIRALCEL OD for the resolution of citalopram enantiomers and its demethylated and propionic acid metabolites in human plasma samples [171]. Rosetti et al. reported the separation of paroxetine under RP conditions on CHIRALPAK IA-3 [172]. Zukowski et al. developed an HPLC method for paroxetine on CHIRALPAK AD [173]. Vivekanand et al. expanded upon this separation and described other separations on CHIRALCEL OD and OJ [174]. Su et al. reported similar separations of paroxetine on CHIRALPAK AD but under SFC conditions [175].

6.6.1.1.6 Cannabinoids

Until more recently (as of the writing of this chapter), the application of CSPs to the separation of cannabinoids has been mostly overlooked. This is mostly due to the achiral nature of many cannabinoid mixtures and analyses that were initially of interest after the 2018 Farm Bill was passed in the United States. With this bill lifting the regulations on the controlled production of hemp, the legalization of products containing cannabidiol (CBD) closely followed. The caveat with this production is/was that the CBD products had to contain less than 0.3% by weight of tetrahydrocannabinol (THC). The resolution of CBD and THC does not require enantiorecognition; therefore, less expensive C_{18} or other commercially available achiral phases have been used so far. However, the screening of CSPs is yielding additional insights on chiral and achiral composition of different hemp samples [176, 177].

Recently, Onishi et al. reported the separation of a 10 cannabinoid mixture on multiple immobilized sub-2 μm CSPs under NP and RP conditions (see Figure 6.12) [178]. The 10 cannabinoids included tetrahydrocannabinolic acid A (THCA-A), cannabidiolic acid (CBDA), delta-8-tetrahydrocannabinol (Δ^8-THC), cannabidiol (CBD), (±)-cannabichromene (CBC), cannabinol (CBN), delta-9-tetrahydrocannabinol (Δ^9-THC), and cannabigerol (CBG). Although the mixture contains 10 unique cannabinoids, CBC is actually a chiral cannabinoid, so the methods were able to separate all 11 peaks in the mixture. Denicola investigated the separation of the same 10 cannabinoid mixture on CHIRALPAK IB N-5 under SFC conditions using methanol as a co-solvent [179, 180]. In this application, two peaks are also observed for the CBC racemate in the sample.

More recently, many are finding that the cannabinoid mixtures do contain chiral mixtures/components (aside from CBC). While Δ^9 THC naturally occurs as the single (−) enantiomer, the non-naturally occurring (+) enantiomer can be detected, particularly from synthetically created samples. While Δ^8 THC is not naturally occurring as either the (−) or (+)

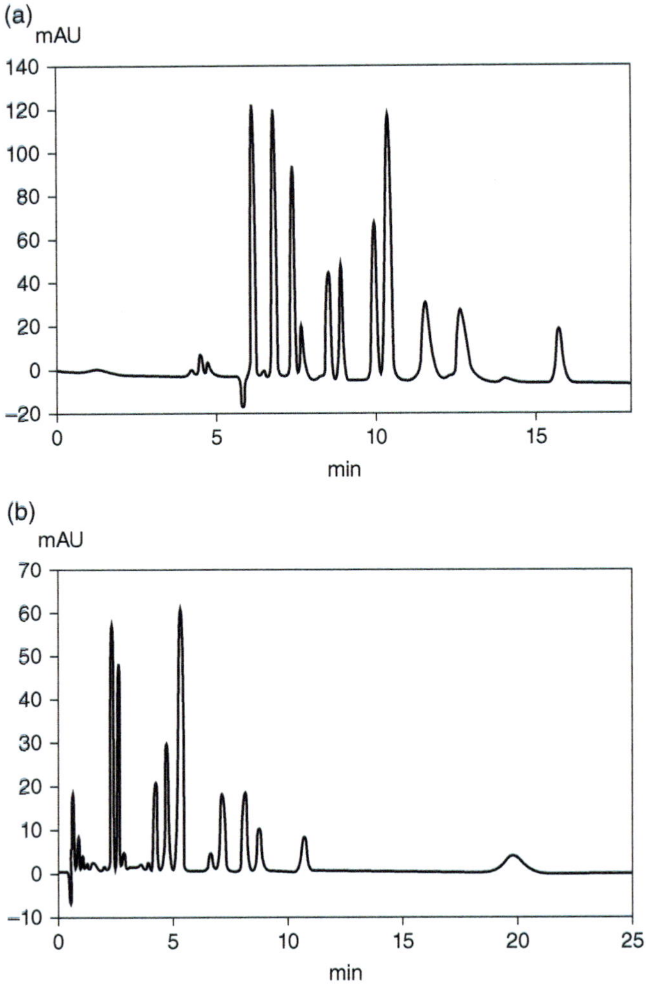

Figure 6.12 Separation of 10 cannabinoid mixtures with (a) CHIRALPAK IB-U in hexane/isopropanol/ethanol/TFA 96/3/1/0.1 and (b) CHIRALPAK IG-U in water/acetonitrile/TFA 45/55/0.1.

enantiomers, both can also be detected from synthetic sources of THC. This achiral/chiral separation of Δ^8 and Δ^9 THC was reported on CHIRALPAK IF under NP conditions with *n*-hexane/2-propanol (95/5, v/v) [181]. Other isomers of THC, like Δ^6 and Δ^{10}, can exist in similar pairs as Δ^8 and Δ^9 THC. This separation was reported by Umstead recently in 2021 under NP conditions on CHIRALPAK IB N-3 and CHIRALPAK IG-3 [182]. The separation of several other chiral cannabinoids, namely, CBC and its

UV-initiated reaction product cannabicyclol (CBL), were reported in the same work on CHIRALPAK IA-3, IB N-3, IC-3, and IG-3.

6.6.1.1.7 Pharmacopeia Applications

As pharmaceuticals are released to the open market, it is critical (as mentioned several times to this point) that accurate quantification of all compounds in the final formulation or delivery form are documented, their effects known, and their allowable limits set. Toward these goals, global pharmacopeia organizations like the United States Pharmacopeia (USP) or European Pharmacopeia (EP) publish monographs that contain a range of validated tests for said pharmaceuticals. One aspect these monographs cover is the determination of enantiomeric or isomeric impurities in drug substances based on methods proposed by pharma companies or research teams. Those methods show somehow also the state-of-the-art in the field.

In such context, polysaccharide-derived columns appear in a number of monographs and receive codes based on their composition to allow the independent choice of the final user, if several column suppliers are available [183]. Logically the first column described was OD-type (USP Code L40), followed by other coated CSPs. Today the immobilized versions are gaining applications and are adopted for validated methods [184, 185]. Some examples of APIs controlled with polysaccharide-derived columns are efavirenz (*R*)-enantiomer (Sustiva), voriconazole-related compound B (VFEND), atomoxetine (*S*)-enantiomer (Strattera), valsartan-related compound A (Diovan), irinotecan hydrochloride enantiomer (Camptosar), abacavir enantiomer (Ziagen), levetiracetam (*R*)-enantiomer (Keppra), tolterodine (*S*)-enantiomer (Detrol), atorvastatin 3-(*S*),5-(*S*)-steriomer (Lipitor), diastereoisomeric ratio of Fulvestrant (Faslodex), paroxetine-related compound C (Paxil), brinzolamide-related compound A (Azopt), or the enantiomeric purity of darifenacin hydrobromide (Enablex).

6.6.1.1.8 Bioanalytical and Clinical Applications

The demand for the determination of enantiomeric excess in bioanalytical applications has encountered a major growth in the last few decades and still continues to expand. Several major reasons are behind such trend.

First of all, a number of enantiomerically pure molecules are commercially available and became off-patent during that period. Generic drug producers appeared all over the world and decided to enter such market, being obliged in every case to demonstrate not only the purity of their APIs but also the bioavailability of their formulations compared with the initial patented originals. Therefore, multiplying the need for mass spectrometry (MS) and RP-compatible chiral methods.

Second, innovative pharma companies have been always sensitive to the biotransformation of the APIs in the body, but have also kept on enhancing their research in the metabolic transformations, the transport, the excretion, the toxicity, etc. of the active ingredients and their metabolites and, therefore, focus as well their interest on chiral interconversions.

Moreover, with the increasing importance of areas like metabolomics, lipidomics, foodomics, etc. in clinical applications and diagnostics, the variable of chirality cannot be excluded from the investigations. Otherwise, conclusions could be biased.

Nevertheless, the complexity of certain biological matrices and the number of samples and of molecules to be identified per sample in single runs require extensive developmental work, often needing LC–MS/MS and 2D chiral and achiral configurations. Such applications are already a reality with most advanced CSPs (e.g. sub-2 μm particle columns), including polysaccharide-derived selectors. Excellent examples in metabolomics [186], constitutional lipidic isomer separations [187], or fatty acid determinations in lipidomics [137] are already published. More would certainly follow with the advent of new innovation in the small particle phases and/or the development of coreshell alternatives, which are not yet commercially available.

6.6.1.2 Agrochemicals

The class of compounds known as agrochemicals has been one that was historically overlooked for enantioselective separations. While many of the chemicals contain stereogenic centers, racemates were manufactured and used ubiquitously without much consideration of their impact. However, much like the case of pharmaceutical active principles, it is not a colossal leap to make the assumption that the application of racemic agrochemicals can have a similar effect on wildlife (e.g. toxicity for bees and other insects) and the surrounding environment. Recently, this has begun to change as legislation from the EU and other governmental agencies has been passed to ensure these impacts are properly assessed.

Ellington et al. reported the separation of crotoxyphos, dialifor, fonofos, fenamiphos, fensulfothion, isofenphos, malathion, methamidophos, profenofos, crufomate, prothiophos, and trichloronate under NP LC conditions on the classic four coated CSPs [188]. Many other publications have also been devoted to the separation of the same compounds with different end applications (RP or environmental testing for instance) [189–198]. More recent works have focused on the separations with immobilized phases, particularly for those compounds not well resolved on coated-type CSPs. For example, 18 chiral pesticides, herbicides, and insecticides were separated by Zhao et al. on CHIRALPAK IG under RP conditions [199]. Merino et al. reported separations on CHIRALPAK IG on 30 pesticides, herbicides, and fungicides under RP and polar organic conditions, 17 of which were improvements over previously reported separations [200]. In 2022, Zhang et al. described the separation of the insecticide fipronil, its metabolites, and photodegradation products on CHIRALPAK IG-3 under MS-compatible mobile phase conditions [138].

In 2011, Jin et al. reported the SFC separation of five chiral pesticides, diclofop-methyl, benalaxyl, acetofenate, myclobutanil, and difenoconazole on CHIRALPAK IB with a differing modifier type and concentration [201]. Przybyciel et al. screened for the chiral separation of the fungicide diniconazole on a series of coated CSPs, reporting amylose *tris*(3,5-dimethylphenylcarbamate) to be the optimal CSP [202]. Iijima described the separation of the fungicide bromuconazole enantiomers on several coated and immobilized-type CSPs, with the best separation conditions found with CHIRALPAK AS using 10% ethanol as a modifier [203]. This was a more complicated example as bromuconazole contains two stereogenic centers. In 2020, Chang et al. developed a method for the insecticide etoxazole on CHIRALPAK IG with 30% methanol as a modifier [204]. Tao et al. achieved the SFC resolution of flutriafol enantiomers on CHIRALPAK IA-3 [205].

6.6.1.3 Food Analysis

The analysis of enantiomeric and isomeric compositions in food allows the traceability of nutritional integrity and potential identification of adulteration. The complexity of such samples in terms of number of components to be analyzed but also matrix interferences make the method development particularly challenging and require preliminary work before application in real samples. Some examples are presented here.

Schaffrath et al. reported the SFC separation of a series of spiro theaspiranes under SFC conditions with CHIRALPAK IA through IF [206]. Baranowska et al. discriminated flavanone, naringenin, and hesperetin under RP conditions on CHIRALPAK AD-3R [207]. Martinez et al. described the RP separation of 6-prenyl-naringenin on CHIRALPAK AD-RH, which is helpful for the discrimination of its phytoestrogen isomer 8-prenyl-naringenin [208]. Ahmed et al. reported the separation of a number of compounds on CHIRALPAK ID-3 under NP conditions, including flavanone and 6-hydroxy-flavanone [209]. Yan et al. investigated the resolution of several diketopiperazine indole alkaloids from hemp seed with CHIRALPAK AS-H [210]. Prado-Cabrero et al. published the separation of lutein, zeaxanthin, and meso-zeaxanthin on CHIRALPAK AD-3 [211]. Park et al. reported the SFC separation of syringaresinal extracted from Ginseng berries on CHIRALPAK IB [212]. Qiao et al. developed a method for the SFC separation of 25 racemic ergostane triterpenoids on CHIRALCEL OJ-H and CHIRALPAK IC [213].

6.6.2 Preparative Applications

There are a number of considerations that need to be made when attempting a preparative separation of a chiral molecule. Chapter 5 has a thorough discussion of the topic, but a few points are mentioned here: solubility of the sample in the mobile phase and loadability of the separation. Because of their increased versatility toward a wide range of solvents (in particular the immobilized CSPs), and their higher saturation capacity compared to other types of CSPs (proteins, cyclodextrins, and antibiotics, for instance), polysaccharide-based CSPs have traditionally and are currently well suited for preparative applications [39, 40, 45, 52, 55, 214–217].

Given the length of time that polysaccharide-based CSPs have been around and are commercialized, there are more examples in the literature than can be sufficiently captured in this chapter [218–227]. However, some of the more impactful and insightful examples will be presented here.

Miller et al. reported the usefulness of chiral HPLC separations as early as 1989 [228]. The group at Searle used preparative chromatography with polysaccharide-based CSPs to purify milligram to kilogram quantities of materials, on the order of 50 projects a year. This allowed for the quick biological screening of their potential targets to either advance or fail more quickly than developing individualized asymmetric syntheses.

In 2008, Cephalon used a continuous chiral chromatographic process to produce multi-kilogram scale quantities of (*R*)-modafinil or armodafinil required for toxicological and clinical studies. LC (and eventually SMB) conditions were developed on coated amylose *tris*(3,5-dimethylphenylcarbamate) with 100% methanol, which resulted in productivity of 0.48 kg of racemate kg^{-1} of CSP per day. In total, more than 600 kg of racemate was processed via this process, which allowed Cephalon's armodafinil to make it through Phase III trials [229].

Several other commercial-scale APIs have been or are manufactured with the use of chiral LC. Sertraline (Zoloft) can be resolved under *n*-hexane/2-propanol conditions with coated amylose *tris*(3,5-dimethylphenylcarbamate) [230]; escitalopram (Lexapro) and levetiracetam (Keppra) are commercially produced with coated amylose *tris*(3,5-dimethylphenylcarbamate) or coated cellulose *tris*(3,5-dimethylphenylcarbamate) [231–233]; and lastly radafaxine, which is a potent metabolite of bupropion [234].

The preparative HPLC separation of the enantiomers of bicalutamide, a nonsteroidal antiandrogen used for the treatment of prostate cancer, was performed with immobilized amylose *tris*(3,5-dimethylphenylcarbamate), using an extended range set of mobile phase conditions, that being *n*-hexane/ethyl acetate/ethanol 100/30/5 (v/v/v). On a 1 cm inner diameter (i.d.) semipreparative column, the authors were able to isolate 960 mg of the active (*R*)-enantiomer [132].

The preparative LC separation of several cathinone compounds has been reported [235]. The first being 3,4-methylenedioxypyrovalerone (MDPV) with NP conditions – using a mixed bed of both coated amylose and cellulose *tris*(3,5-dimethylphenylcarbamate). This resulted in the multi-gram preparative separation of material for initial biological testing. Another example was the separation of pentedrone and methylone at a multi-milligram scale using coated amylose *tris*[(*S*)-α-methylbenzylcarbamate] [236].

A very innovative preparative application was recently published for the isolation of an imaging probe for the detection and stratification of various cancers [136]. The ^{18}F-labeled chiral molecule was isolated from the reaction mixture using 2D chromatography, where the first dimension used was an achiral LC RP separation, and the second dimension a chiral LC separation using a cellulose-based chiral column. The target was isolated with very high enantiomeric purity.

At preparative level, SFC has found a wide adoption in discovery chemistry groups, especially for its ability to provide milligram quantities of materials quickly for initial biological testing. The low viscosity of the mobile phases and the increased flow rate (about three times faster than typical LC conditions) have a number of advantages, and the reduced use of solvent is attractive for waste disposal and overall cost. The advantages of SFC have been discussed in Chapter 3.

The separation principles between LC and SFC on an analytical scale are the same and so are the ones that govern them on a preparative dimension. In contrast, separations found in one or another mode are not necessarily transposable in the same column, despite the correlations made sometimes between the properties of alkanes and supercritical CO_2 [110, 237]. This can be seen for the separation of tofisopam isomers in Figure 6.13.

Much like as the case with LC applications, there have been many reported applications of SFC for semipreparative and preparative scale-up. While it would be comprehensive to cover them all, it would be far too ambitious. A number of selected examples will be shared in this section to demonstrate the usefulness of polysaccharide derivatives under SFC conditions for the separation of different chemical classes.

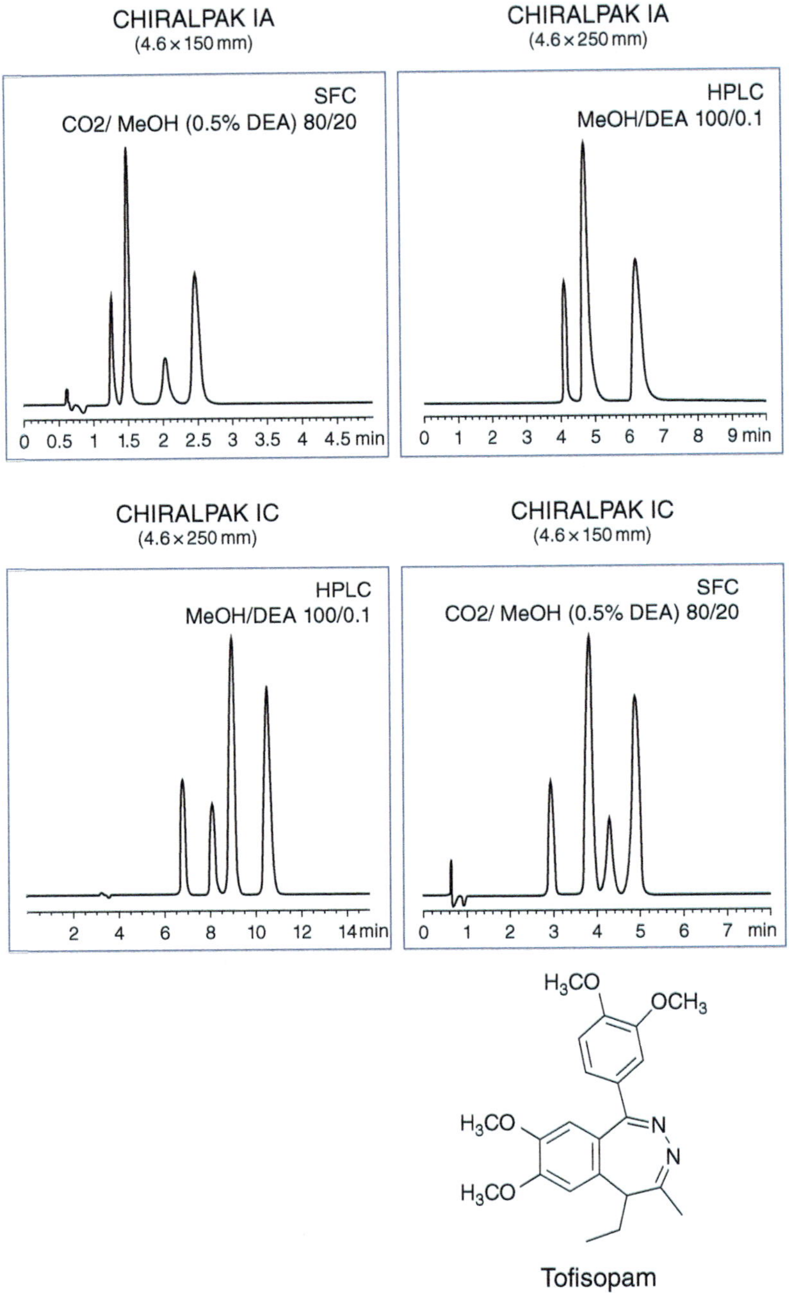

Figure 6.13 Separation of tofisopam isomers in LC and SFC mode. *Source:* Adapted from Zhang et al. [110].

In 2006, Toribio et al. reported the successful SFC preparative separation of omeprazole on coated amylose *tris*(3,5-dimethylphenylcarbamate) [238]. Interesting was the observation that the use of 2-propanol and ethanol as modifiers resulted in reversal of elution order. This same group followed up this work with similar semipreparative separations of lansoprazole, pantoprazole, and rabeprazole using similar approaches.

Li et al. developed a preparative-scale SFC separation for their BMS-962212, a parenteral Factor XIa inhibitor [239]. This was a unique application as it involved a two-step chiral and achiral separation to isolate the final API. The authors were able to accomplish this with coated cellulose *tris*(3,5-dimethylphenylcarbamate) with methanol/acetonitrile as a modifier in CO_2 to purify more than 230 g of the racemic API in >99% enantiomeric purity.

Lee et al. developed a preparative SFC method for the separation of a Pfizer research compound, using CHIRALPAK ID and a non-traditional modifier combination of CO_2 with 40% (90% THF and 10% MeOH with 0.2% isopropylamine). The access to a wider range of solvents for the immobilized CSPs was critical in this example, as the THF was crucial in improving the solubility of the compounds in the injection solvent [117].

Lin et al. developed the semipreparative SFC separation of dihydromyricetin, a naturally occurring compound with a number of reported health benefits, including as an anti-inflammatory agent [240]. Prior to their work, minimal chiral separation records were available. They were successful in separating the enantiomers on amylose *tris*(3,5-dimethylphenylcarbamate) coated CSP with CO_2/methanol (60/40, v/v).

Wu et al. developed a preparative SFC method for the separation of a series of racemic 2,2-dimethyl-3-aryl-propanoic acids [241]. Being such strongly acidic compounds, propanoic acids and their derivatives would generally need an acidic additive like trifluoroacetic acid (TFA) under LC conditions to improve peak shape. However, the slightly acidic nature of CO_2 (equilibrium of carbonic acid with water and carbon dioxide) provided for conditions where this could be avoided. The work was done primarily on coated amylose *tris*(3,5-dimethylphenylcarbamate) (CHIRALPAK AD-H) and cellulose *tris*(4-methylbenzoate) (CHIRALCEL OJ-H) and provided simplified separation conditions from what was previously available.

Dai et al. developed preparative SFC conditions for the separation of a series of atropisomers on coated immobilized CSPs [242]. The authors were successful in separating not only three atropisomers but also other rotational forms (two freely rotating bonds for each atropisomer).

Zehani et al. developed semipreparative SFC conditions for the separation of four 3-carboxamido-5-aryl isoxazole molecules, ligands of the CB2 cannabinoid receptors [243]. A screening with a series of coated amylose and cellulose-derived columns found CHIRALPAK AD-H to be the most productive.

In addition to these specific applications, there have been numerous reviews of the topic published in the literature that can offer a more thorough understanding of the breadth of polysaccharide applications under SFC conditions ([215, 244] as examples); see also Chapter 5.

6.7 Summation

The nearly 50 years of research and commercialization of polysaccharide-based CSPs have made them an invaluable analytical and preparative tool for compounds in the pharmaceutical, agricultural, and numerous other industries for the analysis and separation of enantiomers and other isomeric mixtures. The earliest applications with simple models of Troger's base and TSO have grown into more complex compounds and commercial applications. Initially developed coated CSPs, while still very applicable today, have advanced into immobilized CSPs, providing additional robustness and solvent compatibility. The technology advancement-, also diversifying the particle sizes available, has provided an impetus to the field of chiral separation by chromatography.

Their utility with SFC and LC instrumentation, under a varying range of mobile phase conditions (NP, RP, polar organic mode, etc.), also explains their extensive use. Methods can be somehow "designed" in connection with the instrument availability, the application purpose, and specific aspects of the sample source and sample properties.

Fast or ultrafast chiral analysis can be already achieved today by running short columns packed with 3 or sub-2 μm particles on optimized chromatographic systems such as UHPLC or ultra-high-performance SFC, even in 2D-configurations. Nevertheless, innovation dealing with new selectors and associated technological advances should not cease to undertake the challenges of future therapeutic options and medical advances.

References

1 Hesse, G. and Hagel, R. (1973). Eine vollständige Recemattennung durch eluitons-chromagographie an cellulose-tri-acetat. *Chromatographia* 6 (6): 277–280.

2 Rimbock, K., Kastner, A., and Mannschreck, A. (1986). Microcrystalline tribenzoylcellulose: a high-performance liquid chromatographic sorbent for the separation of enantiomers. *Journal of Chromatography* 351: 346–350.
3 Francotte, E. and Wolf, R.M. (1991). Benzoyl cellulose beads in the pure polymeric form as a new powerful sorbent for the chromatographic resolution of racemates. *Chirality* 3 (1): 43–55.
4 Francotte, E. and Wolf, R.M. (1992). Chromatographic resolution on methylbenzoylcellulose beads: modulation of the chiral recognition by variation of the position of the methyl group on the aromatic ring. *Journal of Chromatography* 595 (1–2): 63–75.
5 Okamoto, Y., Kawashima, M., Yamamoto, K., and Hatada, K. (1984). Useful chiral packing materials for high-performance liquid chromatographic resolution. Cellulose triacetate and tribenzoate coated on macroporous silica gel. *Chemistry Letters* 13 (5): 739–742.
6 Okamoto, Y., Kawashima, M., and Hatada, K. (1984). Chromatographic resolution. 7. Useful chiral packing materials for high-performance liquid chromatographic resolution of enantiomers: phenylcarbamates of polysaccharides coated on silica gel. *Journal of the American Chemical Society* 106 (18): 5357–5359.
7 Ichida, A., Shibata, T., Okamoto, I. et al. (1984). Resolution of enantiomers by HPLC on cellulose derivatives. *Chromatographia* 19: 280–284.
8 Yashima, E. and Okamoto, Y. (1995). Chiral discrimination on polysaccharides derivatives. *Bulletin of the Chemical Society of Japan* 68 (12): 3289–3307.
9 Yashima, E., Yamamoto, C., and Okamoto, Y. (1998). Polysaccharide-based chiral LC columns. *Synlett* 1998 (4): 344–360.
10 Okamoto, Y. and Yashima, E. (1998). Polysaccharide derivatives for chromatographic separation of enantiomers. *Angewandte Chemie International Edition* 37 (8): 1020–1043.
11 Okamoto, Y., Aburatani, R., and Hatada, K. (1987). Chromatographic chiral resolution: XIV. Cellulose tribenzoate derivatives as chiral stationary phases for high-performance liquid chromatography. *Journal of Chromatography* 389: 95–102.
12 Okamoto, Y., Kawashima, M., and Hatada, K. (1986). Chromatographic resolution: XI. Controlled chiral recognition of cellulose triphenylcarbamate derivatives supported on silica gel. *Journal of Chromatography A* 363 (2): 173–186.
13 Berthod, A. (ed.) (2010). *Chiral Recognition in Separation Methods*. Berlin Heidelberg: Springer-Verlag.
14 Barreiro, J.C., Paixão, M.W., Lourenço, T.C. et al. (2016). A high-resolution magic angle spinning NMR study of the enantiodiscrimination of

3,4-methylenedioxymethamphetamine (MDMA) by an immobilized polysaccharide-based chiral phase. *PLoS ONE* 11 (9): e0162892. https://doi.org/10.1371/journal.pone.0162892.

15 Zhang, T., Franco, P., Nguyen, D. et al. (2012). Complementary enantiorecognition patterns and specific method optimization aspects on immobilized polysaccharide-derived chiral stationary phases. *Journal of Chromatography A* 1269: 178–188.

16 Yamamoto, C. and Okamoto, Y. (2007). *Enantiomer Separation: Fundamentals and Practical Methods*. Springer Dordrecht. https://link.springer.com/book/10.1007/978-1-4020-2337-8.

17 Matlin, S.A., Tiritan, M.E., Cass, Q.B., and Boyd, D.R. (1996). Enantiomeric resolution of chiral sulfoxides on polysaccharide phases by HPLC. *Chirality* 8 (1): 147–152.

18 Tiritan, M.E., Cass, Q.B., Del Alamo, A. et al. (1998). Enantiomeric resolution of chiral sulfoxides on polysaccharide phases by HPLC. *Chirality* 10 (7): 573–577.

19 Okamoto, Y., Aburatani, R., Kawashima, M. et al. (1986). Resolution of 4-hydroxy-2-cyclopentenone derivatives by HPLC on cellulose triphenylcarbamate derivatives. *Chemistry Letters* 15 (10): 1767–1770.

20 Okamoto, Y., Sakamoto, H., Hatada, K., and Irie, M. (1986). Resolution of enantiomers by HPLC on cellulose trans- and cis-tris(4-phenylazophenylcarbamate). *Chemistry Letters* 15 (6): 983–986.

21 Okamoto, Y., Kawashima, M., Aburatani, R. et al. (1986). Optical resolution of β-blockers by HPLC on cellulose triphenylcarbamate derivatives. *Chemistry Letters* 15 (7): 1237–1240.

22 Blaschke, G. (1986). Chromatographic resolution of chiral drugs on polyamides and cellulose triacetate. *Journal of Liquid Chromatography* 9 (2–3): 341–368.

23 Francotte, E. and Lohmann, D. (1987). Analytic and preparative resolution of recemic γ- and δ-lactones by chromatography on cellulose triacetate. Relationship between elution order and absolute configuration. *Helvetica Chimica Acta* 70 (6): 1569–1582.

24 Wainer, I.W., Stiffin, R.M., and Shibata, T. (1987). Resolution of enantiomeric aromatic alcohols on a cellulose tribenzoate high-performance liquid chromatography chiral stationary phase: a proposed chiral recognition mechanism. *Journal of Chromatography A* 411 (1–2): 139–151.

25 Krstulovic, A.M., Fouchet, M.H., Burke, J.T. et al. (1988). Direct enantiomeric separation of betaxolol with applications to analysis of bulk drug and biological samples. *Journal of Chromatography A* 452 (1–2): 477–483.

26 Aboul-Enein, H.Y. and Islam, M.R. (1989). Direct high-performance liquid chromatographic separation of penbutolol enantiomers on a cellulose tris-3,5-dimethylphenyl carbamate chiral stationary phase. *Chirality* 1 (4): 301–304.

27 Gaffney, M.H., Stiffin, R.M., and Wainer, I.W. (1989). The effect of alcoholic mobile phase modifiers on retention and stereoselectivity on a commercially available cellulose-based HPLC chiral stationary phase: an unexpected reversal in enantiometric elution order. *Chromatographia* 27 (1): 15–18.

28 Aboul-Enein, H.Y. and Islam, M.R. (1990). Direct separation and optimization of timolol enantiomers on a cellulose tris-3,5-dimethylphenylcarbamate high-performance liquid chromatographic chiral stationary phase. *Journal of Chromatography A* 511 (1–2): 109–114.

29 Hauck, C. and Schildknecht, H. (1990). Separation of enantiomers of the germination stimulant strigol on cellulose triacetate and determination of their biological activity. *Journal of Plant Physiology* 136 (1): 126–128.

30 Aboul-Enein, H.Y. and Islam, M.R. (1990). Direct HPLC separation of carazolol enantiomers on a cellulose tris-3,5-dimethylphenyl carbamate column. *Analytical Letters* 23 (6): 973–980.

31 Straka, R.J., Johnson, K.A., Marshall, P.S., and Remmel, R.P. (1990). Analysis of metoprolol enantiomers in human serum by liquid chromatography on a cellulose-based chiral stationary phase. *Journal of Chromatography B: Biomedical Sciences and Applications* 530 (1–2): 83–93.

32 Aboul-Enein, H.Y. and Islam, M.R. (1990). Direct HPLC separation and optimization of celiprolol enantiomers. *Analytical Letters* 23 (1): 83–91.

33 Aboul-Enein, H.Y. and Islam, M.R. (1990). Direct chromatographic resolution of racemic aminoglutethimide and its acetylated metabolite using different cellulose based chiral stationary phases. *Chromatographia* 30 (3): 223–227.

34 Okamoto, M. and Nakazawa, H. (1991). Reversal of elution order during direct enantiomeric separation of pyriproxyfen on a cellulose-based chiral stationary phase. *Journal of Chromatography A* 588 (1–2): 177–180.

35 Hollenhorst, T. and Blaschke, G. (1991). Direct separation of the enantiomers of propafenone, diprafenone and their major metabolites by high-performance liquid chromatography on modified cellulose and amylose chiral stationary phases. *Journal of Chromatography A* 585 (1–2): 329–332.

36 Okamoto, Y., Aburatani, R., Fukumoto, T., and Hatada, K. (1987). Useful chiral stationary phases for HPLC. Amylose tris (3,5-dimethylphenylcarbamate) and tris (3,5-dichlorophenylcarbamate) supported on silica gel. *Chemistry Letters* 16 (9): 1857–1860.

37 Okamoto, Y., Cao, Z.-K., Aburatani, R., and Hatada, K. (1987). Optical resolution of alcohols as carbamates by HPLC on cellulose tris (phenylcarbamate) derivatives. *Bulletin of the Chemical Society of Japan* 60 (11): 3999–4003.

38 Okamoto, Y., Aburantani, R., Kaida, Y., and Hatada, K. (1988). Direct optical resolution of carboxylic acids by chiral HPLC on tris (3,5-dimethylphenylcarbamate) s of cellulose and amylose. *Chemistry Letters* 17 (7): 1125–1128.

39 Dingenen, J. (1994). Polysaccharide phases in enantioseparations. In: *A Practical Approach to Chiral Separations by Liquid Chromatography* (ed. G. Subramanian). Weinheim: VCH, Chapter 6.

40 Francotte, E. (1997). Chromatography as a separation tool for the preparative resolution of racemic compounds. In: *Chiral Separations, Applications and Technology* (ed. S. Ahuga). Washington: American Chemical Society, Chapter 10.

41 Okamoto, Y. and Kaida, Y. (1994). Resolution by high-performance liquid chromatography using polysaccharide carbamates and benzoates as chiral stationary phases. *Journal of Chromatography A* 666 (1–2): 403–419.

42 Francotte, E. (1994). Contribution of preparative chromatographic resolution to the investigation of chiral phenomena. *Journal of Chromatography A* 666 (1–2): 565–601.

43 Yashima, E. (2001). Polysaccharide-based chiral stationary phases for high-performance liquid chromatographic enantioseparation. *Journal of Chromatography A* 906 (1–2): 105–125.

44 Tachibana, K. and Ohnishi, A. (2001). Reversed-phase liquid chromatographic separation of enantiomers on polysaccharide type chiral stationary phases. *Journal of Chromatography A* 906 (1–2): 127–154.

45 Francotte, E. (2001). Enantioselective chromatography as a powerful alternative for the preparation of drug enantiomers. *Journal of Chromatography A* 906 (1–2): 379–397.

46 Aboul-Enein, H.Y. (2001). High-performance liquid chromatographic enantioseparation of drugs containing multiple chiral centers on polysaccharide-type chiral stationary phases. *Journal of Chromatography A* 906 (1–2): 185–193.

47 de la Puente, M.L., White, C.T., Rivera-Sagredo, A. et al. (2003). Impact of normal-phase gradient elution in chiral chromatography: a novel, robust, efficient and rapid chiral screening procedure. *Journal of Chromatography A* 983 (1–2): 101–114.

48 de la Puente, M.L. (2004). Highly sensitive and rapid normal-phase chiral screen using high-performance liquid chromatography–atmospheric pressure ionization tandem mass spectrometry (HPLC/MS). *Journal of Chromatography A* 1055 (1–2): 55–62.

49 Andersson, M.E., Aslan, D., Clark, A. et al. (2003). Evaluation of generic chiral liquid chromatography screens for pharmaceutical analysis. *Journal of Chromatography A* 1005: 83–101.

50 Cox, G. (ed.) (2005). *Preparative Enantioselective Chromatography*. Oxford: Blackwell Publishing.

51 Wetli, H.A. and Francotte, E. (2007). Automated screening platform with isochronal-parallel analysis and conditioning for rapid method development of chiral separations. *Journal of Separation Science* 30: 1255–1261.

52 Zhang, T. and Franco, P. (2007). Analytical and preparative potential of immobilized polysaccharide-derived chiral stationary phases. In: *Chiral Separation Techniques* (3rd revised and updated edition) (ed. G. Subramanian), 99–134. Weinheim: Wiley-VCH, Chapter 3.

53 Franco, P. and Zhang, T. (2013). Common screening approaches for efficient analytical method development in LC and SFC on columns packed with immobilized polysaccharide-derived chiral stationary phases. In: *Chiral Separations, Methods and Protocols*, 2e (ed. G. Scriba), 113–126. Humana Press, Chapter 6.

54 Bléhaut, J., Franco, P., Zhang, T. et al. (2012). Industrial applications of chiral chromatography. In: *Comprehensive Chirality*, vol. 9 (ed. E.M. Carreira and H. Yamamoto), 400–456. Elsevier Publisher, Chapter 6.

55 Zhang, T. and Franco, P. (2019). Polysaccharide-derived chiral stationary phases for the separation of enantiomers. In: *Chromatographic Method Development*, Jenny Stanford Series on Pharmaceutical Analytics, vol. 3 (ed. G.K. Webster and L. Kott), 116–155. Jenny Stanford Publishing, Chapter 6.

56 Zhang, G.H., Fu, K.Q., Xi, J.B. et al. (2019). Structure screening and performance restoration of chiral separation materials based on chitosan derivatives. *Carbohydrate Polymers* 214: 259–268.

57 Zhang, J., Wang, X.C., Chen, W., and Bai, Z.-W. (2016). Synthesis of substituted phenylcarbamates of N-cyclobutylformylated chitosan and their application as chiral selectors in enantioseparation. *Analyst* 4 (14): 4470–4480.

58 Tang, S., Liu, J.-D., Chen, W. et al. (2018). Performance comparison of chiral separation materials derived from N-cyclohexylcarbonyl and N-hexanoyl chitosans. *Journal of Chromatography A* 1532: 112–123.

59 Wang, J., Huang, S.-H., Chen, W., and Bai, Z.-W. (2016). Eluent tolerance and enantioseparation recovery of chiral packing materials based on chitosan bis(phenylcarbamate)-(n-octyl urea)s for high performance liquid chromatography. *Molecules* 12 (21): 1528.

60 Cass, Q.B., Bassi, A.L., and Matlin, S.A. (1996). Chiral discrimination by HPLC on aryl carbamate derivatives of chitin coated onto microporous aminopropyl silica. *Chirality* 8 (1): 131–135.

61 Ribeiro, J., Tiritan, M., Pinto, M., and Fernandes, C. (2017). Chiral stationary phases for liquid chromatography based on chitin-and chitosan-derived marine polysaccharides. *Symmetry* 9 (9): 190.

62 Toga, Y., Ichida, A., Shibata, T. et al. (2004). Chiral recognition ability of curdlan triacetate: solvent and temperature effects. *Chirality* 16 (4): 272–276.

63 Li, G., Shen, J., and Okamoto, Y. (2015). Synthesis and enantioseparation ability of xylan bisphenylcarbamate derivatives as chiral stationary phases in HPLC. *Chirality* 27 (8): 518–522.

64 Shen, J. and Okamoto, Y. (2016). Efficient separation of enantiomers using stereoregular chiral polymers. *Chemical Reviews* 116: 1094–1138.

65 Franco, P., Senso, A., Oliveros, L., and Minguillón, C. (2001). Covalently bonded polysaccharide derivatives as chiral stationary phases in high-performance liquid chromatography. *Journal of Chromatography A* 906 (1–2): 155–170.

66 Okamoto, Y., Aburatani, R., Miura, S., and Hatada, K. (1987). Chiral stationary phases for HPLC: cellulose tris (3,5-dimethylphenylcarbamate) and tris (3,5-dichlorophenylcarbamate) chemically bonded to silica gel. *Journal of Liquid Chromatography* 10 (8–9): 1613–1628.

67 Yashima, E., Fukaya, H., and Okamoto, Y. (1994). 3,5-Dimethylphenylcarbamates of cellulose and amylose regioselectively bonded to silica gel as chiral stationary phases for high-performance liquid chromatography. *Journal of Chromatography A* 677 (1): 11–19.

68 Enomoto, N., Furukawa, S., Osagawara, Y. et al. (1996). Preparation of silica gel-bonded amylose through enzyme-catalyzed polymerization and chiral recognition ability of its phenylcarbamate derivative in HPLC. *Analytical Chemistry* 68 (17): 2798–2804.

69 Francotte, E. (1996). Photochemically cross-linked polysaccharide derivatives as supports for the chromatographic separation of enantiomers. PCT WO 96/27615.

70 Francotte, E. and Zhang, T. (1997). Photochemically cross-linked polysaccharide derivatives having no photopolymerisable functional groups. PCT WO 97/004011.

71 Francotte, E. (1997). Thermally immobilized polysaccharide derivatives. PCT WO 97/49733.

72 Francotte, E. and Huynh, D. (2022). Immobilization of 3,5-dimethylphenyl carbamate of cellulose and amylose on silica by photochemical and thermal radical processes. *Chirality* https://doi.org/10.1002/chir.23426.

73 Santarelli, X., Muller, D., and Jozefonvicz, J. (1988). Dextran-coated silica packings for high-performance size-exclusion chromatography of proteins. *Journal of Chromatography* 443: 55–62.

74 Kimata, K. and Tsuboi, R. (1994). Optical resolution agent and process for the preparation of the same. US Patent 5 302 633.

75 Oliveros, L., Lopez, P., Minguillon, C., and Franco, P. (1995). Chiral chromatographic discrimination ability of a cellulose 3,5-dimethylphenylcarbamate/10-undecenoate mixed derivative fixed on several chromatographic matrices. *Journal of Liquid Chromatography* 18 (8): 1521–1532.

76 Minguillon, C., Franco, P., Oliveros, L., and Lopez, P. (1996). Bonded cellulose-derived high-performance liquid chromatography chiral stationary phases I. Influence of the degree of fixation on selectivity. *Journal of Chromatography A* 728 (1–2): 407–414.

77 Minguillon, C., Franco, P., and Oliveros, L. (1996). Bonded cellulose-derived high-performance liquid chromatography chiral stationary phases I. Influence of the degree of fixation on selectivity. *Journal of Chromatography A* 728 (1–2): 415–422.

78 Kimata, K., Tsuboi, R., Hosoya, K., and Tanaka, N. (1993). Chemically bonded chiral stationary phase prepared by the polymerization of cellulose p-vinylbenzoate. *Analytical Methods and Instrumentation* 1: 23–29.

79 Oliveros, L., Minguillion, C., and Lopez, P. (1994). Polysaccharide derivatives and their use for preparing stationary chiral phases suitable for separating isomers. French Patent 2 714 671.

80 Garces, J., Franco, P., Oliveros, L., and Minguillon, C. (2003). Mixed cellulose-derived benzoates bonded on allylsilica gel as HPLC chiral stationary phases: influence of the introduction of an aromatic moiety in the fixation substituent. *Tetrahedron: Asymmetry* 14 (9): 1179–1185.

81 Kubota, T., Yamamoto, C., and Okamoto, Y. (2004). Phenylcarbamate derivatives of cellulose and amylose immobilized onto silica gel as chiral stationary phases for high-performance liquid chromatography. *Journal of Polymer Science, Part A: Polymer Chemistry* 42 (18): 4704–4710.

82 Chen, X.M., Yamamoto, C., and Okamoto, Y. (2006). Influence of vinyl monomers and temperature on immobilization of cellulose 3,5-dimethylphenylcarbamate onto silica gel as chiral stationary phases for high-performance liquid chromatography. *Journal of Chromatography A* 1104 (1–2): 62–68.

83 Zhang, S., Ong, T.-T., Ng, S.-C., and On Chan, H.S. (2007). Chemical immobilization of azido cellulose phenylcarbamate onto silica gel via Staudinger reaction and its application as a chiral stationary phase for HPLC. *Tetrahedron Letters* 48 (31): 5487–5490.

84 Peng, G.M., Wu, S.Q., Fang, S.L. et al. (2012). Preparation and chiral separation of a novel immobilized cellulose-based chiral stationary phase in high-performance liquid chromatography. *Journal of Chromatographic Science* 50: 516–522.

85 Chen, X.M., Liu, Y., Kong, I., and Zou, H. (2003). Synthesis of covalently bonded cellulose derivative chiral stationary phases with a bifunctional reagent of 3-(triethoxysilyl) propyl isocyanate. *Journal of Chromatography A* 1010 (2): 185–194.

86 Shen, J., Ikai, T., and Okamoto, Y. (2014). Synthesis and application of immobilized polysaccharide-based chiral stationary phases for enantioseparation by high-performance liquid chromatography. *Journal of Chromatography A* 1363: 51–61.

87 Zhang, T., Kientzy, C., Franco, P. et al. (2005). Solvent versatility of immobilized 3,5-dimethylphenylcarbamate of amylose in enantiomeric separations by HPLC. *Journal of Chromatography A* 1075 (1–2): 65–75.

88 Zhang, T., Schaeffer, M., and Franco, P. (2005). Optimization of the chiral separation of a Ca-sensitizing drug on an immobilized polysaccharide-based chiral stationary phase: case study with a preparative perspective. *Journal of Chromatography A* 1083 (1–2): 96–101.

89 Cirilli, R., Ferretti, R., Gallinilla, B. et al. (2005). Analytical and semipreparative high performance liquid chromatography separation of stereoisomers of novel 3,4-dihydropyrimidin-4(3H)-one derivatives on the immobilized amylose-based Chiralpak IA chiral stationary phase. *Journal of Separation Science* 29: 1399–1406.

90 Zhang, T., Nguyen, D., Franco, P. et al. (2006). Cellulose 3,5-dimethylphenylcarbamate immobilized on silica: a new chiral stationary phase for the analysis of enantiomers. *Analytica Chimica Acta* 557 (1–2): 221–228.

91 Ghanem, A. and Naim, L. (2006). Immobilized versus coated amylose tris(3,5-dimethylphenylcarbamate) chiral stationary phases for the enantioselective separation of cyclopropane derivatives by liquid chromatography. *Journal of Chromatography A* 1101 (1–2): 171–178.

92 Ali, I., Naim, L., Ghanem, A., and Aboul Enein, H.Y. (2006). Chiral separations of piperidine-2,6-dione analogues on Chiralpak IA and Chiralpak IB columns by using HPLC. *Talanta* 69 (4): 1013–1017.

93 Cirilli, R., Simonelli, A., Ferretti, R. et al. (2006). Analytical and semipreparative high performance liquid chromatography enantioseparation of new substituted 1-thiocarbamoyl-3,5-diaryl-4,5-dihydro-(1H)-pyrazoles on polysaccharide-based chiral stationary phases in normal-phase, polar organic and reversed-phase conditions. *Journal of Chromatography A* 1101: 198–203.

94 Cirilli, R., Orlando, V., Ferretti, R. et al. (2006). Direct HPLC enantioseparation of chiral aptazepine derivatives on coated and immobilized polysaccharide-based chiral stationary phases. *Chirality* 18 (8): 621–632.

95 Andersson, S., sin Subramanian, G., ed., (2007) Chiral Separation Techniques: *A Practical Approach* (3rd Revised and updated Edition), VCH, Weinheim, Chapter 17.

96 Cirilli, R., Ferretti, R., Gallinella, B. et al. (2008). High-performance liquid chromatography enantioseparation of proton pump inhibitors using the immobilized amylose-based Chiralpak IA chiral stationary phase in normal-phase, polar organic and reversed-phase conditions. *Journal of Chromatography A* 1177 (1): 105–113.

97 Cirilli, R., Ferretti, R., Gallinella, B. et al. (2008). Enantioseparation of kavain on Chiralpak IA under normal-phase, polar organic and reversed-phase conditions. *Journal of Separation Science* 31: 2206–2210.

98 Cirilli, R., Ferretti, R., De Santis, E. et al. (2008). High-performance liquid chromatography separation of enantiomers of flavanone and 2′-hydroxychalcone under reversed-phase conditions. *Journal of Chromatography A* 1190 (1–2): 95–101.

99 Franco, P., Nguyen, D., and Zhang, T. (2008). Enantiomer resolution screening strategy using multiple immobilised polysaccharide-based chiral stationary phases. *Journal of Chromatography A* 1191 (1–2): 214–222.

100 Zhang, T., Nguyen, D., Franco, P. et al. (2008). Cellulose tris (3,5-dichlorophenylcarbamate) immobilised on silica: a novel chiral stationary phase for resolution of enantiomers. *Journal of Pharmaceutical and Biomedical Analysis* 46 (5): 882–891.

101 Franco, P. and Zhang, T. (2008). Common approaches for efficient method development with immobilised polysaccharide-derived chiral stationary phases. *Journal of Chromatography B* 875 (1): 48–56.

102 Franco, P. and Zhang, T. (2008). 3 μm particle-based chiral stationary phases for the fast and efficient resolution of enantiomers. *LC GC Europe* 21 (9): 430–437.

103 Wang, C., Armstrong, D.W., and Chang, C.D. (2008). Rapid baseline separation of enantiomers and a mesoform of all-trans-astaxanthin, 13-cis-astaxanthin, adonirubin, and adonixanthin in standards and commercial supplements. *Journal of Chromatography A* 1194 (2): 172–177.

104 Ferretti, R., Gallinella, B., La Torre, F. et al. (2009). Direct high-performance liquid chromatography enantioseparation of terazosin on an immobilised polysaccharide-based chiral stationary phase under polar organic and reversed-phase conditions. *Journal of Chromatography A* 1216 (28): 5385–5390.

105 Pawar, R.S., Grundel, E., Mazzola, E. et al. (2010). Preparative chiral separation of the pyrrolizidine alkaloids intermedine and lycopsamine on immobilized amylose and cellulose based stationary phases. *Planta Medica* 75 (5): 55.

106 Franco, P. and Zhang, T. (2010). Finding the best separation for enantiomeric mixtures. *LC GC North America* 23 (9): 818–822.

107 Caille, S., Boni, J., Cox, G.B. et al. (2010). Comparison of large-scale routes to manufacture chiral exo-2-norbornyl thiourea. *Organic Process Research and Development* 14: 133–141.

108 Zhang, T., Nguyen, D., and Franco, P. (2010). Reversed-phase screening strategies for liquid chromatography on polysaccharide-derived chiral stationary phases. *Journal of Chromatography A* 1217 (7): 1048–1055.

109 García, P., Franco, P., Alvarez, R., and De Lera, U.R. (2011). Separation of regioisomers and enantiomers of underivatized saturated and unsaturated fatty acid monoacylglycerols using enantioselective HPLC. *Journal of Separation Science* 34 (9): 999–1003.

110 Zhang, T., Nguyen, N., Franco, P., and Vollmer, M. (2011). Separation of Enantiomers and Conformers of Tofisopam. Agilent Application Note. http://www.chem.agilent.com/Library/applications/5990-9315EN.pdf (accessed 23 March 2022).

111 Zhang, T., Nguyen, N., Franco, P., and Vollmer, M. (2011). Enantiomer Separation of Nonsteroidal Anti-Inflammatory Drugs. Agilent Application Note http://www.chem.agilent.com/Library/applications/5990-9459EN.pdf (accessed 23 March 2022).

112 Ashok, S., Varma, M.S., and Swaminathan, S. (2012). A validated LC method for the determination of the enantiomeric purity of aliskiren hemifumarate in bulk drug samples. *Journal of Chromatography Science* 50: 799–802.

113 Jie, Z., Qiuzheng, D., Suzhen, Z. et al. (2015). Enantioseparation of three important intermediates of tanikolide with immobilized cellulose chiral stationary phase. *Journal of Chromatography Science* 53: 959–962.

114 Geryk, R., Kalíková, K., Vozka, J. et al. (2014). Enantioselective potential of chiral stationary phases based on immobilized polysaccharides in reversed phase mode. *Journal of Chromatography A* 1363: 155–161.

115 Peluso, P., Mamane, V., Aubert, E., and Cossu, S. (2014). High-performance liquid chromatography enantioseparation of polyhalogenated 4,4′-bipyridines on polysaccharide-based chiral stationary phases under multimodal elution. *Journal of Separation Science* 37 (18): 2481–2489.

116 Harrison, A., Melchionna, M., Franco, P., and Hlavac, J. (2014). Solid-phase synthesis and analysis of 3,6-dihydro-2H-1,2-oxazines in their stereo- and regioisomer mixtures. *New Journal of Chemistry* 38 (11): 5491–5499.

117 Lee, J., Lee, J.T., Watts, W.L. et al. (2014). On the method development of immobilized polysaccharide chiral stationary phases in supercritical fluid chromatography using an extended range of modifiers. *Journal of Chromatography A* 1374: 238–246.

118 Rizzo, S., Menta, S., Benincori, T. et al. (2015). Determination of the enantiomerization barrier of the residual enantiomers of C3-symmetric tris[3-(1-methyl-2-alkyl)indolyl]phosphane oxides: case study of a multitasking HPLC investigation based on an immobilized polysaccharide stationary phase. *Chirality* 27 (12): 888–899.

119 Geryk, R., Kalíková, K., Vozka, J., and Tesařová, E. (2015). Immobilized polysaccharide-based stationary phases for enantioseparation in normal versus reversed phase HPLC. *Chromatographia* 78: 909–915.

120 Liu, N., Dong, F., Xu, J. et al. (2015). Stereoselective determination of tebuconazole in water and zebrafish by supercritical fluid chromatography tandem mass spectrometry. *Journal of Agricultural and Food Chemistry* 63: 6297–6303.

121 Islam, M.F. and Lee, W. (2015). Liquid chromatographic enantiomer separation of α-amino acid esters as nitrobenzoxadiazole derivatives using polysaccharide-derived chiral stationary phases. *Journal of the Chosun Natural Science* 8 (2): 111–116.

122 Rao, R.N. and Kumar, K.N. (2015). Thermodynamic evaluation of immobilized cellulose tris(3,5-dichlorophenylcarbamate) as a stationary phase for liquid chromatographic separation of darunavir enantiomers. *Journal of Chromatography Science* 53: 295–301.

123 Menta, S., Carradori, S., Secci, D. et al. (2015). The anancomeric character of the pharmacophore 1,3,4-thiadiazoline framework in chiral spiro-cyclohexyl derivatives: effects on stereochemistry and spiro-junction lability. Thermodynamic aspects. *Journal of Organic Chemistry* 80 (24): 11932–11940.

124 López-Ram-de-Víu, P., Gálvez, J., and Díaz-de-Villegas, M.D. (2015). High-performance liquid chromatographic enantioseparation of unusual amino acid derivatives with axial chirality on polysaccharide-based chiral stationary phases. *Journal of Chromatography A* 1390: 78–85.

125 Qian, Y., Zhao, X.Q., Zhao, L. et al. (2015). Analysis of stereochemistry and biosynthesis of epicatechin in tea plants by chiral phase high performance liquid chromatography. *Journal of Chromatography B* 1006: 1–7.

126 Cirilli, R., Ferretti, R., Gallinella, B. et al. (2009). Development and validation of an enantioselective and chemoselective HPLC method using a Chiralpak IA column to simultaneously quantify (R)-(+)- and (S)-(−)-lansoprazole enantiomers and related impurities. *Journal of Pharmaceutical and Biomedical Analysis* 50 (1): 9–14.

127 Zanitti, L., Ferretti, R., Gallinella, B. et al. (2010). Direct HPLC enantioseparation of omeprazole and its chiral impurities: application to the determination of enantiomeric purity of esomeprazole magnesium trihydrate. *Journal of Pharmaceutical and Biomedical Analysis* 52 (5): 665–671.

128 Sola, I., Viayna, E., Gómez, T. et al. (2015). Multigram synthesis and in vivo efficacy studies of a novel multitarget anti-Alzheimer's compound. *Molecules* 20 (3): 4492–4515.

129 Moldovan, R.C., Dascal, G.S., Mirel, V. et al. (2015). Chiral separation of 16 beta-blockers on immobilized polysaccharide chiral stationary phases. *Farmácia* 63 (6): 909–912.

130 Adhikari, S., Kang, J.S., and Lee, W. (2016). A convenient and validated enantiomer separation of chiral aliphatic amines as nitrobenzoxadiazole derivatives on polysaccharide-derived chiral stationary phases under simultaneous ultraviolet and fluorescence detection. *Chirality* 28 (12): 789–794.

131 Srinivasu, G., Nagesh Kumar, K., Thirupathi, C. et al. (2016). Development and validation of the chiral HPLC method for daclatasvir in gradient elution mode on amylose-based immobilized chiral stationary phase. *Chromatographia* 79 (21): 1457–1467.

132 Saduttoa, D., Ferretti, R., Zanitti, L. et al. (2016). Analytical and semipreparative high performance liquid chromatography enantioseparation of bicalutamide and its chiral impurities on an immobilized polysaccharide-based chiral stationary phase. *Journal of Chromatography A* 1445: 166–171.

133 Gallinella, B., Ferretti, R., Zanitti, L. et al. (2016). Comparison of reversed-phase enantioselective HPLC methods for determining the enantiomeric purity of (S)-omeprazole in the presence of its related substances. *Journal of Pharmaceutical Analysis* 6 (2): 132–136.

134 Kim, S.J., Nam, K.W., Park, B. et al. (2016). A newly developed analytical and semi-preparative enantiomer separation of fluoxetine using polysaccharide-derived chiral stationary phases by high performance liquid chromatography. *KSBB Journal* 31 (3): 186–191.

135 Lotter, C., Poehler, E., Heiland, J. et al. (2016). Enantioselective reaction monitoring utilizing two-dimensional heart-cut liquid chromatography on an integrated microfluidic chip. *Lab on a Chip* 16 (24): 4648–4652.

136 Bowden, G.D., Stotz, S., Kinzler, J. et al. (2021). DoE optimization empowers the automated preparation of enantiomerically pure [18F]talazoparib and its in vivo evaluation as a PARP radiotracer. *Journal of Medicinal Chemistry* 64 (21): 15690–15701.

137 Geibel, C., Zhang, L., Serafimov, K. et al. (2022). Towards enantioselective ultrahigh performance liquid chromatography–mass spectrometry-based metabolomics of branched-chain fatty acids and anteiso-fatty acids under reversed-phase conditions using sub-2-μm amylose- and cellulose-derived chiral stationary phases. *Chirality* 34: 484–497.

138 Paolini, L., Hausser, N., and Zhang, T. (2022). Chiral resolution of the insecticide fipronil enantiomers and the simultaneous determination of its major transformation products by high-performance liquid chromatography interfaced with mass spectrometry. *Chirality* 34: 473–483.

139 Polavarapu, P.L. (ed.) (2018). *Chiral Analysis*. Amsterdam: Elsevier.

140 Batra, S. (ed.) (2014). *Chromatographic Enantioseparation of Pharmaceutical Compounds*. Scholar's Press.

141 Scriba, G.E.K. (ed.) (2019). *Chiral Separations: Methods and Protocols*. Humana Press.

142 Carreira, E.M. and Yamamoto, H. (ed.) (2012). *Comprehensive Chirality*. Amsterdam: Elsevier.

143 Wilson, W.H. (1994). Direct enantiomeric resolution of ibuprofen and flurbiprofen by packed column SFC. *Chirality* 6 (3): 216–219.

144 Johannsen, M. (2001). Separation of enantiomers of ibuprofen on chiral stationary phases by packed column supercritical fluid chromatography. *Journal of Chromatography A* 937 (1–2): 135–138.

145 Wenda, C. and Rajendran, A. (2009). Enantioseparation of flurbiprofen on amylose-derived chiral stationary phase by supercritical fluid chromatography. *Journal of Chromatography A* 1216 (50): 8750–8758.

146 Gyllenhaal, O. and Stefansson, M. (2005). Reversal of elution order for profen acid enantiomers in packed-column SFC on Chiralpak AD. *Chirality* 17 (5): 257–265.

147 Okamoto, Y., Aburatani, R., Kaida, Y. et al. (1989). Direct chromatographic separation of 2-arylpropionic acid enantiomers using tris(3,5-dimethylphenylcarbamate)s of cellulose and amylose as chiral stationary phases. *Chirality* 1 (3): 239–242.

148 He, Y., Wu, Y., Cheng, L. et al. (2018). Separation of ketorolac enantiomers on polysaccharide-based chiral stationary phases using a polar organic mobile phase. *Separation Science Plus* 1 (5): 351–358.

149 Ing-Lorenzini, K.R., Desmeules, J.A., Besson, M. et al. (2009). Two-dimensional liquid chromatography–ion trap mass spectrometry for the simultaneous determination of ketorolac enantiomers and paracetamol in human plasma: application to a pharmacokinetic study. *Journal of Chromatography A* 1216 (18): 3851–3856.

150 Caccamese, S. (1993). Direct high-performance liquid chromatography (HPLC) separation of etodolac enantiomers using chiral stationary phases. *Chirality* 5 (3): 164–167.

151 Pandya, P.A., Shah, P.A., and Shrivastav, P.S. (2021). Simultaneous enantioseparation and simulation studies of atenolol, metoprolol and propranolol on Chiralpak® IG column using supercritical fluid chromatography. *Journal of Pharmaceutical Analysis* 11: 746–756.

152 Garzotti, M. and Hamdan, M. (2002). Supercritical fluid chromatography coupled to electrospray mass spectrometry: a powerful tool for the analysis of chiral mixtures. *Journal of Chromatography B* 770 (1–2): 53–61.

153 Maftouh, M., Granier-Loyaux, C., Chavana, E. et al. (2005). Screening approach for chiral separation of pharmaceuticals: part III. Supercritical fluid chromatography for analysis and purification in drug discovery. *Journal of Chromatography A* 1088 (1–2): 67–81.

154 Chennuru, L.N., Choppari, T., Duvvuri, S., and Dubey, P.K. (2013). Enantiomeric separation of proton pump inhibitors on new generation chiral columns using LC and supercritical fluid chromatography. *Journal of Separation Science* 36 (18): 3004–3010.

155 Rahman, A., Haque, M.R., Sultan, Z. et al. (2017). Enantiomeric determination of omeprazole and esomeprazole by a developed and validated chiral HPLC method and stability studies by microthermal analysis. *Journal of Pharmaceutical Sciences* 16: 221–233.

156 Nozal, M.J., Toribio, J., Bernal, J.L. et al. (2004). Chiral separation of omeprazole and several related benzimidazoles using supercritical fluid chromatography. *Journal of Separation Science* 27 (12): 1023–1029.

157 Kim, M., Yu, S.K., Truong, Q.-K. et al. (2017). Determination of rabeprazole enantiomers in commercial tablets using immobilized cellulose-based stationary phase. *Archives of Pharmaceutical Research* 40: 373–381.

158 Wang, H., Sun, Y., Meng, X. et al. (2015). Determination of lansoprazole enantiomers in dog plasma by column-switching liquid chromatography with tandem mass spectrometry and its application to a preclinical pharmacokinetic study. *Journal of Separation Science* 38: 2960–2970.

159 Sun, L., Cao, Y., Jiao, H. et al. (2015). Enantioselective determination of (R)- and (S)-lansoprazole in human plasma by chiral liquid chromatography with mass spectrometry and its application to a stereoselective pharmacokinetic study. *Journal of Separation Science* 38: 3696–3703.

160 Eom, H.Y., Kang, M., Kang, S.W. et al. (2016). Rapid chiral separation of racemic cetirizine in human plasma using subcritical fluid chromatography-tandem mass spectrometry. *Journal of Pharmaceutical and Biomedical Analysis* 117: 380–389.

161 Kannappan, V. and Kanthiah, S. (2020). Development and optimization of stereoselective liquid chromatographic method for chiral separation of (±)-cetirizine and enantiopurity assessment of R-levocetirizine. *Journal of Analytical Chemistry* 75: 349–357.

162 Ali, I., Al-Othman, Z.A., Al-Warthan, A. et al. (2014). Enantiomeric separation and simulation studies of pheniramine, oxybutynin, cetirizine, and brinzolamide chiral drugs on amylose-based columns. *Chirality* 26 (3): 136–143.

163 Toribio, L., del Nozal, M.J., Bernal, J.L. et al. (2006). Study of the enantiomeric separation of oxfendazole and cetirizine using subcritical fluid chromatography on an amylose-based column. *Journal of Chromatography A* 1121 (2): 268–273.

164 Liu, Q., Zhang, Z., Bo, H., and Sheldon, R.A. (2002). Direct separation of the enantiomers of cetirizine and related compounds by reversed-phase chiral HPLC. *Chromatographia* 56: 233–235.

165 Liu, Y., Wang, X., Yu, J., and Guo, X. (2021). Chiral separation and molecular simulation study of six antihistamine agents on a coated cellulose tri-(3,5-dimethylphenycarbamate) column (Chiralcel OD-RH) and its recognition mechanisms. *Electrophoresis* 42: 1461–1472.

166 Rosetti, A., Ferretti, R., Zanitti, L. et al. (2020). Single-run reversed-phase HPLC method for determining sertraline content, enantiomeric purity, and related substances in drug substance and finished product. *Journal of Pharmaceutical Analysis* 10: 610–616.

167 Jiang, C.-J., Chen, Z.-R., and Zeng, H. (2007). Chiral separation of sertraline hydrochloride by HPLC. *Chinese Journal of Pharmaceutical Analysis* 27 (7): 997–998.

168 Olsen, B.A., Wirth, D.D., and Larew, J.S. (1998). Determination of fluoxetine hydrochloride enantiomeric excess using high-performance liquid chromatography with chiral stationary phases. *Journal of Pharmaceutical and Biomedical Analysis* 17 (4–5): 623–630.

169 Umstead, W.J. (2022). The chiral resolution of fluoxetine. Chiral Technologies Application Note.

170 Geryk, R., Vozka, J., Kalikova, K., and Tesarova, E. (2013). HPLC method for chiral separation and quantification of antidepressant citalopram and its precursor citadiol. *Chromatographia* 76: 483–489.

171 Rochat, B., Amey, M., Van Gelderen, H. et al. (1995). Determination of the enantiomers of citalopram, its demethylated and propionic acid metabolites in human plasma by chiral HPLC. *Chirality* 7 (6): 389–395.

172 Rosetti, A., Ferretti, R., Villani, C. et al. (2021). Simultaneous enantio- and diastereo-selective high-performance liquid chromatography separation of paroxetine on an immobilized amylose-based chiral stationary phase under green reversed-phase conditions. *Journal of Chromatography A* 1653: 462406.

173 Zukowski, J., Brightwell, M., and De Biasi, V. (2003). Chiral HPLC method for chiral purity determination of paroxetine drug substance. *Chirality* 15 (7): 600–604.

174 Vivekanand, V.V., Kumar, V.R., Mohakhud, P.K., and Reddy, G.O. (2003). Enantiomeric separation of the key intermediate of paroxetine using chiral chromatography. *Journal of Pharmaceutical and Biomedical Analysis* 33 (4): 803–809.

175 Su, B., Bao, Z., Xing, H. et al. (2009). Enantioseparation of paroxetine intermediate on an amylose-derived chiral stationary phase by supercritical fluid chromatography. *Journal of Chromatography A* 1216 (26): 5140–5146.

176 Mazzoccanti, G., Ismail, O.H., D'Acquarica, I. et al. (2017). Cannabis through the looking glass: chemo- and enantio-selective separation of phytocannabinoids by enantioselective ultra high performance supercritical fluid chromatography. *Chemical Communications* 91 (53): 12262–12265.

177 Catani, M., Felletti, S., Buratti, A. et al. (2022). Perspectives and pitfalls in potency testing of cannabinoids by high performance liquid chromatography (HPLC). *LCGC North America* 40: 78–81.

178 Onishi, T. and Umstead, W.J. (2021). The separation of cannabinoids on Sub-2 μm immobilized polysaccharide chiral stationary phases. *Pharmaceuticals* 14 (12): 1250.

179 Denicola, C. and Barendt, J. (2018). Streamlined cannabinoid and potency assays utilizing HPLC and SFC. Cannabis Sciences Conference Poster.

180 Denicola, C. and Barendt, J. (2018). Cannabinoid isolation models utilizing immobilized chiral stationary phases and SFC. Emerald Conference Poster.

181 Chiral Technologies. (2018). Application Note – Separation of the Enantiomers of (+/−) Δ8-THC and (+/−) Δ9-THC

182 Umstead, W.J. (2021). The separation of several minor cannabinoids via chiral HPLC. *Cannabis Science and Technology* 4 (6): 44–51.

183 Chromatographic database of USP. https://www.usp.org/resources/chromatographic-columns (accessed 13 May 2022).

184 Chiral Technologies USP list of HPLC columns. https://chiraltech.com/wp-content/uploads/2020/07/USP-Chart-July-2020.pdf (accessed 13 May 2022).

185 HPLC column selection by USP. https://www.glsciences.com/product/lc_method/r3d80e000000jn1w-att/usplist.pdf (accessed 13 May 2022).

186 Cebo, M., Fu, X., Gawaz, M. et al. (2020). Enantioselective ultra-high performance liquid chromatography-tandem mass spectrometry method based on sub-2μm particle polysaccharide column for chiral separation of oxylipins and its application for the analysis of autoxidized fatty acids and platelet releasates. *Journal of Chromatography A* 1624: 461206.

187 Li, P. and Lämmerhofer, M. (2021). Isomer selective comprehensive lipidomics analysis of phosphoinositides in biological samples by liquid chromatography with data independent acquisition tandem mass spectrometry. *Analytical Chemistry* 93: 9583–9592.

188 Ellington, J., Evans, J.J., Prickett, K.B., and Champion, W.L. (2001). High-performance liquid chromatographic separation of the enantiomers of organophosphorus pesticides on polysaccharide chiral stationary phases. *Journal of Chromatography A* 928 (2): 145–154.

189 Li, L., Zhou, S., Jin, L. et al. (2010). Enantiomeric separation of organophosphorus pesticides by high-performance liquid chromatography, gas chromatography and capillary electrophoresis and their applications to environmental fate and toxicity assays. *Journal of Chromatography B* 878 (17–18): 1264–1276.

190 Gadepalli, R.S.V.S., Rimoldi, J.M., Fronczek, F.R. et al. (2007). Synthesis of fenthion sulfoxide and fenoxon sulfoxide enantiomers: effect of sulfur chirality on acetylcholinesterase activity. *Chemical Research in Toxicology* 20 (2): 257–262.

191 Liu, W., Lin, K., and Gan, J. (2006). Separation and aquatic toxicity of enantiomers of the organophosphorus insecticide trichloronate. *Chirality* 18 (9): 713–716.

192 Wang, Y., Tai, K., and Yen, J. (2004). Separation, bioactivity, and dissipation of enantiomers of the organophosphorus insecticide fenamiphos. *Ecotoxicology and Environmental Safety* 57 (3): 346–353.

193 Petrie, B. and Camacho-Munoz, D. (2021). Environmentally friendly analytical method to assess enantioselective behaviour of pharmaceuticals and pesticides in river waters. *Sustainable Chemistry and Pharmacy* 24: 1–13.

194 Wang, P., Jiang, S., Liu, D. et al. (2006). Enantiomeric resolution of chiral pesticides by high-performance liquid chromatograph. *Journal of Agricultural and Food Chemistry* 54 (5): 1577–1583.

195 Hua, Y., Zhou, Q., Wang, P. et al. (2020). En0y high-performance liquid chromatography. *Chirality* 32 (8): 1053–1061.

196 Ye, J., Wu, J., and Liu, W. (2009). Enantioselective separation and analysis of chiral pesticides by high-performance liquid chromatography. *Trends in Analytical Chemistry* 28 (10): 1148–1163.

197 Zhang, H., Chen, S., and Zhou, S. (2012). Enantiomeric separation and toxicity of an organophosporus insecticide, pyraclofos. *Journal of Agricultural and Food Chemistry* 60 (28): 6953–6959.

198 Lin, K., Zhou, S., Xu, C., and Liu, W. (2006). Enantiomeric resolution and biotoxicity of methamidophos. *Journal of Agricultural and Food Chemistry* 54 (21): 8134–8138.

199 Zhao, P., Wang, Z., Li, K. et al. (2018). Multi-residue enantiomeric analysis of 18 chiral pesticides in water, soil and river sediment using magnetic solid-phase extraction based on amino modified multiwalled carbon nanotubes and chiral liquid chromatography coupled with tandem mass spectrometry. *Journal of Chromatography A* 1568: 8–21.

200 Diaz Merino, M.E., Echevarria, R.N., Lubomirsky, E. et al. (2019). Enantioseparation of the racemates of a number of pesticides on a silica-based column with immobilized amylose tris (3-chloro-5-methylphenylcarbamate). *Microchemistry Journal* 149: 103970.

201 Jin, L., Gao, W., Yang, H. et al. (2011). Enantiomeric resolution of five chiral pesticides on a chiralpak IB-H column by SFC. *Journal of Chromatographic Science* 49 (9): 739–743.

202 Przybyciel, M. andKohler, D. (2018). Chiral separation of the pesticide diniconazole by supercritical fluid chromatography (SFC). Sigma Aldrich Application Note, 8.

203 Iijima, S. (2020). Highly sensitive and comprehensive detection for chiral separation of pesticide using HPLC-CD-MS. Jasco Application Note.

204 Chang, W., Nie, J., Geng, Y. et al. (2020). Etoxazole stereoselective determination, bioaccumulation, and resulting oxidative stress in Danio rerio (zebrafish). *Ecotoxicology and Environmental Safety* 192: 110287.

205 Tao, Y., Dong, F., Xu, J. et al. (2014). Green and sensitive supercritical fluid chromatographic–tandem mass spectrometric method for the separation and determination of flutriafol enantiomers in vegetables, fruits, and soil. *Journal of Agricultural and Food Chemistry* 62 (47): 11457–11464.

206 Schaffrath, M., Weidmann, V., and Maison, W. (2014). Enantioselective high performance liquid chromatography and supercritical fluid chromatography separation of spirocyclic terpenoid flavor compounds. *Journal of Chromatography A* 1363: 270–277.

207 Baranowska, I., Hejniak, J., and Magiera, S. (2016). Development and validation of a RP-UHPLC-ESI-MS/MS method for the chiral separation and determination of flavanone, naringenin and hesperetin enantiomers. *Talanta* 159: 181–188.

208 Martinez, S.E. and Davies, N.M. (2015). Stereospecific quantitation of 6-prenylnaringenin in commercially available H. lupulus-containing natural health products and dietary supplements. *Research in Pharmaceutical Sciences* 10 (3): 182–191.

209 Ahmed, M., Gwairgi, M., and Ghanem, A. (2014). Conventional Chiralpak ID vs. capillary Chiralpak ID-3 amylose tris-(3-chlorophenylcarbamate)-based chiral stationary phase columns for the enantioselective HPLC separation of pharmaceutical racemates. *Chirality* 26 (11): 677–682.

210 Yan, X., Zhou, Y., Tang, J. et al. (2016). Diketopiperazine indole alkaloids from hemp seed. *Phytochemistry Letters* 18: 77–82.

211 Prado-Cabrero, A., Beatty, S., Stack, J. et al. (2016). Quantification of zeaxanthin stereoisomers and lutein in trout flesh using chiral high-performance liquid chromatography-diode array detection. *Journal of Food Composition and Analysis* 50: 19–22.

212 Park, H.W., Cho, S.Y., Kim, H.H. et al. (2015). Enantioselective induction of SIRT1 gene by syringaresinol from Panax ginseng berry and Acanthopanax senticosus harms stem. *Bioorganic and Medicinal Chemistry Letters* 25 (2): 307–309.

213 Qiao, X., An, R., Huang, Y. et al. (2014). Separation of 25R/S-ergostane triterpenoids in the medicinal mushroom Antrodia camphorata using analytical supercritical-fluid chromatography. *Journal of Chromatography A* 1358: 252–260.

214 Speybrouck, D. and Lipka, E. (2016). Preparative supercritical fluid chromatography: a powerful tool for chiral separations. *Journal of Chromatography A* 1467: 33–55.

215 De Klerck, K., Mangelings, D., and Vander Hayden, Y. (2012). Supercritical fluid chromatography for the enantioseparation of pharmaceuticals. *Journal of Pharmaceutical and Biomedical Analysis* 69: 77–92.

216 Sardella, R., Ianni, F., Marinozzi, M. et al. (2017). Laboratory-scale preparative enantioseparations of pharmaceutically relevant compounds on commercially available chiral stationary phases for HPLC. *Current Medicinal Chemistry* 24 (8): 796–817.

217 Grybinik, S. and Bosakova, Z. (2021). An overview of chiral separations of pharmaceutically active substances by HPLC (2018–2020). *Monatshefte für Chemie* 152: 1033–1043.

218 Miller, L. (2012). Preparative enantioseparations using supercritical fluid chromatography. *Journal of Chromatography A* 1250: 250–255.

219 White, C. (2005). Integration of supercritical fluid chromatography into drug discovery as a routine support tool: part I. Fast chiral screening and purification. *Journal of Chromatography A* 1074 (1–2): 163–173.

220 Miller, L. and Potter, M. (2008). Preparative chromatographic resolution of racemates using HPLC and SFC in a pharmaceutical discovery environment. *Journal of Chromatography B* 875 (1): 230–236.

221 Ventura, M., Murphy, B., and Goetzinger, W. (2012). Ammonia as a preferred additive in chiral and achiral applications of supercritical fluid chromatography for small, drug-like molecules. *Journal of Chromatography A* 1220: 147–155.

222 Hamman, C., Schmidt, D.E. Jr., Wong, M., and Hayes, M. (2011). The use of ammonium hydroxide as an additive in supercritical fluid chromatography for achiral and chiral separations and purifications of small, basic medicinal molecules. *Journal of Chromatography A* 1218 (43): 7886–7894.

223 Hamman, C., Wong, M., Hayes, M., and Gibbons, P. (2011). A high throughput approach to purifying chiral molecules using 3 μm analytical chiral stationary phases via supercritical fluid chromatography. *Journal of Chromatography A* 1218 (22): 3529–3536.

224 Nguyen, S.T., Butler, M.M., Varady, L. et al. (2010). A concise, total synthesis and antibacterial evaluation of 2-hydroxy-1-(1H-indol-3-yl)-4-methylpentan-3-one. *Bioorganic and Medicinal Chemistry Letters* 20 (19): 5739–5742.

225 Nogle, L.M., Mann, C.W., Watts, W.L. Jr., and Zhang, Y. (2006). Preparative separation and identification of derivatized β-methylphenylalanine enantiomers by chiral SFC, HPLC and NMR for development of new peptide ligand mimetics in drug discovery. *Journal of Pharmaceutical and Biomedical Analysis* 40 (4): 901–909.

226 Toribo, L., Nozal, M.J., Bernal, J.L., and Nieto, E.M. (2003). Use of semipreparative supercritical fluid chromatography to obtain small quantities of the albendazole sulfoxide enantiomers. *Journal of Chromatography A* 1011 (1–2): 155–161.

227 Wang, Z., Jonca, M., Lambros, T. et al. (2007). Exploration of liquid and supercritical fluid chromatographic chiral separation and purification of Nutlin-3 – a small molecule antagonist of MDM2. *Journal of Pharmaceutical and Biomedical Analysis* 45 (5): 720–729.

228 Miller, L. and Bush, H. (1989). Preparative resolution of enantiomers of prostaglandin precursors by liquid chromatography on a chiral stationary phase. *Journal of Chromatography A* 484 (22): 337–345.

229 Hauck, W., Adam, P., Bobier, C., and Landmesser, N. (2008). Use of large-scale chromatography in the preparation of armodafinil. *Chirality* 20 (8): 896–899.

230 Quallich, G.J. (2005). Development of the commercial process for Zoloft®/sertraline. *Chirality* 17 (S1): S120–S126.

231 Ates, C., Surtees, J., Burteau, A.-C. et al. (2009). Oxopyrrolidine compounds, preparation of said compounds, and their use in the manufacturing of levetiracetam and analogues. US Patent 7,563,912.

232 Futagawa, T., Canvat, J.-P., Deleers, M. et al. (2000). Process for the preparation of levetiracetam. US Patent 6,107,492.

233 Bech Sommer, M., Nielsen, O., Petersen, H., et al. (2003). Method for the preparation of escitalopram. World Patent Appl. WO2003/006449.

234 Grinter, T. (2010). The development of an environmentally sustainable process for radafaxine. In: *Green Chemistry in the Pharmaceutical Industry* (ed. P. Dunn, A. Wells and M.T. Williams), 197–220. Weinheim: Wiley-VCH, Chapter 10.

235 Silva, B., Fernandes, C., Tiritan, M.E. et al. (2016). Chiral enantioresolution of cathinone derivatives present in "legal highs", and enantioselectivity evaluation on cytotoxicity of 3,4-methylenedioxypyrovalerone (MDPV). *Forensic Toxicology* 34 (2): 372–385.

236 Silva, B., Pereira, J.A., Cravo, S. et al. (2018). Multi-milligram resolution and determination of absolute configuration of pentedrone and methylone enantiomers. *Journal of Chromatography B* 1100–1101: 158–164.

237 Khater, S., Lozac'h, M.-A., Adam, I. et al. (2016). Comparison of liquid and supercritical fluid chromatography mobile phases for enantioselective separations on polysaccharide stationary phases. *Journal of Chromatography A* 1467: 463–472.

238 Toribio, L., Alonso, C., del Nozal, M.J. et al. (2006). Semipreparative enantiomeric separation of omeprazole by supercritical fluid chromatography. *Journal of Chromatography A* 1137 (1): 30–35.

239 Li, P., Wu, D.-R., Yip, H. et al. (2021). The effect of water on the large-scale supercritical fluid chromatography purification of two factor XIa active pharmaceutical ingredients. *Journal of Chromatography A* 1651: 462318.

240 Lin, Y., Fan, J., Ruan, L. et al. (2019). Semi-preparative separation of dihydromyricetin enantiomers by supercritical fluid chromatography and determination of anti-inflammatory activities. *Journal of Chromatography A* 1606: 460386.

241 Wu, D.-R., Yip, S.H., Li, P. et al. (2016). Additive free preparative chiral SFC separations of 2,2-dimethyl-3-aryl-propanoic acids. *Journal of Pharmaceutical and Biomedical Analysis* 131: 54–63.

242 Dai, J., Wang, C., Traeger, S.C. et al. (2017). The role of chromatographic and chiroptical spectroscopic techniques and methodologies in support of drug discovery for atropisomeric drug inhibitors of Bruton's tyrosine kinase. *Journal of Chromatography A* 1487: 116–128.

243 Zehani, Y., Lemaire, L., Millet, R., and Lipka, E. (2017). Small scale separation of isoxazole structurally related analogues by chiral supercritical fluid chromatography. *Journal of Chromatography A* 1505: 106–113.

244 Speybrouck, D., Howsam, M., and Lipka, E. (2020). Recent developments in preparative-scale supercritical fluid-and liquid chromatography for chiral separations. *Trends in Analytical Chemistry* 133: 116090.

7

Macrocyclic Antibiotics and Cyclofructans

Saba Aslani[1], *Alain Berthod*[2], *and Daniel W. Armstrong*[1]

[1] *Department of Chemistry and Biochemistry, University of Texas, Arlington, TX, USA*
[2] *Institute of Analytical Sciences, University of Lyon 1, CNRS, Villeurbanne, France*

7.1 Introduction

In the 1980s, the advent of commercial high-performance liquid chromatography (HPLC) chiral stationary phases (CSPs) ushered in a scientific/technological revolution that changed the pharmaceutical/medicinal industry, academic research, and virtually all other scientific fields where enantiomeric molecules were important [1–5]. Indeed, this early work provided the impetus for the Food & Drug Administration's implementation of new guidelines for the development of stereoisomeric drugs [6]. However, much of the early commercialized HPLC CSP work was not created in an intellectual vacuum. There were earlier reports using classic column chromatography and even early HPLC to separate enantiomers using immobilized amino acids, "pi-complex" (a.k.a. charge transfer) stationary phases [7–11], natural polymers like starch, cellulose, etc., and synthetic polymers with appended chiral amides, amino acids, and crown ethers [12, 13]. Many of the early chiral selectors were superseded by successive generations and more advanced types of CSPs. Only a few of the original CSPs are still used to any extent today [14–17]. Currently, the two CSP classes that have the broadest selectivity and are used more extensively are (a) percarbamylated derivatives of cellulose and amylose and (b) the macrocyclic glycopeptides [18–20]. This chapter will examine the macrocyclic glycopeptide (antibiotic) class of CSPs as well as a new class of macrocyclic CSPs based on cyclofructans.

Chiral Separations and Stereochemical Elucidation: Fundamentals, Methods, and Applications, First Edition. Edited by Quezia Bezerra Cass, Maria Elizabeth Tiritan, João Marcos Batista Junior, and Juliana Cristina Barreiro.

By the early 1990s, D.W. Armstrong was well aware that certain macrocyclic glycopeptide antibiotics were active against Gram-positive bacteria by binding to a D-alanyl-D-alanine peptide that was essential in bacterial cell wall formation [21, 22]. He thought that this macrocyclic glycopeptide could be a powerful chiral selector and designed and patented in a reliable way to bind it to spherical porous silica used to pack HPLC columns [23–25]. Under the trade name Chirobiotic, the macrocyclic glycopeptide-based chiral columns were successfully commercialized by the Advanced Separation Technologies, Inc. that subsequently was bought in 2006 by Sigma-Aldrich (Supelco) that became MilliporeSigma (Burlington, MA, USA) when Merck acquired Sigma in 2015. The MilliporeSigma company still markets the CHIROBIOTIC™ columns today [26].

By 2014, the macrocyclic glycopeptide chiral selectors were attached to 2.7 μm diameter superficially porous particles (SPPs) [27]. These were patented and marketed by a new company, AZYP, LLC. The SPP columns had shorter retention times, much higher efficiencies, and produced more effective chiral separations. Separations could be done in seconds and even sub-second [28, 29]. Furthermore, a new semisynthetic macrocyclic glycopeptide was introduced (i.e. the NicoShell column) that could do unparalleled chiral separations unmatched by any other known CSP [15, 19].

At the turn of the century, Armstrong developed another class of chiral selectors: the cyclofructans, a variety of cyclic oligosaccharides not related to the macrocyclic glycopeptide family [30–32]. This chapter will describe the structures and properties of both the commercially available macrocyclic glycopeptide selectors and the available cyclofructan-based CSPs. Since these two classes of chiral columns are mostly used in HPLC and also in SFC, many application examples will be found in other chapters describing enantiomeric separations by LC and SFC.

7.2 Macrocyclic Glycopeptides Physicochemical Properties

MilliporeSigma, Agilent, and AZYP LLC are the three commercial companies marketing chiral columns based on macrocyclic glycopeptide selectors. MilliporeSigma markets a variety of CHIROBIOTIC™ columns of different geometries, but all based on fully porous particle (FPP) bonded with four different macrocyclic glycopeptides selectors: ristocetin A (R), teicoplanin (T), vancomycin (V), and teicoplanin aglycone (TAG), the properties of which are listed in Table 7.1. Agilent and AZYP LLC market chiral columns filled with SPPs (or core-shell) based on the most powerful chiral selectors:

Table 7.1 Physicochemical structures and properties of commercially available macrocyclic glycopeptides chiral selector.

Compound name	Ristocetin A	Vancomycin	Teicoplanin	Teicoplanin aglycone
Formula	$C_{95}H_{110}N_8O_{44}$	$C_{66}H_{75}N_9O_{24}Cl_2$	$C_{88}H_{97}N_9O_{33}Cl_2$	$C_{58}H_{43}N_7O_{18}Cl_2$
m.w.	2066	1449	1878	1196
Produced by	*Nocardia lurida*	*Streptomyces orientalis*	*Actinoplanes teicomyceticus*	Chemical hydrolysis of teicoplanin
pK_a[a]	7.0; 7.3; 8.0; 8.3; 10.2; 10.8	2.9; 6.8; 7.7; 8.5; 10.4; 11.7	2.8; 7.6; 8.5; 8.9; 9.3; 11.0	2.7; 7.3; 7.8; 8.3; 8.9; 9.2; 9.8; 11.1
Isoelectric point	7.6	7.2	5.6	6.2
—OH groups (phenols)	21 (4)	9 (3)	14 (4)	7 (6)
Sugar moieties	2	2	3	0
Trade names[b] MilliporeSigma Astec	CHIROBIOTIC R	CHIROBIOTIC V and V2	CHIROBIOTIC T and T2	CHIROBIOTIC TAG
Trade names[c] Agilent	—	InfinityLab Poroshell 120 Chiral-V	InfinityLab Poroshell 120 Chiral-T	—
Trade names[d] AZYP LLC	—	VancoShell	TeicoShell	TagShell

[a] pK_a of ionizable groups including one —COOH (except ristocetin) and one $—NH_2$ groups (two in ristocetin A), plus phenol groups, and a secondary amine in vancomycin (see structures in Figure 7.1).
[b] The MilliporeSigma CHIROBIOTIC columns are filled with 5 μm fully porous particles. MilliporeSigma, 400 summit drive, Burlington, MA 01803, USA. https://www.sigmaaldrich.com/US/en
[c] The Agilent columns are filled with 2.7 μm superficially porous (core-shell) particles. Agilent, 5301 Stevens Creek Blvd, Santa Clara, CA 95051, USA. https://www.agilent.com/en/product/small-molecule-columns/chiral-columns/infinitylab-poroshell-120-chiral
[d] The AZYP LLC chiral columns are filled with 2.7 μm superficially porous (core-shell) particles. AZYP LLC. Planetarium place, CRB 313, Arlington, TX 76019, USA. https://www.azypusa.com/chiral-spp/

vancomycin, teicoplanin, teicoplanin aglycone, and the semisynthetic glycopeptide: NicoShell (Table 7.1).

Figure 7.1 shows the structural similarities observed in the effective macrocyclic glycopeptide selectors. The selector teicoplanin aglycone (TAG) was

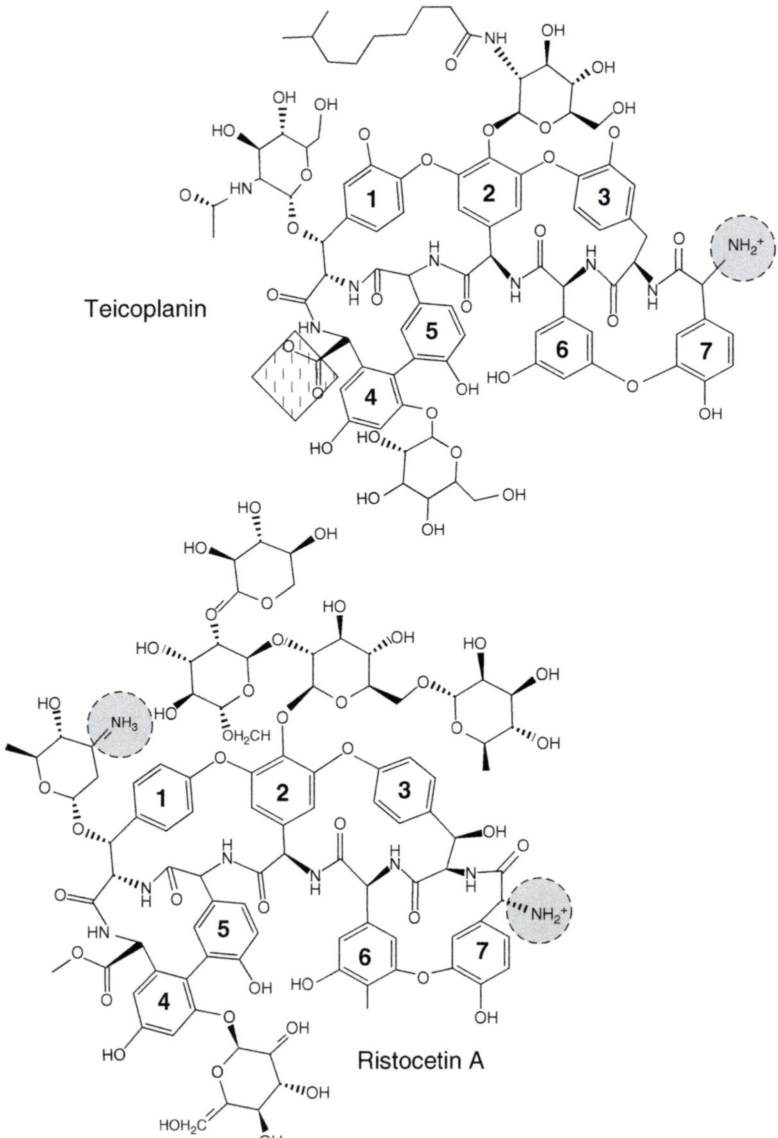

Figure 7.1 Structural similarities between macrocyclic glycopeptides. The numbered phenyl rings belong to the basket structure. The circled amine groups bear a positive charge. The dotted lozenges point to the negatively charged carboxylic groups.

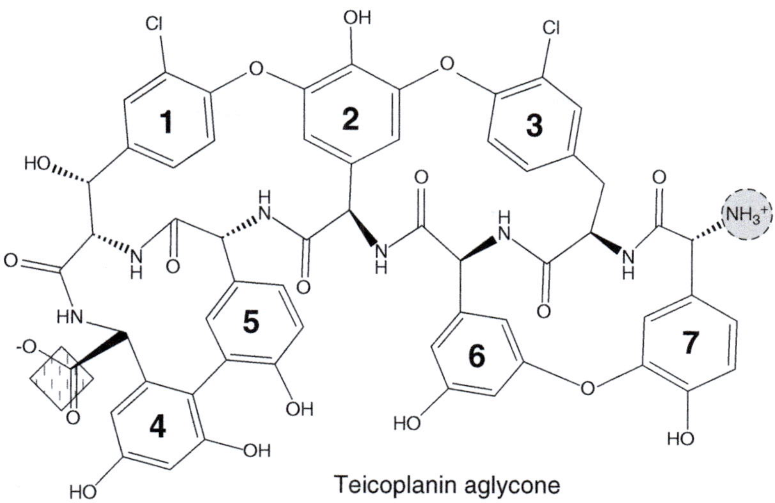

Teicoplanin aglycone

Vancomycin

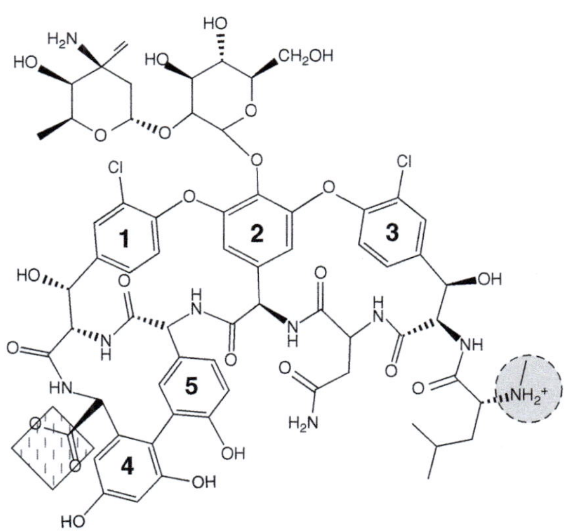

Figure 7.1 (Continued)

obtained by chemically removing the three moieties attached to teicoplanin. A three-dimensional basket shape is formed by the peptide chain linking the numbered phenyl rings (Figure 7.1). This TAG basket structure is obviously found in the original teicoplanin structure, but also exists in analogous form in the ristocetin A and vancomycin structures coming from different bacterial strains (Table 7.1). Phenyl rings 6 and 7 are open in the vancomycin structure. The TAG selector had chiral selectivities different from those of the original teicoplanin showing that, with macrocyclic glycopeptides chiral selectors, the sugar units attached to the central glycopeptide basket may have beneficial or deleterious effects on enantiomeric recognition [33].

An important physicochemical property of the macrocyclic glycopeptides is their charged state that changes widely with the environmental pH. Figure 7.2 shows the charge borne by the teicoplanin chiral selector when the bulk aqueous pH change from pH 1, strongly acidic, to pH 13, highly basic. Table 7.1 lists the six calculated pK_a values of teicoplanin: the lowest pK_a corresponds to the carboxylic acid groups, next are the four phenolic groups, and the last one is the primary amine pK_a. These pK_a values are marked by arrows in Figure 7.2 explaining the change in the teicoplanin electrical global charge. The isoelectric point, around pH 5.6, corresponds to a global neutrality of the molecule that is actually in a zwitterionic state with both a negatively charged carboxylate group and a positively charged amine group. Above the isoelectric point, the phenol groups start to ionize producing a negative global charge up to −5 reached when the amine group releases its proton above pH 11. Analogous behavior can be observed with all macrocyclic glycopeptide chiral selectors. This explains why the mobile phase pH has such a tremendous effect on chiral separations performed using these selectors.

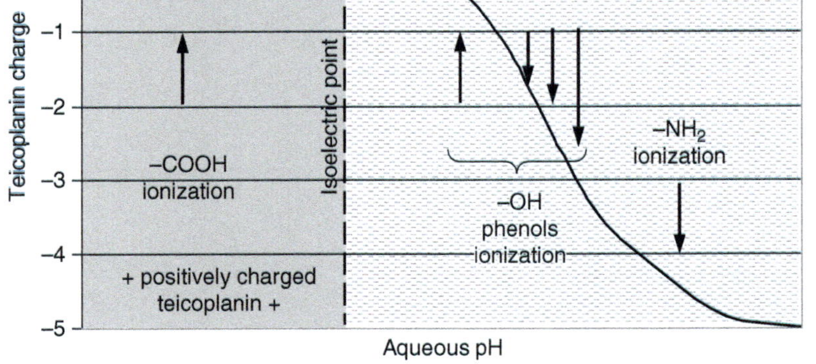

Figure 7.2 Evolution of the teicoplanin global charge with aqueous pH. The arrows locate the pK_a values of the different ionizable groups as listed in Table 7.1. The vertical dotted line is the isoelectric point, pK_i = 5.6, where teicoplanin has a net zero charge.

7.3 Using the Chiral Macrocyclic Glycopeptides Stationary Phases

7.3.1 Mobile Phases and Chromatographic Modes

Five different liquid chromatographic modes can be used with the antibiotic-based columns, namely (i) the reversed phase mode (RPLC); (ii) the normal phase mode (NPLC); (iii) the hydrophilic interaction chromatography mode (HILIC); (iv) the polar organic mode (POM); and (v) the polar ionic mode (PIM). The HILIC mode, which works with acetonitrile containing less than 30% water or buffer mobile phases, is not very useful for chiral separations with macrocyclic CSPs. The PIM mode is almost specific to this CSP class. The described changes in ionization state of the chiral selectors (Figure 7.2) are responsible for important changes in their enantio-recognition capabilities. Macrocyclic glycopeptide CSPs also can be effectively used in supercritical fluid chromatography (SFC). At least 5% and up to 40% organic modifier as well as traces of acids and bases must be added to CO_2 to obtain enantiomeric SFC separations [34].

The macrocyclic CSPs have intermediate polar character. The RPLC mode uses polar mobile phases associated with apolar stationary phases. Methanol/water or acetonitrile/water mixtures are the main RPLC mobile phases. The aqueous part must actually be a buffer of adjusted pH. The stationary phase/solute interactions can be coulombic, H-bonding, inclusion in the macrocyclic glycopeptide cavities, and steric exclusion effects. RPLC can be performed in the isocratic mode with a constant mobile phase composition or in gradient modes in which the proportion of the stronger organic solvent is increased as the chromatogram develops, providing better elution of the most retained compounds.

The NPLC mode is used with a polar CSP and apolar mobile phases. Alkanes such as hexane or heptane are mixed with small amounts of ethanol, or n- or iso-propanol. Steric effects, H-bonding, π–π stacking, and dipole–dipole interactions are involved in the chiral recognition of nonionizable molecules. It was recently demonstrated that trace amounts of water added to alkane–alcohol mobiles phases hydrated the macrocyclic glycopeptide CSPs so that the enantioselectivity was maintained or even improved, but with narrower peaks, showing dramatically increased efficiencies and faster separations [35].

The POM uses nonaqueous polar organic mobile phases, mainly acetonitrile, ethanol, or methanol mobile phases with small amounts of organic acids and bases. The solute separation is primarily adjusted based on solute solubility and H-bond interactions, which can be adjusted by adding less than 10% v/v of another polar solvent, e.g. protic methanol in aprotic acetonitrile or less-polar ethanol in methanol and adjusting acid/base additives. This mode works with ionizable and neutral solutes. Solute-CSP coulombic and H-bonding interactions first create diastereomeric intermediates that will allow for enantiomer differentiation. Also, π–π stacking, dipole–dipole, and steric interactions are involved.

The PIM was especially designed for the macrocyclic antibiotic class of CSPs optimizing drug analysis and clinical assays using mass spectrometric (MS) detection and using the tunable macrocyclic glycopeptide ionization state (Figure 7.2). The PIM mode uses anhydrous polar mobile phases, most often 100% methanol, with small amount of organic acids and bases added to optimize the ionization state of both the solutes and the stationary phase. The PIM mode is exclusively used with ionizable compounds either acids or bases, providing coulombic interactions but also H-bonding, π–π stacking, and steric interactions. The preparation of PIM mobile phases must be correctly done remembering that solvent volumes *versus* concentrations may greatly differ. For example, formic acid (FA) is easily volatile and desirable for MS detection. It has a molecular weight of 46, density of 1.22 $g\,mol^{-1}$, and molar volume of 37.7 $cm^3\,mol^{-1}$. When it is used with triethylamine (TEA), having molecular weight 101.2, density 0.727 $g\,mol^{-1}$, and molar volume 139.2 $cm^3\,mol^{-1}$, an FA/TEA 1/1 v/v mixture contains 7.2 mmol of triethylammonium formate (TEAF) and 19.3 mmol of FA. The 100/0.1/0.1 methanol/TEA/FA PIM mobile phase is actually 0.19 $mol\,l^{-1}$ FA solution containing 72 $mmol\,l^{-1}$ TEAF salt. An aqueous solution with such an acid–base composition would have a pH of about 3. Table 7.2 lists the composition of a variety of possible PIM mobile phases made with TEA and acetic or formic acid. It indicates the very different level of salt and acidity obtained when changing from acetic acid to formic acid. One ml of formic acid contains 26.5 mmol of acid, while one ml of acetic acid contains only 17.4 mmol of acid.

7.3.2 Chromatographic Enantioseparations

7.3.2.1 Amino Acids and Peptides

The macrocyclic glycopeptide CSPs are very effective in separating native amino acids and underivatized small peptides in the RPLC mode [36]. The TAG selector is the most effective in differentiating underivatized amino acid enantiomers. It was demonstrated that the sugar units attached to teicoplanin were actually hindering, somewhat, the approach of amino acids to the core of the selector [33]. Derivatization is used to enhance detection of amino acids at trace levels. Chromophores, such as 9-fluorenylmethyl chloroformate (FMOC), 3,5-dinitrobenzoyl (3,5-DNB), *tert*-butoxycarbonyl (*t*-Boc),

Table 7.2 Composition of polar ionic mobile phases used with antibiotic chiral stationary phases.

	Reagent physicochemical properties				
Compound	**M.W. ($g\,mol^{-1}$)**	**d ($g\,cm^{-3}$)**	**Molar volume ($cm^3\,mol^{-1}$)**	**mmol in a ml**	**pK_a in water**
Formic acid, FA	46	1.220	37.7	26.5	3.75
Acetic acid, AA	60	1.045	57.4	17.4	4.77
Triethylamine, TEA	101.2	0.727	139.2	7.2	10.75
Mobile phase composition v/v/v[a]	**Acid/base v:v**	**Acid/base, mmol/mmol**	**Excess acid, mmol**	**Salt TEAF or TEAA, mmol**	**Excess base, mmol**
PIM mobile phase compositions with formic acid and TEA					
100 : 0.05 : 0.3	1 : 6	1.33/2.16	0	1.33	0.83
100 : 0.05 : 0.15	1 : 3	1.33/1.08	0.25	1.08	0
100 : 0.05 : 0.1	1 : 2	1.33/0.72	0.61	0.72	0
100 : 0.05 : 0.05	1 : 1	1.33/0.36	0.97	0.36	0
100 : 0.1 : 0.4	1 : 4	2.65/2.87	0	2.65	0.22
100 : 0.1 : 0.2	1 : 2	2.65/1.44	1.22	1.44	0
100 : 0.1 : 0.1	1 : 1	2.65/0.72	1.93	0.72	0
100 : 0.2 : 0.6	1 : 3	5.30/4.31	1.00	4.31	0
100 : 0.2 : 0.3	2 : 3	5.30/2.16	3.15	2.16	0

(*Continued*)

Table 7.2 (Continued)

	Reagent physicochemical properties				
Mobile phase composition v/v/v[a]	Acid/base v:v	Acid/base, mmol/mmol	Excess acid, mmol	Salt TEAF or TEAA, mmol	Excess base, mmol
PIM mobile phase compositions with acetic acid and TEA					
100 : 0.05 : 0.3	1 : 6	0.87/2.16	0	0.87	1.28
100 : 0.05 : 0.15	1 : 3	0.87/1.08	0	0.87	0.21
100 : 0.05 : 0.1	1 : 2	0.87/0.72	0.15	0.72	0
100 : 0.05 : 0.05	1 : 1	0.87/0.36	0.51	0.36	0
100 : 0.1 : 0.4	1 : 4	1.74/2.87	0	2.87	1.13
100 : 0.1 : 0.2	1 : 2	1.74/1.44	0.30	1.44	0
100 : 0.1 : 0.1	1 : 1	1.74/0.72	1.02	0.72	0
100 : 0.2 : 0.6	1 : 3	3.48/4.31	0	4.31	0.83
100 : 0.2 : 0.3	2 : 3	3.48/2.16	1.33	2.16	0

TEAA, triethylammonium acetate; TEAF, triethylammonium formate.
The shaded compositions are for basic mobile phases. All other compositions are acidic.
[a] Acetonitrile or methanol can be used as solvent.

or 5-dimethylamino-1-sulfonyl (dansyl) derivatives, are added for enhanced UV detection. However, only the amine side of amino acids can be derivatized. The carboxylic acid group must be free as it provides the primary docking interaction with the positively charged amine of the macrocyclic selectors. Derivatizing the carboxylic acid of an amino acid completely negated its enantiomer recognition [37].

7.3.2.2 Chiral Compounds

Macrocyclic glycopeptide CSPs are very effective in separating enantiomers containing acidic or basic functionalities using coulombic interactions, aromatic rings for π–π interactions, polar groups, or dipolar groups for H-bond interactions. A complementary effect was observed with the different CSPs and especially between the teicoplanin and ristocetin A CSPs. Considering the structural similarities observed between these chiral selectors (Figure 7.1), it is not surprising that a partial enantiomeric separation of a compound on a given selector can provide a complete separation on another one. Table 7.3 lists the most effective mobile phase compositions for the different classes of analytes: neutral, acids, or bases. It is pointed out that the ristocetin A chiral selector (CHIROBIOTIC R columns) is unable to separate bases but is very effective in separating chiral acids.

With UV detection, it is recommended to start with the RPLC mode that is effective with all macrocyclic glycopeptide CSPs. In the RPLC mode, the recommended initial pH 4 water-rich composition may produce long solute retention times. Adjusting the methanol proportion, in either isocratic or gradient mode, will permit one to obtain the desired solute retention. When mass spectrometry is used for detection, the most effective mode with ionizable compounds is the PIM mode starting with the acidic 100/0.1/0.1 composition (Table 7.2). The POM mode also is useful with MS detection and neutral molecules. The NPLC mode provides only 10–15% of the published chiral separations performed on antibiotic CSPs.

As a general comment, macrocyclic glycopeptides CSPs must be part of the set of chiral columns available for assays in any laboratory involved in enantiomer separations. Macrocyclic glycopeptide CSPs have supplanted protein-based column having similar enantioselectivity as they are much more stable and versatile working in different LC modes or even with SFC mobile phases.

7.3.2.3 Particle Structure

As has been explained in the dedicated chapter on chiral separations by LC, recent developments in fast chromatographic separations were possible using smaller particles, short columns, and SPP (a.k.a. core-shell)

Table 7.3 Mobile phase compositions for optimal enantiomeric separations on the four macrocyclic glycopeptides selectors.

Compound type	Reversed phase	Normal phase	Polar organic	Polar ionic
Vancomycin				
Neutral	+	+	+	
Acids	+			
Bases	+			+
Ristocetin A				
Neutral	+	+		+
Acids	+		+	
Bases	no —COOH group, cannot separate bases			
Teicoplanin				
Neutral	+	+	+	
Acids	+			+
Bases	+			+
Teicoplanin aglycone				
Neutral	+	+	+	
Acids	+			+
Bases	+			+

Best starting RPLC composition: 20/80 v/v methanol/20 mmol l^{-1} ammonium formate buffer pH 4.
Best starting NPLC mobile phase composition: 80/20 v/v heptane/ethanol.
Best starting POM mobile phase composition: 100% pure methanol.
Best starting PIM: composition: 100/0.1/0.1 v/v methanol/acetic acid/triethylamine.

particles [15, 19, 27–29]. The great advantage of the SPP particles is that they give similar chromatography performance to FPP of much smaller diameters, but with much less driving pressure. The CHIROBIOTIC columns are packed with 5 μm FPP particles. The AZYP and Agilent chiral columns are packed with 2.7 μm SPP particles having a solid core of 1.7 μm and a shell thickness of 0.5 μm (Table 7.1). The enantiorecognition capability of the two types of particles is the same if they are bonded with the same selector.

Figure 7.3 compares the separation of the α-methyl-benzylamine enantiomers on a 25-cm CHIROBIOTIC V2 column [38] and a 10-cm VancoShell column. The CHIROBIOTIC V2 column, packed with 5 μm FPP vancomycin CSP, gives the highest enantioresolution factor Rs = 2.1 in 18 minutes

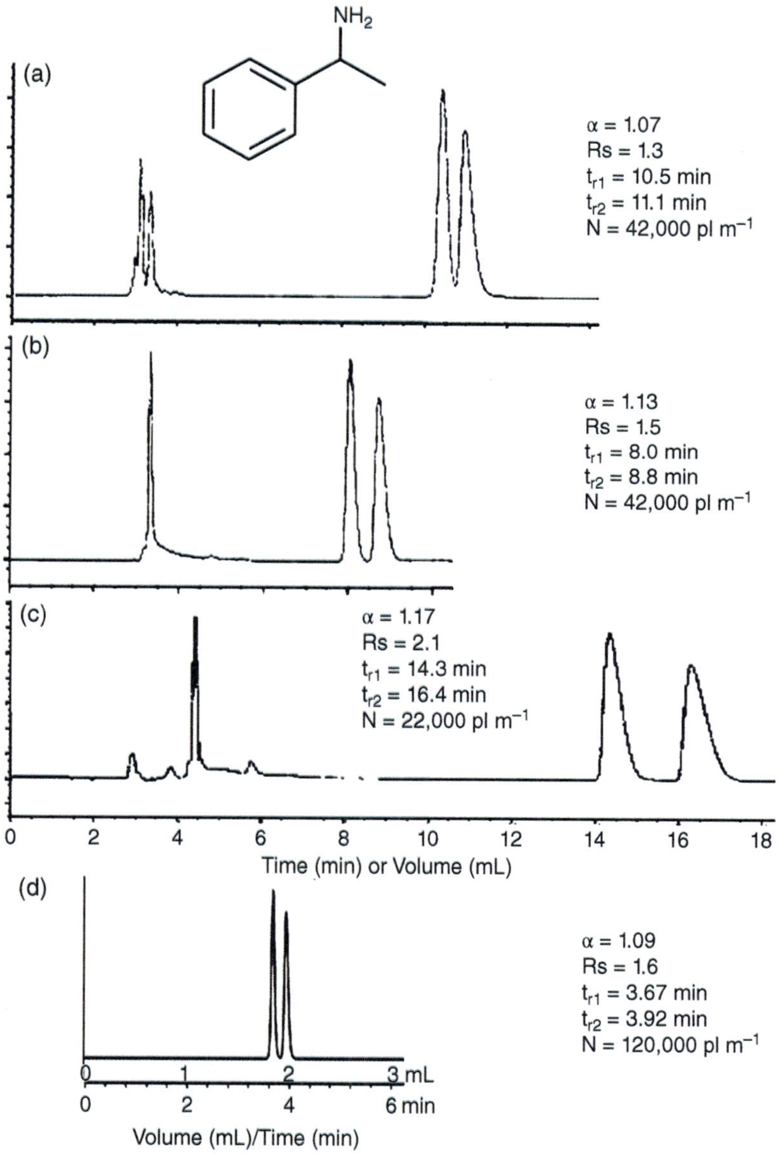

Figure 7.3 Separation of the enantiomers of α-methyl-benzylamine on a 25 cm × 0.46 cm 5 μm FPP CHIROBIOTIC V2 column (a, b, and c), and a 10 × 0.46 cm 2.7 μm SPP VancoShell column (d). Mobile phases: (a) PIM 100/0.05/0.05 methanol/NH_3/TFA, 1 ml min^{-1}, P = 50 bar; UV detection 254 nm. (b) RPLC 75/25 methanol/buffer NH_4TFA 0.15 mol l^{-1} pH 3.6, 1 ml min^{-1}, P = 110 bar; (c) RPLC 50/50 methanol/buffer NH_4TFA 0.15 mol l^{-1} pH 3.6, 1 ml min^{-1}, P = 140 bar; (d) RPLC 90/10 methanol/buffer NH_4TFA 0.15 mol l^{-1} pH 3.6, 0.5 ml min^{-1}, P = 90 bar.

and 18 ml of RP mobile phase (Chromatogram 7.3c). The highest resolution factor is obtained at the cost of long experiment duration. Increasing the organic modifier content in the RPLC mobile phase to 75% reduces the peak retentions (Figure 7.3b). The enantioresolution factor is also reduced. Figure 7.3a shows that the chromatogram obtained using a PIM mobile phase produced a very slight peak overlap. Figure 7.3d shows the chromatogram obtained in similar conditions with a modern VancoShell SPP column. The baseline separation is obtained in less than 4 minutes using about 2 ml of mobile phase. The SPP column efficiency is three to four times higher than that of the FPP column, partly due to the smaller 2.7 μm particles and also to the core-shell structure. The price is the very high driving pressure of the mobile phase. At 0.5 ml min^{-1} and with a much shorter 10-cm length, the 2.7 μm SPP column needs a driving pressure of 90 bar (1300 p.s.i) comparable to what was observed with the 5 μm FPP 25-cm columns (Figure 7.3). A similar 25-cm SPP column operated with the same mobile phase at 1 ml min^{-1} would need a huge 450 bar (6500 psi) pressure, four times higher than that of the FPP 5 μm pressure with the same operating conditions.

7.4 Using and Protecting Macrocyclic Glycopeptide Chiral Columns

7.4.1 Operating Conditions

Depending on the detection method used, it is always best to inject minimum amounts of sample. Typical injection volumes on 10–15 cm analytical columns are between 1 and 5 μl of 1 mg ml^{-1} solute concentration. Depending on the mobile phase viscosity, the typical operating pressures for a 10- x0.46-cm CHIROBIOTIC column operated at 1 ml min^{-1} flow are 25–35 bar (~400 psi), 30–40 bar (~500 psi), and 50–80 bar (~700–1100 psi) for the PIM or POM, the NPLC, and the RPLC modes, respectively. For a 2.7 μm column of the same geometry with the same mobile phase composition and flow rate, the operating pressure will be three to four times higher. The operating pressure is proportional to the column length and mobile phase viscosity. It is inversely proportional to the square of the particle diameter (Figure 7.3).

Care should always be exercised in carefully filtering the mobile phases since the column entrance frit acts as a filter. With poorly filtered mobile phases, microparticles will accumulate on the column head, producing an increased operating pressure. When such increased operating pressure is observed, the chiral columns can be operated temporarily in the reverse direction with an open outlet to help washing and removing accumulated material.

7.4.2 Storage

After RPLC use with buffers, the chiral column must be rinsed with 100% water to eliminate all traces of salts. Then it should be flushed with several column volumes of pure methanol or ethanol for storage. When not used, chiral columns should be thoroughly rinsed with pure methanol or ethanol and stored with both ends tightly sealed by end plugs.

7.5 Cyclofructans

7.5.1 Cyclofructan Structure and Properties

Cyclofructans are cyclic oligosaccharides consisting of β2-1 linked D-cyclofuranose units (Figure 7.4). They are produced by fermentation of inulin, a starchy polysaccharide, using an extracellular enzyme from a strain of *Bacillus circulans* OKUMZ31B or MCI-2554 [39]. Cyclofructans with degrees of polymerization from six to eight, abbreviated as CF6, CF7, and CF8, respectively, are found in commercially available cyclofructan raw material. Figure 7.4a reveals the inner part of the cyclofructan CF6 structure, which is an 18-crown-6 ether. Figure 7.4b shows that one side of the crown ether opening is completely blocked being covered by three primary alcohol groups of the attached cyclofuranoses tightly linked together by H-bonds.

Cyclofructans were first used as chiral agents in fast atom bombardment MS [40]. However, the native cyclofructans were ineffective and consequently methylated cyclofructan CF6 was used. Molecular inclusion in the hydrophilic crown-ether ring is a needed interaction for efficient enantiomeric recognition. As shown by Figure 7.4b (i.e. the hydrophilic CF6 side), the cyclofructan crown ether is not accessible in its native state. The primary alcohols of three cyclofuranose rings are tightly linked by H-bonds forcing them to cover the folded crown-ether cavity. The other side of the CF6 unit is more hydrophobic, but with ethylene groups covering the crown-ether ring. The primary cyclofuranose alcohols are also the most reactive CF6 groups for derivatization being the first to be methylated or derivatized. Such derivatization negates H-bonds allowing the methylated cyclofuranose units to open and making the hydrophilic crown ether ring

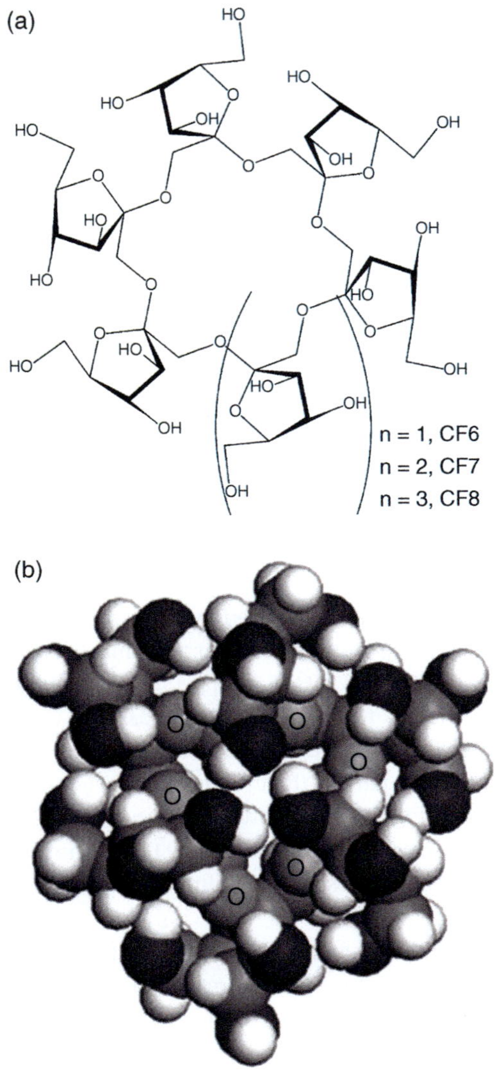

Figure 7.4 Cyclofructan structure. (a) Molecular structure of CF6. CF7 and CF8 would have one or two more cyclofuranose units, respectively. (b) Space filling model of CF6 showing how the internal crown-ether ring is covered by three primary alcohol groups tightly held by H-bonds. Atom shades: hydrogen, white; carbon, gray; ether oxygen of the crown ring, gray with a black circle; other oxygen, black.

accessible. Derivatized cyclofructans are efficient chiral selectors used in LC, GC, SFC, and CE [30–32, 39].

Table 7.4 lists the commercially available chiral columns based on cyclofructan-6 derivatized by isopropyl, noted P, or R-naphthyl-ethyl carbamate, noted RN. They can be packed with either FPP 5 μm particles with the trade name Larihc or with SPP 2.7 μm particles under the trade name LarihcShell. Native cyclofructan-6 was also bonded to FPP or SPP silica particles and

Table 7.4 Physicochemical structures and properties of commercially available cyclofructan 6-based chiral columns.

Derivative	Isopropyl	*R*-naphthyl-ethyl	Benzoic acid	Propyl sulfonic acid	None
Formula	—CO—NH—C$(CH_3)_2$	—CO—NH—C(CH_3)—naphthyl	—CH_2—phi—COOH	—$(CH_2)_3$—SO_3H	—
Number of substitution	4	6	4	4	0
Trade name AZYP LLC[a]	Larihc CF6-P	Larihc CF6-RN	—	—	Frulic-N
Particle diameter (μm)	5	5	—	—	5
Particle nature	FPP	FPP	—	—	FPP
Trade name AZYP LLC[a]	LarihcShell P	LarihcShell RN	FructoShell-B	FructoShell-S	FructoShell-N
Trade name Agilent[b]	InfinityLab Poroshell 120-CF	—	—	—	—
Particle diameter (μm)	2.7	2.7	2.7	2.7	2.7
Particle nature	SPP	SPP	SPP	SPP	SPP

FPP: fully porous particles; phi: phenyl ring; SPP superficially porous particles (core-shell). Shell thickness 0.5 μm.
[a] AZYP LLC, Planetarium place, Arlington, TX76019, USA. https://www.azypusa.com/
[b] Agilent, 5301 Stevens Creek Blvd, Santa Clara, CA 95051, USA. https://www.agilent.com/store/

packed in columns offered as Frulic-N and FructoShell-N, respectively, and is dedicated to HILIC separations. Similarly, cyclofructan-6 was derivatized with moderately acidic benzoic acid, noted B, and strongly acidic propyl sulfonic acid, noted S, but are only available as SPP columns, respectively, named FructoShell-B and FructoShell-S, for HILIC separations [41]. It will be shown that all Table 7.4 columns can be used with CO_2/methanol SFC mobile phases as well.

7.5.2 Chiral Separations with Cyclofructan-Based Stationary Phases

The cyclofructan CSPs were found most useful to separate the enantiomers of chiral amines [39, 42]. It is known that the hydrophilic cavity of the 18-crown-6 is an exact match for the 0.151 nm potassium cation. The ammonium cation has comparable ionic radii of 0.152 nm. It is then not surprising that the open crown ether cavity of the bonded cyclofructan is a very good primary docking station for ionized organic primary amines. The hydrophilic —OH groups, hydrophobic methylene groups, and cyclofuranose units of cyclofructan make different secondary and tertiary interactions with the two enantiomers of the chiral amine, allowing them to be separated. The enantiomers of secondary amines and even tertiary amines were also well separated by cyclofructan CSPs [42]. Unlike other chiral crown ether CSPs, the analyte amine group does not have to be in its cationic form. However, moderately acidic mobile phases, with pH between 3.5 and 5.5, are effective. With MS detection, the RPLC mobile phases can be made with acetate or formate triethyl ammonium buffers with pH between 4 and 5. The acetonitrile/methanol POM mobile phases must be acidic meaning that the proportion of acetic or formic acid must be similar to that of triethylamine (Table 7.2). The heptane/ethanol NPLC mobile phases must contain a small amount, e.g. 0.1%, of acetic or trifluoroacetic acid [42].

Beside amines, the cyclofructan CSPs separated a variety of neutral, acidic, and amphiprotic chiral compounds showing selectivity comparable to that of cyclodextrin CSP and approaching that of polysaccharide-based CSPs [43]. Atropisomers were also well separated by cyclofructan CSPs [44].

7.5.3 Cyclofructan Stationary Phases Used in the HILIC Mode

The HILIC mode separates polar and hydrophilic solutes using a polar stationary phase swept by an acetonitrile-rich mobile phase [45]. Native CF6 and cyclofructan-6 derivatized with anionic moieties were designed to be used in the HILIC mode and showed great capabilities [41]. The nucleic acids uracil and cytosine are, respectively, very poorly and moderately retained in HILIC mode. Comparing their retentions on different stationary phases allows estimating their hydrophobic/hydrophilic character. Similarly, the benzyl-trimethyl ammonium chloride solute is a salt whose positive cation retention is an indicator of the electrostatic character or ion-exchange capability of a stationary phase [41]. Figure 7.5 shows the separation of these three test solutes on the three commercially available FructoShell columns and compared to a simple bare silica column. The chromatograms show that the FructoShell-S column is the most polar, with all negatively charged sulfonic groups highly retaining the positively charged ammonium compound. The FructoShell-B is less polar, showing less retention of both the cationic solute 3 and cytosine. This HILIC column compares well to the bare silica column for hydrophilicity and surface charge. The unique character of the FructiShell-N column is evidenced by Figure 7.5. With the cytosine retention slightly higher than that on both bare silica and FructoShell-B, FructoShell-N shows little retention of the cationic ammonium solute 3. It means that there are no surface charges or accessible silanols on this stationary phase. The cyclofructan units cover the silica surface by a polar but molecular layer hindering access to the underlying residual silanols and making it an excellent HILIC stationary phase [41].

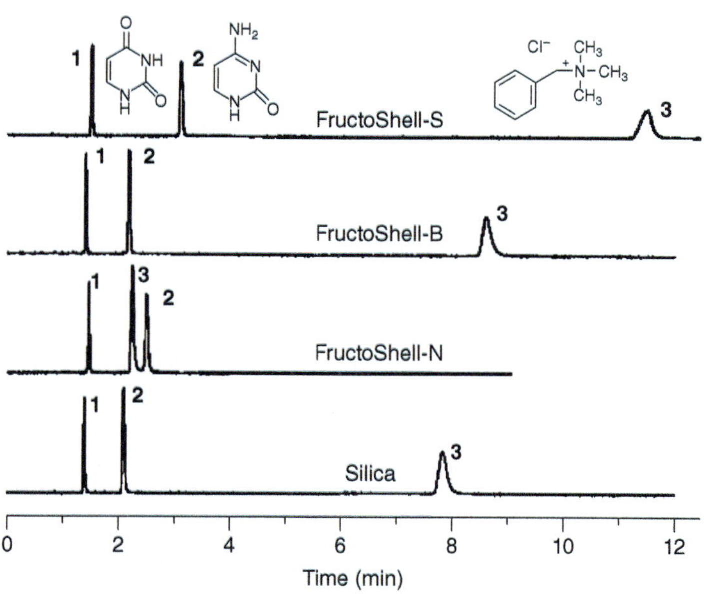

Figure 7.5 Assessing stationary phase polarity by separating 1-uracil, 2-cytosine, and 3-benzyl trimethyl ammonium chloride on the indicated 10 × 0.46 cm columns packed with 2.7 μm SPPs, HILIC mobile phase: acetonitrile/20 mmol l^{-1} ammonium acetate buffer pH 6.8, 80/20 v/v, 1 ml min^{-1}, UV detection at 220 nm.
Source: Adapted from Spudeit et al. [14].

Cyclofructan columns were successfully used in separating polar compounds in the HILIC mode. Compounds families included β-blockers, xanthines, polar acidic or basic drugs, nucleic acid bases, and nucleosides [41, 46]. β-Blockers are chiral; however, their enantiomers were not resolved: they were separated by the FructoShell-S column showing a single peak containing the two enantiomers [46]. The HILIC FructoShell-N columns could be tested for chiral separations in the HILIC mode since at least two primary alcohol groups of the native cyclofructan-6 unit were used for bonding to silica. This could make the internal crown ether accessible for solute docking.

7.5.4 Cyclofructan Stationary Phases Used in Supercritical Fluid Chromatography

Cyclofructan as well as macrocyclic glycopeptide columns can be used in SFC. Modern SFC is rarely performed using pure supercritical CO_2. Significant proportions of organic modifier, mainly methanol, are added to

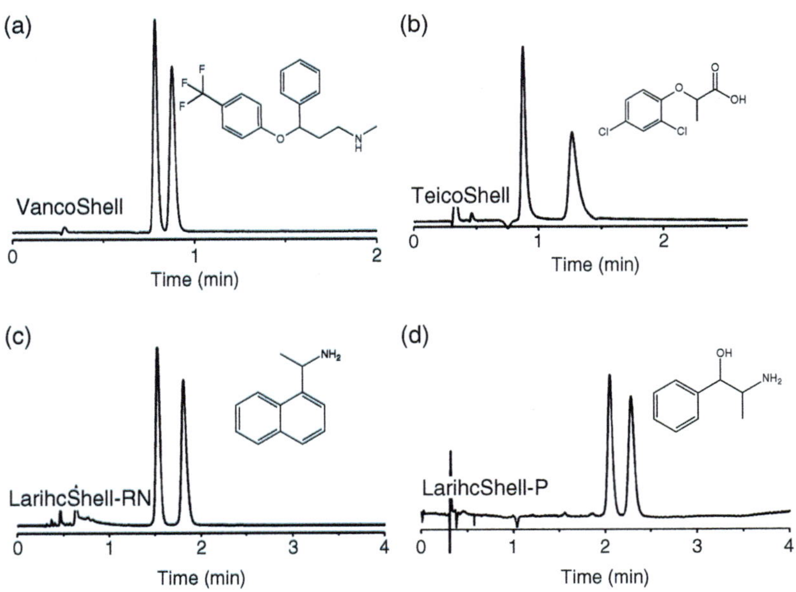

Figure 7.6 Fast supercritical fluid enantiomeric separations on the indicated 10 × 0.46 cm SPP chiral columns. Flow rate 4 ml min^{-1}, temperature 30 °C, backpressure 100 bar, detection UV 254 nm. (a) Fluoxetine. CO_2/methanol/water/TEA/TFA 75/25/3/0.1/0.1; (b) Dichloroprop, CO_2/methanol/TEA/FA 60/40/0.1/0.1; (c) 1-Naphthyl-ethylamine, CO_2/methanol/TEA/TFA 80/20/0.2/0.3; (d) Norephedrine, CO_2/methanol/TEA/TFA 80/20/0.2/0.3.

CO_2 changing the state from supercritical to subcritical but maintaining the SFC advantages of speed, efficiency, environmental friendliness, and ease of recovery of separated compounds. Figure 7.6 presents fast chiral SFC separations performed on a variety of SPP chiral columns [20]. With subcritical CO_2 mobile phases containing between 20 and 40% of methanol and at a flow rate of 4 ml min^{-1}, the presented enantiomers are baseline separated in less than 2 minutes. The fastest SFC chiral separation ever (not shown) was performed on a 5-cm LarihcShell-P column (2.7 μm SPP packing, CO_2 with methanol/TEA/TFA 20/0.2/0.3% v/v), baseline separating the enantiomers of 1,2,2-triphenylethylamine in 0.2 minutes (12 seconds) [20]. The low viscosity of the CO_2/methanol 80/20% with TEA/TFA 0.3/0.2% allowed having a 14 ml min^{-1} flow rate at room temperature with only 100 bar backpressure.

A wide variety of chiral compounds, including basic, neutral, and acid drugs, amino acids, indanes, and xanthanes, were enantioseparated on cyclofructan or antibiotic columns [20, 47]. The addition of small amounts of water (less than 5%) to the methanol organic modifier was found to enhance dramatically the column efficiency improving the amine peak symmetry. The column efficiency could reach 200,000 pl m^{-1} with 2.7 μm SPP core-shell particles corresponding to a reduced plate height of 1.9 particle diameter [47]. These CSPs used in SFC allow greener approaches toward solving separation problems.

7.6 Conclusions

Macrocyclic glycopeptide CSPs have proven to be versatile, broadly selective, sturdy, and efficient. These CSPs can be used in all HPLC modes. The reversed phase mode is preferred with the aqueous mobile phases allowing separation of a wide variety of enantiomers. However, the nonaqueous polar organic and polar ionic modes are also very effective with acidic and basic ionizable compounds. The enantiomers of a variety of chiral analytes also were separated using heptane/ethanol normal mobile phases. These CSPs also can be used in SFC with some polar methanol and acids or bases added to the CO_2 mobile phase. The recent development of SPPs, a.k.a. core-shell particles, bonded with macrocyclic antibiotics permit faster and higher efficiency chiral separations of many chiral compounds. Chiral columns based on macrocyclic antibiotic selectors, either FPP or SPP type, must be part of the column set of any laboratory involved in chiral separations.

The cyclofructan-based CSPs are very useful to separate amines and basic drugs. The active chiral selectors are derivatized cyclofructans so that the internal 18-crown-6 ether is accessible to compounds with amine groups. These cyclofructan CSPs work in RPLC, POM, and NPLC modes. Native cyclofructan bonded phases were very effective in the HILIC mode separating a variety of non-chiral polar compounds. All cyclofructan bonded phases could be used in SFC.

References

1 Pirkle, W.H. and Finn, J.M. (1981). Chiral high-pressure liquid chromatographic stationary phases. 3. General resolution of arylalkycarbionols. *J. Organomet. Chem.* 46 (14): 2935–2938. https://doi.org/10.1021/jo00327a019.

2 Armstrong, D.W. and DeMond, W. (1984). Cyclodextrin bonded phases for liquid chromatographic separation of optical, geometrical, and structural isomers. *J. Chromatogr. Sci.* 22 (9): 411–415. https://doi.org/10.1093/chromsci/22.9.411.

3 Okamoto, Y., Shiro, H., Heimei, Y. et al. (1984). Resolution, absolute configuration, and photoracemization of chiral troponoid acetals, cyclophepta[2,1-b:2,3-b']DI[1,4]benzoxazines. *Chem. Lett.* 13 (7): 1149–1152. https://doi.org/10.1246/cl.1984.1149.

4 Hermansson, J. (1983). Direct liquid chromatographic resolution of racemic drugs using α_1-acid glycoprotein as the chiral stationary phase. *J. Chromatogr. A* 269: 71–80. https://doi.org/10.1016/S0021-9673(01)90787-3.

5 Allenmark, S., Bomgren, B., and Boren, H. (1983). Direct liquid chromatographic separation of enantiomers on immobilized protein stationary phases III. Optical resolution of a series of N-aroyl D,L-amino acids by high-performance liquid chromatography on bovine serum albumin covalently bound to silica. *J. Chromatogr. A* 264: 63–68. https://doi.org/10.1016/S0021-9673(01)95006-X.

6 U.S. Food & Drug Administration Guidance Document (1992). Development of new stereoisomeric drugs.

7 Davankov, V.A., Bochkov, A.S., Kurganov, A.A. et al. (1980). Separation of unmodified α-amino acid enantiomers by reverse phase HPLC. *Chromatographia* 13: 677–685. https://doi.org/10.1007/BF02303437.

8 Klemm, L.H. and Reed, D. (1960). Optical resolution by molecular complexation chromatography. *J. Chromatogr. A* 3: 364–368. https://doi.org/10.1016/S0021-9673(01)97011-6.

9 Mikes, F., Boshart, G., and Gil-Av, E. (1976). Resolution of optical isomers by high-performance liquid chromatography, using coated and bonded chiral charge-transfer complexing agents as stationary phases. *J. Chromatogr. A* 122 (7): 205–221. https://doi.org/10.1016/S0021-9673(00)82245-1.

10 Mikes, F. and Boshart, G. (1978). Binaphtyl-2,2'-diyl hydrogen phosphate. A new chiral atropisomeric selector for the resolution of helicenes using high performance liquid chromatography. *J. Chem. Soc. Chem. Commun.* 4: 173–175. https://doi.org/10.1039/C39780000173.

11 Armstrong, D.W. (1984). Chiral stationary phases for high performance liquid chromatographic separation of enantiomers: a mini-review. *J. Liq. Chromatogr.* 7 (S2): 356–376. https://doi.org/10.1080/01483918408073942.

12 Blaschke, G. (1980). Chromatographic resolution of racemates. New analytical methods (17). *Angew. Chem. Int. Ed.* 19 (1): 13–24. https://doi.org/10.1002/anie.198000131.

13 Sogah, G.D.Y. and Cram, D.J. (1979). Host-guest complexation. 14. Host covalently bound to polystyrene resin for chromatographic resolution of enantiomers of amino acid and ester salts. *J. Am. Chem. Soc.* 101 (11): 3035–3042. https://doi.org/10.1021/ja00505a034.

14 Spudeit, D.A., Breitbach, Z.S., Dolzan, M.D. et al. (2015). Superficially porous particle based hydroxyprolyl-β-cyclodextrin stationary phase for high-efficiency enantiomeric separations. *Chirality* 27 (11): 788–794. https://doi.org/10.1002/chir.22526.

15 Hellinghausen, G., Roy, D., Lee, J.T. et al. (2018). Effective methodologies for enantiomeric separations of 150 pharmacology and toxicity related 1°, 2°, and 3° amines with core-shell chiral stationary phases. *J. Pharm. Biomed. Anal.* 5 (155): 70–81. https://doi.org/10.1016/j.jpba.2018.03.032.

16 Zhang, Y., He, L., Xuezhi, L. et al. (2019). Selective depletion of chiral 4-hydroxypraziquantel metabolites in three types of aquaculture fish by LC-MS/MS. *J. Agric. Food Chem.* 67: 4098–4104. https://doi.org/10.1021/acs.jafc.8b04728.

17 Hellinghausen, G., Readel, E.R., Wahab, M.F. et al. (2019). Mass spectrometry-compatible separations of 100 pesticides using core-shell chiral stationary phases and evaluation of iterative curve fitting models for overlapping peaks. *Chromatographia* 82: 221–233. https://doi.org/10.1007/s10337-018-3604-3.

18 Zhang, J., Sun, J., Liu, Y. et al. (2019). Immobilized cellulose-based chiralpak IC chiral stationary phase for enantioseparation of eight imidazole antifungal drugs in normal-phase conditions using high-performance liquid chromatography. *Chromatographia* 82: 649–660. https://doi.org/10.1007/s10337-019-03688-y.

19 Hellinghausen, G., Roy, D., Wang, Y. et al. (2018). A comprehensive methodology for the chiral separation of 40 tobacco alkaloids and their carcinogenic E/Z-(R,S)-tobacco-specific nitrosamine metabolites. *Talanta* 181: 132–141. https://doi.org/10.1016/j.talanta.2017.12.060.

20 Roy, D. and Armstrong, D.W. (2019). Fast super/subcritical fluid chromatographic enantioseparations on superficially porous particles bonded with broad selectivity chiral selectors relative to fully porous particles. *J. Chromatogr. A* 8 (1650): 360339. https://doi.org/10.1016/j.chroma.2019.06.060.

21 Fesik, S.W., O'Donnell, T.J., Gampe, R.T., and Olejniczak, E.T. (1986). Determining the structure of a glycopeptide-Ac_2-Lys-D-Ala-D-Ala complex using NMR parameters and molecular modeling. *J. Am. Chem. Soc.* 108: 3165–3170. https://doi.org/10.1021/ja00272a002.

22 Somma, S., Gastaldo, L., and Corti, A. (1984). Teicoplanin, a new antibiotic from Actinoplanes teicomyceticus nov. sp. *Antimicrob. Agents Chemother.* 26: 917–923. https://doi.org/10.1128/AAC.26.6.917.

23 Armstrong, D.W., Tang, Y., Chen, S. et al. (1994). Macrocyclic antibiotics as a new class of selectors for liquid chromatography. *Anal. Chem.* 66 (9): 1473–1484. https://doi.org/10.1021/ac00081a019.

24 Armstrong, D.W., Liu, Y., and Ekborgott, H. (1995). A covalently bonded teicoplanin chiral stationary phase for HPLC enantioseparations. *Chirality* 7 (6): 474–497. https://doi.org/10.1002/chir.530070614.

25 Armstrong, D.W. Macrocyclic antibiotics as separation agents. US Patents 08/198,409 February 22, 1994 and 09/187,369 November 16, 1998 and 10/712,249 November 13, 2003.

26 Macrocyclic glycoprotein stationary phases for chiral separations and highly selective achiral separations. https://www.sigmaaldrich.com/US/en/technical-documents/technical-article/analytical-chemistry/large-molecule-hplc/chirobiotic (accessed 16 August 2022).

27 Patel, D.C., Breitbach, Z.S., Wahab, M.F. et al. (2015). Gone in seconds: praxis, performance, and peculiarities of ultrafast chiral liquid chromatography with superficially porous particles. *Anal. Chem.* 87 (18): 9137–9148. https://doi.org/10.1021/acs.analchem.5b00715.

28 Wahab, M.F., Wimalasinghe, R.M., Wang, Y. et al. (2016). Salient sub-second separations. *Anal. Chem.* 88 (17): 8821–8826. https://doi.org/10.1021/acs.analchem.6b02260.

29 Patel, D.C., Wahab, M.F., O'Haver, T.C., and Armstrong, D.W. (2018). Separations at the speed of sensors. *Anal. Chem.* 90 (5): 3349–3356. https://doi.org/10.1021/acs.analchem.7b04944.

30 Sun, P. and Armstrong, D.W. (2010). Effective enantiomeric separations of racemic primary amines by the isopropyl carbamate-cyclofructan-6 chiral stationary phase. *J. Chromatogr. A* 1217 (30): 4904–4918. https://doi.org/10.1016/j.chroma.2010.04.079.

31 Kalikova, K., Janeckova, L., Armstrong, D.W., and Tesarova, E. (2011). Characterization of new *R*-naphthylethyl cyclofructan 6 chiral stationary phase and its comparison with *R*-naphthylethyl β-cyclodextrin-based column. *J. Chromatogr. A* 1218 (10): 1393–1398. https://doi.org/10.1016/j.chroma.2011.01.023.

32 Smuts, J.P., Hao, X.Q., Han, Z. et al. (2014). Enantiomeric separations of chiral sulfonic and phosphoric acids with barium-doped cyclofructan selectors via an ion interaction mechanism. *Anal. Chem.* 82 (2): 1282–1290. https://doi.org/10.1021/ac403686a.

33 Berthod, A., Chen, X., Kullman, J.P. et al. (2000). Role of the carbohydrate moieties in chiral recognition on teicoplanin-based LC stationary phases. *Anal. Chem.* 72 (8): 1767–1780. https://doi.org/10.1021/ac991004t.

34 Liu, Y., Berthod, A., Mitchell, C.R. et al. (2002). Super/subcritical fluid chromatography chiral separations with macrocyclic glycopeptides stationary phases. *J. Chromatogr. A* 978: 185–204. https://doi.org/10.1016/S0021-9673(02)01356-0.

35 Aslani, S., Wahab, F.M., Kenari, M. et al. (2022). An examination of water on normal phase enantioseparations. *Anal. Chim. Acta* 1200: 339608. https://doi.org/10.1016/j.aca.2022.339608.

36 Berthod, A., Liu, Y., Bagwill, C., and Armstrong, D.W. (1996). Facile liquid chromatographic enantioresolution of native amino-acids and peptides using a teicoplanin chiral stationary phase. *J. Chromatogr. A* 731: 123–137. https://doi.org/10.1016/0021-9673(95)01198-6.

37 Berthod, A., Qiu, H.X., Staroverov, S. et al. (2010). Chiral recognition with macrocyclic glycopeptides: mechanisms and applications. In: *Chiral Recognition in Separation Methods* (Ed. A. Berthod), 203–222. Heidelberg: Springer https://doi.org/10.1007/978-3-642-12445-7_7.

38 Beesley, T.E. and Lee, J.T. (2007). Method development and optimization of enantioseparations using macrocyclic glycopeptide chiral stationary phases. In: *Chiral Separation Techniques*, 3rd edn (Ed. G. Subramanian), 1–28. Weinheim: Wiley-VCH.

39 Wang, C., Sun, P., and Armstrong, D.W. (2010). Cyclofructans, a new class of chiral stationary phases. In: *Chiral Recognition in Separation Methods* (ed. A. Berthod), 77–96. Heidelberg: Springer.

40 Sawada, M., Takai, Y., Yamada, H. et al. (1995). Chiral recognition in host – guest complexation determined by the enantiomer-labeled guest method using fast atom bombardment mass spectrometry. *J. Am. Chem. Soc.* 117: 7726–7736. https://doi.org/10.1021/ja00134a017.

41 Yu, J.J., Wey, M., Firooz, S.K., and Armstrong, D.W. (2021). Ionizable cyclofructan 6-based stationary phases for hydrophilic interaction liquid chromatography using superficially porous particles. *Chromatographia* 84: 821–832. https://doi.org/10.1007/s10337-021-04063-6.

42 Padivitage, N.L.T., Dodbiba, E., Breitbach, Z.S., and Armstrong, D.W. (2014). Enantiomeric separations of illicit drugs and controlled substances using cyclofructan-based (LARIHC) and cyclobond I 2000 RSP HPLC chiral stationary phases. *Drug Test. Anal.* 6: 542–551. https://doi.org/10.1002/dta.1534.

43 Agathokleous, E.A., Stavrou, I.J., and Kapnissi-Christodoulou, C. (2022). Comparison of cyclofructan-, cyclodextrin-, and polysaccharide-based chiral stationary phases for the separation of pharmaceuticals. *Anal. Bioanal. Chem.* 414: 1323–1333. https://doi.org/10.1007/s00216-021-03754-1.

44 Frink, L.A., Berthod, A., Xu, Q.L. et al. (2016). Separation of 2-naphthol atropisomers on cyclofructan-based chiral stationary phases. *J. Liq. Chromatogr. Relat. Technol.* 39: 710–717. https://doi.org/10.1080/10826076.2016.1234395.

45 Alpert, A.J. (1990). Hydrophilic interaction chromatography for the separation of peptides, nucleic acids, and other polar compounds. *J. Chromatogr. A* 499: 177–196. https://doi.org/10.1016/S0021-9673(00)96972-3.

46 Padivitage, N.L.T. and Armstrong, D.W. (2011). Sulfonated cyclofructan 6 based stationary phase for hydrophilic interaction chromatography. *J. Sep. Sci.* 34: 1636–1647. https://doi.org/10.1002/jssc.201100121.

47 Firooz, S.K., Wahab, M.F., Yu, J.J., and Armstrong, D.W. (2021). High efficiency functionalized hydrophilic cyclofructans as stationary phases in sub/supercritical fluid chromatography. *Talanta* 232: 122308. https://doi.org/10.1016/j.talanta.2021.122308.

8

Cyclodextrins

Gerhard K. E. Scriba, Mari-Luiza Konjaria, and Sulaiman Krait

Department of Pharmaceutical/Medicinal Chemistry, Friedrich Schiller University Jena, Jena, Germany

8.1 Introduction

Cyclodextrins (CDs) are cyclic oligosaccharides discovered by the French pharmacist and chemist Antoine Villiers in 1891 as crystalline products from incubations of potato starch with *Bacillus amylobacter* [1]. Villiers named the products "*cellosines*" based on the similarity to cellulose, especially to withstand acid hydrolysis. Significant contributions to cyclodextrin chemistry were made by the Austrian chemist Franz Schradinger, who studied the degradation of starch by *Bacillus macerans*. He isolated two crystalline products, which he initially termed "*dextrin A*" and "*dextrin B*" and later "*crystallized dextrin α*" and "*crystallized dextrin β*"; γ-dextrin was first isolated in 1935. Upon determination of their cyclic structure as well as correct number of D-glucopyranosyl residues, the "*cyclohexaamylose*" and "*cycloheptaamylose*" were proposed for α-dextrin and β-dextrin, respectively. The term "*cyclodextrin*" was eventually coined by the German chemist Friedrich Cramer in 1949 [1] and has later been adopted by IUPAC [2]. Systematic names considering the fact that CDs do not have a free anomeric center were also proposed, for example, cycloheptakis-(1,4)-α-D-glycosyl for α-CD [1], but these names are rarely used.

The cyclic nature of the CDs had already been hypothesized by Schradinger. P. Karrer and coworkers were the first to suggest that the D-glucopyranose molecules were connected via α-1,4 glycosidic linkages.

Chiral Separations and Stereochemical Elucidation: Fundamentals, Methods, and Applications, First Edition. Edited by Quezia Bezerra Cass, Maria Elizabeth Tiritan, João Marcos Batista Junior, and Juliana Cristina Barreiro.

The cyclic structure was proven by K. Freudenberg and F. Cramer using optical rotation data as well as by D. French and coworkers and W. Borchert by X-ray crystallography in the late 1940s [1]. French also determined the correct number of D-glycosyl residues in α-CD and β-CD.

It was also recognized that CDs possess the structure of a hollow torus with a hydrophilic outside and a hydrophobic cavity allowing the formation of inclusion complexes with a large number of molecules. This property plus the fact that large quantities of the CDs can be produced industrially for a reasonable price resulted in numerous applications and the practical use in many industrial sectors such as pharmaceuticals, biotechnology, chemistry including catalysis, agrochemicals, cosmetics, food, toiletries, and textiles [3]. Upon derivatization of the hydroxy groups, many CD derivatives can be obtained with different properties compared to the native CDs. Furthermore, it was noted that CDs, being chiral compounds, can complex analytes in a stereospecific manner, which has led to their use as chiral selectors in the separation of stereoisomers, especially enantiomers. This chapter will highlight the use of CDs in the major separation techniques for stereoisomer analysis, i.e. chromatography including gas chromatography (GC), thin-layer chromatography (TLC), high-performance liquid chromatography (HPLC) as well as supercritical fluid chromatography (SFC) and especially electrophoretic techniques such as capillary electrophoresis (CE), where CDs have become the most often applied chiral selectors. In addition, further applications such as in membrane technologies as well as aspects of the structures of analyte-CD complexes will be addressed.

8.2 Structure and Properties

CDs are cyclic oligosaccharides composed of D-glucopyranose units linked by α-1,4 glycosidic bonds (Figure 8.1). They are produced by digestion of starch by the enzyme cyclodextrin glycosyl transferase (CGTase, EC 2.4.1.19) from various *Bacillus* strains and other bacteria [4]. The most abundant CDs, α-CD, β-CD, and γ-CD differ in the number of glucose units, i.e. six D-glucosyl molecules (α-CD), seven molecules (β-CD), and eight molecules (γ-CD) (Figure 8.1). CDs containing nine and more residues (typically up to about 30 glucose molecules although structures containing up to 100 glucosyl units have also been reported) are called "large ring CDs" [5]. These CDs are rarely applied in separation sciences and will not be covered in this chapter.

In native α-CD, β-CD, and γ-CD, the D-glucopyranose units are in the $^4C_1$ chair conformation resulting in a relatively rigid almost circular structure

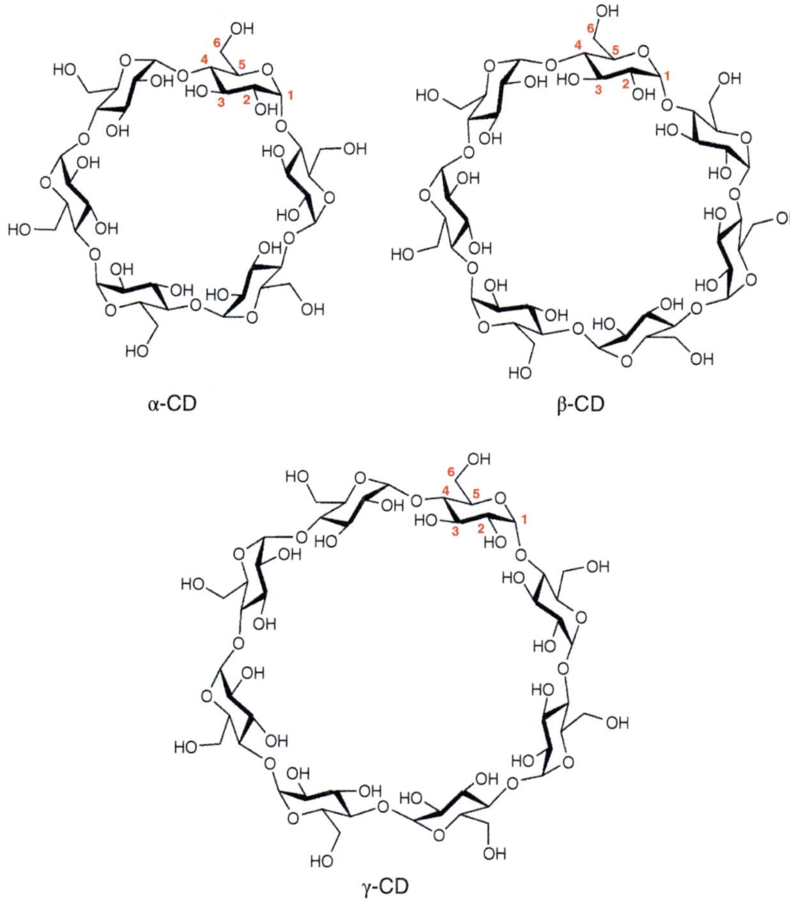

Figure 8.1 Structures of α-CD, β-CD, and γ-CD.

typically described as a hollow torus, i.e. a truncated conical cylinder with a hydrophilic outside surface and a lipophilic cavity [6]. The wider rim contains the secondary 2- and 3-hydroxy groups and is often termed the "secondary rim," while the narrower rim is formed by the primary 6-hydroxy groups, also called the "primary rim." The 2-OH group points toward the opening, while the 3-OH group points away from the rim. Furthermore, there are dynamical flip-flop hydrogen bonds between the 3-OH and the 2-OH group of adjacent D-glucose units, which impart structural rigidity and are considered as the reason for the much lower aqueous solubility of β-CD compared to α-CD and γ-CD [7]. γ-CD is more flexible than the other two CDs. In α-CD, the hydroxymethyl groups (C(6)-OH) tend to arrange in the

Table 8.1 Properties of native cyclodextrins.

	α-CD	β-CD	γ-CD
D-Glucopyranose units	6	7	8
Molecular formula (anhydrous)	$C_{36}H_{60}O_{30}$	$C_{42}H_{70}O_{35}$	$C_{48}H_{80}O_{40}$
Molecular weight (g mol^{-1})	972.8	1135.0	1297.1
Approximate cavity diameter (Å)	4.7–5.3	6.0–6.5	7.5–8.3
Approximate outer diameter (Å)	14.6	15.4	17.5
Approximate height (Å)	7.9	7.9	7.9
Approximate volume of cavity (Å^3)	174	262	427
Solubility in water (g/100 ml)	14.5	1.85	23.2
Approximate pK_a (25 °C)	12.33	12.20	12.08

Source: Adapted from Szejtli [6].

gauche-gauche (gg) conformation pointing outward of the macrocycle in order to avoid steric crowding. With the increase in size, the proportion of hydroxymethyl groups in the gauche-trans (gt) conformation increases so that the population of 6-OH groups pointing toward the opening increases and becomes dominant in γ-CD. The interior cavity is lined by the glycosidic oxygen atoms as well as the axial C(3)-H and C(5)-H rendering it lipophilic. The glycosidic oxygens are protruding into the cavity, so that the diameter of the rim formed by the glycosidic oxygens is smaller than the outer primary or secondary rims. Thus, the interior is not exactly a torus but has rather the form of a truncated hourglass. The distance from the cavity center formed by the glycosidic oxygens to the rim formed by the primary 6-OH groups is larger (approximately 0.31 nm) compared to the rim formed by the secondary OH groups (approximately 0.24 nm) [7]. In contrast to the cavity, the outside surface of the CDs is hydrophilic due to the hydroxy groups, which makes CDs water-soluble, although the solubility of β-CD is about 8 and 12 times lower than that of α-CD and γ-CD, respectively. Further properties as well as dimensions of the three native CDs are summarized in Table 8.1.

The hydroxy groups display different reactivities and selection of various regio- and site-specific chemical reactions including specific activation or protecting groups enable the synthesis of many derivatives. For example, the 2-OH group is the most acidic one, while the 6-hydroxy group is the least acidic on one hand, but the most accessible especially for spacious reagents. Substitution at the 3-OH group is the most difficult reaction. Consequently, a large number of so-called randomly substituted and single isomer CDs have been prepared [8–11]. Furthermore, it is possible to selectively derivatize an individual OH group of a single glucose residue of a CD. While the single isomer CDs as well as the CDs bearing a substituent at a single glucose molecule are defined compounds, randomly substituted CDs are a mixture of compounds with a different degree of substitution as well positional isomers. Nonetheless, companies are able to produce such CDs in a reproducible way with a comparable composition of the batches. Examples of commercially available CD derivatives, which are used as chiral selectors in CE have been compiled in Table 8.2. These comprise neutral derivatives as well as CDs containing chargeable or permanently charged substituents. Derivatization also increases the relatively low water-solubility of β-CD in most cases.

Table 8.2 Examples of commercially available cyclodextrin derivatives.

Derivative	Substituents
Neutral CDs	
Methyl-α-CD	CH_3, randomly substituted
Methyl-β-CD	CH_3, randomly substituted
Methyl-γ-CD	CH_3, randomly substituted
Heptakis(2,6-di-*O*-methyl)-β-CD	CH_3 in positions 2 and 6
Heptakis(2,3,6-tri-*O*-methyl)-β-CD	CH_3 in positions 2, 3, and 6
2-Hydroxypropyl-α-CD	CH_2—CH(OH)—CH_3, randomly substituted
2-Hydroxypropyl-β-CD	CH_2—CH(OH)—CH_3, randomly substituted
2-Hydroxypropyl-γ-CD	CH_2—CH(OH)—CH_3, randomly substituted
Negatively charged CDs	
Carboxymethyl-β-CD	CH_2—COONa, randomly substituted
Sulfated α-CD	SO_3Na, randomly substituted
Sulfated β-CD	SO_3Na, randomly substituted
Sulfated γ-CD	SO_3Na, randomly substituted
Sulfobutylether-β-CD	CH_2—CH_2—CH_2—CH_2—SO_3Na, randomly substituted
Succinyl-β-CD	CO—CH_2—CH_2—COOH, randomly substituted
Heptakis(6-sulfo)-β-CD	SO_3Na in position 6

(*Continued*)

Table 8.2 (Continued)

Derivative	Substituents
Heptakis(2,3-di-*O*-acetyl-6-sulfo)-β-CD	CH_3CO in positions 2 and 3, SO_3Na in position 6
Hexakis(2,3-di-*O*-methyl-6-sulfo)-α-CD	CH_3 in positions 2 and 3, SO_3Na in position 6
Heptakis(2,3-di-*O*-methyl-6-sulfo)-β-CD	CH_3 in positions 2 and 3, SO_3Na in position 6
Octakis(2,3-di-*O*-methyl-6-sulfo)-γ-CD	CH_3 in positions 2 and 3, SO_3Na in position 6
Positively charged CDs	
(2-Hydroxy-3-trimethylammoniopropyl)-α-CD	CH_2—CH(OH)—CH_2—$N(CH_3)_3Cl$, randomly substituted
(2-Hydroxy-3-trimethylammoniopropyl)-β-CD	CH_2—CH(OH)—CH_2—$N(CH_3)_3Cl$, randomly substituted
(2-Hydroxy-3-trimethylammoniopropyl)-γ-CD	CH_2—CH(OH)—CH_2—$N(CH_3)_3Cl$, randomly substituted
6-Monodeoxy-6-monoamino-β-CD	NH_2 instead of 6-OH group on one D-glucose unit
Heptakis(6-deoxy-6-amino)-β-CD	NH_2 in position 6

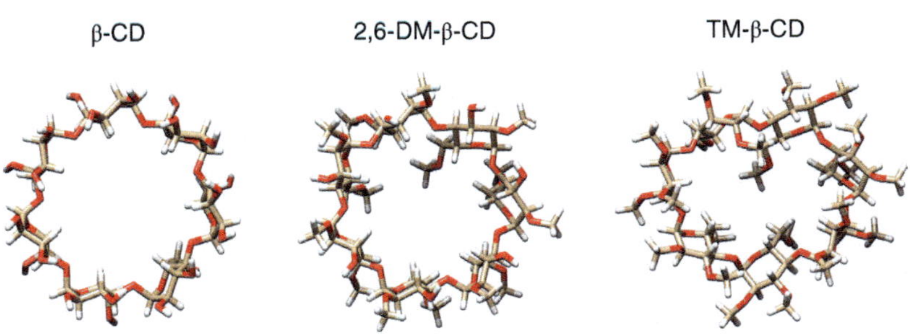

Figure 8.2 X-ray structures of β-CD, 2,6-DM-β-CD, and TM-β-CD (structures released from the Cambridge Structural Database entries AGAZOX [12], DEZMIE10 [13], GELKEN10 [14]).

Depending on the substitution, the round shape of the CD may become distorted. This is especially true for substitution of the 3-OH group. Figure 8.2 compares the molecular modeling structures of β-CD, *heptakis*(2,6-di-*O*-methyl)-β-CD (2,6-DM-β-CD), and *heptakis*(2,3,6-tri-*O*-methyl)-β-CD (TM-β-CD). Methylation of the 2-OH groups does not affect the formation of O(3)···O(2) hydrogen bonds which stabilizes the round structure of CDs. Thus, 2,6-DM-β-CD also has an almost round shape. However, methylation of the 3-hydroxy group increases the distance between O(3) and O(2), weakening the hydrogen bonds. Moreover, it has been shown by X-ray crystallography that one of the D-glucopyranose molecules undergoes a conformational change to the $^{0}S_{2}$-twist boat conformation in complexes [14], consequently, the shape of TM-β-CD is elliptic.

8.3 Cyclodextrin Complexes

The complexation ability of CDs already noted early on by Schradinger has been the foundation of the numerous applications of CDs not only in industrial technologies including pharmaceuticals, biotechnology, chemistry, agrochemicals, cosmetics, food, toiletries, and textiles but also in separation sciences. The technological utilizations rely on the inclusion of lipophilic guest molecules in the cavity of the CDs. However, as it will be discussed below in CE enantioseparations, there have also been cases where complexes of the external type were formed.

Due to the orientation of the D-glucopyranose molecules in the CDs, the cavity has a hydrophobic surface because it is lined by C—H bonds and the glycosidic oxygen bonds. Thus, inclusion complex formation is mediated via hydrophobic interactions within the cavity as well as hydrogen bonds with the polar groups on the CD rims [15]. Several studies have also shown a significant contribution of van der Waals interactions in the complex formation. Furthermore, inside the cavity, CDs have high electron density and Lewis-base character because the non-bonding electron pairs of the glycosidic oxygen atoms are directed toward the inside of the cavity. This enables dipole–dipole interactions between guest molecules and the CD as well as π–π interactions in case of aromatic analytes. Charged CDs can also engage in ionic (electrostatic) interactions. Furthermore, a conformational change upon complex formation ("induced fit") can contribute to CD complexation [15].

In addition to intermolecular interactions between CDs and analytes, the expulsion of so-called high-energy water from the cavity appears to play a role in inclusion complex formation [7]. In the absence of a guest molecule, the cavities of the CDs are occupied by water molecules. The number depends on the size, and a statistical occupancy of 2.6, 6.5, and 8.8 water molecules inside the cavity has been reported based on X-ray diffraction data for α-CD, β-CD, and γ-CD, respectively, but a higher number can be accommodated (Table 8.1). These water molecules are "unorganized" because only a limited number of hydrogen bonds are possible, less than the average of about four interactions per molecule as in bulk water. Thus, as the water molecules inside the cavity cannot form a stable hydrogen-bonded network, they are "energetically frustrated." Upon inclusion of a molecule, these water molecules are expelled from the cavity into the bulk phase, where they can form more hydrogen bonds. As a result, CD-solute complexation is mainly enthalpically driven unlike the classical entropically driven complexation process based on hydrophobic interactions [7]. In this context, the desolvation of the analyte from the (aqueous) solution or other interactions of the analyte molecule must also be considered.

The complexation of guest molecules by CDs has been studied extensively by spectroscopic as well as separation techniques so that a fairly good understanding of the structures of the complexes exist. A large number of reports on X-ray crystallographic data for the characterization of CD-solute complexes have been published providing direct evidence of the inclusion of guest molecules in the cavity. However, such structures may not reflect the exact structure of a given complex in solution. Therefore, an advantage is the fact that CDs are soluble chiral selectors. Thus, they can be studied in solution by spectroscopic methods such as NMR spectroscopy comparable to the separation conditions especially those applied in CE so that direct correlations between complex structures and enantioseparations can be drawn. Especially NMR techniques such as nuclear Overhauser enhancement spectroscopy (NOESY) and rotating frame Overhauser enhancement spectroscopy (ROESY) have contributed to the understanding because these methods allow the evaluation of the spatial proximity of atoms or functional groups of guest and host molecules [16–18]. Furthermore, molecular modeling and molecular dynamics simulation approaches in combination with enantioseparation studies have been performed in order to understand the chiral recognition mechanisms and to analyze the binding thermodynamics of complex formation [19, 20].

The combination of CE with NMR spectroscopy (and in some cases also molecular modeling) has been especially illustrative for enantioseparations in CE given the fact that essentially the same experimental conditions can be applied for the spectroscopic characterization of the complexes and the enantioseparation experiment as mentioned above. This is more difficult in chromatographic techniques where the CD is immobilized to a solid support so that the entry of the solute into the cavity may not be possible from either side. In fact, it has been reported that different results were obtained for CD-mediated enantioseparations in HPLC depending on the attachment of the CD to the solid support via the primary or secondary side. No clear evidence can be derived if an inclusion complex is formed or if the separation is based on superficial complexes due to interactions of the analytes with the outside of the CD rims.

Many publications describing the determination of the structure of the CD-analyte complexes by NMR spectroscopy have been performed in combination with CE [16, 17]. Thus, correlations between enantioseparations and complex structures could be analyzed. In most cases, formation of 1 : 1 complexes between analytes and CDs has been observed but higher order structures were also described. Complexation of guest molecules occurred by inclusion of lipophilic moieties of the analytes either via the narrower primary side or (more often) via the wider, secondary side of the torus. The structure of the CD-analyte complex depended on the type of CD as well as experimental conditions. This may also affect the chiral recognition of enantiomers by the different CDs, which can result in a different enantiomer migration order in CE. An example is illustrated in Figure 8.3 showing schematically the inclusion complexes formed between ketoprofen and *hexakis*(2,3,6-tri-*O*-methyl)-α-CD (TM-α-CD), TM-β-CD, and *octakis*(2,3,6-tri-*O*-methyl)-γ-CD (TM-γ-CD), respectively, as well as the corresponding electropherograms obtained with the CDs as chiral selectors in a sodium acetate buffer, pH 5.0, as background electrolyte (BGE) [21]. The structures of the complexes were derived from NMR data including ROESY measurements. The data indicated that the ketoprofen entered the cavity of TM-α-CD from the wider secondary side with the benzoyl moiety located deep inside the cavity, while in case of the other CDs, the drug was included via the narrower, primary side. In the complex involving TM-β-CD, the benzoyl ring was situated inside the cavity, while the phenyl ring was found inside the cavity of TM-γ-CD with the benzoyl moiety located outside the torus. Opposite migration order of the ketoprofen enantiomers was observed between TM-γ-CD and the other two CDs. In contrast, despite considerable structural differences between the ketoprofen complexes with TM-α-CD and TM-β-CD, identical enantiomer migration order was observed in CE. However, there are also several cases described in the literature when an opposite migration sequence of enantiomers could also be observed when only very minor differences in the structures of the complexes were concluded from NMR data [16, 22, 23]. Thus, the CE enantiomer migration order cannot be concluded from the structures of the complexes a priori, and, furthermore, small differences such as the depth of inclusion may result in opposite migration order.

Inclusion of lipophilic moieties of the guest molecules into the cavity is considered the primary way of complex formation in case of CDs and, consequently, the basis for enantioseparations. However, it has been shown that at least in CE inclusion complex formation is not a prerequisite for

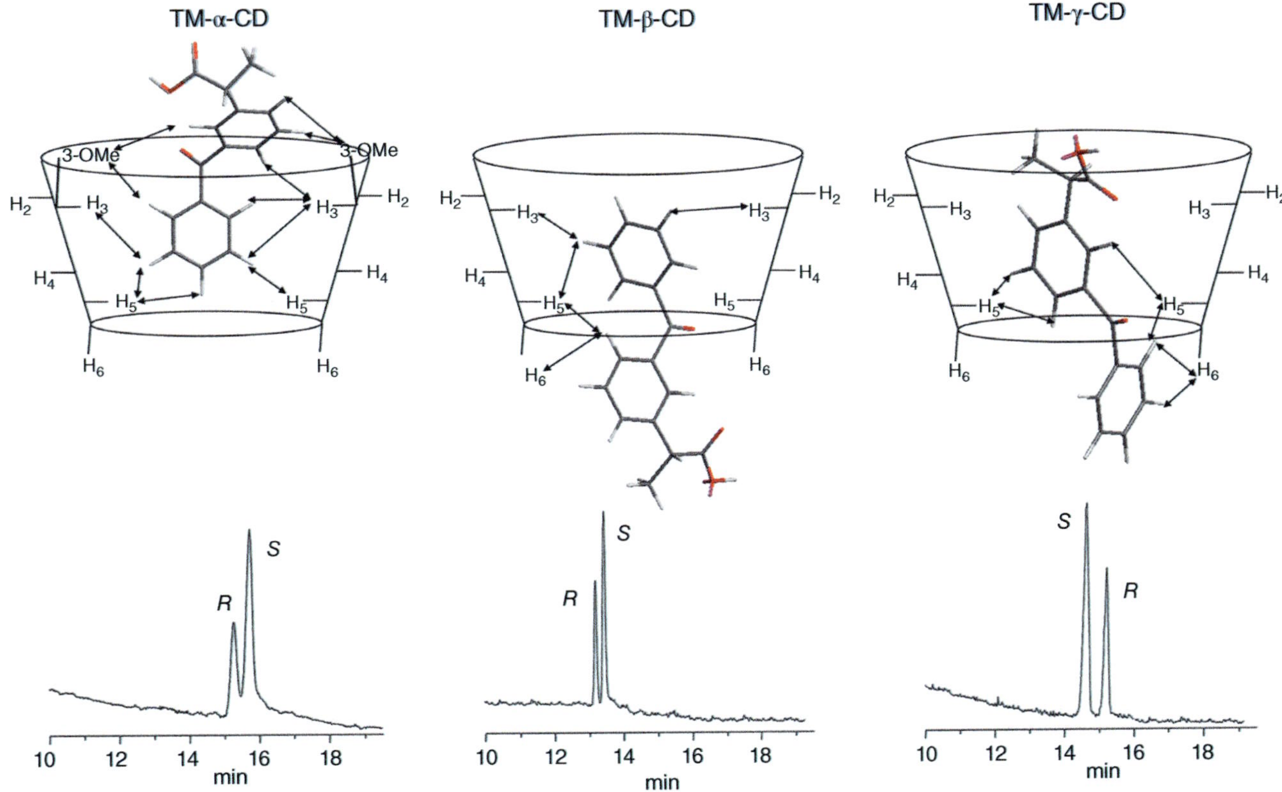

Figure 8.3 Schematic structures of the complexes formed between ketoprofen and *hexakis*(2,3,6-tri-*O*-methyl)-α-CD (TM-α-CD), *heptakis*(2,3,6-tri-*O*-methyl)-β-CD (TM-β-CD), and *octakis*(2,3,6-tri-*O*-methyl)-γ-CD (TM-γ-CD) (top) and electropherograms of the CE enantioseparations (bottom). The arrows indicate intermolecular NOEs between ketoprofen and the CDs. CE conditions: 50/60.2 cm, 50 μm ID fused-silica capillary; 60 mM sodium acetate, pH 5.0, containing 75 mM TM-α-CD, 50 mM TM-β-CD or 50 mM TM-γ-CD; 35 °C; 30 kV; detection at 254 nm. *Source:* Adapted from Samakaschvili et al. [21] with permission of Wiley.

chiral separations. External complexes displaying primarily the interactions between the analyte and the CD rims or only very weak NOE interactions of the analytes with the protons inside the cavity were derived by NMR spectroscopy, while effective enantioseparations were observed in CE. Such external complexes have been primarily observed either with sulfated CDs or in non-aqueous electrolytes [20, 22, 23], although the formation of an external complex between native α-CD and the drug terbutaline in aqueous media has been reported recently [24]. Moreover, formation of "shallow" complexes, where the analyte resides on top of the CD like a lid without entering deeply into the cavity has been described. The structure of a complex can also depend on the experimental conditions. An example is the complex of talinolol with *heptakis*(2,3-di-*O*-acetyl-6-*O*-sulfo)-β-CD (HDAS-β-CD) in aqueous and non-aqueous electrolyte solutions [25]. In the aqueous solution, ROESY NMR data indicated the inclusion of the phenyl ring into the cavity of the CD, while in a methanolic solution an inclusion complex could not be derived and only interactions of the *N*-isobutyl group with the wider rim of HDAS-β-CD were detected (Figure 8.4). Despite these apparent structural differences, identical enantiomer migration order was observed in the CE experiments.

The fact that a compound is complexed by different CDs in various ways is illustrated for terbutaline in Figure 8.5. Inclusion complexes are formed in most cases but in case of α-CD an external complex was observed. The interaction with γ-CD resulted in a "shallow" complex where terbutaline only interacted with the wider rim of the CD. Such behavior has not been

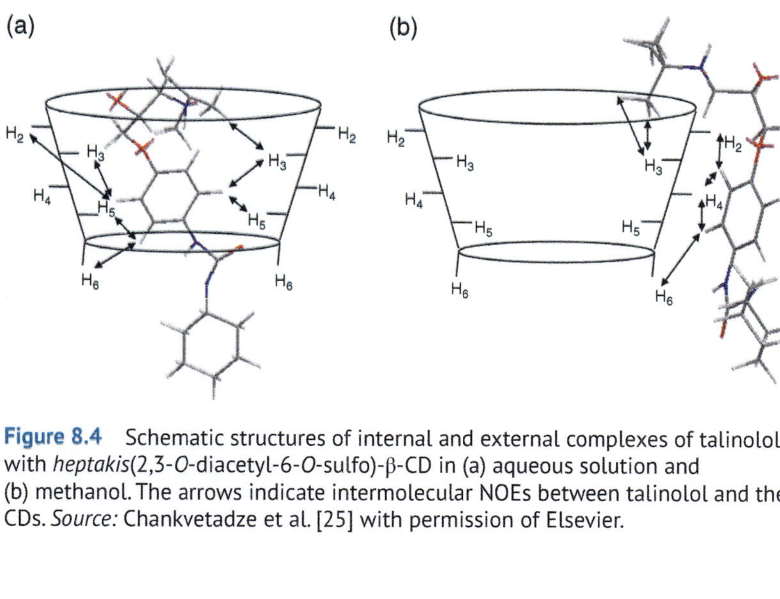

Figure 8.4 Schematic structures of internal and external complexes of talinolol with *heptakis*(2,3-*O*-diacetyl-6-*O*-sulfo)-β-CD in (a) aqueous solution and (b) methanol. The arrows indicate intermolecular NOEs between talinolol and the CDs. *Source:* Chankvetadze et al. [25] with permission of Elsevier.

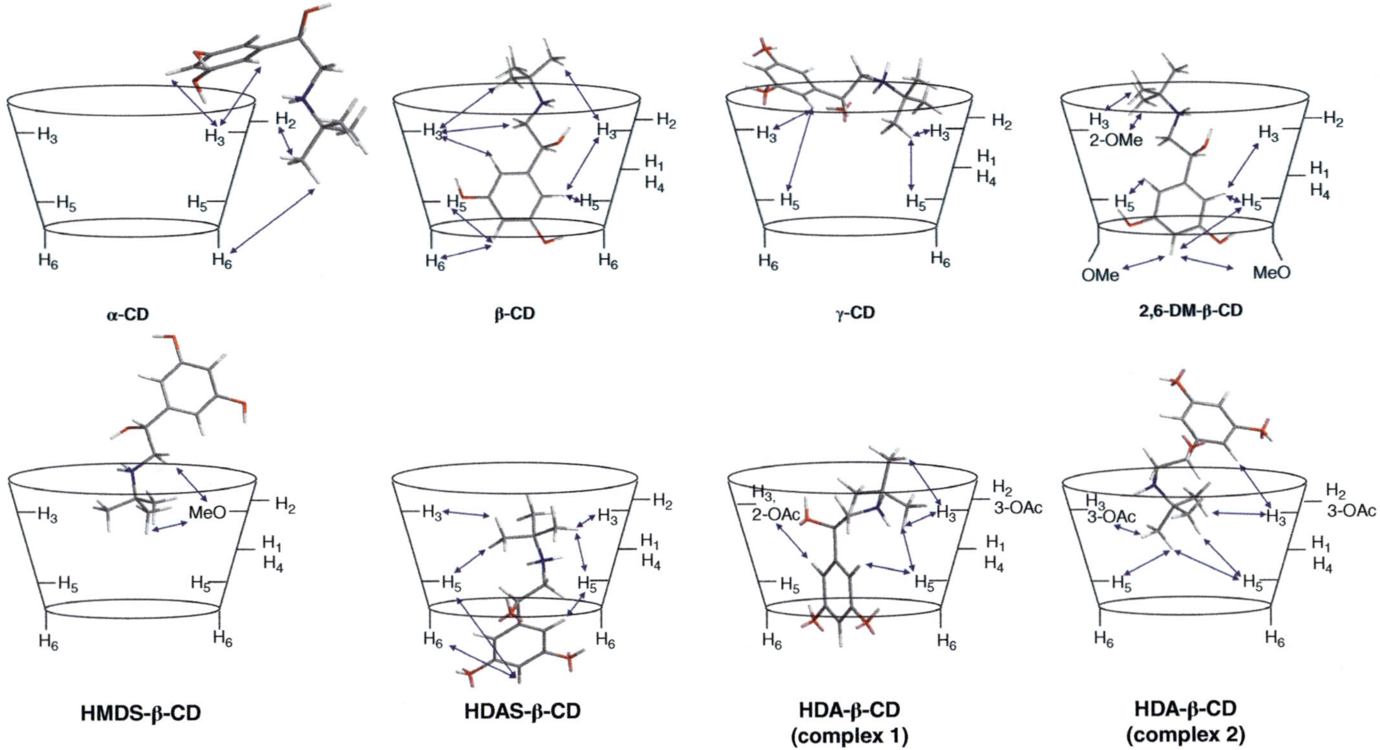

Figure 8.5 Schematic structures of the complexes of terbutaline with CDs. Complexes with α-CD, β-CD, γ-CD, and *heptakis*(2,3-di-*O*-acetyl)-β-CD (HDA-β-CD) are reprinted from Gogolashvili et al. [24] with permission of Wiley-VCH and complexes with *heptakis*(2-*O*-methyl-3,6-di-*O*-sulfo)-β-CD (HMDS-β-CD) and *heptakis*(2,3-*O*-diacetyl-6-*O*-sulfo)-β-CD (HDAS-β-CD) from Gogolashvili et al. [26] with permission of Elsevier. The structure of the 2,6-DM-β-CD/terbutaline complex was drawn from the data presented in Varga et al. [27]. The arrows indicate intermolecular NOEs between terbutaline and the CDs.

frequently observed in CE but is discussed as a mechanism in GC and HPLC. Inclusion complexes with the aromatic ring entering from the wider side of the CD were derived for β-CD and 2,6-DM-β-CD. In case of *heptakis*(2-*O*-methyl-3,6-di-*O*-sulfo)-β-CD (HDMS-β-CD) and *heptakis*(2,3-*O*-diacetyl-6-*O*-sulfo)-β-CD (HDAS-β-CD), the *N*-*tert*-butyl moiety was included into the cavity for the former via the secondary side not penetrating too far into the cavity, while this moiety entered via the primary side in case of HDAS-β-CD and was inserted more deeply. Data for *heptakis* (2,3-di-*O*-acetyl)-β-CD (HDA-β-CD) were inconclusive and could be interpreted as an inclusion complex resulting from the insertion of the aromatic moiety of the drug (complex 1) or insertion of the *tert*-butyl residue (complex 2).

Moreover, different complexes may be formed simultaneously between an analyte and a CD. This has been shown for the chiral separation of the dibenzoxepinopyrrole derivative asenapine by β-CD [28]. It was concluded from ROESY measurements that two different inclusion complexes were present in solution, one with the phenyl moiety of the drug inside the β-CD cavity and one with the chlorophenyl ring inside the cavity. In both cases, the compound entered the CD from the wider secondary side. Consequently, the observed CE enantioseparation was due to the average formation of both types of complexes.

As stated earlier, most often 1 : 1 complexes between CDs and solutes have been observed. Higher order complexes refer in most cases to the association of more than one CD or analyte molecule in a complex. Apart from 1 : 1 complexes, the composition 1:2 (analyte:CD) is the second most abundant [29], while a stoichiometry of 1:3, 2:1, or 2:2 is seldom observed. Moreover, ternary complexes including buffer additives such as surfactants like SDS or (chiral) ionic liquids could be observed. In most cases, the stoichiometry of higher order complexes has been derived from Job's plots and molecular modeling, but, unless NMR has been employed, the true nature of the complex cannot be reliably proven. Furthermore, complexes with different stoichiometry may be present simultaneously in a solution. The rare example of the formation of a 2 : 1 complex has been described between γ-CD and daclatasvir (DCV) [30]. 1D and 2D ROESY measurements provided evidence for the simultaneous presence 1 : 1 and 2 : 1 complexes illustrated in Figure 8.6. The NMR spectrum revealed the presence of three sets of signals for DCV (illustrated by the insert in Figure 8.6a) indicating three "species" of the molecule in the mixture. In addition, two "species" of γ-CD appeared to be present. NOEs observed between the aromatic H-2 and H-6 protons of one of the DCV "species" upon irradiation at the resonance of the same protons of another DCV "species" (shown as green arrows in Figure 8.6b) as well as the integration of the aromatic DCV signals and the

(a)

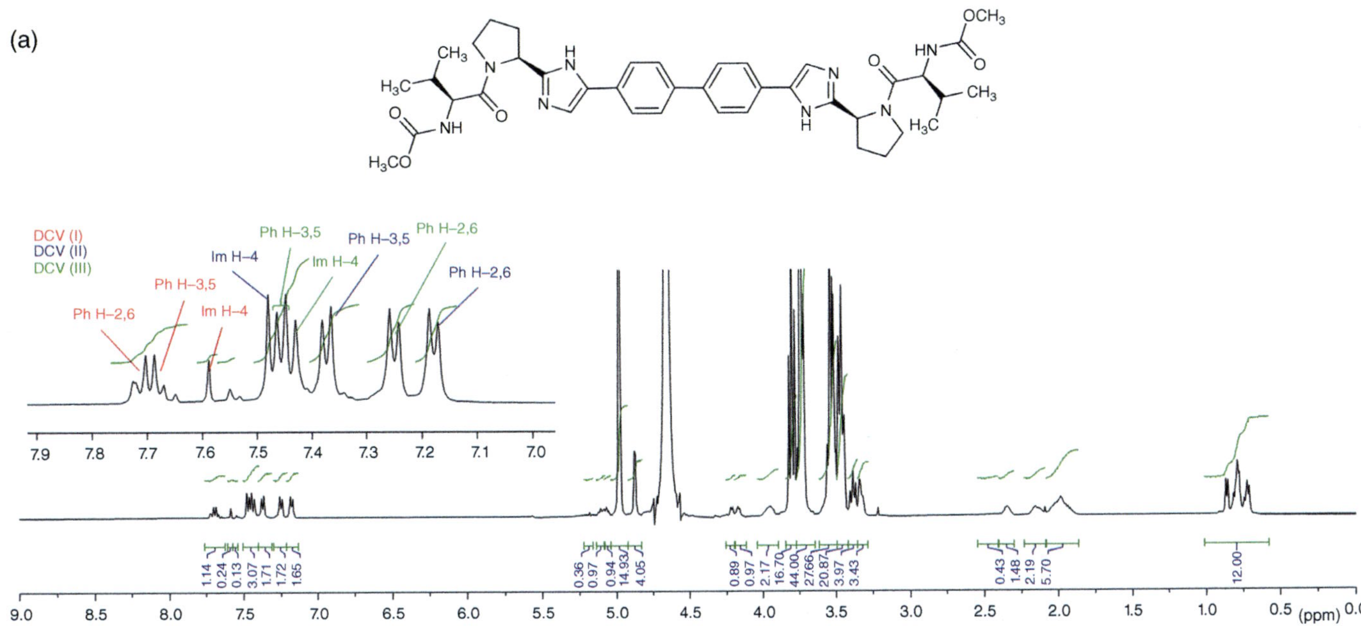

Figure 8.6 ^{1}H-NMR spectrum of a mixture of DCV and γ-CD in deuterated phosphate buffer in D_2O, apparent pH 2.5 (a) and schematic representations of the 1 : 1 and 2 : 1 complexes derived from ROESY experiments (b). The inset in the ^{1}H-NMR spectrum shows the expanded region of the aromatic protons. The arrows indicate intramolecular NOEs (red), intermolecular NOEs between DCV and γ-CD (blue), and intermolecular NOEs between DCV molecules in the 2 : 1 complex (green). *Source:* Adapted from Krait et al. [30] with permission of Elsevier.

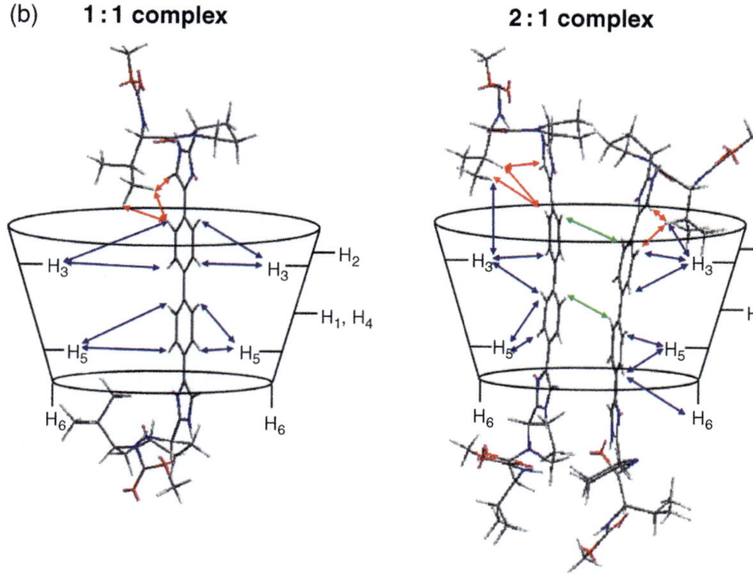

Figure 8.6 (Continued)

signal of the anomeric CD protons led to the conclusion that two DCV molecules are stacked inside the CD cavity with the aromatic rings in close vicinity to each other. NOEs were also observed between the "dual" DCV and one of the γ-CD species. The third set of DCV signals only displayed NOEs with the other γ-CD species. From further ROESY data, it was concluded that in both complexes the biphenyl moiety of DCV is included in the CD cavity with the MOC-Val side chains protruding on both sides (Figure 8.6b). Mass spectrometry confirmed the simultaneous presence of both complexes in solution. In CE experiments, two peaks with a plateau were observed when analyzing DCV in the presence of γ-CD in the BGE. However, the two peaks did not correspond to an enantioseparation but rather the slow complexation equilibrium of the drug. One peak represented free DCV, while the second one contained both the 1 : 1 and the 2 : 1 complexes, which co-migrated under the experimental conditions. Surprisingly, the complexes displayed the higher charge density as they migrated faster than the free, non-complexed solute. The case of DCV complexation by γ-CD is also an example that complexation does not necessarily translate into an enantioseparation.

As can be concluded from the published data on CD-analyte complexes, knowledge of the complex structure cannot be used for the prediction of the enantiomer migration sequence in CE or the elution order in chromatographic techniques but are useful for the understanding of the elution or migration behavior observed under the respective experimental conditions. For further examples and discussion of the structure of CD-analyte complexes as well as their effect on chiral separations, the reader is also referred to, for example, [16, 17, 20, 22, 23].

8.4 Application in Separation Science

Apart from technological use by the pharmaceutical, chemical, cosmetics, or textile industry, CDs have been frequently applied in separation sciences by chromatographic and electrophoretic methods. The premier application has been the separation of chiral isomers, especially enantiomers. CDs are composed of D-glucosyl residues so that they are chiral compounds. β-CD, for example, contains 35 stereogenic centers with a defined configuration. Consequently, CDs can bind analyte enantiomers in a stereospecific manner. The resulting diastereomeric complexes typically differ in the binding thermodynamics, which can be exploited for enantioseparations. Most often, chiral selectors are applied in separation sciences for enantioseparations (stereoisomer separations) due to their enantioselectivity (stereoselectivity). However, it should be kept in mind that very often the selectors also display chemoselectivity because binding thermodynamics also differ for structurally related or unrelated compounds. Thus, chiral selectors may enable the simultaneous separation of stereoisomers and other compounds.

8.4.1 Gas Chromatography

GC using capillary columns is a fast, sensitive, and high-resolution technique due to the narrow peak width, enabling the simultaneous separation of many analytes. A limitation of GC is the restriction to volatile and thermally stable analytes or their derivatives. Because separations cannot be affected by modifications of the polarity of the mobile phase as in other chromatographic techniques, (enantio)selectivity of the chiral stationary phase (CSP) becomes a predominant factor.

Although research on the application of CDs as CSPs in GC had been studied earlier, their use as chiral selectors was published for the first time in 1983 by Kościelkski and coworkers, who reported the enantioseparation of terpene hydrocarbons such as α- and β-pinene [31]. Further development of CDs as CSPs in GC occurred in the late 1980s to early 1990s and to date, CDs represent by far the dominant class of chiral selectors based on commercially available columns. In fact, there is only one popular non-CD chiral column, known as Chirasil-Val, which is based on valine-*tert*-butylamide. The selector is available with the L- or D-configuration and is widely used for enantioseparations of amino acids, α-amino alcohols, or α-hydroxy carboxylic acids.

8.4.1.1 Types of Cyclodextrins

Most of the CDs used as chiral selectors in GC are lipophilic derivatives in contrast to liquid-based enantioseparation techniques where hydrophilic CDs dominate, especially in CE. There are essentially two chemical types of substituents in the CD derivatives, alkyl groups such as methyl, pentyl, or *tert*-butyldimethylsilyl and more polar substituents such as acetyl, trifluoroacetyl, or butyryl groups [32–34]. Initially, peralkylated CDs such as *heptakis*-(2,3,6-tri-*O*-methyl)-β-CD or *heptakis*-(2,3,6-tri-*O*-pentyl)-β-CD as well as the corresponding α-CD and γ-CD analogs were applied. Subsequently, CDs containing different substituents at individual hydroxy groups of the D-glucopyranose units were developed including 2,3-di-*O*-acetyl-6-*O*-*tert*-but-yldimethylsilyl-β-cyclodextrin, 2,3-di-*O*-methyl-6-*O*-*tert*-but-yldimethylsilyl-β-CD, or 2,6-di-*O*-pentyl-3-*O*-trifluoroacetyl-β-CD as well as analogous α-CD or γ-CD derivatives. With regard to cavity size, β-CD derivatives are more frequently applied than γ-CD, while only few columns are based on α-CD derivatives. More than 30 CD-based chiral GC columns have been commercialized, examples of selectors have been compiled in Table 8.3. Further developments of CDs as chiral selectors in GC included the synthesis of CD-based ionic liquids by introducing permanently positively charged butylimidazolinium or trialkylphosphonium substituents [35]. These selectors showed good enantioselectivity and chromatographic performance, but they have not been commercialized to date.

8.4.1.2 Types of Columns

Several approaches have been realized in the manufacture of chiral CD-based GC columns. The first approach utilizes CD derivatives with low melting points. These comprise partially or fully pentylated derivatives, which are viscous liquids with good wetting ability so that they can be directly coated onto the inner surface of fused-silica capillaries. This is referred to as undiluted selector columns. The method was first applied by W. König and coworkers in 1988 coating per-*O*-pentylated α-CD and 3-*O*-acetyl-2,6-di-*O*-pentyl-α-CD or -β-CD to glass capillaries [36]. Another example is 2,6-di-*O*-pentyl-3-*O*-trifluroacetyl-α-CD, -β-CD, or -γ-CD introduced by D. W. Armstrong and colleagues in 1990 [37]. The undiluted selector columns exhibit enantioselectivity toward a wide range of analytes and are thermally stable up to 180–200 °C.

Table 8.3 Examples of CDs used as chiral selectors in commercial GC columns. The information was compiled from manufacturer websites.[a]

Cyclodextrin	Type[b]	Dominant mechanism	Examples of compound classes
2,3,6-Tri-*O*-methyl-β-CD	I, II, III	Inclusion	Alcohols, diols, esters, lactones, ketones, olefins, and hydrocarbons
2,3,6-Tri-*O*-pentyl-β-CD[c]	I	Inclusion	Alcohols, olefins, esters, and alkyl halogenides
2,6-Di-*O*-pentyl-3-*O*-methyl-β-CD[d]	I	Inclusion	Heterocycles, amines, amino acid derivatives, and epoxides
2,6-Di-*O*-methyl-6-*O*-*tert*-butyldimethylsilyl-β-CD[e]	I, II	Inclusion and surface	Amino acids, (aromatic) acids, amines, amino alcohols, alcohols, diols, esters, epoxides, and ketones
2,3-Di-*O*-acetyl-6-*O*-*tert*-butyldimethylsilyl-β-CD[e]	II	Surface and inclusion	Alcohols, esters, lactones, ketones, aldehydes, and lactones
2,6-Di-*O*-pentyl-3-trifluoroacetyl-γ-CD[f]	I	Surface	Amino acids, acids, amines, alcohols, polyols, esters, lactones, epoxides, and halogenated compounds
2,6-Di-*O*-pentyl-3-propionyl-γ-CD	I	Surface	Amines, esters, lactones, epoxides, higher alcohols, and diols

[a] Merck/Suppelco (http://www.sigmaaldrich.com/analytical-chromatography/chiral-chromatography), Machery-Nagel (http://www.mn-net.com/de/chromatographie/gaschromatographie-gc), and Restek (http://www.restek.com/row/products/columns/gc-columns).
[b] I, undiluted, coated to capillary surface; II, embedded in polysiloxane coated to capillary surface; III, covalently bound.
[c] α-CD derivative also available.
[d] α-CD and γ-CD derivatives also available.
[e] γ-CD derivative also available.
[f] α-CD and β-CD derivatives also available.

The second type of chiral columns pioneered by V. Schurig and coworkers [38] has been referred to as diluted selectors because the selector is dissolved in a polysiloxane, and this mixture is coated onto the capillary wall, thus combining the stereoselectivity of the CDs with the properties of the polysiloxanes. This results in columns with high efficiency and resolution and thermal stability of up to 230 °C, in some cases up to even higher temperatures. The CD derivatives often (but not exclusively) feature a *tert*-butyldimethylsilyl substituent in position 6 of the D-glucopyranoses, which improves the solubility of the CD in the polysiloxane matrix. It has been recognized that a large and bulky substituent in position 6 has little effect on the chiral selectivity of the CDs so that this site is used for the introduction of the *tert*-butyldimethylsilyl substituent. As polysiloxane, low- to mid-polar phases such as DB 5 (5% phenyl 95% methylpolysiloxane), DB 35 (35% diphenyl 65% methylpolysiloxane), or DB-1701 (14% cyanopropylphenyl 86% methylpolysiloxane) are used. In the case of diluted selector columns, the separation factor (α) of an enantioseparation depends on the concentration of the CD in the polysiloxane. It has been shown experimentally and theoretically that α does not increase linearly because enantioselectivity improves with increasing the CD concentrations, but efficiency decreases because of poor mass transfer [39, 40]. Thus, the optimum of the selectivity is reached at relatively low CD concentrations. Apart from polysiloxanes, ionic liquids served as solvents for CDs. Although such columns also showed good enantioselectivity, they were not commercialized.

CSPs have also been prepared by covalent attachments of the CDs to the polysiloxane, the so-called grafted selectors. The advantage of this approach is the fact that apolar siloxanes can be used so that the analysis time of polar analytes at low temperatures can be reduced. Moreover, the concentration of the selector is not limited by its solubility. However, thermal stability is only increased moderately and some loss of enantioselectivity is observed. In the commercial product (Chirasil-β-Dex™), permethylated β-CD is attached via a C11-linker to dimethylpolysiloxane.

In addition to single selectors, the mixed binary selector approach combines the enantioselectivity of different types of chiral selectors based on their unique interactions with analytes in chiral separations. Such binary CSPs can be obtained by (i) simultaneous covalent attachment of (two) different selectors to a polysiloxane, (ii) dissolving one selector in a polysiloxane to which the other selector is bound, and (iii) by directly linking one selector to the other such as attaching *N*-acetyl-L-valine-*tert*-butylamide to methylated β-CD. Due to the different chiral recognition mechanism of the selectors, the enantioseparation of an analyte may be enhanced compared to the individual selector or deteriorated in case of counteracting enantiorecognition. Currently, columns with binary selectors are not commercially available.

8.4.1.3 Separation Mechanisms

In CD-mediated enantioseparations in GC, two mechanisms are considered: inclusion complex formation and the much looser surface interaction with the outside of the CDs [34, 41]. Hydrogen bonding, van der Waals or dipole–dipole as well as steric interactions may be involved. The dominant mechanism for a given CD can be derived from thermodynamic and sample loading studies. Enantiomers having relatively high enthalpy ($\Delta H°$) and entropy ($\Delta S°$) values also display low loading capacity, suggesting a dominant role of inclusion complexation. In contrast, low $\Delta H°$ and $\Delta S°$ values typically coincide with high sample loading capacity. This indicates that surface interaction is the predominant mechanism [42].

The interaction type as well as the enantioselectivity of the CDs is controlled by the cavity size and the type and position of the substituents, especially the type of substituent at the 3-OH group, which appears to partially control the inclusion complex formation [41, 42]. Thus, upon derivatization of the 3-OH group with more polar substituents such as trifluoroacetic acid, which enables dipole–dipole interactions, the chiral separation mechanism does not appear to involve inclusion complex formation. Such selectors display higher efficiencies and faster separations compared to CDs relying primarily on inclusion phenomena. Therefore, the available CD CSPs differ in their enantioselectivity toward a given set of analytes and may also display alternative enantiomer elution order.

In contrast to liquid phase separation techniques like HPLC and CE, there are no competing solvent molecules in the stationary phase in GC, which allows the association between molecules even at elevated temperatures. This allows enantioseparations of racemic analytes that cannot be achieved by liquid phase enantioseparations. Examples are hydrocarbons which lack functional groups so that chiral recognition by a selector is essentially based on weak van der Waals interactions [43]. Combined with the high efficiency of enantioselective GC, the unique cavity structure of CDs often allows the enantioseparation of chiral hydrocarbon compounds as illustrated in Figure 8.7.

Enantioselectivities of the CD derivatives depend on the position and polarity of the substituents, which govern surface interaction versus inclusion complexation as addressed earlier [34]. Polar substituents enable dipole–dipole interactions and may favor surface interaction (especially when placed in position 3 of the D-glucopyranose units), while inclusion complexation recognizes size, shape, and functionality of analytes. Moreover, steric effects may play a role as does the size of the cavity. All these factors affect interactions between the selector and the analytes. Thus, the various selectors may display complementary selectivities toward a given set of analytes and reversal of the enantiomer elution order may be observed in the case of a certain selector when switching from the β-CD derivative to the γ-CD derivative or vice versa. This phenomenon is also observed in liquid-based enantioseparation techniques

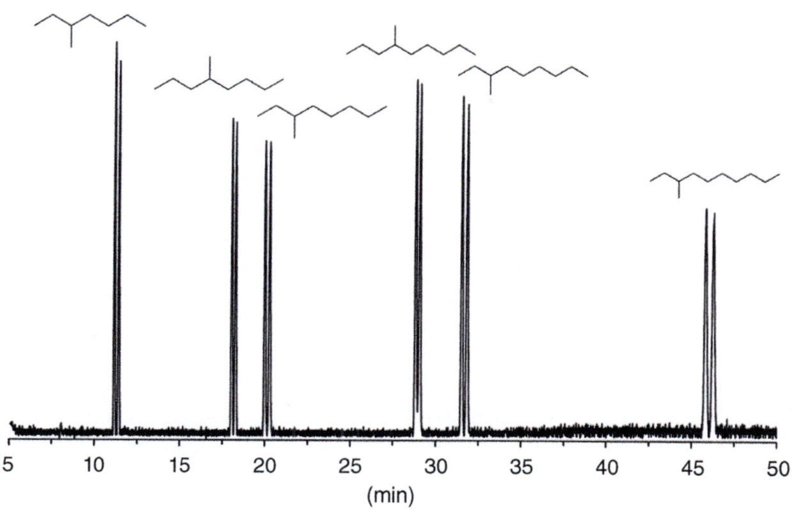

Figure 8.7 Gas chromatogram of resolved hydrocarbon enantiomers 3-methylheptane, 4-methyloctane, 3-methyloctane, 4-methylnonane, 3-methylnonane, and 3-methyldecane separated on a capillary column coated with permethylated β-cyclodextrin (Chirasil-Dex CB, 25 m, 0.25 mm inner diameter, film thickness 0.25 μm, Varian-Chrompack). Split injection 1:50 at injector temperature of 230 °C, 1.3 ml min^{-1} constant flow of He (99.9995%) carrier gas, oven temperature programmed 3 minutes at 35 °C, 1 °Cmin^{-1} to 15 minutes at 70 °C. *Source:* Adapted from Meierhenrich et al. [44] with permission of Wiley.

such as HPLC or CE. Different selectivity including reversal of the enantiomer elution order may also be encountered when changing from surface interaction to inclusion mechanisms. Furthermore, the type of column, i.e. undiluted, diluted, or grafted, may affect selectivities so that one type of column may be preferable for a certain group of analytes. In case of derivatization of the compounds prior to GC analysis, the separation efficiency may also be modified by the type of the derivative.

8.4.1.4 Applications

Many separations of chiral analytes by GC on CD-based CSPs have been reported in the literature, as summarized, for example, in [32–34, 39, 41]. Table 8.3 lists classes of compounds that have been successfully separated by the respective CD selectors. GC is primarily applied to the enantioseparation of volatile compounds of biological origin including terpenes or pheromones and can also be used in case of polar compounds such as amino acids upon derivatization.

Judged from the number of applications, 2,3-di-*O*-methyl-6-*O*-*tert*-butyl dimethylsilyl-β-CD appears to be the most frequently applied chiral selector. This CD has largely replaced the initially often used 2,3,6-tri-*O*-methyl-β-CD because of higher peak efficiency and greater selectivity. The CD is diluted with a polysiloxane carrier, but the composition of the carrier seemed to have little influence on the selectivity. Another selector with a broad applicability is 2,6-di-*O*-pentyl-3-trifluoroacetyl-γ-CD, which is considered by some authors as the most versatile chiral selector in GC. The trifluoroacetyl group enables strong dipole–dipole interactions so that this phase was highly selective not only for oxygen-containing compounds including alcohols, diols, aldehydes, ketones, acids, esters, or lactones but also for halogenated compounds.

8.4.2 Thin-Layer Chromatography

Despite the experimental simplicity of the technique, enantioseparations by TLC have not been often applied because of the much lower separation efficiency of TLC compared to high resolution techniques such as GC, HPLC, or CE. Chiral TLC plates containing a CD as chiral selector are not commercially available. In the published literature, CDs have been most often used as chiral mobile phase additives in combination with silica gel or aluminum oxide TLC plates [45, 46]. Another option was the impregnation (coating) of the silica gel layer with a CD or immobilization of a CD to the solid support. In recent years, polymeric layers containing CDs have been developed as TLC stationary phases for enantioseparations. In contrast to the other analytical separation technologies, where all three native CDs or various derivatives of them have been utilized, essentially only β-CD or simple derivatives such as 2-hydroxypropyl-β-CD (HP-β-CD) have been employed in TLC methods. Only few analytes have been separated by such CD-based methods including amino acids and their derivatives, β-blockers, local anesthetics, ibuprofen, ofloxacin, citalopram, or clopidogrel [46, 47].

8.4.3 High-Performance Liquid Chromatography

HPLC is the most often applied separation technique for non-volatile analytes. This refers to achiral as well as chiral separations. Although the vast majority of enantioselective liquid chromatography separations is performed on polysaccharide-based CSPs, CDs are nonetheless important chiral selectors in HPLC.

Initially, CDs were used as mobile phase additives as in TLC. The first report on a CD-based column dates to 1978, when Harada and colleagues reported the enantioresolution of mandelic acid derivatives on crosslinked CD gels [48]. Upon covalent attachment to silica gel, the application in

HPLC became available. While the publications utilizing such columns reported their application to the separation of positional isomers in 1983 [49, 50], the first enantioseparation was reported in 1984 by Armstrong and DeMond [51]. Since then, many CD derivatives and CSPs have been synthesized [8, 52–54] and columns have been commercialized.

8.4.3.1 Types of Columns

Several binding chemistries have been realized for the covalent attachment of the CDs to the silica gel support, typically via a spacer consisting of 8–12 atoms. In most cases, the CD is attached via the primary 6-position, but methods for linking to the secondary side have also been developed. The type of linker and the binding to either the primary or the secondary side affect the chromatographic performance as well as enantioselectivity of the columns [55]. Two general strategies have been employed for attachment of CDs to the solid support. In the first approach, the column is obtained by immobilizing the CD to functionalized silica gel. Post immobilization derivatization of the CD may apply. Alternatively, a CD derivative featuring a linker with a ω-triethyloxysilyl group is synthesized, which is subsequently bound to native silica gel. Single isomer CDs with a defined position and number of the substituents can be immobilized in a specific manner by this approach.

Common linkers utilized for the preparation of CD-based chiral columns are summarized in Figure 8.8a. The amine linker was the first type to be employed for bonding CDs to silica gel [49, 50]. This linkage is obtained by reacting amino-functionalized silica with a 6-tosyl-CD. The ether linkage, also referred to as the epoxide linkage, was introduced by Armstrong and coworkers [51] by reaction of glycidyl epoxide-functionalized silica gel with a CD. The advantage of this linker is the absence of nitrogen as a hydrogen bond-forming group. Many commercially available columns are based on this linkage such as the Cyclobond™ columns by Astec. Another linker present in commercial ChiraDex™ (Merck) or UltronES-CD™ (Shinwa) columns is the carbamate (urethane) linker, which can be obtained by reaction of isocyanate silica with the CD. Furthermore, CDs have been linked to silica supports via amide bonds or urea linkages in several studies [8]. A novel frequently evaluated bonding chemistry in research papers is based on a triazole synthesized by reaction of a ω-alkyne-modified silica gel and an azido-CD in the presence of Cu^+ ions (click chemistry) [8, 52]. Further binding chemistries as well as crosslinked selectors with multiple attachments of the CDs to a polymer network have also been synthesized [8, 52] and multiple layered CD CSPs have been explored [56]. Commercial columns are based on particulate silica, but analogous bonding chemistries have been used for the synthesis of

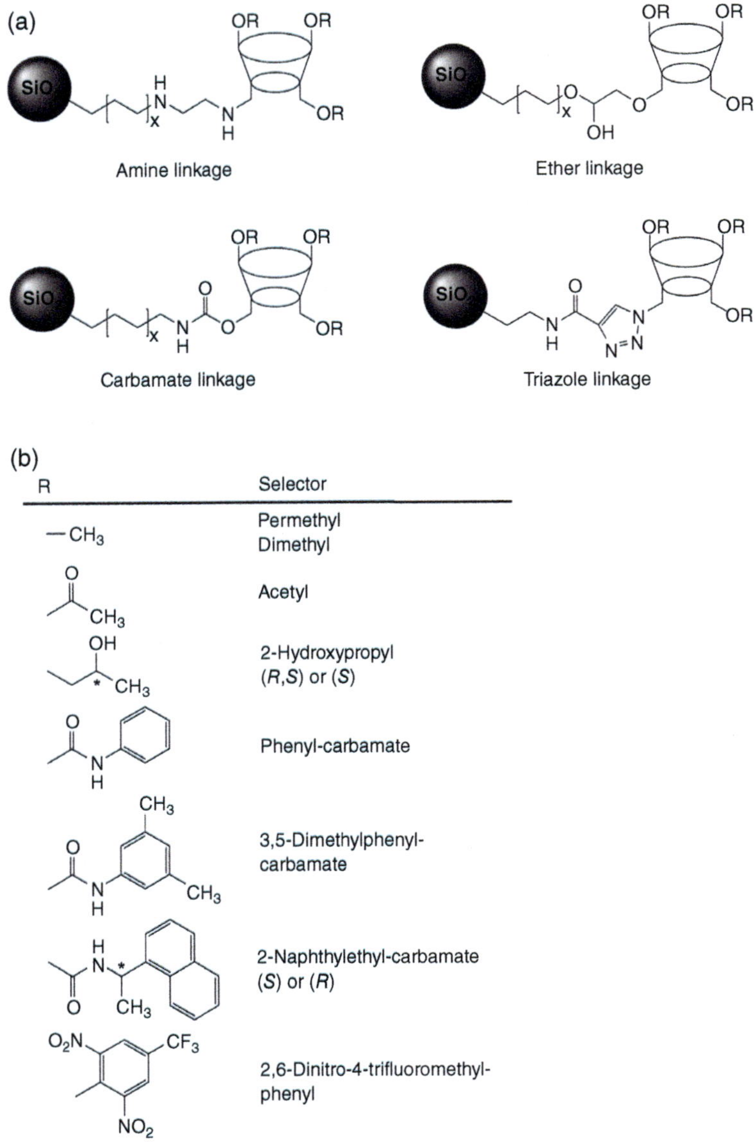

Figure 8.8 Schematic representation of (a) types of linkers and (b) substituents of the CD derivatives employed as chiral stationary phases in HPLC.

CD-functionalized monoliths [57, 58]. For high-speed and high-efficiency enantioseparations, 2-hydroxypropyl-β-CD has been immobilized to 2.7 μm superficially porous particles (core-shell materials) [59]. The column has been commercialized under the trade name CDShell-RSP™ (Azyp, LLC).

8.4.3.2 Types of Cyclodextrins

Native α-CD, β-CD, and γ-CD have been immobilized to silica gel particles, but with regard to CD derivatives, most commercial columns are based on β-CD because this CD showed the broadest applicability for small molecules. The exceptions are (per)methylated as well as acetylated α-CD and γ-CD. Common substituents are listed in Figure 8.8b. The substituents modify interactions or offer additional interaction modes, thus modifying the enantioselectivity of the derivatives compared to the native CDs. Moreover, enantioselectivity may also depend on the operation mode, i.e. normal phase, reversed phase, or polar organic mode.

Aliphatic (non-aromatic) substituents comprise methyl, acetyl, or 2-hydroxypropyl groups. Methylated CDs may come in two variants depending on the manufacturer: permethylated CDs, where all OH groups have been derivatized and 2,3-dimethyl-CDs featuring methyl groups selectively in positions 2 and 3 of the D-glucopyranose units. Methylation disables the formation of hydrogen bonds so that less compounds were enantioresolved under identical conditions compared to native CDs. Acetylation changes the hydrogen bond-donating character of hydroxy groups to hydrogen bond-accepting groups. Moreover, the cavity of the CDs is extended upon der derivatization of the secondary hydroxy groups so that accommodation of larger molecules is possible. This applies to all CD derivatives. 2-Hydroxypropyl-β-CD based stationary phases displayed the broadest applicability among the non-aromatic CD derivatives. The 2-hydroxypropyl group is either racemic or (*S*)-configured. Typically, seven moieties are bound to the CD resulting in an increased size of the cavity. Moreover, the hydrogen bond-donating hydroxy groups are extended in the cavity which appears to be advantageous for the chiral recognition. Although it was envisioned that 2-(*S*)-hydroxypropyl-β-CD would show increased enantioselectivity compared to its racemic counterpart, this could not be observed in practice. In fact, most analytes were separated with comparable enantioselectivities and only in a few cases superior resolution was observed with 2-(*S*)-hydroxypropyl-β-CD.

Aromatic substituents are phenyl carbamate, 3,5-dimethylphenyl carbamate, 2-(*R*)- or 2-(*S*)-naphthylethyl carbamate as well as 2,6-dinitro-4-trifluoromethyl-phenyl ether groups (Figure 8.8b). Aromatic substituents increase the cavity size of the CD enabling the accommodation of larger molecules. Moreover, additional intermolecular interactions such as π–π- or dipole–dipole interactions can contribute to enantioselective recognition. The carbamate groups are both, hydrogen bond donating (N—H) as well as hydrogen bond accepting (C=O) and provide additional dipole–dipole interaction sites. Moreover, these chiral columns can be operated in the normal phase as well as the reversed phase and polar organic mode, while aliphatic substituted CD columns are not effective under normal phase conditions. The 2-naphthylethyl-carbamate-β-CDs are the most versatile CD-based selectors in HPLC. On average about six moieties are bound to the CD. Under normal phase conditions, reversal of the enantiomer elution order of the enantiomers of many compounds were observed changing from the (*S*)-configured to the (*R*)-configured selector [60], while the former proved to be more effective resolving the enantiomers of the test analytes. The 3,5-dimethylphenyl moiety and the 2,6-dinitro-4-trifluoromethyl phenyl residue are complementary with regard to their π-interaction abilities. The latter is π electron basic due to the electron withdrawing substituents, while the former is π electron acidic because of the electron-donating effects of the methyl groups. Moreover, the linkage to β-CD is different, a carbamate in one case and an ether in the other case. Consequently, different enantioselectivities were observed. 2,6-Dinitro-4-trifluoromethylphenyl-β-CD was especially effective resolving aromatic analytes under reversed phase conditions.

In research papers, many further CD derivatives have been explored as chiral selectors in HPLC [52, 54, 61]. These include, for example, cationic CD derivatives or phenylcarbamate-derivatized CDs bearing 3-chloro-4-methylphenyl substituents [62].

8.4.3.3 Separation Mechanisms

Four main factors affect the performance of chiral CD columns in HPLC. First, the binding chemistry used to immobilize the CD on the solid support, i.e. the spacer as well as the type and position of the linkage. This may alter the accessibility of the CD and enable interactions such as hydrogen bonds, which can be beneficial or detrimental for a chiral separation. As stated earlier, binding of the CD to either the primary or the secondary side can determine the chromatographic performance as well as enantioselectivity [55]. The second factor is the coverage of the selector, i.e. the "concentration" on the support, which is affected by the nature of the support, especially the pore size and surface area apart from the bonding chemistry. The third factor is the size of the CD, that is whether the selector is based on α-CD, β-CD, or γ-CD. In case of the formation of inclusion complexes, the size of the cavity determines the type of analytes that can be accommodated and has an impact on the stability of the diastereomeric CD-analyte complexes. However, cavity size does not seem to be of utmost importance because most commercial columns are based on β-CD. Finally, the fourth and probably the most important factor is the type of the derivative groups of the selector as well as their linkages with the CD. This governs the usage of the column under the various elution modes and, more importantly, increases the size of the CD and can also provide additional interaction sites as well as types of interactions that may be established. For example, as described earlier, aromatic substituents enable intermolecular π–π and dipole–dipole interactions, while carbamate linkages can establish hydrogen bonds.

A chromatographic separation is always governed by the interplay between stationary phase and mobile phase. Therefore, in chiral separations, selectivity may also be affected by the mobile phase despite the fact that it is not chiral (at least not in the direct separation mode). However, the nature of the mobile phase determines the strength and type of interactions that can be formed between analyte and selector and thereby the overall performance of the system. CD-based chiral columns have been operated in the normal phase, reversed phase, and polar organic mode depending on the CD derivative. Switching from one mode to another affects the enantioselectivity and opposite elution order may be observed. Native and aliphatic derivatized CDs are very effective chiral selectors when operated in the reversed phase and polar organic mode, while they are much less useful under normal phase conditions. This has been attributed to the fact that hydrophobic inclusion complexes are not formed so that analytes can only interact with the surface of the CD, thereby limiting the usefulness of these columns. In contrast, aromatic CD derivatives are effective selectors also under normal phase conditions because they provide additional interaction sites. Derivatization also increases the cavity enabling the accommodation of larger molecules.

As in GC, two basic chiral recognition mechanisms are discussed in HPLC, inclusion complex formation and surface interaction. Inclusion complexation is driven by the displacement of water molecules from the CD cavity and further stabilization of the complex is mediated via van der Waals interactions as well as additional interactions with the CD rim or substituents on the rim including hydrogen bonds or dipole–dipole interactions as well as π–π interactions in case of aromatic substituents. Steric repulsion may also contribute. Inclusion complex formation is considered the dominant mechanism under reversed phase conditions. In the normal-phase or polar organic mode, the cavity of the CDs is occupied by solvent molecules that cannot be expelled so that inclusion complexation does not occur. Moreover, hydrophobic interactions between analytes and the interior of the CD cavity are prevented. Thus, enantioseparations are mediated via surface interaction primarily at the wider secondary side of the CDs. In the polar organic mode, polar interactions between the hydroxy groups or polar substituents of the CDs and functional groups of the analytes such as hydrogen bonding or dipolar interactions dominate the separation mechanism in addition to steric repulsion. In contrast, dipole–dipole and especially π–π interactions become more pronounced in apolar eluents used in normal-phase chromatography. Consequently, only aromatic analytes can be enantioresolved in the normal-phase mode.

8.4.3.4 Applications

Although CDs are not the most frequently used chiral selectors in HPLC, they represent an important class as can be judged from the number of commercialized columns. Apart from enantiomers, CD-based chiral columns have been shown to effectively separate diastereomers and structural isomers. The enantioseparations of many chiral analytes using CD-based columns have been summarized [8, 52–54]. Information can also be found in manufacturer handbooks, websites, and application databases. Chiral analytes included (un)derivatized amino acids, aromatic alcohols, flavones, aryl-alkylamines, carboxylic acids, hydantoins, dihydrobenzofuranes, or sulfoxides as well as pharmaceutical drugs including β-blockers, barbiturates, benzodiazepines, nonsteroidal anti-inflammatory drugs, thiazide diuretics, opioids, and alkaloids. The chiral columns can be operated under different elution modes, so that different enantioselectivities may result by selection of a given CD selector and/or the elution mode.

8.4.4 Supercritical Fluid Chromatography

SFC has gained popularity in recent years also in the field of enantioseparations [63]. As discussed in detail in Chapter 3, high separation efficiencies can be obtained making it an ideal technique for enantioseparations. In SFC, the same columns as in HPLC can be used. Nowadays, carbon dioxide (CO_2) is the preferred mobile phase because of the low critical temperature (31 °C) and pressure (7.3 MPa) and advantages such as inertness, availability in large quantities, and low toxicity. Moreover, CO_2 evaporates easily so that only a low amount of waste is generated from the mobile phase. Thus, SFC is often considered a green technology. The ease of removal of most of the mobile phase by evaporation also makes SFC an ideal technique for preparative enantioseparations. However, CO_2 has a low polarity comparable to *n*-hexane. To elute polar compounds, it is necessary to add co-solvents, often short-chain alcohols such as methanol, ethanol, and 2-propanol or acetonitrile typically in concentrations around 20% (up to 60%). The co-solvent can also adsorb to the stationary phase changing its polarity. Moreover, 0.1–1.0% of an organic acid and/or an organic base may be added to improve peak shape and resolution. Small amounts of water in the organic co-solvent can also increase separation efficiency [8, 64, 65].

The combination of CO_2 as a non-polar solvent and a polar co-solvent renders the mobile phase comparable to phases employed in normal-phase HPLC. However, under these conditions, inclusion complexation by native CDs is disfavored. Consequently, the use of CDs in SFC is limited as shown by the fact that these selectors were applied in less than 5% of the published enantioseparations, while polysaccharide-based CSPs predominate in SFC as in HPLC [64]. This limitation of neutral CDs has been tried to overcome by CSPs featuring cationic CDs, which allow ionic interactions [66]. These CSPs effectively separated the enantiomers of amino acid derivatives, flavones, or thiazide diuretics. Nonetheless, many chiral amino acid derivatives, pharmaceuticals, or pesticides have been enantioseparated in SFC on CD-based stationary phases [8].

8.4.5 Capillary Electromigration Techniques

Within the last three decades, CE has matured to a valuable liquid separation technique competing with HPLC in many areas. Although lower precision and reproducibility are often cited as disadvantages of CE, advantages include flexibility as well as the low consumption of sample and chemicals as well as the fact that organic solvents are rarely applied. Probably the most prominent advantage is the high efficiency and resolving power. A selectivity factor α of 1.01 typically corresponds to a baseline resolution, whereas in HPLC α values of at least 1.02 are required. For further details on chiral CE, see Chapter 4.

8.4.5.1 Types of Cyclodextrins

In CE, native α-CD, β-CD, and γ-CD as well as their derivatives have been utilized as chiral selectors. In fact, numerous CD derivatives are commercially available (e.g. from CycloLab, Cyclodextrin Shop, or Merck). CDs are by far the most often applied chiral selectors in CE. Statistical evaluations of the published literature have indicated that about 2/3 of the publications report enantioseparations mediated by CDs. Often applied CD derivatives have been compiled in Table 8.2. Derivatization increases the cavity of a CD and also enables additional interactions with the analytes. Moreover, compared to native β-CD, its derivatives display higher aqueous solubility although derivatives with very low aqueous solubility ($< 0.1\,\text{mg}\,\text{ml}^{-1}$) are also known, which prevents their use as chiral selectors in CE.

Apart from grouping according to cavity size, CDs are typically divided based on the charge, i.e. into neutral and charged CDs, which are further subdivided into negatively charged (anionic) and positively charged (cationic) CDs (Table 8.2). Native and neutral CD derivatives do not possess self-mobility and cannot be used for the separation of uncharged analytes because there is a difference neither in the effective mobilities of the diastereomeric derivatives nor in the mobilities of the free analyte and the analyte-CD complex. Thus, their use is limited to charged analytes unless other modes such as micellar electrokinetic chromatography (MEKC) are employed. In contrast, charged CDs can also be applied to the separation of uncharged (as well as charged) compounds due to the inherent self-mobility of the CDs, so that the free analytes possess a different mobility than that of the analyte-CD complexes (see also Section 8.4.5.2).

A further distinction refers to randomly substituted CDs versus single isomer CDs. Randomly substituted CDs are a mixture of a (large) number of CD isomers differing in the number as well as in the position of the substituents. Randomly substituted CDs are characterized by the degree of substitution, which is the average number of substituents either per glucose molecule or per CD molecule. They can be further characterized by the percentage of substituents in positions 2, 3, or 6, but this is not often determined. The synthesis of randomly substituted CDs is relatively simple, but they will differ from manufacturer to manufacturer and even from batch to batch for the same vendor. As a consequence, the enantioseparation selectivity in a given assay may vary depending on the source of the CD. This fact has to be considered when developing a validated method for routine use. Single isomer CDs are composed of essentially only one isomer. Their synthesis is more tedious including protection group chemistry [8, 10, 11, 67], but batches are basically identical so that the same separation selectivity can be expected when switching from one batch to another. Therefore, single isomer CDs may be preferable in validated routine assays. On the other hand, a higher enantioseparation selectivity is often observed using randomly substituted as compared to the analogous single isomer CDs. This is due to the fact that the individual isomers present in randomly substituted CDs differing in the number and positions of the substituents all contribute to the given enantioseparation. In case of a single isomer CD, only a specific isomer mediates the enantioresolution.

8.4.5.2 Separation Mechanisms

In CE, the chiral selector is added to the BGE so that it is mobile in contrast to the use of a CSP in chromatographic techniques. Thus, the complex formed between selector and selectand is also mobile, so that the effective mobility of an analyte interacting with a chiral selector is a function of the mobility of the free analyte, the mobility of the complex, and the strength of the interaction expressed as the stability (complexation) constant. The simplest model for CE enantioseparations has been derived by Wren and

Rowe [68]. An enantioseparation is observed when the effective mobilities of the (*S*)- and the (*R*)-enantiomer, μ_{eff}^{S} and μ_{eff}^{R}, differ:

$$\Delta\mu = \mu_{\mathrm{eff}}^{S} - \mu_{\mathrm{eff}}^{R} = \frac{\mu_{\mathrm{f}} + \mu_{\mathrm{cplx}}^{S} K_S[C]}{1 + K_S[C]} - \frac{\mu_{\mathrm{f}} + \mu_{\mathrm{cplx}}^{R} K_R[C]}{1 + K_R[C]} \tag{8.1}$$

μ_{f} is the mobility of the free analyte, μ_{cplx}^{S} and μ_{cplx}^{R} the mobilities of the complexes formed between the selector and the analyte enantiomers, and K_S and K_R the respective complexation constants and $[C]$ the concentration of the chiral selector. A prerequisite is that the mobilities of the complexes differ from the mobility of the free analyte ($\mu_{\mathrm{f}} \neq \mu_{\mathrm{cplx}}$). From Eq. (8.1), it is evident that an enantioseparation is achieved in case of different affinities of the enantiomers toward the selector expressed as complexation constants of the enantiomers, i.e. $K_S \neq K_R$, or in case of different mobilities of the enantiomer-selector complexes, i.e. $\mu_{\mathrm{cplx}}^{S} \neq \mu_{\mathrm{cplx}}^{R}$. The first mechanism is also referred to as the chromatographic enantioselective mechanism or the thermodynamic mechanism. With regard to this mechanism, there is no conceptual difference between CE and chromatographic techniques using CSPs. Therefore, enantioselective CE is also referred to as chiral electrokinetic chromatography (EKC) reflecting the fact that the enantioseparation is based on a chromatographic principle, while the movement of the analytes is based on electrokinetic phenomena. The second mechanism based on differences of the mobilities of the complexes is termed the electrophoretic enantioselective mechanism, which is not feasible with a selector fixed to a solid support. It may be caused by differences in the hydrodynamic radii of the complexes or by differences of the pK_{a} values of the diastereomeric complexes. Both mechanisms can contribute simultaneously but the chromatographic mechanism is typically the dominant mechanism because the hydrodynamic radii of the two enantiomer-selector complexes do not differ significantly and the effective charges of the two complexes are identical. However, a striking difference between enantioseparations in pressure-driven chromatographic systems and systems based on electrophoretic phenomena is the fact that enantiomers can also be separated in the case of equal binding constants ($K_S = K_R$) solely based on differences in the electrophoretic mobilities of the diastereomeric complexes [69]. This separation mechanism is not possible with the immobilized selectors in chromatography. Another advantage of chiral CE is the fact that different selectors can be easily combined to achieve an enantioseparation.

The simple mathematical model for CE enantioseparations according to Eq. (8.1) has several disadvantages such as the fact that it does not account for the protonation equilibria of the enantiomers in their free and complexed forms. Thus, different charge densities of the diastereomeric complexes may result at a given pH of the BGE based on a complexation-induced pK_a shift leading to different mobilities of the diastereomeric selector-enantiomer complexes. For a discussion of further complex models including the protonation equilibria or multiple equilibria, see, for example, [70–72].

Complex formation between CDs and analyte enantiomers occurs primarily via inclusion of lipophilic moieties of the solutes into the toroidal cavity either from the wider or from the narrower side displacing water molecules from the cavity as one of the driving forces of the process. However, aggregates with only shallow insertion of the molecules into the cavity, analogous to surface interaction mechanism in chromatography, or formation of external complexes have also been described as discussed in detail in Section 8.3. Hydrophobic and van der Waals interactions as well as hydrogen bond formation are primarily involved in complex formation and stabilization. Steric factors as well as ionic interactions in the case of charged CDs may be present as well.

8.4.5.3 Migration Modes and Enantiomer Migration Order Using CDs as Selectors

In CE, the chiral selector as well as the solute may be neutral, anionic, or cationic. In addition to buffer additives (other than the chiral selector) and the nature of the capillary wall, which may both affect the electroosmotic flow (EOF), the charge of the analyte, and the chiral selector determine the mechanism and direction of the migration in CE. Therefore, the nature of the chiral additive contributes not only to the separation selectivity but, in case of charged selectors, also to the migration direction and magnitude. Consequently, various scenarios can be applied to alter the effective mobilities of the enantiomers in order to obtain a certain migration order [70, 73, 74]. Considering the effect of the chromatographic principle reflected by the complexation of the solutes and the electrophoretic principle on the effective mobilities of the enantiomers, the following mechanisms may affect the enantiomer migration order (combinations may apply):

- the strength of the complexation of the enantiomers by the selector
- the direction and magnitude of the EOF
- the direction and magnitude of the mobility of the free analyte
- the direction and magnitude of the mobility of the chiral selector
- the direction and magnitude of the mobility of the diastereomeric selector-enantiomer complexes

The scenarios are also possible because detection can be performed at the cathodic end (normal polarity of the applied voltage) as well as the anodic end of the capillary (reversed polarity of the applied voltage) so that cathodic as well as anodic mobility can be exploited. Selected migration modes, considering charged or neutral analytes as well as neutral and charged chiral selectors, are schematically shown in Figure 8.9. Selection of appropriate experimental conditions can result in a reversal of the enantiomer migration order, which may be important for the determination of the enantiomeric excess when the minor enantiomer has to be determined in the presence of a large excess of the major enantiomer. For example, it may be desirable to determine the minor enantiomer in front of a large tailing peak. Further experimental conditions resulting in a different enantiomer migration order can be envisioned, so that the scenarios discussed are not comprehensive. For example, only single selector systems as well as neutral and negatively charged selectors have been considered here.

The enantiomers of basic compounds can be separated at low pH using uncharged selectors and detection at the cathode (Figure 8.9a). Under these conditions, the analyte is positively charged and migrates to the cathode. The weaker complexed enantiomer is detected first because it possesses the higher effective mobility. Negatively charged analytes can be separated using neutral selectors under high EOF (Figure 8.9b) or low EOF conditions (Figure 8.9c). The negatively charged enantiomers migrate to the anode. In the presence of a high EOF, overall mobility is directed to the cathode so that the stronger complexed enantiomer will migrate first. In contrast, it will be detected second under low EOF conditions and detection at the anode because the weaker complexed enantiomer possesses a relatively higher mobility toward the anode. Using a negatively charged selector for the enantioseparation of a cationic compound, the weaker complexed enantiomer will be detected first at the cathodic end of the capillary at a low selector concentration (Figure 8.9d). At a high concentration, the carrier ability of the charged selector can be exploited detecting the analyte at the anode. Thus, upon reversal of the polarity of the applied voltage, the stronger complexed analyte will be transported faster to the detector at the anodic end of the capillary (Figure 8.9e). Because of their electrophoretic mobility, charged selectors can also be used for the enantioseparation of neutral analytes as outlined for the combination of an uncharged analyte and a negatively charged selector in Figure 8.9f and g. The weaker bound enantiomer is detected first at the cathode at low selector concentrations in the presence of a sufficiently high EOF (Figure 8.9f). In contrast, under reversed polarity of the applied voltage exploiting the carrying ability of the selector present at high concentrations in the BGE, the stronger complexed enantiomer will be detected first at the anodic end of the capillary (Figure 8.9g).

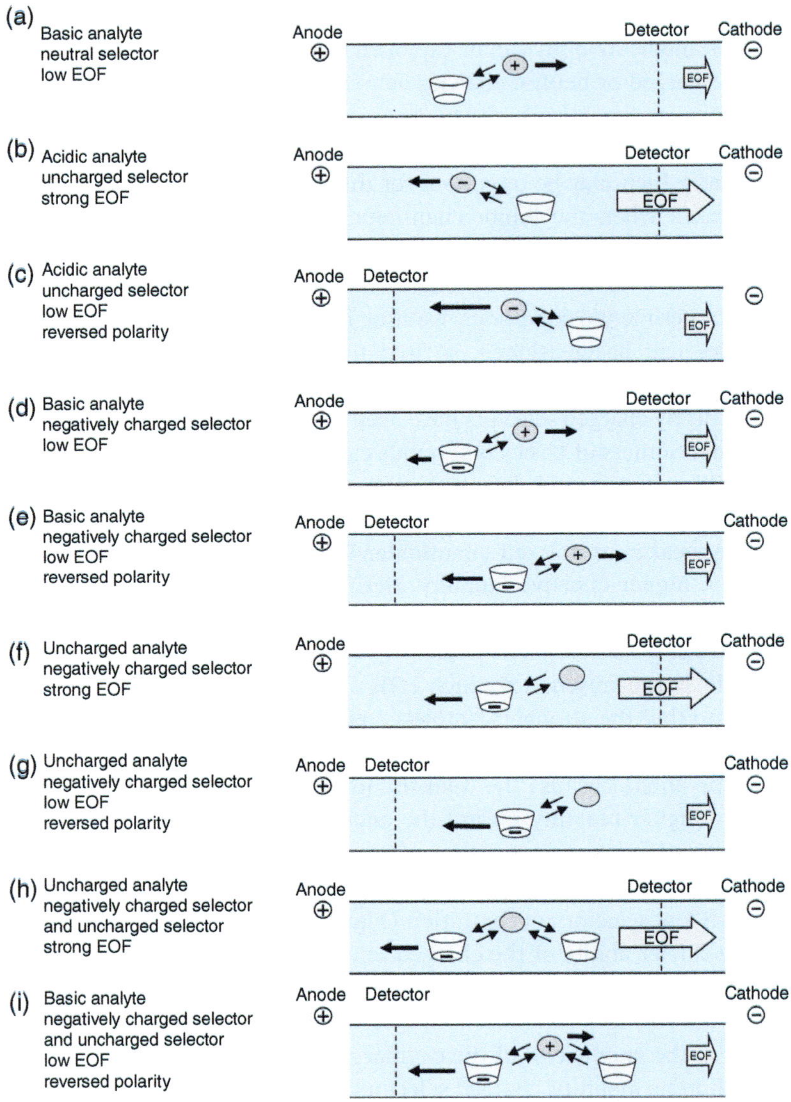

Figure 8.9 Schematic representation of migration modes in capillary electrophoresis enantioseparations. The oval represents the analyte with a charge indicated where appropriate. The basket represents the CD. A charged selector is indicated by the additional minus sign. The arrows represent the electrophoretic mobility of analyte and selector, respectively. The overall mobility in the different scenarios is always directed toward the detector. For details of modes (a) to (i) see text.

It is apparent from these examples that a designed enantiomer migration order can be achieved for a given compound by selection of the appropriate selector and conditions. Because each selector has its own enantioselective recognition toward analytes, the migration order will be primarily affected by the selector. For example, it has been shown that the enantiomer migration order can depend on the size of the CD cavity so that different enantiomer migration orders were observed for α-, β-, and γ-CDs. Moreover, different derivatives of a CD may also display opposite migration order. Furthermore, reversal of the enantiomer migration order may be achieved by the selection of the appropriate experimental conditions. It is obvious that the scenarios schematically shown in Figure 8.9b and c, Figure 8.9d and e as well as Figure 8.9f and g will lead to a different enantiomer migration order if the compound can be analyzed under the specific experimental conditions. For example, opposite migration order of the enantiomers of an anionic analyte using an uncharged CD can be observed under low EOF conditions (mechanism Figure 8.9c) versus high EOF conditions (mechanism Figure 8.9b) as illustrated for 1,1′-binaphthyl-2,2′-diyl hydrogen phosphate in the presence of β-CD as shown in Figure 8.10. At low pH, the EOF is very low so that the anionic analyte migrates toward the anode under reversed polarity of the voltage. Therefore, the weaker complexed (*R*)-(−)-enantiomer is detected first because of the higher effective mobility. In the presence of a high EOF at higher pH, the analytes can be detected at the cathode applying normal polarity of the voltage. Because the stronger bound (*S*)-(+)-enantiomer possesses a lower overall mobility toward the anode, it is now detected first at the cathodic end of the capillary.

The reversal of the enantiomer migration order utilizing the carrier ability of a selector is illustrated in Figure 8.11 by the example of carboxymethyl-β-CD for the simultaneous separation of isomers and enantiomers of the dipeptide α/β-Asp-PheOMe in pH 5.25 BGE. At this pH, the peptides are essentially uncharged, while the CD bears negative charges. At low CD concentration, the weaker complexed enantiomer is detected first at the cathode if a sufficiently high EOF exists (mechanism Figure 8.9f) resulting in the electropherogram shown in Figure 8.11a. In contrast, at a high CD concentration using a polyacrylamide-coated capillary for suppression of the EOF, the CD acts as a carrier for the analytes to the anode (mechanism Figure 8.9g). In the example, the migration order of each enantiomeric pair is opposite at high concentrations of carboxymethyl-β-CD under reversed polarity of the applied voltage (Figure 8.11b). Furthermore, if an amphoteric compound is analyzed as positively charged or as negatively charged species, the migration order of the enantiomers may change provided that

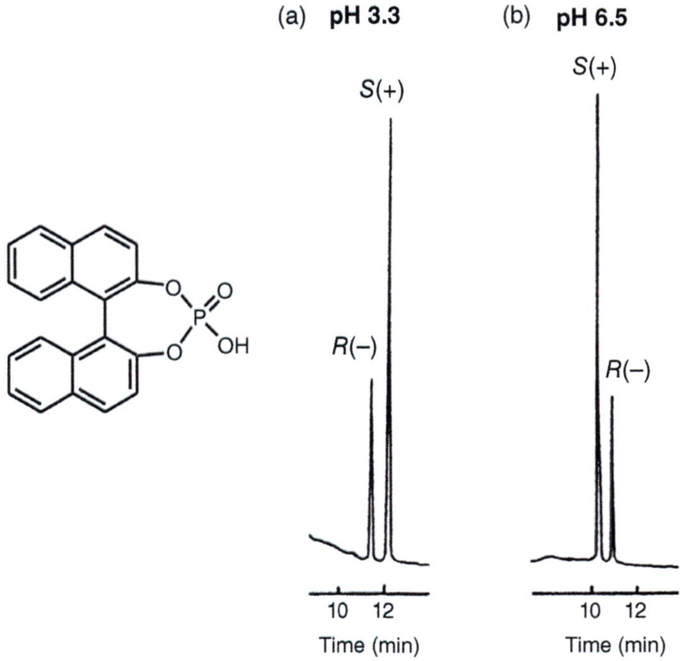

Figure 8.10 Enantioseparation of 1,1′-binaphthyl-2,2′-diyl hydrogen phosphate (*S/R* ratio 2 : 1) using 2.5 mg ml^{-1} β-CD as chiral selector in 50 mM phosphate buffer (a) pH 3.3 and negative polarity of the applied voltage (−400 V cm^{-1}), and (b) pH 6.5 and normal polarity of the applied voltage (+400 V cm^{-1}). Experimental conditions: 50/60 cm, 50 μm ID fused-silica capillary, 20 °C, UV detection at 210 nm. *Source:* Reproduced from Chankvetadze et al. [75] with permission of Elsevier.

the chiral recognition of the selector toward the enantiomers does not change as a function of the charge of the analyte (Figure 8.9a and c).

In CE, the combination of chiral selectors can also be easily applied. For useful combinations, the selectors must possess opposite chiral recognition toward the enantiomers or different mobilities in case the chiral recognition is the same but with a different magnitude. In the first case, the migration order is a function of the CD used at the higher concentration in the binary mixture. This mechanism is essentially the same as using the individual selector alone. In the second case, typically a charged and neutral selector are combined. The migration order of the enantiomers in the absence of the neutral selector results from the complexation by the charged complexation agent depending on the fact whether the analyte itself is charged or uncharged and detection is subsequently carried out at the anodic or the cathodic end of the capillary. For example, when using a negatively charged

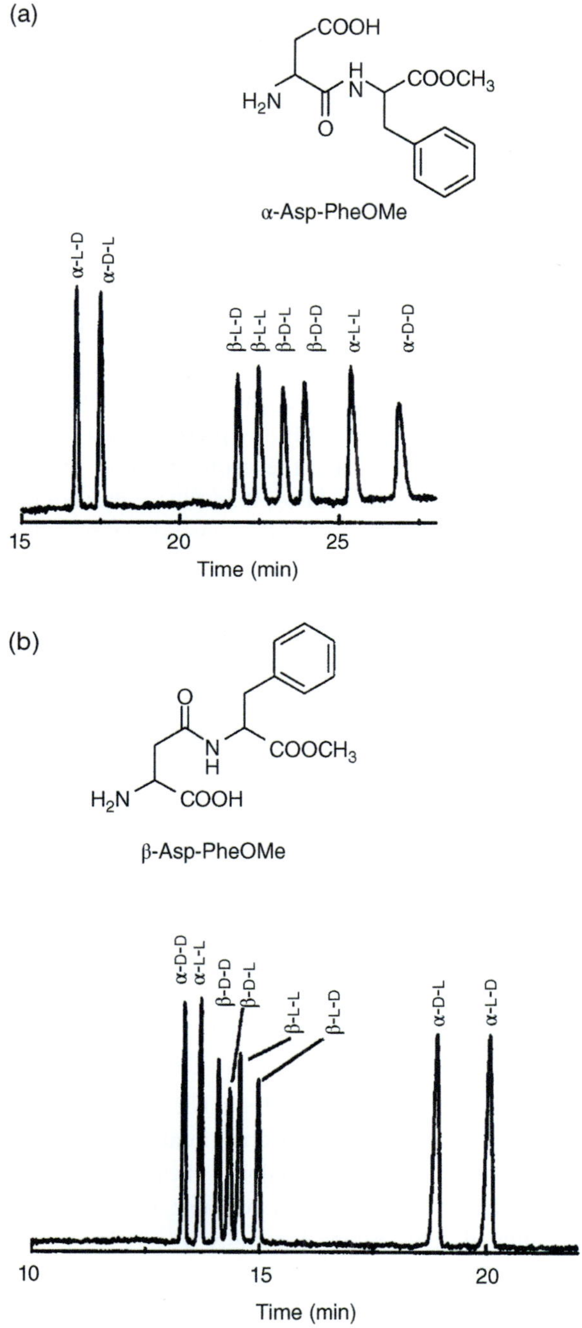

Figure 8.11 Separation of enantiomers of the α/β-isomers of the dipeptides α/β-Asp-PheOMe under (a) normal polarity of the applied voltage (detection at the cathode) and (b) exploiting the carrier ability of the selector under reversed polarity of the applied voltage (detection at the anode). Experimental conditions: (a) 40/47 cm, 50 μm ID fused-silica capillary, 50 mM sodium phosphate buffer, pH 5.25, 15 mg ml^{-1} carboxymethyl-β-CD, +20 kV, 20 °C, UV detection at 215 nm and (b) 40/47 cm, 50 μm ID polyacrylamide-coated capillary, 50 mM sodium phosphate buffer, pH 5.25, 60 mg ml^{-1} carboxymethyl-β-CD, −20 kV, 20 °C, UV detection at 215 nm. *Source:* Reprinted from Sabah and Scriba [76] with permission of Elsevier.

CD for the separation of a neutral analyte with detection at the cathode, the weaker complexed enantiomer will migrate first because it is carried to the detector by the EOF, while the analyte-selector complex migrates toward the anode (Figure 8.9f).

In CE, the combination of CDs as chiral selectors is feasible. Typically, a neutral and a charged CD are selected. By adding a neutral CD with the same chiral recognition but stronger analyte complexation compared to the charged CD, the competition between the selectors for the analyte enantiomers results in a lower overall binding of the stronger complexed enantiomer by the negatively charged selector. In the presence of a strong EOF, the complex between the neutral CD and the stronger bound enantiomer is carried to the detector at the cathode faster than the weaker bound enantiomer (Figure 8.9h) and reversal of the enantiomer migration order occurs compared to the sole presence of the negatively charged CD. A similar observation results for protonated analytes in combination with a negatively charged CD under reversed polarity of the applied voltage detecting at the anodic end of the capillary (Figure 8.9i). In this set-up, the stronger complexed enantiomer is detected first when only working with the negatively charged selector. Adding an uncharged selector with the same chiral recognition but stronger analyte complexation compared to the charged selector competition between the selectors for the analyte enantiomers results in a lower overall binding of the stronger complexed enantiomer by the negatively charged selector as stated earlier. As a consequence, the weaker bound enantiomer is less retarded by the uncharged selector increasing its binding to the charged selector so that it now reaches the detector at the anode first. For a more comprehensive discussion of the enantiomer migration order in CE, see also [70, 73, 74].

8.4.5.4 Applications

Based on the number of publications, the use of CDs in CE enantioseparations is one of the fastest growing areas in stereoisomer analysis. As stated earlier, CDs are applied as chiral selectors in the majority of cases. Although quite a number of CD derivatives are commercially available, novel CDs are still synthesized and evaluated. This refers especially to single isomer CDs [8, 11, 67, 77]. Moreover, CE has often been combined with NMR spectroscopy as well as molecular modeling in mechanistic studies in order to understand the complexation behavior of CDs toward analytes, the structures of the CD-analyte complexes and/or to derive the rationale for migration phenomena observed in CE as discussed in Section 8.3.

CE enantioseparations using CDs as chiral selectors have been reported in many fields. Many pharmaceutical applications refer to the chiral analysis of drugs in bulk ware or finished products most often in the form of assays for the enantiomeric purity of a drug [77–80] or to determine the enantiomeric excess following organic synthesis. Such determinations can be applied to analytes with multiple chiral centers as illustrated for the drug tapentadol in Figure 8.12. The separation of all four stereoisomers was achieved using sulfated α-CD as chiral selector within less than 10 minutes analysis time and the stereoisomeric impurities in (*R*,*R*)-tapentadol could be determined at the 0.15% level [81]. Pharmaceutical drugs are also typically used for the evaluation of novel CDs because of their availability and the structural diversity. Furthermore, bioanalytical methods have been developed in order to study the stereoselective in vivo and in vitro metabolism of drugs. In forensic and toxicological analyses, chiral methods have been used, for example, for the analysis of illegal substances. In food analysis, chiral CE methods were applied for the analysis of enantiomerically pure compounds in foods and food supplements [82], for example, for distinguishing between the natural or synthetic origin of a compound. In the latter case, the racemate is typically found. CD-mediated enantioseparation is also a topic in environmental analysis for the determination of drugs in water samples or pesticides in soil or water samples [83]. Another application is the determination of the configuration of amino acids in various samples including food and environmental samples or in the search for

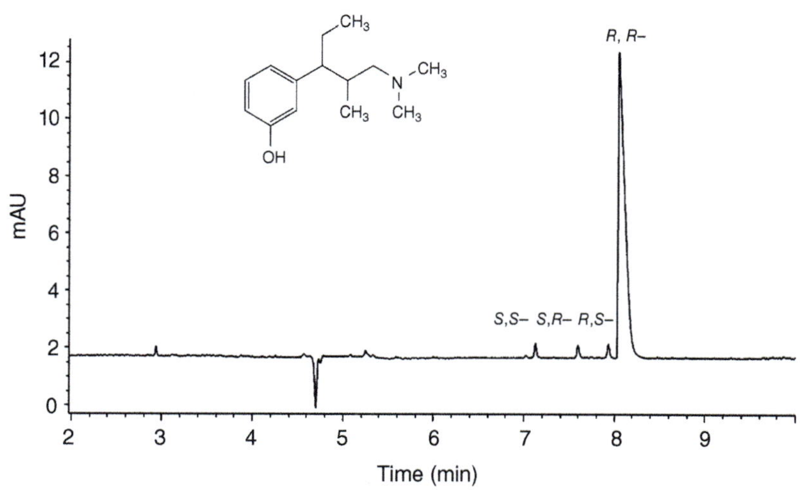

Figure 8.12 CE separation of the tapentadol stereoisomers for the determination of the enantiomeric purity of (*R*,*R*)-tapentadol containing the stereoisomeric impurities at the 0.15% level. Experimental conditions: 56/64.5 cm, 50 μm ID fused-silica capillary; 100 mM sodium borate pH 9.5 with 1.0% (w/v) of sulfated α-CD; 15 °C; 25 kV. *Source:* Reprinted from Znaleziona et al. [81] with permission of Elsevier.

extraterrestrial life, for example, by analyzing meteor samples or as part of space missions. In fact, chiral CE systems are implemented in the Organic Capillary Electrophoresis Analysis System (OCEANS) or the Mars Organic Analyzer (MOA), which are under development for future Mars and other space missions [84].

Chiral CE separations were typically performed in the EKC mode using a single CD as selector. However, combinations of CDs (so-called dual selector systems) or, more recently, combinations of CDs with chiral ionic liquids have also been applied. Moreover, in case of uncharged analytes or in order to increase separation selectivity, MEKC or microemulsion electrokinetic chromatography (MEEKC) methods were reported [77, 78, 85].

8.4.6 Membrane Technologies

In recent years, membrane technologies for enantioseparations have emerged as a platform for large-scale enantioseparations. Advantages include low energy consumption, continuous operation, and high processing capacity as well as the possibility of simple scale-up and integration into other separation processes [86]. According to the status of the membrane phase, chiral membranes can be divided into liquid membranes and solid membranes. In liquid membranes, the selector is dissolved in the liquid phase. These can be further subdivided into bulk liquid membranes (Figure 8.13a), emulsion liquid membranes (Figure 8.13b), and supported liquid membranes (Figure 8.13c) [87, 88]. In the first case, the membrane consists of an organic water-immiscible solvent between aqueous donating and receiving phases. Emulsion membranes are double emulsions composed of an emulsion of the aqueous receiving phase in an organic phase, which is dispersed in the aqueous donating phase. In most supported liquid membranes, the liquid phase is held within the pores of a porous membrane. Liquid membranes are characterized by ease of design and fast mass transfer, but their use is hampered by their low long-term stability. CDs are not frequently found as components of liquid membranes, but β-CD and 2,3,6-tri-*O*-acetyl-β-CD have been used in bulk liquid membranes [86, 88].

Solid membranes either are inherently chiral or consist of an achiral structure, which has been functionalized by immobilization of the chiral selector, for example, by impregnation, esterification, amidation, thiol-ene click reactions, or grafting. Inherently, chiral membranes are prepared from chiral polymers such as poly-L-glutamates, alginate, chitosan, cellulose, and their derivatives. Because such membranes often display somewhat limited chiral separation efficiency, they are further functionalized by immobilization of chiral selectors. For example, β-CD has been incorporated in

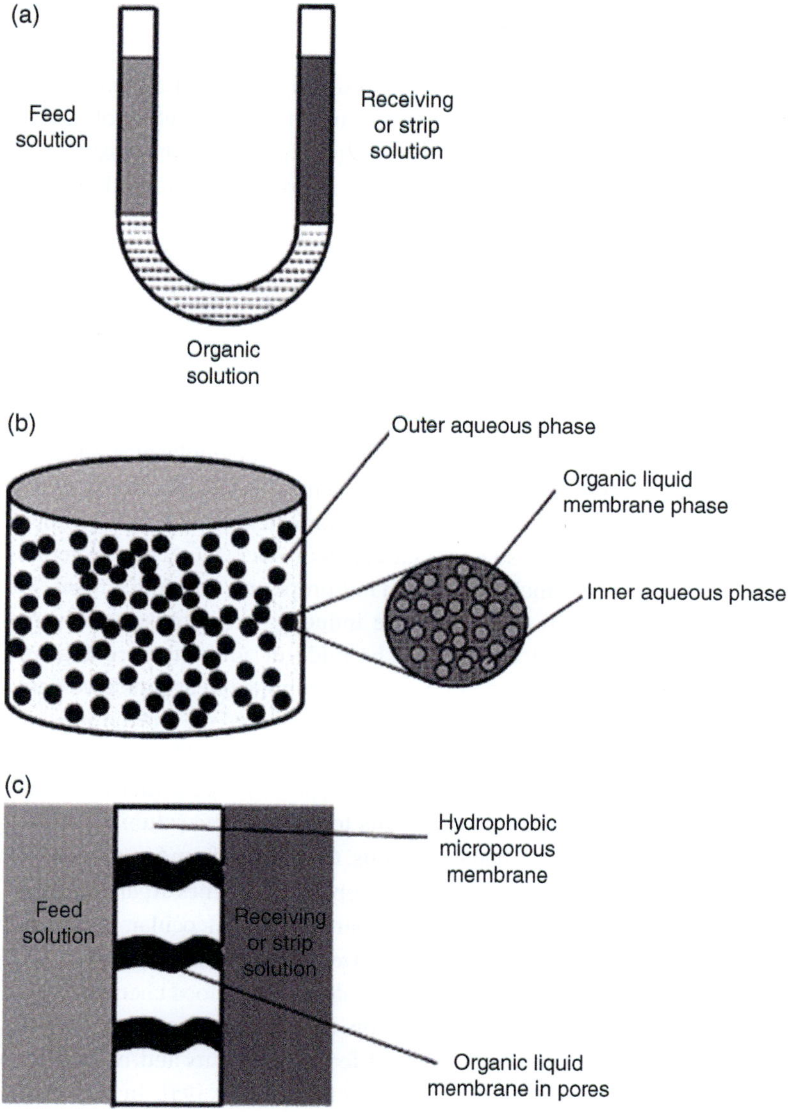

Figure 8.13 Types of liquid membranes: (a) bulk liquid membrane, (b) emulsion liquid membrane, and (c) supported liquid membrane. *Source:* Clark et al. [87] with permission of Elsevier.

membranes based on cellulose acetate [89]. Achiral supports used for the preparation of chiral solid membranes are quite diverse and include polymers such as nylon or polysulfone, metal organic frameworks, covalent organic frameworks, porous organic cages as well as carbon nanotubes, ceramics, zeolite, or silica gel. In most cases, β-CD and derivatives not only were immobilized to such supports but γ-CD has also been employed for chiral functionalization [86, 89, 90].

So far only few analytes have been separated by using chiral membranes containing CDs including model analytes such as 1-phenylethanol, amino acid derivatives, such as phenylalanine, tryptophan, and *p*-hydroxyphenyl glycine as well as the pharmaceutical drugs naproxen, chlorthalidone, and propranolol [86, 88–90].

8.5 Miscellaneous Applications

Apart from analytical separations sciences, CDs are applied in many other fields, a few of which will be briefly addressed. In many of these applications, the stereoselectivity of the complexation of target molecules is not the primary issue, except in organic reactions or sensor-based detection systems. An important area is the pharmaceutical/biomedical field. CD derivatives are approved drugs such as sugammadex *octakis*(6-deoxy-6-(2-carboxyethyl) thio)-γ-CD, Bridion®), which is approved in Europe and the United States for the reversal of neuromuscular blockade induced by the muscle relaxants rocuronium bromide and vecuronium bromide in adults undergoing surgery [91]. CD-based therapies as cholesterol-complexing agents have demonstrated benefits of methylated β-CD derivatives in the prevention and treatment of atherosclerosis, while 2-hydroxypropyl-β-CD (Trappsol® Cyclo) is in clinical trials for Niemann-Pick disease Type C1 [92]. Furthermore, CDs are used in pharmaceuticals as excipients to increase the solubility and/or stability of drugs or to mask bad taste. Thus, CDs can be found in many Food and Drug Administration (FDA) and European Medicines Agency (EMA) approved drugs or clinical candidates for parenteral, oral, ocular, intranasal, pulmonary, topical, vaginal, rectal, or transdermal administration [93, 94]. CDs for site-specific delivery have also been designed. In food chemistry, CDs have been employed in a similar fashion as in pharmaceutical analysis, i.e. the solubilization and stabilization of food ingredients and/or additives as well as for taste masking or improvement of flavor [95]. In the textile industry, CDs have been incorporated into the fibers or used in odor-eliminating sprays as a tool to remove odors by encapsulating components of sweat or cigarette smoke [96]. Furthermore, CDs have been employed in electrochemical [97] and fluorescence-based sensors [3, 98], as components of liquid crystals [3], or as catalysts in organic reactions [99].

Apart from monomeric CDs, CD polymers have been increasingly developed in technological applications. Many types of CD polymers have been prepared including linear polymers, crosslinked polymers, and hyperbranched polymers [100]. Sometimes, nanoparticle-sized CD crosslinked polymers are also referred to as CD nanosponges because of their sponge-like structure formed by the three-dimensional network. As in the case of monomeric CDs, applications of the polymers were based on the inclusion complex formation ability. Consequently, CD polymers have been explored as drug delivery systems in pharmaceutical sciences and biomedical applications or as additives in the food industry. Interesting applications also refer to the complexation of various species by CDs in solution, which is exploited for the removal of toxic compounds. A biomedical example is the clearance of toxic compounds in the bloodstream [101]. CDs have also been part of polymers utilized for the removal of inorganic and organic micropollutants such as heavy metal ions, endocrine-disrupting chemicals, or pharmaceutical drugs in wastewater treatment [3, 100].

8.6 Conclusions and Outlook

Due to their ability of (stereospecific) complexation of a large variety of compounds, CDs have found numerous applications in analytical chemistry as well as many other technologies including pharmacy, chemistry, food, textiles, or environmental technologies to name just a few. With regard to analytical separation sciences, CDs will continue to play an important role in enantioseparations especially in GC and CE. New CDs will be developed as chiral selectors not just for CE but also for HPLC or SFC and further applications for the analysis and determination of the chiral purity of drugs can be expected. This is in contrast to GC, where CD-based selectors seem to have matured beyond further development.

Apart from analytical applications, CDs have enormous potential in further technological applications in the pharmaceutical, chemical, or food industry as well as in environmental sciences, although these have not been the focus of the chapter. Especially, the importance of novel systems for drug delivery and stabilization of pharmaceutical formulations will drive research in this field using not only monomeric CDs but increasingly also CD polymers. The same applies to membrane technologies and polymers, for example, for wastewater treatment.

References

1 Crini, G., Fourmentin, S., and Lichtfouse, E. (ed.) (2020). *The History of Cyclodextrins*. Cham, Switzerland: Springer-Nature.

2 Moss, G.P., Smith, P.A.S., and Tavernier, D. (1995). Glossary of class names of organic compounds and reactive intermediates based on structure. *Pure Appl. Chem.* 67: 1307–1375.

3 Fourmentin, S., Crini, G., and Lichtfouse, E. (ed.) (2018). *Cyclodextrin Applications in Medicine, Food, Environment and Liquid Crystals*. Cham, Switzerland: Springer-Nature.

4 Biwer, A., Antranikian, G., and Heinzle, E. (2002). Enzymatic production of cyclodextrins. *Appl. Microbiol. Biotechnol.* 59: 609–617.

5 Larson, K.L. (2002). Large cyclodextrins. *J. Incl. Phenom. Macrocycl. Chem.* 43: 1–13.

6 Szejtli, J. (1998). Introduction and general overview of cyclodextrin chemistry. *Chem. Rev.* 98: 1743–1753.

7 Sandilya, A.A., Natarajan, U., and Priya, M.H. (2020). Molecular view into cyclodextrin cavity: structure and hydration. *ACS Omega* 5: 25655–25667.

8 Tang, W., Ng, S.-C., and Sun, D. (ed.) (2013). *Modified Cyclodextrins in Chiral Separations*. Heidelberg: Springer.

9 Rezanka, M. (2019). Synthesis of substituted cyclodextrins. *Environ. Chem. Lett.* 17: 49–63.

10 Rezanka, M. (2018). Synthesis of cyclodextrin derivatives. In: *Cyclodextrin Fundamentals, Reactivity and Analysis* (ed. S. Fourmentin, G. Crini and E. Lichtfouse), 57–103. Cham, Switzerland: Springer-Nature.

11 Guo, C. and Xiao, Y. (2021). Negatively charged cyclodextrins: synthesis and applications in chiral analysis – a review. *Carbohydr. Polym.* 256: 117517.

12 Alexander, J.M., Clark, J.L., Brett, T.J., and Stezowski, J.J. (2002). Chiral discrimination in cyclodextrin complexes of amino acid derivatives: β-cyclodextrin/N-acetyl-l-phenylalanine and N-acetyl-D-phenylalanine complexes. *Proc. Natl Acad. Sci. USA* 99: 5115–5120.

13 Harata, K. (1988). The structure of the cyclodextrin complex. XXI. Crystal structures of *heptakis*(2,6-di-O-methyl)-β-cyclodextrin complexes with *p*-iodophenol and *p*-nitrophenol. *Bull. Chem. Soc. Jpn* 61: 1939–1944.

14 Harata, K., Hirayama, F., Arima, H. et al. (1992). Crystal structure of *heptakis*(2,3,6-tri-O-methyl)-β-cyclodextrin complexes with *m*-iodophenol and 4-biphenylacetic acid. Guest-induced conformational change of a pyranose ring. *J. Chem. Soc., Perkin Trans. 2* 1159–1166.

15 Liu, L. and Guo, Q.-X. (2002). The driving forces in the inclusion complexation of cyclodextrins. *J. Incl. Phenom. Macrocycl. Chem.* 42: 1–14.

16 Salgado, A. and Chankvetadze, B. (2016). Applications of nuclear magnetic resonance spectroscopy for the understanding of enantiomer separation mechanisms in capillary electrophoresis. *J. Chromatogr. A* 1467: 95–144.

17 Chankvetadze, B. (2004). Combined approach using capillary electrophoresis and NMR spectroscopy for an understanding of enantioselective recognition mechanisms by cyclodextrins. *Chem. Soc. Rev.* 33: 337–347.

18 Dodziuk, H., Kozinsky, W., and Ejchart, A. (2004). NMR studies of chiral recognition by cyclodextrins. *Chirality* 16: 90–105.

19 Peluso, P., Dessì, A., Dallocchio, R. et al. (2019). Recent studies of docking and molecular dynamics simulation for liquid-phase enantioseparations. *Electrophoresis* 40: 1881–1896.

20 Peluso, P. and Chankvetadze, B. (2021). Native and substituted cyclodextrins as chiral selectors for capillary electrophoresis enantioseparations: structures, features, applications and molecular modeling. *Electrophoresis* 42: 1676–1708.

21 Samakaschvili, S., Salgado, A., Scriba, G.K.E., and Chankvetadze, B. (2013). Comparative enantioseparation of ketoprofen with trimethylated α-, β-, and γ-cyclodextrin in capillary electrophoresis and study of related selector-selectand interactions using nuclear magnetic resonance spectroscopy. *Chirality* 25: 79–88.

22 Scriba, G.K.E. (2016). Chiral recognition in separation science – an update. *J. Chromatogr. A* 1467: 56–78.

23 Scriba, G.K.E. (2020). Chiral recognition in separation sciences. Part I: polysaccharide and cyclodextrin selectors. *Trends Anal. Chem.* 120: 115639.

24 Gogolashvili, A., Chankvetadze, L., Takaishvili, N. et al. (2020). Separation of terbutaline enantiomers in capillary electrophoresis with neutral cyclodextrin-type chiral selectors and investigation of the structure of selector-selectand complexes using nuclear magnetic resonance spectroscopy. *Electrophoresis* 41: 1023–1030.

25 Chankvetadze, L., Servais, A.C., Fillet, M. et al. (2012). Comparative enantioseparation of talinolol in aqueous and non-aqueous capillary electrophoresis and study of related selector-selectand interactions by nuclear magnetic resonance spectroscopy. *J. Chromatogr. A* 1267: 206–216.

26 Gogolashvili, A., Tatunashvili, E., Chankvetadze, L. et al. (2018). Separation of terbutaline enantiomers in capillary electrophoresis with cyclodextrin-type chiral selectors and investigation of structure of selector-selectand complexes. *J. Chromatogr. A* 1571: 231–239.

27 Varga, E., Benkovics, G., Darcsi, A. et al. (2019). Comparative analysis of the full set of methylated β-cyclodextrins as chiral selectors in capillary electrophoresis. *Electrophoresis* 40: 2789–2798.

28 Szabo, Z.-I., Toth, G., Vöglyi, G. et al. (2016). Chiral separation of asenapine enantiomers by capillary electrophoresis and characterization of cyclodextrin complexes by NMR spectroscopy, mass spectrometry and molecular modeling. *J. Pharm. Biomed. Anal.* 117: 398–404.

29 Connors, K.A. (1997). The stability of cyclodextrin complexes in solution. *Chem. Rev.* 97: 1325–1358.

30 Krait, S., Salgado, A., Villani, C. et al. (2020). Unusual complexation behavior between daclatasvir and γ-cyclodextrin a multiplatform study. *J. Chromatogr. A* 1628: 461448.

31 Kościelkski, T., Sybilska, D., and Jurczak, J. (1983). Separation of α- and β-pinene into enantiomers in gas-liquid chromatography systems via α-cyclodextrin inclusion complexes. *J. Chromatogr.* 280: 131–134.

32 Juza, M. and Schurig, V. (2014). The analytical separation of enantiomers by gas chromatography on chiral stationary phases. In: *Practical Gas Chromatography* (ed. K. Dettmer-Wilde and W. Engewald), 529–575. Heidelberg: Springer.

33 Cagliero, C., Sgorbini, B., Cordero, C. et al. (2021). Separation of stereoisomers by gas chromatography. In: *Gas Chromatography* (ed. C. Poole), 581–614. Amsterdam: Elsevier.

34 He, L. and Beesley, T.E. (2005). Application of enantiomeric gas chromatography: a review. *J. Liq. Chromatogr. Relat. Technol.* 28: 1075–1114.

35 Huang, K., Zhang, X., and Armstrong, D.W. (2010). Ionic cyclodextrins in ionic liquid matrices as chiral stationary phases for gas chromatography. *J. Chromatogr. A* 1217: 5261–5273.

36 König, W.A., Lutz, S., Mischnick-Lübbecke, P. et al. (1988). Cyclodextrins as chiral stationary phases in capillary gas chromatography – part I: pentylated α-cyclodextrin. *J. Chromatogr.* 447: 193–197.

37 Li, W.Y., Jin, H.L., and Armstrong, D.W. (1990). 2,6-Di-O-pentyl-3-O-trifluoroacetyl cyclodextrin liquid stationary phases for capillary gas chromatographic separation of enantiomers. *J. Chromatogr.* 509: 303–324.

38 Schurig, V. and Novotny, H.P. (1988). Separation of enantiomers on diluted permetylated β-cyclodextrin by high-resolution gas chromatography. *J. Chromatogr.* 441: 155–163.

39 Schurig, V. (2005). Contributions to the theory and practice of the chromatographic separation of enantiomers. *Chirality* 17: S205–S226.

40 Pino, V., Lantz, A.W., Anderson, J.L. et al. (2006). Theory and use of the pseudophase model in gas-liquid chromatographic enantiomeric separations. *Anal. Chem.* 78: 113–119.

41 Patil, R., Weatherly, C.A., and Armstrong, D.W. (2018). Chiral gas chromatography. In: *Chiral Analysis – Advances in Spectroscopy, Chromatography and Emerging Methods*, 2e (ed. P.L. Polavarapu), 469–506. Amsterdam: Elsevier.

42 Berthod, A., Li, W., and Armstrong, D.W. (1992). Multiple enantioselective retention mechanisms on derivatized cyclodextrin gas chromatographic chiral stationary phases. *Anal. Chem.* 64: 873–879.

43 Sicoli, G., Kreidler, D., Czesla, H. et al. (2009). Gas chromatographic enantioseparation of unfunctionalized chiral alkanes: a challenge in separation science. *Chirality* 21: 183–198.

44 Meierhenrich, U., Nguyen, M.-J., Barbier, B. et al. (2003). Gas chromatographic separation of saturated aliphatic hydrocarbon enantiomers on permethylated β-cyclodextrin. *Chirality* 15: S13–S16.

45 Kowalska, T. and Sherma, J. (ed.) (2007). *Thin Layer Chromatography in Chiral Separations and Analysis*. Boca Raton, FL: CRC Press.

46 Bereznitski, Y., Thompson, R., O'Neil, E., and Grinberg, N. (2001). Thin-layer chromatography – a useful technique for the separation of enantiomers. *J. AOAC Int.* 84: 1242–1251.

47 Günther, K., Richter, P., and Möller, K. (2004). Separation of enantiomers by thin-layer chromatography. In: *Chiral Separations – Methods and Protocols* (ed. G. Gübitz and M.G. Schmid), 29–59. Totowa, NJ: Humana Press.

48 Harada, A., Furue, M., and Nozakura, S. (1978). Optical resolution of mandelic acid derivatives by column chromatography on crosslinked cyclodextrin gels. *J. Polym. Sci., Polym. Chem. Ed.* 16: 189–196.

49 Fujimura, K., Veda, T., and Ando, T. (1983). Retention behavior of some aromatic compounds on chemically bonded cyclodextrin silica stationary phase in liquid chromatography. *Anal. Chem.* 55: 446–450.

50 Kawaguchi, Y., Tanaka, M., Nakae, M. et al. (1983). Chemically bonded cyclodextrin stationary phase for liquid chromatographic separation of aromatic compounds. *Anal. Chem.* 55: 1852–1857.

51 Armstrong, D.W. and DeMond, W. (1984). Cyclodextrin bonded phases for the liquid chromatographic separation of optical, geometrical, and structural isomers. *J. Chromatogr. Sci.* 22: 411–415.

52 Xiao, Y., Ng, S.-C., Tan, W.T.T.Y., and Wang, Y. (2012). Recent development of cyclodextrin chiral stationary phases and their applications in chromatography. *J. Chromatogr. A* 1269: 52–68.

53 Mitchell, C.R. and Armstrong, D.W. (2004). Cyclodextrin-based chiral stationary phases for liquid chromatography: a twenty-year overview. In: *Chiral Separations – Methods and Protocols* (ed. G. Gübitz and M.G. Schmid), 61–112. Totowa, NJ: Humana Press.

54 Zhou, J., Tang, J., and Tang, W. (2015). Recent development of cationic cyclodextrins for chiral separations. *Trends Anal. Chem.* 65: 22–29.

55 Li, X., Yao, X., Xiao, Y., and Wang, Y. (2017). Enantioseparation of single layer native cyclodextrin chiral stationary phases: effect of cyclodextrin orientation and a modeling study. *Anal. Chim. Acta* 990: 174–184.

56 Zhou, J., Ang, B., Tang, J., and Tang, W. (2018). A cationic cyclodextrin clicked bilayer chiral stationary phase for versatile chiral separation in HPLC. *New J. Chem.* 42: 3526–3533.

57 Adly, F.G., Antwi, N.Y., and Ghanem, A. (2016). Cyclodextrin-functionalized monolithic capillary columns: preparation and chiral applications. *Chirality* 28: 97–109.

58 Guo, J., Lin, Y., Xiao, Y. et al. (2016). Recent developments on cyclodextrin functionalized monolithic columns for the enantioseparation of chiral drugs. *J. Pharm. Biomed. Anal.* 130: 110–125.

59 Spudeit, D.A., Breitbach, Z.S., Dolzan, M.D. et al. (2015). Superficially porous particle based hydroxypropyl-β-cyclodextrin stationary phase for high-efficiency enantiomeric separations. *Chirality* 27: 788–794.

60 Berthod, A., Chang, S.-C., and Armstrong, D.W. (1992). Empirical procedure that uses molecular structure to predict enantioselectivity of chiral stationary phases. *Anal. Chem.* 64: 395–404.

61 Teixeira, J., Tiritan, M.E., Pinto, M.M.M., and Fernandes, C. (2019). Chiral stationary phases for liquid chromatography: recent developments. *Molecules* 24: 865.

62 Zhou, J., Yang, B., Tang, J., and Tang, W. (2016). Cationic cyclodextrin clicked chiral stationary phase for versatile enantioseparations in high-performance liquid chromatography. *J. Chromatogr. A* 1467: 169–177.

63 Felletti, S., Ismail, O.H., De Luca, C. et al. (2019). Recent achievements and future challenges in supercritical fluid chromatography for the enantioselective separation of chiral pharmaceuticals. *Chromatographia* 82: 65–75.

64 West, C. (2014). Enantioselective separations with supercritical fluids – review. *Curr. Anal. Chem.* 10: 99–120.

65 Galea, C.M., Vander Heyden, Y., and Mangelings, D. (2017). Separation of stereoisomers. In: *Supercritical Fluid Chromatography* (ed. C.F. Poole), 345–379. Amsterdam: Elsevier.

66 Tang, R.-Q., Ong, T.-T., and Ng, S.-C. (2012). Chemically bonded cationic β-cyclodextrin derivatives and their applications in supercritical fluid chromatography. *J. Chromatogr. A* 1224: 97–103.

67 Fejös, I., Kalydi, E., Malanga, M. et al. (2020). Single isomer cyclodextrins as chiral selectors in capillary electrophoresis. *J. Chromatogr. A* 1627: 461375.

68 Wren, S.A.C. and Rowe, R.C. (1992). Theoretical aspects of chiral separation in capillary electrophoresis. I. Initial evaluation of a model. *J. Chromatogr. A* 603: 235–241.

69 Chankvetadze, B., Lindner, W., and Scriba, G.K.E. (2004). Enantiomer separations in capillary electrophoresis in the case of equal binding constants of the enantiomer with a chiral selector: commentary on the feasibility of the concept. *Anal. Chem.* 76: 4256–4260.

70 Chankvetadze, B. (2018). Contemporary theory of capillary electrophoresis. *J. Chromatogr. A* 1567: 2–25.

71 Rizzi, A. (2001). Fundamental aspects of chiral separations by capillary electrophoresis. *Electrophoresis* 22: 3079–3106.

72 Müllerova, L., Dubský, P., and Gaš, B. (2014). Twenty years of development of dual and multi-selector models in capillary electrophoresis: a review. *Electrophoresis* 35: 2688–2700.

73 Chankvetadze, B., Schulte, G., and Blaschke, G. (1997). Nature and design of enantiomer migration order in chiral capillary electrophoresis. *Enantiomer* 2: 157–179.

74 Chankvetadze, B. (2002). Enantiomer migration order in chiral capillary electrophoresis. *Electrophoresis* 23: 4022–4035.

75 Chankvetadze, B., Schulte, G., and Blaschke, G. (1996). Reversal of the enantiomer elution order in capillary electrophoresis using charged and neutral cyclodextrins. *J. Chromatogr. A* 732: 183–187.

76 Sabah, S. and Scriba, G.K.E. (1998). Electrophoretic stereoisomer separation of aspartyl dipeptides and tripeptides in untreated fused-silica and polyacrylamide-coated capillaries using charged cyclodextrins. *J. Chromatogr. A* 822: 137–145.

77 Zhu, Q. and Scriba, G.K.E. (2016). Advances in the use of cyclodextrins as chiral selectors in capillary electromigration techniques – fundamentals and applications. *Chromatographia* 79: 1403–1435.

78 Krait, S., Konjaria, M.-L., and Scriba, G.K.E. (2021). Advances of capillary electrophoresis enantioseparations in pharmaceutical analysis (2017–2020). *Electrophoresis* 42: 1709–1725.

79 El Deeb, S., Wätzig, H., Abd El-Hady, D. et al. (2016). Recent advances in capillary electrophoretic migration techniques for pharmaceutical analysis (2013–2015). *Electrophoresis* 37: 1591–1608.

80 Saz, J.M. and Marina, M.L. (2016). Recent advances on the use of cyclodextrins in the chiral analysis of drugs by capillary electrophoresis. *J. Chromatogr. A* 1467: 79–94.

81 Znaleziona, J., Fejös, I., Sevcik, J. et al. (2015). Enantiomeric separation of tapentadol by capillary electrophoresis – study of chiral selectivity manipulation by various types of cyclodextrins. *J. Pharm. Biomed. Anal.* 105: 10–16.

82 Herrero, M., Simo, C., Garcia-Canas, V. et al. (2010). Chiral capillary electrophoresis in food analysis. *Electrophoresis* 31: 2106–2114.

83 Hernandez-Borges, J., Rodrigues-Delgado, M.A., Garcia-Montelongo, F.J., and Cifuentes, A. (2005). Chiral analysis of pollutants and their metabolites by capillary electromigration methods. *Electrophoresis* 26: 3799–3813.

84 Willis, P.A., Creamer, J.S., and Mora, M.F. (2015). Implementation of microchip electrophoresis instrumentation for future spaceflight missions. *Anal. Bioanal. Chem.* 407: 6939–6963.

85 Salido-Fortuna, S., Castro-Puyana, M., and Marina, M.L. (2020). Chiral micellar electrokinetic chromatography. *J. Chromatogr. A* 1626: 461383.

86 Xie, R., Chu, L.-Y., and Deng, J.-G. (2008). Membranes and membrane processes for chiral resolution. *Chem. Soc. Rev.* 37: 1243–1263.

87 Clark, J.D., Han, B., Bhown, A.S., and Wickramasinghe, S.R. (2005). Amino acid resolution using supported liquid membranes. *Sep. Purif. Technol.* 42: 201–211.

88 Fernandes, C., Tiritan, M.E., and Pinto, M.M.M. (2017). Chiral separation in preparative scale: a brief overview of membranes as tools for enantiomeric separation. *Symmetry* 9: 206.

89 Liu, T., Li, Z., Wang, J. et al. (2021). Solid membranes for chiral separation: a review. *Chem. Eng. J.* 410: 128247.

90 Choi, H.-J., Ahn, Y.-H., and Koh, D.-Y. (2021). Enantioselective mixed matrix membranes for chiral resolution. *Membranes* 11: 279.

91 Welliver, M., McDonough, J., Kalynych, N., and Redfern, R. (2008). Discovery, development, and clinical application of sugammadex sodium, a selective relaxant binding agent. *Drug Des. Devel. Ther.* 2: 49–59.

92 Coisne, C., Tilloy, S., Monflier, E. et al. (2016). Cyclodextrins as emerging therapeutic tools in the treatment of cholesterol-associated vascular and neurodegenerative diseases. *Molecules* 21: 1748.

93 Rincon-Lopez, J., Almanza-Arjona, Y.C., Riascos, A.P., and Rojas-Aguirre, Y. (2021). Technological evolution of cyclodextrins in the pharmaceutical field. *J. Drug Delivery Sci. Technol.* 61: 102156.

94 Conceicao, J., Adeoye, O., Cabral-Marques, H.M., and Sousa Lobo, J.M. (2018). Cyclodextrins as drug carriers in pharmaceutical technology: the state of art. *Curr. Pharm. Des.* 24: 1405–1433.

95 Gonzalez Pereira, A., Carpena, M., Garcia Oliveira, P. et al. (2021). Main applications of cyclodextrins in the food industry as the compounds of choice to form host-guest complexes. *Int. J. Mol. Sci.* 22: 1339.

96 McQueen, R.H. and Vaezafshar, S. (2020). Odor in textiles: a review of evaluation methods, fabric characteristics, and odor control technologies. *Text. Res. J.* 90: 1157–1173.

97 Healy, B., Yu, T., da Silva Alves, D.C. et al. (2021). Cyclodextrins as supramolecular recognition systems: applications in the fabrication of electrochemical sensors. *Materials* 14: 1668.

98 Ogoshi, T. and Harada, A. (2008). Chemical sensors based on cyclodextrin derivatives. *Sensors* 8: 4961–4982.

99 Bai, C.C., Tian, B.R., Zhao, T. et al. (2017). Cyclodextrin-catalyzed organic synthesis: reactions, mechanisms, and applications. *Molecules* 22: 1475.

100 Liu, Y., Lin, T., Cheng, C. et al. (2021). Research progress on synthesis and application of cyclodextrin polymers. *Molecules* 26: 1090.

101 Varan, C., Anceschi, A., Sevli, S. et al. (2020). Preparation and characterization of cyclodextrin nanosponges for organic toxic molecule removal. *Int. J. Pharm.* 585: 119485.

9

Pirkle Type

Maria Elizabeth Tiritan, Madalena Pinto, and Carla Fernandes

Laboratório de Química Orgânica e Farmacêutica, Departamento de Ciências Químicas, Faculdade de Farmácia da Universidade do Porto, Porto, Portugal
Centro Interdisciplinar de Investigação Marinha e Ambiental (CIIMAR), Matosinhos, Portugal

9.1 Introduction

Pirkle-type or brush-type chiral stationary phases (CSPs) are based in low-molecular-size chiral selectors covalently bonded to the chromatographic support via a spacer [1]. Although the chromatographic selectors are small molecules, they can establish multiple interactions with the analytes through chiral discrimination leading to enantioselectivity. This event showed in CSPs generally arises from a combination of hydrogen bond, dipole–dipole, π–π interactions, ionic, and van der Waals attractions and repulsions. Usually, the Pirkle-type CSPs comprise π-donor and/or π-acceptor moieties as attractive π–π interaction sites [2, 3].

One of the main advantages of Pirkle-type CSPs is the facility of preparation processes by a rational strategy planning [2]. In fact, the selectors can be designed to create interactions with a specific chiral analyte through predictable enantiorecognition mechanisms since Pirkle-type CSPs comprise small molecules as chiral selectors [4, 5]. The expectedness enantiorecognition is one of the driving forces for the planning/design, development, and improvement of various Pirkle-type CSPs. These studies are also essential to clarify the type of interactions between a CSP and the enantiomers to estimate the enantioselectivity and the elution order [2, 3].

Chiral Separations and Stereochemical Elucidation: Fundamentals, Methods, and Applications, First Edition. Edited by Quezia Bezerra Cass, Maria Elizabeth Tiritan, João Marcos Batista Junior, and Juliana Cristina Barreiro.

The design and development of this type of CSPs started in the 1970s by Professor William H. Pirkle (1934–2018) and coworkers resulting in the first commercial chiral column launched in 1981. Then, successive generations of CSPs were developed by Pirkle's group for high-performance liquid chromatography (HPLC) [2, 3], based on the principle of reciprocity [6, 7] and also on studies of chiral recognition phenomena, established by the analysis of chromatographic [8, 9] and spectroscopic data [10, 11]. The development of efficient enantioselective Pirkle-type chiral selectors using in silico tools was also reported, exploring chemical combinatorial libraries together with high throughput screening [12, 13]. Over the years, computational studies of chromatographic separations have become a necessary tool for elucidation of the chiral recognition mechanisms for diverse Pirkle-type CSPs [14], particularly Whelk-O1® CSP, one of the most useful and applied CSPs, which has been more investigated [15–22].

The development of several generations of CSPs developed by Pirkle and coworkers was an inspiration to other research groups to dive into this fascinating field [1, 23]. To follow the constant challenges on different areas, such as medicinal chemistry [24], food [25], biomedical and environmental analysis [26], and forensic and toxicological sciences [27, 28], the development of new Pirkle-CSPs has been a continuous and evolutionary subject [29]. The strategies encompassed the development of new chiral selectors, introduction of new chromatographic supports, and application of different synthetic approaches for preparation of CSPs [30]. The referred approaches also considered the follow-up of the instrumental developments and technical progress associated with chromatography separation techniques.

Currently, some companies commercialize Pirkle-type CSPs, but Regis Technologies is the leading manufacturer [31]. Designed in partnership with Professor William Pirkle, and with nearly 50 years of experience, nine CSPs (Phenylglycine®, β-Gem 1®, Leucine®, Whelk-O1®, Whelk-O2®, Pirkle 1-J®, α-Burke 2®, ULMO®, and DACH-DNB®) are commercially available in a broad range of particle sizes and dimensions for analytical to preparative scale separations. All of them can be used not only for liquid chromatography (LC) but also for supercritical fluid chromatography (SFC) applications, offering a range of selectivity and distinct advantages for broadening the scope of applications.

But where did Pirkle's "inspiration" to develop CSPs come from? In the following sections, the answer to this question will be given and a summary of the extraordinary work developed by Pirkle and coworkers will be presented.

9.2 CSPs Developed by Pirkle's Group: Chronological Evolution

In Figure 9.1, some of the most important landmarks of the evolution of the successive generations of CSPs developed by Pirkle's group are briefly summarized. Nuclear magnetic resonance (NMR) studies with chiral

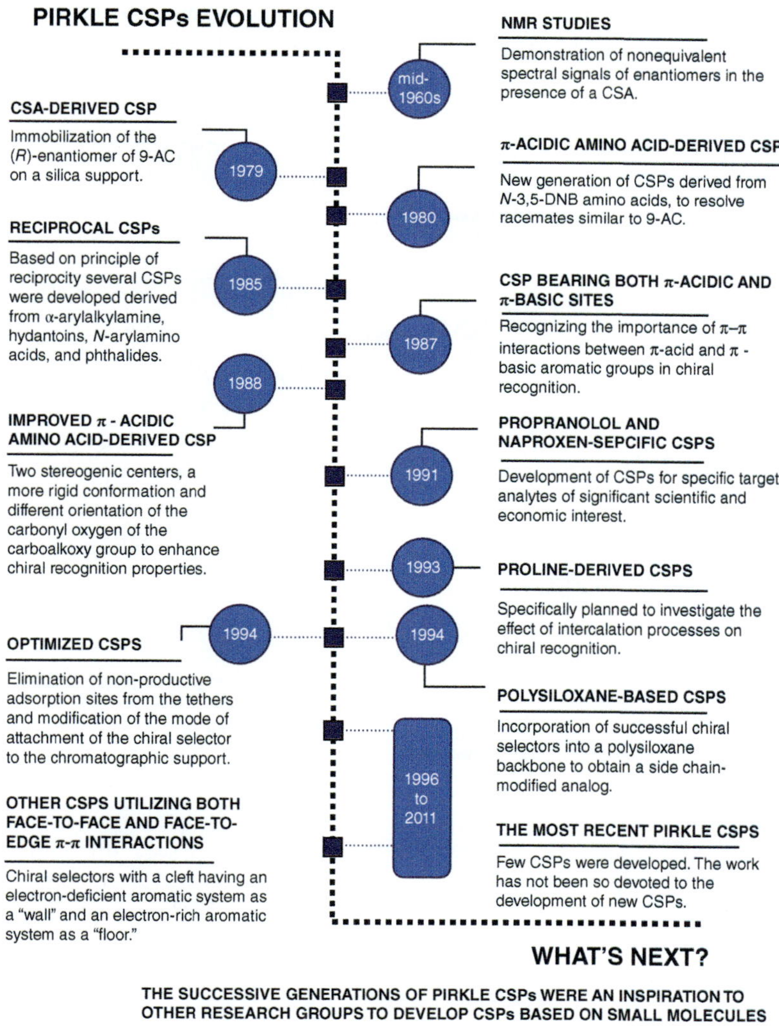

Figure 9.1 Timeline of events related to the history of CSPs developed by Pirkle's group.

solvating agents (CSAs) were the impulse for planning the first CSPs. In the mid-1960s, Pirkle and coworkers demonstrated non-equivalent NMR spectral signals of enantiomers in the presence of a CSAs [32–47]. To elucidate the differences observed on spectral data, they performed molecular models [46, 48].

In 1979, they immobilized the (*R*)-enantiomer of the most broadly useful CSA, 2,2,2-trifluoro-1-(9-anthryl)ethanol (9-AC) [45–47], on a silica support to develop a CSP (Figure 9.2, CSP-1). This CSP has shown capability to enantioseparate a wide variety of π-acidic racemates [3, 49–51], such as

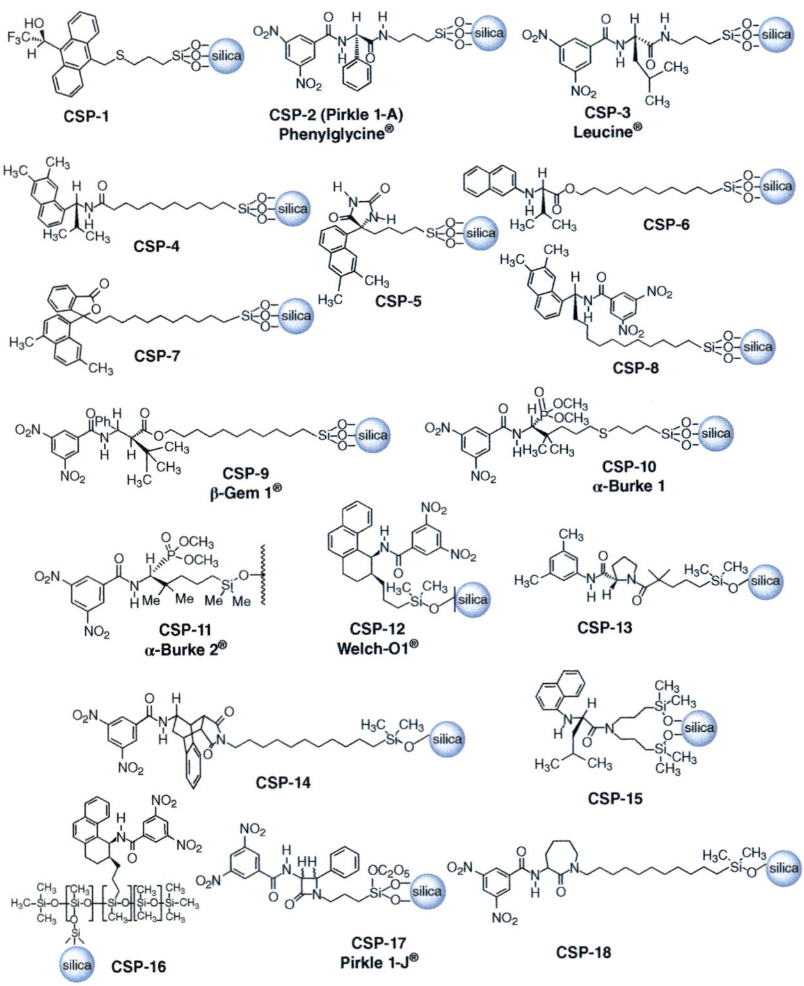

Figure 9.2 Examples of CSPs developed by Pirkle's group.

N-3,5-dinitrobenzoyl (*N*-3,5-DNB) derivative of phenylglycine with a good separation factor ($\alpha = 1.78$) [49].

In 1980, regarding the chromatographic results obtained with **CSP-1**, a new generation of CSPs derived from π-acidic amino acids was developed and evaluated to resolve racemates like the alcohol 9-AC. One of the most successful CSPs was derived from *N*-3,5-DNB phenylglycine (Figure 9.2, **CSP-2**), which was able to separate a broad spectrum of racemates, including drugs, on both analytical and preparative scales [3, 6, 8, 49, 51–65]. Concomitantly with the development and evaluation of these CSPs, Pirkle and coworkers proposed chiral recognition models to investigate the mechanisms concerning enantioselectivity. One of these models was based on the chromatographic resolution of *N*-acylated 1-aryl-1-aminoalkanes by **CSP-2** [53], and a representation of the molecular recognition is shown in Figure 9.3.

CSP-2 was commercialized in 1981 as the Pirkle 1A column, the first commercially available HPLC column for the chromatographic separation of enantiomers. Nowadays, it is commercially available as Phenylglycine. Another analog CSP comprising *N*-3,5-DNB leucine as chiral selector (**CSP-3**) was also commercialized as Leucine [31].

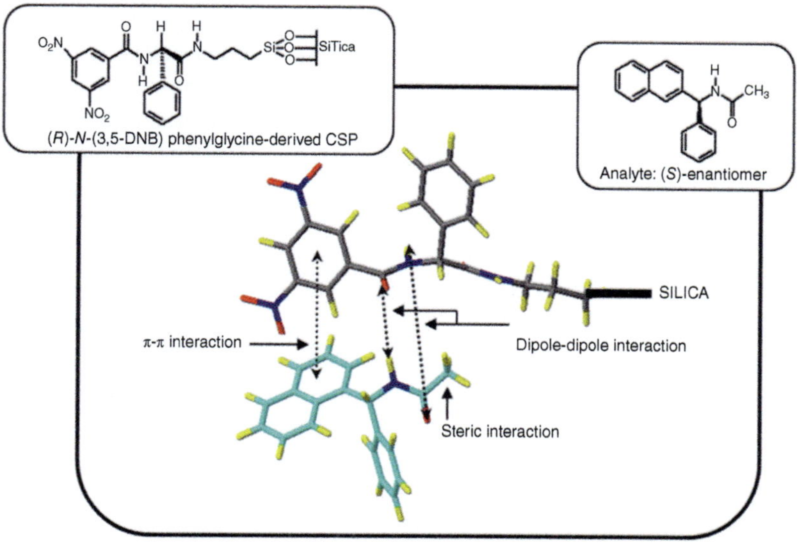

Figure 9.3 Chiral recognition model involving the interactions between **CSP-2** and the most retained enantiomer of an analyte. *Source:* Reproduced from Fernandes et al. [2]/with permission of Springer.

Then, based on the principle of reciprocity, a significant number of CSPs, were developed derived from α-arylalkylamine [66], hydantoins [9], *N*-aryl amino acids [67], among others. The concept of reciprocity predicts that if an enantiomer of A, immobilized on an inert support, can distinguish between the enantiomers of B, therefore an immobilized enantiomer of B should, using similar interactions, be able to distinguish between the enantiomers of A [6]. The development and evaluation of reciprocal α-arylalkylamine-derived CSPs, such as **CSP-4** (Figure 9.2), allowed the study of some aspects of chiral recognition, namely, the influence of the nature of the linkage of the selector to the tether [66], and the effect of the inter-strand distance [68]. Some rational improvements were also performed on the development of hydantoin-derived CSPs to increase the enantiomeric resolution [9]. For example, it was found that the electron donating ability of the methyl groups of **CSP-5** (Figure 9.2), increased the π-basicity of the aromatic ring and, as a consequence, the strength of π–π interactions [9]. CSPs derived from *N*-aryl-α-amino acids, such as *N*-(2-naphthyl)valine (**CSP-6**, Figure 9.2) [67], include a combination of moieties such as π-donor, basic and acidic sites organized in space allowing multiple and complementary interactions with analytes for an efficient enantioresolution [67]. Also based on studies of chiral recognition mechanisms [61], a phthalide-derived CSP (**CSP-7**, Figure 9.2) was prepared and evaluated, embodying a bulky alkyl group and a methyl-substituted naphthyl ring system, providing good enantioselectivity [69].

At that stage, Pirkle and coworkers had already a great deal of knowledge of the mechanisms of interactions among the developed CSPs and several analytes. By recognizing the importance of π–π interactions between π-acid and π-basic aromatic groups in chiral recognition, a CSP was developed in 1987 which bears both π-acid *N*-3,5-DNB group and π-basic 1-(6,7-dimethyl) naphthyl group (**CSP-8**, Figure 9.2). **CSP-8** was able to separate the enantiomers of both π-acid or π-basic analytes [70].

Besides, constant improvements and refinements were carried out considering both the rational planning as well as the knowledge of chiral recognition mechanisms involved on CSPs. Such kind of improvement upon the widely studied *N*-3,5-DNB phenylglycine-derived CSP (**CSP-2**) led to the development of **CSP-9** (Figure 9.2), in 1988, comprising two stereogenic centers, a more rigid conformation and a different orientation of the carbonyl oxygen of the carboalkoxy group [62]. In some cases, a better enantioresolution was achieved compared to that of **CSP-2**. A remarkable result was the separation of the enantiomers of underivatized naproxen [71]. **CSP-9** was commercialized as β-Gem1®.

Just till 1989, the strategy used by Pirkle's group for the planning of CSPs was mainly focused on mechanistic studies, which allowed the development of several CSPs with great applicability. Some of the rational principles were sufficiently understood to make possible the development of target-directed CSPs. Thus, diverse chiral selectors for specific target analytes of high scientific and economic interest were synthetized, specifically for propranolol (β-blocker drug) and naproxen, a non-steroidal anti-inflammatory drug (NSAID) [72]. The efforts made to develop CSPs for propranolol and other β-blockers drugs were a success [73]. **CSP-10**, named as α-Burke 1, comprising geminal dimethyl groups responsible for a steric barrier (Figure 9.2), is one example [74]. **CSP-11** (Figure 9.2), derived from **CSP-10**, was commercialized as α-Burke 2, and it is not restricted to the enantioseparation of β-blockers, many other enantiomers can be separated on this π-acceptor CSP [31, 75].

Regarding naproxen-directed CSPs, first, the (*S*)-enantiomer was immobilized on a support of silica to screen candidates for naproxen selectors. A number of candidates were found to display a usable level of enantioselectivity. A mechanistic hypothesis concerning the enantiorecognition of naproxen was lately formulated and used to design an improved naproxen selector [72, 76]. The strategy applied by Pirkle's group was named "immobilized guest method" and was an extension of many studies carried out with reciprocal CSPs. By analysis of the chiral recognition mechanisms and the structural features relevant for enantioselectivity [76], **CSP-12** (Figure 9.2) was developed, which provided excellent enantioresolution of naproxen and other NSAIDs [77]. This CSP is commercially available as Welch-O1® CSP and is the Pirkle-type CSP that, still nowadays, has the broadest application in academic and industrial laboratories. As a general rule, enantiomers that exhibit a hydrogen bond acceptor and an aromatic moiety close to the stereogenic center tend to be well separated on this CSP [78]. Welch-O1 is widely used for enantioseparation due to its semi-rigid framework (cleft type) holding two aromatic moieties: a π-acidic 3,5-dinitrobenzamide group, as a "wall," and perpendicular to a π-polynuclear aromatic group, as a "floor" (Figure 9.4). The amide NH in the cleft formed by the two aromatic systems works as a hydrogen bond donor [79]. These structural features allow at least three simultaneous interactions to occur: hydrogen bonding, face-to-edge π–π, and face-to-face π–π interactions [77].

Recently, a covalent trifunctional version of the Whelk-O1 CSP was commercialized as Whelk-O2. This CSP retains the same chiral selector but incorporates a trifunctional linkage to the silica support. Whelk-O2 was designed to improve the resistance of the CSP to hydrolysis while using strong organic modifiers such as trifluoroacetic acid. It is ideal for preparative separations since the material is bonded on 10 μm, 100 Å spherical Kromasil silica [31].

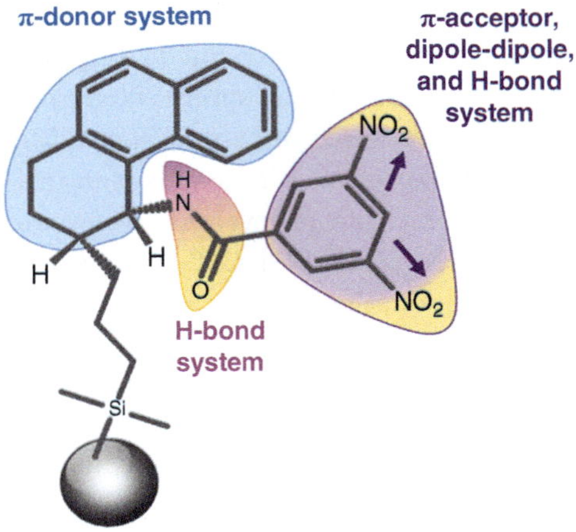

Figure 9.4 Structure of the Welch-O1 CSP and the features of the different fragments of the chiral selector. *Source:* Reproduced from West and Khater [79]/ with permission of Elsevier.

In 1993, new CSPs derived from L-proline were also developed (for example **CSP-13**, Figure 9.2) for investigation of the effect of intercalation processes on chiral recognition [80].

Considering the success of Welch-O1, another stage on the evolution of Pirkle's work started in 1994 through the development of other CSPs with structural features allowing both face-to-face and face-to-edge π–π interactions [3]. All chiral selectors presented a cleft having an electron-deficient aromatic system as a "wall" and an electron-rich aromatic system as a "floor." One example is the **CSP-14** (Figure 9.2), comprising π-basic and π-acidic rings perpendicular oriented that increase the rigidity of the global structure [81].

Also in 1994, Pirkle and coworkers began to develop another generation of CSPs, after optimization and modification of existing CSPs. In general, the optimized CSPs showed higher enantioselectivity [3]. One of the improvements implicated the elimination of non-productive adsorption sites from the tethers of the CSPs, which were responsible for "unnecessary" interactions on chiral recognition that increased the analyte retention without distinguishing between enantiomers [82]. The modification of the mode of attachment of the chiral selector to the chromatographic support was also a progress [83]. **CSP-15** (Figure 9.2), in which the chiral selector was connected by a doubly tethered tertiary amide-link, is a representative example [84]. The incorporation of successful chiral selectors into a polysiloxane backbone was another strategy (for example **CSP-16**, Figure 9.2) [85].

Although having essentially the same scope and attributes, in general, the polysiloxane-based CSPs afforded much less retention and demonstrated greater enantioselectivity and resolution. The increase in enantioselectivity was attributed, in part, to the reduction of undesired non-specific analyte interactions with the support silanols owing to effective shielding of the surface by the polymer [85].

From 1996 to 2011, Pirkle's group research was not so much devoted to the development of new CSPs; to the best of our knowledge, only three new CSPs were developed [86–88].

CSP-17 (Figure 9.2), derived from a conformational rigid α-amino β-lactam system, was developed for the resolution of the enantiomers of several β-blockers and analogs [88]. Its unusual β-lactam structure significantly alters its molecular recognition properties. **CSP-17** afforded improved resolution compared to the less rigid **CSP-10** (α-Burke 1) derived from an α-amino phosphonate [89] and was commercialized as Pirkle 1-J.

CSP-18 (Figure 9.2) was the latest CSP described by Pirkle and coworkers for the separation of a series of π-basic racemic analytes. Along with other CSPs, including **CSP-3**, it was used to resolve the enantiomers of their precursors to explore the chiral self-recognition phenomenon, in addition to other analytes such as some drugs [90].

Over decades, Pirkle and coworkers studied a variety of NMR CSAs, planned and developed many new CSPs, and performed a series of chromatographic enantioresolution studies. The strategy applied to design new CSPs emphasizes the perception of the chiral molecular mechanisms of recognition. Pirkle's work was a pivotal source to provide efficient chiral columns that are still nowadays one of the most used in both analytical and preparative chromatographic applications.

Pirkle's CSPs were first designed mainly for enantioseparation of analytes comprising aromatic moieties and to operate under normal elution conditions. However, these CSPs are able to be operated in polar, ionic, and reversed elution modes. Many important applications were allowed by the possibility to use reversed mode of elution [91–93].

Pirkle's studies gave a great input for the understanding of the chiral recognition phenomena with contributions that paved the way for the subsequent advances in the discovery, development, and manufacture of enantiopure pharmaceuticals, agrochemicals, and chemicals. Moreover, his research group at the University of Illinois became a key source of other researchers that are a reference in enantioseparation field, who grow up under his guidance. Part of the story of Pirkle's life and work, including some professional and personal memories, was recently collected by Christopher J. Welch as a tribute to his mentor and friend [94].

9.3 Pirkle-Type CSPs Developed by Other Research Groups

The history of planning and development of CSPs by Pirkle's group is recognized as a great inspiration to other researchers to generate new CSPs. There are some fundamental reviews that describe in detail the vast repertory of new CSPs obtained by covalent linkage of low-molecular-mass selectors to the chromatographic support [1, 23]. Among them, it is of note the review by Gasparrini et al. [23], which extensively revised the achievements on this type of CSPs until 2000, with emphasis on those obtained by total synthesis. Furthermore, a literature survey covering the report on Pirkle-type CSPs from January 2000 to March 2017 was reported by Fernandes et al. [1], that emphasizes the chemical nature of the new chiral selectors, new insights in the development strategies, and their applications in LC.

In fact, since the introduction of the first CSP, a wide number of new Pirkle-type CSPs has been developed by different research groups and their applications in several areas of research have grown exponentially. Some of Pirkle-type CSPs were closely related to the originally described CSPs, comprising amino acids or analogs (amines or amino alcohol) with π-basic and π-acidic aromatic moieties (Figure 9.5). Among them, **CSP-19** is of remark, which was developed by the Austrian researchers Uray, Lindner, and Maier [95, 96] and commercialized as ULMO CSP. The ULMO CSP is based on a 3,5-DNB derivative of diphenylethylenediamine and has a general ability to enantioseparate many classes of compounds, particularly the enantiomers of aryl carbinols [31]. Another relevant example is **CSP-20**, designed in the 1990s by the Italian chemists Gasparrini, Misiti, and Villani [97, 98] and commercialized as DACH-DNB CSP. The DACH-DNB CSP contains the 3,5-DNB derivative of *trans*-(1*R*,2*R*)-diaminocyclohexane (DACH) and proved to be very effective to resolve a broad range of racemate classes [23, 99].

Some research groups followed the same approach of Pirkle's group by developing CSPs without "superfluous" adsorption sites that can be responsible for increased retention and reduced enantioselectivity [82]. A series of CSPs based on the original Pirkle *N*-(3,5-DNB)-L-leucine CSP are one example of this strategy [100]. The leucine-derived **CSP-21** (Figure 9.5), containing a 3,5-dimethylphenyl substituent in the amide-attaching tether, demonstrated the best performance.

CSPs comprising various amino acids, such **CSP-22** (Figure 9.5), have been prepared and evaluated [101]. Other CSPs comprising linear [102–104] and cyclopeptide-bonded CSPs [105] were also reported. Wu et al. described

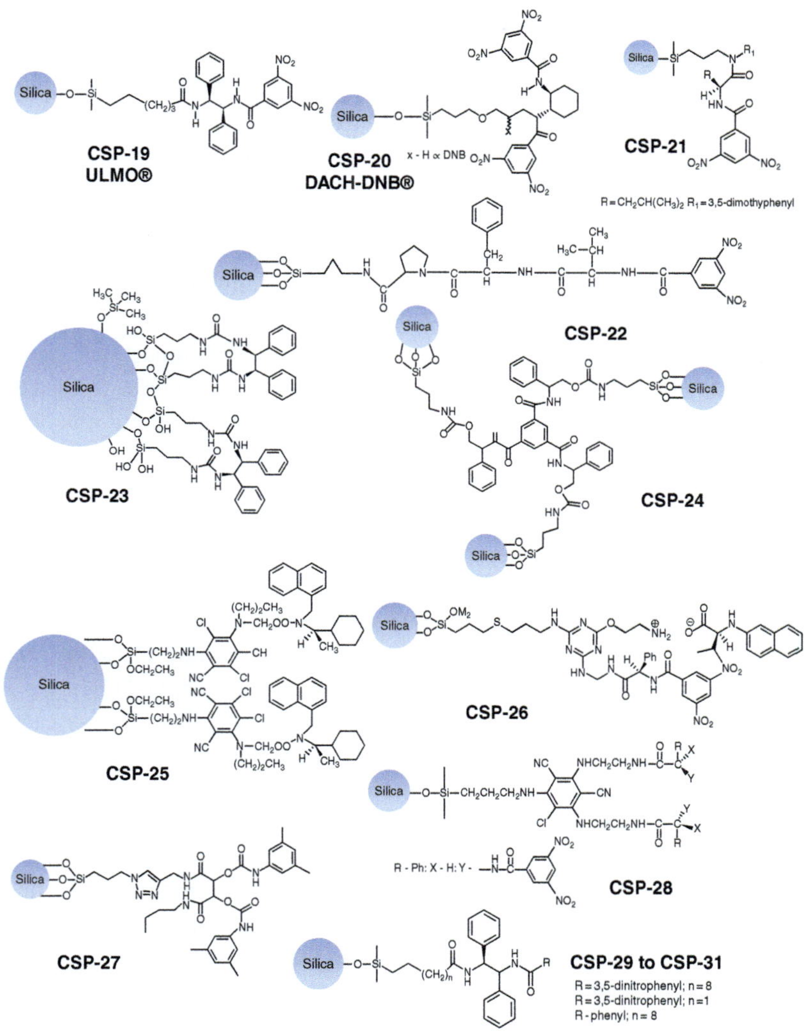

Figure 9.5 Examples of CSPs comprising selectors related to those developed by Pirkle's group.

several CSPs based on poly-*N*-substituted glycines, also called peptoids, the first ones based on peptoid oligomers with α-chiral aromatic side chains [106]. They studied several features including the influence of the chain length [106], terminal group of peptoid selector [107, 108], and configuration of peptoid units [109, 110] on enantiomeric separation performance.

Even though they were chemically similar to the CSPs from Pirkle's group, several of the Pirkle-type CSPs developed were structurally organized in a different way. For example, the "crab-like" **CSP-23** has been successfully prepared from 1,2-diphenyl 1,2-ethylenediamine as chiral selector covalently bonded to silica gel through a bidentate tether [111]. Other examples are the development of C3 symmetric CSPs [112, 113], such as the (*R*)-phenylglycinol *N*-1,3,5-benzenetricarboxylic acid-derived CSP (**CSP-24**) (Figure 9.5) [113].

Another interesting and efficient strategy used to increase enantioselectivity involved the development of CSPs comprising a biselector or combinatorial selectors (mixture of chiral selectors in different proportions), such as **CSP-25** [114] and **CSP-26** (Figure 9.5) [115], which comprise the selectors of two original Pirkle's CSPs [49, 116]. The enantiorecognition performance exhibited by the biselector CSPs when compared with the monoselector CSPs demonstrated the versatility of this approach [115].

Research to obtain new CSPs followed several synthetic strategies including non-conventional synthetic approaches, such as click chemistry, solid-phase, and in situ procedures [30]. Through click chemistry, selectors based on *O,O′*-diaryl tartardiamide and *O,O′*-bis-(arylcarbamoyl) tartardiamine were synthesized and immobilized on silica gel, affording CSPs such as **CSP-27** [117]. By solid-phase synthesis, the amino acid-derived CSPs were developed, including **CSP-28**, doubling the number of chiral units per each binding site to the silica, resulting in an increased number of recognition sites [118]. A versatile, innovative, and *in situ* procedure was followed to prepare new acylated Pirkle-type CSPs, such as **CSP-29** to **CSP-31** [119]. The key step of this method was the preparation of *N*-hydroxysuccinimide activated silica, followed by the immobilization of the enantiomeric pure diamine with only one single amino group through a simple substitution reaction. Then, the free second amino group was acylated with different acyl halogenides, affording the CSPs [119].

Like Pirkle's approach to develop CSPs for specific target analytes, such as β-blockers [72–74], NSAIDs [72, 77], and thiazide diuretics [87], some research groups followed the same concept to design new CSPs (Figure 9.6). For example, **CSP32** and **CSP-33** were developed for the specific resolution of racemic 2-aryloxypropionic acids [120], while **CSP-34** to **CSP-38** for separation of BINAPO enantiomers [121], and **CSP-39** [122] along with three CSP analogs (**CSP-40** to **CSP-42**) [123] to separate pyrethroid insecticides.

Another strategy was the design of CSPs from drugs and analogs as chiral selectors (Figure 9.6). **CSP-43** and **CSP-44**, based on the cephalosporin antibiotic cefaclor [124, 125], and **CSP-45** and **CSP-46**, based on the NSAID (*S*)-naproxen and its analog, are illustrative examples [121].

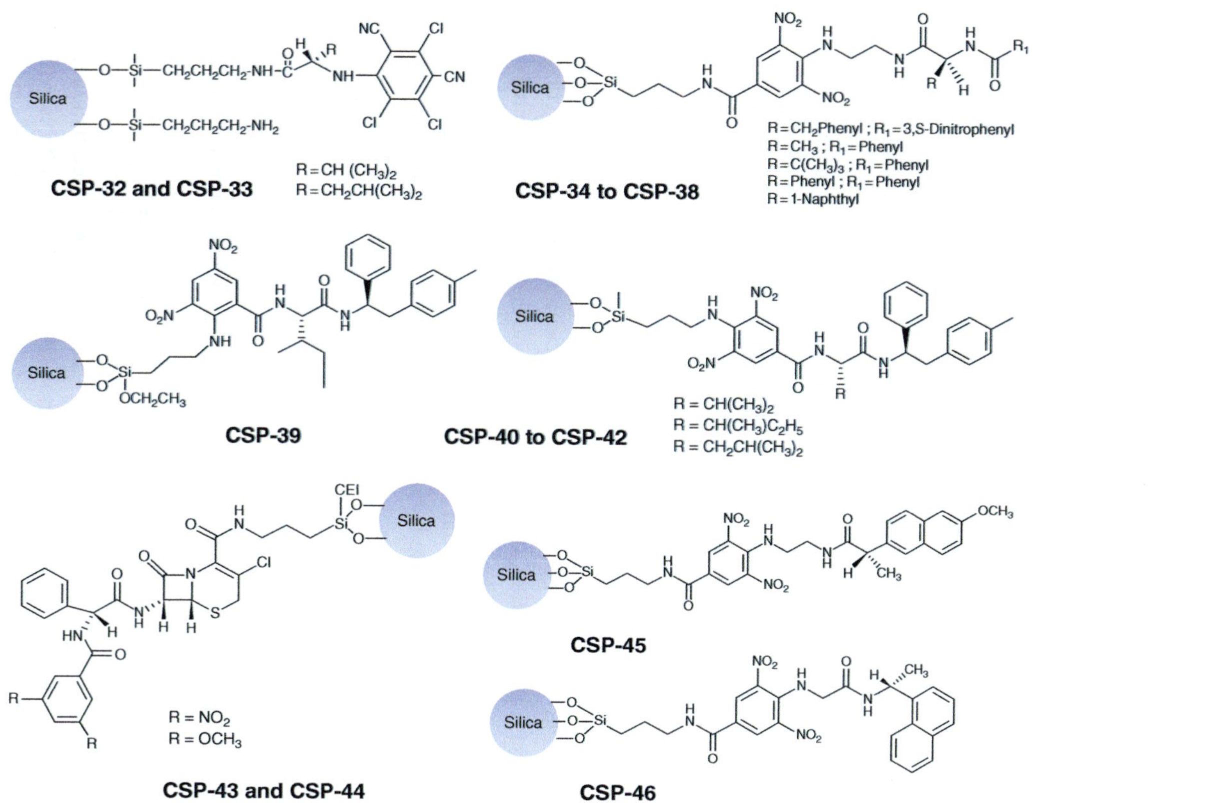

Figure 9.6 Example of CSPs design for specific target analytes (**CSP-32** to **CSP-42**) and from drugs and analogs as chiral selectors (**CSP-43** to **CSP-46**).

Other designed Pirkle-type CSPs were chemically different from those "classically" developed by Pirkle and coworkers (Figure 9.7). Some of them have chiral selectors obtained from natural sources or analogs, such as the dehydroabietic acid and friedelin (**CSP-47** to **CSP-49**) [126] or a terpenoid compound (**CSP-50**) [127]. Chiral derivatives of xanthones (CDXs), in addition to interesting biological activities [128], were bound to a chromatographic support and explored as CSPs, by our group [129, 130]. Some examples are displayed in Figure 9.7 (**CSP-51** to **CSP-54**). In general, they showed high specificity for the enantioseparation of CDXs mainly under normal-phase elution conditions.

Several chiral ion-exchange CSPs, based on the cinchona alkaloids quinine and quinidine, have been developed by Lindner's group (see Chapter 11). Considered as a subset of Pirkle-type CSPs, these CSPs comprise ionizable moieties, being the ionic interactions between the oppositely charged selectors and the analytes pivotal for retention and enantiodiscrimination [14].

Innovation was given not only to the chiral selectors but also to the chromatographic support of the CSPs. Therefore, new approaches to improve column efficiency were explored, including the reduction of the particle size [131], the introduction of core-shell particles [132, 133], and other support materials, such as monolithic [134, 135].

The reduction of the particle size of the chromatographic support was considered an important breakthrough in the field of chromatography as it allowed an improvement on the kinetic performance of the column, an improvement on its response signal and in the sensibility of analytical methods [131]. In addition, this approach met the demand for faster analysis with high column efficiency, associated to a decrease of solvent consumption and amount of chiral selector needed for the preparation of the CSP [136]. The first CSP based on sub-2 μm silica particles was prepared by Gasparrini's group through covalent attachment of the selector of Whelk-O1 CSP to spherical, high-surface area 1.8 μm porous silica particles [131, 137]. The new sub-2-μm Whelk-O1 CSP presented high stability and a wide range of applications for several compounds in considerably short analysis times [138, 139]. A sub-2 μm version of Whelk-O1 CSP is commercially available. The same group also immobilized the chiral selector of DACH-DNB CSP to 1.9 μm silica particles [131, 137].

The introduction of monoliths appeared as a feasible alternative to some traditional particle-based packing materials [140]. A monolithic Pirkle-type CSP was obtained via an "in-column modification" process which included immobilization of a proline-derived chiral selector using click chemistry [134]. In another example, an "in-column" preparation of a new

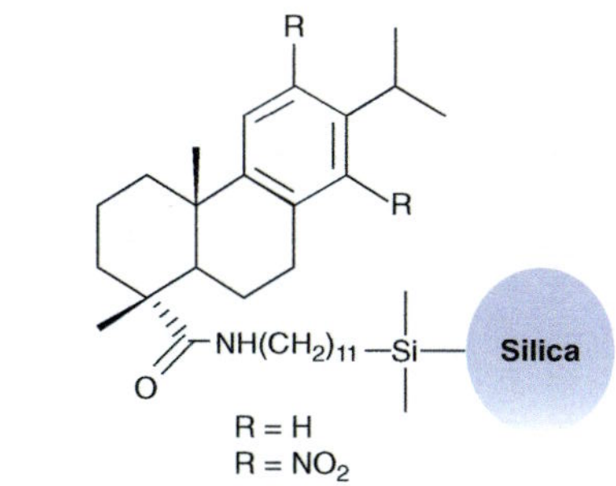

CSP-47 and **CSP-48**

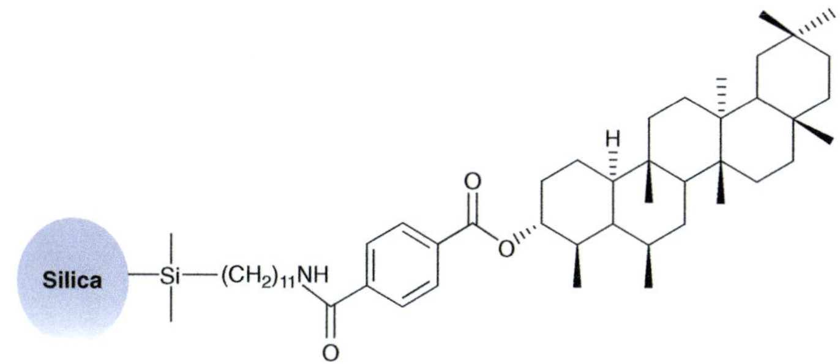

CSP-49

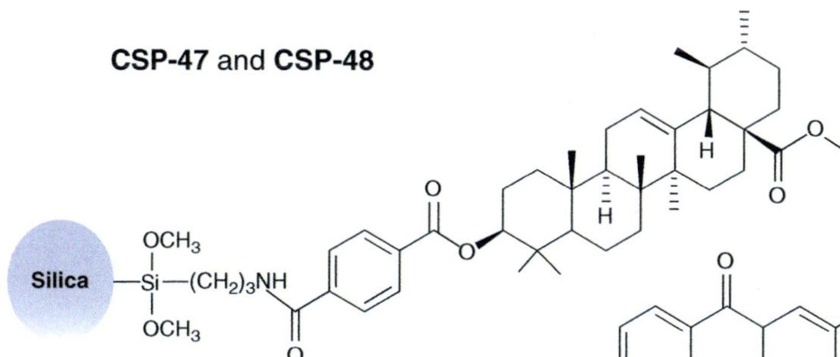

CSP-50

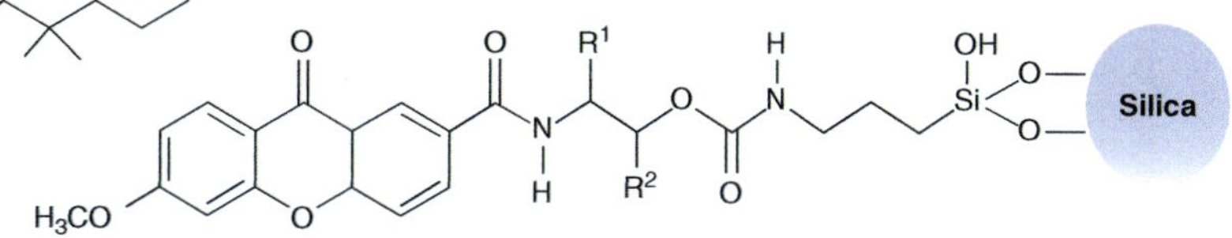

CSP-51 to **CSP-54**

(*S*) R^1 = isopropyl R^2 = H
(*R*) R^1 = phenyl R^2 = H
(*S*) R^1 = phenyl R^2 = H
(*R,S*) R^1 = phenyl R^2 = phenyl

Figure 9.7 Example of CSPs chemically different to those "classically" developed by Pirkle and coworkers.

monolithic Pirkle-type CSP containing (*R*)-acryloyloxy-β-β-dimethyl-γ-butyrolactone as chiral selector was described [135].

The Pirkle-type Whelk-O1 CSP was also prepared on 2.6 μm core-shell particles [132]. In addition, Regis commercialized Whelk-O1 core CSP bonded to 2.7 μm core-shell particles for faster, highly efficient separations without increasing system back-pressure [141]. A CSP obtained from chiral core-shell silica microspheres with DACH was synthesized using the layer-by-layer method [142]. Rapid and efficient enantioresolutions of several chiral analytes with C2 symmetrical axes, including binaphthyl derivatives and biphenanthrol, were obtained.

9.4 Example of Applications in Analytical and Preparative Scales

Compared to other CSPs, that are frequently employed, such as polysaccharides (Chapter 6), macrocyclic antibiotics (Chapter 7), cyclodextrins (Chapter 8), or protein-based (Chapter 10), the great advantage of Pirkle-type phases is the possibility to control the configuration of the chiral selectors and the option to have the chiral selectors in both enantiomeric forms, as the example of Whelk-O1 which is commercially available as (*R,R*) and (*S,S*) configurations. This feature gives the opportunity to control the elution order of the enantiomers by choosing the adequate configuration of the chiral selector. The control of the elution order of the enantiomers is very useful in analytical applications as, for example, in determination of the enantiomeric purity because the trace enantiomer is preferably eluted before the major one, thus avoiding being misplaced by the tailing peak of the main enantiomer [143]. For preparative scale applications, the desired enantiomer is better to elute first because it can be obtained with higher enantiomeric purity than the second eluted enantiomer (see Chapter 5). Although their loadability is not as high as that reported for polysaccharide CSPs (Chapter 6), Pirkle-type CSPs have higher loadability than the CSPs based on proteins, cyclodextrins, and macrocyclic glycopeptides, being suitable for enantioseparation in preparative scale applications by LC [24], SFC, and also chiral membranes [144]. Moreover, this type of CSPs has other advantages, including universal solvent compatibility, and stability at broader temperature and pressure ranges [129]. Their main field of application is in LC and more recently in SFC [79, 145]. The enantioseparation and the range of the different class of compounds are always described when a new CSP is reported, but practical applications are restricted to few commercialized columns. Herein, selected examples of enantiomeric purity evaluation of bioactive substances and few biomedical and environmental applications are described to give an overview of the scope of the applications.

9.4.1 Analytical Applications

Enantioselective LC methods for enantiomeric ratio determination are essential for the evaluation of the enantiomeric purity of synthetic, semi-synthetic, or natural biological active compounds. Concerning synthetic compounds with potential biological activity, a series of CDXs, prepared by the chiral pool approach, was evaluated in L-Phenylglycine® and (*S,S*)-Whelk-O1® columns using multimodal elution conditions [15]. As expected, the capability of L-Phenylglycine to resolve the CDXs series was much lower than that of (*S,S*)-Whelk-O1. Polar-organic conditions presented the best chromatographic parameters for analytical applications allowing excellent resolutions with lower run time than both normal and reversed elution conditions. (*S,S*)-Whelk-O1, under polar-organic conditions, demonstrated high resolution for the CDXs possessing an aromatic moiety linked to the stereogenic center with enantioselectivity (α) and resolution (Rs) ranging from 1.91 to 7.55 and from 6.71 to 24.16, respectively. Docking studies considering the results for (*S,S*)-Whelk-O1 yielded scores in accordance with the chromatographic results regarding α and the enantiomeric elution order. Figure 9.8 shows the most stable docked conformations of both enantiomers of 6-methoxy-9-oxo-*N*-(1-(p-tolyl) ethyl)-9*H*-xanthene-2-carboxamide (CDX-3) complexed to the (*S,S*)-Whelk-O1 selector. Later, the enantiomeric purity of the series of CDXs was evaluated by (*S,S*)-Whelk-O1 columns under polar-organic conditions [128].

The determination of enantiomeric purity in enantioselective synthesis is not problematic when both enantiomers are available, and the chromatographic conditions allow good α and Rs values. The enantiomeric purity of a series of 32 CDXs was evaluated by comparing the retention time of the target enantiomer with the chromatograms of the enantiomeric mixtures. The chromatogram of the "pure" enantiomer, obtained by the chiral pool synthesis process, with 1% solution of the opposite enantiomer was used to calculate the enantiomeric ratio as illustrated in Figure 9.9. The enantiomeric purity of all 32 CDXs was close to 100%.

The determination of the enantiomeric ratio can be far more complicated either when the substance is within a complex mixture from a natural source, the α value and Rs are low and reference samples of the minor enantiomer, or of the racemate, are not easily available [143]. Detection of minor enantiomer in the absence of reference samples was reported by the

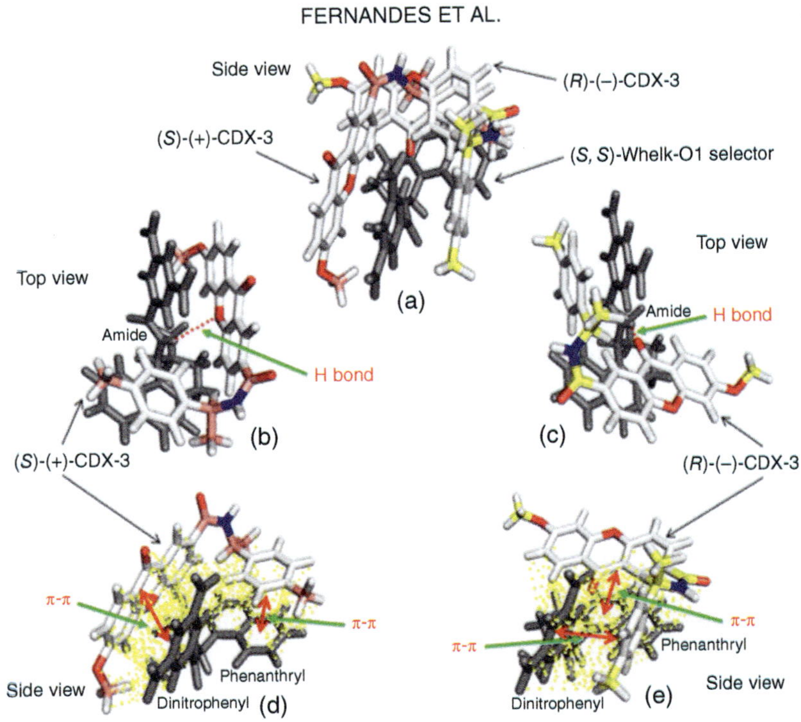

Figure 9.8 The most stable docked conformations of both enantiomers of 6-methoxy-9-oxo-*N*-(1-(p-tolyl) ethyl)-9*H*-xanthene-2-carboxamide (CDX-3) complexed to (*S*,*S*)-Whelk-O1 selector (gray): (a) superimposed CDX–CSP complexes, H-bonding interactions of (b) (*S*)-(+)-CDX-3 and (c) (*R*)-(−)-CDX-3, and π−π interactions of (d) (*S*)-(+)-CDX-3 and (e) (*R*)-(−)-CDX-3. The non-aromatic carbon atoms of (*S*)-(+)-CDX-3 and of (*R*)-(−)-CDX-3 are represented in pink and yellow, respectively; red dashes represent H-bonding interactions; yellow dashes represent all the other interactions; π−π interactions are highlighted with a red double edge arrow. *Source:* Adapted from Fernandes et al. [15].

combination of LC "Inverted Chirality Columns Approach" and MS/MS detection in the evaluation of semi-synthetic derivatives of camptothecin using the (*S*,*S*)- and (*R*,*R*)-DACH-DNB® columns (Figure 9.10).

The analysis of gimatecan and other semi-synthetic derivatives demonstrated that it was possible to determine the minor enantiomer even when only one standard enantiomer is available by calculating the retention and α values for the virtual racemate and also by choosing the configuration of the CSP that allows the trace enantiomer to elute first [143]. It can be straightforward if both enantiomeric forms of the CSP are available by switching

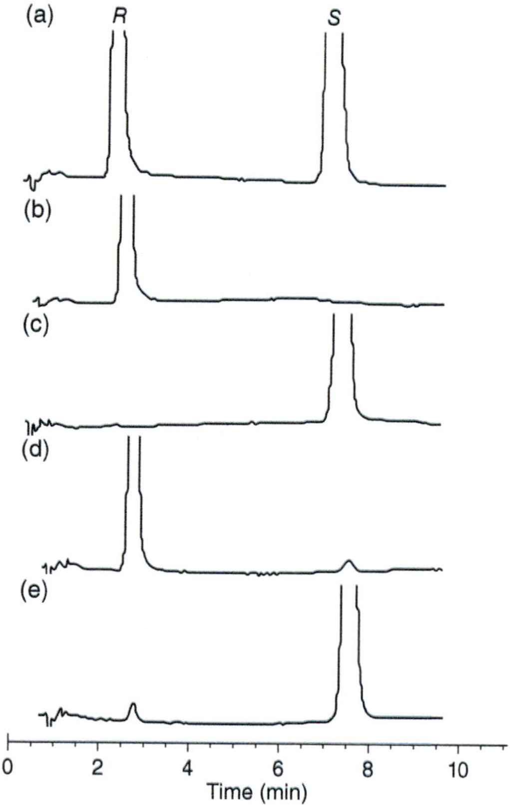

Figure 9.9 (a) Chromatogram of enantiomeric mixture, (b) chromatogram of (*R*)-(−)-CDX, (c) chromatogram of (*S*)-(+)-CDX, (d) chromatogram of (*R*)-(−)-CDX spiked with 1% of (*S*)-(+)-CDX, (e) chromatogram of (*S*)-(+)-CDX spiked with 1% of (*R*)-(−)-CDX. Chromatographic conditions: MeOH/ACN (50 : 50 v/v), flow rate 1.0 ml min^{-1}. Column: (*S,S*)-Whelk-O1 CSP and detection wavelength 254 nm. *Source:* Adapted from Carraro et al. [128].

between the two CSPs to reverse the elution order of the enantiomeric pair, according to the reciprocal principle of selectand–selector-systems. This strategy enables a precise and accurate quantification by peak area integration of the trace enantiomer (Figure 9.10). The same strategy was also followed for trace quantification of the (+)-enantiomer of Δ^9-*trans*-tetrahydrocannabinol in ethanol extracts from medicinal marijuana (Bedrocan®) using the (*S,S*)- and (*R,R*)-Whelk-O1 CSPs in the sub-2 μm format by SFC (example showed in Figure 9.11) [146]. The (+)-enantiomer was found in 0.135%, which should not be negligible in bioactivity evaluation.

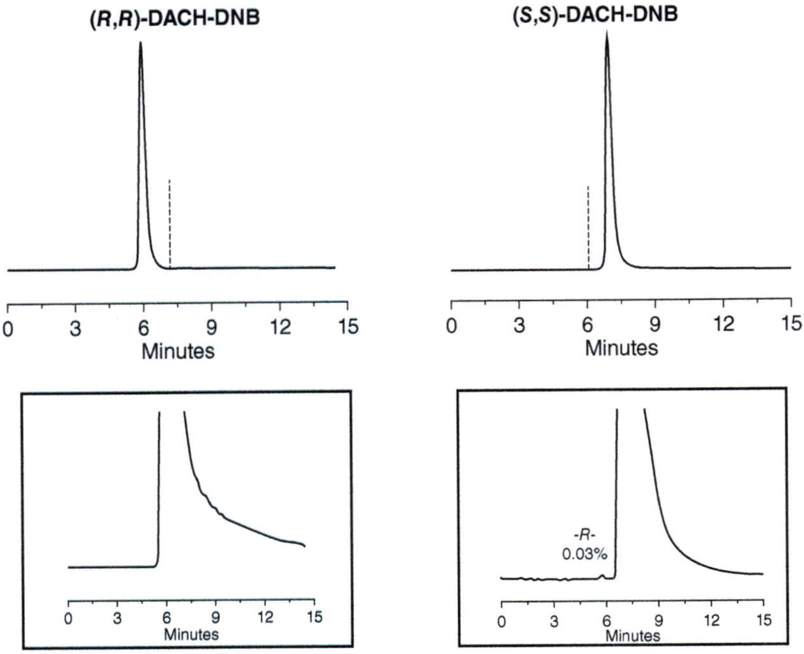

Figure 9.10 Separation of (*S*)-gimatecan from its (*R*)-enantiomer with low concentration of the (*R*)-enantiomer using the DACH-DNB columns and MS detection. *Source:* Reproduced from Carraro et al. [128]/with permission of American Chemical Society.

Most practical applications use the famous Whelk-O1 CSPs, but the first generation of Pirkle columns has also been applied in many recent works for quantification of amino acids after previous derivatization. Regarding biomedical applications, metabolomics for novel biomarker screening based on D-amino acids have been reported with 2D-LC systems with a Pirkle-type *N*-3,5-DNB derivative of amino acids column in the second dimension. D-Amino acids have been recognized as physiologically active molecules and have been proposed as biomarkers of different and relevant human pathologic states. The trace amounts of D-amino acids were screened for prognosis of chronic kidney disease [147]. This study examined whether multiple chiral amino acids would be associated with kidney functions, comorbidities, and prognosis of chronic kidney disease. For that, the amino acids were derivatized with 4-fluoro-7-nitro-2,1,3-benzoxadiazole (NBD-F), separated in a reversed-phase column (monolithic ODS) in the first dimension and the target amino acids were automatically collected using a multi-loop valve, and transferred to the enantioselective KSAACSP-001S or

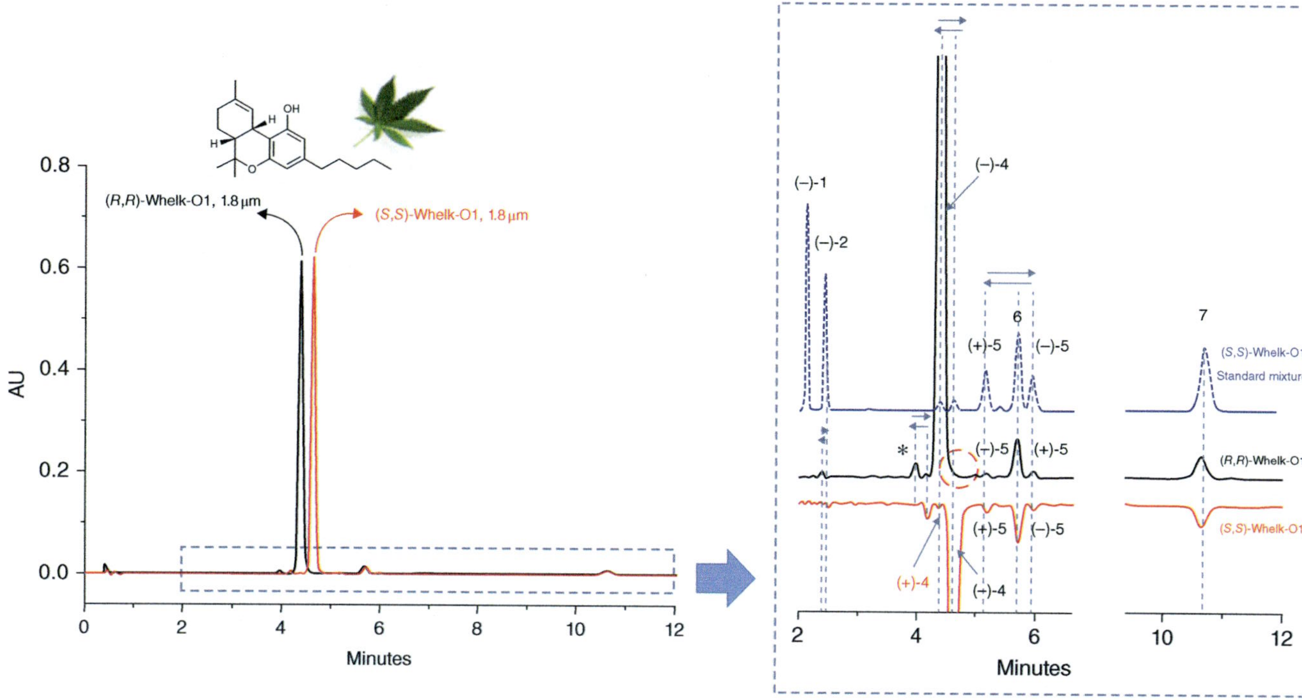

Figure 9.11 Bedrocan ethanol extract analyzed by applying the "Inverted Chirality Columns Approach" protocol. For peak identification, the chromatogram of a six-component cannabinoid standard mixture is shown in the inset as a dotted line trace. The asterisk denotes a chiral unknown impurity. *Source:* Reproduced from Carraro et al. [128]/Royal Society of Chemistry.

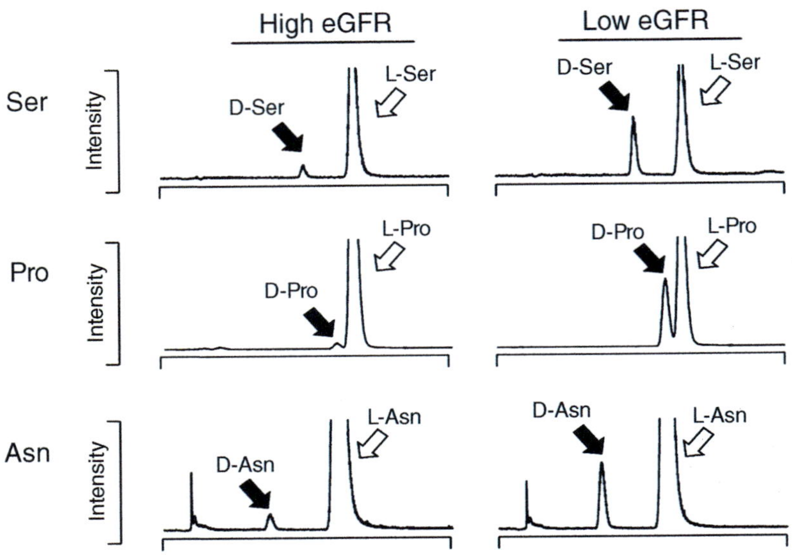

Figure 9.12 Detection of 4-fluoro-7-nitro-2,1,3-benzoxadiazole (NBD)-labeled amino acid enantiomers in the plasma of patients with chronic kidney disease using a 2D-LC-MS/MS system. eGFR: estimated glomerular filtration ratio. *Source:* Reproduced from Kimura et al. [147]/with permission of Springer/Public Domain CC BY 4.0.

a Sumichiral OA-3200 columns (*N*-3,5-DNB derivative of L-leucine and *tert*-leucine, respectively) in the second dimension (see also Chapter 1). The MS/MS detection was employed to confirm the presence of D-amino acids (Figure 9.12); 16 out of 21 D-amino acids were detected in plasma from 108 patients in a longitudinal cohort. Here, the selection of the precise configuration of the CSP is important to guarantee that the trace D-amino acid will be the first enantiomer eluted.

In another study with 2D-LC to quantify hydrophilic amino acids in mammalian tissues and physiological fluids, a microbore-ODS column was used in the first dimension and the narrowbore Sumichiral OA-2500S (*N*-3,5-DNB derivative of 1-naphthylglycine) was used in the second dimension to separate NBD-labeled amino acids [148] (Figure 9.13).

Despite the possibility of enantioseparation of amino acids without any previous derivatization, the selectivity can be improved by derivatization and also to expand the choices of detection methods. Recently, Prof. Kenji Hamse and coworkers reported the quantification of D-phenylglicine in plasma and urine samples using 2D-LC [149]. In another study, a three-dimensional (3D)-LC method was also established for the determination of

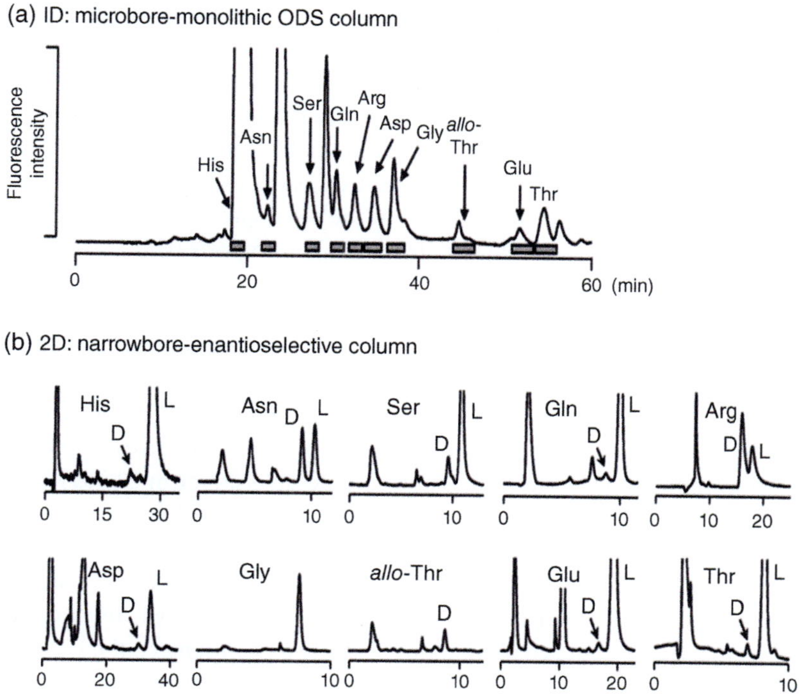

Figure 9.13 Separation of NBD-derivatized hydrophilic amino acids in rat urine using the described 2D-LC system: a microbore-ODS column (a) and the continuously interlinked enantiomer separation using a narrowbore-Sumichiral OA-2500S column (b). The fractions indicated by gray bars are on-line collected to a multi-loop device and successively transferred to the enantioselective column. *Source:* Reproduced from Hamase et al. [148]/with permission of Elsevier.

the D-amino acids of asparagine, serine, alanine, and proline in the plasma of patients with chronic kidney disease [150] and also to quantify the enantiomeric ratio of alanine, 2-aminobutyric acid, valine, norvaline, and isovaline in extraterrestrial samples [151], all in combination with fluorescence derivatization with NBD-F. The NBD-aminoacids were enantioseparated in the second and third dimensions by a Pirkle-type KSAACSP-001S column and monitored by fluorescence detectors [149–151].

Even though there are a variety of analytical methodologies to monitor the quality of foods and to evaluate the level of environmental contaminations, chiral analyses is needed to assure the safety and the correct risk assessment since they allow the quantification of both enantiomers. However, the development of enantioselective analytical methods to quantify residues of chiral pollutants in environmental and food matrices is a challenging task for analytical chemists due to the complexity of the matrices and the need to develop multi-target methods [26]. The enantiomeric fraction (EF) [152] is a key parameter to assess the environmental fate and distribution of chiral pollutants since its variation depends on biological processes, such as biosorption, biodegradation, biotransformation, uptake, and bioaccumulation, and/or has impact on toxicological effects. In this sense, many efforts have been applied in the development of enantioselective analyses for environmental matrices [26] but reports with Pirkle-type columns are scarce. Under our knowledge, only one LC–MS/MS method has been described with Pirkle-type CSPs to monitor surface water. (*S*,*S*)-Whelk-O®1 was applied to enantioseparate profens and warfarin in a dual multi-target analytical method that included different therapeutic classes of chiral pharmaceuticals (beta-blockers, antidepressants, one beta-2-adrenergic agonist, NSAIDs and stimulants) and illicit drugs, such as cocaine and its metabolites, and amphetamines in a survey in Douro River, Portugal [93]. The enantioseparation of the pesticide haloxyfop has also been reported with (*R*,*R*)-Whelk-O®1 by LC in normal elution mode to evaluate the behavior of haloxyfop in four soil samples. Enantioselective degradation was found with EF values ranging from 0.058 to 0.61. In order to determine the absolute configurations of the two enantiomers, the UV and CD chromatograms of the racemate and (*R*)-haloxyfop were compared at 225 nm. The results indicated the first eluted enantiomer was *S*-(−)-haloxyfop and the second was *R*-(+)-haloxyfop [153].

Regarding food analysis, besides the importance of chiral analysis to investigate toxicity and bioaccumulation paths of environmental contaminants (e.g. pharmaceuticals and pesticides), it is a valuable tool to evaluate food quality and genuineness, that is, to determine the geographical origin of a certain sample aiming to find out fraud or adulteration [25]. D-Amino acids have been identified in various foods, particularly in fermentation systems, such as vinegar, beer, and wine. The presence of D-amino acids can be associated to inadequate food processing under conditions of fermentation or adulteration that can induce the racemization of amino acids. The same strategy used in biomedical applications with NBD-amino acids can be used in food analyses. For instance, the determination of amino acids in various fermented products using a 2D-LC method with MS/MS detection and the KSAACSP-001S column to enantioseparate the NBD-amino acids in the second dimension was reported [154]. A method for quality control of a L-cysteine in dietary supplements also used NBD-labeled amino acids, but here the (*R*,*R*)-Whelk-O1® column was applied for the enantioseparation under reserved elution mode. Baseline enantioseparation was achieved in less than 3 minutes. The derivatives were then detected using negative

ESI-MS in selected reaction monitoring mode. The identity of the reaction product was confirmed by high resolution mass spectrometry (HRMS). The analysis and quantification of native cysteine remain a critical concern since its thiol group is highly reactive and easily evolves to the disulfide cystine derivative through oxidation reactions [155].

9.4.2 Preparative Applications

The majority of preparative enantioseparations are performed using polysaccharide-based or Pirkle-type phases (see Chapter 5). For small scale (milligram to gram), only a limited number of CSPs and mobile phases are selected for screening. Previous experience demonstrated that the use of three different polysaccharide-based CSPs and the Whelk-O1 CSP enable resolution of 80% of the compounds in an efficient way and only the remaining 20% was subjected for further screenings [156].

The membrane technologies for enantioseparations have also emerged as a platform for large-scale enantioseparations. Advantages include low energy consumption, continuous operation, and high processing capacity as well as the possibility of simple scale-up and integration into other separation processes. Recently, chiral resolution capacity of mixed matrix membranes (MMMs) containing silicon-based mesoporous array (named MCM-41) and grafted with the Pirkle-type chiral selector D-(3,5-dinitrobenzoyl)-α-phenylglycine was reported [144]. The performance of the MMMs was evaluated through filtration tests with the analyte 9-AC (Pirkle's alcohol) with good enantioselectivity ($\alpha = 1.28$).

9.5 Conclusions and Perspectives

Pirkle-type CSPs are based in low-molecular-size chiral selectors that can establish enantioselective interactions with the analytes through a combination of hydrogen bond, dipole–dipole, π–π interactions, ionic, and van der Waals attractions and repulsions. Usually, the Pirkle-type CSPs comprise π-donor and/or π-acceptor moieties as attractive π–π interaction sites. The main advantages of Pirkle-type CSPs are the possibility to design the selector by a rational strategy planning due to the predictable chiral recognition mechanisms. This strategy is crucial for the planning/design, development, and improvement of CSPs for target analytes, which motivated many researchers to develop new CSPs based on Pirkle-type CSPs. The possibility to control the configuration of the selectors and to have them in both enantiomeric forms, as the example of the famous Whelk-O1 CSPs, available as the (*R*,*R*) and (*S*,*S*) configurations, is a crucial feature to control the elution order of the enantiomers, which is very useful in both analytical and preparative scale applications.

Most practical applications employed the Whelk-O1 CSPs, but the first generation of Pirkle columns has also been applied in many recent works for quantification of amino acids after previous derivatization.

All the aforementioned data are an inspiration to carry out new research in an area with an increasingly societal impact.

References

1 Fernandes, C., Phyo, Y.Z., Silva, A.S. et al. (2018). Chiral stationary phases based on small molecules: an update of the last 17 years. *Separation and Purification Reviews* 47 (2): 89–123.

2 Fernandes, C., Tiritan, M.E., and Pinto, M. (2013). Small molecules as chromatographic tools for HPLC enantiomeric resolution: Pirkle-type chiral stationary phases evolution. *Chromatographia* 76 (15–16): 871–897.

3 Welch, C.J. (1994). Evolution of chiral stationary phase design in the Pirkle laboratories. *Journal of Chromatography A* 666 (1–2): 3–26.

4 Scriba, G.K.E. (2012). Chiral recognition mechanisms in analytical separation sciences. *Chromatographia* 75 (15–16): 815–838.

5 Scriba, G.K.E. (2016). Chiral recognition in separation science – an update. *Journal of Chromatography A* 1467: 56–78.

6 Pirkle, W.H. and Däppen, R. (1987). Reciprocity in chiral recognition. Comparison of several chiral stationary phases. *Journal of Chromatography A* 404 (C): 107–115.

7 Schurig, V. (2016). The reciprocal principle of selectand-selector-systems in supramolecular chromatography. *Molecules* 21 (11): 1535.

8 Pirkle, W.H., Hyun, M.H., Tsipouras, A. et al. (1984). A rational approach to the design of highly effective chiral stationary phases for the liquid chromatographic separation of enantiomers. *Journal of Pharmaceutical and Biomedical Analysis* 2 (2): 173–181.

9 Pirkle, W.H. and Hyun, M.H. (1985). Preparation and use of hydantion-based chiral stationary phases. *Journal of Chromatography A* 322 (C): 309–320.

10 Pirkle, W.H., Murray, P.G., and Wilson, S.R. (1996). X-ray crystallographic evidence in support of a proposed chiral recognition mechanism. *Journal of Organic Chemistry* 61 (14): 4775–4777.

11 Pirkle, W.H. and Selness, S.R. (1995). Chiral recognition studies: intra- and intermolecular 1H{1H}-nuclear overhauser effects as effective tools in the study of bimolecular complexes. *The Journal of Organic Chemistry* 60 (10): 3252–3256.

12 Blodgett, J., Wang, Y., Li, T. et al. (2002). Resolution of tert-butyl-1-(2-methylnaphthyl)phosphine oxide using selectors identified from a chemical combinatorial library. *Analytical Chemistry* 74 (20): 5212–5216.

13 Brahmachary, E., Ling, F.H., Svec, F., and Fréchet, J.M.J. (2003). Chiral recognition: design and preparation of chiral stationary phases using selectors derived from Ugi multicomponent condensation reactions and a combinatorial approach. *Journal of Combinatorial Chemistry* 5 (4): 441–450.

14 Lämmerhofer, M. (2010). Chiral recognition by enantioselective liquid chromatography: mechanisms and modern chiral stationary phases. *Journal of Chromatography A* 1217 (6): 814–856.

15 Fernandes, C., Palmeira, A., Santos, A. et al. (2013). Enantioresolution of chiral derivatives of xanthones on (S,S)-Whelk-O1 and l -phenylglycine stationary phases and chiral recognition mechanism by docking approach for (S,S)-Whelk-O1. *Chirality* 25 (2): 89–100.

16 Lipkowitz, K.B. (2001). Atomistic modeling of enantioselection in chromatography. *Journal of Chromatography A* 906 (1–2): 417–442.

17 Del Rio, A., Hayes, J.M., Stein, M. et al. (2004). Theoretical reassessment of whelk-O1 as an enantioselective receptor for 1-(4-halogeno-phenyl)-1-ethylamine derivatives. *Chirality* 16 (suppl): S1–S11.

18 Del Rio, A., Piras, P., and Roussel, C. (2005). Data mining and enantiophore studies on chiral stationary phases used in HPLC separation. *Chirality* 17 (suppl): S74–S83.

19 Del Rio, A., Piras, P., and Roussel, C. (2006). Enantiophore modeling in 3D-QSAR. A data mining application on Whelk-O1 chiral stationary phase. *Chirality* 18 (7): 498–508.

20 Zhao, C. and Cann, N.M. (2006). Solvation of the Whelk-O1 chiral stationary phase: a molecular dynamics study. *Journal of Chromatography A* 1131 (1–2): 110–129.

21 Zhao, C. and Cann, N.M. (2007). The docking of chiral epoxides on the Whelk-O1 stationary phase: a molecular dynamics study. *Journal of Chromatography A* 1149 (2): 197–218.

22 Zhao, C.F., Diemert, S., and Cann, N.M. (2009). Rational optimization of the Whelk-O1 chiral stationary phase using molecular dynamics simulations. *Journal of Chromatography A* 1216 (32): 5968–5978.

23 Gasparrini, F., Misiti, D., and Villani, C. (2001). High-performance liquid chromatography chiral stationary phases based on low-molecular-mass selectors. *Journal of Chromatography A* 906 (1–2): 35–50.

24 Sardella, R., Ianni, F., Marinozzi, M. et al. (2017). Laboratory-scale preparative enantioseparations of pharmaceutically relevant compounds on commercially available chiral stationary phases for HPLC. *Current Medicinal Chemistry* 24 (8): 796–817.

25 Alvarez-Rivera, G., Bueno, M., Ballesteros-Vivas, D., and Cifuentes, A. (2020). Chiral analysis in food science. *TrAC Trends in Analytical Chemistry* 123: 115761.

26 Barreiro, J.C., Tiritan, M.E., and Cass, Q.B. (2021). Challenges and innovations in chiral drugs in an environmental and bioanalysis perspective. *TrAC Trends in Analytical Chemistry* 116326.

27 Coelho, M.M., Fernandes, C., Remião, F., and Tiritan, M.E. (2021). Enantioselectivity in drug pharmacokinetics and toxicity: pharmacological relevance and analytical methods. *Molecules* 26 (11): 3113.

28 Ribeiro, A.R.L., Maia, A.S., Ribeiro, C., and Tiritan, M.E. (2020). Analysis of chiral drugs in environmental matrices: current knowledge and trends in environmental, biodegradation and forensic fields. *TrAC Trends in Analytical Chemistry* 124: 115783.

29 Teixeira, J., Tiritan, M.E., Pinto, M.M., and Fernandes, C. (2019). Chiral stationary phases for liquid chromatography: recent developments. *Molecules* 24 (5): 865.

30 Fernandes, C., Teixeira, J., Pinto, M.M.M., and Tiritan, M.E. (2021). Strategies for preparation of chiral stationary phases: progress on coating and immobilization methods. *Molecules* 26 (18).

31 Regis Technologies, Inc. https://www.registech.com/other-chiral-phases (accessed 3 June 2022).

32 Pirkle, W.H. (1966). The nonequivalence of physical properties of enantiomers in optically active solvents. Differences in nuclear magnetic resonance spectra. I. *Journal of the American Chemical Society* 88 (8): 1837.

33 Pirkle, W.H. and Beare, S.D. (1967). Nonequivalence of the nuclear magnetic resonance spectra of enantiomers in optically active solvents. IV. Assignment of absolute configuration. *Journal of the American Chemical Society* 89 (21): 5485–5487.

34 Pirkle, W.H. and Burlingame, T.G. (1967). Nonequivalence of the nuclear magnetic reasonance spectra of enantiomers in optically active solvents III. *Tetrahedron Letters* 8 (41): 4039–4042.

35 Pirkle, W.H., Burlingame, T.G., and Beare, S.D. (1968). Optically active NMR solvents VI. The determination of optical purity and absolute configuration of amines. *Tetrahedron Letters* 9 (56): 5849–5852.

36 Pirkle, W.H., Beare, S.D., and Muntz, R.L. (1974). Assignment of absolute configuration of sulfoxides by NMR. A solvation model. *Tetrahedron Letters* 15 (26): 2295–2298.

37 Pirkle, W.H., Beare, S.D., and Muntz, R.L. (1969). Optically active solvents for nuclear magnetic resonance. X. Enantiomeric nonequivalence of sulfinamides, sulfinates, sulfites, thiosulfinates, phosphine oxides, and amine oxides. *Journal of the American Chemical Society* 91 (16): 4575.

38 Pirkle, W.H. and Hoekstra, M.S. (1976). Chiral nuclear magnetic resonance solvating agents. Resolution, determination of enantiomeric purity, and assignment of absolute configuration of cyclic and acyclic sulfinate esters. *Journal of the American Chemical Society* 98 (7): 1832–1839.

39 Pirkle, W.H., Muntz, R.L., and Paul, I.C. (1971). Chiral nuclear magnetic resonance solvents. XI. A method for determining the absolute configuration of chiral N,N-dialkylarylamine oxides. *Journal of the American Chemical Society* 93 (11): 2817–2819.

40 Pirkle, W.H. and Beare, S.D. (1969). Optically active solvents in nuclear magnetic resonance spectroscopy. IX. Direct determinations of optical purities and correlations of absolute configurations of α-amino acids. *Journal of the American Chemical Society* 91 (18): 5150–5155.

41 Pirkle, W.H. and Rinaldi, P.L. (1977). Nuclear magnetic resonance determination of enantiomeric compositions of oxaziridines using chiral solvating agents. *Journal of Organic Chemistry* 42 (20): 3217–3219.

42 Pirkle, W.H. and Sikkenga, D.L. (1977). The use of chiral solvating agents for nuclear magnetic resonance determination of enantiomeric purity and absolute configuration of lactones. Consequences of three-point interactions. *Journal of Organic Chemistry* 42 (8): 1370–1374.

43 Pirkle, W.H. and Boeder, C.W. (1977). Estimation of allene optical purities by nuclear magnetic resonance. *Journal of Organic Chemistry* 42 (23): 3697–3700.

44 Pirkle, W.H. and Rinaldi, P.L. (1978). Erratum: nuclear magnetic resonance determination of enantiomeric compositions of oxaziridines using chiral solvating agents (*Journal of Organic Chemistry* (1977) 42, (3217)). *Journal of Organic Chemistry* 43 (26): 5027.

45 Pirkle, W.H. and Hoekstra, M.S. (1974). An example of automated liquid chromatography. Synthesis of a broad-spectrum resolving agent and resolution of 1-(1-naphthyl)-2,2,2-trifluoroethanol. *Journal of Organic Chemistry* 39 (26): 3904–3906.

46 Pirkle, W.H. and Sikkenga, D.L. (1975). Use of achiral shift reagents to indicate relative stabilities of diastereomeric solvates. *Journal of Organic Chemistry* 40 (23): 3430–3434.

47 Pirkle, W.H., Hoekstra, M.S., and Miller, W.H. (1976). Electronic effects in asymmetric induction. Reaction of para-substituted phenylsulfinyl chlorides with S-(+)-1-(1-naphthyl)-2,2,2-trifluoroethanol. *Tetrahedron Letters* 17 (25): 2109–2112.

48 Pirkle, W.H. and Sikkenga, D.L. (1976). Resolution of optical isomers by liquid chromatography. *Journal of Chromatography A* 123 (2): 400–404.

49 Pirkle, W.H., House, D.W., and Finn, J.M. (1980). Broad spectrum resolution of optical isomers using chiral high-performance liquid chromatographic bonded phases. *Journal of Chromatography A* 192 (1): 143–158.

50 Pirkle, W.H. and House, D.W. (1979). Chiral high-pressure liquid chromatographic stationary phases. 1. Separation of the enantiomers of sulfoxides, amines, amino acids, alcohols, hydroxy acids, lactones, and mercaptans. *Journal of Organic Chemistry* 44 (12): 1957–1960.

51 Welch, C.J. (1995). Crawling out of the chiral pool: the evolution of Pirkle-type chiral stationary phases. In: *Advances in Chromatography*, vol. 35 (ed. P.R. Brown and E. Grushka), 171–197. New York: Marcel Dekker.

52 Pirkle, W.H., Tsipouras, A., and Sowin, T.J. (1985). Preparative separation of enantiomers by flash chromatography. *Journal of Chromatography A* 319 (C): 392–395.

53 Pirkle, W.H., Welch, C.J., and Hyun, M.H. (1983). A chiral recognition model for the chromatographic resolution of N-acylated 1-aryl-1-aminoalkanes. *Journal of Organic Chemistry* 48 (25): 5022–5026.

54 Pirkle, W.H. and Welch, C.J. (1984). Chromatographic separation of the enantiomers of acylated amines on chiral stationary phases. *Journal of Organic Chemistry* 49 (1): 138–140.

55 Pirkle, W.H., Welch, C.J., Mahler, G.S. et al. (1984). Chromatographic separation of the enantiomers of N-acylated heterocyclic amines. *Journal of Organic Chemistry* 49 (13): 2504–2506.

56 Pirkle, W.H. and Tsipouras, A. (1984). Direct liquid chromatographic separation of benzodiazepinone enantiomers. *Journal of Chromatography A* 291: 291–298.

57 Pirkle, W.H. and McCune, J.E. (1989). Separation of the enantiomers of N-protected alpha- amino acids as anilide and 3,5-dimethylanilide derivatives. *Journal of Chromatography A* 479 (2): 419–423.

58 Pirkle, W.H. and Hamper, B.C. (1988). Chromatographic separation of the enantiomers of 1,3-dithiolane-1-oxides. *Journal of Chromatography A* 450 (2): 199–210.

59 Pirkle, W.H., Pochapsky, T.C., Mahler, G.S., and Field, R.E. (1985). Chromatographic separation of the enantiomers of 2-carboalkoxyindolines and N-aryl-α-amino esters on chiral stationary phases derived from N-(3,5-dinitrobenzoyl)-α-amino acids. *Journal of Chromatography A* 348: 89–96.

60 Pirkle, W.H. and Burke, J.A. (1992). Separation of the enantiomers of the 3,5-dinitrobenzamide derivatives of alpha-amino phosphonates on four chiral stationary phases. *Journal of Chromatography A* 598 (2): 159–167.

61 Pirkle, W.H. and Sowin, T.J. (1987). Direct liquid chromatographic separation of phthalide enantiomers. *Journal of Chromatography A* 387 (C): 313–321.

62 Pirkle, W.H. and McCune, J.E. (1988). An improved chiral stationary phase for the facile separation of enantiomers. *Journal of Chromatography A* 441 (2): 311–322.

63 Pirkle, W.H. and McCune, J.E. (1989). Improved chiral stationary phase for the separation of the enantiomers of chiral acids as their anilide derivatives. *Journal of Chromatography A* 471 (C): 271–281.

64 Pirkle, W.H. and Murray, P.G. (1990). The separation of the enantiomers of a variety of non-steroidal anti-inflammatory drugs (NSAIDS) as their anilide derivatives using a chiral stationary phase. *Journal of Liquid Chromatography* 13 (11): 2123–2134.

65 Pirkle, W.H., Welch, C.J., and Zych, A.J. (1993). Chromatographic investigation of the slowly interconverting atropisomers of hindered naphthamides. *Journal of Chromatography A* 648 (1): 101–109.

66 Pirkle, W.H. and Hyun, M.H. (1985). α-arylalkylamine-derived chiral stationary phases. Evaluation of urea linkages. *Journal of Chromatography A* 322 (C): 295–307.

67 Pirkle, W.H. and Pochapsky, T.C. (1986). Erratum: a new, easily accessible reciprocal chiral stationary phase for the chromatographic separation of enantiomers (*Journal of the American Chemical Society* (1986) 108 (352–354)). *Journal of the American Chemical Society* 108 (9): 2492.

68 Pirkle, W.H. and Hyun, M.H. (1985). Effect of interstrand distance upon chiral recognition by a chiral stationary phase. *Journal of Chromatography A* 328: 1–9.

69 Pirkle, W.H. and Sowin, T.J. (1987). Design, preparation and performance of a phthalide-based chiral stationary phase. *Journal of Chromatography A* 396 (C): 83–92.

70 Hyun, M.H. and Pirkle, W.H. (1987). Preparation and evaluation of a chiral stationary phase bearing both π-acidic and -basic sites. *Journal of Chromatography A* 393 (2): 357–365.

71 Pirkle, W.H. and Welch, C.J. (1991). Chromatographic separation of underivatized naproxen enantiomers. *Journal of Liquid Chromatography* 14 (18): 3387–3396.

72 Pirkle, W.H., Welch, C.J., Burke, J.A., and Lamm, B. (1992). Target-directed design of chiral stationary phases. *Analytical Proceedings* 29: 225–227.

73 Pirkle, W.H. and Burke Iii, J.A. (1991). Chiral stationary phase designed for β-blockers. *Journal of Chromatography A* 557 (1–2): 173–185.

74 Pirkle, W.H. and Burke, J.A. (1989). Preparation of a chiral stationary phase from an α-amino phosphonate. *Chirality* 1: 57–62.

75 Welch, C.J. and Perrin, S.R. (1995). Improved chiral stationary phase for ß-blocker enantioseparations. *Journal of Chromatography A* 690 (2): 218–225.

76 Pirkle, W.H., Welch, C.J., and Lamm, B. (1992). Design, synthesis, and evaluation of an improved enantioselective naproxen selector. *Journal of Organic Chemistry* 57 (14): 3854–3860.

77 Pirkle, W.H. and Welch, C.J. (1992). An improved chiral stationary phase for the chromatographic separation of underivatized naproxen enantiomers. *Journal of Liquid Chromatography* 15 (11): 1947–1955.

78 Pirkle, W.H. and Welch, C.J. (1994). Use of simultaneous face to face and face to edge π–π interactions to facilitate chiral recognition. *Tetrahedron: Asymmetry* 5 (5): 777–780.

79 West, C. and Khater, S. (2020). Characterization of retention and separation mechanisms with Pirkle-type enantioselective stationary phases in supercritical fluid chromatography. *Journal of Chromatography A* 1626: 461352.

80 Pirkle, W.H. and Murray, P.G. (1993). Chiral stationary phase design. Use of intercalative effects to enhance enantioselectivity. *Journal of Chromatography A* 641 (1): 11–19.

81 Pirkle, W.H. and Liu, Y. (1994). Design, synthesis, resolution, determination of absolute configuration, and evaluation of a chiral naproxen selector. *Journal of Organic Chemistry* 59 (23): 6911–6916.

82 Pirkle, W.H. and Welch, C.J. (1992). Effect of superfluous remote polar functionality on chiral recognition. *Journal of Chromatography A* 589 (1–2): 45–51.

83 Pirkle, W.H., Murray, P.G., and Burke, J.A. (1993). Use of homologous series of analytes as mechanistic probes to investigate the origins of enantioselectivity on two chiral stationary phases. *Journal of Chromatography A* 641 (1): 21–29.

84 Pirkle, W.H. and Bowen, W.E. (1994). Chiral stationary phase design: a study in optimization. *Journal of High Resolution Chromatography* 17: 629–633.

85 Schleimer, L.M., Pirkle, W.H., and Schurig, V. (1994). Enantiomer separation by high-performance liquid chromatography on polysiloxane-based chiral stationary phases. *Journal of Chromatography A* 679 (1): 23–34.

86 Pirkle, W.H. and Liu, Y. (1996). Incremental development of chiral selectors for underivatized profens. *Journal of Chromatography A* 736 (1–2): 31–38.

87 Hyun, M.H. and Pirkle, W.H. (2000). Liquid chromatographic separation of the stereoisomers of thiazide diuretics. *Journal of Chromatography A* 876 (1–2): 221–227.

88 Pirkle, W.H. and Koscho, M.E. (1999). Structural optimization of a chiral selector for use in preparative enantioselective chromatography. *Journal of Chromatography A* 840 (2): 151–158.

89 Pirkle, W.H. and Lee, W. (2010). Separation of the enantiomers of β-blockers using brush type chiral stationary phase derived from conformationally rigid α-amino β-lactam. *Bulletin of the Korean Chemical Society* 31 (3): 620–623.

90 Lee, W., Snyder, S.E., Volkers, P.I. et al. (2011). Assessing chiral self-recognition using chiral stationary phases. *Tetrahedron* 67 (37): 7143–7147.

91 Asnin, L., Gritti, F., Kaczmarski, K., and Guiochon, G. (2010). Features of the adsorption of Naproxen on the chiral stationary phase (S,S)-Whelk-O1 under reversed-phase conditions. *Journal of Chromatography A* 1217 (3): 264–275.

92 Asnin, L.D. and Guiochon, G. (2010). Retention of Naproxen enantiomers on the chiral stationary phase Whelk-O1 under reversed-phase conditions. A reconsideration of the adsorption mechanism in the light of new experimental data. *Journal of Chromatography A* 1217 (10): 1709–1711.

93 Coelho, M.M., Lado Ribeiro, A.R., Sousa, J.C.G. et al. (2019). Dual enantioselective LC–MS/MS method to analyse chiral drugs in surface water: monitoring in Douro River estuary. *Journal of Pharmaceutical and Biomedical Analysis* 170: 89–101.

94 Welch, C.J. and William, H. (2020). Pirkle: stereochemistry pioneer. *Chirality* 32 (7): 961–974.

95 Maier, N.M. and Uray, G. (1996). Diphenylethanediamine (DPEDA) as chiral selector: VII. Efficient HPLC resolution of a series of underivatized diarylcarbinols on a chiral stationary phase based on C5-amide-bonded N-3,5-dinitrobenzoyl-DPEDA. *Chirality* 8 (7): 490–493.

96 Maier, N.M. and Uray, G. (1996). Diphenylethanediamine derivatives as chiral selectors V. Efficient normal-phase high-performance liquid chromatographic enantioseparation of underivatized chiral arylalcohols on four differently linked 3,5-dinitrobenzoyldiphenylethanediamine-derived chiral stationary phases. *Journal of Chromatography A* 732 (2): 215–230.

97 Gasparrini, F., Misiti, D., Villani, C., and La Torre, F. (1991). Chromatographic resolution of 1,2-amino alcohols on a chiral stationary phase containing N,N′-(3,5-dinitrobenzoyl)-trans-1,2-diaminocyclohexane. Theoretical and practical aspects. *Journal of Chromatography A* 539 (1): 25–36.

98 Galli, B., Gasparrini, F., Misiti, D. et al. (1992). HPLC resolution of atropoisomeric compounds on a CSP derived from (1R;2R)-diaminocyclohexane: thermodynamic data from variable temperature chromatography. *Chirality* 4 (6): 384–388.

99 Cancelliere, G., D'Acquarica, I., Gasparrini, F. et al. (2006). Twenty years of research on silica-based chiral stationary phases. *Journal of Separation Science* 29 (6): 770–781.

100 Forjan, D.M., Kontrec, D., and Vinković, V. (2006). Performance of brush-type HPLC chiral stationary phases with tertiary amide in the connecting tether. *Chirality* 18 (10): 857–869.

101 Li, Y., Jiang, D., Huang, D. et al. (2015). Preparation and characterization of tripeptide chiral stationary phases with varying amino acid sequences and terminal groups. *Analytical Methods* 7 (9): 3772–3778.

102 Lao, W. and Gan, J. (2009). Evaluation of triproline and tri-α-methylproline chiral stationary phases. Retention and enantioseparation associated with hydrogen bonding. *Journal of Chromatography A* 1216 (25): 5020–5029.

103 Huang, J., Chen, H., and Li, T. (2006). Improvement of proline chiral stationary phases by varying peptide length and linker. *Journal of Chromatography A* 1113 (1–2): 109–115.

104 Huang, J., Chen, H., Zhang, P., and Li, T. (2006). Improvement of proline enantioselective stationary phases by replacing the 9-fluorenylmethoxycarbonyl group. *Journal of Chromatography A* 1109 (2): 307–311.

105 Li, J., Li, Y., Chen, T. et al. (2013). Preparation, chromatographic evaluation and comparison between linear peptide- and cyclopeptide-bonded stationary phases. *Talanta* 109: 152–159.

106 Wu, H., Liang, T., Yin, C. et al. (2011). Enantiorecognition ability of peptoids with α-chiral, aromatic side chains. *The Analyst* 136 (21): 4409–4411.

107 Wu, H., Su, X., Li, K. et al. (2012). Improvement of peptoid chiral stationary phases by modifying the terminal group of selector. *Journal of Chromatography A* 1265: 181–185.

108 Wu, H., Song, G., Liang, X., and Ke, Y. (2012). Investigation of peptoid chiral stationary phases terminated with N′-substituted phenyl-L-proline/leucine amide. *Chinese Journal of Chemistry* 30 (12): 2791–2797.

109 Wu, H., Li, K., Yu, H. et al. (2013). Investigation of peptoid chiral stationary phases varied in absolute configuration. *Journal of Chromatography A* 1281: 155–159.

110 Wu, H., Song, G., Wang, D. et al. (2013). Study of stereomeric peptoid chiral stationary phases containing different chiral side chains. *Journal of Chromatography A* 1298: 152–156.

111 Kotoni, D., Villani, C., Bell, D.S. et al. (2013). Bidentate urea-based chiral selectors for enantioselective high performance liquid chromatography: synthesis and evaluation of "Crab-like" stationary phases. *Journal of Chromatography A* 1297: 157–167.

112 Yu, J., Armstrong, D.W., and Ryoo, J.J. (2018). Synthesis of new C3 symmetric amino acid- and aminoalcohol-containing chiral stationary phases and application to HPLC enantioseparations. *Chirality* 30 (1): 74–84.

113 Yu, J., Ryoo, D.H., Lee, J.M., and Ryoo, J.J. (2016). Synthesis and application of C2 and C3 symmetric (R)-phenylglycinol-derived chiral stationary phases. *Chirality* 28 (3): 186–191.

114 Kontrec, D., Vinković, V., and Šunjić, V. (2000). Preparation and evaluation of chiral stationary phases based on N,N-2,4- (or 4,6)-disubstituted 4,5-(or 2,5)-dichloro-1,3-dicyanobenzene. *Chirality* 12 (2): 63–70.

115 Lecci, C. and Iuliano, A. (2005). Synthesis and evaluation of a new biselector s-triazine based chiral stationary phase for enantioselective HPLC: potentiality of the approach and perspectives. *Biomedical Chromatography* 19 (6): 439–446.

116 Pirkle, W.H. and Pochapsky, T.C. (1986). A new, easily accessible reciprocal chiral stationary phase for the chromatographic separation of enantiomers. *Journal of the American Chemical Society* 108 (2): 352–354.

117 Wu, H., Ji, S., Yang, B. et al. (2012). Investigation of brush-type chiral stationary phases based on O,O′-diaroyl tartardiamide and O,O′-bis-(arylcarbamoyl) tartardiamide. *Journal of Separation Science* 35 (3): 351–358.

118 Kontrec, D., Abatangelo, A., Vinković, V., and Šunjić, V. (2001). Solid-phase synthesis of chiral stationary phases based on 2,4,5,6-tetrachloro-1,3-dicyanobenzene derivatives spaced from N-3,5-dinitrobenzoyl α-amino acids: comparative study of their resolution efficacy. *Chirality* 13 (6): 294–301.

119 Kosjek, B. and Uray, G. (2001). Immobilization of difunctional building blocks on hydroxysuccinimide activated silica: versatile in situ preparation of chiral stationary phases. *Chirality* 13 (10): 657–667.

120 Vinkovic, V., Kontrec, D., Sunjic, V. et al. (2001). Mechanism of chiral recognition in the enantioseparation of 2-aryloxypropionic acids on new brush-type chiral stationary phases. *Chirality* 13 (9): 581–587.

121 Vinković, V., Roje, M., Kontrec, D., and Šunjić, V. (2008). Separation of BINAPO enantiomers on brush-type chiral stationary phases. *Arkivoc* 2008 (8): 242–253.

122 Tan, X., Hou, S., and Wang, M. (2007). Enantioselective and diastereoselective separation of synthetic pyrethroid insecticides on a novel chiral stationary phase by high-performance liquid chromatography. *Chirality* 19 (7): 574–580.

123 Tan, X.L., Hou, S.C., Bian, Q.H., and Wang, M. (2007). Synthesis of novel chiral stationary phases for high-performance liquid chromatography. *Chinese Chemical Letters* 18 (4): 461–464.

124 Hwan, S.C., Hyun, J.H., Hee, J.C. et al. (2008). Preparation and application of a new liquid chromatographic chiral stationary phase based on cefaclor. *Bulletin of the Korean Chemical Society* 29 (11): 2135–2139.

125 Choi, H.J., Ha, H.J., Shin, M.S. et al. (2009). Preparation of a new π-basic chiral stationary phase based on cefaclor and the liquid chromatographic resolution of-acidic analytes. *Journal of Liquid Chromatography and Related Technologies* 32 (13): 1879–1890.

126 Moiteiro, C., Fonseca, N., Curto, M.J.M. et al. (2006). Development of novel brush-type chiral stationary phases based on terpenoid selectors: HPLC evaluation and theoretical investigation of enantioselective binding interactions. *Tetrahedron: Asymmetry* 17 (23): 3248–3264.

127 Wang, D.C., Luo, J., Yao, S., and Song, H. (ed.) (2016). The synthesis and characterization of novel brush-type chiral stationary phase based on terpenoid selector for resolution of chiral drugs. In: *2nd International Conference on Chemical and Material Engineering, ICCME 2015*. EDP Sciences.

128 Carraro, M.L., Palmeira, A., Tiritan, M.E. et al. (2017). Resolution, determination of enantiomeric purity and chiral recognition mechanism of new xanthone derivatives on (S,S)-whelk-O1 stationary phase. *Chirality* 29 (6): 247–256.

129 Fernandes, C., Tiritan, M.E., Cravo, S. et al. (2017). New chiral stationary phases based on xanthone derivatives for liquid chromatography. *Chirality* 29: 430–442.

130 Phyo, Y.Z., Teixeira, J., Tiritan, M.E. et al. (2020). New chiral stationary phases for liquid chromatography based on small molecules: development, enantioresolution evaluation and chiral recognition mechanisms. *Chirality* 32 (1): 81–97.

131 Cancelliere, G., Ciogli, A., D'Acquarica, I. et al. (2010). Transition from enantioselective high performance to ultra-high performance liquid chromatography: a case study of a brush-type chiral stationary phase based on sub-5-micron to sub-2-micron silica particles. *Journal of Chromatography. A* 1217 (7): 990–999.

132 Ismail, O.H., Pasti, L., Ciogli, A. et al. (2016). Pirkle-type chiral stationary phase on core–shell and fully porous particles: are superficially porous particles always the better choice toward ultrafast high-performance enantioseparations? *Journal of Chromatography. A* 1466: 96–104.

133 Waterlot, C., Ghinet, A., and Lipka, E. (2018). Core-shell particles: a way to greening liquid chromatography in environmental applications. *Current Chromatography* 5 (2): 78–90.

134 Slater, M.D., Fréchet, J.M.J., and Svec, F. (2009). In-column preparation of a brush-type chiral stationary phase using click chemistry and a silica monolith. *Journal of Separation Science* 32 (1): 21–28.

135 Ghanem, A., Ikegami, T., and Tanaka, N. (2011). New silica monolith bonded chiral (R)-γ butyrolactone for enantioselective micro high-performance liquid chromatography. *Chirality* 23 (10): 887–890.

136 Cielecka-Piontek, J., Zalewski, P., Jelińska, A., and Garbacki, P. (2013). UHPLC: the greening face of liquid chromatography. *Chromatographia* 76: 1429–1437.

137 Kotoni, D., Ciogli, A., Molinaro, C. et al. (2012). Introducing enantioselective ultrahigh-pressure liquid chromatography (eUHPLC): theoretical inspections and ultrafast separations on a new sub-2-$^{\mu}$m Whelk-O1 stationary phase. *Analytical Chemistry* 84 (15): 6805–6813.

138 Kotoni, D., Ciogli, A., D'Acquarica, I. et al. (2012). Enantioselective ultra-high and high performance liquid chromatography: a comparative study of columns based on the Whelk-O1 selector. *Journal of Chromatography A* 1269: 226–241.

139 Wang, Y., Ai, F., Ng, S.C., and Tan, T.T.Y. (2012). Sub-2μm porous silica materials for enhanced separation performance in liquid chromatography. *Journal of Chromatography A* 1228: 99–109.

140 Khoo, H.T. and Leow, C.H. (2021). Advancements in the preparation and application of monolithic silica columns for efficient separation in liquid chromatography. *Talanta* 224: 121777.

141 Group. R (2021). Whelk-O1 ® (Chiral Core-Shell). http://www.registech.com/whelk-o1 (accessed 3 June 2022).

142 Wu, X., You, L., Di, B. et al. (2013). Novel chiral core-shell silica microspheres with trans-(1R,2R)-diaminocyclohexane bridged in the mesoporous shell: synthesis, characterization and application in high performance liquid chromatography. *Journal of Chromatography A* 1299: 78–84.

143 Badaloni, E., Cabri, W., Ciogli, A. et al. (2007). Combination of HPLC "inverted chirality columns approach" and MS/MS detection for extreme enantiomeric excess determination even in absence of reference samples. Application to camptothecin derivatives. *Analytical Chemistry* 79 (15): 6013–6019.

144 Vedovello, P., Costa, J.A.S., Fernandes, C. et al. (2022). Evaluation of chiral separation by Pirkle-type chiral selector based mixed matrix membranes. *Separation and Purification Technology* 289: 120722.

145 Mazzoccanti, G., Manetto, S., Ciogli, A. et al. (2022). A perspective on enantioselective chromatography by comparing ultra-high performance supercritical fluid chromatography and normal-phase liquid chromatography through the use of a Pirkle-type stationary phase. *TrAC Trends in Analytical Chemistry* 147: 116511.

146 Mazzoccanti, G., Ismail, O.H., D'Acquarica, I. et al. (2017). Cannabis through the looking glass: chemo- and enantio-selective separation of phytocannabinoids by enantioselective ultra high performance supercritical fluid chromatography. *Chemical Communications* 53 (91): 12262–12265.

147 Kimura, T., Hamase, K., Miyoshi, Y. et al. (2016). Chiral amino acid metabolomics for novel biomarker screening in the prognosis of chronic kidney disease. *Scientific Reports* 6: 26137.

148 Hamase, K., Miyoshi, Y., Ueno, K. et al. (2010). Simultaneous determination of hydrophilic amino acid enantiomers in mammalian tissues and physiological fluids applying a fully automated micro-two-dimensional high-performance liquid chromatographic concept. *Journal of Chromatography A* 1217 (7): 1056–1062.

149 Hsiao, S.W., Ishii, C., Furusho, A. et al. (2021). Determination of phenylalanine enantiomers in the plasma and urine of mammals and D-amino acid oxidase deficient rodents using two-dimensional high-performance liquid chromatography. *Biochimica et Biophysica Acta. Proteins and Proteomics* 1869 (1): 140540.

150 Furusho, A., Koga, R., Akita, T. et al. (2019). Three-dimensional high-performance liquid chromatographic determination of Asn, Ser, Ala, and Pro enantiomers in the plasma of patients with chronic kidney disease. *Analytical Chemistry* 91 (18): 11569–11575.

151 Furusho A, Akita T, Mita M, Naraoka H, Hamase K. Three-dimensional high-performance liquid chromatographic analysis of chiral amino acids in carbonaceous chondrites. *Journal of Chromatography A* 2020;1625 461255.

152 Tiritan, M.E., Fernandes, C., Maia, A.S. et al. (2018). Enantiomeric ratios: why so many notations? *Journal of Chromatography A* 1569: 1–7.

153 Sun, M., Liu, D., Shen, Z. et al. (2015). Stereoselective quantitation of haloxyfop in environment samples and enantioselective degradation in soils. *Chemosphere* 119: 583–589.

154 Ishii, C., Akita, T., Nagano, M. et al. (2019). Determination of chiral amino acids in various fermented products using a two-dimensional HPLC-MS/MS system. *Chromatography* 40 (2): 83–87.

155 Pucciarini, L., Saluti, G., Galarini, R. et al. (2020). Optimized one-pot derivatization and enantioseparation of cysteine: application to the study of a dietary supplement. *Journal of Pharmaceutical and Biomedical Analysis* 180 (113066): 1–8.

156 Leek, H., Thunberg, L., Jonson, A.C. et al. (2017). Strategy for large-scale isolation of enantiomers in drug discovery. *Drug Discovery Today* 22 (1): 133–139.

10

Proteins

Jun Haginaka

Institute for Biosciences, Mukogawa Women's University, Koshien Kyuban-cho, Nishinomiya, Japan

10.1 Introduction

In the early 1950s, the binding of the enantiomers of the anionic azo-dye, α-[[4-[2-[4-(dimethylamino)phenyl]diazenyl]benzoyl]amino]benzeneacetic acid, to bovine serum albumin (BSA) [1] or human serum albumin (HSA) [2] was examined. Binding of the (−)-form to BSA with an affinity of about twofold higher than that of the (+)-form was observed, while the difference in the binding to HSA was almost insignificant. Furthermore, it was reported that L-tryptophan (Trp) was bound to serum albumin (bovine mercaptoalbumin or HSA) with a higher affinity than the D-form, which had almost no binding to serum albumins [3]. L-Trp was bound to serum albumin predominantly at one site in a highly stereospecific manner: the indole ring, carboxy group, and α-hydrogen of the L-Trp interacted with the protein. Especially, a "close fit" to the protein at the α-hydrogen was important [3]. These studies clearly indicated that serum albumins could differentiate enantiomers of some ligands, and that they could be used as chiral selectors for the enantioseparations. In 1973, BSA-succinoylaminoethyl-agarose beads were used for the separation of Trp enantiomers. On the BSA-agarose beads, D- and L-Trp were clearly resolved, and the D-enantiomer eluted first. This result was consistent with the bindings of Trp enantiomers to BSA in the solution described above. This was the first report on the use of a protein-based chiral stationary phase (CSP) for enantioseparation purposes in liquid chromatography (LC) [4].

Chiral Separations and Stereochemical Elucidation: Fundamentals, Methods, and Applications, First Edition. Edited by Quezia Bezerra Cass, Maria Elizabeth Tiritan, João Marcos Batista Junior, and Juliana Cristina Barreiro.

Protein- or glycoprotein-immobilized CSPs have the enantioselectivity for a wide range of compounds because of multiple binding sites on the surfaces of proteins or glycoproteins and multiple binding interactions between proteins or glycoproteins and ligands. However, in general they are unstable against heat, acids, organic modifiers, and so on because those induce reversible or irreversible changes in their conformation.

Proteins so far employed have included BSA [5], HSA [6], trypsin [7], α-chymotrypsin [8], lysozyme [9], pepsin [10], fatty acid-binding protein (FABP) [11], penicillin G acylase (PGA) [12], streptavidin [13], and lipase B [14], while glycoproteins have included human α_1-acid glycoprotein (AGP, hAGP) [15], ovomucoid from chicken egg whites (OMCHI) [16], ovoglycoprotein from chicken egg whites (OGCHI) [17] (now, it is termed chicken AGP [cAGP] [18]), avidin [19], riboflavin-binding protein (RfBP) [20], ovotransferrin (conalbumin) [21], cellobiohydrolase [22], glucoamylase [23], antibody (immunoglobulin G) [24], $\alpha_3\beta_4$-nicotinic acetylcholine receptor (nAChR) [25], and human liver organic cation transporter (hOCT) [26]. Physical properties of these proteins and glycoproteins used as CSPs are shown in Table 10.1.

Proteins and glycoproteins that were immobilized and adsorbed onto supports were used as CSPs in LC. Proteins and glycoproteins were also used as chiral selectors in capillary electrophoresis (CE) after being immobilized, adsorbed, and encapsulated onto the capillary or the CSPs being packed into the capillary. These methods used in CE are termed affinity capillary electrochromatography (CEC). It is also possible to dissolve proteins and glycoproteins in the mobile phase and the running buffers, respectively, in chiral LC and CE. However, these methods will not be covered in this chapter.

A lot of book chapters [27–30] and review articles [31–36] dealing with proteins or glycoproteins as chiral selectors in LC or CE have been published. This chapter deals with preparation of protein- and glycoprotein-based CSPs, the types of protein and glycoprotein used in LC and CEC, and the chiral recognition mechanism on proteins and glycoproteins.

10.2 Preparation of Protein- and Glycoprotein-Based Chiral Stationary Phases

Silicas, zirconias, and polymers have been used as supports for the immobilization of a protein or glycoprotein [33]. Silica-based materials were used as a particle, capillary, and monolith. The disadvantage of silica-based materials is that the mobile phase pH used is limited to the ranges of 2–8 [32, 33].

Table 10.1 Physical properties of proteins and glycoproteins used for CSPs.

Protein	Molecular mass (Da)	Carbohydrate content (%)	Isoelectric point	Origin
		Protein		
Bovine serum albumin (BSA)	66,000	—[a]	4.7	Bovine serum
Human serum albumin (HSA)	66,000	—	4.7	Human serum
Trypsin	23,000	—	10.1–10.8	Pancreas
α-Chymotrypsin	25,000	—	8.1–8.6	Pancreas
Lysozyme	14,300	—	10.5	Egg white
Pepsin	34,600	—	<1	Porcine stomach
Fatty acid binding protein (FABP)	14,000	—	9.0	Chicken liver
Penicillin G acylase (PGA)	85,000	—	8.1	Bacterium
Streptavidin	53,000	—	5	Bacterium
Lipase B	33,000		6.0	Fungus
		Glycoprotein		
α_1-Acid glycoprotein (AGP, hAGP)	41,000–43,000	45	2.7–3.8	Human serum
	33,000	35	2.7–3.8	
Ovomucoid (OMCHI)	28,000	30	4.1	Egg white
Ovoglycoprotein (OGCHI) (Chicken α_1-acid glycoprotein [cAGP])	30,000	25	4.1	Egg white

(*Continued*)

Table 10.1 (Continued)

Protein	Molecular mass (Da)	Carbohydrate content (%)	Isoelectric point	Origin
Avidin	66,000	7	10	Egg white
Riboflavin binding protein (RfBP)	32,000–36,000	14	4	Egg white
Ovotransferrin (Conalbumin)	77,000	2.6	6.1	Egg white
Cellulase				Fungus
Cellobiohydrolase I (CBH I, Cel7A)	65,000	6	3.9	
Cellobiohydrolase II (CBH II, Cel6A)	53,000		5.9	
Cellobiohydrolase 58 (CBH 58, Cel6B)	60,000		3.8	
Glucoamylase				Fungus
Glucoamylase G1	94,000	30–35		
Glucoamylase G2	85,000			
Antibody (Immunoglobulin G)	150,000	2–3	6–8	Vertebrate
$\alpha_3\beta_4$-Nicotinic acetylcholine receptor (nAChR)	309,000			Cell line
Human liver organic cation transporter (hOCT)	61,000			Cell line

[a] No sugar moieties.

Zirconias and polymers have been introduced to overcome pH limitations of silica materials [33]. The former is used as a particle, and the latter is used as a particle, capillary, and monolith.

There are two methods for the preparation of protein- and glycoprotein-immobilized materials: noncovalent and covalent immobilization onto materials. The physical adsorption has been first carried out using bare or anion-exchange silicas [32, 33]. BSA could be cross-linked with glutaraldehyde after entrapped on the pore of silica particles [37], while avidin, which has a high isoelectric point (pI) value, was adsorbed onto a fused-silica capillary [38] or silica monolith [39]. BSA [40, 41] or HSA [42] was adsorbed onto anion-exchange silicas by ionic interactions. Furthermore, proteins or glycoproteins that were permeated into the phospholipid bilayer membrane were coated on a fused-silica capillary [43, 44].

Proteins and glycoproteins could be immobilized onto functionalized supports (silicas and polymers) via their amino, carboxy, and sulfhydryl groups [45]. Functionalized supports, epoxy-, diol-, or amino-supports, were used for the immobilization of proteins or glycoproteins via their amino groups. Side chain amino group(s) of a protein or glycoprotein such as Lys, Arg, and/or an *N*-terminal amino group could be responsible for the reaction.

In the case of epoxy-supports, a protein or glycoprotein was directly reacted without activation of the supports [46, 47]. This involves the nucleophilic attack on an epoxy group on the supports by amine groups on a protein or glycoprotein, yielding a stable secondary amine linkage.

Diol-supports were activated with 1,1′-carbonyldiimidazole or *N,N′*-disuccinimidylcarbonate (DSC) used for the preparation of protein- and glycoprotein-immobilized materials, where a carbamate bond could be formed [47]. Moreover, diol-supports were activated with tresyl chloride and reacted with a protein or glycoprotein, where it was immobilized to the supports via a secondary amine linkage [46]. Furthermore, diol-supports were oxidized to their aldehyde forms by periodic acid. A protein or glycoprotein was immobilized to the aldehyde-supports by Schiff base formation. This Schiff base was reduced to a secondary amine using sodium cyanoborohydride [46, 47].

With regard to immobilization of a protein or glycoprotein to amino-supports, the method includes the activation of porous amino-supports by DSC followed by reaction with a protein or glycoprotein [17], where the urea bond could be formed. Instead of DSC, *N,N′*-disuccinimidylsuberate (DSS) was used for the activation of amino-supports [48]. In addition, proteins or glycoproteins can be bound to amino-silicas or amino-polymers using glutaraldehyde as a cross-linker, resulting in cross-linking by Schiff base formation followed by reduction with sodium cyanoborohydride [32].

Proteins and glycoproteins can be bound to amino-silicas via their carboxy groups by using water-soluble carbodiimide (for example, 1-ethyl-3-(3′-dimethylaminopropyl)carbodiimide [EDC]) and *N*-hydroxysulfosuccinimide (HSSI) [49], where a protein or glycoprotein was immobilized to the base materials via an amide bond. Side chain carboxy group(s) of a protein or glycoprotein such as Glu, Asp, and/or a *C*-terminal carboxy group could be responsible for the reaction.

Moreover, proteins or glycoproteins were immobilized onto amino-supports via their sulfhydryl groups. Their sulfhydryl groups were reacted with amino-silicas activated by succinimidyl 4-(*N*-maleimidomethyl) cyclohexane-1-carboxylate (SMCC) [50]. Similarly, amino-supports were activated by succinimidyl iodoacetate (SIC) for reaction with a sulfhydryl group [50]. Though this method is site selective, it could be only applicable to a protein or glycoprotein, which has a free sulfhydryl group.

Glycoproteins have been immobilized onto hydrazide-activated supports (silicas or polymers) in the following steps: (i) oxidation of hydroxyl groups of sugar moieties into aldehyde groups by periodic acid, (ii) preparation of hydrazide-supports by reaction of oxalic dihydrazide with aldehyde-supports, and (iii) immobilization of oxidized glycoproteins to the activated supports [51].

Membrane proteins such as $\alpha_3\beta_4$-nAChR [25] and hOCT [26] were immobilized on an artificial membrane, whose surface was formed by covalent binding of membrane-forming phospholipids to silicas.

10.3 Types of Protein- and Glycoprotein-Based Chiral Stationary Phases

10.3.1 Proteins

10.3.1.1 Bovine Serum Albumin

BSA, which has a molecular mass of 66 kDa, consists of 583 amino acid residues. BSA-immobilized agarose was used for the enantioseparation of Trp: D- and L-Trp were clearly resolved using the CSP as described in the Introduction [4]. For the first time, BSA was immobilized onto silicas by Allenmark et al. [5]. Though silicas were mainly used as the base materials, polymers also have been used for the immobilization of BSA in LC. The polymers used include hydroxy-ethylmethacrylate, polystyrene-divinylbenzene perfusion beads, and polyethylene hollow-fiber membranes [52]. Nakamura et al. [52] prepared a BSA-multilayer-adsorbed porous hollow-fiber membrane as a CSP. BSA-four-layered-adsorbed porous hollow-fiber membranes gave higher resolution for Trp enantiomers than BSA-monolayered ones. CSPs based on BSA were used for the separations of a variety of acidic and neutral enantiomers such as *N*-derivatized amino acids, aromatic amino acids, uncharged solutes, sulfoxides, and sulfoximine derivatives in LC [32, 33]. An amine, prilocaine, was separated using the mobile phase whose pH used is 8.9 [53].

In CEC, capillaries filled with silica particles consisting of BSA cross-linked with glutaraldehyde have been prepared for the resolution of Trp enantiomers [54]. A CEC column was prepared with physically adsorbed BSA on strong anion-exchange packings [41]. BSA was bound to the capillary wall [55]: the method included etching the capillary wall with sodium hydroxide, epoxy-diol coating with 3-glycidoxypropyltrimethoxysil ane followed by hydrolysis with hydrochloric acid, activation with tresyl chloride, and BSA coupling. Dinitrophenyl-amino acids and 3-hydroxy-1,4-benzodiazepines were enantioseparated. The capillary walls were coated with a phospholipid coating with BSA [43] as a chiral selector permeated into the phospholipid bilayer membrane. 1-(4-Iodobutyl)-1,4-dimethylpiperazin-1-ium iodide was applied as a first coating layer in the capillary to effectively suppress the electroosmotic flow and to stabilize the phospholipid coating of BSA. Carboxylated single-walled carbon nanotubes (c-SWNTs) were cut into short pipes by chemical oxidation in a mixture of concentrated sulfuric and nitric acids. BSA was bound to the shortened c-SWNTs, which bear carboxy groups on their surfaces, by using water-soluble carbodiimide (for example, EDC) via amide bonds. Chiral separation was realized utilizing c-SWNT-BSA conjugates immobilized on microchannel for microchip CEC [56]. Successful separation of Trp enantiomers was achieved in less than 70 s. BSA was immobilized to diol- or amino-silica monoliths via Schiff base formation followed by reduction with sodium cyanoborohydride and used for separation of Trp enantiomers in CEC [57]. BSA was encapsulated in silica monoliths using a sol–gel method by three steps: (i) formation of sol particles during hydrolysis and condensation, (ii) addition of BSA into the sol, and (iii) trapping BSA in the growing silicate network [58]. The prepared monoliths were used for enantioseparation of Trp in CEC.

It was initially reported that CSPs based on a BSA fragment(s) had lower capacity and enantioselectivity than that of the intact BSA-based CSPs, being less stable [59, 60]. CSPs based on the *N*-terminal half of a BSA fragment(s), which was mainly cleaved between Asp307 and Phe308 by pepsin and whose molecular masses were ca. 35,000 Da, were prepared [61, 62]. The intact and fragmented BSA were immobilized to amino-silicas activated by DSC. The bound amounts of the BSA fragment were 2.2–2.7 times higher than that of the intact BSA. Chiral resolution of 2-arylpropionic acid derivatives, benzodiazepines, warfarin, and benzoin was attained with the BSA fragment columns in LC. The BSA fragment-based CSPs gave higher enantioselectivity for lorazepam and benzoin because of the higher bound amounts, and lower enantioselectivity for other compounds tested, compared to the intact BSA-based CSPs [62].

10.3.1.2 Human Serum Albumin

HSA, which has a molecular mass of 66 kDa, consists of 585 amino acid residues. HSA-based CSPs were first immobilized onto diol-silica particles by Domenici et al. [6] and used in LC. Recently, Hage and his group immobilized HSA to polymer monoliths based on glycidyl methacrylate (GMA) and ethylene glycol dimethacrylate (EDMA) by several methods (epoxy, CDI, DSS, and Schiff base methods) as described in Section 10.2 [47]. Diol-silica monoliths and polymer monoliths based on GMA and EDMA were oxidized by using periodic acid to yield aldehyde groups. HSA was immobilized to the aldehyde-supports via Schiff base followed by reduction with sodium cyanoborohydride [63]. The Schiff base method gave the best resolution in chiral separations of (*R/S*)-warfarin and D/L-Trp. The HSA-immobilized silica monoliths gave higher retention and higher or comparable resolution and efficiency when compared to HSA-immobilized silica particles or HSA-immobilized GMA/EDMA monoliths. It was concluded that silica monoliths can be valuable alternatives to silica particles or GMA/EDMA monoliths when used with immobilized HSA as a CSP in LC [63]. Furthermore, HSA was immobilized via its sulfhydryl groups [50]. It was found that the SMCC and succinimidyl iodoacetate (SIA) methods gave HSA-based CSPs with comparable or improved enantioselectivity and stability compared to those prepared by the Schiff base method. Figure 10.1 shows enantioseparations of Trp, warfarin, and ibuprofen on HSA-based CSPs prepared by the SMCC and SIA methods [50]. The SMCC method gave higher enantioselectivity for Trp, warfarin, and ibuprofen than the SIA method in LC.

A variety of weakly acidic and neutral compounds, which include 2-arylpropionic acid derivatives, such as naproxen, flurbiprofen, ibuprofen, ketoprofen, and fenoprofen, reduced folates, such as leucovorin and 5-methyltetrahydrofolate, and benzodiazepines, such as oxazepam, lorazepam, and temazepam, were resolved on HSA-based CSPs [32].

In CEC, HSA-immobilized LC silica packing materials were packed into fused-silica capillaries [64]. HSA was coated onto the capillary, about 0.7 monolayer of HSA adsorbed on a typical uncoated-silica capillary at pH 7.4 [65]. Enantioseparation of warfarin was attained using the preceding method showing strong bindings to HSA (assocation constant [*K*a] $\geq 10^5\,M^{-1}$) but no

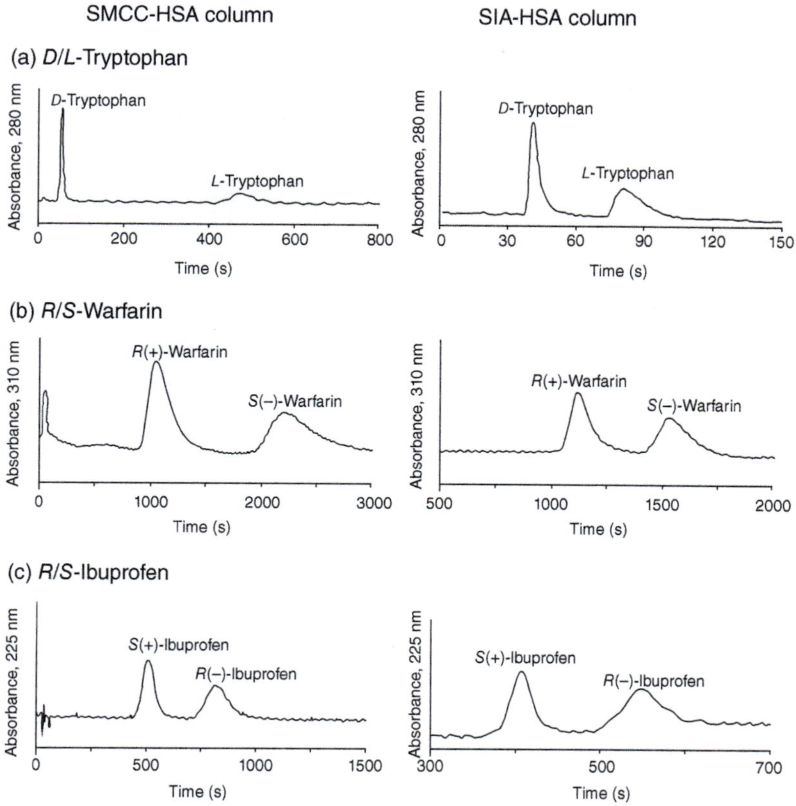

Figure 10.1 Enantioseparations of (a) Trp, (b) warfarin, and (c) ibuprofen on HSA-based CSPs prepared by the SMCC and SIA methods. The LC conditions were as follows: sample concentration, 20 mmol l^{-1} Trp, ibuprofen, or warfarin; sample volume, 20 μl; mobile phase for Trp, pH 7.4, 0.067 mol l^{-1} potassium phosphate buffer; mobile phase for ibuprofen, pH 7.0, 0.067 mol l^{-1} potassium phosphate buffer containing 8 vol% 2-propanol and 1 mmol l^{-1} octanoic acid; mobile phase for warfarin, pH 7.0, 0.067 mol l^{-1} potassium phosphate buffer containing 5 vol% 2-propanol and 1 mmol l^{-1} octanoic acid; flow rate for the SMCC HSA column, 1.5 ml min^{-1} for Trp, 1.0 ml min^{-1} for ibuprofen and warfarin; flow rate for the SIA HSA column, 0.3 ml min^{-1} for Trp and warfarin, 0.5 ml min^{-1} for ibuprofen; column size, 5 cm × 4.6 mm I.D.; temperature, 25 °C. *Source:* Reproduced from Mallik et al. [50]/with permission of American Chemical Society.

enantioseparation of Trp showing weaker bindings to HSA ($Ka \leq 10^4 M^{-1}$). This is due to the low effective protein concentration on the capillary wall. Graphene oxide (GO)-modified silica monoliths were prepared using three types of amine spacers for GO immobilization, including ammonia, ethanediamine (EDA), and polyethyleneimine [66]. HSA-GO-EDA-based silica monoliths exhibited the best chiral recognition ability for phenylalanine (Phe), Trp, warfarin, ibuprofen, azelastine, salbutamol, nefopam, propranolol, and chlortrimeton enantiomers among the prepared HSA monoliths in CEC. Furthermore, HSA and cellulase were co-immobilized on a poly(GMA-*co*-EDMA) monolith. CSPs based on the mixed selectors exhibited a broader range of enantioselectivity than the single selector in CEC mode [67].

CSPs based on the *N*-terminal half of an HSA fragment, whose molecular mass is about 35,000 Da, were prepared [68]. The HSA fragment was immobilized to amino-silicas activated by DSC. CSPs based on an HSA fragment had lower enantioselectivity than the intact HSA-based CSPs for all racemates (2-arylpropionic acid derivatives, benzodiazepines, and warfarin) tested. It was reported that the primary binding sites of warfarin were on an *N*-terminal half of HSA, and those of 2-arylpropionic acid derivatives and benzodiazepines were on the other half of HSA [68]. However, as will be described in Section 10.4, the two enantiomers of warfarin adopt very similar conformations from a co-crystal structure of (*R*)- or (*S*)-warfarin with HSA and make many of the same specific contacts with amino acid side chains at the binding site, thus accounting for the relative lack of stereospecificity of the HSA–warfarin interaction [69]. Therefore, it is questionable whether the primary binding sites might be the same as the enantioselective binding sites on the protein.

10.3.1.3 Trypsin and α-Chymotrypsin

Both trypsin and α-chymotrypsin are a family of serine proteases and catalyze the enantioselective hydrolysis of amide and esters. CSPs based on trypsin were introduced by Thelohan et al. [7] and those based on α-chymotrypsin were by Wainer et al. [8]. Trypsin-based CSPs can resolve *O*-, *N,O*-derivatized amino acids that are substrates of the enzyme [7]. This means that chiral separations are due to the activity of the enzyme, and that the chiral recognition site could be on the enzyme activity site. α-Chymotrypsin-based CSPs can enantioseparate amino acids, amino acid derivatives, dipeptides, and other compounds such as naproxen and aryloxypropionic acids [32].

10.3.1.4 Lysozyme and Pepsin

CSPs based on lysozyme and pepsin were developed by Haginaka et al. [9, 10]. Basic (chlorpheniramine, alprenolol, and tolperisone) and uncharged (benzoin and lorazepam) enantiomers were resolved in LC on lysozyme-based CSPs, while no resolution of acidic enantiomers was observed. Similarly, basic (homochlorcyclizine, verapamil, and alprenolol) and uncharged (oxazepam) enantiomers were separated on pepsin-based CSPs, while no resolution of acidic enantiomers was observed [10].

In CEC based on lysozyme, the fused-silica capillary walls were coated with a multilayer assembly of exfoliated γ-zyrconium phosphate and lysozyme, and Trp enantiomers were separated [70]. Furthermore, the capillary walls were coated with a phospholipid coating with lysozyme [71] as the chiral selector permeated into the phospholipid bilayer membrane as described in Section 10.3.1.1 for BSA. The chiral separation of D- and L-Trp was attained with the phospholipid-lysozyme membrane.

In CEC based on pepsin, silica monoliths functionalized with 3-aminopropyl trimethoxysilane were reacted with glutaraldehyde and pepsin. The resulting Schiff bases and residual aldehyde groups were reduced to secondary amines and hydroxyl groups, respectively, using sodium cyanoborohydride [72] and used for the enantioseparation of nefopam hydrochloride. c-SWNTs were incorporated into poly(GMA-*co*-EDMA) monoliths and reacted with pepsin via conversion of epoxide groups to amine groups using glutaraldehyde as the cross-linker [73]. The resulting Schiff bases were reduced to secondary amines as described earlier. The pepsin-modified poly(GMA-*co*-EDMA) monoliths with c-SWNTs exhibited much higher enantioselectivity than those without c-SWNTs for basic drugs such as nefopam, clenbuterol, amlodipine, and propranolol. Furthermore, pepsin-GO-EDA-based silica monoliths were prepared [66] with the same method with HSA-GO-EDA-based silica monoliths. Furthermore, instead of c-SWNTs and GO, amino-modified mesoporous silica nanoparticles were incorporated into poly(GMA-*co*-EDMA) monoliths [74]. In this case, it was supposed that pepsin could be immobilized to amino-modified mesoporous silica nanoparticles and polymer monoliths. Similarly, amino-modified mesoporous silica nanoparticles contributed to excellent enantioselectivity for basic drugs (azelastine, salmeterol, clenbuterol, propranolol, nefopam, ritodrine, and so on) as shown in Figure 10.2.

10.3.1.5 Fatty Acid-Binding Protein

FABPs belong to a family of low molecular mass proteins that bind fatty acids exhibiting high affinity binding constants for small endogenous and exogenous lipophilic ligands [11]. A FABP from chicken liver was immobilized onto aminopropyl-silicas activated by DSC. CSPs based on FABP were successfully applied to enantioseparations of some aryl- and aryloxy-propionic acids (e.g. suprofen, flurbiprofen, 2-(3-ethylphenoxy)propionic acid, 2-(3-phenylphenoxy)propionic acid, 2-(3-phenoxyphenoxy)propionic acid, and 2-(2,6-dimethylphenoxy)propionic acid) [11].

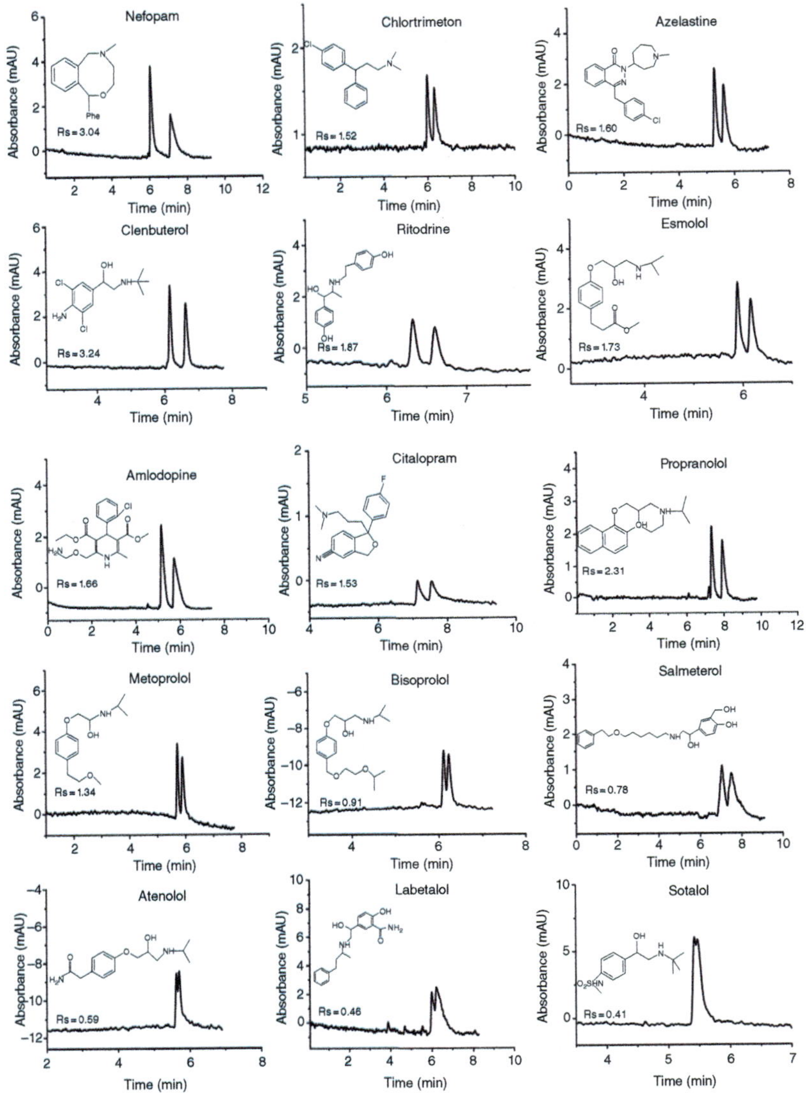

Figure 10.2 Enantioseparation of basic chiral drugs on pepsin-based poly(GMA-EDMA-NH2-MSN) monoliths by CEC. Experimental conditions: running buffer, 15 mmol l^{-1} ammonium acetate-acetic acid buffer (pH 4.2 for azelastine and amlodipine, and pH 5.8 for others); applied voltage, 20 kV for azelastine and amlodipine, 15 kV for salmeterol and labetalol, and 10 kV for others; injection, 10 kV × 1 s for salmeterol, citalopram, and amlodipine, and 1 kV × 1 s for others; detection, 210 nm for clenbuterol, salmeterol, and labetalol, 215 nm for azelastine and propranolol, 220 nm for ritodrine and metoprolol, 240 nm for citalopram and amlodipine, and 225 nm for others; temperature, 20 °C. *Source:* Reproduced from Xu et al. [74]/with permission of Elsevier.

10.3.1.6 Penicillin G Acylase

PGA from *Escherichia coli* catalyzes the hydrolysis of penicillin G to phenylacetic acid and 6-aminopenicillanic acid. PGA was immobilized via its amino or carboxy groups on silica particles [12]. The former was attained using amino-silicas activated by DSC and epoxy-silicas without activation, and the latter using amino-silicas with EDC and HSSI. The best results in terms of bound amounts of PGA, enzymatic activity, and enantioselectivity were obtained with PGA immobilized on epoxy-silicas. 2-Arylpropionic acid derivatives, such as ketoprofen, suprofen and fenoprofen, and phenoxypropionic acids, such as 2-(4-phenoxyphenoxy)propionic acid and 2-(4-benzylphenoxy)propionic acid were enantioseparated on PGA-based CSPs [12]. Furthermore, PGA was immobilized on epoxy-silica monoliths [75]. PGA-immobilized silica particles and monoliths were compared with respect to their enantioselectivity [75]. The former gave higher resolution for the enantiomers tested than the latter, despite the lower bound amount of PGA on the former than that on the latter. The worse resolution values on the latter are the consequence of the lower enantioselectivity rather than the lower column efficiency. A possible explanation could be that the high protein coverage on the PGA monoliths results in an increase in the non-specific interaction, while, on the other hand, hindering the access to the specific catalytic site, i.e. chiral binding site.

Furthermore, PGA immobilized onto epoxy-silica monoliths were used for enantioseparation of 2-arylpropionic acids in both capillary LC and electrically assisted capillary LC [76] using commercial CE instrumentation. However, the tested electrically assisted capillary LC conditions did not provide any significant improvement of both enantioselectivity and column efficiency.

10.3.1.7 Streptavidin

Avidin and streptavidin have very similar primary structures and are tetrameric proteins, being isolated from hen egg white and from the bacterium *Streptomyces avidinii*, respectively [13]. Their isoelectric points (pIs) are very different, 10 and 5, respectively. Streptavidin POROS, poly(styrene-divinylbenzene) porous perfusion beads, were packed into a microbore column. The streptavidin-based CSPs were used for the separation of adenosine enantiomers [13]. D-Adenosine was retained more strongly than the L-enantiomer and bound to the streptavidin-specific sites. By formation of streptavidin–biotin complex, streptavidin-based CSPs resulted in a drastic decrease in the retentivity and enantioselectivity via a direct and/or indirect effect. The direct effect is that some solutes have high affinity for the biotin binding sites, and the indirect one is that the strong interaction between the avidin and biotin may cause a conformational change of streptavidin.

10.3.1.8 Lipase

Candida antarctica lipase B is a family of enzymes that hydrolyze triglycerides. Lipase B has a molecular mass of 33 kDa and a pI of 6.0 [77]. Lipase B was immobilized to diol-silicas and amino-silicas [14]. Diol-silicas were oxidized to the aldehyde-silicas by periodic acid. Lipase B has been immobilized by Schiff base formation either to the aldehyde-silicas or to amino-silicas using in the latter case glutaraldehyde. In both methods, the formed Schiff bases were reduced to a secondary amine. CSPs based on lipase B were used for separation of several aromatic alcohols and diniconazole enantiomers in LC [14]. Lipase B was also immobilized or encapsulated to poly(GMA-*co*-EDMA) monoliths [78] and directly immobilized to epoxy-groups of the monoliths. The encapsulated monoliths showed better enantioselective separations for most of the investigated compounds [78]. A variety of drug classes, such as β-blockers, α-blockers, anti-inflammatory drugs, antifungal drugs, dopamine antagonists, norepinephrine-dopamine reuptake inhibitors, catecholamines, sedative hypnotics, diuretics, antihistaminics, anticancer, and antiarrhythmic drugs were enantioseparated in LC [78].

10.3.2 Glycoproteins

10.3.2.1 Human α_1-Acid Glycoprotein

AGP (orosomucoid [ORM]), a member of the lipocalin family, is a 41–43 kDa glycoprotein with pI of 2.7–3.8. The peptide moiety is a single chain of 183 amino acids for AGP. The carbohydrate content represents 45% of the molecular weight attached in the form of five to six highly sialylated complex-type-*N*-linked glycans [32]. Recently, it was reported that the average molecular mass of AGP could be ca. 33 kDa [79] by matrix-assisted laser-desorption ionization time-of-flight mass spectrometry, and that the sugar content of AGP could be estimated to be about 35% [79]. Besides the high heterogeneity of glycans, the protein part has also been found to show polymorphism. The variants are encoded by two different genes: the F1*S variant (ORM1) is encoded by the alleles of the same gene, while the A variant (ORM2) is encoded by a different gene [80]. There is a difference of at least 22 amino acid residues between the F1*S and A variants, while F1 and S forms differ only in a few residues.

CSPs based on AGP were developed by Hermansson [15]. In general, AGP was bound to silicas via its amino group. Recently, Hage and his group bound AGP via its carbohydrate chains after periodate oxidation to hydrazide-activated supports [51]. Silica particles, silica monoliths, and GMA/EDMA monoliths were used as the support and their retention and enantioselective properties were compared [51]. The higher surface area of the silica monolith gave materials that contained 1.5- to 3.6-times more immobilized protein per unit volume when compared to a silica particle or GMA/EDMA monolith. The AGP-immobilized silica monoliths gave higher retention as well as better resolution and efficiency than the AGP-immobilized silica particles or GMA/EDMA monoliths [51].

A wide range of basic, neutral, and acidic enantiomers were separated using AGP-based CSPs in LC [32, 81]. Various factors, such as mobile phase pH, type, and concentration of organic modifier and charged modifier; ionic strength; and temperature, affect the retentivity and enantioselectivity of solutes on AGP-based CSPs [81]. The optimization procedure for AGP-based CSPs was proposed by changing various factors described above, with the premise that the method can be used in LC–MS [82]. The hydrogen bonding properties and hydrophobicity of the uncharged organic modifier significantly influence the enantioselectivity of the solutes: verapamil was not enantioseparated with a mobile phase containing propan-1-ol, but was completely resolved with a mobile phase containing acetonitrile [81]. Enantioseparation of methylphenobarbital with a pure phosphate buffer (pH 7.0) as the mobile phase resulted in no resolution at all, while addition of only 2 vol% propan-2-ol to the mobile phase resulted in an induction of the enantioselectivity and complete resolution of the enantiomers [81]. Furthermore, addition of charged modifiers such as *N,N*-dimethyl-*n*-octylamine (DMOA), tetrapropylammonium bromide, tetrabutylammonium bromide, and sparteine affected the retentivity and enantioselectivity of uncharged, anionic, and cationic solutes [81]. These modifiers compete with the solute enantiomers at a binding site(s) or bound to an allosteric site(s). Furthermore, the adsorption of these modifiers on the protein induces a reversible change in its conformation.

AGP-immobilized LC silica particles [18] were packed into fused-silica capillaries and used in CEC mode.

10.3.2.2 Chicken Ovomucoid

OMCHI is a glycoprotein in egg whites and has a trypsin inhibitory activity [83]. OMCHI-based CSPs were developed by Miwa et al. [16]. OMCHI was immobilized to aminopropyl-silica particles [16] or amino-polymer particles [84], both of which were activated by DSC. The OMCHI-based CSPs were utilized for the resolution of acidic, basic, and neutral enantiomers in bulk drugs and formulations and for the assay of enantiomers in biological fluids [32] in LC. Kirkland et al. [85] reported that the OMCHI column (commercially available as ovomucoid [OVM] column) provided better long-term stability for repetitive injections compared to the commercially available AGP column, and that the column stability proved superior to other protein-based columns tested. Two columns were also compared: the OMCHI column tended to separate large molecules better than the AGP column [86].

In CEC, OMCHI was encapsulated in silica monoliths and used for enantioseparation of benzoin, eperisone, and chlorpheniramine [58].

The OMCHI-based CSPs were prepared from crude OMCHI, which includes ca. 10% OGCHI. As described in the Section 10.3.2.3., the chiral recognition ability of OMCHI reported previously [16] came from OGCHI, and OMCHI had no appreciable chiral recognition ability. Therefore, pure OMCHI-based CSPs cannot separate a lot of enantiomers.

10.3.2.3 Chicken α_1-Acid Glycoprotein

Recently, Haginaka et al. isolated and characterized a new protein from chicken egg whites [17]. It was first termed OGCHI as reported by Ketterer [87]. In addition, it was found that 10% of OGCHI was included in crude OMCHI preparations [17]. The pI values of OMCHI and OGCHI are the same, as shown in Table 10.1. Therefore, OGCHI was included in OMCHI preparations as an impurity. OGCHI was immobilized onto aminopropyl-silica particles activated with DSC like crude OMCHI. The OGCHI-based CSPs gave much higher chiral recognition ability than the crude OMCHI-based CSPs. Moreover, OMCHI- and OGCHI-based CSPs were made from isolated, pure proteins and were compared with regard to chiral recognition abilities. Furthermore, it was found that OGCHI is preferentially bound to DSC-activated amino-silicas compared to OMCHI [88]. This is why CSPs based on crude OMCHI, which are now commercially available as an OVM column, show moderate chiral recognition ability. Chiral recognition properties of OGCHI and ovoglycoprotein from Japanese quail egg whites (OGJPQ) were compared [89]. The OGCHI-based CSPs are suitable for chiral resolution of basic compounds, while the OGJPQ-based CSPs are suitable for that of acidic compounds. The results obtained revealed that the chiral recognition of various solutes should be efficiently attained by complementary use of both CSPs.

OGCHI was immobilized onto aminopropyl-silica particles by reaction with its amino and carboxy groups [90]. In the former, OGCHI was bound to DSC-activated amino-silicas, while in the latter OGCHI activated by EDC and HSSI was bound to amino-silicas. Figure 10.3a and b shows chromatograms of 2-phenylpropionic acid and naproxen on the OGCHI-based CSPs immobilized via amino and carboxy groups of the protein, respectively. The former could be suitable for enantioseparations of basic ligands, while the latter is suitable for enantioseparations of acidic ligands [90]. These results

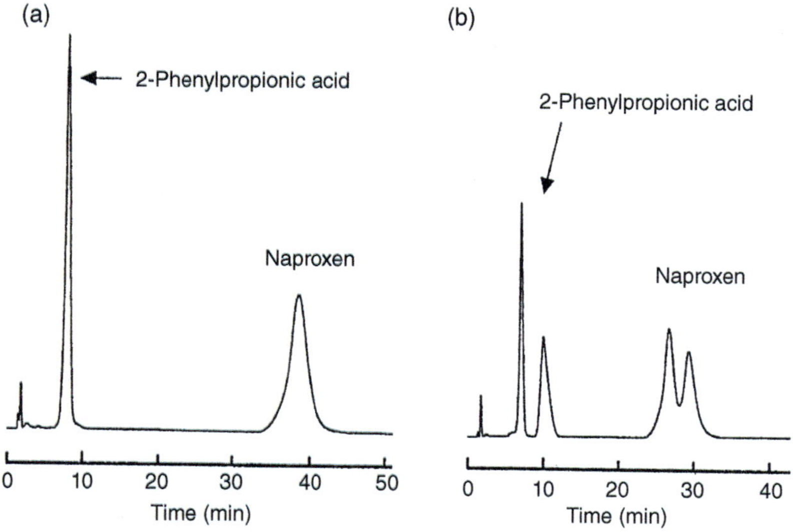

Figure 10.3 Chromatograms of 2-phenylpropionic acid and naproxen on OGCHI (cAGP)-based CSPs prepared via (a) amino and (b) carboxy groups of the protein. LC conditions: column, 100 mm × 2.0 mm I.D.; mobile phase, 20 mmol l^{-1} phosphate buffer (pH 4.0)–ethanol (90 : 10, v/v); column temperature, 25 °C; flow rate, 0.2 ml min^{-1}; detection, 220 nm. *Source:* Reproduced from Haginaka et al. [90]/with permission of Elsevier.

suggest that electrostatic interactions between amino and carboxy groups of OGCHI with acidic and basic ligands, respectively, should work for their retentions and enantioseparations [88], and that non-specific immobilization of OGCHI via amino and carboxy groups could interfere with nonselective and selective interactions based on electrostatic interactions.

Recently, a cDNA clone encoding OGCHI was isolated and the amino acid sequence of OGCHI was clarified [18]. OGCHI consisted of 203 amino acids including a predictable signal peptide of 20 amino acids. The mature OGCHI showed 31–32% identities to rabbit and human AGPs. Thus, OGCHI should be chicken AGP (cAGP), a member of the lipocalin family.

cAGP was immobilized onto aminopropyl-silica particles, whose nominal particle diameters were 5, 3, and 2.1 μm, activated with DSC [91]. There were not so much differences in the retention and enantioseparation factors among three cAGP columns, while their resolution values were in the order of 2.1 μm > 3 μm > 5 μm silica particles and their height equivalent of a theoretical plate values were in the reversed order. These results indicate that 2.1-μm silica particles are useful for the preparation of cAGP- or protein-based CSPs.

10.3.2.4 Avidin

Avidin is a basic glycoprotein found in egg whites, with a pI of 10.0, that strongly binds biotin with *Ka* of ~10^{15} M^{-1} [92]. Avidin was immobilized to DSC-activated amino-silicas and applied to the enantioseparation of acidic compounds, such as 2-arylpropionic acid derivatives in LC [19]. Furthermore, avidin was immobilized to amino-silicas activated with DSS [48]. Avidin-based CSPs prepared by activation with DSS gave longer retentivity and higher enantioselectivity than those with DSC [48]. Therefore, the former could enantioseparate acidic compounds efficiently, while basic and neutral compounds were separated with lower enantioseparation factors than those for acidic compounds [32]. This could be due to the fact that DSS works as a spacer for immobilization of a protein, thus allowing an increased flexibility and availability of the protein. Furthermore, avidin-based CSPs prepared by activation with DSS were used for the direct serum injection assays of drug enantiomers in biological fluids [29]. This could be due to the fact that the outer surface of avidin was hydrophilic, while the internal surface worked as enantioselective sites.

In CEC, avidin was adsorbed onto capillary walls and used for enantioseparation of 2-arylpropionic acids, warfarin, abscisic acid (ABA), and dansyl amino acids [38]. The physical adsorption method of avidin to silica monoliths was successfully applied to enantioseparations in LC and CEC modes [39]. Furthermore, avidin in phospholipid membrane was coated on fused-silica capillaries using biotinylated phospholipids and employed as a chiral selector for the separation of amino acids such as Trp and phenylthiohydantoin-Ser and -Thr [44].

Not only was enantioselectivity but also the interaction with racemates significantly lost by the formation of avidin–biotin complex [19], as in the case of the formation of streptavidin–biotin complex.

10.3.2.5 Riboflavin-Binding Protein and Ovotransferrin

CSPs based on RfBP from chicken egg white were introduced by Mano et al. for the enantioseparation of acidic, basic, and neutral compounds [20]. On the other hand, Massolini et al. [93] described the use of RfBP from chicken egg yolk as the CSPs. Egg white and yolk RfBPs seem to be the product of the same gene but to have undergone different post-translational modifications. Therefore, the amino acid sequences of both RfBPs are the same, but the latter lacks in the last 11 to 13 amino acids by proteolytical cleavage, in addition to differences in carbohydrate chains [93]. Both CSPs based on chicken egg white and yolk RfBPs were immobilized on amino-silicas activated by DSC and gave similar enantioselectivity. Furthermore, CSPs based on quail egg white and yolk RfBPs were developed by Lorenzi et al. [94].

CSPs based on ovotransferrin from chicken egg white were immobilized to amino-silicas activated by DSC [20]. The CSPs were utilized for chiral resolution of a basic compound, azelastine [32]. Ovotransferrin was stable to heat when combined with iron, copper, manganese, and zinc. Furthermore, ovotransferrin was further stabilized by immobilization to supports.

10.3.2.6 Cellobiohydrolase

Cellulases are cellulose-hydrolyzing enzymes. The fungus *Trichoderma reesei* produces four major cellulases: two cellobiohydrolases, CBH I (Cel7A, 64 kDa, pI 3.9) and CBH II (Cel6A, 53 kDa, pI 5.9), and two endoglucanases, EG I (Cel7B, 55 kDa, pI 4.5) and EG II (Cel6B, 48 kDa, pI 5.5) [95]. They are all acidic glycoproteins and have a common structural organization with a binding domain connected to the rest of the enzyme (i.e. the core) through a flexible arm [96]. The interconnecting region is rich in Ser, Thr, and Pro residues and is highly glycosylated [96]. The core is enzymatically active. CSPs based on CBH I (Cel7A), which can resolve acidic, basic, and uncharged enantiomers, have been the most extensively investigated among CSPs based on cellulases [97]. CBH I was immobilized to aldehyde-silicas followed by Schiff base formation and reduction. Especially, high enantioselectivity was obtained for separations of β-blockers such as propranolol, oxprenolol, and metoprolol [97]. Contrary to CBH I (Cel7A), CBH II (Cel6B) could enantioseparate only a few β-blockers, in addition to some racemates, e.g. mexiletine, ibuprofen, chlorthalidone, and pentobarbital [98]. Cellobiohydrolase 58 (CBH 58, Cel7D) from *Phanerochaete chrysosporium*, which is the counterpart to Cel7A from *T. reesei*, was immobilized to silica supports. CBH 58 (Cel7D, 60 kDa, pI 3.8), similar to CBH I (Cel7A), is an excellent chiral selector for β-blockers and even expresses a broader enantioselectivity for other basic compounds than CBH I [99].

It is interesting that at mobile phase pH 5.5 the retention time of the less retained (*R*)-propranolol enantiomer decreases with increasing temperature on CBH I (Cel7A)-based CSPs, while that of the (*S*)-enantiomer increases, causing a large increase in the enantioseparation factor when the temperature is raised from 5 to 45 °C as shown in Figure 10.4 [100]. It was found that the adsorption of the more retained enantiomer, (*S*)-propranolol, is endothermic while that of (*R*)-enantiomer is exothermic [100]. This is why unusual chiral separation of propranolol occurs on the CBH I (Cel7A)-based CSPs. Fornstedt et al. [101] reported that a temperature rise from 15 to 25 °C gave no significant difference in the circular dichroism (CD) spectra of CBH I (Cel7A), but that at 45 °C the CD spectra indicated a conformational change of the protein.

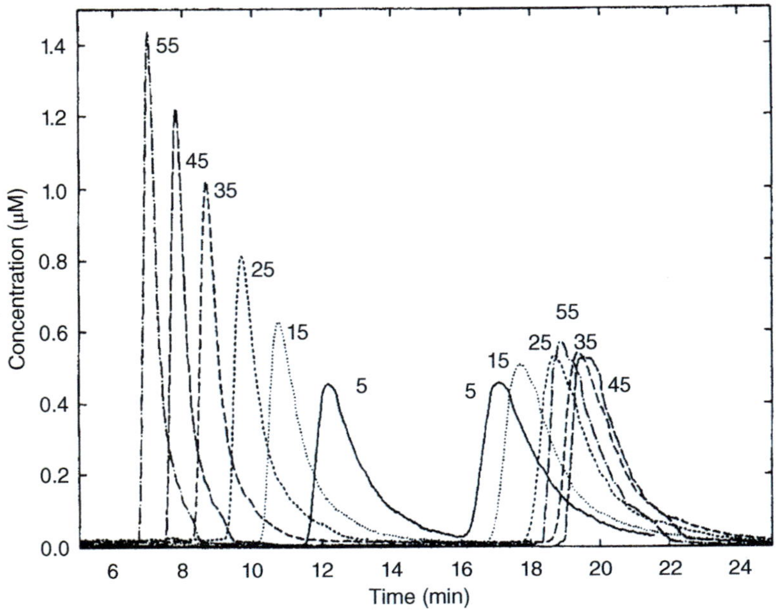

Figure 10.4 Elution profiles of (*R*)- and (*S*)-propranolol at different temperatures. (*S*)-Propranolol is the more retained enantiomer and the one showing the unusual temperature dependence at this mobile phase pH value. Conditions: column, 100 mm × 4.6 mm I.D.; stationary phase, immobilized CBH I (Cel7A) on silicas; mobile phase, acetate buffer at pH 5.47, $I = 0.02$; mobile phase flow rate, 0.8 ml min^{-1}; sample, 10 μl of a 0.05 mmol l^{-1} solution of (*R*)- and (*S*)-propranolol, each, in the acetate buffer. *Source:* Reproduced from Fornstedt et al. [100]/with permission of American Chemical Society.

Furthermore, an increase in both retentivity and enantioselectivity for some β-blockers was observed when exchanging potassium with sodium ion in the buffer used as the mobile phase [102]. The CD spectra of CBH I dissolved in potassium or sodium phosphate buffers displayed a small difference. This can be explained by an altered protein conformation, which might cause an altered retention behavior as well as a different enantioselectivity for β-blockers.

Cellulase from *T. reesei* ATCC 26921 (Cel) was immobilized onto aminopropyl-silica particles via their amino and carboxy groups using DSC and EDC and HSSI, respectively [103]. They were termed *N*-Cel and *C*-Cel, respectively. Despite their smaller retention factors on a *C*-Cel column, the enantioseparation factors and resolution values of β-blockers were similar using either *N*- or *C*-Cel columns. Furthermore, the former showed lower column efficiency than the latter. This could be due to the fact that strong non-specific interactions between amine groups of β-blockers with carboxy groups of *C*-Cel could work for the retention of β-blockers, and that the carboxyl groups of Cel in enantioselective binding sites could not be involved in the immobilization of Cel to aminopropyl-silica particles. In addition, *C*-Cel was prepared using aminopropyl-silica particles with nominal particle diameters of 5, 3, and 2.1 μm, respectively. A *C*-Cel column prepared with 2.1-μm aminopropyl-silica particles gave the highest enantioselectivity and column efficiency among three *C*-Cel columns as in the case of the cAGP column.

In CEC, capillaries filled with silica particles consisting of mixed proteins, CBH I and BSA, were prepared [104]. Since it was difficult to obtain stable silica particles with CBH I alone, it was copolymerized with BSA. The enantioseparation of β-adrenergic antagonists (acebutolol, alprenolol, atenolol, metoprolol, pindolol, prenalterol, and propranolol) was attained. As described in Section 10.3.1.2, cellulase was immobilized with HSA on polymer monoliths [67]. CSPs based on the mixed proteins gave a broad range of enantioselectivity.

10.3.2.7 Glucoamylase

Glucoamylase (amyloglycosidase) is an enzyme used to break down starch to mono- and di-saccharides. CSPs based on glucoamylases G1 and G2 from *Aspergillus niger* were introduced by Karlsson and his coworkers [23, 105, 106]. Glucoamylase G1 consists of three parts: the first 440 amino acids contain the catalytic domain, the 441–511 amino acids contain a highly *O*-glycosylated linker segment, and the 512–616 amino acids contain a *C*-terminal domain responsible for raw starch binding and dispensable for activity. Glucoamylase G2 consists of 512 amino acids, including the catalytic domain and linker segment without a *C*-terminal domain. First, glucoamylase G1 was immobilized to amino-silicas activated by DSC. However, the stability of the CSPs produced was very poor and their enantioselectivity was lost after passing less than 1000 column volumes through the columns. Next, glucoamylase G1 was immobilized on oxidized diol-silicas (aldehyde-silicas) by reductive amination. The diol-silicas with 50 nm pore sizes gave the highest enantioselectivity and stability among the diol-silicas tested. On this CSP, amino alcohols were separated with high separation factors and high column efficiencies up to 30,000 m^{-1}, while acidic enantiomers could not be separated [23, 105]. Furthermore, glucoamylase G2 was immobilized using the same method applied for glucoamylase G1 [106]. Two glucoamylase-based CSPs were compared with regard to retentive and enantioselective properties for amino alcohols. The retention factors on the glucoamylase G2-based CSP were higher than those on the glucoamylase G1-based CSP, while the enantioseparation factors were similar. The higher retention factor is due to the fact that the amounts of immobilized glucoamylase G2 on the supports are about three times more than those of glucoamylase G1 [106]. Addition of acarbose, which binds to the catalytic domain and inhibits the glucoamylase activity, to the mobile phase resulted in total loss of enantioselectivity of the glucoamylase G2-based CSP [106]. These results indicate that the catalytic domain is responsible for the chiral recognition.

The enantioselective retention could be controlled by several mobile phase parameters, e.g. mobile phase pH, type, and concentration of organic modifier and column temperature [105]. A drastic increase in the retention factor was observed with an increase in the mobile phase pH from 6.0 to 7.6, while the enantioselectivity remained unchanged on both glucoamylases G1- and G2-based CSPs [105, 106]. The effects of column temperature and concentration of propan-2-ol on the enantioseparation of alprenolol are shown in Figure 10.5 [105]. A drastic increase in the enantioselectivity of

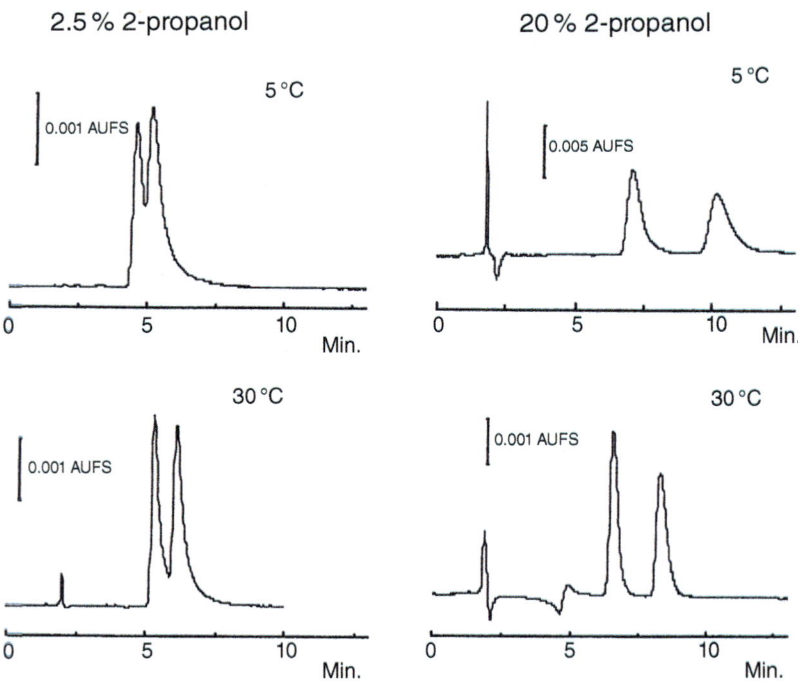

Figure 10.5 Effects of propan-2-ol and column temperature on the enantiomeric resolution of alprenolol. Mobile phase: phosphate buffer (pH 7.0, $I = 0.01$) and propan-2-ol. Solute: (*R*)- and (*S*)-alprenolol. *Source:* Reproduced from Strandberg et al. [105]/with permission of Springer Nature.

alprenolol by increasing the content of propan-2-ol in the mobile phase as well as an increase in the enantioselectivity by elevating the column temperature are shown. A further thermodynamic study indicated that the bi-Langmuir model (containing nonselective and enantioselective sites) was appropriate to describe the system [107].

10.3.2.8 Antibody (Immunoglobulin G)

In 1982, Mertens et al. raised polyclonal antibodies for (+)- and (−)-ABA in rabbit sera and prepared their immunoglobulin G (IgG) fractions. Then, the IgG antibodies immobilized onto cyanogen bromide-activated Sepharose 4B were utilized for the isolation of ^{3}H-(+)-ABA as well as ^{3}H-(−)-ABA [24]. The antibody-based CSPs could be reused at least six to eight times without loss of capacity. Similarly, an antibody for (+)-ABA was produced using ABA-4′-*p*-aminobenzoyl-hydrazone coupled to keyhole limpet hemocyanin (KLH) as an immunogene [108]. The obtained monoclonal antibodies were immobilized to cyanogen bromide-activated Sepharose 4B. The target enantiomer was retained by the antibody-based CSPs and eluted second, while the opposite enantiomer eluted with the void volume. The CSPs were used 10 times without loss of capacity. The major disadvantage of antibody-based CSPs is that a strong mobile phase is required for elution of a more retained enantiomer, and that they are unstable for repeated uses.

Hofstetter et al. raised stereospecific antibodies against a broad class of substances, α-amino acids, and applied them to the enantioseparations of α-amino acids [109–111]. *p*-Amino-D- and *p*-amino-L-Phe, respectively, were coupled to KLH or BSA via the *p*-amino group by diazotization of its tyramine residues, and the resulting conjugates, *p*-azo-D-Phe-KLH or -BSA and *p*-azo-L-Phe-KLH or -BSA, were used as the immunogenes for rabbits. The produced monoclonal antibodies were bound to Sepharose 4B, POROS-OH (poly(styrene-divinylbenzene) porous perfusion beads) and diol-silica particles activated with DSC. Enantioseparations of α-amino acids could be achieved on antibody-based CSPs by using the latter two supports under LC conditions. Figure 10.6 shows enantioseparation of Tyr, Phe, and *p*-aminoPhe on a monoclonal antibody-based CSP against L-amino acids [110]. The obtained CSP gave the enantioselectivity for a wide variety of α-amino acids: the higher values for aromatic and bulky side chain amino acids such as Phe, Trp, Tyr, and histidine (His) and lower ones for aliphatic amino acids [110]. Using an anti-D-amino acid antibody-based CSP, the L-enantiomers eluted with the void volume, while the D-enantiomers retained well. Inverted elution orders were obtained on CSPs prepared from an anti-L-amino acid antibody [110]. These results indicate that the elution orders of α-amino acids can be predicted by using anti-D- or L-amino acid antibody-based CSPs. The effect of the mobile phase parameters

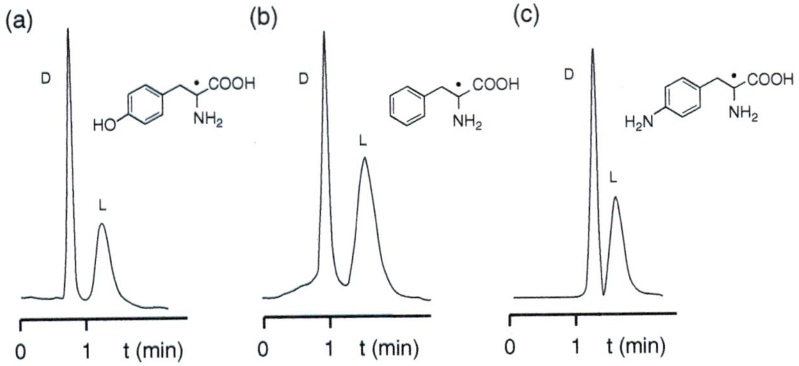

Figure 10.6 Enantioseparations of (a) Tyr, (b) Phe, and (c) *p*-aminoPhe on an anti-L-amino acid antibody-based CSP. LC conditions: column, 250 mm × 4.6 mm I.D.; mobile phase, phosphate buffered saline, pH 7.4; flow rates of (a and b) 4 and (c) 2 ml min^{-1}. *Source:* Reproduced from Hofstetter et al. [111]/with permission of Elsevier.

such as flow rate, temperature, and organic modifier on the enantioseparations of various aromatic amino acids was examined [111]. The retention factor of the second eluted enantiomer is dependent on the affinity between the analyte and the immobilized antibody, while that was independent of the flow rate. With an increase in column temperature, the retention factor of a solute decreased. However, the van't Hoff plot was not linear. This means that at an elevated temperature, conformational changes of the immobilized antibody could occur [111]. Furthermore, the addition of an organic modifier did not improve the separation.

Hofstetter and his group prepared CSPs based on anti-α-hydroxy acids [112]. *p*-Amino-D- and *p*-amino-L-phenylacetic acid, respectively, were coupled to KLH via the *p*-amino group by diazotization of its tyrosyl residues and were immunized to rabbits, as described for *p*-aminoPhe [109]. The prepared monoclonal antibodies against D-α-hydroxy acids were immobilized to diol-silica particles or POROS-OH activated with DSC. Chiral separations of several aliphatic and aromatic members of this class of compounds were achieved under mild isocratic buffer conditions using phosphate buffered saline, pH 7.4, as the mobile phase. Enantioseparations of D,L-4-hydroxymandelic acid and D,L-vanillomandelic acid on a monoclonal antibody-based CSP against D-hydroxy acid [112] were attained using phosphate buffer saline as the mobile phase. As described above, the D-enantiomers of 4-hydroxymandelic acid and vanillomandelic acid eluted second. No significant changes of column performance were observed after 300 injections [112].

Anti-D-methamphetamine monoclonal antibodies were immobilized onto Sepharose 4B activated by cyanogen bromide or amino-silicas modified with glutaraldehyde [113]. The antibody-based CSPs were applied to the enantioseparation of amphetamine and methamphetamine. Anti-3,3′,5-triiodo-L-thyronine (anti-L-T_3) polyclonal antibodies were immobilized onto amino-silicas modified with glutaraldehyde [114]. The enantiomers of T3 showed baseline separation under mild isocratic mobile phase conditions using phosphate buffer, pH 7.4.

10.3.2.9 Nicotinic Acetylcholine Receptor and Human Liver Organic Cation Transporter

Recently, membrane-bound proteins such as receptors, ion channels and drug transporters (for example, $\alpha_3\beta_4$-nAChR [25] and and hOCT [26]), have been incorporated into chromatographic systems. The $\alpha_3\beta_4$-nAChR and hOCT were immobilized on immobilized artificial membrane stationary phases. CSPs based on $\alpha_3\beta_4$-nAChR could separate dextromethorphan (DM) and levomethorphan (LM), while those based on hOCT could separate verapamil, propranolol, atenolol, and pseudoephedrine enantiomers.

10.4 Chiral Recognition Mechanisms on Protein- and Glycoprotein-Based Chiral Stationary Phases

10.4.1 Human Serum Albumin

HSA has three homologous domains (I–III), and each of them is composed of two subdomains (A and B). There are two principal drug-binding sites (sites I and II) in subdomains IIA and IIIA, respectively (Figure 10.7) [115]. Sites I and II are referred to as warfarin-azapropazone and indole-benzodiazepine sites, respectively. In addition, other minor binding sites were proposed to elucidate bindings of a given drug, whose bindings could not be explained by the principal drug-binding sites [116]. Selective modification of some amino acid residues of HSA was tried to clarify their roles in enantioselective bindings. Tyr411 (located in Site II) or Lys199 (located in Site I) was selectively acetylated [117, 118]. In most cases, the acetylation of Tyr411 resulted in a decrease in the retentivity and enantioselectivity of most drugs tested in LC, while their increases were observed with some drugs such as benoxaprofen, temazepam, and oxazepam hemisuccinate, but not other profens and benzodiazepines, which bind to site II. These results suggest that site II should be considered as a large and flexible area composed of several subsites [117]. The acetylation of Lys199 affected the

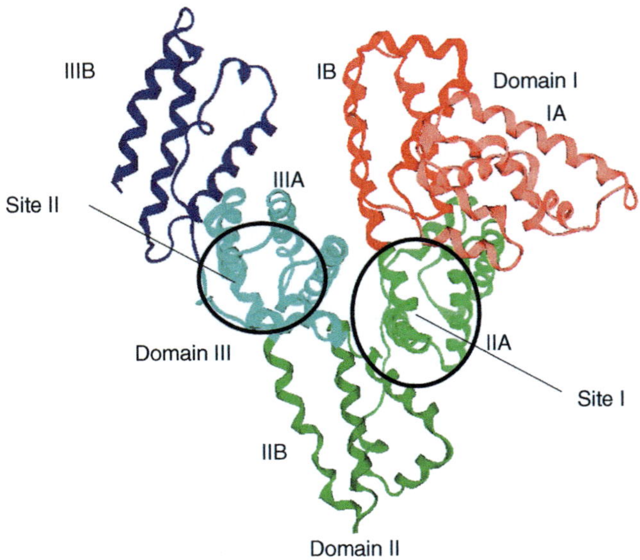

Figure 10.7 Crystal structure of HSA. The subdivision of HSA into domain (I–III) and subdomains is indicated, and approximate locations of sites I and II are shown. *Source:* Reproduced from Chuang and Otagiri [69]/with permission of John Wiley & Sons.

bindings of drugs, which bind to both sites I and II of HSA. It is interesting that an increase in enantioseparation factors in LC was observed with drugs, which exclusively bind to site II [118]. The modification of Cys34 of HSA with ethacrylic acid was found to improve the retentivity and enantioselectivity of a solute on the HSA-based CSPs in LC [119]. The retention times are significantly shorter for most of the drugs tested, and the enantioselectivity was enhanced in many cases. Furthermore, it was found that selective modification of Trp214 (located on site I) with *o*-nitrophenylsulfenyl chloride gave the same number of binding sites as intact HSA for (*R*)-warfarin and L-Trp (i.e. probes for sites I and II, respectively) but lower association equilibrium constants for both of these solutes [120]. It was suggested that the weaker bindings of (*R*)-warfarin could be due to a direct blocking of site I by modified Trp214, and that those of L-Trp could be due to an allosteric-induced change in Site II.

He and Carter determined the three-dimensional structure of HSA, which shows that the binding sites I and II are located in hydrophobic cavities in the subdomains IIA and IIIA, respectively [121]. In addition, the crystal structure of HSA-myristate complexed with (*R*)- and (*S*)-warfarin was determined [122]. The structures confirm that warfarin binds to subdomain IIA (site I) in the presence of myristate and reveal the molecular details of HSA–drug interactions. The experimental difference of the electron density map for *R*-(+)- and *S*-(−)-warfarin in subdomain IIA from the HSA-myristate structure [122] was compared. The (*R*)- and (*S*)-enantiomers bind in the pocket in almost identical conformations. Additionally, the two enantiomers of warfarin adopt very similar conformations when bound to HSA and make many of the same specific contacts with amino acid residues at the binding site, thus resulting in the relative lack of enantioselective bindings of the HSA–warfarin complex [122]. Similarly, dockings of the ketoprofen enantiomers to the model structures of HSA gave similar interactions with the carboxyl and aromatic benzoyl groups of both enantiomers. These results indicated that the subdomain IIIA (site II) might not be the major location for the stereoselective binding of ketoprofen [69]. It was supposed that the enantioselective binding sites exist at the interface of the other subdomains of HSA, not subdomains IIA and IIIA (sites I and II).

10.4.2 Penicillin G Acylase

It has been reported that the binding site of PGA consists of three major regions responsible for the ligand recognition by the enzyme: the catalytic residue SerB1; the oxyanion hole (stabilizing the negative charge present on the ligand carboxylate group) formed by GlnB23, AlaB69, AsnB241; and a hydrophobic pocket that is able to accommodate lipophilic groups [123]. Chiral recognition mechanism on PGA was investigated using molecular modeling and docking studies [124, 125]. As shown in Figure 10.8a, the binding mode for (*S*)-2-(4-chlorophenyl)-2-phenoxyacetic acid is characterized by the presence of numerous hydrogen bonding interactions involving the carboxy group with both the backbone NH and the OH groups of SerB386 [124]. An additional hydrogen bonding is established between the ether oxygen of the ligand and the ArgB263 side chain. Moreover, an ionic interaction is formed between the negatively charged carboxy group and the positively charged *N*-terminal SerB1. A set of charge-transfer interactions is also found between the ligand phenoxy and phenyl moieties and the PheB24, PheA146, PheB71, and PheB256 aromatic rings. With regard to (*R*)-2-(4-chlorophenyl)-2-phenoxyacetic acid, the calculated snapshot is similar to that found for the (*S*)-enantiomer, even if the hydrogen bonding interaction between the ether oxygen and ArgB263 side chain is absent and the phenoxy moiety is placed in the same position occupied by the phenyl ring in the (*S*)-enantiomer (Figure 10.8b) [124]. The absence of the hydrogen bonding interaction with ArgB263 would suggest that the

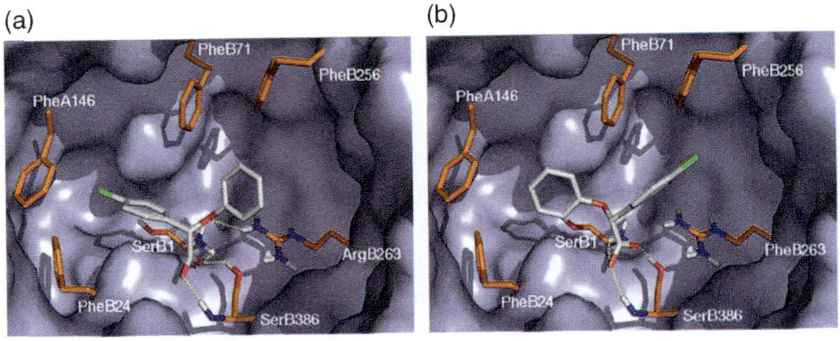

Figure 10.8 Binding mode of (*S*)-2-(4-chlorophenyl)-2-phenoxyacetic acid (a) and (*R*)-2-(4-chlorophenyl)-2-phenoxyacetic acid (b) within PGA. For clarity reasons only interacting residues are displayed. Hydrogen bonds between ligand and protein are shown as dashed yellow lines. Ligand (white) and interacting key residues (orange) are represented as stick models, while the protein as a light gray Connolly surface. *Source:* Adapted from Lavecchia et al. [124]/with permission of Elsevier.

(*R*)-enantiomer forms a less stable complex with PGA. This simulation result well coincides with the elution order of (*R*)/(*S*) of 2-(4-chlorophenyl)-2-phenoxyacetic acid in LC.

10.4.3 Human α_1-Acid Glycoprotein

It was originally thought that drug bindings to hAGP occurred at a single hydrophobic pocket or cleft within the protein domain of the molecule or at more than one binding site [126]. Furthermore, the hydrophobic, electrostatic, and hydrogen bonding interactions could play an important role in the retentivity and enantioselectivity of a solute on an AGP-based CSP [81]. Native hAGP consists of about 70% F1*S variant (ORM1) and 30% A variant (ORM2). The ORM1 has lobs I, II, and III as the ligand-binding sites, while ORM2 has only lobs I and II [127]. Figure 10.9 shows crystal structures of F1*S (left, ORM1) and A (center, ORM2) variants of hAGP [128]. It was reported that many ligands bound almost to the same sites on hAGP, that is, lobes I–III [127].

Selective binding of coumarin enantiomers (warfarin, phenprocoumon, and acenocoumarol) to hAGP genetic variants, ORM1 and ORM2, was investigated. All investigated compounds bound stronger to ORM1 than to ORM2 [80]. ORM1 and native hAGP preferred the binding of (*S*)-enantiomers of warfarin and acenocoumarol, while no enantioselectivity was observed in bindings of phenprocoumon. Furthermore, a new homology model of hAGP was built and the models of ORM1 and ORM2

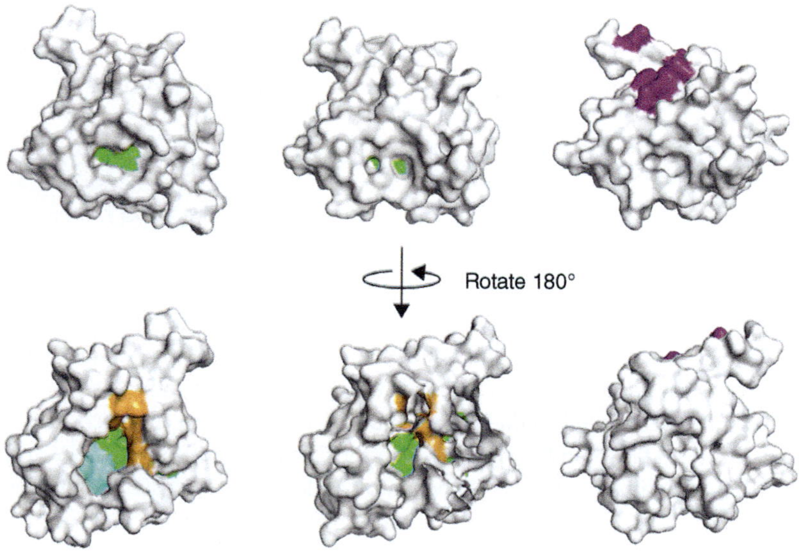

Figure 10.9 Crystal structures of F1*S (left) and A (middle) variants of hAGP and a built model structure of cAGP (right). The A variant of hAGP was only shown by one of subunits. The cavities in hAGP were called as lobes, highlighted with green for lobe I, orange for lobe II, and cyan for lobe III. The amino acids, which consist of feasible ligand-binding sites for cAGP, were colored in magenta.
Source: Reproduced from Haginaka et al. [128]/MDPI/Public Domain CC BY 4.0.

suggested that the binding cavity, including Trp122, for ORM 1 was the same with that for ORM2, and that the difference in binding to hAGP genetic variants could be caused by steric factors; ORM2 formed a smaller, more hydrophobic cavity as compared to ORM1 [80]. Comparison of dockings of (*R*)- and (*S*)-acenocoumarol to ORM1 and ORM2 models revealed that dockings to ORM1 resulted in a much lower intermolecular energy than dockings to ORM2; although binding to both variants was possible, ORM1 binding was more favorable. Energy differences between (*R*)- and (*S*)-enantiomers are not significant and show a slight preference for (*S*)-enantiomers in the case of both ORM1 and ORM2 [80]. However, these simulation results could not clearly explain the enantioselective bindings of coumarin enantiomers to ORM1 and ORM2.

Ligand-binding properties of hAGP were investigated by using CD methods [129]. The induced CD spectra of drug–hAGP complexes were observed with many classes of drugs. Results of additional CD experiments performed by using recombinant hAGP mutants showed no changes in the ligand binding ability of Trp122Ala in sharp contrast with the Trp25Ala, which was unable to induce extrinsic CD signal with either ligand. These findings suggest that, likely via π–π stacking mechanism, Trp25 is essentially involved in the hAGP binding of the drugs studied (chlorpromazine, dpyridamole, terazosin, propranolol, primaquine, and carbamazepine) [129].

10.4.4 Turkey Ovomucoid

Various ovomucoids, such as ovomucoid from turkey egg whites (OMTKY) and OMCHI, exist as three tandem, independent domains [32]. Each domain and combination domains (first and second domains, and second and third domains) were isolated, purified, and characterized [130]. Furthermore, CSPs were made with purified OMTKY and OMCHI domains to test chiral recognition properties [130]. The third domain of OMTKY and OMCHI consisted of glycosylated (OMTKY3S and OMCHI3S) and unglycosylated domains (OMTKY3 and OMCHI3). The third domains of the OMTKY and OMCHI domains were found to be enantioselective to at least two classes of compounds, benzodiazepines and 2-arylpropionic acid derivatives. Glycosylation of the third domain did not affect chiral recognition. However, either the first, second, or a combination domain of OMTKY gave no appreciable chiral recognition ability [130]. One might ask why whole, intact OMTKY and OMCHI do not have chiral recognition ability, but the third domains of OMTKY and OMCHI do. A possible explanation is that ligands cannot reach the chiral binding sites on whole, intact OMTKY and OMCHI because of steric hindrance, but they can reach the chiral binding sites on the third domains after cleavage of the connection peptide of each domain.

The chiral recognition mechanism of OMTKY3 was elucidated by using nuclear magnetic resonance (NMR) measurements, molecular modeling, and ligand docking studies [130]. On the surface of OMTKY3, there are two distinct binding sites (nonselective and enantioselective binding sites): in the former, hydrophobic interactions mainly work for the binding, while in the latter, hydrophobic, electrostatic, and hydrogen bonding interactions play important roles. The enantioselective binding model for (*R*)- and (*S*)-U-80413, which is one of 2-arylpropionic acid derivatives, with OMTKY3 showed similarities and differences in orientation and intermolecular interactions between (*R*)- and (*S*)-U-80413 [130]. The carboxy groups of each enantiomer engage in electrostatic interactions with the positive charge on Arg21. The carbonyl group on U-80413's central ring shares a hydrogen bond with NH_3^+ group of Lys34. The distinguishing difference between the enantiomers is the proximity of the phenyl group of (*R*)-U-80413 and Phe53. The elution of U-80413 on OMTKY3-based CSPs in LC was in the order of (*S*) and (*R*). The simulation result was consistent with the elution order of U-80413 enantiomers in LC.

10.4.5 Chicken α_1-Acid Glycoprotein

The recombinant cAGP was prepared by the *E. coli* expression system, and its chiral recognition ability was confirmed by CE [18]. Since proteins expressed in *E. coli* are not modified by any sugar moieties, this result shows that the protein domain of the cAGP could be responsible for the chiral recognition [18]. cAGP consists of 183 amino acid residues and has only one Trp residue at the 26 position [18]. The Trp26 residue was modified with 2-nitrophenylsulfenyl chloride and chiral separations of neutral, acidic, and basic compounds were examined on cAGP- and Trp-modified cAGP-based CSPs [131]. The chiral separation ability for propranolol, alprenolol, and oxprenolol was lost on the Trp-modified cAGP-based CSP, while chlorpheniramine, ketoprofen, and benzoin were still enantioseparated on the Trp-modified cAGP-based CSP, despite lower enantioselectivity than that on the cAGP-based CSP. These results suggest that the Trp26 residue could be responsible for chiral recognition of these compounds. Competition studies using DMOA as a competitor indicated that propranolol, alprenolol, and oxprenolol competed with DMOA on a single binding site near the Trp26 region and that further bindings of chlorpheniramine, ketoprofen, and benzoin occurred at the secondary binding site in a noncompetitive fashion with DMOA [131]. Furthermore, ligand-binding properties of cAGP were investigated by using CD methods. Analysis of the extrinsic CD spectra employing the Trp26-modified protein and CD displacement experiments revealed that a single Trp26 residue of cAGP, conserved in the whole lipocalin family, is part of the binding site, and that it is essentially involved in the ligand-binding process via π–π stacking interaction. This results in the appearance of strong induced CD bands due to the non-degenerate intermolecular exciton coupling between the π–π^* transitions of the stacked indole ring-ligand chromophores, such as in chlorpromazine, (*R,S*)-primaquine, (*R,S*)-propranolol, tamoxifen, diclofenac, (*R,S*)-ketoprofen, (*S*)-ketoprofen, and diazepam [132].

Recently, the model structure of cAGP was generated using hAGP (ORM2) as a template [128]. Subsequently, each enantiomer of benzoin, chlorpheniramine, and propranolol was simulated and docked to a certain cavity on the generated model structure of cAGP [128]. The enantiomers of benzoin and chlorpheniramine were located onto a similar position of the surface; however, the docked propranolol enantiomers slightly shifted to the relatively small cliff. As shown in Figure 10.10, a carboxy group of E168 is close to an amino group of (*R*)-chlorpheniramine at a distance of 2.6 Å, while an amino group of (*S*)-chlorpheniramine could interact with a carboxy group of D161 within 3.0 Å instead of E168 for (*R*)-chlorpheniramine. Electrostatic

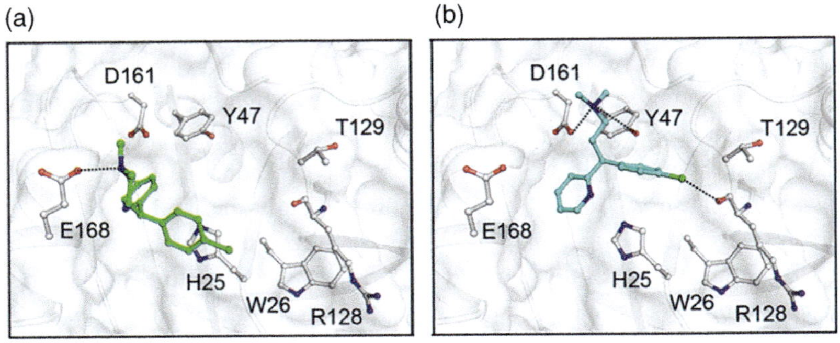

Figure 10.10 Simulated docking structures of cAGP with (a) (*R*)-chlorpheniramine and (b) (*S*)-chlorpheniramine. The docked compounds were represented in either green or blue for carbon atoms. Besides, nitrogen and oxygen atoms were colored in blue and red, in each. The remarked amino acid residues were labeled and shown in gray for carbon atoms, and feasible interactions were depicted as dotted lines. *Source:* Reproduced from Haginaka et al. [128]/MDPI/Public Domain CC BY 4.0.

interactions between amino groups of (*R*)- and (*S*)-chlorpheniramine and carboxy groups of E168 and D161, respectively, could work for the recognition. In addition, as shown in Figure 10.10b, Y47 was located near the bound (*S*)-chlorpheniramine with a distance of 3.9 Å. Furthermore, the chlorine atom of (*S*)-chlorpheniramine was close to a carbonyl group for the main chain between R128 and T129 with a distance of 3.6 Å. It implied that (*S*)-chlorpheniramine could interact with the main chain via halogen bonding interactions. These results indicate that (*S*)-chlorpheniramine interacts with cAGP more tightly than (*R*)-chlorpheniramine. Furthermore, (*R*)- and (*S*)-propranolol were docked to the generated model structure of cAGP (Figure 10.11) [128]. Interestingly, the docked propranolol was slightly shifted toward H25 and W26. Although Y47 was again suggested to be involved in the binding of (*R*)-propranolol, with a distance of 4.7 Å, it could be weak to anchor propranolol. A carbonyl group for the main chain between R128 and T129 was remarkably located to 2.7 and 2.8 Å for (*R*)-propranolol and (*S*)-propranolol, respectively. The former hydrogen bonding interactions were with an amino group of (*R*)-propranolol, while the latter hydrogen bonding interactions were with a hydroxyl group of (*S*)-propranolol. These results indicate that the (*R*)-propranolol binding to cAGP is more favorable than that of the (*S*)-enantiomer. The elution orders of chlorpheniramine and propranolol enantiomers on cAGP-based CSPs in LC were consistent with the simulation results on the model structure of cAGP.

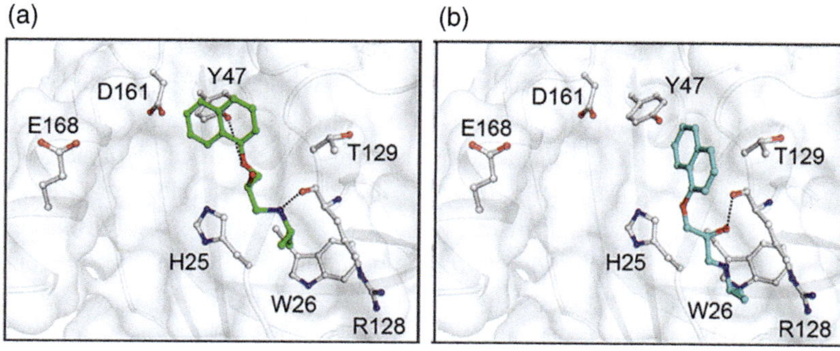

Figure 10.11 Simulated docking structures of cAGP with (a) (*R*)-propranolol and (b) (*S*)-propranolol. The docked compounds were represented as in Figure 10.10. Besides, the same colors were used for nitrogen and oxygen atoms, and the remarked amino acid residues as in Figure 10.10. Feasible interactions were depicted as dotted lines. *Source:* Reproduced from Haginaka et al. [128]/MDPI/ Public Domain CC BY 4.0.

As described in Section 10.4.3, the ligand binding sites of hAGP were totally different from those of cAGP in this study: Figure 10.9 shows crystal structures of F1*S (left, ORM1) and A (center, ORM2) variants of hAGP and a built model structure of cAGP (right) [128]. These structures indicate that the ligand-binding sites of hAGP and cAGP are totally different. As described in Section 10.4.3, a new homology model of hAGP could not explain enantioselective bindings of warfarin derivatives on lobes I–III of hAGP [80]. On the other hand, it was reported that Trp25 was essentially involved in the hAGP binding of drugs by CD methods [129]. Taking into account CD studies on hAGP and ligand bindings on cAGP, there might be enantioselective binding sites near Trp25 on hAGP and near Trp26 on cAGP.

10.4.6 Cellobiohydrolase

CBH I and CBH II (Cel7A and Cel6A), which contain core (catalytic domain), flexible region and binding domain, were enzymatically degraded into two fragments: core and flexible region + binding domain [98]. The dominating enantioselective site of propranolol and other solutes was located on the core, the main part of the enzyme [98]. This result is in good agreement with those of X-ray crystallography and chromatography studies.

The three-dimensional structure of the catalytic domain of Cel7A was determined by X-ray crystallography [133]. In addition, the catalytic domain of Cel7A was co-crystallized with (*S*)-propranolol, and the three-dimensional structure of the complex was determined [134]. It has been shown that the binding site is a ~4.5 nm long tunnel, as shown in Figure 10.11a [135]. Regarding CSPs based on the mutated Cel7A, only the Asp214Asn of the Cel7A mutant retained low, while the enantioselectivity was completely lost for the Glu212Gln and Glu217Gln mutants [134]. The loss of enantioselectivity was accompanied by the loss of catalytic activity for the mutants. These results revealed that carboxy functions of Glu212 and Glu217 were essential for catalysis as well as chiral recognition, and that the exchange of either one impaired both the activity and chiral recognition. Asp214 appears to be involved in chiral recognition but with less importance [134]. Thus, the carboxy groups of Glu212 and Glu217 interact with the nitrogen group of propranolol, and Try376 interacts with the naphthyl moiety (Figure 10.12b) [134].

Furthermore, from the structure of a complex of Cel7A with (*S*)-propranolol, it was found that (*S*)-propranolol bound at the active site: the catalytic residues Glu212 and Glu217 made tight salt links with the secondary amino group of (*S*)-propranolol, as shown in Figure 10.13. The oxygen atom attached to the chiral center of (*S*)-propranolol forms hydrogen bonds to the nucleophile Glu212 and to Gln175, whereas the aromatic naphthyl moiety stacks with the indole ring of Trp376. Co-crystallization of Cel7A with the (*R*)-propranolol failed. It was hypothesized that (*R*)-enantiomer could interact with Cel7A in the same binding site with (*S*)-enantiomer, resulting in a poorer fit of the (*R*)-enantiomer due to steric hindrance with the hydroxyl group, rather than weaker binding due to the loss of hydrogen bonding [134].

The addition of cellobiose or lactose, inhibitors of CBH, to the mobile phase resulted in the decreased retention of β-blockers, such as propranolol and atenolol, indicating that these drugs bind to the same site on a CBH 58 (Cel7D)-based CSPs [52]. These results were further confirmed with covalent modification of the carboxy groups of CBH I (Cel7A) [136] to amide groups. The covalent modification was carried out either in the presence or absence of cellobiose, which has proven to inhibit the enzymatic activity. The enantioseparation factors of propranolol on CSPs based on CBH I were almost unchanged during the reaction periods in the presence of cellobiose in LC, while they decreased rapidly without the inhibitor.

10.4.7 Antibody

As described in Section 10.3.2.8, CSPs based on antibodies against D- or L-Phe were useful for the enantioseparations of a broad number of α-amino acids. The enantioselective binding of D-amino acids to the mouse recombinant anti-D-amino acid antibody (clone 67.36) (the anti-D-AA

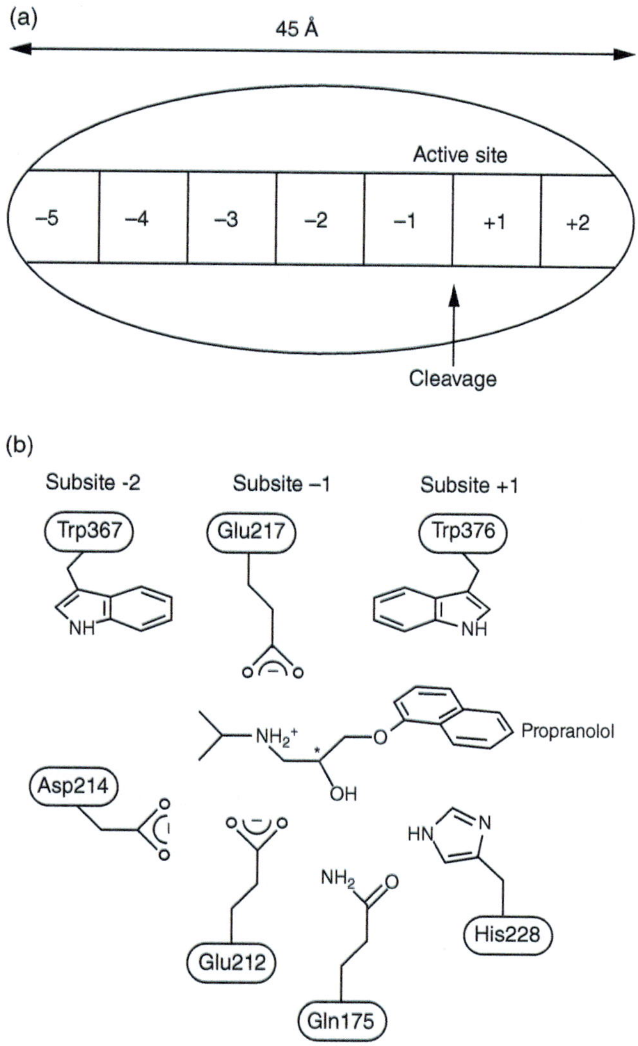

Figure 10.12 (a) Schematic representation of the cellulose-binding tunnel in CBH I (Cel7A). The tunnel is ~4.5 nm long and contains at least seven subsites for binding of glucosyl residues. The active site, where glycosidic-bond cleavage occurs, is located near the exit of the tunnel between subsites −1 and+1. (b) Schematic model for the binding of (*R*)/(*S*)-propranolol in the active site of CBH I (Cel7A). Residues in or near the active site are shown, including the proposed acid/base catalyst, Glu217, and the nucleophile, Glu212. All acidic residues are shown in their deprotonated state and propranolol in its protonated state (p*K*a, ~9.5). The chiral center is marked with an asterisk. The model was built manually, and the best fit of both forms of propranolol was obtained when the carboxy groups of Glu212 and Glu217 interact with the nitrogen group and Trp376 stacks with the naphthyl moiety of the ligand. *Source:* Adapted from Ståhlberg et al. [134]/with permission of Elsevier.

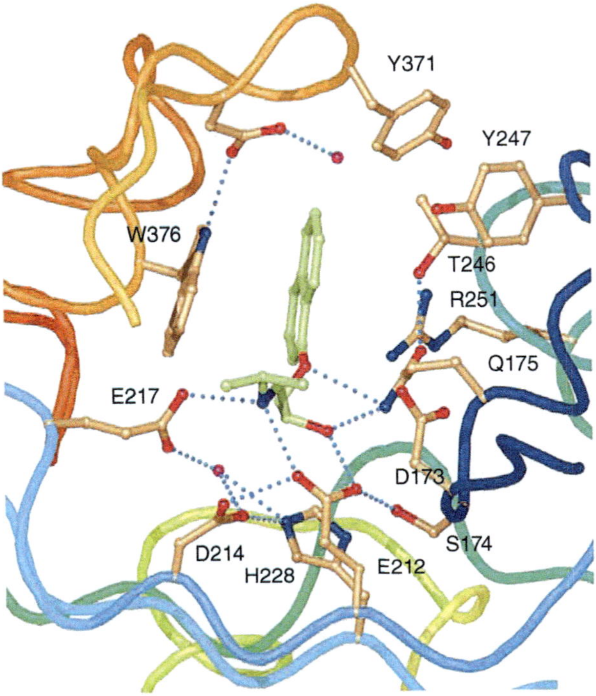

Figure 10.13 Ball and stick representation of the active site with the (*S*)-propranolol molecule (green C atoms), protein residues (light-brown C) with atoms within 4Å distance from the ligand, and two water molecules (magenta). Probable hydrogen bonds are indicated with blue dots and relevant parts of the C_a backbone are shown, color ramped from blue at residue 171 to red at the C terminus, residue 434. The main interactions are the hydrophobic stacking of the naphthyl group with Trp376, the bidentate salt-link interaction between the positively charged secondary amine and the catalytic residues Glu212 and Glu217, and hydrogen bonding of the chiral hydroxy group with Gln175 and Glu212. Two residues in contact with the isopropyl moiety, Tyr145 and Tyr171, were omitted for clarity. *Source:* Reproduced from Ståhlberg et al. [134]/with permission of John Wiley & Sons.

67.36_rotamer_Y40L-L137L structure) was investigated using molecular modeling and ligand docking studies [137]. The model ligand used was the *p*-aminoPhe portion linked to tyramine via diazotization, which was used as a hapten for raising an antibody. The results indicate that in addition to four hydrogen bonds, formed between amino acid residues in the binding site and the ligand, a number of hydrophobic interactions are involved in the formation of the antibody–ligand complex as shown in Figure 10.14 [137]. The aromatic side chain of the ligand interacts with Trp and Tyr residues in

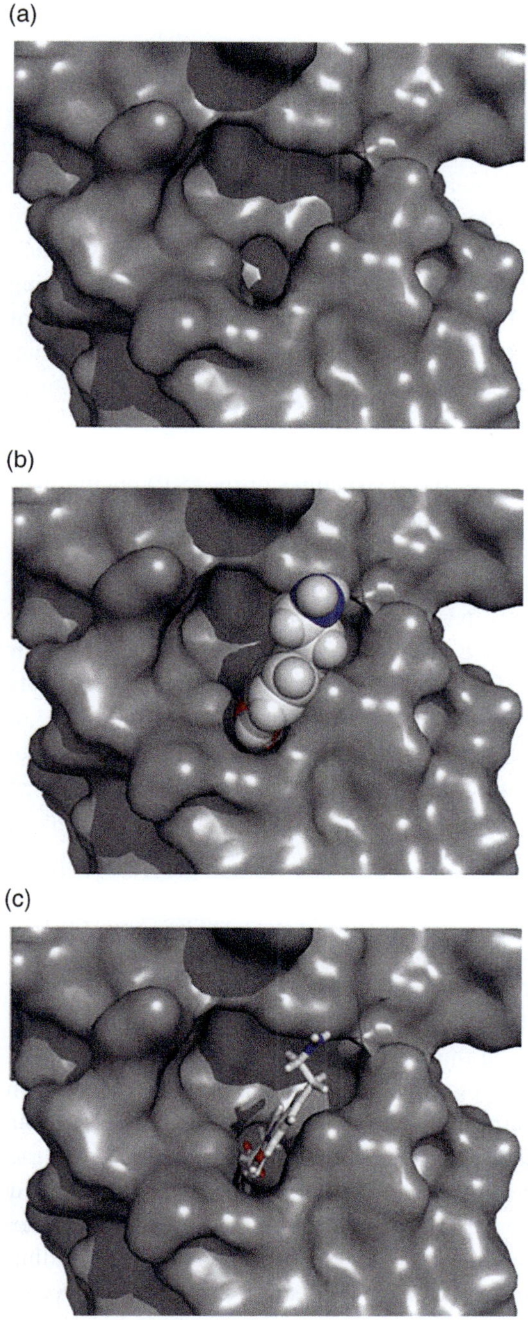

Figure 10.14 Surface contour images of the modeled anti-D-amino acid 67.36_rotamer_Y40L-L137L structure (a) without the ligand, (b) with the docked ligand in a spherical representation, and (c) with the docked ligand in a stick representation. *Source:* Reproduced from Ranieri et al. [137]/with permission of John Wiley & Sons.

the binding site through π–π stacking. Docking experiments performed with D-Phe and D-norvaline showed that these ligands are bound to the antibody in a way analogous to that of the D-enantiomer of the model ligand. However, the L-enantiomer of the model ligand cannot access the binding site due to steric hindrance.

Recombinant antibody fragments (antigen-binding fragment, Fab) have been prepared for *SR* (a) finrozole, which has two chiral centers, using genetic engineering techniques [138]. The obtained Fab fragments, which have a molecular weight of about 30 kDa, were bound to Chelating Sepharose Fast Flow loaded with copper ions. Furthermore, single mutant (Tyr96Val) or double mutants (Tyr96Val/Trp33Ala mutants) of the Fab fragments were prepared to decrease the affinity by one or more orders of magnitude. These resulted in an increase of the lifetime of the antibody fragment-based CSPs [139] and in the use of a lower concentration of organic solvent for elution. The crystal structures of the (*S*,*R*)-enantiomer-specific Fab fragment, ENA11His, are determined in the absence or the presence of the hapten [140]. The hapten molecule was tightly bound in a deep cleft between the light and heavy chains of the Fab fragment. From the complex structure, it was also possible to describe the molecular basis for enantioselectivity and to deduce the absolute configurations of all the four different stereoisomers of finrozole. The antibody fragment selectively binds the (*S*,*R*)-enantiomer from the racemic mixture. Asp95 and Asn35 of the H-chain in the antibody seem to provide this specificity through hydrogen bonding interactions [140].

10.4.8 Nicotinic Acetylcholine Receptor and Human Liver Organic Cation Transporter

The enantioselective bindings of DM and LM to $\alpha_3\beta_4$-nAChR were investigated using molecular modeling and docking studies. The lowest-energy docked conformations of the DM and LM complexes were both located at the Val/Phe ring and involved the insertion of the hydrophobic portion of both molecules into the hydrophobic cleft found at this position, as shown in Figure 10.15 [141]. The mirror image relationship between the two enantiomers and their lack of conformational mobility produce two unique orientations, which result in distinctly different interactions with nearby amino acid moieties comprising the lumen. This initial binding interaction of DM and LM with the hydrophobic pocket is not enantioselective, but it tethers the molecules to the receptor. The second step is the configurationally defining hydrogen bonding interactions. These two steps in the binding process are interconnected and produce a dynamic enantioselective binding [142].

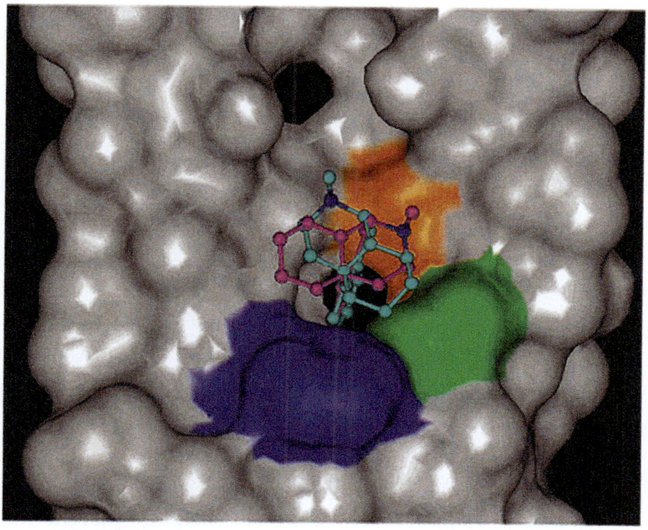

Figure 10.15 Overlay of the most stable docked orientations of DM (cyan) and LM (magenta) complexes. Nitrogen atoms of both enantiomers are colored in blue. The binding pocket formed between β_4 and α_3 helices is shown in detail to highlight the interactions leading to an enantioselectivity. The hydrophobic part of the $\alpha_3\beta_4$-nAChR channel with depiction of the cleft formed between Phe (blue) and Val (green) and Ser (orange). *Source:* Reproduced from Jozwiak et al. [141]/with permission of American Chemical Society.

With regard to the modeling of enantioselective binding to hOCT, the resulting model contained four binding sites, a positive ion interaction site, a hydrophobic interaction site, and two hydrogen bond acceptor sites [143]. When (*R*)-verapamil was fit to the proposed pharmacophore, all the relevant functional groups of the molecule matched the hypothesis, whereas (*S*)-verapamil could be mapped to only three of the model feature sites. These data suggest that for (*R*)- and (*S*)-verapamil, chiral recognition is a multistep process involving an initial tethering of the selectant to the selector, most probably occurring at the positive ion interaction site, followed by conformational adjustments that produce the optimum interactions. This process results in a distribution of selectant–selector complexes of varying relative stabilities and the observed enantioselectivity.

10.5 Conclusions

The advantages of protein- or glycoprotein-based CSPs are that they have enantioselectivity for a wide range of compounds because of multiple binding sites on the surfaces of proteins or glycoproteins and multiple binding interactions between proteins or glycoproteins and ligands. In general, they are unstable against heat, acids, and organic modifiers because those induce reversible or irreversible changes in their conformation.

Protein- and glycoprotein-based CSPs have been prepared and used for the enantioseparations of various compounds in LC and CEC. The supports used were mainly silicas and polymers, which are used as a particle, capillary, and monolith. Proteins so far used were BSA and HSA, trypsin, α-chymotrypsin, lysozyme, pepsin, FABP, PGA, streptavidin, and lipase, while glycoproteins so far used were AGP (hAGP), OGCHI (now termed cAGP), avidin, RfBP, ovotransferrin, cellobiohydrolase, glucoamylase, antibody, $\alpha_3\beta_4$-nAChR, and hOCT. It is interesting that the chiral recognition ability of OMCHI reported previously came from cAGP, a member of the lipocalin family, and that OMCHI had no appreciable chiral recognition ability. OMCHI-based CSPs, which are commercially available, were made from crude OMCHI including cAGP. Furthermore, antibodies against α-amino acids and α-hydroxy acids were produced so that the antibody-based CSPs could be utilized for enantioseparations of members of these classes of compounds.

The chiral recognition sites of some proteins or glycoproteins were identified and their chiral recognition mechanisms were elucidated by X-ray crystallography, NMR measurements, and molecular modeling and ligand docking studies.

CPSs of proteins and glycoproteins can be utilized for the separations of not only enantiomers but also diastereomers, including drugs and agrochemicals. In the future, a new protein or glycoprotein that has excellent chiral recognition properties can be found and introduced as CSPs. By genetic technology it is possible to prepare a protein that is stable and has excellent chiral recognition abilities. Furthermore, the domain or fragment of a protein, which has excellent chiral recognition abilities, will be prepared by genetic technology.

Furthermore, protein- and glycoprotein-based CSPs will be useful for examining the occurrence, fate, and transport of chiral compounds in various environmental matrices.

References

1 Karush, F. (1952). The interaction of optically isomeric dyes with bovine serum albumin. *The Journal of Physical Chemistry* 56: 70–77.

2 Karush, F. (1954). The interaction of optically isomeric dyes with human serum albumin. *Journal of the American Chemical Society* 76: 5536–5542.

3 Mcmenamy, R.H. and Oncley, J.L. (1958). The specific binding of L-tryptophan to serum albumin. *The Journal of Biological Chemistry* 233: 1436–1447.

4 Stewart, K.K. and Doherty, R.F. (1973). Resolution of DL-tryptophan by affinity chromatography on bovine-serum albumin-agarose columns. *Proceedings. National Academy of Sciences. United States of America* 70: 2850–2852.

5 Allenmark, S., Bomgren, B., and Borén, H. (1983). Direct liquid chromatographic separation of enantiomers on immobilized protein stationary phases: III. Optical resolution of a series of *N*-aroyl *d,l*-amino acids by high-performance liquid chromatography on bovine serum albumin covalently bound to silica. *Journal of Chromatography A* 264: 63–68.

6 Domenici, E., Bertucci, C., Salvadori, P. et al. (1990). Synthesis and chromatographic properties of an HPLC chiral stationary phase based upon human serum albumin. *Chromatographia* 29: 170–176.

7 Thelohan, S., Jadaud, P., and Wainer, I.W. (1989). Immobilized enzymes as chromatographic phases for HPLC: the chromatography of free and derivatized amino acids on immobilized trypsin. *Chromatographia* 28: 551–555.

8 Wainer, I.W., Jadaud, P., Schombaum, G.R. et al. (1988). Enzymes as HPLC stationary phases for chiral resolutions: initial investigations with α-chymotrypsin. *Chromatographia* 25: 903–907.

9 Haginaka, J., Murashima, T., and Seyama, C. (1994). Separation of enantiomers on a lysozyme-bonded silica column. *Journal of Chromatography A* 666: 203–210.

10 Haginaka, J., Miyano, Y., Saizen, Y. et al. (1995). Separation of enantiomers on a pepsin-bonded column. *Journal of Chromatography A* 708: 161–168.

11 Massolini, G., De Lorenzi, E., Calleri, E. et al. (2001). Properties of a stationary phase based on immobilised chicken liver basic fatty acid-binding protein. *Journal of Chromatography B* 751: 117–130.

12 Massolini, G., Calleri, E., de Lorenzi, E. et al. (2001). Immobilized penicillin G acylase as reactor and chiral selector in liquid chromatography. *Journal of Chromatography A* 921: 147–160.

13 Ravelet, C., Mchaud, M., Ravel, A. et al. (2004). Streptavidin chiral stationary phase for the separation of adenosine enantiomers. *Journal of Chromatography A* 1036: 155–160.

14 He, Z., Lv, C., Fan, X., and Zhou, Z. (2011). A novel lipase-based stationary phase in liquid chromatography. *Analytica Chimica Acta* 689: 143–148.

15 Hermansson, J. (1983). Direct liquid chromatographic resolution of racemic drugs using α_1-acid glycoprotein as the chiral stationary phase. *Journal of Chromatography A* 269: 71–80.

16 Miwa, T., Ichikawa, M., Tsuno, M. et al. (1987). Direct liquid chromatographic resolution of racemic compounds. Use of ovomucoid as a column ligand. *Chemical and Pharmaceutical Bulletin* 35: 682–686.

17 Haginaka, J., Seyama, C., and Kanasugi, N. (1995). The absence of chiral recognition ability in ovomucoid: ovoglycoprotein-bonded HPLC stationary phases for chiral recognition. *Analytical Chemistry* 67: 2539–2547.

18 Sadakane, Y., Matsunaga, H., Nakagomi, K. et al. (2002). Protein domain of chicken α_1-acid glycoprotein is responsible for chiral recognition. *Biochemical and Biophysical Research Communications* 295: 587–590.

19 Miwa, T., Miyakawa, T., and Miyake, Y. (1988). Characteristics of an avidin-conjugated column in direct liquid chromatographic resolution of racemic compounds. *Journal of Chromatography A* 457: 227–233.

20 Mano, N., Oda, Y., Asakawa, N. et al. (1992). Development of a flavoprotein column for chiral separation by high-performance liquid chromatography. *Journal of Chromatography A* 623: 221–228.

21 Mano, N., Oda, Y., Asakawa, N. et al. (1992). Conalbumin-conjugated silica gel, a new chiral stationary phase for high-performance liquid chromatography. *Journal of Chromatography A* 603: 105–109.

22 Erlandsson, P., Marle, I., Hansson, L. et al. (1990). Immobilized cellulase (CBH I) as a chiral stationary phase for direct resolution of enantiomers. *Journal of the American Chemical Society* 112: 4573–4574.

23 Nyström, A., Strandberg, A., Aspergren, A. et al. (1999). Use of immobilized amyloglucosidase as chiral selector in chromatography. Immobilization and performance in liquid chromatography. *Chromatographia* 50: 209–214.

24 Mertens, R., Stuning, M., and Weiler, E.W. (1982). Metabolism of tritiated enantiomers of abscisic acid prepared by immunoaffinity chromatography. *Die Naturwissenschaften* 69: 595–597.

25 Jozwiak, K., Hernandez, S.C., Kellar, K.J., and Wainer, I.W. (2003). Enantioselective interactions of dextromethorphan and levomethorphan with the alpha 3 beta 4-nicotinic acetylcholine receptor: comparison of chromatographic and functional data. *Journal of Chromatography B* 797: 373–379.

26 Moaddel, R., Patel, S., Jozwiak, K. et al. (2005). Enantioselective binding to the human organic cation transporter-1 (hOCT1) determined using an immobilized hOCT1 liquid chromatographic stationary phase. *Chirality* 17: 501–506.

27 Haginaka, J. (2010). Chiral separations using proteins and peptides as chiral selectors. In: *Chiral Separations by Capillary Electrophoresis* (ed. A. Van Eeckhaut and Y. Michotte). Boca Raton, FL: CRC Press.

28 Haginaka, J. (2011). Mechanistic aspects of chiral recognition on protein-based stationary phases. In: *Advances in Chromatography*, vol. 49 (ed. E. Grushka and N. Grinberg). Boca Raton, FL: CRC Press.

29 Haginaka, J. (2012). Chromatographic separations and analysis: protein and glycoprotein stationary phases. In: *Comprehensive Chirality*, vol. 8 (ed. E. Carreira and H. Yamamoto). Amsterdam: Elsevier.

30 Scriba, G.K.E. (2019). Chiral recognition in separation science: an overview. In: *Chiral Separations. Methods and Protocols* (ed. G.K.E. Scriba). Totowa, NJ: Humana Press.

31 Haginaka, J. (2000). Enantiomer separation of drugs by capillary electrophoresis using proteins as chiral selectors. *Journal of Chromatography A* 875: 235–254.

32 Haginaka, J. (2001). Protein-based chiral stationary phases for high-performance liquid chromatography enantioseparations. *Journal of Chromatography A* 906: 253–273.

33 Millot, M.C. (2003). Separation of drug enantiomers by liquid chromatography and capillary electrophoresis, using immobilized proteins as chiral selectors. *Journal of Chromatography B* 797: 131–159.

34 Haginaka, J. (2008). Recent progresses in protein-based chiral stationary phases for enantioseparations in liquid chromatography. *Journal of Chromatography B* 875: 12–19.

35 Bi, C., Zheng, X., Azaria, S. et al. (2016). Chromatographic studies of protein-based chiral separations. *Separations* 3: 27.

36 Cardoso, T., Almeida, A.S., Remião, F., and Fernandes, C. (2021). Enantioresolution and binding affinity studies on human serum albumin: recent applications and trends. *Chemosensors* 9: 304.

37 Thompson, R.A., Andersson, S., and Allenmark, S. (1989). Direct liquid chromatographic separation of enantiomers on immobilized protein stationary phases: VII. Sorbents obtained by entrapment of cross-linked bovine serum albumin in silica. *Journal of Chromatography A* 465: 263–270.

38 Liu, Z., Otsuka, K., and Terabe, S. (2001). Chiral separation by open tubular capillary electrochromatography with adsorbed avidin as a stationary phase. *Journal of Separation Science* 24: 17–26.

39 Liu, Z., Otsuka, K., Terabe, S. et al. (2002). Physically adsorbed chiral stationary phase of avidin on monolithic silica column for capillary electrochromatography and capillary liquid chromatography. *Electrophoresis* 23: 2973–2981.

40 Jacobson, S.C. and Guiochon, G. (1992). Contribution of ionically immobilized bovine serum albumin to the retention of enantiomers. *Journal of Chromatography A* 600: 37–42.

41 Ye, M., Zou, H., Liu, Z. et al. (2002). Study of competitive binding of enantiomers to protein by affinity capillary electrochromatography. *Journal of Pharmaceutical and Biomedical Analysis* 27: 651–660.

42 Millot, M.C., Taleb, N.L., and Sebille, B. (2002). Binding of human serum albumin to silica particles by means of polymers: a liquid chromatographic study of the selectivity of resulting chiral stationary phases. *Journal of Chromatography B* 768: 157–166.

43 Wiedmer, S.K., Bo, T., and Riekkola, M.L. (2008). Phospholipid-protein coatings for chiral capillary electrochromatography. *Analytical Biochemistry* 373: 26–33.

44 Han, N.Y., Hautala, J.T., Bo, T. et al. (2006). Immobilization of phospholipid-avidin on fused-silica capillaries for chiral separation in open-tubular capillary electrochromatography. *Electrophoresis* 27: 1502–1509.

45 Xuan, H. and Hage, D.S. (2005). Immobilization of α_1-acid glycoprotein for chromatographic studies of drug–protein binding. *Analytical Biochemistry* 346: 300–310.

46 Marle, I., Karlsson, A., and Pettersson, C. (1992). Separation of enantiomers using α-chymotrypsin-silica as a chiral stationary phase. *Journal of Chromatography A* 604: 185–196.

47 Mallik, R., Jiang, T., and Hage, D.S. (2004). High-performance affinity monolith chromatography: development and evaluation of human serum albumin columns. *Analytical Chemistry* 76: 7013–7012.

48 Oda, Y., Asakawa, N., Abe, S. et al. (1991). Avidin protein-conjugated column for direct injection analysis of drug enantiomers in plasma by high-performance liquid chromatography. *Journal of Chromatography A* 572: 133–141.

49 Marle, I., Jönsson, S., Isaksson, R. et al. (1993). Chiral stationary phases based on intact and fragmented cellobiohydrolase I immobilized on silica. *Journal of Chromatography A* 648: 333–347.

50 Mallik, R., Wa, C., and Hage, D.S. (2007). Development of sulfhydryl-reactive silica for protein immobilization in high-performance affinity chromatography. *Analytical Chemistry* 79: 1411–1424.

51 Mallik, R., Xuan, H., and Hage, D.S. (2007). Development of an affinity silica monolith containing alpha1-acid glycoprotein for chiral separations. *Journal of Chromatography A* 1149: 294–304.

52 Nakamura, M., Kiyohara, S., Saito, K. et al. (1999). High resolution of DL-tryptophan at high flow rates using a bovine serum albumin-multilayered porous hollow-fiber membrane. *Analytical Chemistry* 71: 1323–1325.

53 Allenmark, S. (1986). Optical resolution by liquid chromatography on immobilized bovine serum albumin. *Journal of Liquid Chromatography* 9: 425–442.

54 Birnbaum, S. and Nilsson S. (1992). Protein-based capillary affinity gel electrophoresis for the separation of optical isomers. *Analytical Chemistry* 64: 2872–2874.

55 Hofstetter, H., Hofstetter, O., and Schurig, V. (1998). Enantiomer separation using BSA as chiral stationary phase in affinity OTEC and OTLC. *Journal of Microcolumn Separations* 10: 287–291.

56 Weng, X., Bi, H., Liu, B., and Kong, J. (2006). On-chip chiral separation based on bovine serum albumin-conjugated carbon nanotube as stationary phase in a microchannel. *Electrophoresis* 27: 3129–3315.

57 Hong, T., Zheng, Y., Hu, W., and Ji, Y. (2014). Preparation and evaluation of bovine serum albumin immobilized chiral monolithic column for affinity capillary electrochromatography. *Analytical Biochemistry* 464: 43–50.

58 Kato, M., Sakai-Kato, K., Matsumoto, N., and Toyo'oka, T. (2002). A protein-encapsulation technique by the sol-gel method for the preparation of monolithic columns for capillary electrochromatography. *Analytical Chemistry* 74: 1915–1921.

59 Erlandsson, P. and Nilsson, S. (1989). Use of a fragment of bovine serum albumin as a chiral stationary phase in liquid chromatography. *Journal of Chromatography A* 482: 35–51.

60 Andersson, S., Allenmark, S., Erlandsson, P., and Staffan, N.S. (1990). Direct liquid chromatographic separation of enantiomers on immobilized protein stationary phases: VIII. A comparison of a series of sorbents based on bovine serum albumin and its fragments. *Journal of Chromatography A* 498: 81–91.

61 Haginaka, J. and Kanasugi, N. (1995). Enantioselectivity of bovine serum albumin-bonded columns produced with isolated protein fragments. *Journal of Chromatography A* 694: 71–80.

62 Haginaka, J. and Kanasugi, N. (1997). Enantioselectivity of bovin serum albumin-bonded columns produced with isolated protein fragments. II. Characterization of protein fragments. *Journal of Chromatography A* 769: 215–223.

63 Mallik, R. and Hage, D.S. (2008). Development of an affinity silica monolith containing human serum albumin for chiral separations. *Journal of Pharmaceutical and Biomedical Analysis* 46: 820–830.

64 Lloyd, D.K., Li, S., and Ryan, P. (1995). Protein chiral selectors in free-solution capillary electrophoresis and packed-capillary electrochromatography. *Journal of Chromatography A* 694: 285–296.

65 Yang, J. and Hage, D.S. (1994). Chiral separation in capillary electrophoresis using human serum albumin as buffer additive. *Analytical Chemistry* 66: 2719–2725.

66 Hong, T., Chen, X., Xu, Y. et al. (2016). Preparation of graphene oxide-modified affinity capillary monoliths based on three types of amino donor for chiral separation and proteolysis. *Journal of Chromatography A* 1456: 249–256.

67 Xu, S., Wang, Y., Tang, Y., and Ji, Y. (2018). A protein-based mixed selector chiral monolithic stationary phase in capillary electrochromatography. *New Journal of Chemistry* 42: 13520.

68 Matsunaga, H., Qiang, F.Q., and Haginaka, J. (2002). HPLC chiral stationary phases produced with isolated human serum albumin fragments. *Analytical Sciences* 18: 27–30.

69 Chuang, V.T. and Otagiri, M. (2006). Stereoselective binding of human serum albumin. *Chirality* 18: 159–166.

70 Geng, L., Bo, T., Liu, H. et al. (2004). Capillary coated layer-by-layer assembly of γ-zirconium phosphate/lysozyme nanocomposite film for open tubular capillary electrochromatography. *Chromatographia* 59: 65–70.

71 Bo, T., Wiedmer, S.K., and Riekkola, M.L. (2004). Phopholipid-lysozyme coating for chiral separation in capillary electrophoresis. *Electrophoresis* 25: 1784–1791.

72 Hong, T., Chi, C., and Ji, Y. (2014). Pepsin-modified chiral monolithic column for affinity capillary electrochromatography. *Journal of Separation Science* 37: 3377–3383.

73 Miao, C., Bai, R., Xu, S. et al. (2017). Carboxylated single-walled carbon nanotube-functionalized chiral polymer monoliths for affinity capillary electrochromatography. *Journal of Chromatography A* 1487: 227–234.

74 Xu, S., Mo, R., Jin, C. et al. (2017). Mesoporous silica nanoparticles incorporated hybrid monolithic stationary phase immobilized with pepsin for enantioseparation by capillary electrochromatography. *Journal of Pharmaceutical and Biomedical Analysis* 140: 190–198.

75 Calleri, E., Massolini, G., Lubda, D. et al. (2004). Evaluation of a monolithic epoxy silica support for penicillin G acylase immobilizatio. *Journal of Chromatography A* 1031: 93–100.

76 Gotti, R., Fiori, J., Calleri, E. et al. (2012). Chiral capillary liquid chromatography based on penicillin G acylase immobilized on monolithic epoxy silica column. *Journal of Chromatography A* 1234: 45–49.

77 Uppenberg, J., Hansen, M.T., Patkar, S., and Jones, T.A. (1994). The sequence, crystal structure determination and refinement of two crystal forms of lipase B from Candida antarctica. *Structure* 2: 293–308.

78 Ahmed, M. and Ghanem, A. (2014). Enantioselective nano liquid chromatographic separation of racemic pharmaceuticals: a facile one-pot in situ preparation of lipase-based polymer monoliths in capillary format. *Chirality* 26: 754–763.

79 Haginaka, J. and Matsunaga, H. (2000). Separation of enantiomers on HPLC chiral stationary phases based on human plasma alpha1-acid glycoprotein: effect of sugar moiety on chiral recognition ability. *Enantiomer* 5: 37–45.

80 Hazai, E., Visy, J., Fitos, I. et al. (2006). Selective binding of coumarin enantiomers to human α_1-acid glycoprotein genetic variants. *Bioorganic and Medicinal Chemistry* 4: 1959–1965.

81 Hermansson, J. (1989). Enantiomeric separation of drugs and related compounds based on their interaction with α_1-acid glycoprotein. *Trends in Analytical Chemistry* 8: 251–259.

82 Michishita, T., Franco, P., and Zhang, T. (2010). New approaches of LC-MS compatible method development on α1-acid glycoprotein-based stationary phase for resolution of enantiomers by HPLC. *Journal of Separation Science* 33: 3627–3637.

83 Linweaver, H. and Murray, C.W. (1947). Identification of the trypsin inhibitor of egg white with ovomucoid. *The Journal of Biological Chemistry* 171: 565–581.

84 Miwa, T., Sakashita, S., Ozawa, H. et al. (1991). Application of an ovomucoid-conjugated polymer column for the enantiospecific determination of chlorprenaline concentrations in plasma. *Journal of Chromatography A* 566: 163–171.

85 Kirkland, K.M., Neilson, K.L., and McCombs, D.A. (1991). Comparison of a new ovomucoid and a second-generation α_1-acid glycoprotein-based chiral column for the direct high-performance liquid chromatography resolution of drug enantiomers. *Journal of Chromatography A* 545: 43–58.

86 Zhou, L., Mao, B., and Ge, Z. (2008). Comparative study of immobilized alpha 1 acid glycoprotein and ovomucoid protein stationary phases for the enantiomeric separation of pharmaceutical compounds. *Journal of Pharmaceutical and Biomedical Analysis* 46: 898–906.

87 Ketterer, B. (1965). Ovoglycoprotein, a protein of hen's-egg white. *The Biochemical Journal* 96: 372–376.

88 Haginaka, J. and Takehira, H. (1997). Separation of enantiomers on a chiral stationary phase based on ovoglycoprotein: II. Comparison of chiral recognition properties with crude ovomucoid. *Journal of Chromatography A* 777: 241–247.

89 Haginaka, J., Kagawa, C., and Matsunaga, H. (1999). Separation of enantiomers on a chiral stationary phase based on ovoglycoprotein: VII. Comparison of chiral recognition ability of ovoglycoprotein from chicken and Japanese quail egg whites. *Journal of Chromatography A* 858: 155–165.

90 Haginaka, J., Okazaki, Y., and Matsunaga, H. (1999). Separation of enantiomers on a chiral stationary phase based on ovoglycoprotein: V. Influence of immobilization method on chiral resolution. *Journal of Chromatography A* 840: 171–181.

91 Matsunaga, H. and Haginaka, J. (2014). Separation of enantiomers on chiral stationary phase based on chicken α-acid glycoprotein: effect of silica particle diameters on column performance. *Journal of Chromatography A* 1363: 96–100.

92 Bayer, E.A. and Wilchek, M. (1990). Application of avidin-biotin technology to affinity-based separations. *Journal of Chromatography A* 510: 3–11.

93 Massolini, G., De Lorenzi, E., Ponci, M.C. et al. (1995). Egg yolk riboflavin binding protein as a new chiral stationary phase in high-performance liquid chromatography. *Journal of Chromatography A* 704: 55–65.

94 De Lorenzi, E., Massolini, G., Lloyd, D.K. et al. (1997). Evaluation of quail egg white riboflavin binding protein as a chiral selector in high-performance liquid chromatography and capillary electrophoresis. *Journal of Chromatography A* 790: 47–64.

95 Uzcategui, E., Ruiz, A., Montesino, R. et al. (1991). The 1,4-β-d-glucan cellobiohydrolases from *Phanerochaete chrysosporium*. I. A system of synergistically acting enzymes homologous to *Trichoderma reesei*. *Journal of Biotechnology* 19: 271–286.

96 Tomme, P., Van Tilbeurgh, H., Pettersson, G. et al. (1988). Studies of the cellulolytic system of *Trichoderma reesei* QM 9414. *European Journal of Biochemistry* 170: 575–581.

97 Marle, I., Erlandsson, P., Hansson, L. et al. (1991). Separation of enantiomers using cellulase (CBH I) silica as a chiral stationary phase. *Journal of Chromatography A* 586: 233–248.

98 Henriksson, H., Jönsson, S., Isaksson, R., and Pettersson, G. (1995). Chiral separation based on immobilized intact and fragmented cellobiohydrolase II (CBH II): a comparison with cellobiohydrolase I (CBH I). *Chirality* 7: 415–424.

99 Henriksson, H., Munoz, I.G., Isaksson, R. et al. (2000). Cellobiohydrolase 58 (P.c. Cel 7D) is complementary to the homologous CBH I (T.r. Cel 7A) in enantioseparations. *Journal of Chromatography A* 898: 63–74.

100 Fornstedt, T., Sajonz, P., and Guiochon, G. (1997). Thermodynamic study of an unusual chiral separation. Propranolol enantiomers on an immobilized cellulase. *Journal of the American Chemical Society* 119: 1254–1264.

101 Fornstedt, T., Götmar, G., Anderson, M., and Guiochon, G. (1999). Dependence on the mobile-phase pH of the adsorption behavior of propranolol enantiomers on a cellulase protein used as the chiral selector. *Journal of the American Chemical Society* 121: 1164–1174.

102 Hedeland, M., Jönsson, S., Isaksson, R., and Pettersson, C. (1998). Unexpected difference in enantioselective retention on cellulase (CBH I) silica stationary phase caused by exchange of potassium for sodium ion in the mobile phase. *Chirality* 10: 513–518.

103 Matsunaga, H. and Haginaka, J. (2016). Separation of enantiomers on chiral stationary phase based on cellulase: effect of preparation method and silica particle diameters on chiral recognition ability. *Journal of Chromatography A* 1467: 155–162.

104 Nilsson, S., Schweitz, L., and Petersson, M. (1997). Three approaches to enantiomer separation of beta-adrenergic antagonists by capillary electrochromatography. *Electrophoresis* 18: 884–890.

105 Strandberg, A., Nyström, A., Behr, S., and Karlsson, A. (1999). Use of immobilized amyloglucosidase as chiral selector in chromatography. Immobilization and performance in liquid chromatography. *Chromatographia* 50: 215–222.

106 Karlsson, A., Arfwidsson, I., Husovic, Z., and Svensson, B. (2001). Use of immobilised glucoamylase G2 for separation of enantiomers. *Chromatographia* 54: 717–723.

107 Götmar, G., Samuelsson, J., Karlsson, A., and Fornstedt, T. (2007). Thermodynamic characterization of the adsorption of selected chiral compounds on immobilized amyloglucosidase in liquid chromatography. *Journal of Chromatography A* 1156: 3–13.

108 Quarrie, S.A. and Galfre, G. (1985). Use of different hapten-protein conjugates immobilized on nitrocellulose to screen monoclonal antibodies to abscisic acid. *Analytical Biochemistry* 151: 389–399.

109 Hofstetter, O., Hofstetter, H., Wilchek, M. et al. (2000). Production and applications of antibodies directed against the chiral center of α-amino acids. *International Journal of Bio-Chromatography* 5: 165–174.

110 Hofstetter, O., Lindstrom, H., and Hofstetter, H. (2002). Direct resolution of enantiomers in high-performance immunoaffinity chromatography under isocratic conditions. *Analytical Chemistry* 74: 2119–2125.

111 Hofstetter, O., Lindstrom, H., and Hofstetter, H. (2004). Effect of the mobile phase on antibody-based enantiomer separations of amino acids in high-performance liquid chromatography. *Journal of Chromatography A* 1049: 85–95.

112 Franco, E.J., Hofstetter, H., and Hofstetter, O. (2008). Enantiomer separation of α-hydroxy acids in high-performance immunoaffinity chromatography. *Journal of Pharmaceutical and Biomedical Analysis* 46: 907–913.

113 Lua, A.C. and Chou, T.Y. (2002). Preparation of immunoaffinity columns for direct enantiomeric separation of amphetamine and/or methamphetamine. *Journal of Chromatography A* 967: 191–199.

114 Koidl, J., Hödl, H., Schmid, M.G. et al. (2006). Chiral separation of T3 enantiomers using stereoselective antibodies as a selector in micro-HPLC. *Journal of Biochemical and Biophysical Methods* 69: 33–42.

115 Lämmerhofer, M. (2010). Chiral recognition by enantioselective liquid chromatography: mechanisms and modern chiral stationary phases. *Journal of Chromatography A* 1217: 814–856.

116 Hage, D.S., Noctor, T.A.G., and Wainer, I.W. (1995). Characterization of the protein binding of chiral drugs by high-performance affinity chromatography interactions of R- and S-ibuprofen with human serum albumin. *Journal of Chromatography A* 693: 23–32.

117 Noctor, T.A.G. and Wainer, I.W. (1992). The in situ acetylation of an immobilized human serum albumin chiral stationary phase for high-performance liquid chromatography in the examination of drug–protein binding phenomena. *Pharmaceutical Research* 9: 480–484.

118 Bertucci, C., Canepa, A., and Altemura, P. (1997). Acetylated human albumin as chiral selector in HPLC: resolution of racemic drugs and protein binding studies. *Annali di Chimica (Rome)* 87: 45–52.

119 Bertucci, C. and Wainer, I.W. (1997). Improved chromatographic performance of a modified human albumin based stationary phase. *Chirality* 9: 335–340.

120 Chattopadhyay, A., Tian, T., Kortum, L., and Hage, D.S. (1998). Development of tryptophan-modified human serum albumin columns for site-specific studies of drug–protein interactions by high-performance affinity chromatography. *Journal of Chromatography B* 715: 183–190.

121 He, X.M. and Carter, D.C. (1992). Atomic structure and chemistry of human serum albumin. *Nature* 358: 209–215.

122 Petitpas, I., Bhattacharya, A.B., Twine, S. et al. (2001). Crystal structure analysis of warfarin binding to human serum albumin: anatomy of drug site I. *The Journal of Biological Chemistry* 276: 22804–22809.

123 Duggleby, H.J., Tolley, S.P., Hill, C.P. et al. (1995). Penicillin acylase has a single-amino-acid catalytic centre. *Nature* 373: 264–268.

124 Lavecchia, A., Cosconati, S., Novellino, E. et al. (2007). Exploring the molecular basis of the enantioselective binding of penicillin G acylase towards a series of 2-aryloxyalkanoic acids: a docking and molecular dynamics study. *Journal of Molecular Graphics and Modelling* 25: 773–783.

125 Temporini, C., Calleri, E., Fracchiolla, G. et al. (2007). Enantiomeric separation of 2-aryloxyalkyl- and 2-arylalkyl-2-aryloxyacetic acids on a penicillin G acylase-based chiral stationary phase: influence of the chemical structure on retention and enantioselectivity. *Journal of Pharmaceutical and Biomedical Analysis* 45: 211–218.

126 Kremer, J.M., Wilting, J., and Janssen, L.H. (1988). Drug binding to human alpha-1-acid glycoprotein in health and disease. *Pharmacological Reviews* 40: 1–47.

127 Nishi, K., Ono, T., Nakamura, T. et al. (2011). Structural insights into differences in drug-binding selectivity between two forms of human alpha1-acid glycoprotein genetic variants, the A and F1*S forms. *The Journal of Biological Chemistry* 286: 14427–14434.

128 Haginaka, J., Yamashita, T., Tsujino, H., and Arisawa, M. (2021). Revisiting chiral recognition mechanism on chicken alpha 1-acid gycoprotein: location of chiral binding sites and insight into chiral binding mechanism. *Separations* 8: 73.

129 Zsila, F. and Iwao, Y. (2007). The drug binding site of human alpha1-acid glycoprotein: insight from induced circular dichroism and electronic absorption spectra. *Biochimica et Biophysica Acta* 1770: 797–809.

130 Pinkerton, T.C., Howe, W.J., Urlich, E.L. et al. (1995). Protein binding chiral discrimination of HPLC stationary phases made with whole, fragmented, and third domain Turkey ovomucoid. *Analytical Chemistry* 67: 2354–2367.

131 Matsunaga, H. and Haginaka, J. (2006). Investigation of chiral recognition mechanism on chicken α_1-acid glycoprotein using separation system. *Journal of Chromatography A* 1106: 124–130.

132 Zsila, F., Matsunaga, H., Bikádi, H., and Haginaka, J. (2006). Multiple ligand-binding properties of the lipocalin member chicken α_1-acid glycoprotein studied by circular dichroism and electronic absorption spectroscopy: the essential role of the conserved tryptophan residue. *Biochimica et Biophysica Acta* 1760: 1248–1273.

133 Divne, C., Ståhlberg, J., Reinikainen, T. et al. (1994). The three-dimensional crystal structure of the catalytic core of cellobiohydrolase I from *Trichoderma reesei*. *Science* 265: 524–528.

134 Ståhlberg, J., Henriksson, H., Divne, C. et al. (2001). Structural basis for enantiomer binding and separation of a common β-blocker: crystal structure of cellobiohydrolase Cel7A with bound (*S*)-propranolol at 1.9 Å resolution. *Journal of Molecular Biology* 305: 79–93.

135 Henriksson, H., Ståhlberg, J., Koivula, A. et al. (1997). The catalytic amino-acid residues in the active site of cellobiohydrolase 1 are involved in chiral recognition. *Journal of Biotechnology* 57: 115–125.

136 Hedeland, M., Henriksson, H., Isaksson, R., and Pettersson, C. (1998). Studies on the enantioselective retention mechanisms of cellobiohydrolase I (CBH I) by covalent modification of the intact and fragmented protein. *Chirality* 10: 760–769.

137 Ranieri, D.I., Corgliano, D.M., Franco, E.J. et al. (2008). Investigation of the stereoselectivity of an anti-amino acid antibody using molecular modeling and ligand docking. *Chirality* 20: 559–570.

138 Nevanen, T.K., Söderholm, L., Kukkonen, K. et al. (2001). Efficient enantioselective separation of drug enantiomers by immobilised antibody fragments. *Journal of Chromatography A* 925: 89–97.

139 Nevanen, T.K., Hellman, M.L., Munck, N. et al. (2003). Model-based mutagenesis to improve the enantioselective fractionation properties of an antibody. *Protein Engineering* 16: 1089–1097.

140 Parkkinen, T., Nevanen, T.K., Koivula, A., and Rouvinen, J. (2006). Crystal structures of an enantioselective Fab-fragment in free and complex forms. *Journal of Molecular Biology* 357: 471–480.

141 Jozwiak, K., Ravichandran, S., Collins, J.R., and Wainer, I.W. (2004). Interaction of noncompetitive inhibitors with an immobilized alpha3beta4 nicotinic acetylcholine receptor investigated by affinity chromatography, quantitative-structure activity relationship analysis, and molecular docking. *Journal of Medicinal Chemistry* 47: 4008–4021.

142 Jozwiak, K., Moaddel, R., Ravichandran, S. et al. (2008). Exploring enantiospecific ligand–protein interactions using cellular membrane affinity chromatography: chiral recognition as a dynamic process. *Journal of Chromatography B* 875: 200–207.

143 Moaddel, R., Ravichandran, S., Bighi, F. et al. (2007). Pharmacophore modelling of stereoselective binding to the human organic cation transporter (hOCT1). *British Journal of Pharmacology* 151: 1305–1314.

11

Chiral Stationary Phases Derived from Cinchona Alkaloids

Michael Lämmerhofer[1] *and Wolfgang Lindner*[2]

[1] *Institute of Pharmaceutical Sciences, University of Tübingen, Tübingen, Germany*
[2] *Institute of Analytical Chemistry, University of Vienna, Vienna, Austria*

11.1 Introduction

Cinchona alkaloids are widely utilized chiral auxiliaries in organic chemistry and chiral technologies with a long tradition in classical fields, such as fractionated crystallization and asymmetric synthesis. Their unique structures with multiple functional groups (comprising the tertiary amine of the bulky quinuclidine ring, a planar electron-rich quinoline ring representing a π-base, a secondary alcohol, and a vinyl group), a plethora of stereogenic centers (N_1, C_3, C_4, C_8, and C_9), and the fact that they are readily available natural products shifted them into the focus of interest in the field of liquid chromatographic (LC) enantiomer separation (Figure 11.1). Among them, quinine (QN) and quinidine (QD) are more powerful chiral selectors than their epimers as well as cinchonidine (CD) and cinchonine (CN) (Figure 11.1). The functional groups, in particular the C_9 hydroxyl moiety, facilitate the chemical derivatization to enhance their chiral recognition capabilities. They also offer several choices for linkage to a chromatographic support such as silica or allow the introduction of a polymerizable group for the synthesis of macroporous polymer monoliths or of other polymeric phases [1]. In the majority of cases, the exocyclic vinyl group attached to the quinuclidine ring is used for immobilization to silica; it is located remote from the primary chiral recognition site (vide infra), and therefore this surface anchoring strategy does not negatively influence the chiral recognition

Chiral Separations and Stereochemical Elucidation: Fundamentals, Methods, and Applications, First Edition. Edited by Quezia Bezerra Cass, Maria Elizabeth Tiritan, João Marcos Batista Junior, and Juliana Cristina Barreiro.

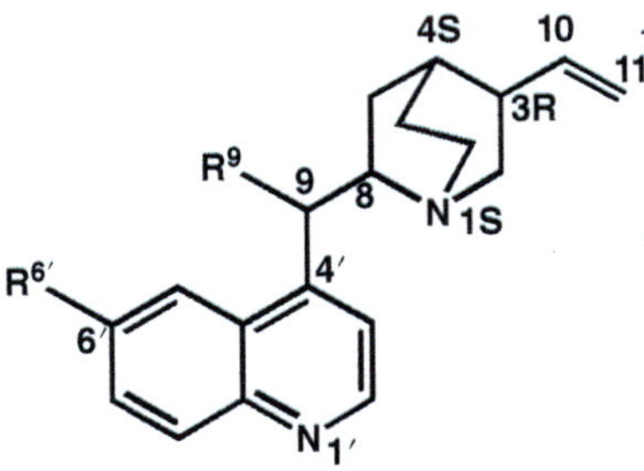

	$R^{6'} = OCH_3$	$R^9 = OH$
QN	(8*S*,9*R*)	Quinine
QD	(8*R*,9*S*)	Quinidine
EQN	(8*S*,9*S*)	Epiquinine
EQD	(8*R*,9*R*)	Epiquinidine
	$R^{6'} = H$	$R^9 = OH$
CD	(8*S*,9*R*)	Cinchonidine
CN	(8*R*,9*S*)	Cinchonine
ECD	(8*S*,9*S*)	Epicinchonidine
ECN	(8*R*,9*R*)	Epicinchonine

Figure 11.1 Cinchona alkaloids, their stereochemistry, substitution pattern, and abbreviations.

process. Moreover, it opens the possibility of further structural modifications at the C_9 hydroxyl group to modulate the chiral recognition process.

This chapter is focused on the use of cinchona alkaloids as chiral selectors in (U)HPLC. It is emphasized, however, that the chiral recognition capabilities can be preserved and exploited in many other separation technologies such as supercritical fluid chromatography (SFC) [2, 3], capillary electrophoresis (CE) [4], capillary electrochromatography (CEC) [5], centrifugal partition chromatography (CPC) [6], and supported liquid membrane separations [7]. This summary aims at providing an overview on the most popular cinchona alkaloid-derived chiral stationary phases (CSPs), their characteristics, insights into chiral recognition, separation mechanisms in LC, method development, and trends on modern supports such as superficially porous particle (SPP) designs, and new applications. This chapter builds on previous review articles [8, 9] and describes selected aspects.

11.2 Cinchona Alkaloid-Derived Chiral Stationary Phases

Early attempts on the use of cinchona alkaloids immobilized to resins for enantiomer separation date back to the 1950s [10]. Later, cinchona alkaloids have been successfully incorporated into mobile phases as counterions for ion-pair enantiomer separation of chiral acids [11]. The first silica-based quinine-bonded CSP was finally reported in the mid-1980s [12]. This CSP with immobilized quinine having free C_9-hydroxyl group, yet, showed limited enantioseparation capability. The potential and scope of applicability for HPLC enantiomer separation of cinchona alkaloid-derived CSPs were greatly enhanced with the introduction of carbamoylated cinchonan derivatives [13].

Subsequently, a large number of distinct structural variants were evaluated regarding their enantiorecognition capability [8]. It turned out that *O*-9-(*tert*-butylcarbamoyl)quinine and quinidine-based CSPs, both chemically bonded to thiol-modified silica, were a good compromise in terms of broad enantiomer selectivity [14, 15]. These CSPs are commercially available with tradenames CHIRALPAK QN-AX and CHIRALPAK QD-AX (Figure 11.2a). QN and QD refer to quinine- and quinidine-derived CSPs and AX indicate their primary mode of operation as weak chiral anion-exchangers mainly used for the enantiomer separation of chiral acids, such as *N*-derivatized amino acids [13, 15, 16] and peptides, aryloxy [13] and arylcarboxylic acids [13] as well as hydroxyl alkanoic acids [17] and various other chiral acids (vide infra).

As chiral anion-exchangers have limited applicability scope, attempts to combine their chiral recognition principles with those of chiral cation-exchange materials [18] finally led to the development of zwitterionic chiral ion-exchangers [19]. Structural variants with sulfohexyl-carbamate moiety emerged as favorable selectors and are available as CHIRALPAK ZWIX(+), quinine-derived with 1″*S* and 2″*S* configuration in the cation-exchange site, and ZWIX(−), quinidine-derived with 1″*R* and 2″*R* configuration in the cation-exchanger moiety (Figure 11.2b). As will be outlined in the following text, these hybrid selectors with chiral weak anion-exchange (WAX) and chiral strong cation-exchange (SCX) moieties combine the applicability spectra of chiral anion-exchangers and cation-exchangers comprising chiral acids and chiral bases, respectively. Furthermore, the two primary ion-exchange binding sites act in a synergistic manner for chiral recognition of amphoteric compounds including α-, β- and γ-amino acids [19, 20], primary

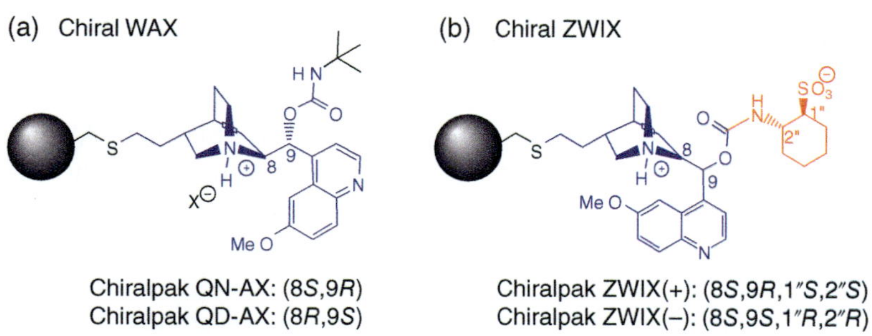

Figure 11.2 Commercially available cinchona alkaloid-derived CSPs. (a) Chiral weak anion-exchangers and (b) zwitterionic chiral ion-exchangers.

and secondary amino acids [19], aliphatic and aromatic as well as cyclic and acyclic amino acids [19, 21], amino acids with apolar, polar, or charged side chains [19], amino acids containing one or two chiral centers [21], amino carboxylic acids [19], amino sulfonic acids [22], amino phosphonic [22] and phosphinic acids [22], as well as small peptides [20, 23, 24]. ZWIX CSPs have been applied for the enantioresolution of many specialty amino acids as discussed in detail by Ilisz et al. [25].

Of particular practical impact is the pseudo-enantiomeric behavior of QN and corresponding QD CSPs which can be widely observed for both QN-AX and QD-AX as well as ZWIX(+) and ZWIX(−) CSPs (Figure 11.3). As can be readily derived from Figure 11.1, quinine and quinidine are diastereomers (configurations in position 1, 3, and 4 are identical and opposite at carbon 8 and 9). From an experimental viewpoint, the corresponding QN and QD selectors behave like enantiomers because the spatial orientation of the functional groups and primary interaction sites at C_9 is decisive for the chiral recognition mechanism. Here, QN and QD have opposite configurations leading to enantiomer-like receptor surface (cf. Figure 11.4a and b). It leads to opposite enantiomer affinities on QN- and QD-derived CSPs and reversed enantiomer elution orders, both on AX (Figure 11.3a and b) and on ZWIX CSPs (Figure 11.3c and d). This is a real advantage over CSPs derived from other natural products such as polysaccharide, protein, macrocyclic antibiotic CSPs which exist only in one enantiomeric form. Due to the diastereomeric instead of enantiomeric stereochemical relationship, the inversion upon switching from QN- to QD-derived CSPs is not perfect, i.e. the separation factors may be significantly different and not exactly the same as in enantiomeric CSPs with truly opposite configurations. It is dependent on the analyte and chiral recognition mechanism whether a perfect reversal with very similar separation factors can be achieved or not. In certain cases, a reversal of elution order is not observed. For instance, a reversed elution order is observed for the enantiomers of 2-hydroxybutyric acid by exchanging the QD-AX for a QN-AX column, which is not the case for ZWIX(+) and ZWIX(−) [29]. On the contrary, no inversion of elution order is obtained for tartaric acid upon switching from QN- to QD-derived CSPs (both AX and ZWIX) [29]. It is due to an altered chiral recognition mechanism on the corresponding pseudo-enantiomeric CSPs. However, a reversal of elution order was found for tartaric acid by changing from the AX to the corresponding ZWIX column [29]. As can be seen, there are some caveats to be obeyed in experimental practice of this highly useful molecular recognition principle.

Except for specific discussions, the current chapter concentrates on the commercially available QN-AX, QD-AX, ZWIX(+), and ZWIX(−) CSPs.

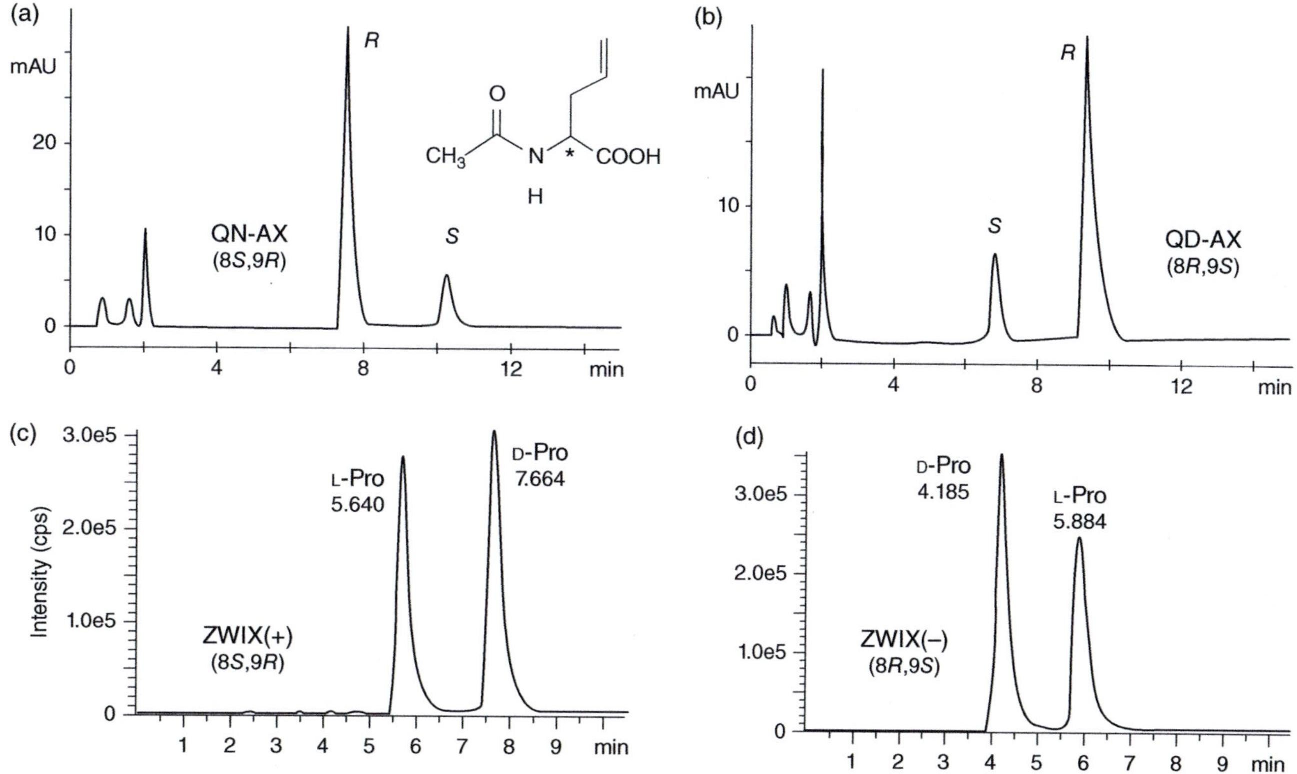

Figure 11.3 Pseudo-enantiomeric character and reversal of elution orders on quinine and quinidine carbamate CSPs. (a) *N*-acetyl-2-allylglycine enantiomer separation on QN-AX, and (b) on QD-AX. (c) Proline (Pro) enantiomer separation on ZWIX(+) and (d) ZWIX(−). *Sources:* (a and b) Reproduced from Lämmerhofer and Lindner [8]/with permission of Taylor and Francis Group. (c and d) Reproduced from Horak and Lämmerhofer [115]/with permission of Elsevier.

11.3 Chiral Recognition

In the commercial QN/QD-AX and ZWIX(+)/(−) CSPs, the cinchonan-based selectors are attached to thiol silica via the vinyl group at C_3. As can be seen from the X-ray crystal structures depicted in Figure 11.4, it is located on the back side of the receptor-like chiral selector exposing the binding pocket to the surface being therefore optimally accessible for the solutes. From numerous studies (NMR, molecular modeling, and X-ray), it is known that quinine and quinidine carbamates adopt preferentially an *anti-open* conformation, like shown in Figure 11.4, upon protonation of the quinuclidine nitrogen, which in LC occurs with common weakly acidic mobile phases [27, 30]. It means that the selector adopts a conformational arrangement with the torsional angle $C_4'{-}C_9{-}C_8{-}N_1$ of +180° in the *open* conformation and the torsion $C_3'{-}C_4'{-}C_9{-}C_8$ of +90° in the *anti*-conformation [27]. This presents the binding site in an open conformation as depicted in Figure 11.4a–d with the protonated quinuclidine in a central position and the bulky methoxyquinoline and alkyl carbamate lined up to form a binding cleft for analyte insertion driven by the primary ionic interaction between quinuclidinium and carboxylate or other anions. Supported by hydrogen bonds (with carbamate) and π–π-interaction (with electron-rich quinoline ring) as well as hydrophobic and/or steric interactions at the carbamate residue, enantiomer recognition can be quite efficient for many acidic analytes, in particular *N*-derivatized amino acids (Figure 11.4a and b), and *N*-derivatized peptides (Figure 11.4c) [26]. Both the quinoline ring and carbamate moiety can dynamically adjust their conformations to maximize interactions (cf. *N*-derivatized amino acid vs. peptide binding in Figure 11.4c) [26]. Even when the analyte looks like loosely attached to the quinuclidine ring by nonspecific ionic interactions [28], steric interactions in the binding pocket guided by bulky methoxyquinoline and steric carbamate residues lead to reasonable enantioselectivity (Figure 11.4d [27] and 11.4e [28]).

Information on molecular recognition mechanisms of various types of solutes by zwitterionic cinchonan-based CSPs originate mainly from chromatographic and molecular modeling investigations. In several articles, A. Carotti and coworkers published molecular dynamics (MD) studies performed with the zwitterionic selectors and in complexes with various types of analytes ranging from acidic, basic to zwitterionic

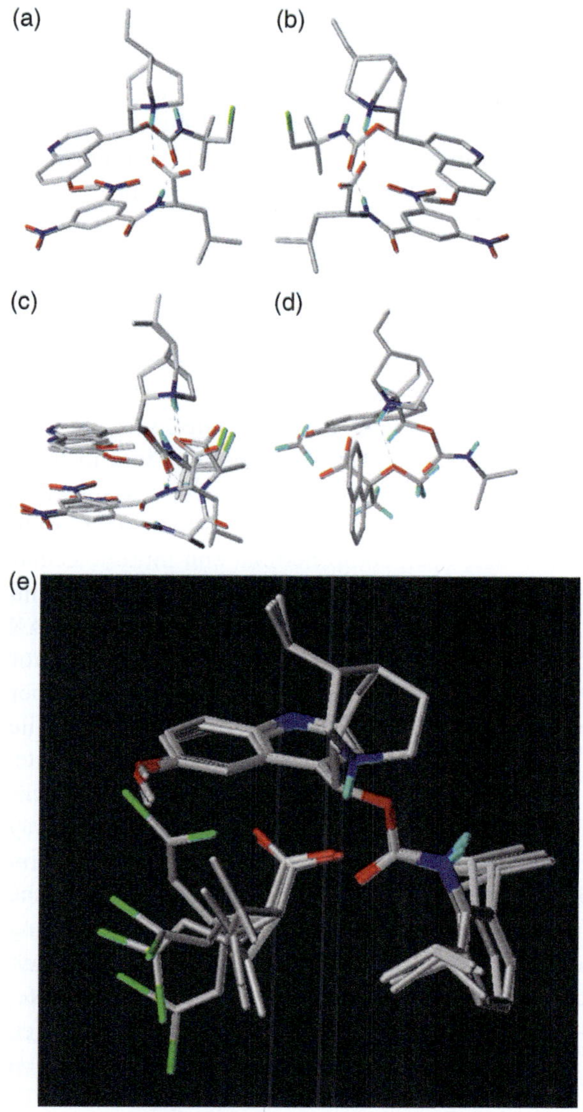

Figure 11.4 X-ray crystal structures of selector-selectand complexes (ion-pairs): (a) O_9-(β-chloro-tert-butylcarbamoyl)quinine with *N*-(3,5-dinitrobenzoyl)-(*S*)-leucine, (b) the pseudo-enantiomeric complex of O_9-(β-chloro-*tert*-butylcarbamoyl)quinidine with *N*-(3,5-dinitrobenzoyl)-(*R*)-leucine, (c) superposition of complexes between O_9-(β-chloro-*tert*-butylcarbamoyl)quinine with *N*-(3,5-dinitrobenzoyl)-(*S*)-leucine and *N*-(3,5-dinitrobenzoyl)-(*S*)-alanyl-(*S*)-alanine, and (d) O_9-(*tert*-butylcarbamoyl)quinine and (*S*)-2-methoxy-2-(1-naphthyl)propionic acid. (e) Overlay of single crystal X-ray structures of the four complexes of O_9-(2,6-diisopropylphenylcarbamoyl)quinine and permethrinic acid ((*1R,3R*)-cis-, (*1S,3S*)-cis-, (*1R,3S*)-trans, and (*1S,3R*)-trans-). Superposition by fit of quinuclidine and C_9 carbon atoms. Most hydrogens have been omitted for the purpose of clarity. *Sources:* (a–c) Reproduced from Czerwenka et al. [26]/with permission of American Chemical Society. (d) Reproduced from Akasaka et al. [27]/with permission of John Wiley & Sons. (e) Reproduced from Bicker et al. [28]/with permission of Elsevier.

analytes [24, 31–35]. In these modeling studies, the linker and silica surface are realistically reconstructed, e.g. by considering four 3-mercaptopropyl-functionalized silanols (~1.97 μmol m^{-2}), 8 free silanols (~8.0 μmol m^{-2}), and 44 silicon atoms for each grafted selector unit (~0.5 μmol m^{-2}), at the base of a cubic box of 30 Å side lengths. Silicon atoms and their bonded hydrogen atoms in the base layer are usually set frozen during the molecular dynamics. During the MD simulations, several thousands of frames can be produced during ms-time scale dynamics which should allow a realistic sampling of favorable conformations and selector-solute binding states. Conformational analysis of free selectors provided favorable minimal energy conformations as depicted in Figure 11.5a and b for ZWIX(+) and ZWIX(−), respectively [31]. It can be seen that in these models, generated with acetonitrile/water (90 : 10; v/v) as solvation model, self-protonation of the quinuclidine by the sulfonic acid group occurs leading to the favorable open conformation, and intramolecular ionic interaction appears to be possible due to its long-range nature and spatial closeness in the favorable conformation. Like the chiral WAX selectors, also the ZWIX selectors present favorable binding clefts for analyte insertion (Figure 11.5a and b). Analyte binding, on the other hand, seems still driven by counterionic electrostatic interactions, while the analyte appears to be repelled from the co-ionic ion-exchange site. The acidic analytes in Figure 11.5c and d are attached to the quinuclidinium moiety through the carboxylate groups which are oriented away from the sulfonate moiety of the ZWIX selectors. The latter acts as intramolecular counterion (vide supra). For efficient chiral recognition, the primary attractive electrostatic interactions must be accompanied by any other supportive interactions (e.g. H-bonding, π–π-interaction with quinoline, hydrophobic or steric interaction) that are typically analyte-dependent, of short-range nature and may occur enantioselectively (cf. Figure 11.5c and d). For basic analytes, the cation-exchange interaction is the driving force for binding to the sulfonate moiety of the ZWIX selector (Figure 11.5e and f) [33]. The enantiomer which is effectively supported by additional binding increments is more strongly attached to the ZWIX selector (Figure 11.5f), while nonspecific electrostatic interaction alone does not provide high affinity binding (Figure 11.5e) and hence this enantiomer is less retained. The synergistic effect of anion-exchange and cation-exchange sites by simultaneous double ion-pairing enables chiral recognition and separation of the stereoisomers of zwitterionic analytes, like free amino acids and peptides (such as shown in

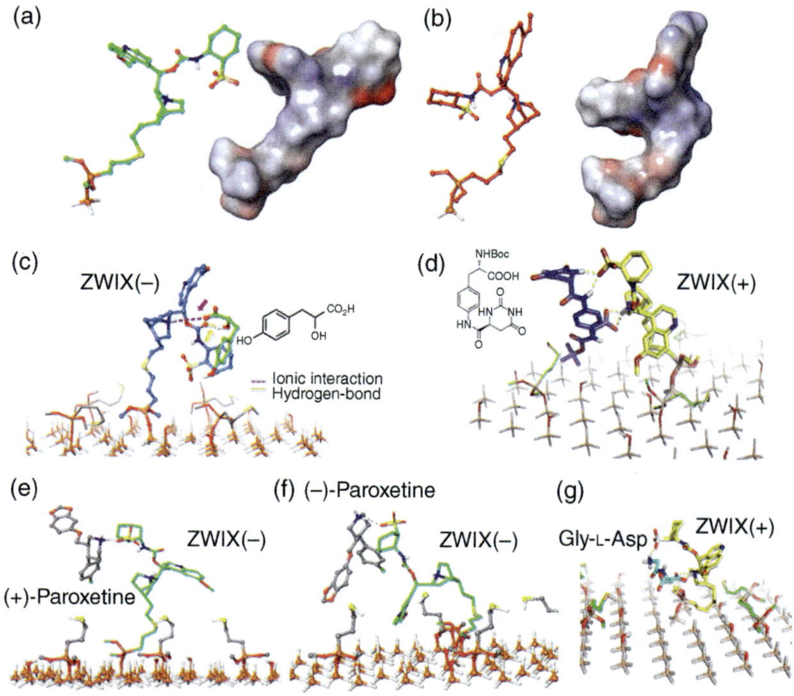

Figure 11.5 Preferential conformations of ZWIX selectors and tentative binding models of various analytes. Minima conformations of ZWIX(+) (a) and ZWIX(−) (b). Favorable docking poses from molecular dynamics simulations of (c) 3-(4-hydroxyphenyl)lactic acid on ZWIX(−), (d) a BOC-amino acid nucleic acid on ZWIX(+), (e) (+)-paroxetine on ZWIX(−), (f) (−)-paroxetine on ZWIX(−), and (g) Gly-L-Asp on ZWIX(+). *Sources:* (a and b) Reproduced from Bäurer et al. [31]/with permission of Elsevier. (c) Reproduced from Varfaj et al. [35]/with permission of Elsevier. (d) Reproduced from Ianni et al. [32]/with permission of Elsevier. (e and f) Reproduced from Grecso et al. [33]/with permission of Elsevier. (g) Adapted from Ianni et al. [24].

Figure 11.5g for the dipeptide glycyl-L-aspartic acid, Gly-L-Asp) [24]. While these are just a few simple exemplary models of selector-analyte interactions, they give an idea about the mechanistic peculiarities of how the WAX-type and zwitterionic-type cinchonan-based CSPs can recognize and distinguish enantiomers of chiral solutes. Support for these tentative binding models comes from various mechanistic chromatographic studies (vide infra).

11.4 Chromatographic Retention Mechanisms

11.4.1 Multimodal Applicability

Covalent attachment of the cinchona alkaloid-derived chiral selectors to silica via a reactive silane makes these stationary phases compatible with essentially all types of mobile phases commonly used in LC. Applicable organic solvents range from very apolar (heptane, hexane), over mid-polar (tetrahydrofuran, *tert*-butyl methylether, chloroform, dichloromethane), to highly polar (acetonitrile, methanol, water) and define the primary chromatographic elution mode. Under aqueous conditions, pH values must remain in a range which does not catalyze hydrolysis of the siloxane bonding and silica backbone, i.e. within pH 2–8. Under these circumstances, the cinchona alkaloid-derived CSPs truly show multimodal applicability, like many other CSP classes. They can be applied with

1) reversed-phase elution conditions using mobile phases composed of methanol-buffer or acetonitrile-buffer mixtures (RP mode) [36],
2) polar organic elution conditions composed of nonaqueous polar organic solvents, such as methanol, acetonitrile, or mixtures thereof with acids (formic, acetic, rarely trifluoroacetic acid) [37, 38], bases (ammonia, diethylamine, triethylamine) [38, 39], or both as additives (PO mode),
3) hydrophilic interaction chromatography conditions with acetonitrile-rich (>50%, typically 90–98%) aqueous mobile phases containing volatile organic buffers (HILIC mode) [40, 41],
4) normal-phase elution conditions using alkane-based eluents with polar organic modifiers such as 2-propanol and organic acids/bases as additives (NP mode) [42].
5) Furthermore, supercritical fluids ($scCO_2$) with polar modifiers and additives (SFC mode) have also been used with QN/QD-AX [2] and ZWIX(+)/(−) [43] CSPs.

Through these distinct elution conditions, it becomes possible to strengthen or weaken specific types of non-covalent bonds and interactions, respectively, and hence balance specific and nonspecific analyte–adsorbent interactions, thus modulating enantioselectivity and overall retentivity as will be discussed in the following text.

11.4.2 Surface Charge of Cinchonan-Based CSPs

For the chromatographic separation of charged analytes on cinchona alkaloid-derived CSPs, their surface charge plays an important role. Knowledge on how the surface charge changes with altered conditions is therefore of great utility to understand the retention behavior of analytes on these CSPs. Thereby, the net surface charge is dependent on the ionization degree and the respective surface concentration of both the chiral selectors and residual silanols of supporting silica particles.

From a practical viewpoint, it is convenient to characterize the net surface charge of CSP particles by pH-dependent ζ-potential measurements using electrophoretic light scattering [31, 44]. For this purpose, the particles are suspended at a concentration of 0.2 mg ml^{-1} in an electrolyte solution (here 2 mM buffer of defined pH in 10 mM KCl to keep the ionic strength constant throughout the series of experiments). A double layer forms on the charged surface with a fixed layer of counterions directly attached to the charged surface and an adjacent diffusive layer enriched with counterions. Upon application of an electric field, the charged particles start to migrate according to their electrophoretic mobility which depends on the potential at the shear surface, i.e. the ζ-potential of the particles under given conditions (pH, ionic strength, temperature, and dielectric permittivity of the medium).

The ζ-potential of commonly employed fully porous plain silica particles typically varies between −10 mV at pH 3.5 and −40 mV at pH 9.5 under above specified conditions [44]. Modification with a reactive 3-mercaptopropylsilane layer reduces the ζ-potential by around 5–10 mV but remains negative throughout the entire tested pH range due to residual silanols. In contrast, *tert*-butylcarbamoyl quinine-modified silica particles (Chiralpak QN-AX) exhibit a positive ζ-potential at low pH of 3.5 due to protonation of the chiral selector (pK_a of quinuclidine = 8.08, pK_a of quinoline = 4.05 according to calculations by Marvin) (Figure 11.6) [31]. The ζ-potential only slightly declines from +40 mV at pH 3.5 to +30 mV at pH 5.5, spanning the useful pH range with optimal anion-exchange capacity (most acidic solutes are at least partially dissociated in this pH range and thus can be retained by ionic interactions). By further increasing the pH, the surface charge drops significantly but remains positive up to pH 7.5. In this pH range (between 5.5 and 7.5), the QN-AX and QD-AX CSPs have weak effective anion-exchange capacity, hence less retention while enantioselectivity is still available. Above pH 7.5, charge reversal from positive to negative ζ-potentials occurs (Umpolung) due to reduced dissociation of the chiral selector and dominating effect of dissociated residual silanols (Figure 11.6). Under such repulsive electrostatic effects, retention of anionic analytes on QN/QD-AX will be low or they will elute even before t_0.

The ζ-potential characteristics of the zwitterionic CSPs (Chiralpak ZWIX(+) and ZWIX(−)) is dominated by the sulfonic acid moiety (pK_a = −0.97, calculated with Marvin Sketch). Its presence results in an

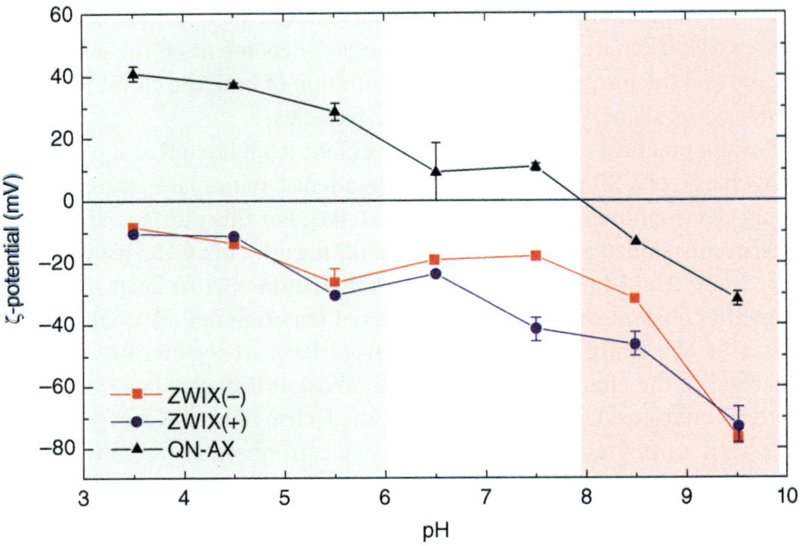

Figure 11.6 pH-dependent ζ-potential measurements by electrophoretic light scattering on a Zetasizer Nano ZS of QN-AX, ZWIX(+), and ZWIX(−) CSP particles dispersed at 0.2 mg ml^{-1} in 1 mM buffer containing 10 mM KCl. Green shadow indicates pH range with favorable ion-exchange capacity, light grey shadow pH range with minor ion-exchange capacity, and red shadow pH range without significant ion-exchange capacity (AX CSPs) and less useful pH range (ZWIX CSPs). *Source:* Reproduced from Baeurer et al. [31]/with permission of Elsevier.

offset of around −50 mV for the zwitterionic CSPs relative to the AX CSPs (Figure 11.6) [31]. It has the consequence that the surface of the ZWIX phases is net negatively charged over the entire pH range. Evidently, there is no simple 1:1 charge balance from the two oppositely charged ion-exchange sites for reasons that are discussed in detail elsewhere [31]. Like for the AX phases, for the ZWIX CSPs the favorable pH range also spans in most cases weakly acidic conditions (indicated in Figure 11.6 by a green shadow): the SCX site is dissociated due to its strongly acidic character and the AX site is protonated also offering suitable anion-exchange capacity; weakly acidic analytes are at least partially dissociated, basic solutes are protonated, and hence the retention of acids by anion-exchange, of bases by cation-exchange, and of amphoteric analytes by simultaneous anion- and cation-exchange is possible. Optimized mobile phase conditions for many enantiomer separations reported in the literature use weakly acidic conditions and underpin this interpretation.

11.4.3 Retention Mechanisms and Models, and Method Development on Chiral WAX CSPs

The prime scope of application of QN-AX and QD-AX CSPs is for enantiomer separation of chiral acids. Only a few studies evaluated their applicability for other analytes, e.g. neutral or basic analytes adopting normal-phase (e.g. 2-substituted imidazo[1,5-*b*]-quinazoline-1,5-diones [42] or arylalcohols [45]) or reversed-phase (e.g. 2-substituted imidazo[1,5-*b*]-quinazoline-1,5-diones [36]) elution conditions, but they have little practical importance. In a few reports, the enantiomer separation of zwitterionic compounds on AX CSPs could be accomplished. Thereby, the primary ionic interaction of the solutes with the cinchonan-based selector was supported by additional strong solute–adsorbent interactions due to aromatic moieties (thyroxine [46], phenylalanine [47]) or acidic amino acid side chains (e.g. 1-aminoindan-1, 5-dicarboxylic acid [48]).

Regardless of elution conditions (RP, PO, HILIC, and NP), the separations of chiral acids on QN/QD-AX follow an anion-exchange principle. The stoichiometric displacement model [49] is a simple retention model for such an ion-exchange retention process. This model treats ion-exchange as a stoichiometric process in which one solute ion is displaced from the ion-exchange site by one counterion *C*, and vice versa. Consequently, plots of log *k* versus log [*C*] are linear in accordance to Eq. (11.1)

$$\log k = \log K_z - Z \cdot \log [C] \tag{11.1}$$

wherein k is the retention factor, $[C]$ the molar concentration of the counterion in the eluent, Z is the slope of the linear regression line, and $\log K_z$ the intercept. In Eq. (11.1), K_z represents a system-specific constant that is related to the ion-exchange equilibrium constant K (in L mol^{-1}), the surface area S (in m^2 g^{-1}), the charge density on the surface i.e. the number of ion-exchange sites q_x available for adsorption (in mol m^{-2}), and the mobile phase volume V_0 (in L) in the column (Eq. 11.2)

$$K_z = \frac{K \cdot S \cdot (q_x)^Z}{V_0} \tag{11.2}$$

As discussed earlier, q_x changes with the pH and K also depends on the affinity of the solute toward the ion-exchange site. The slope Z depends on the ratio of the effective charge numbers [50] of solute ($z_{\text{eff,S}}$) and counterion ($z_{\text{eff,C}}$) ($Z = z_{\text{eff,S}}/z_{\text{eff,C}}$). This explains the rapid increase of retention for multiply charged analytes which can be set off by use of multiply charged counterions like citric acid in the eluent. Thus, counterion concentration and type are prime experimental variables to adjust retention with typically no significant effect on enantioselectivity. As can be seen in Figure 11.7a, retention decreases with increase of counterions in the mobile phase, with multiply charged counterions being stronger competitors, which may guide the selection of elution conditions. Besides, the pH is another major experimental variable that governs retention in ion-exchange. As pointed out earlier, it regulates the dissociation state of the CSPs and also that of analytes and additives (co- and counterions) of the mobile phase. According to an ion-exchange model reported by Sellergren and Shea [52] (Eq. 11.3), the highest retention should be obtained under pH conditions at which the product of dissociation degree of ion-exchange selector (here quinine carbamate QN) α^*_{QN} and analyte α^*_{SA} is maximal (ϕ) represents the phase ratio, K the distribution coefficient, and $[\mathrm{QN}]_{\mathrm{tot}}$ the total ion-exchange capacity, i.e. here the concentration of immobilized quinine carbamate selector.

$$k = \phi \cdot \frac{1}{[C]} \cdot K \cdot \alpha^*_{\mathrm{SA}} \cdot \alpha^*_{\mathrm{QN}} \cdot [\mathrm{QN}]_{\mathrm{tot}} \tag{11.3}$$

From these considerations, it can be derived that the WAX-type QN-AX and QD-AX CSPs possess their highest ion-exchange capacity for chiral carboxylic acids in the pH range between 3 and 5, as shown arlier. More acidic analytes may have stronger retention at even lower pH. They exhibit still reasonable anion-exchange capacities at pH values between 5 and 6, which significantly drops between 6 and 7.5. Above pH 8 electrostatic repulsion is dominating for anionic solutes unless they are overcome by other strong analyte adsorbent interactions.

Under reversed-phase elution conditions (RP mode), the strength of the counterion increases in the order acetate < formate < phosphate < citrate. Other counterions have been rarely used in RP mode. A reasonable guiding principle for the selection of counterion type can be similarity considerations: for carboxylic acid type analytes, acetate or formate may be preferred unless the high UV cutoff is prohibitive, e.g. for analytes without strong chromophoric group; for phosphoric and phosphonic acid analytes, phosphate buffer may be advantageous. Criteria for the selection of the RP mode for chiral separations of chiral acids can be solubility reasons, e.g. if analytes are poorly soluble under nonaqueous (PO, NP) and HILIC conditions, and when hydrophobic interactions are desirable, e.g. due to lack of other additional interaction sites besides the primary anionic moiety or as an additional binding increment to resolve homologs. A linear solvent strength

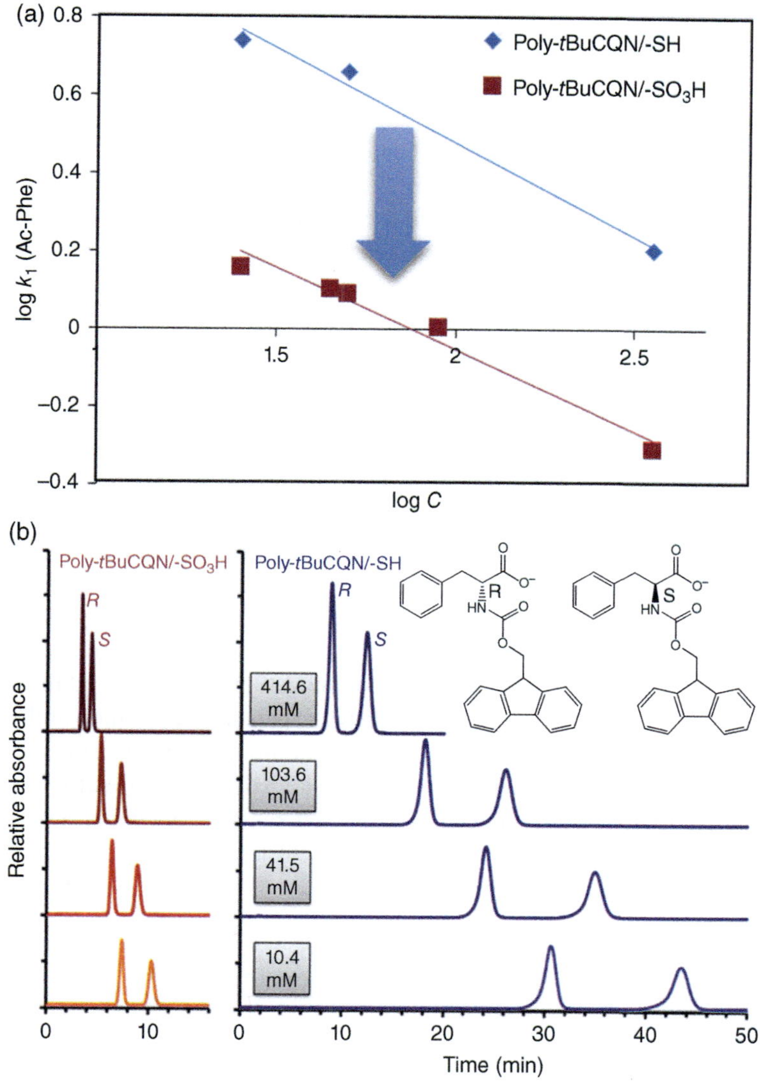

Figure 11.7 Counterion effect in accordance to stoichiometric displacement model and retention shift of first eluted D-enantiomer of *N*-acetyl-phenylalanine due to surface-anchored counterions (a). Surface-anchored counterions obtained by oxidation of residual thiols to sulfonic acid moieties in a polymeric *t*BuCQN-bonded CSP allows decreasing the counterion concentration in the mobile phase by factor of around 10 which makes mobile phases more ESI-MS friendly (b). *Sources:* (a) Adapted from Woiwode et al. [51]. (b) Reproduced from Woiwode et al. [51]/with permission of Elsevier.

dependency (Eq. 11.4) can be typically observed if ionic strength is kept constant, similar to what is observed in RPLC.

$$\log k = \log k_0 - S \cdot \phi \tag{11.4}$$

wherein ϕ is the organic modifier content and S is a solute-dependent parameter related to its hydrophobic contact area, k_0 the (extrapolated) retention factor at 100% aqueous mobile phase. The relative importance of the experimental factors with decreasing effect on enantioselectivity in RP mode is as follows: pH, organic modifier type, counterion type, temperature, and co-ion (has little effect). A good starting point for method development can be a mobile phase with 20% ammonium acetate buffer and 80% methanol (pH of mixture adjusted to 6) or a mixed buffer-methanol gradient thereof. If retention is too short, the buffer concentration can be reduced, followed by optimization of pH and test of acetonitrile as organic modifier.

In the nonaqueous PO mode, the pH of the eluent is adjusted by the acid–base ratio of the additives to either plain methanolic eluent or methanol-acetonitrile mixtures. Acetic acid, formic acid and ammonia, and diethylamine are the most common additives. Usually, the acidic component is added in (three to fourfold) excess to achieve adequate anion-exchange capacity. Acid/base ratio and methanol/acetonitrile percentage are the two most influential factors for optimization. The acidic additive represents the counterion and can be adjusted to balance the strength of the ionic interaction and retention according to an anion-exchange principle (vide supra). An anion-exchange principle can even persist in normal-phase elution mode for acidic analytes which then requires adequate acidic additive concentrations to balance the strong electrostatic interaction and retention in such apolar medium [53].

11.4.4 Retention Mechanisms and Method Development on ZWIX CSPs

Retention principles on ZWIX CSPs can be discussed by the nature of the chiral analytes, i.e. acidic, basic, or zwitterionic. Most of the time, polar organic conditions were applied for the ZWIX CSPs.

For acidic solutes, an anion-exchange principle dominates retention and the above-described rationale for WAX-type CSPs can be applied to ZWIX phases as well. Although the ZWIX-CSP surface is net negatively charged, the acidic analytes are guided by long-range electrostatic attraction to the respective AX-site of the selector (i.e. the quinuclidinium group), as illustrated in Figure 11.5c and d. However, due to the existence of the intramolecular counterion in the ZWIX selector, retention is much lower than that on the corresponding WAX-type CSP at equal conditions (e.g. factor 4 for 3-hydroxyalkanoic acids) [54]. This may be favorable for multiply charged analytes, such as dicarboxylic acids like tartaric or isocitric acid with factor 10 lower retention times on the ZWIX CSPs than the corresponding AX CSPs [29]. In practice, the concentration of the counterions in the mobile phase can be substantially reduced with ZWIX-CSPs as compared to the QN/QD-AX CSPs, which is favorable for ESI-MS hyphenation where high ionic strength may lead to significant ion suppression and sensitivity loss.

For basic analytes, a cation-exchange principle is dominating. The general principles are the same but the charge sign of the fixed charge of the ion-exchanger, counter- and co-ion, and analyte are reversed. As shown earlier, the driving force for solute–sorbent interaction for basic analytes is the electrostatic interaction between the protonated base and the SCX moiety (Figure 11.5e and f). A variety of chiral amino alcohols [19] (e.g. β-sympathomimetics [19], β-blockers [19], mefloquine [19, 55]) and a number of other chiral bases (flecainide [19], paroxetine [33]) have been separated into enantiomers on ZWIX CSPs. Usually, polar organic mode with methanol-acetonitrile mixtures as eluents were employed and the acidic additive (mostly formic acid) was present in two to threefold excess to protonate the basic analyte. In this cation-exchange mode, the elution strength dictated by the counterion increases from tertiary amine (e.g. diisopropylethylamine) over secondary amine (e.g. diethylamine) to ammonia [38]. A good choice of mobile phase is a mixture of methanol, acetonitrile, and water (49 : 49 : 2; v/v) with 25 mM formic acid and 12.5 mM ammonium formate, as exploited in the analysis of mefloquine by LC–MS [55]. Although trifluoroacetic acid as a co-ion is often neglected and not recommended, it cannot be completely ruled out: for instance, equimolar concentrations of trifluoroacetic acid and ammonia as counterion (25 mM) provided fast and highly efficient enantiomer separation for salbutamol [56]. It appears of importance that the ionic interactions and also the intramolecular counterion effect are finely tuned to be neither too weak nor too strong.

The retention mechanism of zwitterionic chiral molecules has been investigated in numerous works [15, 38, 41]. The polar organic mode is recommended with typically twofold molar excess of acidic additive (formic acid preferred) as compared to basic additive (diethylamine or ammonia) [38]. With trifluoroacetic acid, retention is usually too short and acetic acid gives lower resolution [38]. The above (FA excess) conditions assure double ion-pairing for zwitterionic solutes combining anion-exchange and cation-exchange at the quinuclidinium and sulfonate moieties, respectively, as illustrated in Figure 11.5g [24]. Although the guidelines of the column supplier suggest relatively high additive concentrations to balance the ionic interactions (e.g. methanol/acetonitrile/water, 49:49:2 (v/v/v) containing 50 mM formic acid and 25 mM diethylamine), it has been demonstrated that the ZWIX phases can be operated even without additives due to the intramolecular counterion effect that seems, as the sole displacer of the analytes from the zwitterionic ion-exchange selector, sufficient for effective elution of the zwitterionic analytes [41]. For a more in-depth explanation, an electrostatic attraction–repulsion model was postulated (Figure 11.8a) [41]. Indeed, a simple intramolecular counterion effect with the sulfonate being the counterion at the AX site and the quinuclidinium at the SCX site would not sufficiently explain the elution ability with eluents completely devoid of electrolytes. For such conditions, multivalent binding and mutual charge saturation by this double ion pairing process would expect a very strong (even infinitely long) adsorption for the zwitterionic analyte if only electrostatic attraction is considered, which means no elution would occur. On the other hand, if only electrostatic repulsion between the ionic groups of analyte and selector with equal charge sign is considered, the zwitterionic analyte should elute before the void volume marker, i.e. before t_0. In fact, neither of the two scenarios was observed experimentally (Figure 11.8b). U-shaped retention profiles were observed both with methanol- and acetonitrile-based electrolyte-free aqueous-organic mobile phases. In organic-rich eluents, (attractive) electrostatics dominate retention due to thin solvation shells resulting from the organic solvents. As the water content increases, the ionic interaction sites get strongly solvated and electrostatic interactions are effectively disrupted by the water shielding effect. After this so-called "balanced region," hydrophobic interactions gained increasing importance with water-rich mobile phases, thus promoting again increasing retention. Enantioselectivity was largest in the balanced region. An electrostatically driven "attraction–repulsion model," which claims balanced electrostatic interactions due to simultaneously active attractive and repulsive electrostatic interactions, was postulated to explain this favorable characteristic allowing elution of zwitterionic solutes on zwitterionic CSPs without counterions in the mobile phase. As attractive electrostatic interactions are enhanced due to less polar solvents, electrostatic repulsion is also strengthened, largely compensating the effect from the former. Likewise, if a strongly polar solvent reduces electrostatic attraction by thick solvation shells, repulsive electrostatic interaction is affected similarly, which leads to a well-balanced interaction throughout various conditions. However, although separation factors were similar with and without eluent electrolytes, the

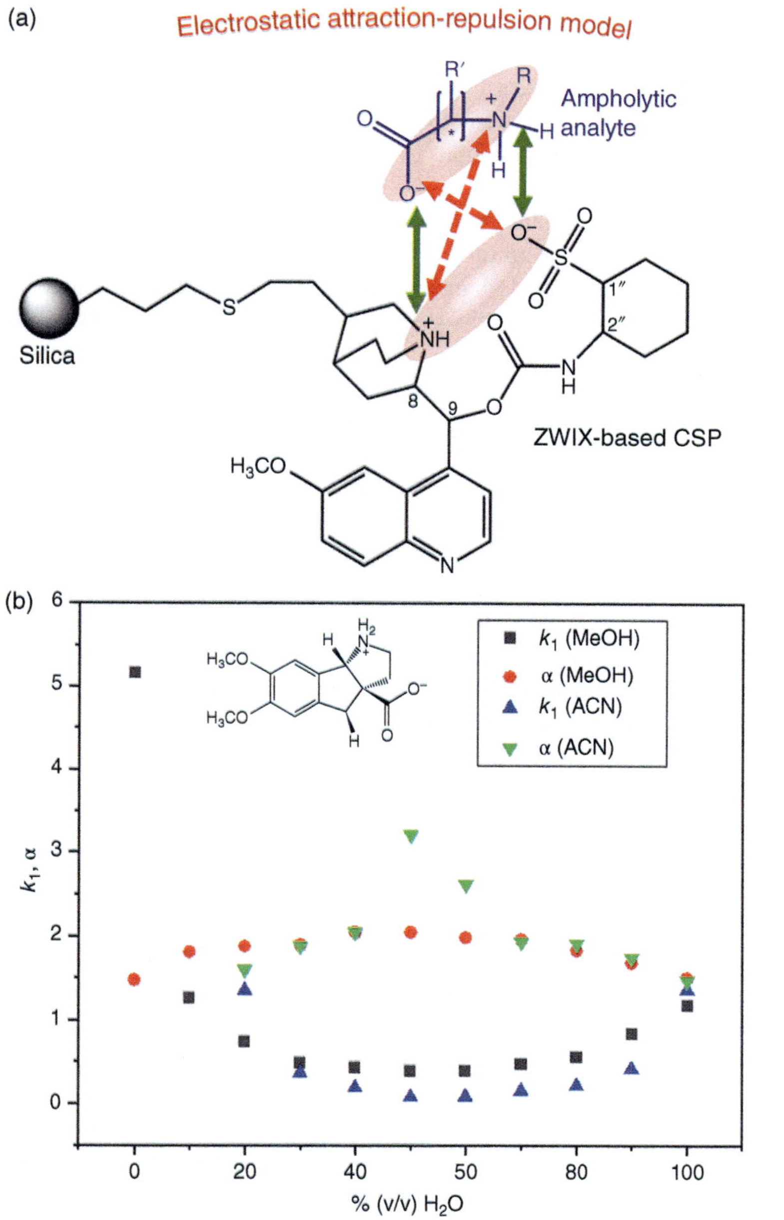

Figure 11.8 Electrostatic attraction–repulsion model (a) and solvent effects for amino acid enantiomer separation with counterion-free mobile phases (b). *Sources:* (a) Reproduced from Mimini et al. [41]/with permission of Elsevier. (b) Adapted from Mimini et al. [41].

presence of acid/base additives in the mobile phase provided better resolution values due to enhanced mobile-stationary phase exchange rates and resultant higher efficiencies [41]. From a practical viewpoint, a mobile phase with a low concentration of buffer additive should also lead to more robust chromatographic conditions and is therefore usually preferred over eluents devoid of any additives.

A detailed systematic study on method development and optimization of mobile phase factors for amino acids and other zwitterionic analytes are reported in Refs. [20, 57]. Methanol as a protic solvent has been found to be a good choice as the basis for the mobile phase and the addition of a small percentage of water (e.g. 2%) provided beneficial performance [20]. It increases elution strength, improves the efficiency, reduces peak tailing, enhances solubility of analyte and additives in the mobile phase, and is also favorable for ESI stability in MS detection. High percentages of water typically lead to too fast elution and loss of enantioselectivity and resolution. On the contrary, if stronger retention is needed, a part (e.g. 50%) of the methanol can be replaced by acetonitrile. In some instances, tetrahydrofuran instead of acetonitrile was favorable for the separation, e.g. for peptide separations [20]. As pointed out earlier, the presence of co- and counterions in the mobile phase ensures proper counterbalancing of the intramolecular counterion effects and regulates the selector analyte interactions to finely tune their strength (Figure 11.9) [20]. Addition of just basic additives, e.g. 25 mM diethylamine (Figure 11.9a) or excess of basic additives (Figure 11.9c), is unfavorable because it leads to loss of ionic character of the amino group of the analyte and the ZWIX phase (corresponding to the unfavorable red region in Figure 11.6). It leads to insufficient retention due to poor actual ion-exchange capacity. In contrast, acidic additives alone, e.g. formic acid (Figure 11.9e) and acetic acid (Figure 11.9b), provide satisfactory retention and good selectivity but poor peak shapes. A mixture of acidic and basic additives with the former in excess, which corresponds to the green pH region in Figure 11.6, gives good enantioselectivity and peak shapes (Figure 11.9d). If longer retention is required, some of the methanol can be replaced by acetonitrile (Figure 11.9g). If MS compatibility is required, ammonia instead of diethylamine is a good choice with slightly higher elution strength but similar enantioselectivity and resolution (Figure 11.9f). Overall, 50 mM formic acid and 25 mM diethylamine (or ammonium hydroxide) in methanol/acetonitrile/water (49 : 49 : 2; v/v/v) is recommended as the starting mobile phases for method development [20]. Instead, methanol/tetrahydrofuran/water at the same ratio and with the same additives can be a good alternative in some instances and provides better selectivity, e.g. for peptides. If stronger elution is required,

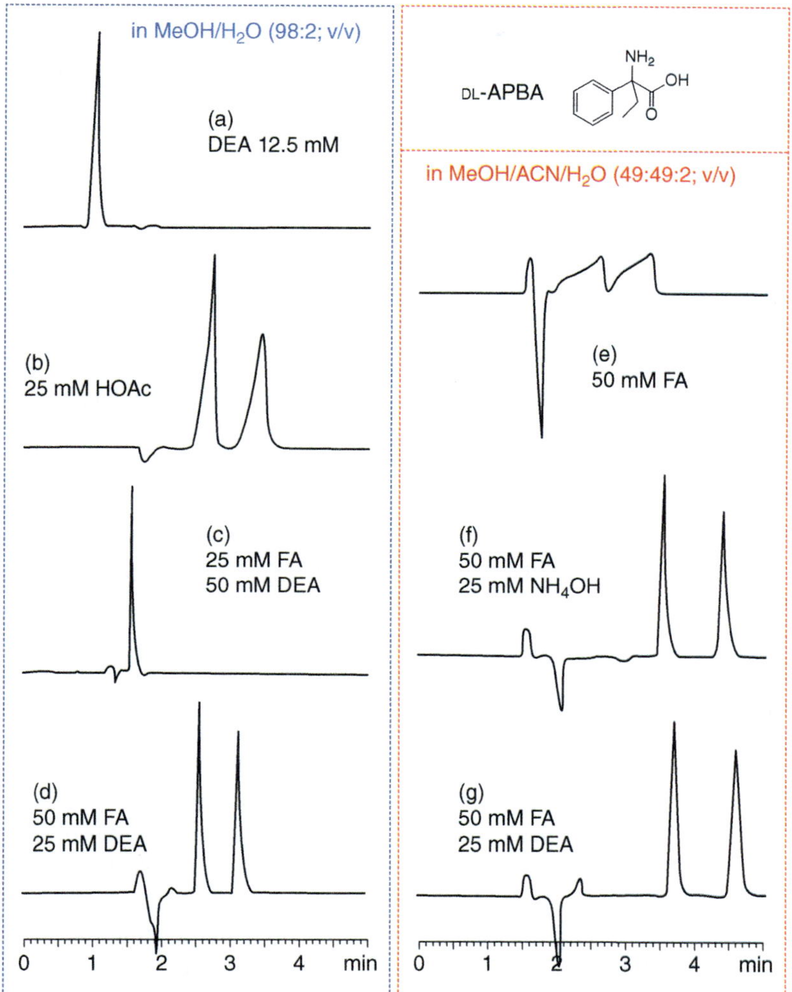

Figure 11.9 Optimization of mobile phase composition for enantiomer separation of amino acids (exemplified by 2-amino-2-phenylbutyric acid, APBA) on ZWIX phases. *Source:* Reproduced from Zhang et al. [20]/with permission of Elsevier.

methanol/water (98 : 2; v/v) and/or higher additive concentrations can make the separation faster. For MS with ZWIX phases, a mobile phase composed of 25 mM formic acid and 12.5 mM ammonium formate in methanol/acetonitrile/water or 25 mM formic acid and 25 mM ammonium formate in methanol/water 98 : 2 (v/v) may be the first choice for the start of method development.

11.5 Structural Variants of Cinchona Alkaloid CSPs and Immobilization Chemistries

Over the years, numerous structural variants of cinchonan-type selectors have been synthesized to investigate structure–enantioselectivity relationships, optimize enantioselectivity, minimize or modulate secondary interactions with the supporting particles, improve immobilization chemistry and stability, and so forth. It became rapidly evident that immobilization of free quinine (or quinidine) with free C_9-hydroxyl group exhibits some but only limited enantioselectivity [14, 58]. The introduction of a carbamate moiety greatly enhanced enantioselectivity for a wide variety of chiral acids [14]. Moreover, carbamates also showed favorable enantioselectivity levels over more rigid amide and urea analogs [53], especially for *N*-derivatized amino acids, although urea-type cinchonan CSPs revealed elevated enantioselectivities for specific analytes, like *N*-derivatized secondary amino acids (e.g. Pro) [53] and *N*-derivatized peptides [26]. These findings can be readily rationalized by the binding pose presented in Figure 11.4a. By removing the *tert*-butyl carbamate moiety, the quinine selector gets more conformational flexibility, and largely unhindered free accessibility of the anion-exchange site becomes possible for both enantiomers. Insignificant differences in the diastereomeric binding energetics lead to very similar separation factors for the two enantiomers of chiral acids on the free quinine CSP. The preferential binding mode also gets perturbed when epimeric forms of the cinchona alkaloids are utilized as selectors [15]. Consistently, only small α-values were obtained for *N*-acylated amino acids on the C_9-epimeric forms of QN-AX and QD-AX CSPs. This can be readily anticipated if we look at the X-ray crystal structures (Figure 11.4a and b): if we exchange the hydrogen and the carbamate residue, and thus the configuration at C_9, the binding site at N_1 loses its favorable pocket-like structure and also leads to altered conformations. Interestingly, the elution order is controlled by the absolute configuration of the C9-stereogenic center, i.e. QN- and *epi*-QD-derived CSPs, both with (9*R*)-configuration, show stronger retention for the (*S*)-enantiomer of *N*-acyl amino acids, while the elution order is reversed for QD- and *epi*-QN-derived CSPs, both with (9*S*)-configuration and the same elution order [15]. It indicates that the C_9-carbamate together with the WAX site dominate the binding mechanism, while simultaneous interaction with the quinoline is a supporting binding increment. The 6′-residue also plays a significant role [26]. *tert*-Butylcarbamate of CD and CN with hydrogen in 6′ position exhibit lower enantioselectivity for *N*-acylamino acids compared to their quinine and quinidine counterparts (see e.g. [26]). It seems that the methoxy substituent shields the electrostatic interaction of the analyte at the

WAX site and contributes to the formation of a deeper binding pocket. Consequently, further 6′-modifications have been investigated and it could be demonstrated that bulkier neopentyl and 1-adamantylmethyl (instead of methyl) further enhance this effect leading to exceptional separation factors (e.g. α =32.6 for 3,5-dinitrobenzoyl(DNB)-leucine on a CSP with 6′-neopentoxy and *tert*-butylcarbamate residue) [26]. A similar effect can be achieved with modulation of the carbamate residue [59]. Increasing the bulkiness of the carbamate residue leads to increased α-values for *N*-acyl-amino acids. Steric interactions between carbamate residue and amino acid side chains (as evident from their spatial proximity in the X-ray crystal structures of Figure 11.4a–c) become more important with bulkier aliphatic groups at those sites. However, once they are too voluminous, steric repulsion effects may lead to a decline of separation factors upon further increasing the size of the carbamate residue. For this reason, *tert*-butyl was found to be a good compromise as a carbamate residue. Together with the 6′-residue of quinine and quinidine, *tert*-butylcarbamates form relatively deep, yet still accessible binding pockets for average organic acids and in particular *N*-derivatized amino acids. Aromatic carbamate residues such as 2,6-diisopropylphenyl, 3,5-dinitrophenyl, 3,4-dichlorophenyl, 3,5-bis(trifluoromethyl)phenyl, and triphenylmethyl (trityl) can further enhance the steric effect and have shown complementary enantioselectivity profiles as well as advantageous enantioselectivity for small chiral acids and arylcarboxylic acids that lack a polar substituent at the stereogenic center [58]. This is exemplified in Figure 11.4e which shows the overlay of X-ray crystal structures of the four permethrinic acid stereoisomers in complex with 2,6-diisopropylphenylcarbamoylquinine [28]. The 2,6-diisopropylphenyl residue provides significant space filling close to the WAX site and hence a steric/van der Waals type interaction site that induces enhanced stereoselectivity. It showed significantly improved separation factors over the *tert*-butylcarbamate CSP. Ketoprofen enantiomers were successfully separated on the 2,6-diisopropylphenylcarbamoyl quinine CSP (α =1.09; R_S =1.60) in the reversed-phase mode, while the corresponding *tert*-butylcarbamate CSP did not afford such a separation [60]. Evidently, optimal carbamate and 6′-residue must be seen in relation to the analyte which needs to be enantioresolved. If all these binding increments at 6′-position, carbamate residue and analyte residues are optimized, α-values up to 100 and even higher can be achieved, like exemplified by O_9-1-adamantylcarbamoyl-6′-neopentyloxycinchonidine CSP (quinine analog with 6′-neopentoxy and 1-adamantyl carbamate), which exhibited exceptionally high enantioselectivity for the enantioseparation of *N*-3,5-DNB-α-amino acids, e.g. α =88.5 for (*R*,*S*)-*N*-3,5-DNB-4-methylleucine [61].

In this context, another aspect of high relevance for LC enantiomer separation is the enantiomeric purity of the chiral selector that is immobilized. It has been generally anticipated that an enantiomeric impurity of a chiral selector leads to a reduction of its enantioselectivity. Its first experimental verification by pseudo-enantiomeric cinchonidine – (with QN configurations) and CN-derived CSPs (having QD configurations) with 6′-neopentoxy and 1-adamantyl carbamate residues revealed that for highly enantioselective chiral selectors even trace levels of the opposite enantiomer of the selector immobilized on the stationary phase can lead to a dramatic loss of enantioselectivity [61]. For example, the separation factor of *N*-3,5-DNB-4-methylleucine dropped from 88.5 to 51.4 and to 17 when controlled quantities of 1 and 5% of opposite (pseudo-enantiomeric) selector were admixed, respectively [61]. Hence, racemization in a selector could be a veritable problem, but multiple chiral centers may alleviate the problem which seems to be also a principle of biological systems in which pharmacological receptors are proteinogenic containing multiple chiral centers in homochiral form [61].

To generate even more advanced receptor-like chiral selectors, quinine and quinidine carbamates as well as ethers have been assembled to dimers (e.g. with 1,6-hexamethylene, 1,2-cyclohexylene, 1,3-adamantylene, phthalazine-1,4-diyl spacer [6]) [26, 62]. For specific analytes, such dimeric selectors might present some unique binding features and even allow for multivalent binding. In some instances, this has led to enhanced enantioselectivity, although it is not a general principle [26]. To further corroborate the possibility of cooperative binding by two independent receptor units, cinchonan-type chiral selectors with other receptor-type substructure moieties comprising calixarene [53, 63], crown ether [64, 65], peptoid [66], and cyclodextrins [67, 68] have been synthesized and evaluated for LC enantiomer separation. Calix[4]arene-cinchona carbamate hybrid type receptors covalently immobilized to silica did not exhibit the expected cooperativity effect between the calix[4]arene module and the cinchonan units in terms of chiral recognition [63]. It was concluded that the carbamate linker allows too much conformational flexibility which poses the two units in unfavorable spatial orientation. For this reason, the carbamate linker has been replaced by a rigid urea functionality [53]. By this partial freezing of the unfavorable conformational flexibility in the calixarene-cinchona assembly, spatially better-defined binding domains were expected. Indeed, at least for certain analytes, cooperativity of the two individual receptor units was proposed with calix[4]arene module providing an accessible binding domain with improved analyte shape complementarity [53]. In another study, an achiral benzo-18-crown-6 moiety was linked via carbamate functionality with quinine to introduce a binding site for primary amines which can complex ammonium ions by triple hydrogen bonding [64, 65]. While the enantioselectivity of *N*-derivatized amino acids was typically lower compared to that of a QN-AX benchmark, a few primary amino acids could be resolved on the crown ether-hybrid CSPs. The fact that the attached crown ether was achiral, along with both its distance from the cinchona moiety and flexibility, evidently limited the chiral recognition capability of the crown ether-quinine carbamate hybrid selectors. Quinine and quinidine selectors have been also connected via the carbamate linker with peptoid chains, which have shown some utility as CSPs before, to create new hybrid CSPs [66]. Peptoids are a class of compounds that refer to *N*-substituted glycine oligomers that may be substituted at the glycine-nitrogen, in this particular application, by chiral moieties. The columns were tested in normal-phase and polar organic modes against the individual generic CSPs. The peptoid-quinine carbamate hybrid CSPs benefitted to a certain extent from combined application spectra. Although in the majority of tests the generic CSPs showed higher α-values than the hybrid CSPs, in some instances the latter exhibited even larger enantioselectivity indicating some cooperative effects in chiral recognition. Recently, cyclodextrin derivatives have also been merged with quinine/quinidine [67, 68]. 3,5-Dimethylphenylcarbamoylated β-cyclodextrin was connected via a single 6-imidazolium-propylcarbamate linker to quinine or quinidine moieties [68]. Again, the hybrid CSPs showed mostly lower enantioselectivity than the generic CSPs with one selector unit. However, enantioselectivity could be essentially maintained in spite of low selector surface coverages (e.g. 0.023 and 0.027 $mmol\,g^{-1}$ in the hybrid CSPs, which is around factor 10 lower than on quinine carbamate CSPs). In an advanced version, the quinine carbamate was linked to silica at the C_9-carbamate residue and the C_{11}-position was utilized to attach the cyclodextrin at a single 6-position via a thioether group [67]. In one embodiment, the free hydroxyls of the cyclodextrin moiety were further derivatized with *p*-tolylisocyanate to the corresponding carbamates. It seemed that the two complementary selectors, quinine and functional cyclodextrin, retained their enantioselectivity which widened the enantioseparation applicability spectra of the hybrid CSPs.

The above concepts of merging two chiral selectors into a combined synergistic selector motif that extends the scope of applicability followed the basic idea of the ZWIX selectors, which have been obtained by hybridizing quinine or quinidine carbamate-based chiral anion-exchangers with cyclohexanesulfonic acid-based strong chiral cation-exchangers [19]. Also, structural variants of the ZWIX selectors have been described, including achiral β-alanine and taurine cation-exchanger moieties [19, 69], chiral amino acid and amino sulfonic acid carbamate residues [70–72], 6′-ω-carboxyalkyl) moieties of distinct alkyl length [69], and last but not least ZWIX analogs with 2-aminocyclohexanesulfonic acid with stereochemistry distinct to the ones of the commercial ZWIX CSPs (i.e. 1″*S*,2″*R* and 1″*R*,2″*S* instead of 1″*S*,2″*S* and 1″*R*,2″*R*) [33]. While the variants with achiral cation-exchanger site showed decent enantioselectivity for acidic analytes, their chiral recognition ability for bases and amino acids was fairly modest [19]. This is not surprising for basic analytes, because their chiral recognition takes place at the SCX site and stereogenic centers close to this primary interaction site are considered favorable or even a requirement for effective chiral recognition (see Figure 11.5e and f). Also, for zwitterionic analytes, the SCX moiety is tentatively involved in binding and chirality close to the binding can be favorable (Figure 11.5 g). Like short and long achiral alkyl-chained cation-exchanger carbamate moieties, 6′-carboxyalkyl substituents mainly act as intramolecular counterions, having neither extremely detrimental nor favorable effect on separation factors [69]. The chiral cation-exchange selector units of the commercial ZWIX phases have been carefully selected as a reasonable compromise over a large set of chiral analytes [19, 69].

QN/QD-AX and ZWIX(+)/(−) CSPs have the selector tethered to silica via a thioether group at C_{11} position. It is the result of thiol-ene click chemistry (radical addition reaction) involving the vinyl double bond of the selector and 3-mercaptopropylsilica. This reaction can be performed under mild conditions and in various solvents [15], either thermally initiated or by photoinitiation [73]. It allows control of the surface coverage by charging the reaction mixture with dedicated quantities of chiral selector and 3-mercaptopropyl-silica [74]. Thiol-ene click immobilization has also been pursued at the C_9-carbamate residue, e.g. allylcarbamate [13]. An alternative, yet straightforward chemistry, utilizes a triethoxysilylpropylisocyanate for carbamoylation followed by direct linkage to silica through a silanization reaction [13]. This chemistry, unfortunately, is less flexible in terms of choice of carbamate residues. Hence, Kacprzak and Lindner introduced the 1,3-dipolar Huisgen cycloaddition click reaction to allow for efficient immobilization of CSPs with functional groups that are incompatible with radical addition (thiol click) reactions [75]. Along this line, 3,5-dinitrophenylcarbamoylquinine was immobilized by the 1,3-dipolar Huisgen cycloaddition click reaction via a triazolo-linker. It featured enhancement of the selectivity toward specific π-donating analytes such as aryloxypropionic acids and profens [75]. Kohout et al. compared systematically the effect of different immobilization strategies on chiral recognition by thiol-ene (radical) or an azide-yne (copper(I)-catalyzed) click reaction [76]. Enantioselectivity can be modulated by different immobilization chemistry, either through direct effect of chiral recognition at the selector moiety or through modulation of nonspecific binding at the linker (triazolo- vs. thioether), or by the supporting particle backbone which utilizes different activation in the two approaches, viz. 3-mercaptopropyl-silica vs. 3-azidopropyl silica (see also Ref. [77]). An elegant approach for the deconvolution of the extent of nonspecific binding contributions to overall enantioselectivity was proposed by Maier et al. [74]. Reduction of nonspecific binding to the CSP surface at achiral sites usually reduces enantioselectivity. It can be even reverted, e.g. if nonspecific binding is made repulsive, such as by the introduction of immobilized counterions (which can be achieved e.g. by acidic endcappings) [51]. Through long-ranged electrostatic interactions, these surface-anchored counterions act like the above-mentioned intramolecular counterions in ZWIX phases and significantly reduce retention without significant detrimental effect on enantioselectivity (see Figure 11.7b).

The above-described attachment of cinchona alkaloid-derived selectors by thiol-ene or azide-yne click reactions involves modified-silica reactive intermediates (3-mercapto- or 3-azidopropyl-silica), which are bonded to silica via a hydrosilylation reaction of trialkoxysilanes. It typically leads to a distribution of mono-, bi-, and trifunctional siloxane linkages with a population maximum for the bifunctional bonding as proven by solid-state ^{29}Si cross-polarization/magic angle spinning (CP/MAS) NMR [78]. Recently, it could be demonstrated that this bonding chemistry can be shifted to primarily trifunctional siloxane bonds by use of 3-mercaptopropylsilatrane as the silanization agent instead of 3-mercaptopropyltrimethoxysilane [79]. The silatrane reagent further showed faster reaction kinetics and also led to a more efficient CSP (factor 3 lower C-term). It has been ascribed to a more homogeneous monolayer as compared to heterogeneous oligomeric network structures that result with trialkoxysilanes on the silica surface which suffer from poor stationary phase mass transfer. A highly stabilized surface bonding can be created by the coating of poly(3-mercaptopropyl) methylsiloxane onto vinylsilica and subsequent double thiol-ene click reaction for simultaneous immobilization of the admixed chiral *tert*-butylcarbamoyl quinine selector on the polythiol and crosslinking of the polythiol film to the vinylsilica surface [80–82]. It can be attractive if highly stable CSPs are needed for LC with harsh conditions (e.g. high concentration of phosphate buffers) or to minimize background signal in highly sensitive MS detection.

11.6 Cinchonan-Based UHPLC Column Technologies

Attempts to improve chromatographic efficiency by altering the morphology of chromatographic supports date back to the early days of chromatography. In the 1990s, small-sized silica was already commercially available in the form of nonporous silica particles with diameters of 1.5 and 1.0 μm (MICRA NPS-1.5 and Kovasil NPS-1.0). The former was evaluated as support for quinine carbamate-based CSPs [83]. Up to 70,000 theoretical plates m^{-1} could be reached with such material on standard HPLC equipment using short columns (33 mm length). Due to the low specific surface (~3 $m^2\ g^{-1}$) available in the column (about a factor 100 lower than on 100 Å fully porous silica particles), fast separations within 2 minutes analysis time could be obtained, yet the applicability was limited. For instance, the enantiomers of selected analytes, e.g. 3,5-DNB-leucine, could be baseline resolved within 1.5 minutes. The extremely high efficiencies expected from the small particle diameter could not be realized, among other things, due to lack of adequate UHPLC equipment at that time.

In the early 2000s, monolithic silica columns with 2 μm macropores and 120 nm mesopores were introduced (Chromolith technology). Its adaptation for chiral separation by chemically bonding the *tert*-butylcarbamoylquinine chiral anion-exchanger selector as a stationary phase by a flow through approach onto the surface of 3-mercaptopropyl-modified silica monolith resulted in useful columns with improved efficiency compared to the corresponding 5 μm fully porous silica particle CSP [84]. H/u curves were significantly flatter for the monolith than for the 5 μm particle column, overall comparable to that of 3 μm particles but with much better permeability. A variety of chiral probes could be resolved with similar enantioselectivity to that of the 5 μm particle benchmark reaching 35–90,000 plates m^{-1} (compared to 30–40,000 plates m^{-1} for 5 μm QN-AX CSP). Due to the high permeability of the monolithic bed, 6 monolithic columns each of 10 cm length could be coupled in-line by zero-dead volume connectors providing a 60-cm-long column. With this column, the back pressure was still only 4.5 MPa. A plate count of 30,000 per column could be thus accomplished which is exceptional for a chiral column and could be useful for challenging separations, of course at the expense of analysis times on such a long column.

With the introduction of sub-2 μm particle columns and UHPLC in 2004 as well as the renaissance of SPP columns in 2006 [85], LC entered into a new era. It took a while until the new chromatographic support technology also found its way into the chiral separation field [86]. Gasparini and coworkers were the first to adopt the sub-2 μm particle technology for Pirkle-type chiral columns [87]. The first ever enantiomer separation on a SPP column was reported in 2011 by Lindner and coworkers [88]. Core-shell silica (Halo) with particle diameter of 2.7 μm, composed of 1.7 μm impermeable nonporous core and 0.5 μm porous shell with mean pore diameter of 90 Å, was modified with the chemistry of the QD-AX CSP. Twenty proteinogenic amino acids were resolved as *N*-(3-ferrocenylpropionyl) derivatives on *tert*-butyl quinidine carbamate-based core-shell particle column within 10 minutes by LC-ESI-MS/MS which indicated the great potential of SPP CSPs for fast enantiomer separations.

A more detailed investigation into the performance and speed of the quinine carbamate-based core-shell particle column was then performed by Armstrong and coworkers [89]. In this work, *tert*-butylcarbamoylquinine was immobilized onto 3-mercaptopropyl-modified 2.7 μm superficially porous silica (Poroshell 120 Å, 1.7 μm impermeable core with 0.5 porous silica shell, 120 $m^2\,g^{-1}$; selector coverage 1.61 $\mu mol\,m^{-2}$) (QShell). A hydrosilylated analog with direct silica linkage at the vinyl group yielded factor 3 lower surface coverage. A theoretical plate number of 136,000 m^{-1} (around factor 3 more than for commercial 5 μm benchmark) and reduced plate height of 2.6 could be achieved for dichloroprop using RP mode (80/20 methanol/100 mM ammonium acetate; pH_a 6.0) with a UHPLC instrument [89]. Similar resolutions could be obtained on a 5 cm long SPP column as for a 15 cm long 5 μm QN-AX column at significantly lower run times. H/u curves revealed lower C-term contribution to peak broadening for the 2.7 μm core-shell column as compared to 5 μm fully porous QN-AX. A direct comparison of HPLC and SFC van Deemter plots showed that the H/u curves are slightly shifted to higher velocities but their slopes (indicative for the mass-transfer resistance i.e. the C-term) were fairly comparable, meaning no gain by SFC, due to the high methanol percentage (40%) in the scCO2 which resulted in subcritical conditions [89]. In addition, also the slow adsorption/desorption kinetics of ion-exchange processes, which is a significant contribution to the C-term, is expected to be essentially the same in SFC and HPLC. For this reason, the expected improvement in mass transfer by switching from HPLC to SFC was not realized.

To address the question of which modern particle platform provides the best performance in terms of efficiency and speed, QN-AX [90] and ZWIX(+) [91] selectors were immobilized on different particles including various fully porous particles (FPPs) of distinct diameters (1.7, 3, and 5 μm) and pore size (120 and 200 Å) as well as SPP of different size (2 and 2.7 μm) and pore size (90 and 160 Å). The key findings of these studies are summarized in Figure 11.10. Both new particle types (1.7 μm FPP and 2.7 μm SPP) outperformed the 5 μm FPP QN-AX benchmark (N/m ~ 30,000; h_{min} ~ 3.7) in terms of speed and efficiency (Figure 11.10a), with wider pore

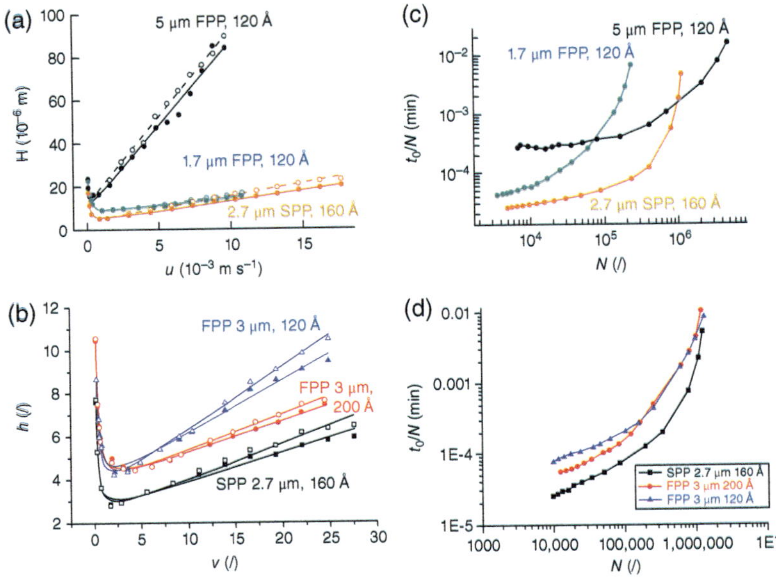

Figure 11.10 Comparison of chromatographic performance of different particle morphologies and pore size of QN-AX and ZWIX phases. (a) H/u curves of 1.7 μm fully porous particle (FPP) column (120 Å) and 2.7 μm superficially porous particle (SPP) column (160 Å) with 5 μm FPP benchmark (120 Å). (b) H/u curves of 2.7 μm SPP column (160 Å) and 3 μm FPP column (200 Å) with 3 μm FPP benchmark (120 Å). (c and d) Corresponding kinetic plots. Analytes: (a and c) FMOC-phenylalanine and (b and d) 2-amino-2-phenylbutyric acid. *Sources:* (a and c) Reproduced from Schmitt et al. [90]/with permission of Elsevier. (b and d) Reproduced from Geibel et al. [91]/with permission of Elsevier.

materials (160 or 200 Å) being advantageous (over 90 or 120 Å) [90]. As can be seen in Figure 11.10a, the 160 Å SPP QN-AX CSP with 2.7 μm particle diameter (1.7 μm nonpermeable core and 0.5 μm porous shell) has a very similar mass transfer coefficient (C-term), as indicated by the steepness of the H/u curves, compared to the 120 Å 1.7 μm FPP QN-AX CSP [90]. The SPP CSP provides even lower theoretical plate heights at low flow rates and equal plate heights at high velocity. Due to its higher permeability resulting from the larger particle diameter, the flow range in which it can be adopted is significantly extended. Hence, it was concluded that the 160 Å 2.7 μm SPP (N/m ~ 160,000; h_{min} ~ 1.7) outperforms the 120 Å 1.7 μm FPP column (N/m ~ 104,000; h_{min} ~ 4.3). A further reduction of the particle diameter to 2.0 μm 160 Å SPP QN-AX CSP did not bring a further advantage (maybe due to suboptimal packing and/or loss by extra-column effects) [90]. Hence, the 2.7 μm 160 Å SPP CSP appeared to be a good compromise. The corresponding 90 Å material showed a worse mass transfer [90]. Probably, the surface bonding leads to further reduction of the pore size which seems then too narrow having a negative effect on pore diffusion. In a follow-up study, the SPP technology was also adopted for the ZWIX(+) phase [91]. H/u curves of the 2.7 µm 160 Å SPP ZWIX CSP (N/m ~ 145,000; h_{min} ~ 2.6) were compared to that of 3 µm 120 Å FPP ZWIX CSP benchmark (N/m ~ 75,000; h_{min} ~ 4.2) and a corresponding 200 Å wide-pore material (Figure 11.10b). This work confirms that wider pores (200 Å instead of 120 Å) are favorable for the mass transfer (as indicated by the flatter H/u curves of the 200 vs. 120 Å FPP in Figure 11.10b) probably due to less restricted pore diffusion. Furthermore, most striking in this comparison is the significantly lower A-term of the SPP CSP in spite of comparable particle diameter (~3 µm). It is known that core-shell particles have a narrower particle diameter distribution which may lead to more homogenous packings [85].

The individual contributions to peak broadening were deconvoluted for the different particle morphologies by use of the second-order effective medium theory [90, 91] in accordance to Eq. (11.5)

$$h = h_a + h_b + h_{c,m} + h_{c,s} + h_{c,ads} \tag{11.5}$$

wherein h, h_a, h_b, $h_{c,m}$, $h_{c,s}$, and $h_{c,ads}$ are the total reduced plate height ($h = H/d_p$), the eddy diffusion (A-term), longitudinal diffusion (B-term), mobile phase and stationary phase mass transfer resistance (C-term) as well as mass transfer resistance term (C-term) due to slow adsorption–desorption, all in reduced coordinates [91]. It was shown that the biggest contribution to peak dispersion in the 5 µm 120 Å FPP QN-AX benchmark column originates from the slow adsorption–desorption kinetics (i.e. the exchange kinetics on the CSP surface due to slow ion-exchange process) followed by the eddy diffusion term [90]. A more thorough deconvolution for the ZWIX CSPs revealed that the better performance of the 2.7 µm 160 Å SPP column is due to (i) lower b-term, (ii) lower c,s-term, (iii) lower c,ads-term, and (iv) favorable a-term, while c,m-term was comparable for the FPP columns (having different pore size) with similar particle diameter (3 µm) [91]. In the ZWIX core-shell phase, the largest contribution seems to arise from the eddy diffusion followed by the slow adsorption–desorption kinetics.

SPP CSPs can be regarded as the best compromise between performance, speed, and backpressure [85]. As can be seen from Figure 11.10c and d, for both QN-AX and ZWIX the SPP particle technology provided the lowest x-axis asymptote indicative of faster separations at comparable efficiencies (plate numbers N) and the same pressure drop [90, 91]. As a matter of consequence, SPP CSPs have been promoted for fast (sub-minute [90, 91]) and superfast (sub-second [89, 92]) separations. For instance, the *tert*-butylquinine carbamate core-shell columns (5 cm long) enable enantiomer separations in the sub-minute time scale [47]. Using even shorter core-shell columns (e.g. with 1 cm length), analysis times of <1.5 seconds could be achieved by UHPLC at high flow rates, with optimized UHPLC instrumentation having reduced extra-column volume and enhanced detector sampling rates [89, 92]. Such fast enantiomer separations on SPP CSPs open up new avenues, e.g. their application as second dimension chiral separation in enantioselective 2D-LC [93–95].

11.7 Applications

An extensive overview of the applicability of cinchona alkaloid-derived CSPs along with optimization strategies of experimental conditions has been published recently [25]. Herein, only selected examples of pharmaceutical and biomedical applications are briefly described to give an idea of the scope of application of QN-AX and QD-AX as well as ZWIX(+) and ZWIX(−) columns.

11.7.1 Pharmaceutical and Biotechnological Applications

A classical pharmaceutical application of cinchonan-based CSPs is for enantiomeric purity determination of drug substances, intermediates, and drug products of chiral pharmaceuticals. In one recently reported example, ZWIX(+) was used for the quantification of elagolix enantiomers in a drug product, marketed as Orilissa [96]. Elagolix is a gonadotropin-releasing hormone (GnRH) antagonist, which is therapeutically used to treat pain in premenopausal women accompanying endometriosis. The IUPAC name is 4-[[(1*R*)-2-[5-(2-fluoro-3-methoxyphenyl)-3-[[2-fluoro-6-(trifluoromethyl) phenyl]methyl]-4methyl-2,6-dioxopyrimidin-1-yl]-1-phenylethyl]amino]butanoic acid, a zwitterionic compound. Only the *R*-enantiomer has the therapeutic effect. Thus, the *S*-enantiomer has to be considered as an enantiomeric impurity and was controlled in bulk active pharmaceutical ingredient and finished product. A large number of CSPs including 16 polysaccharide CSPs (coated and immobilized) under polar organic, reversed-phase and normal-phase modes were evaluated; however, Chiralpak ZWIX(+) showed the highest enantioselectivity with favorable elution of the enantiomeric impurity before the main peak. The optimized method could then be utilized for its intended purpose in pharmaceutical quality control. In a similar application, the *R*-enantiomer impurity of pregabalin,

(*S*)-3-(aminomethyl)-5-methylhexanoic acid), a γ-amino acid derivative which is an antiepileptic and analgesic drug could be analyzed on ZWIX(+) as the first eluting enantiomer [97]. These examples clearly demonstrate the superior performance of ZWIX CSPs over many other CSPs for chiral zwitterionic drugs.

N-Derivatized amino acids are important building blocks in peptide synthesis. The most widespread protocols for peptide synthesis nowadays are based on 9-fluorenylmethoxycarbonyl (FMOC)-amino acid building blocks. A high chemical and enantiomeric purity of the building blocks is of utmost importance and must be rigorously controlled. It is well known that higher purity leads to purer products, and the subsequent chromatographic purification (usually performed by RP HPLC) poses less challenges if the purity of the starting materials is better. The longer the peptide the more critical are impurities in the building blocks as the purity of the product decreases with each coupling step. D-Amino acids in FMOC building blocks should be avoided, and their presence must be controlled as they lead to diastereomeric peptide impurities which not only lead to rapid decrease in the target peptide yield but may also be difficult to remove during purification. It was demonstrated that the enantiomers of FMOC-amino acids used as building blocks for peptide synthesis can be all separated on Chiralpak QN/QD-AX and ZWIX(+) and ZWIX(−) [3, 39, 93, 98, 99]. The former chiral anion-exchangers, however, gave much better resolutions and are preferred, especially the QN-AX on which the enantiomeric impurities were eluted first in front of the main L-enantiomer. However, the reversed elution orders on QD-AX CSPs can be quite useful for validation of results or for preparative isolation of specialty amino acids that are synthesized as racemates (the first eluted enantiomer can be produced with higher enantiomeric purity [100, 101]). Also on ZWIX(+) the D-enantiomers elute first [99] with reversed order on ZWIX(−) [3]. It was demonstrated that even traces of enantiomeric impurity down to 0.01% related to the main L-enantiomer can be analyzed [98]. Fluorescence detection could further enhance detection sensitivity and specificity. A shortcoming is the limited enantioselectivity for secondary amino acids like FMOC-Pro (proline). An interesting alternative option could be SFC which has shown enhanced resolution for many FMOC-amino acids [3]. Other peptide building blocks like *tert*-butyloxycarbonyl (BOC)- and benzyloxycarbonyl (Z)-amino acids can be resolved as well on cinchona alkaloid CSPs.

Not only is the stereointegrity control of synthesized peptides a requirement in the field of quality control of therapeutic peptides but also the structural elucidation of the constituent amino acids of natural peptide products. From an analytical perspective, it involves the hydrolytic cleavage of the peptide by hydrochloric acid followed by the enantiomer separation of the released amino acids. In a number of studies, cinchona alkaloid CSPs were employed in such applications for elucidation of the amino acid configurations of peptides from biotechnological production. For example, the cyclic lipopeptide poaeamide from *Pseudomonas poae* contained several D-amino acids, which could be identified by precolumn derivatization of the hydrolysate with Sanger's reagent (2,4-dinitrofluorobenzene) and analysis on the ZWIX phases [102]. The results were validated by GC–MS with Chirasil Val. The fatty acid side chain of this lipopeptide contained a 3-hydroxydecanoic acid, the absolute configuration of which could be determined on the ZWIX phases as well [54, 102]. This work already indicated the challenges and problems that arise from isobaric analogs of threonine (Thr) (allo-Thr, homo-serine (homo-Ser)) and leucine (Leu)/isoleucine (Ile) (allo-Ile) even when tandem-MS detection is used. Thus, for similar applications, multiple heart-cutting enantioselective [94, 95] and full comprehensive 2D-LC methods [93] were developed for absolute configuration determination in bacitracin and gramicidin as well as other non-ribosomal peptides. In one study, the amino acids from the peptide hydrolysate were derivatized with Sanger's reagent, leading to 2,4-dinitrophenyl-derivatized (DNP) amino acids which were separated on an achiral sub-2 μm C18 RP column in the first dimension, followed by enantioselective analysis of the transferred multiple heart cuts on a core-shell *tert*-butylcarbamoyl quinine-based chiral weak anion-exchanger column (2.7 μm Halo 90 Å as supporting particles) [94]. Both primary and secondary amino acids could be well resolved into enantiomers on QN-AX and QD-AX. The interface consisted of an 8-port/2-position valve connected to two-loop decks each equipped with six 40 μl loops for fraction storage. The second dimension run could be finished within 210 seconds. This way the DNP-derivatives of 25 amino acids (20 proteinogenic plus β-alanine (β-Ala), allo-Thr, homo-Ser, allo-Ile, and ornithine (Orn)) were separated with a total run time of 130 minutes (first and second dimension together) transferring 29 fractions into the second dimension in a time-based manner. The practical applicability was documented by the determination of the absolute configurations of the amino acids in the non-ribosomal therapeutic peptides bacitracin and gramicidin after their hydrolysis [94]. Recently, this multiple heart-cutting 2D-LC platform could be further advanced through the acceleration of the second dimension separation by using a tandem column consisting of a 5 cm long 160 Å 2.7 μm core-shell *tert*-butylcarbamoylquinine column coupled to a 2 cm long 3 μm fully porous ZWIX(+) precolumn for enantioselective separation of all proteinogenic amino acids derivatized with AccQ reagent in 68 seconds [95]. It allowed to shorten the total run time to

45 minutes which, together with MS detection, greatly improved the practical applicability as a generic enantioselective amino acid analysis platform.

For impurity profiling, it may be desirable to control the stereoisomeric purity also on the intact peptide level. While it is not possible to obtain and separate all possible stereoisomers of such complex peptides mentioned earlier, it is readily possible to fully separate di- and tripeptide stereoisomers (both enantiomers and diastereomers), on cinchona alkaloid-derived CSPs ether in free [23, 24] or *N*-derivatized forms [103, 104]. The ZWIX CSPs have great capability to resolve not only the enantiomers but also diastereomers without derivatization. Optimized conditions for peptide stereoisomer separations have been reported by Zhang et al. [20]. In this application, it is remarkable that the ZWIX phases also have sequence selectivity. For example, it was shown for a tripeptide (Gly-Leu-Ala) with two chiral centers that all four-possible stereoisomers could be fully resolved as well as those of the corresponding retro-peptide stereoisomer Ala-Leu-Gly (Figure 11.11).

However, the full resolution of the stereoisomers of pharmaceuticals with multiple chiral centers like peptides is often hampered by the limited peak capacity of chiral columns, as compared to RP columns, as well as their insufficient chemoselectivity. For this reason, it may happen that enantiomers are well separated but diastereomers coelute in a complex mixture of all stereoisomers. Two-dimensional LC approaches with an achiral RP column for diastereomer separation and a chiral column for enantiomer separation can readily deal with such problems. For example, an achiral-chiral heart-cutting LC–MS method with direct transfer was reported for peptide stereoisomer separation including enantiomers [103]. Diastereomers of di- to tetrapeptides were separated on an RP column and enantiomeric separation on a quinine-based column finished off the resolution of the complex mixtures with analytes having multiple stereogenic centers.

Enantioselective amino acid analysis is also of relevance in food chemistry. For example, rapid enantioselective amino acid analysis of a complex L-amino acid supplement (containing L-His (histidine), L-Ile, L-Leu, L-Lys (lysine), L-Met (methionine), L-Phe (phenylalanine), L-Thr, L-Trp (tryptophan), and L-Val (valine)) could be achieved within 2.5 minutes implementing AccQ-tagging and UHPLC separation on a core-shell QN-AX 2.7 μm, 160 Å (50×3 mm ID) column with ESI-QTOF-MS/MS detection in data-dependent acquisition mode [105]. A uniformly-^{13}C,^{15}N-labeled amino acid internal standard mixture, and high-resolution MS detection allowed the unequivocal identification and assignment of the L-amino acids. Significant D-amino acid quantities could be found for Thr (0.71%) and Val (1.32%). Their identity could be verified by the racemized internal standard.

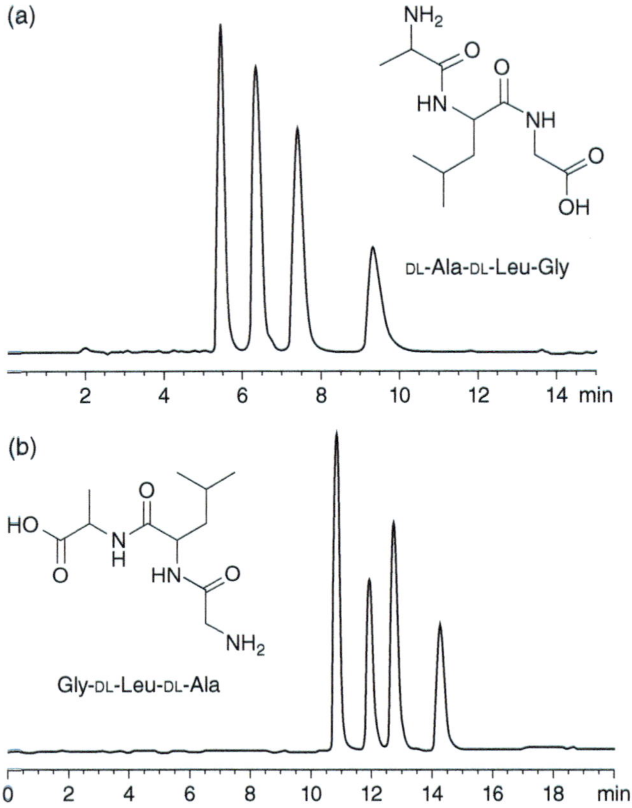

Figure 11.11 Peptide stereoisomer separation of tripeptides with reversed sequence on ZWIX. Chromatographic conditions: Column, Chiralpak ZWIX(+) 250 × 4 mm id for (a), ZWIX(−) 150 × 3 mm id for (b); mobile phase, methanol/water (98 : 2; v/v) (a); methanol/tetrahydrofuran/water (49 : 49 : 2; v/v/v) (b), detection, evaporative light scattering detector (ELSD). *Source:* Reproduced from Zhang et al. [57]/with permission of John Wiley & Sons.

To deal with a more complex real sample, namely honey samples collected from different regions of Argentina, a chiral × achiral 2D-LC with a quinine-bonded column with free C_9-hydroxyl group (bonded to 3-mercaptopropyl-silica) in the first dimension and a short 3 μm C18 or phenylbutyl stationary phase in the second was established [106]. The method involved precolumn derivatization with Sanger's reagent. While multiple heart-cutting 2D-LC methods target specific analytes, this fully comprehensive enantioselective assay is truly untargeted. The challenge here is that the second dimension has to be extremely fast (typically 1 minutes or less). This is hard to accomplish with most chiral columns.

The reversed order chiral × achiral is easier to realize. In the mentioned application, the 8-port/2-position interface was equipped with two 20 μl sampling loops that were alternatingly used for sample collection and injection; while one loop was filled, the other one was analyzed. While the analysis suffered from lack of MS detection, it can be regarded as a successful proof of principle of chiral × achiral 2D-LC. Also chiral × chiral 2D-LC has been realized for amino acid analysis, either with complementary quinine carbamate CSPs having distinct carbamate residues (*tert*-butyl vs. 2,6-diisopropylphenyl) [58] or with CSPs with complementary configurations, viz. core-shell quinine and quinidine carbamate CSPs, in first and second dimensions, respectively (see Figure 11.12) [93]. The latter approach allowed to realize specific peak patterns in which the amino acid configuration can be simply recognized by their positioning above or below the diagonal parity line of the 2D-contour plots, which indicates equal retention in both dimensions, such as obtained for achiral components (achiral amino acids and reagent peaks).

The QN-AX CSP also has the capability to separate the enantiomeric impurity of L-thyroxine (L-T4) besides the enantiomers of the structurally related triiodothyronine (T3) [46]. These thyroid hormones are essential for normal growth and development of organs as well as of the central nervous system in infants. They need to be supplemented in case of endogenous deficiencies. L-T4 tablets of four suppliers have been analyzed on QN-AX and the D-T4 levels typically ranged between 0.1 and 1.6%, for one supplier even 7.3%; also D- and L-T3 could be detected and quantified as impurities simultaneously.

Cinchona alkaloid CSPs have also been applied in SFC mode with up to 40 or even 50% methanol in $scCO_2$ with buffer additives (acetic acid, formic acid, ammonia, etc.) which results in subcritical conditions. For this reason, the expected gain in efficiency compared to LC could not be realized but similar performance was typically achieved. QN-AX and QD-AX CSPs were employed for SFC enantiomer separation of chiral sulfonates [2]. ZWIX phases showed great utility for SFC enantiomer separation in anion-exchange mode [43], cation-exchange mode [107, 108], and also for zwitterionic analytes [109]. Due to the intramolecular counterion effect and lower overall retentivity, there is more flexibility available for tuning conditions and optimizing enantioresolutions. FMOC-amino acids were separated in anion-exchange mode by SFC on ZWIX(+) and ZWIX(−) with up to 40% methanol as protic cosolvent [43]. Basic chiral compounds, e.g. indole-type basic compounds [107] and β-carboline, tetrahydroisoquinoline and benzazepine analogs [108], could be resolved in cation-exchange mode on ZWIX phases by SFC. SFC enantiomer separations of zwitterionic analytes (alanine, glutamine) were reported as well [109]. The SFC elution mode may be particularly appealing for preparative applications.

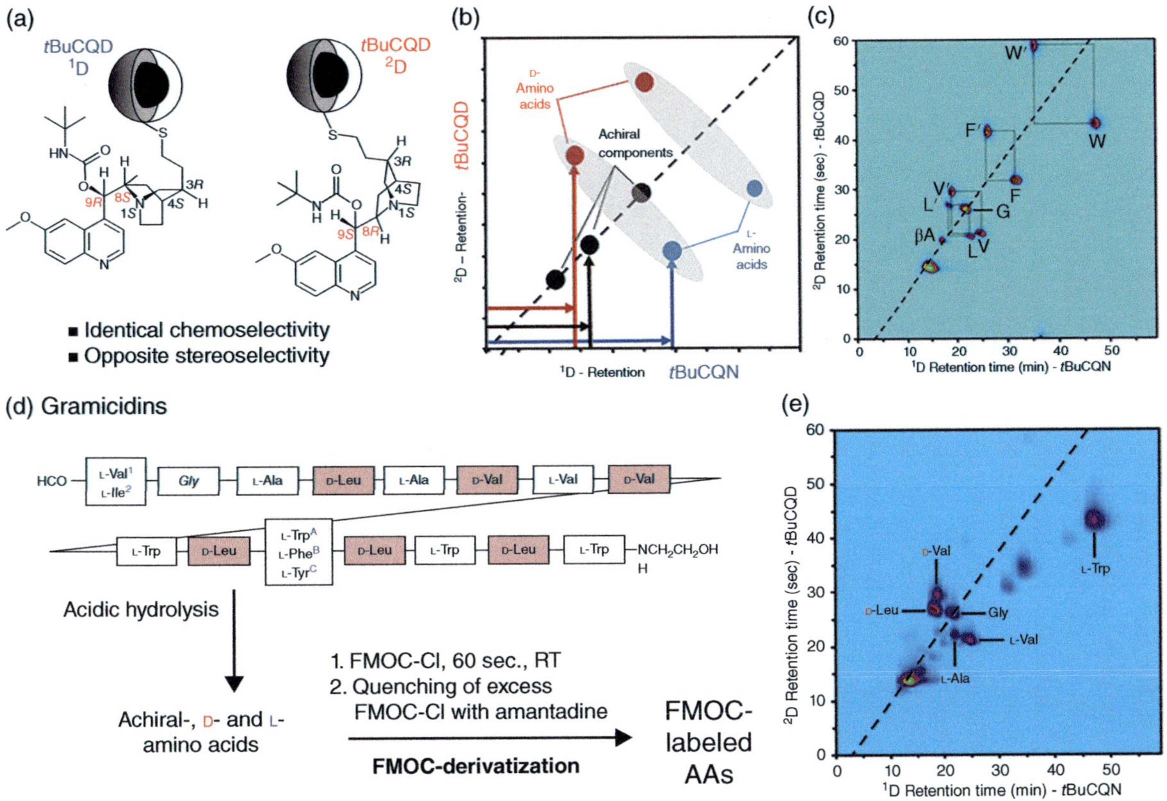

Figure 11.12 Inverted chiral column approach by chiral x chiral 2D-LC. (a) *t*BuCQN (^{1}D) and *t*BuCQD (^{2}D) SPP columns are chemically identical, but have opposite configurations at C_8 and C_9, thus opposite enantiomer affinity and reversed elution order. (b) Achiral amino acids and reagent peaks show similar retention on ^{1}D and ^{2}D columns due their equal chemoselectivity; therefore, they are lying on a parity line in the 2D-contour plot. FMOC-L-amino acids are stronger retained on the ^{1}D (*t*BuCQN) column, therefore on the contour plot below the diagonal parity line, and FMOC-D-amino acids on the ^{2}D (*t*BuCQD) column, therefore above the diagonal parity line. (c) Experimental realization of the concept from (b). (d) Structure of gramicidin and sample preparation protocol for peptide hydrolysis and amino acid derivatization. (e) Determination of the amino acid configurations in the gramicidin peptide hydrolysate: D-Leu and D-Val could be detected (above diagonal line) while other amino acids had L-configuration, as expected. L, leucine (Leu); V, valine (Val); F, phenylalanine (Phe); W, Tryptophan (Trp); βA, β-alanine (β-Ala); G, Glycine (Gly); βA and G are achiral; dashed letters in (c) indicate D-configuration. *Sources:* (a–c, e) Reproduced from Woiwode et al. [93]/with permission of American Chemical Society. (d) Adapted from Woiwode et al. [93].

11.7.2 Biomedical Applications

In a large-scale population pharmacokinetics study of more than 2000 blood samples of pregnant African females who received racemic mefloquine for chemoprophylaxis against malaria, cinchona-based chiral columns demonstrated their suitability in bioanalytical applications [55]. Mefloquine is a racemic mixture of (+)-(11*S*,12*R*)- and (−)-(11*R*,12*S*)-enantiomers. Although it has been introduced for treatment and prophylaxis of malaria almost 30 years ago, information about the stereoselectivity of biotransformation was still contradictory, probably due to the only recently clarified absolute configurations. It is known that the cytochrome P450 3A subfamily metabolizes mefloquine to 4-carboxymefloquine, which shows no antimalarial activity in vitro. To shed light on this issue, dried blood spot sampling was combined with enantioselective separation on a quinidine-based zwitterionic CSP column, Chiralpak ZWIX(−) (3.0 × 150 mm, 3 μm), using isocratic elution with methanol/acetonitrile/water (49 : 49 : 2; v/v) containing 25 mM formic acid and 12.5 mM ammonium formate. Enantiomers and primary metabolites of mefloquine could be simultaneously analyzed within 10 minutes using tandem mass spectrometry in selected reaction monitoring (SRM) acquisition mode with high sensitivity and excellent robustness. The entire study was concluded on a single analytical column which tolerated the injection of more than 2000 clinical samples (solvent extract of dried blood spots) along with calibrants and quality controls [55].

In a number of papers, the potential of QN-AX, QD-AX, ZWIX(+), and ZWIX(−) for enantioselective disease biomarker studies of short chiral hydroxyalkanoic acids has been indicated (Figure 11.13) [17, 29, 110]. This includes chiral 2-hydroxycarboxylic acids that are associated with inherited metabolic diseases, such as lactic acid, 2-hydroxy-3-methylbutyric acid, 2-hydroxy-3-methylpentanoic acid, 2-hydroxyisocaproic acid, 2-hydroxyadipic acid, and also 2-hydroxyglutaric acid that are monitored in organic acidurias. 2-Hydroxyisocaproic acid is of interest in Zellweger syndrome, and 2-hydroxycaproic acid was tested as marker of *Nocardia* infections in cerebrospinal fluid of patients with meningitis. 2-Hydroxybutyric acid in the plasma has been reported as a suitable marker for early stage type II diabetes and 2-isopropylmalic acid for asthma. Among all of them, 2-hydroxyglutaric acid (2-HG) has reached significant popularity in recent years. D-2-HG is an oncometabolite that can serve as a marker for tumors with mutant isocitrate dehydrogenase 1/2. These hydroxy acid compounds have some tendency to chelate with metal ions and may have a poor peak shape for this reason. To alleviate this problem, ethylenediaminetetraacetic acid (EDTA) is sometimes added to the mobile phase as a metal chelator. Although the hyphenation with ESI-MS has not been reported yet, the employed polar organic or

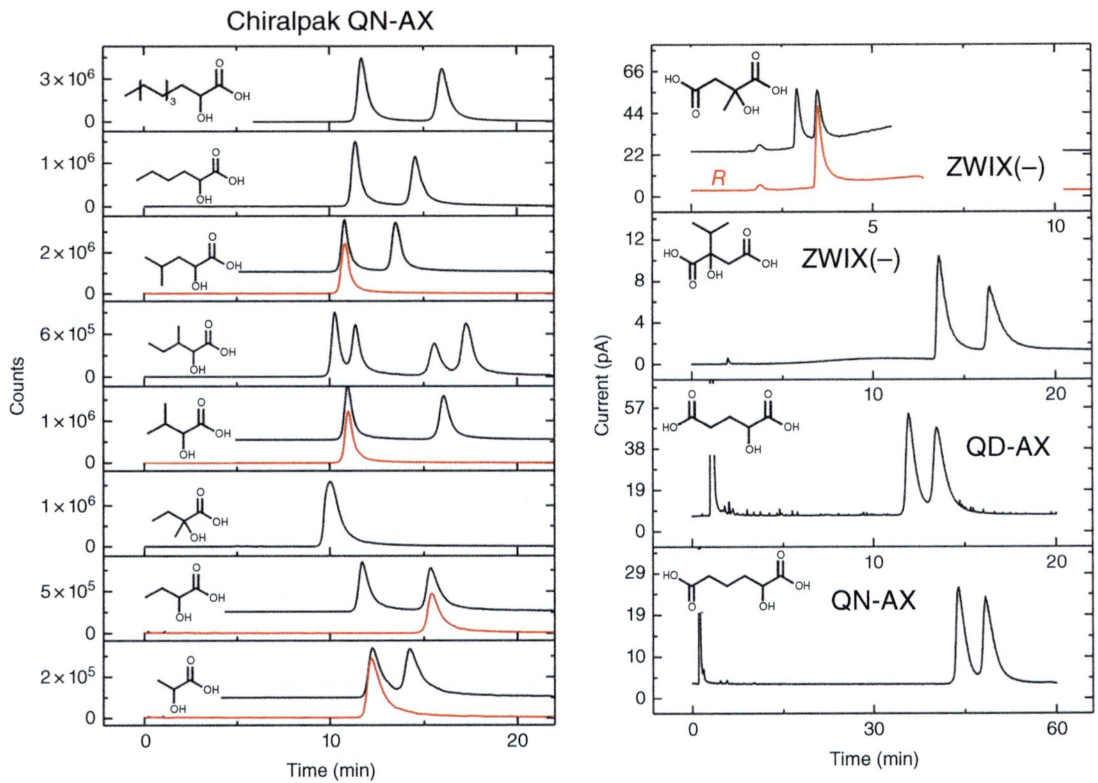

Figure 11.13 Hydroxy alkanoic acid enantiomer separations [17]. Chromatographic conditions: left panel: Chiralpak QN-AX column; mobile phase acetonitrile/methanol (50 : 50; v/v) containing 15 mM formic acid and 2.5 mM ammonia, 10 °C, 1.0 ml min^{-1}; right panel (from top to bottom): Chiralpak ZWIX(–), acetonitrile/methanol (50 : 50; v/v) containing 30 mM formic acid, 10 °C, 1.0 ml min^{-1}; QD-AX, methanol 100% containing 40 mM formic acid and 5 mM ammonia, at 1.0 ml min^{-1} and 10 °C; QN-AX, acetonitrile/methanol (50 : 50; v/v) containing 15 mM formic acid and 2.5 mM ammonia, 10 °C, 1.0 ml min^{-1}. *Source:* Reproduced from Calderon et al. [17]/with permission of John Wiley & Sons.

reversed-phase mobile phase conditions are ESI compatible and hence applicability for bioanalysis can be envisioned.

Several studies addressed the analysis of D-amino acids in biological fluids and tissues as potential biomarkers of diseases [88, 111–114]. A sensitive and specific enantioselective LC-mass spectrometric assay based on tandem MS with selective reaction monitoring acquisition and precolumn derivatization was reported for the simultaneous analysis of D-alanine (Ala), D-aspartic acid (Asp), and D-serine (Ser) in rat plasma and tissues [112]. The biological roles of these amino acids are partly known. D-Ser and D-Ala act at the *N*-methyl-D-aspartic acid (NMDA) receptor and D-Asp can be found in various endocrine tissues. Two distinct tagging strategies were evaluated in this work, viz. use of precolumn derivatization reagents 6-aminoquinolyl-*N*-hydroxysuccinimidylcarbamate (AccQ) or *p*-*N,N,N*-trimethylammonioanilyl *N′*-hydroxysuccinimidyl carbamate iodide (TAHS). Enantiomer separation was accomplished by a ZWIX(+) column within 20 minutes. Both mass tags introduce favorable groups both for chiral recognition (carbamoyl NH as H-donor, C=O as H-acceptor, aromatic groups for π–π-interaction with quinoline of the chiral selector) as well as chemical functionalities with either high proton affinity (AccQ) or permanent charge (TAHS) for enhancing electrospray ionization efficiency and MS detection sensitivity in positive ESI mode. It has been shown that AccQ-tagging, that leads to anionic derivatives with commonly employed eluents, allows the enantiomer separation of both primary and secondary, neutral, aromatic, polar, and apolar amino acids on ZWIX(+), with D-enantiomers eluting first except for Pro (Figure 11.14) [115, 116]. In the work by Hamase and coworkers, it was demonstrated that TAHS-derivatization, which yields zwitterionic derivatives, affords even higher resolutions for the three targeted amino acids: Ala, Ser, and Asp [112]. In terms of sensitivity, the two reagents were largely comparable with similar lower limits of quantification (LLOQs) (ca. 0.05–0.1 nmol ml^{-1}). TAHS derivatization was then applied due to resulting higher resolutions for the determination of the D-amino acid levels (all first eluted) besides the L-amino acids in plasma and various tissues of three rats. D-Forms could be detected in all samples.

The L- and D-amino acid levels in the brain of wild-type and mutant mice with D-amino acid oxidase (DAO) knockout were analyzed by Armstrong and coworkers [113]. In this work, a quinine carbamate core-shell column (2.7 μm, 50×4.6 mm ID) was employed in gradient elution mode with 100 mM ammonium formate/methanol (pH 6 and 5, respectively, in channel A and B). Tandem MS with a triple quadrupole instrument in SRM mode allowed sensitive detection of the AccQ-derivatized amino acids including Arg (arginine), His, Pro, and Ile/Leu isobaric amino acids in

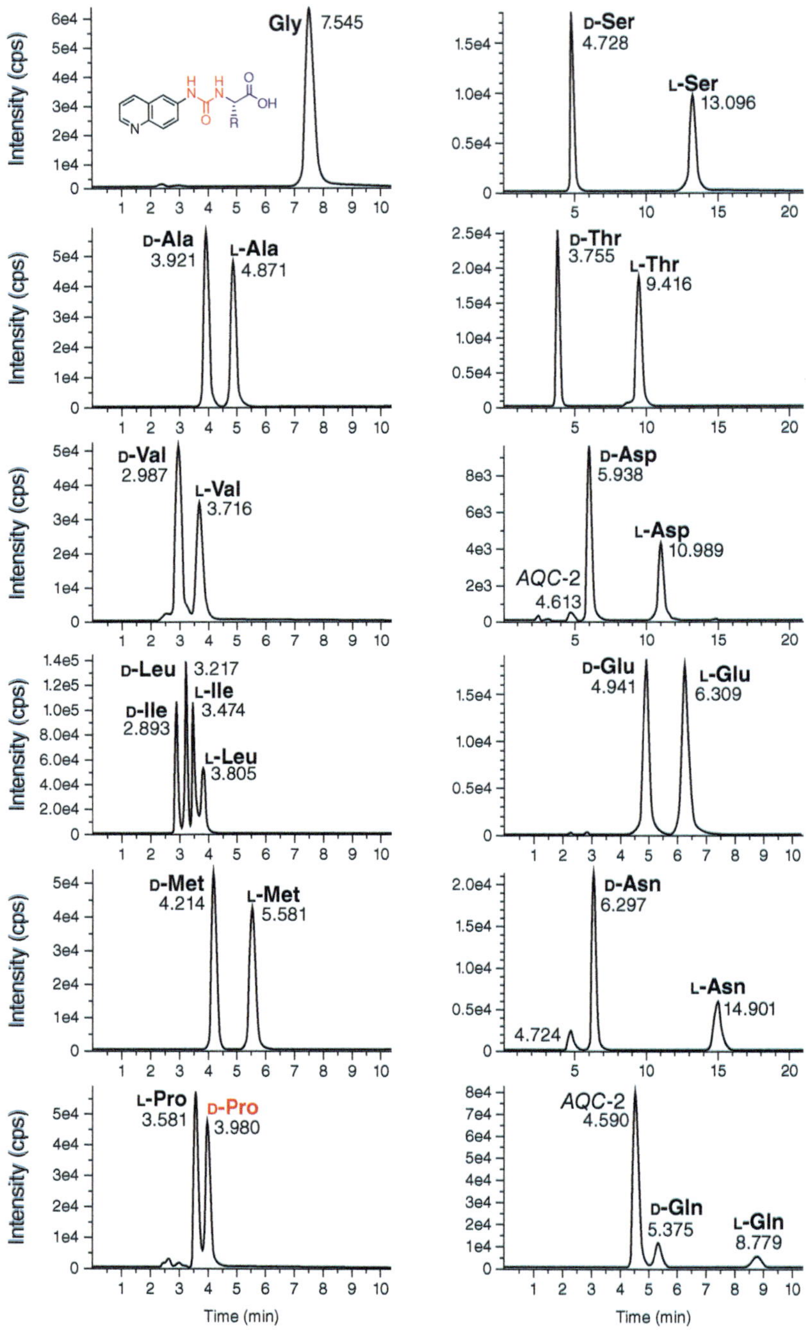

Figure 11.14 Stereoselective separation of the AccQ-derivatized amino acid mixture containing DL-amino acids (0.025 mM) and U-$^{13}C^{15}N$-L-amino acids (0.0125 mM; without Trp, Gln, and Asn) on Chiralpak ZWIX(+) and ZWIX(−) column (150 × 4 mm, 3 μm) using gradient elution combined with HPLC-ESI-QTOF-MS detection in positive ionization mode. *Source:* Reproduced from Horak and Lämmerhofer [115]/with permission of Elsevier.

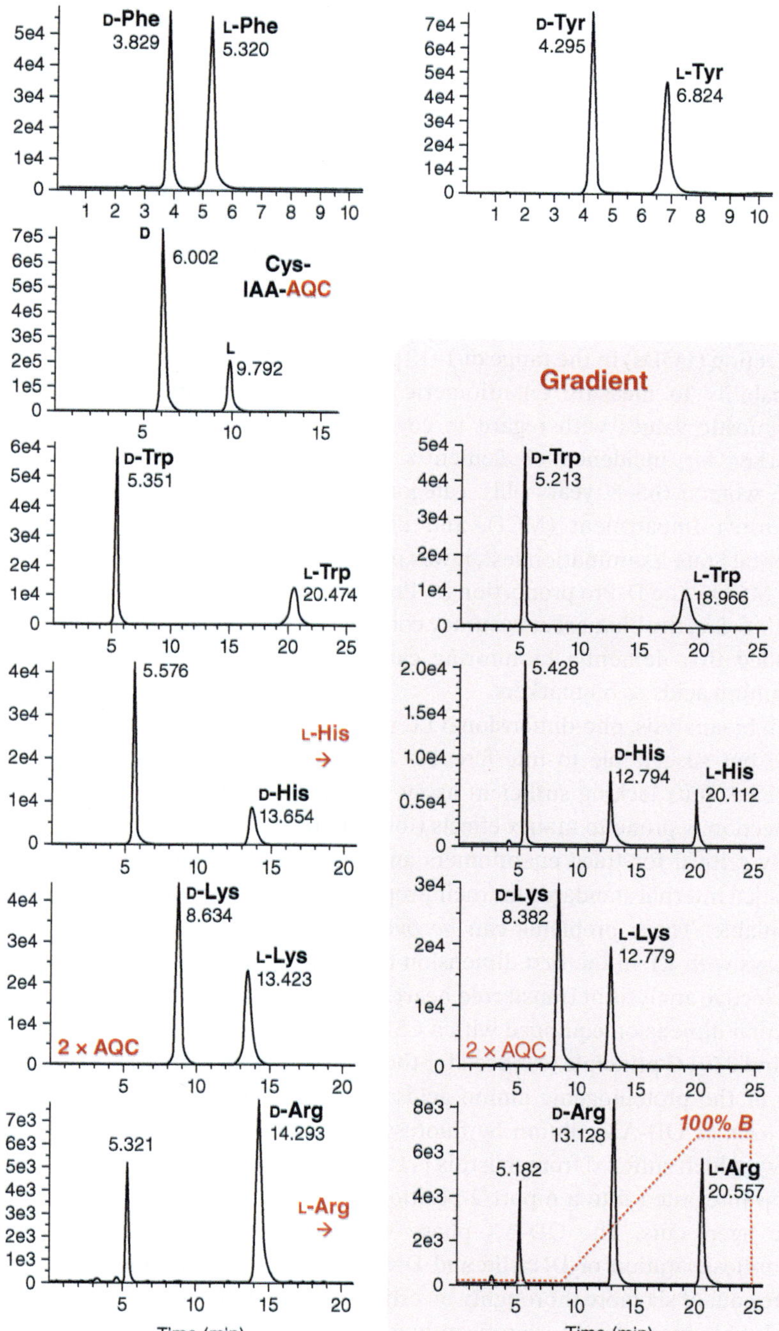

Figure 11.14 (Continued)

16 minutes. Elevated levels of D-amino acids in the brain of DAO knockout mice could be primarily detected for Leu, Ile, Ser, Ala, and Arg, while other amino acids were not significantly affected.

Kimura et al. utilized tandem columns of either QN-AX and ZWIX(+) or QD-AX and ZWIX(−) (the latter combination shows reversed elution order compared to the former tandem column) with tandem mass spectrometry in SRM acquisition mode to separate the amino acids as their AccQ-derivatives in ESI(+) mode [114]. The AccQ-tagged amino acids were eluted in around 40 minutes on the tandem column (consisting of two 150 mm long chiral columns). The validated method with limits of detection (LODs) in the range of 1–10 pmol ml^{-1} was then evaluated for its capability to measure enantiomeric ratios in serum and assess their diagnostic values with regard to cognitive decline as a predictive biomarker for incidence of dementia. In a cross-sectional study with 305 women (65–80 years old), categorized into 3 groups, control, mild-cognitive-Impairment (MCI), and dementia groups, using the Mini-Mental State Examination test, it was possible to derive a useful biomarker for MCI by the D-Pro proportion (D-Pro/(D-Pro + L-Pro)). By combination with D-Ser, the biomarker accuracy could be further increased. It was concluded that dementia monitoring can be improved by including trace D-amino acids as biomarkers.

In bioanalysis, one-dimensional LC with fluorescence detection is sensitive but susceptible to interferences by other metabolites that are also labeled, thus lacking sufficient assay specificity. On the other hand, MS detection is prone to matrix effects (ion suppression) that may be particularly critical for trace enantiomers and requires adequate stable isotope labeled internal standards for their proper correction, which are not always available. These problems can be overcome by enantioselective 2D-LC assays with RP in the first dimension to remove interferences and enantioselective analysis of transferred heart cuts of targeted analyte peaks into a second dimension equipped with a CSP. Such a multiple heart-cutting RP-chiral 2D-LC principle was used for the enantioselective amino acid analysis of the proteinogenic amino acids as dansyl (DNS) derivatives using QN-AX or QD-AX column by fluorescence detection in milk samples of cows which suffered from mastitis [117]. A single 100 μl fraction collection loop integrated into a 6-port/2-position valve was utilized for transfer of the heart cuts. The QD-AX phase was preferred, which allowed the enantioseparation of DNS-Ile and DNS-Leu. The D-enantiomer levels of three out of six more thoroughly investigated compounds, namely, Asp, Ile, and Val, were found significantly elevated in milk from cows with mastitis [117]. A similar approach with single-loop interface was employed for the analysis of Pro and hydroxyproline enantiomers in mammals after their derivatization by NBD-F (4-fluoro-7-nitro-2,1,3-benzoxadiazole) with 3,5-diisopropylphenylcarbamoyl or *tert*-butylcarbamoylquinine/quinidine as the second dimension CSP [118]. The analysis of branched amino acids in several tissues and physiological fluids in rat, dog, and mouse with QN-AX in the second dimension allowed detection of various D-amino acids (D-Leu in all rat tissues and physiological fluids tested; D-allo-Ile in rat, dog, and mouse urine) [111].

A multi-loop interface approach, which provides more flexibility to store fractions from the first dimension until the second dimension is ready for analysis, turned out to be even more powerful [111]. Such a multi-loop multiple heart-cutting 2D-LC approach was used in combination with NBD-F precolumn derivatization (yielding low fmol LODs) and a 3 μm octadecylsilica (ODS) particle based or monolithic microbore column (1 mm ID). This 2D-LC system was employed for the analysis of D-amino acids as biomarkers in various matrices with either QN-AX (D-amino acids eluted first) or QD-AX (D-amino acids eluted after the L-amino acids) as the enantioselective column in the second dimension. For example, branched chain amino acids (Val, Leu, Ile, and allo-Ile) were analyzed in biological samples taken from mammals, and a high amount of D-allo-Ile was found in the urine of healthy rats, whereas the L-form was not detected [111]. In another study, *N*-methyl-D-aspartic acid (NMDA) analogs were determined by this 2D-LC assay in various tissues and plasma including Wistar rat [119]. Various D-amino acids (Ala, Asp, Glu [glutamic acid], Ser, Leu, and allo-Ile) could also be detected by 2D-LC in foodstuff and beverages using QN-AX and other CSPs in the second dimension. For the enantioselective analysis of lactate and 3-hydroxybutyrate enantiomers, which are important biomarkers in clinical analysis, precolumn derivatizations with the fluorescent reagent 4-(*N*-chloroformylmethyl-*N*-methylamino)-7-nitro-2,1,3-benzoxadi azole (NBD-COCl) was carried out and the derivatized samples separated in the first dimension with a monolithic ODS capillary column [120]. The second dimension was equipped with two distinct chiral columns, QD-AX CSP for lactic acid and a homemade Pirkle-type CSP for 3-hydroxybutyrate enantiomer separation, which were integrated into a 6-port/2-position switching valve behind the two-loop sampling interface. The assay was validated for plasma, urine, and saliva.

There are many more applications in which it could be shown that D-amino acids are useful biomarkers for disease, and hence it is assumed that they will receive further attention in the future in clinical analysis.

11.8 Conclusions

Cinchona alkaloid-derived CSPs are well established as powerful alternatives for the stereoisomer separation of chiral compounds. Anion-exchange type quinine and quinidine carbamate-based CSPs have favorable application scope for chiral acids, while zwitterionic quinine and quinidine carbamate CSPs feature extended applicability spectra comprising chiral acids, bases, and zwitterionic analytes. The pseudo-enantiomeric relationship of the corresponding quinine and quinidine CSPs allows reversal of enantiomer elution orders by switching between them and is a peculiarity not existing for other chiral selectors derived from natural products, such as polysaccharides and macrocyclic antibiotics. These columns possess multimodal applicability in LC, yet can be also applied for supercritical fluid-type enantiomer separations. Due to the requirement of polar modifiers up to 50%, a gain in efficiency by SFC, as compared to LC, cannot be realized, yet other benefits (reduction of organic solvents, straightforward sample recovery in preparative applications) can be materialized. Transfer to 2.7 μm core-shell particle technology enables enantiomer separations in the sub-minute time scale. It seems the best choice regarding the performance-speed-backpressure compromise compared to sub-2 μm FPP UHPLC technology. It is attractive for high throughput enantiomer separations and enantioselective 2D-LC. As the most appealing application, a generic enantioselective amino acid analysis platform using precolumn derivatization and LC–MS detection can be established with quinine/quinidine columns that is of great utility for analysis of the stereointegrity of therapeutic peptides, D-amino acid biomarkers, and other applications in pharmaceutical, biotechnological, and biomedical analysis.

The future will bring more high-throughput applications with cinchona alkaloid-derived core-shell particle columns, integration in multi-column screening and 2D-LC platforms, and broader usage in SFC enantiomer separations. On the other hand, these WAX- and ZWIX-type chiral columns hold great promise for a variety of "off-label" applications, such as achiral analytical applications, in particular in the field of biopharmaceuticals analysis including therapeutic peptides, oligonucleotides, and plasmids.

References

1 Lämmerhofer, M., Peters, E.C., Yu, C. et al. (2000). Chiral monolithic columns for enantioselective capillary electrochromatography prepared by copolymerization of a monomer with quinidine functionality. 1. Optimization of polymerization conditions, porous properties, and chemistry of the stationary phase. *Analytical Chemistry* 72 (19): 4614–4622.

2 Pell, R., Schuster, G., Laemmerhofer, M., and Lindner, W. (2012). Enantioseparation of chiral sulfonates by liquid chromatography and subcritical fluid chromatography. *Journal of Separation Science* 35 (19): 2521–2528.

3 Lajko, G., Grecso, N., Toth, G. et al. (2017). Liquid and subcritical fluid chromatographic enantioseparation of Nα-Fmoc proteinogenic amino acids on quinidine-based zwitterionic and anion-exchanger type chiral stationary phases. A comparative study. *Chirality* 29 (6): 225–238.

4 Lämmerhofer, M. and Lindner, W. (2004). Chiral separations by capillary electrophoresis using cinchona alkaloid derivatives as chiral counter-ions. In: *Chiral Separations* (ed. G. Gübitz and M.G. Schmid). Totowa, NJ: Humana Press Inc.

5 Lämmerhofer, M. (2010). Enantioseparations by capillary electrochromatography. In: *Chiral Separations by Capillary Electrophoresis* (ed. A.V. Eeckhaut and Y. Michotte), 393–457. Boca Raton, FL: CRC Press, Taylor and Francis.

6 Gavioli, E., Maier, N.M., Minguillón, C., and Lindner, W. (2004). Preparative enantiomer separation of dichlorprop with a cinchona-derived chiral selector employing centrifugal partition chromatography and high-performance liquid chromatography: a comparative study. *Analytical Chemistry* 76 (19): 5837–5848.

7 Maximini, A., Chmiel, H., Holdik, H., and Maier, N.W. (2006). Development of a supported liquid membrane process for separating enantiomers of N-protected amino acid derivatives. *Journal of Membrane Science* 276 (1): 221–231.

8 Lämmerhofer, M. and Lindner, W. (2008). Liquid chromatographic enantiomer separation and chiral recognition by cinchona alkaloid-derived enantioselective separation materials. *Advances in Chromatography (Boca Raton, FL, United States)* 46: 1–107.

9 Ilisz, I., Bajtai, A., Lindner, W., and Peter, A. (2018). Liquid chromatographic enantiomer separations applying chiral ion-exchangers based on cinchona alkaloids. *Journal of Pharmaceutical and Biomedical Analysis* 159: 127–152.

10 Grubhofer, N. and Schleith, L. (1953). Modifizierte Ionenaustauscher als spezifische Adsorbentien. *Die Naturwissenschaften* 40 (19): 508.

11 Pettersson, C. (1984). Chromatographic separation of enantiomers of acids with quinine as chiral counter ion. *Journal of Chromatography A* 316: 553–567.

12 Rosini, C., Bertucci, C., Pini, D. et al. (1985). Cinchona alkaloids for preparing new, easily accessible chiral stationary phases.I. 11-(10,11-Dihydro-6′-methoxy-cinchonan-9-ol)-tiopropylsilanized silica. *Tetrahedron Letters* 26 (28): 3361–3364.

13 Lämmerhofer, M. and Lindner, W. (1996). Quinine and quinidine derivatives as chiral selectors I. Brush type chiral stationary phases for

high-performance liquid chromatography based on cinchonan carbamates and their application as chiral anion exchangers. *Journal of Chromatography A* 741 (1): 33–48.

14 Mandl, A., Nicoletti, L., Lämmerhofer, M., and Lindner, W. (1999). Quinine- versus carbamoylated quinine-based chiral anion exchangers: a comparison regarding enantioselectivity for N-protected amino acids and other chiral acids. *Journal of Chromatography A* 858 (1): 1–11.

15 Maier, N.M., Nicoletti, L., Lämmerhofer, M., and Lindner, W. (1999). Enantioselective anion exchangers based on cinchona alkaloid-derived carbamates: influence of C8/C9 stereochemistry on chiral recognition. *Chirality* 11 (7): 522–528.

16 Sievers-Engler, A., Lindner, W., and Lämmerhofer, M. (2014). Ligand-receptor binding increments in enantioselective liquid chromatography. *Journal of Chromatography. A* 1363: 79–88.

17 Calderon, C., Santi, C., and Lämmerhofer, M. (2018). Chiral separation of disease biomarkers with 2-hydroxycarboxylic acid structure. *Journal of Separation Science* 41 (6): 1224–1231.

18 Hoffmann, C.V., Lämmerhofer, M., and Lindner, W. (2007). Novel strong cation-exchange type chiral stationary phase for the enantiomer separation of chiral amines by high-performance liquid chromatography. *Journal of Chromatography. A* 1161 (1–2): 242–251.

19 Hoffmann, C.V., Pell, R., Lämmerhofer, M., and Lindner, W. (2008). Synergistic effects on enantioselectivity of zwitterionic chiral stationary phases for separations of chiral acids, bases, and amino acids by HPLC. *Analytical Chemistry* 80 (22): 8780–8789.

20 Zhang, T., Holder, E., Franco, P., and Lindner, W. (2014). Method development and optimization on cinchona and chiral sulfonic acid-based zwitterionic stationary phases for enantiomer separations of free amino acids by high-performance liquid chromatography. *Journal of Chromatography. A* 1363: 191–199.

21 Orosz, T., Forro, E., Fulop, F. et al. (2018). Effects of N-methylation and amidination of cyclic β-amino acids on enantioselectivity and retention characteristics using cinchona alkaloid- and sulfonic acid-based chiral zwitterionic stationary phases. *Journal of Chromatography. A* 1535: 72–79.

22 Gargano, A.F.G., Kohout, M., Macikova, P. et al. (2013). Direct high-performance liquid chromatographic enantioseparation of free α-, β- and γ-aminophosphonic acids employing cinchona-based chiral zwitterionic ion exchangers. *Analytical and Bioanalytical Chemistry* 405 (25): 8027–8038.

23 Bajtai, A., Ilisz, I., Howan, D.H.O. et al. (2020). Enantioselective resolution of biologically active dipeptide analogs by high-performance liquid chromatography applying cinchona alkaloid-based ion-exchanger chiral stationary phases. *Journal of Chromatography. A* 1611: 460574.

24 Ianni, F., Sardella, R., Carotti, A. et al. (2016). Quinine-based zwitterionic chiral stationary phase as a complementary tool for peptide analysis: mobile phase effects on enantio- and stereoselectivity of underivatized oligopeptides. *Chirality* 28 (1): 5–16.

25 Ilisz, I., Bajtai, A., Lindner, W., and Péter, A. (2018). Liquid chromatographic enantiomer separations applying chiral ion-exchangers based on cinchona alkaloids. *Journal of Pharmaceutical and Biomedical Analysis* 159: 127–152.

26 Czerwenka, C., Lämmerhofer, M., Maier, N.M. et al. (2002). Direct high-performance liquid chromatographic separation of peptide enantiomers: study on chiral recognition by systematic evaluation of the influence of structural features of the chiral selectors on enantioselectivity. *Analytical Chemistry* 74 (21): 5658–5666.

27 Akasaka, K., Gyimesi-Forrás, K., Lämmerhofer, M. et al. (2005). Investigations of molecular recognition aspects related to the enantiomer separation of 2-methoxy-2-(1-naphthyl)propionic acid using quinine carbamate as chiral selector: an NMR and FT-IR spectroscopic as well as X-ray crystallographic study. *Chirality* 17 (9): 544–555.

28 Bicker, W., Chiorescu, I., Arion, V.B. et al. (2008). Contributions to chromatographic chiral recognition of permethrinic acid stereoisomers by a quinine carbamate chiral selector: evidence from X-ray diffraction, DFT computations, 1H NMR, and thermodynamic studies. *Tetrahedron: Asymmetry* 19 (1): 97–110.

29 Calderon, C. and Lämmerhofer, M. (2017). Chiral separation of short chain aliphatic hydroxycarboxylic acids on cinchonan carbamate-based weak chiral anion exchangers and zwitterionic chiral ion exchangers. *Journal of Chromatography. A* 1487: 194–200.

30 Maier, N.M., Schefzick, S., Lombardo, G.M. et al. (2002). Elucidation of the chiral recognition mechanism of cinchona alkaloid carbamate-type receptors for 3,5-dinitrobenzoyl amino acids. *Journal of the American Chemical Society* 124 (29): 8611–8629.

31 Baeurer, S., Ferri, M., Carotti, A. et al. (2020). Mixed-mode chromatography characteristics of chiralpak ZWIX(+) and ZWIX(−) and elucidation of their chromatographic orthogonality for LC × LC application. *Analytica Chimica Acta* 1093: 168–179.

32 Ianni, F., Carotti, A., Marinozzi, M. et al. (2015). Diastereo- and enantioseparation of a Nα-Boc amino acid with a zwitterionic quinine-based stationary phase: focus on the stereorecognition mechanism. *Analytica Chimica Acta* 885: 174–182.

33 Grecso, N., Kohout, M., Carotti, A. et al. (2016). Mechanistic considerations of enantiorecognition on novel cinchona alkaloid-based zwitterionic chiral stationary phases from the aspect of the separation of trans-paroxetine enantiomers as model compounds. *Journal of Pharmaceutical and Biomedical Analysis* 124: 164–173.

34 Sardella, R., Macchiarulo, A., Urbinati, F. et al. (2018). Exploring the enantiorecognition mechanism of cinchona alkaloid-based zwitterionic chiral stationary phases and the basic trans-paroxetine enantiomers. *Journal of Separation Science* 41 (6): 1199–1207.

35 Varfaj, I., Protti, M., Di Michele, A. et al. (2021). Efficient enantioresolution of aromatic α-hydroxy acids with cinchona alkaloid-based zwitterionic stationary phases and volatile polar-ionic eluents. *Analytica Chimica Acta* 1180: 338928.

36 Gyimesi-Forrás, K., Kökösi, J., Szász, G. et al. (2004). Liquid chromatographic enantiomer separations of novel quinazolone derivatives on quinine carbamate based chiral stationary phases using hydro-organic mobile phases. *Journal of Chromatography A* 1047 (1): 59–67.

37 Gyimesi-Forrás, K., Akasaka, K., Lämmerhofer, M. et al. (2005). Enantiomer separation of a powerful chiral auxiliary, 2-methoxy-2-(1-naphthyl)propionic acid by liquid chromatography using chiral anion exchanger-type stationary phases in polar-organic mode; investigation of molecular recognition aspects. *Chirality* 17 (S1): S134–S142.

38 Hoffmann, C.V., Reischl, R., Maier, N.M. et al. (2009). Investigations of mobile phase contributions to enantioselective anion- and zwitterion-exchange modes on quinine-based zwitterionic chiral stationary phases. *Journal of Chromatography. A* 1216 (7): 1157–1166.

39 Lee, M., Jang, M.D., Lee, W., and Park, J.H. (2008). Chiral separation of N-fluorenylmethoxycarbonyl α-amino acids on quinine carbamate-bonded silica by high performance liquid chromatography. *Bulletin of the Korean Chemical Society* 29 (1): 257–260.

40 Tanacs, D., Orosz, T., Ilisz, I. et al. (2021). Unexpected effects of mobile phase solvents and additives on retention and resolution of N-acyl-D,L-leucine applying Cinchonane-based chiral ion exchangers. *Journal of Chromatography. A* 1648: 462212.

41 Mimini, V., Ianni, F., Marini, F. et al. (2019). Electrostatic attraction-repulsion model with cinchona alkaloid-based zwitterionic chiral stationary phases exemplified for zwitterionic analytes. *Analytica Chimica Acta* 1078: 212–220.

42 Gyimesi-Forrás, K., Maier, N.M., Kökösi, J. et al. (2009). Enantiomer separation of imidazo-quinazoline-dione derivatives on quinine carbamate-based chiral stationary phase in normal phase mode. *Chirality* 21 (1): 199–207.

43 Lajko, G., Ilisz, I., Toth, G. et al. (2015). Application of cinchona alkaloid-based zwitterionic chiral stationary phases in supercritical fluid chromatography for the enantioseparation of Nα-protected proteinogenic amino acids. *Journal of Chromatography. A* 1415: 134–145.

44 Sánchez Muñoz, O.L., Hernández, E.P., Lämmerhofer, M. et al. (2003). Estimation and comparison of ζ-potentials of silica-based anion-exchange type porous particles for capillary electrochromatography from electrophoretic and electroosmotic mobility. *Electrophoresis* 24 (3): 390–398.

45 Götmar, G., Asnin, L., and Guiochon, G. (2004). Adsorption of the enantiomers of 2,2,2-trifluoro-1-(9-anthryl)-ethanol on silica-bonded chiral quinidine-carbamate. *Journal of Chromatography A* 1059 (1): 43–52.

46 Gika, H., Lämmerhofer, M., Papadoyannis, I., and Lindner, W. (2004). Direct separation and quantitative analysis of thyroxine and triiodothyronine enantiomers in pharmaceuticals by high-performance liquid chromatography. *Journal of Chromatography B* 800 (1): 193–201.

47 Losacco, G.L., Wang, H., Haidar Ahmad, I.A. et al. (2022). Enantioselective UHPLC screening combined with in Silico modeling for streamlined development of ultrafast enantiopurity assays. *Analytical Chemistry* 94: 1804–1812. https://doi.org/10.1021/acs.analchem.1c04585.

48 Sardella, R., Lämmerhofer, M., Natalini, B., and Lindner, W. (2008). Enantioselective HPLC of potentially CNS-active acidic amino acids with a cinchona carbamate based chiral stationary phase. *Chirality* 20 (3/4): 571–576.

49 Kopaciewicz, W., Rounds, M.A., Fausnaugh, J., and Regnier, F.E. (1983). Retention model for high-performance ion-exchange chromatography. *Journal of Chromatography A* 266: 3–21.

50 Williams, B.A. and Vigh, G. (1997). Dry look at the CHARM (charged resolving agent migration) model of enantiomer separations by capillary electrophoresis. *Journal of Chromatography A* 777 (2): 295–309.

51 Woiwode, U., Sievers-Engler, A., Zimmermann, A. et al. (2017). Surface-anchored counterions on weak chiral anion-exchangers accelerate separations and improve their compatibility for mass-spectrometry-hyphenation. *Journal of Chromatography. A* 1503: 21–31.

52 Sellergren, B. and Shea, K.J. (1993). Chiral ion-exchange chromatography: correlation between solute retention and a theoretical ion-exchange model using imprinted polymers. *Journal of Chromatography A* 654 (1): 17–28.

53 Krawinkler, K.H., Maier, N.M., Sajovic, E., and Lindner, W. (2004). Novel urea-linked cinchona-calixarene hybrid-type receptors for efficient chromatographic enantiomer separation of carbamate-protected cyclic amino acids. *Journal of Chromatography A* 1053 (1): 119–131.

54 Ianni, F., Pataj, Z., Gross, H. et al. (2014). Direct enantioseparation of underivatized aliphatic 3-hydroxyalkanoic acids with a quinine-based zwitterionic chiral stationary phase. *Journal of Chromatography. A* 1363: 101–108.

55 Geditz, M.C.K., Lindner, W., Lämmerhofer, M. et al. (2014). Simultaneous quantification of mefloquine (+)- and (−)-enantiomers and the carboxy metabolite in dried blood spots by liquid chromatography/tandem mass spectrometry. *Journal of Chromatography B: Analytical Technologies in the Biomedical and Life Sciences* 968: 32–39.

56 Hanafi, R.S. and Lämmerhofer, M. (2018). Response surface methodology for the determination of the design space of enantiomeric separations on cinchona-based zwitterionic chiral stationary phases by high performance liquid chromatography. *Journal of Chromatography. A* 1534: 55–63.

57 Zhang, T., Holder, E., Franco, P., and Lindner, W. (2014). Zwitterionic chiral stationary phases based on cinchona and chiral sulfonic acids for the direct stereoselective separation of amino acids and other amphoteric compounds. *Journal of Separation Science* 37 (11): 1237–1247.

58 Woiwode, U., Ferri, M., Maier, N.M. et al. (2018). Complementary enantioselectivity profiles of chiral cinchonan carbamate selectors with distinct carbamate residues and their implementation in enantioselective two-dimensional high-performance liquid chromatography of amino acids. *Journal of Chromatography. A* 1558: 29–36.

59 Lämmerhofer, M., Franco, P., and Lindner, W. (2006). Quinine carbamate chiral stationary phases: systematic optimization of steric selector-selectand binding increments and enantioselectivity by quantitative structure-enantioselectivity relationship studies. *Journal of Separation Science* 29 (10): 1486–1496.

60 Sardella, R., Lisanti, A., Carotti, A. et al. (2014). Ketoprofen enantioseparation with a cinchona alkaloid based stationary phase: enantiorecognition mechanism and release studies. *Journal of Separation Science* 37 (19): 2696–2703.

61 Levkin, P.A., Naier, N.M., Schurig, V., and Lindner, W. (2010). Strong detrimental effect of a minute enantiomeric impurity of a chiral selector on the enantioselectivity factor. *Angewandte Chemie, International Edition* 49 (42): 7742.

62 Franco, P., Lämmerhofer, M., Klaus, P.M., and Lindner, W. (2000). Novel cinchona alkaloid carbamate C9-dimers as chiral anion-exchange type selectors for high-performance liquid chromatography. *Journal of Chromatography A* 869 (1): 111–127.

63 Krawinkler, K.H., Maier, N.M., Ungaro, R. et al. (2003). Novel cinchona carbamate selectors with complementary enantioseparation characteristics for N-acylated amino acids. *Chirality* 15 (S1): S17–S29.

64 Zhao, J., Wu, H., Wang, D. et al. (2015). Improvement of chiral stationary phases based on cinchona alkaloids bonded to crown ethers by chiral modification. *Journal of Separation Science* 38 (22): 3884–3890.

65 Wang, D., Zhao, J., Wu, H. et al. (2015). Preparation and evaluation of novel chiral stationary phases based on quinine derivatives comprising crown ether moieties. *Journal of Separation Science* 38 (2): 205–210.

66 Wu, H., Wang, D., Song, G. et al. (2014). Novel chiral stationary phases based on peptoid combining a quinine/quinidine moiety through a C9-position carbamate group. *Journal of Separation Science* 37 (8): 934–943.

67 Lu, X., Chen, M., Yang, J. et al. (2021). Surface-up construction of quinine bridged functional cyclodextrin for single-column versatile enantioseparation. *Journal of Chromatography A* 462786.

68 Zhu, L., Zhu, J., Sun, X. et al. (2020). Novel chiral stationary phases based on 3,5-dimethyl phenylcarbamoylated β-cyclodextrin combining cinchona alkaloid moiety. *Chirality* 32 (8): 1080–1090.

69 Hoffmann, C.V., Reischl, R., Maier, N.M. et al. (2009). Stationary phase-related investigations of quinine-based zwitterionic chiral stationary phases operated in anion-, cation-, and zwitterion-exchange modes. *Journal of Chromatography. A* 1216 (7): 1147–1156.

70 Pell, R., Sic, S., and Lindner, W. (2012). Mechanistic investigations of cinchona alkaloid-based zwitterionic chiral stationary phases. *Journal of Chromatography. A* 1269: 287–296.

71 Wernisch, S. and Lindner, W. (2012). Versatility of cinchona-based zwitterionic chiral stationary phases: enantiomer and diastereomer separations of non-protected oligopeptides utilizing a multi-modal chiral recognition mechanism. *Journal of Chromatography. A* 1269: 297–307.

72 Wernisch, S., Pell, R., and Lindner, W. (2012). Increments to chiral recognition facilitating enantiomer separations of chiral acids, bases, and ampholytes using cinchona-based zwitterion exchanger chiral stationary phases. *Journal of Separation Science* 35 (13): 1560–1572.

73 Geibel, C., Dittrich, K., Wolter, M., and Lämmerhofer, M. (2020). Thiol-ene photo-click immobilization of a chiral chromatographic ligand on silica particles. *Journal of Chromatography. A* 1622: 461133.

74 Levkin, P., Maier, N.M., Lindner, W., and Schurig, V. (2012). A practical method for the quantitative assessment of non-enantioselective versus enantioselective interactions encountered in liquid chromatography on brush-type chiral stationary phase. *Journal of Chromatography. A* 1269: 270–278.

75 Kacprzak, K.M. and Lindner, W. (2011). Novel Pirkle-type quinine 3,5-dinitrophenylcarbamate chiral stationary phase implementing click chemistry. *Journal of Separation Science* 34 (18): 2391–2396.

76 Kohout, M., Wernisch, S., Tuma, J. et al. (2018). Effect of different immobilization strategies on chiral recognition properties of cinchona-based anion exchangers. *Journal of Separation Science* 41 (6): 1355–1364.

77 Hettegger, H., Kohout, M., Mimini, V., and Lindner, W. (2014). Novel carbamoyl type quinine and quinidine based chiral anion exchangers implementing alkyne-azide cycloaddition immobilization chemistry. *Journal of Chromatography. A* 1337: 85–94.

78 Hellriegel, C., Skogsberg, U., Albert, K. et al. (2004). Characterization of a chiral stationary phase by HR/MAS NMR spectroscopy and investigation of enantioselective interaction with chiral ligates by transferred NOE. *Journal of the American Chemical Society* 126 (12): 3809–3816.

79 Geibel, C., Theiner, J., Wolter, M. et al. (2021). Controllable organosilane monolayer density of surface bonding using silatranes for thiol functionalization of silica particles for liquid chromatography and validation of microanalytical method for elemental composition determination. *Journal of Chromatography. A* 1653: 462418.

80 Wolter, M., Chen, X., Woiwode, U. et al. (2021). Preparation and characterization of poly(3-mercaptopropyl)methylsiloxane functionalized silica particles and their further modification for silver ion chromatography and enantioselective high-performance liquid chromatography. *Journal of Chromatography. A* 1643: 462069.

81 Schmitt, K. and Lämmerhofer, M. (2018). Optimization of the surface modification process of cross-linked polythiol-coated chiral stationary phases synthesized by a two-step thiol-ene click reaction. *Journal of Separation Science* 41 (6): 1338–1345.

82 Zimmermann, A., Horak, J., Sievers-Engler, A. et al. (2016). Surface-crosslinked poly(3-mercaptopropyl)methylsiloxane-coatings on silica as new platform for low-bleed mass spectrometry-compatible functionalized stationary phases synthesized via thiol-ene click reaction. *Journal of Chromatography. A* 1436: 73–83.

83 Piette, V., Lämmerhofer, M., Bischoff, K., and Lindner, W. (1997). High-performance liquid chromatographic enantioseparation of N-protected α-amino acids using nonporous silica modified by a quinine carbamate as chiral stationary phase. *Chirality* 9 (2): 157–161.

84 Lubda, D. and Lindner, W. (2004). Monolithic silica columns with chemically bonded tert-butylcarbamoylquinine chiral anion-exchanger selector as a stationary phase for enantiomer separations. *Journal of Chromatography A* 1036 (2): 135–143.

85 Luo, C., DeStefano, J.J., Langlois, T.J. et al. (2021). Fundamental to achieving fast separations with high efficiency: a review of chromatography with superficially porous particles. *Biomedical Chromatography* 35 (7): e5087.

86 Ciogli, A., Ismail, O.H., Mazzoccanti, G. et al. (2018). Enantioselective ultra high performance liquid and supercritical fluid chromatography: the race to the shortest chromatogram. *Journal of Separation Science* 41 (6): 1307–1318.

87 Cancelliere, G., Ciogli, A., D'Acquarica, I. et al. (2010). Transition from enantioselective high performance to ultra-high performance liquid chromatography: a case study of a brush-type chiral stationary phase based on sub-5-micron to sub-2-micron silica particles. *Journal of Chromatography A* 1217 (7): 990–999.

88 Reischl, R.J., Hartmanova, L., Carrozzo, M. et al. (2011). Chemoselective and enantioselective analysis of proteinogenic amino acids utilizing N-derivatization and 1-D enantioselective anion-exchange chromatography in combination with tandem mass spectrometric detection. *Journal of Chromatography A* 1218 (46): 8379–8387.

89 Patel, D.C., Breitbach, Z.S., Yu, J.J. et al. (2017). Quinine bonded to superficially porous particles for high-efficiency and ultrafast liquid and supercritical fluid chromatography. *Analytica Chimica Acta* 963: 164–174.

90 Schmitt, K., Woiwode, U., Kohout, M. et al. (2018). Comparison of small size fully porous particles and superficially porous particles of chiral anion-exchange type stationary phases in ultra-high performance liquid chromatography: effect of particle and pore size on chromatographic efficiency and kinetic performance. *Journal of Chromatography A* 1569: 149–159.

91 Geibel, C., Dittrich, K., Woiwode, U. et al. (2019). Evaluation of superficially porous particle based zwitterionic chiral ion exchangers against fully porous particle benchmarks for enantioselective ultra-high performance liquid chromatography. *Journal of Chromatography. A* 1603: 130–140.

92 Patel, D.C., Wahab, M.F., O'Haver, T.C., and Armstrong, D.W. (2018). Separations at the speed of sensors. *Analytical Chemistry* 90 (5): 3349–3356.

93 Woiwode, U., Reischl, R.J., Buckenmaier, S. et al. (2018). Imaging peptide and protein chirality via amino acid analysis by chiral × chiral two-dimensional correlation liquid chromatography. *Analytical Chemistry* 90 (13): 7963–7971.

94 Woiwode, U., Neubauer, S., Lindner, W. et al. (2018). Enantioselective multiple heartcut two-dimensional ultra-high-performance liquid chromatography method with a Coreshell chiral stationary phase in the second dimension for analysis of all proteinogenic amino acids in a single run. *Journal of Chromatography. A* 1562: 69–77.

95 Karongo, R., Ge, M., Geibel, C. et al. (2021). Enantioselective multiple heart cutting online two-dimensional liquid chromatography-mass spectrometry of all proteinogenic amino acids with second dimension chiral separations in one-minute time scales on a chiral tandem column. *Analytica Chimica Acta* 1180: 338858.

96 Choppari, T., Gunnam, S., Chennuru, L.N., and Cherla, P.M. (2020). Direct stereoselective method development and validation of elagolix on zwitterionic chiral stationary phase by HPLC. *International Journal of Pharmaceutical Sciences and Drug Research* 12 (5): 517–524.

97 Chennuru, L.N., Choppari, T., Nandula, R.P. et al. (2016). Direct separation of pregabalin enantiomers using a zwitterionic chiral selector by high performance liquid chromatography coupled to mass spectrometry and ultraviolet detection. *Molecules* 21 (11): 1578.

98 Peter, A., Grecso, N., Toth, G. et al. (2016). Ultra-trace analysis of enantiomeric impurities in proteinogenic N-Fmoc-amino-acid samples on cinchona alkaloid-based chiral stationary phases. *Israel Journal of Chemistry* 56 (11–12): 1042–1051.

99 Lajko, G., Grecso, N., Toth, G. et al. (2016). A comparative study of enantioseparations of Nα-Fmoc proteinogenic amino acids on quinine-based zwitterionic and anion exchanger-type chiral stationary phases under hydro-organic liquid and subcritical fluid chromatographic conditions. *Molecules* 21 (11): 1579.

100 Forssen, P., Edstroem, L., Lämmerhofer, M. et al. (2012). Optimization strategies accounting for the additive in preparative chiral liquid chromatography. *Journal of Chromatography. A* 1269: 279–286.

101 Arnell, R., Forssen, P., Fornstedt, T. et al. (2009). Adsorption behaviour of a quinidine carbamate-based chiral stationary phase: role of the additive. *Journal of Chromatography. A* 1216 (16): 3480–3487.

102 Gerhardt, H., Sievers-Engler, A., Jahanshah, G. et al. (2016). Methods for the comprehensive structural elucidation of constitution and stereochemistry of lipopeptides. *Journal of Chromatography A* 1428: 280–291.

103 Czerwenka, C., Maier, N.M., and Lindner, W. (2004). Liquid chromatographic–mass spectrometric separation of oligoalanine peptide stereoisomers: influence of absolute configuration on enantioselectivity and two-dimensional separation of diastereomers and enantiomers. *Journal of Chromatography A* 1038 (1): 85–95.

104 Czerwenka, C., Lämmerhofer, M., and Lindner, W. (2003). Structure-enantioselectivity relationships for the study of chiral recognition in peptide enantiomer separation on cinchona alkaloid-based chiral stationary phases by HPLC: influence of the N-terminal protecting group. *Journal of Separation Science* 26 (17): 1499–1508.

105 Karongo, R., Ge, M., Horak, J. et al. (2021). Rapid enantioselective amino acid analysis by ultra-high performance liquid chromatography-mass spectrometry combining 6-aminoquinolyl-N-hydroxysuccinimidyl carbamate derivatization with core-shell quinine carbamate anion exchanger separation. *Journal of Chromatography Open* 1: 100004.

106 Acquaviva, A., Siano, G., Quintas, P. et al. (2020). Chiral x achiral multidimensional liquid chromatography. Application to the enantioseparation of dintitrophenyl amino acids in honey samples and their fingerprint classification. *Journal of Chromatography A* 1614: 460729.

107 Bajtai, A., Lajko, G., Szatmari, I. et al. (2018). Dedicated comparisons of diverse polysaccharide- and zwitterionic cinchona alkaloid-based chiral stationary phases probed with basic and ampholytic indole analogs in liquid and subcritical fluid chromatography mode. *Journal of Chromatography. A* 1563: 180–190.

108 Ilisz, I., Bajtai, A., Szatmari, I. et al. (2020). Enantioseparation of β-carboline, tetrahydroisoquinoline and benzazepine analogues of pharmaceutical importance: utilization of chiral stationary phases based on polysaccharides and sulfonic acid modified cinchona alkaloids in high-performance liquid and subcritical fluid chromatography. *Journal of Chromatography. A* 1615: 460771.

109 Raimbault, A., Ma, C.M.A., Ferri, M. et al. (2020). Cinchona-based zwitterionic stationary phases: exploring retention and enantioseparation mechanisms in supercritical fluid chromatography with a fragmentation approach. *Journal of Chromatography. A* 1612: 460689.

110 Calderon, C., Horak, J., and Lämmerhofer, M. (2016). Chiral separation of 2-hydroxyglutaric acid on cinchonan carbamate based weak chiral anion exchangers by high-performance liquid chromatography. *Journal of Chromatography. A* 1467: 239–245.

111 Hamase, K., Morikawa, A., Ohgusu, T. et al. (2007). Comprehensive analysis of branched aliphatic d-amino acids in mammals using an integrated multi-loop two-dimensional column-switching high-performance liquid chromatographic system combining reversed-phase and enantioselective columns. *Journal of Chromatography A* 1143 (1): 105–111.

112 Karakawa, S., Shimbo, K., Yamada, N. et al. (2015). Simultaneous analysis of D-alanine, D-aspartic acid, and D-serine using chiral high-performance liquid chromatography-tandem mass spectrometry and its application to the rat plasma and tissues. *Journal of Pharmaceutical and Biomedical Analysis* 115: 123–129.

113 Du, S., Wang, Y., Weatherly, C.A. et al. (2018). Variations of L- and D-amino acid levels in the brain of wild-type and mutant mice lacking D-amino acid oxidase activity. *Analytical and Bioanalytical Chemistry* 410 (12): 2971–2979.

114 Kimura, R., Tsujimura, H., Tsuchiya, M. et al. (2020). Development of a cognitive function marker based on D-amino acid proportions using new chiral tandem LC-MS/MS systems. *Scientific Reports* 10 (1): 804.

115 Horak, J. and Lämmerhofer, M. (2019). Stereoselective separation of underivatized and 6-aminoquinolyl-N-hydroxysuccinimidyl carbamate derivatized amino acids using zwitterionic quinine and quinidine type stationary phases by liquid chromatography-high resolution mass spectrometry. *Journal of Chromatography. A* 1596: 69–78.

116 Hellinger, R., Horak, J., and Lindner, W. (2013). Enantioseparation of 6-aminoquinolyl-N-hydroxysuccinimidyl carbamate tagged amino acids and other zwitterionic compounds on cinchona-based chiral stationary phases. *Analytical and Bioanalytical Chemistry* 405 (25): 8105–8120.

117 Ianni, F., Sardella, R., Lisanti, A. et al. (2015). Achiral-chiral two-dimensional chromatography of free amino acids in milk: a promising tool for detecting different levels of mastitis in cows. *Journal of Pharmaceutical and Biomedical Analysis* 116: 40–46.

118 Tojo, Y., Hamase, K., Nakata, M. et al. (2008). Automated and simultaneous two-dimensional micro-high-performance liquid chromatographic determination of proline and hydroxyproline enantiomers in mammals. *Journal of Chromatography B* 875 (1): 174–179.

119 Koga, R., Miyoshi, Y., Negishi, E. et al. (2012). Enantioselective two-dimensional high-performance liquid chromatographic determination of N-methyl-d-aspartic acid and its analogues in mammals and bivalves. *Journal of Chromatography A* 1269: 255–261.

120 Liu, S.-L., Oyama, T., Miyoshi, Y. et al. (2015). Establishment of a two-dimensional chiral HPLC system for the simultaneous detection of lactate and 3-hydroxybutyrate enantiomers in human clinical samples. *Journal of Pharmaceutical and Biomedical Analysis* 116: 80–85.

Part III

Methods for Stereochemical Elucidation

12

X-Ray Crystallography for Stereochemical Elucidation

Ademir F. Morel and Robert A. Burrow

Departamento de Química, Universidade Federal de Santa Maria, RS, Brazil

12.1 Introduction

Structural determination by means of a single crystal X-ray diffraction experiment is considered an almost fool-proof method, providing a three-dimensional structural model of the measured compound that reveals details such as molecular conformation and relative stereochemistry in the solid state [1, 2]. The accompanying software suites of modern single crystal X-ray diffractometers guide the novice user, steering them away from common pitfalls. Some software suites even have an "automatic" collection, solution, and refinement routine for those with no experience what-so-ever [3]. The recent advances in the equipment, which now features large, sensitive area detectors, an X-ray beam of just about 100 μm in diameter, low-temperature sample cooling apparatus, and quite often two X-ray sources, usually with Mo K_α and Cu K_α radiation, means that even the smallest crystal yields very useable data that often leads to the confident determination of crystal structure, and for chiral compounds its relative stereochemistry, with little doubt, and the absolute stereochemistry with some care. The now almost infallible structure solution algorithms no longer need a precise space group determination to give quite often the correct complete structure in the correct space group. *CheckCIF* [4, 5], an online crystallographic verification service, facilitates the detection of subtle crystallographic errors, for those more experienced. An experienced

Chiral Separations and Stereochemical Elucidation: Fundamentals, Methods, and Applications, First Edition. Edited by Quezia Bezerra Cass, Maria Elizabeth Tiritan, João Marcos Batista Junior, and Juliana Cristina Barreiro.

crystallographer following current best practices extremely rarely will produce an inaccurate crystal structure.

There are alternatives to single crystal X-ray diffraction. Powder X-ray diffraction is now quite advanced and with quite a bit more work from a crystalline powder can produce results rivaling single crystal structures [6]. Electron diffraction is the newest technique, just a few years old at the time of writing, and has allowed the structural determination from sub-micron crystals [7–9]. These techniques can be seen as complementary to single crystal X-ray diffraction and the same principles can be applied to them.

When the subject of the experiment is a crystal of a natural product, the determination of its absolute configuration is highly desired. First, a distinction should be made between relative configuration, absolute configuration, and absolute structure. If a molecule has two or more stereogenic centers, these centers will have a relative configuration, for example (*R**,*R**)/(*S**,*S**) or (*R**,*S**)/(*S**,*R**). The relative configuration is determined with the three-dimensional structure of the crystal contents by the X-ray diffraction experiment. If the molecule crystallizes in a non-Sohnke space group, that is one which has a mirror or inversion symmetry element, the molecule by crystallographic symmetry will exist in the crystal as racemic pairs. On the other hand, if the molecule crystallizes in a Sohnke space group, it will be enantiomerically pure, with the exception being a crystal with inversion twinning. The absolute structure is a crystallographic term that refers to the chirality of the contents of the crystal. Once the absolute structure of a crystal is determined, the absolute configuration – the configurations of all the stereogenic centers of the molecule – is consequently determined. The absolute structure of the crystal must be determined by some means to determine with confidence the absolute configuration of the molecule of the experiment. If the means is mainly by the X-ray diffraction experiment, much more care than for a normal X-ray experiment is essential. One of the best methods of determining the absolute stereochemistry of a natural product is by a well-performed single crystal X-ray diffraction experiment. This chapter will explore these aspects.

12.2 Absolute Structure and Absolute Configuration

If the natural product crystallizes in a Sohnke space group, it will be enantiomerically pure and the absolute structure of the contents of the crystal will determine the absolute configuration of the natural product [10]. The exception to this is when inversion twinning occurs, in which case, the crystal is a solid solution of the enantiomorphs. There are 65 Sohnke space

Table 12.1 Achiral and chiral Sohnke space groups.

Crystal system	Achiral Sohnke space groups	Chiral Sohnke space groups
Triclinic	$P1$ (#1)	
Monoclinic	$P2$ (#3), $P2_1$ (#4), $C2$ (#5)	
Orthorhombic	$P222$ (#16), $P222_1$ (#17), $P2_12_12$ (#18), $P2_12_12_1$ (#19), $C222_1$ (#20), $C222$ (#21), $F222$ (#22), $I222$ (#23), $I2_12_12_1$ (#24)	
Tetragonal	$P4$ (#75), $P4_2$ (#77), $I4$ (#79), $I4_1$ (#80), $P422$ (#89), $P42_12$ (#90), $P4_222$ (#93), $P4_22_12$ (#94), $I422$ (#97), $I4_122$ (#98)	$P4_1$ (#76)/$P4_3$ (#78), $P4_122$ (#91)/$P4_322$ (#95), $P4_12_12$ (#92)/$P4_32_12$ (#96)
Trigonal	$P3$ (#143), R3 (#146), $P312$ (#149), $P321$ (#150), R32 (#155)	$P3_1$ (#144)/$P3_2$ (#145), $P3_112$ (#151)/$P3_212$ (#153), $P3_121$ (#152)/$P3_221$ (#154)
Hexagonal	$P6$ (#168), $P6_3$ (#173), $P622$ (#177), $P6_322$ (#182)	$P6_1$ (#169)/$P6_5$ (#170), $P6_2$ (#171)/$P6_4$ (#172), $P6_122$ (#178)/$P6_522$ (#179), $P6_222$ (#180)/$P6_422$ (#181)
Cubic	$P23$ (#195), $P2_13$ (#198), $F23$ (#196), $I23$ (#197), $I2_13$ (#199), $P432$ (#207), $F432$ (#209), $F4_132$ (#210), $I432$ (#211), $P4_232$ (#208), $I4_132$ (#214)	$P4_332$ (#212)/$P4_132$ (#213)

The International Tables of Crystallography space group number is in parenthesis.

groups (Table 12.1) of the total of 230 total space groups. The Sohnke space groups do not contain a mirror, combination mirror, simple inversion, or combination inversion symmetry operation (called symmetry operators of the second kind), which would generate the enantiomer by the crystallographic symmetry. Note that there are an additional 73 non-centrosymmetric space groups which are not Sohnke space groups since, though they have no inversion or combination inversion symmetry elements, they have mirror or combination mirror symmetry operations that would generate an enantiomorph. It is always important to remember that the combination symmetry elements such as glide planes that contain a mirror symmetry element and rotary-inversion axes that contain an inversion symmetry element also generate enantiomorphs. The Sohnke space groups can be further divided into 43 achiral Sohnke space groups and 11 pairs of chiral Sohnke space groups. The chiral Sohnke space groups are in pairs of groups, which only differ by their chirality from the polarity of the screw axis. When a crystal is found to be in one chiral Sohnke space group, the enantiomeric space group must also be considered, and the best group used for the final refinement.

The scattering of X-rays by a crystal is due to the repeated three-dimensional nature of the electron density of the atoms within the crystal. The unit cell, the three-dimension building block of the crystal, is created by the application of the crystal symmetry to the asymmetric unit that can be viewed as the minimum unit of the crystal. Within the asymmetric unit, for each reflection, designated by its planes (h,k,l), the atoms contribute scattering factors, f, based on their elemental types, which can be summed over inverse space to produce the structure factor for that reflection, $F_{(h,k,l)}$. The structure factor is a complex number consisting of a real part and an imaginary part. For centrosymmetric structures, the scattering factor, f, is a real number and Friedel's law which states that $F_{(h,k,l)} = F_{(-h,-k,-l)}$ holds or that the diffraction pattern is inherently centrosymmetric. The pairs of reflections $F_{(h,k,l)}$ and $F_{(-h,-k,-l)}$ are called Friedel pairs. For non-centrosymmetric structures, anomalous scattering, or resonant scattering, happens due to the excitation of electrons of the atoms by the X-rays, which cause the scattered X-rays to be out of phase with the incident beam, giving rise to two addition components: a real, dispersive component f' and an imaginary, absorptive component f'' to the scattering factor. The scattering factor then can be written as $f = f_0 + f' + if''$, where f_0 is the scattering factor of the atom without resonant scattering effects. The magnitude of f'' generally increases in approximate cubic relationships with atomic number and X-ray wavelength, with deep drops near atomic absorption edges. Figure 12.1 shows the variation of the magnitude of f'' with X-ray wavelength for atoms of carbon, oxygen, and nitrogen, common elements in an organic crystal. The values of f'' for these light elements are extremely small. For comparison, heavier atoms ($Z > 11$) have a magnitude of f'' usually over 1 electron and it can be over 10 electrons. The magnitude of f'' is much larger for X-ray with longer wavelengths. For example, an oxygen atom with Mo K_α radiation has $f'' = 0.0059$ electrons while with Cu K_α radiation $f'' = 0.0289$ electrons, an almost fivefold increase. For non-centrosymmetric structures, imaginary part of the scattering factor causes Friedel's law to no longer hold, so $F_{(h,k,l)} \neq F_{(-h,-k,-l)}$, and the diffraction pattern is no longer centrosymmetric. The nonequivalent pairs $F_{(h,k,l)}$ and $F_{(-h,-k,-l)}$ are called Bijvoet pairs, and the difference between the pair is called the Bijvoet difference. The ratio of Bijvoet difference to the structure factor is the Bijvoet ratio [11]. The Friedif parameter can be calculated to show the magnitude of the resonant scattering effect of a compound for the measurement wavelength; a value greater than 80 means that the absolute structure can be reliably determined [12].

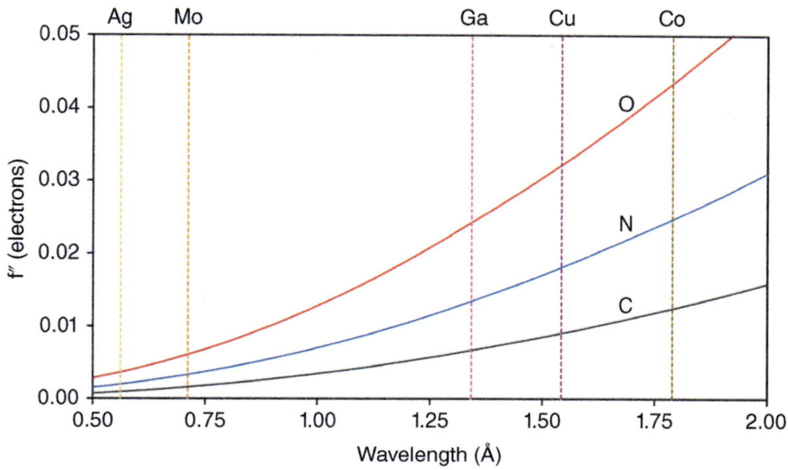

Figure 12.1 The wavelength dependence of the resonant scattering component for atoms of carbon, nitrogen, and oxygen, with the wavelengths for the characteristic K_α radiation for X-ray tubes with Ag, Mo, Ga, Cu, or Co anodes. *Source:* National Institute of Standards and Technology / https://physics.nist.gov/PhysRefData/FFast/html/form.html / last accessed November 22, 2022.

The resonant scattering by the crystal can be used to establish the absolute structure of the contents of the crystal. The absolute configuration of the molecular compound is consequently defined by the absolute structure of the contents of the crystal. The determination of the absolute structure of a chiral compound by a single crystal X-ray diffraction experiment solely by resonant scattering depends on the difference of the magnitude of the Bijvoet pairs $F_{(h,k,l)}$ and $F_{(-h,-k,-l)}$. For structures that contain just light atoms (H—F), due to the very small differences, this difference is just tenths or hundredths of electrons, indicated by values of Friedif in the single digits. This requires extremely careful measurement of the X-ray intensities using a radiation with a longer wavelength such as Cu K_α. Usually wavelengths longer than Cu K_α are not used due to the low resolution of the resulting structure. Use of a low temperature cooling system to cool the crystal to 100–150 K is very useful as the thermal motion will be reduced considerable. This not only usually gives a better-defined structure with smaller uncertainties but also greatly increases the intensities of the higher angle reflections, where there are greater number of reflections, and gives a much better estimate of the Bijvoet difference. Data collection with a high multiplicity of observations (redundancy) of at least four times the number of unique reflections covering at least all the Friedel/Bijvoet pairs and preferably the whole Ewald sphere should be performed. Advertising material from Rigaku boasts that the absolute structure of an organic compound with just H, C, and O atoms was determined using Mo K_α radiation, quite a feat made possible using a photon counting X-ray detector with an extremely low noise floor [13].

The statistical analysis of the Bijvoet differences to determine the absolute structure of a crystal by its resonant scattering has evolved greatly since it was first used for a rubidium sodium tartrate with Zr K_α radiation by Bijvoet [14]. The current state of the art for crystals in Sohnke space groups has the refinement of the handedness of the absolute structure, defined as the Flack parameter *x*, which represents the molar fraction of the inverted structure, together with all the other structural parameters. Rather than using the original structure factors, $F_{(h,k,l)}$, for refinement, the compound expression $(1 - x)|F_{(h,k,l)}|^2 + x|F_{(-h,-k,-l)}|^2$ is used [15]. If the refinement gives the Flack parameter $x \approx 0$, the absolute structure is correct; if the Flack parameter $x \approx 1$, the structure is inverted; and if the Flack parameter $x \approx ½$, the structure exists as an inversion twin or the structure is racemic. For the case where $x > ½$, the structure should be inverted. This is usually a simple inversion symmetry operation. For the Sohnke space groups $I4_1$, $I4_122$, $F4_132$, the inversion center is not at the origin of the cell and the inversion symmetry operation needs to consider the position of the inversion center [16]. As all measured values, the Flack *x* has an uncertainty which, for this case, indicates the success of the X-ray diffraction experiment to determine the absolute structure. When the value of the Flack *x* is reported, after the number the uncertainty is given in parenthesis as an integer referring to the last digits of the Flack *x*. Usually, a tolerance of 0.04 is acceptable [10, 17]. This is because the uncertainty is an estimation of the standard deviation, σ. Statistically, 2.97σ encompasses 95% of the probable values for a normal distribution. Therefore, the value plus or minus three times the uncertainty is estimated to be the probable range of values from the experiment.

To reduce the uncertainly of the Flack parameter, post-refinement estimates can be made. One post-refinement estimate is the use of Parsons' quotients method [18] for the Flack *x* parameter, which gives about a threefold lower uncertainty of the absolute structure. This is calculated automatically in SHELXL97. The Flack *x* can be calculated by three different methods, which give different results: the refinement of *x* together with the full matrix refinement of the structure assuming an inversion twin, and classical method with *x* calculated post-refinement with or without Parsons' quotient method. The best Flack *x* is the first and can be reported as Flack *x* (twin). The second and third are also reported in the literature as the Flack *x*, with the third sometimes reported as Parsons' *z*. The post-refinement determination of Flack *x* can give inconsistent results [19].

For more difficult cases, a more sophisticated analysis can be done using Bayesian statistics [20–22]. This is available as a built-in function of the *PLATON* software package [23]. The *PLATON* software package implements several methods, which should give consistent results for a crystal structure with good data [24, 25]. First, the classical post-refinement Flack x and the Flack x using Parsons' quotient methods, reported as Parson's z, are calculated. Next, a statistical analysis is made, which can either by two-way: probability of the correct absolute structure or inverted structure, or three-way: probability of the correct absolute structure, the inverted structure, or a racemic mixture. The method used depends on the possibility that the structure is racemic. Finally, the Hooft y [20, 21] is calculated from the statistical analysis.

As an example of the improvement using statistical methods, in a recent structure of a very poorly diffracting crystal of a natural product, the refinement results in a Flack $x = 0.140$ with an uncertainty of 0.270, which is insufficient to determine the absolute structure. Using the Parsons' quotient method, $x = 0.036$ with an uncertainty of 0.046. The statistical analysis by *PLATON* has the probability of the correct absolute structure in the two-way analysis as 1.000. For the three-way analysis, probability of the correct absolute structure is again 1.000, while the probability of a racemic twin is 4.0×10^{-26} and inverted structure 2×10^{-115}. The same data set gives a Hooft y of 0.06 with uncertainty of 0.04 and a Parsons' z [24, 26] of 0.05 with an uncertainty of 0.05. These post-refinement methods allow the absolute structure to be more confidently considered correct.

If Cu K_α radiation is not available, there are other means to determine the absolute structure. One way is to determine the chirality of one carbon atom in the structure by other means, such as by a nuclear magnetic resonance experiment. All X-ray diffraction experiments will provide the relative stereochemistry of a chiral compound. The additional information of the absolute configuration of one chiral center, for example by the Horeau or Mosher's method, will therefore establish all the rest. Similarly, if the compound has a component of known chirality, this is enough the give the absolute configuration. Otherwise, a synthetic method can be used to add either a heavy atom from the third or greater period, which will provide a resonant scatterer or a fragment of known chirality. The formation of a salt of the natural product, by methylation, acid, or base chemistry, can add an ion containing a strong resonant scatterer. The formation of co-crystals can be used to crystallize oils and viscous liquids to determine the absolute structure [27]. Sometimes, fortuitously, a solvent molecule with heavy atoms such as chloroform crystallizes together with the chiral compound and thus provides a resonant scatterer [28].

12.3 Best Practices

A recent review by Linden [29] details best practices for the determination of the absolute structure of a crystal structure and hence the absolute configuration of the natural product. Some of the points have been mentioned previously. The very first best practice is to choose the best crystal possible for measurement and not infrequently this means testing several crystals to get one that diffracts well and is not an agglomerate or twinned crystal. The latter can be measured and can give comparable structural parameters as a single crystal. However, such crystals require more work to process and can give poor estimates of Bijvoet differences, hampering the determination of the absolute structure.

The data collection should have a good multiplicity of observations, also called redundancy, which means the unique reflections should each be measured several times in different positions, so the incident and diffracted beams have passed through the crystal by many paths. The estimate of the true intensities of the unique reflections are much better using the mean intensities of an averaged set of a good number of equivalent reflections. Other benefits are better frame scaling and better empirical "multi-scan" absorption correction. A multiplicity of observations for the unique reflections of four or greater is recommended. With the large area detector of modern diffractometers, the multiplicity of observations can easily go over 10 without using too much time for the overall data collection. The unique reflections should be appropriate for the non-centrosymmetric point group of the crystal system. At this stage, the space group has not been determined but should be assumed for a chiral natural product to be one of the Sohnke space groups and therefore a Sohnke point group should be chosen for the data collection strategy, Table 12.2. When there are two possible point groups, the one with the lowest symmetry should be preferred; these are indicated by a box around the point group in the table. The quality of the data also needs to be excellent since the Bijvoet differences are extremely small and can easily be lost in the noise. High multiplicity of observations with weak data will not lead to a good absolute structure. Measuring noise repeatedly will just result in more noise. The time per measured image should produce a large fraction of strong reflections, reflections with their signal more than three times the noise. A percentage of 80–95% strong reflections is desired. Integration of the data after a 100 or so images will give the data statistics before much time is spent and a correction to the data collection can be made, if necessary. At this point, the space group should be able to be determined and the point group adjusted. Some trade-off is required between the multiplicity of observations and time per image for

Table 12.2 Crystal systems and their point groups.

	Point Group		
	Non-centrosymmetric		Centrosymmetric
Crystal system	**Sohnke**	**Non-Sohnke**	
Triclinic	$\boxed{1}$		$\bar{1}$
Monoclinic	$\boxed{2}$	m	$2/m$
Orthorhombic	$\boxed{222}$	$mm2$	mmm
Tetragonal	$\boxed{4}$, 422	$4mm$, $\bar{4}$, $\bar{4}2m$	$4/m$, $4/mmm$
Trigonal	$\boxed{3}$, 32	$3m$	$\bar{3}$, $\bar{3}m$
Hexagonal	$\boxed{6}$, 622	$6mm$, $\bar{6}$, $\bar{6}2m$	$6/m$, $6/mmm$
Cubic	$\boxed{23}$, 432	$43m$, $\bar{4}3m$	$m\,\bar{3}$, $m\,\bar{3}m$

The point groups surrounded by a box are those best used for chiral natural products.

small or otherwise weakly diffracting crystals. With experience, it becomes easier to balance between the two to get measurement times that are of reasonable lengths. Some helpful references are [30–37]. Even with a modern diffractometer with a sensitive area detector and a high-brilliance micro-focus Cu K_α source, a usual natural product crystal will take two to five days for a good data collection with data and multiplicity of observations to determine the absolute structure.

After the data collection, the images are processed. Each image is a two-dimensional slice of reciprocal space, x^*y^*, and the sequence of the images represents the z^* dimension. For each sequence of images, the integration program creates stacks of images to create three-dimensional representations of the inverse space in which the reflections appear as approximate spheres. From these stacks, the crude reflection intensities and their counting statistical errors are determined. Usually, the accurate unit cell parameters are determined as well. The next step is the data reduction: correction of the crude reflections for crystal illumination and incident and diffracted path differences, called scaling, and for X-ray adsorption effects. The usual "multi-scan" absorption correction works adequately, but, if possible, the effort to perform face indexing of the crystal followed by a numeric absorption correction should be made.

Even though the space group, sometimes still not yet determined, is a non-centrosymmetric Sohnke group, the corresponding centrosymmetric point group should be used for both integration and data scaling and absorption correction. A point group of a space group is a group that results when all the translational symmetry elements are removed from the space group. If the space group is already known, its Laue point group – the Sohnke point group with an inversion center added – should be used. Where there are two possible Laue point groups, the one of least symmetry should be used initially. After the determination of the correct space group, if the Laue point group is of higher symmetry, the integration should be redone using the correct Laue point group. Although the use of the Laue point group in place of the Sohnke point group appears counterintuitive, the Bijvoet differences are so small that the benefit of the extra multiplicity of observations is much, much greater than the distinction between the Bijvoet pairs. Also, small systematic errors might occur if the Laue point group is not used. Though small, these errors are sufficient to skew the intensity statistics and distort the absolute structure determination.

The determination of the space group with modern software is usually semiautomatic. Some time ago this was a crucial step, fraught with difficulty, and a wrong choice meant no structure was obtained, especially for a crystal without a heavy atom. The robust dual space routines of the latest crystal structure solution software, notably *SHELXT* [38], no longer need a space group, just the correct crystal system and Bravais lattice to give typically an almost complete structure in the correct space group. At this point, it is wise to verify all possible solutions in all suggested space groups and to rule out the wrong ones. The structure solution programs only indicate which solution is most probable and it is left to the crystallographer to make sure. Sometimes the wrong space group refines adequately. The correct space group will always give better results.

The refinement of the structure is these days straightforward. Occasionally, non-hydrogen atoms in the structure solution will be wrong or missing, and no hydrogen atoms will be present. The missing atoms can be found in the electronic difference map after a few cycles of refinement but need to be attributed correctly. If there is disorder or large thermal motion of a chain of atoms, the missing atoms may need to be found in steps. The electronic density map is generated from the Fourier transform of the structure factors. But only the magnitude of the structure factors have been measured experimentally. The phase has not been measured and is generated by the crystal structure model. Regions in space with missing atoms will have poorly defined phases. As the completion of the structure progresses, the crystal structure model becomes more complete, and the phases better determined. The contrary can happen: if an atom is placed in model in the wrong place, it will generate wrong phases and the crystal structure model will become worse. In this case, the wrongly placed atom is generally attributed a large thermal motion. Simply deleting it will correct the model. More complicated problems such as disorder are covered in reference [39].

The specific example of a wrong atom would be a nitrogen atom changed for a carbon atom or vice versa. Here, a structure determined by a complementary method such as nuclear magnetic resonance is very helpful. The use of difference density maps is very useful to visualize the electron density map to verify the atoms of a structure and to locate possible hydrogen atoms. The software suite that comes with the diffractometer generally has a good option. The *ShelXle* [40] is a good free program that can generate three-dimensional electron density maps, along with a lot of additional functionality. If an atom, say nitrogen, is changed wrongly for a carbon atom, instead of the correct seven electrons at the atomic position, the model will have just six electrons. These electrons are spread out through space, occupying a certain volume. The refinement, to accommodate just six electrons, will reduce the size of the "carbon" atom by reducing its thermal motion so that it envelopes the equivalent of just six electrons. Also, if hydrogen atoms are considered, the "carbon" atom will have one more hydrogen atom to maintain its valency. This produces several artifacts in the model. The "carbon" will have a smaller than expected size, which is usually visible but can be hidden by thermal motion effects. The electronic difference density map will have a very visible positive region around the "carbon" for the one electron that was not modeled. A more subtle change would be a small positive region for the missing nitrogen hydrogen atom and two negative regions where the incorrect carbon hydrogens are placed. The verification programs, such as *CheckCIF*, check for these through the Hirshfeld test and the difference density map. If the incorrect atoms are changed for the correct ones, the fitness (R_1, wR_2) parameters will decrease slightly.

The use of the R values for the correctness of a possible structure is not so simple since excellent data can make a bad structure seem very reasonable. The use of the difference density map as well as expected bond angles and lengths goes a long way to confirm the correctness of a structure. *Mogul*, part of the Cambridge Structural Database (CSD) suite from the Cambridge Crystallographic Data Centre (CCDC), available with Mercury [41], can perform a statistical analysis of the comparison of the bond angles and lengths within a structure to other similar structures found within the CSD.

When the structure is well determined, the Flack x parameter can be checked. If it is close to 1, the structure should be inverted, and the refinement continued with the new configuration. If the value deviates significantly from 0, the structure should be refined as a racemic twin. The post-refinement methods, Flack x with Parsons' quotients (Parsons' z),

Hooft *y*, and the Bayesian statistical methods, should be used to confirm the absolute structure. A special note needs to be made if the natural product is found in an achiral Sohnke space group. Even though the structure may refine nicely and has a well determined (with a low uncertainty) absolute structure, the determined absolute structure might be inverted from the real absolute structure. It is necessary to refine in the enantiomorphic space group, which is inverted to the original space group. The best structure that with the lowest *R* factors and lowest atomic positional uncertainties will be the correct one.

The final step, the generation of the crystallographic information framework (CIF) file is extremely important [42–44]. The diffractometer software suite often can generate a ready CIF file directly from the refinement. The programs *publCIF* [45, 46] from the IUCr and *EnCIFer* [47] from the CCDC can be used to edit and complement the CIF file with additional information. The CIF file is the primary archival means and can be deposited with the CCDC [48, 49] and Crystallographic Open Database [50, 51].

12.4 Structure Validation

The validation of the crystal structure should be done using the crystallographic information framework (CIF) file with the *PLATON* [23] structural verification routine, either with the standalone program or though the *CheckCIF* service [4, 5, 52] of the IUCr. All validation ALERTS (A, B, C, and G levels) should be checked thoroughly. If any crystallographic errors are found, these should be corrected. Any remaining ALERTS should be responded to in the validation response report, which should be added to the CIF file.

The final validation of the crystal structure ideally should be done with all the experimental analyses together by crystallographer and the natural products team. There is always the possibility that the measured single crystal might not represent the natural product being studied but could be the result of an impurity.

12.5 The Absolute Configuration of (+)-Lanatine A

(+)-Lanatine A was isolated from the aerial parts of *Lupinus lanatus* (Fabaceae) from Frederico Westphalen, Rio Grande do Sul, Brazil, in the period from October to December 2008 [53]. Large crystals were grown by slow evaporation. At the time, the X-ray facility had a single crystal X-ray diffractometer with just Mo K_α radiation, so it was used. A low temperature data collection would be ideal; however, the low temperature apparatus at the time was not functioning. The crystal diffracted very well, giving data up to $\theta = 30.5°$. The structure was well determined, Figure 12.2. The resonant scattering from the all-organic structure is extremely low – the short wavelength and only light atoms meant that the Friedel value was just 5 – and the usual absolute structure parameters were not very conclusive. With the use of Parson's quotients, the Flack x was calculated to be −0.1(2). The uncertainty of 0.2 signifies that the true value is 95% within the found parameter plus or minus three times the uncertainty. The Flack $x = -0.1(2)$ would probably be between −0.7 and 0.5, 95% of the time. Remembering that a Flack x of zero means the correct handedness, this is almost enough to consider the handedness of the absolute structure to be correct if there were no possibility of a racemic mixture, which would have a Flack $x = 0.5$. Due to the large number of well-determined Bijvoet pairs, the Bayesian statistics is more decisive. The probability of the correct handedness considering only an enantiomerically pure crystal, P2(true), is 1.000. For the case which allows a racemic mixture, the probability of the correct enantiomorph is P3(true), which in this case is 0.999, and the probability of a racemic twin, P3(rac-twin), is 0.001. The analysis gives a near certain probability that the correct enantiomorph is as refined, Table 12.3. The absolute structure was chosen to be this and the absolute configuration of (+)-lanatine A was reported.

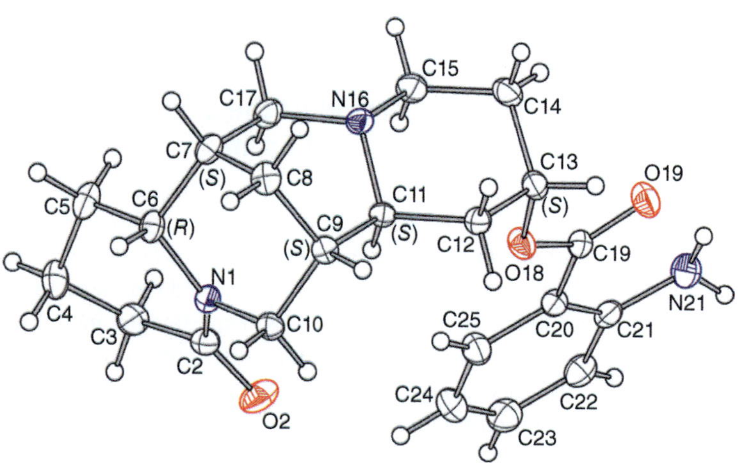

Figure 12.2 Probable absolute configuration of (+)-lanatine A, $C_{22}H_{29}N_3O_3$, determined by the Bayesian analysis of the Bijvoet differences.

Table 12.3 Data for the X-ray crystal structure of (+)-Lanatine A.

Parameter	(+)-Lanatine A
Formula	$C_{22}H_{29}N_3O_3$
Crystal system	Orthorhombic
Space group	$P2_12_12_1$
T	295 K
Radiation	Mo K_α
Reflections	57942
Unique reflections	5973
R_{int}	0.0153
R_1 [$F^2 > 2\sigma(F^2)$]	0.0372
wR_2 [all data]	0.1017
Resonant scattering	0.0006
Friedif	5
Bijvoet pairs	2605 (100%)
Flack x	−0.1(2)
Hooft y	−0.3(2)
Parsons' z	−0.3(2)
Probability P2(true)	1.000
Probability P3(true), P3(rac-twin)	0.999, 0.001

12.6 The Absolute Configuration of the Diacetylated Form of Acrenol and the Acetylated Form of Humirianthol

Acrenol and humirianthol were isolated from the tubers of *Humirianthera ampla* (Miers) collected from Rio Branco, Acre, in Brazil and their structures determined by nuclear magnetic resonance (NMR) experiments, including their relative stereochemistry [54]. The absolute configuration of C-15 of humirianthol was determined using the Horeau method [55] and could be used to estimate the absolute configuration of humirianthol. Due to the similarity between the structure of humirianthol and acrenol, the diterpenoid core of acrenol was presumed to have the same absolute structure. The relative stereochemistry of the C-15 atom of the dihydroxyethyl group of acrenol was not possible to determine by the NMR experiments. None of the natural products isolated from *H. ampla* crystallized. Their acetyl derivatives were synthesized to form good crystals. The diacetyl derivative of acrenol and the acetyl derivative of humirianthol formed crystals from the evaporation of a chloroform solution, and their structures were measured. At the time, in 1993, the use of Cu K_α radiation in single crystal X-ray diffractometers was extremely rare; Mo K_α radiation was used almost exclusively. The crystal of diacetylated acrenol was measured and diffracted very weakly, yielding data only to 20° on θ. When the structure was solved, it was apparent that fortuitously a molecule of chloroform co-crystallized as a solvate together with diacetylated acrenol. The chlorine atoms contributed a high resonant scattering to the diffraction pattern as confirmed by the large Friedif value, enough that the absolute structure of the crystal was established. The absolute configuration was therefore determined unambiguously, and the structure is of (15*R*)-15,16-diacetato-3β,20-epoxy-3α-hydroxy-9-epi-7-pimaren-19,6β-olide chloroform solvate, Figure 12.3a. The stereodescriptors of the stereogenic

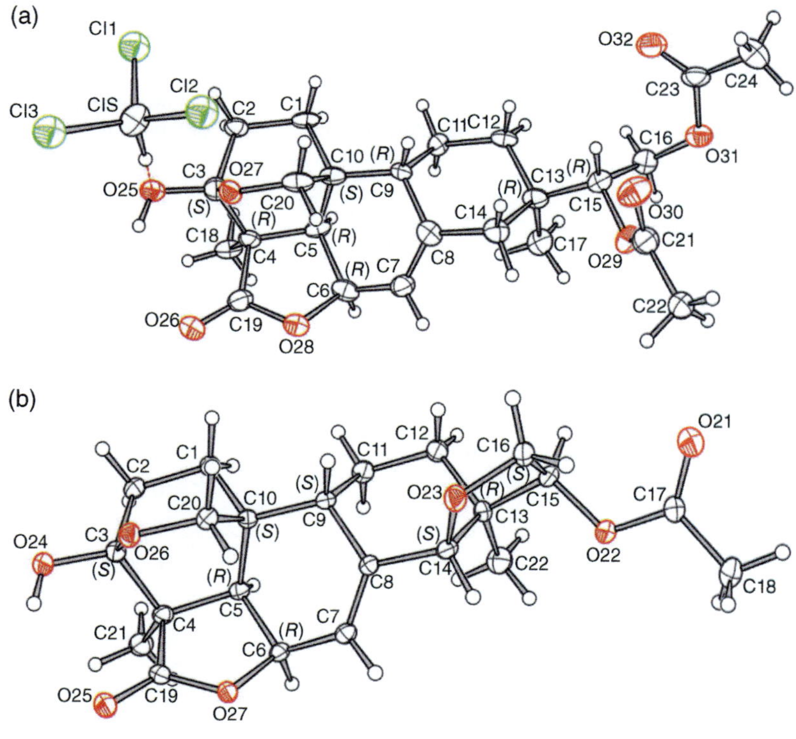

Figure 12.3 Absolute configuration of the diacetylated form of acrenol, $C_{24}H_{32}O_8{\cdot}CHCl_3$, determined by the resonant scattering of the chlorine atoms (a) and absolute structure of the acetylated form of humirianthol, $C_{22}H_{28}O_7$, determined by the absolute (*S*) configuration of C-15 determined by the Horeau method (b).

carbon centers are C-3 (*S*), C-4 (*R*), C-5 (*R*), C-6 (*R*), C-9 (*R*), C-10 (*S*), C-13 (*R*), and C-15 (*R*). The acetylated humirianthol gave crystals with just the natural product [56]. The absolute structure of the crystal could not be determined from the resonant scattering. However, with the knowledge of the absolute configuration of C-15, (*S*), determined experimentally, the absolute configuration of acetylated humirianthol could be established reliably, Figure 12.3b. The stereodescriptors of the stereogenic carbon centers are C-3 (*S*), C-4 (*R*), C-5 (*R*), C-6 (*R*), C-9 (*S*), C-10 (*S*), C-13 (*R*) C-14 (*S*), and C-15 (*S*). The absolute stereochemistries of the substituted pimarene systems for acrenol and humirianthol are determined to be identical. Table 12.4 summarizes the crystal data and the absolute structure parameters for diacetylated acrenol and acetylated humirianthol. No absolute structure parameters are available for the crystal structure of acetylated humirianthol since the Bijvoet pairs were incorrectly merged and the original diffraction images are no longer available.

Table 12.4 Data for the X-ray crystal structures of diacetylated acrenol and acetylated humirianthol.

Parameter	Diacetylated acrenol chloroform solvate	Acetylated humirianthol
Formula	$C_{24}H_{32}O_8 \cdot CHCl_3$	$C_{22}H_{28}O_7$
Crystal system	Orthorhombic	Orthorhombic
Space group	$P2_12_12_1$	$P2_12_12_1$
T	100 K	150 K
Radiation	Mo K_α	Mo K_α
Reflections	8484	10 310
Unique reflections	2410	2039
R_{int}	0.132	0.15
R_1 $[F^2 > 2\sigma(F^2)]$	0.063	0.059
wR_2 (all data)	0.161	0.146
Resonant scattering	0.0081	0.0007
Friedif	121	
Bijvoet pairs	985 (97%)	
Flack x	0.2(2)	
Hooft y	0.18(11)	
Parsons' z	0.19(9)	
Probability P2(true)	1.000	
Probability P3(true), P3(rac-twin)	0.958, 0.042	

12.7 The Absolute Configuration of Ester Form of Clemateol

Clemateol was isolated from the essential oil from the leaves of *Calea clematidea* (Asteraceae), collected in November 2013 in Santana do Livramento, Rio Grande do Sul, Brazil [57, 58]. Its absolute configuration was determined initially by the Horeau method, but the referees of the publication had reservations about the efficacy of the method. Clemateol has two stereogenic carbon centers, C-3 and C-4. A new study used Mosher's method [59] to determine the chirality of C-4 to be (*R*) of its (*R*)- and (*S*)-2-methoxy-2-phenylacetate ester derivatives agreeing with the previous assignment by Horeau's method. The single crystals of the (*R*)-2-methoxy-2-phenylacetate (MPA) ester derivative of clemateol were obtained. The measurement of the crystal on a single crystal X-ray diffraction with Cu K_α radiation at 100 K was graciously done at the Universidade Federal Fluminense. The knowledge of the stereogenic carbon atom of the MPA group would have been sufficient to determine the absolute structure of the crystal and, by that, the absolute structure of the (*R*)-MPA ester of clemateol. The crystal structure was well determined, Figure 12.4, and the

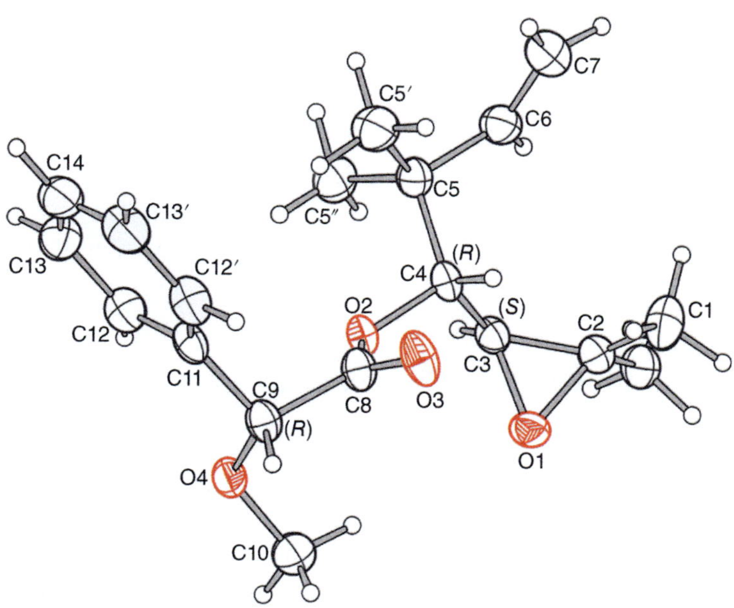

Figure 12.4 The absolute configuration of (*R*)-MPA ester of clemateol, $C_{19}H_{26}O_4$, determined by the resonant scattering effects in the single crystal experiment and by Mosher's method for C-4.

Table 12.5 Data for the X-ray crystal structure of the (*R*)-MPA ester of clemateol.

Parameter	(*R*)-MPA ester of clemateol
Formula	$C_{19}H_{26}O_4$
Crystal system	Orthorhombic
Space group	$P2_12_12_1$
T	100 K
Radiation	Cu K_α
Reflections	7630
Unique reflections	3487
R_{int}	0.0228
R_1 [$F^2 > 2\sigma(F^2)$]	0.0369
wR_2 (all data)	0.1001
Resonant scattering	0.0034
Friedif	33
Bijvoet pairs	1450 (93%)
Flack x	0.04(9)
Hooft y	0.06(8)
Parsons' z	0.07(10)
Probability P2	1.000
Probability P3(true), P3(rac-twin)	1.000, 4×10^{-7}

resonant scattering effects were strong enough, with a Friedif of 33, to give the absolute structure, Table 12.5. The absolute configuration of the (*R*)-MPA ester of clemateol was C-3 (*S*) and C-4 (*R*).

12.8 Relative Configurations of Waltherione A, Waltherione B, and Vanessine

Waltherione A, waltherione B, and vanessine were obtained from the stem bark of the shrub *Waltheria douradinha* (Saint-Hilaire), collected in São Pedro do Sul, Rio Grande do Sul, Brazil [60]. Waltherione A had been the subject of a previous study in which its relative stereochemistry was reported [61]. In the new study, crystals of waltherione A were obtained from the slow evaporation of a methanol: water (1 : 1 v/v) solution. The crystal structure was found in the chiral Sohnke space group $P4_3$. The alternative setting $P4_1$ only differs from $P4_3$ by the polarity of the screw axis. At the time in 2005, no reasonable crystal structure solution was found in the $P4_1$. Now, repeating the effort with *SHELXT* [38], the impressive dual-space crystal-structure determination program, which was not available at the time of the original structure solution and refinement, the only solution was given in $P4_3$. Even forcing *SHELXT* to use just the space group $P4_1$ yielded no acceptable solution in that space group. The only correct space group therefore can only be $P4_3$. In the asymmetric unit of its crystal structure, there are four independent molecules of waltherione A, and, disordered over two sites, one water molecule and one methanol molecule. Just one molecule is shown in Figure 12.5a. As the crystal was measured with Mo K_α radiation, the resonant scattering effects are too weak to reliably determine the absolute structure. The relative configuration of waltherione A was definitively found, and the previous relative configuration was corrected. Crystals of the *N*-methylated form of waltherione A and the *O*-methylated form of waltherione B were each obtained from CH_2Cl_2:MeOH (4 : 1 v/v) solutions. Both crystal structures were found in $P2_12_12_1$. Due to the use of Mo K_α radiation, only their relative configurations can be designated; the Bayesian statistical analysis is not sufficient, Table 12.6. It is obvious that the configurations found for the original waltherione A and its *O*-methylated form are inverted. This suggests that the absolute structure indications for at least one of the structures is wrong. The absolute configuration needs to be determined using Cu K_α radiation to determine the correct absolute structure.

12.9 The Absolute Configuration of Condaline A

Condaline A was isolated from the root bark of the tree *Condalia buxifolia* (Rhamnaceae) collected near Lavras do Sul (Rio Grande do Sul, Brazil) [62]. To form crystals of the natural product, condaline A dissolved in acetone was treated with methyl iodide to synthesize the methylated cation with an iodide anion. Slow evaporation of solution produced yellow crystals suitable for the X-ray diffraction experiment. The single crystal study found two methylated cations, two iodide anions and six solvated water molecules in

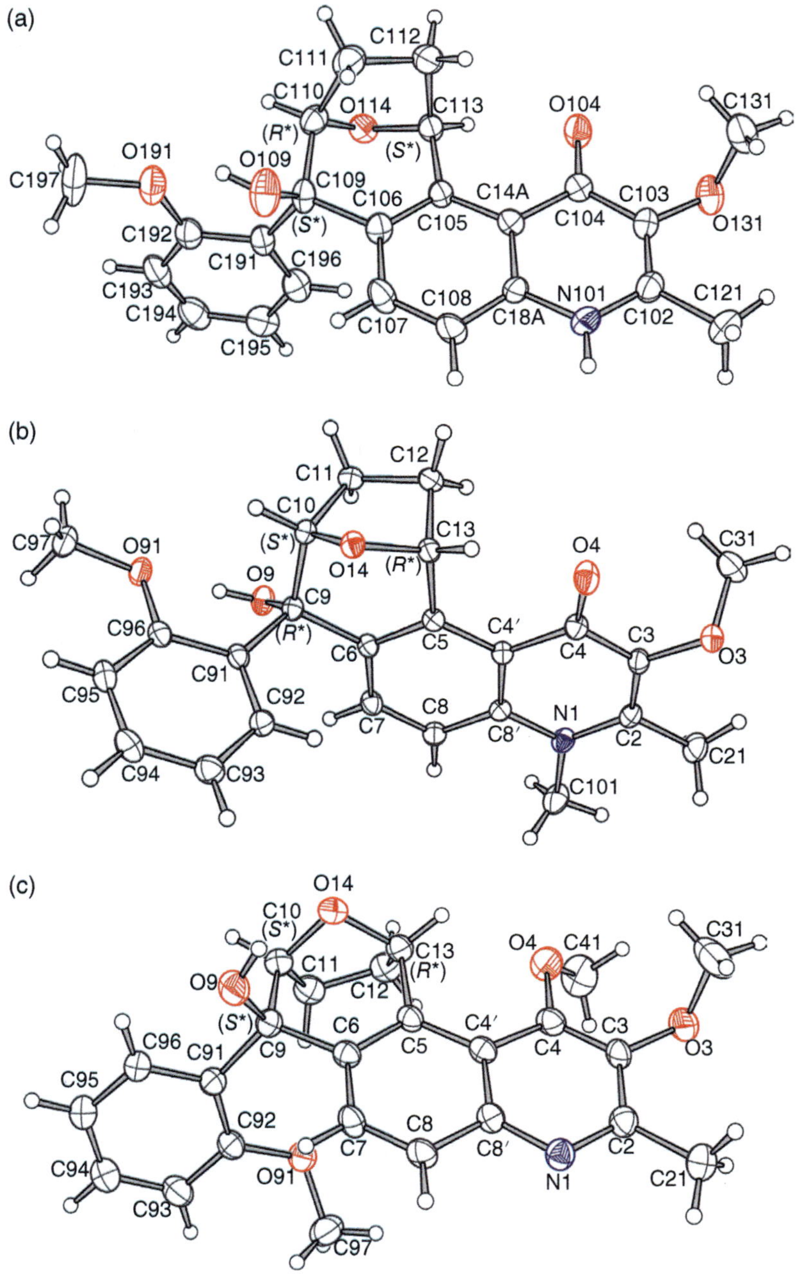

Figure 12.5 Relative configurations of waltherione A, $(C_{23}H_{24}NO_4)_4{\cdot}CH_3OH{\cdot}H_2O$ (a); the *N*-methylated form of waltherione A, $C_{24}H_{25}NO_5$ (b); and the *O*-methylated form waltherione B, $C_{24}H_{25}NO_5$ (c).

Table 12.6 Data for the X-ray crystal structures of waltherione A, the *O*-methylated form of waltherione B, and the *N*-methylated form of vanessine.

Parameter	Waltherione A	*N*-methylated Waltherione A	*O*-methylated Waltherione B
Formula	$(C_{23}H_{24}NO_4)_4{\cdot}CH_3OH{\cdot}H_2O$	$C_{24}H_{25}NO_5$	$C_{24}H_{25}NO_5$
Crystal system	Tetragonal	Orthorhombic	Orthorhombic
Space group	$P4_3$	$P2_12_12_1$	$P2_12_12_1$
T	295 K	295 K	295 K
Radiation	Mo K_α	Mo K_α	Mo K_α
Reflections	50 305	13 172	21 243
Unique reflections	17 055	3 132	4 082
R_{int}	0.1159	0.0422	0.0916
$R_1\ [F^2 > 2\sigma(F^2)]$	0.0592	0.0404	0.0512
wR_2 (all data)	0.1587	0.1103	0.1304
Resonant scattering	0.0007	0.0006	0.0006
Friedif	6	6	6
Bijvoet pairs	7 989 (91%)	2 009 (80%)	1 734 (98%)
Flack x	−0.3(7)	−0.3(8)	−0.3(10)
Hooft y	−0.4(6)	−0.2(6)	−0.4(9)
Parsons' z	−0.5(7)	−0.1(7)	−0.7(12)
Probability P2(true)	0.936	0.868	0.739
Probability P3(true), P3(rac-twin)	0.689, 0.264	0.590, 0.320	0.495, 0.330

the asymmetric unit of the monoclinic $P2_1$ Sohnke space group. Figure 12.6 shows just one methylated condaline A cation, the terminal amine containing N34 was methylated, with one iodide anion. Despite the weak diffraction by the crystal, the pronounced resonant scattering due to the iodide anions, as seen by the large value for the resonant scattering and Friedif, allows the definite determination of the absolute structure. The Flack x of 0.076 had an uncertainty of just 0.012, Table 12.7.

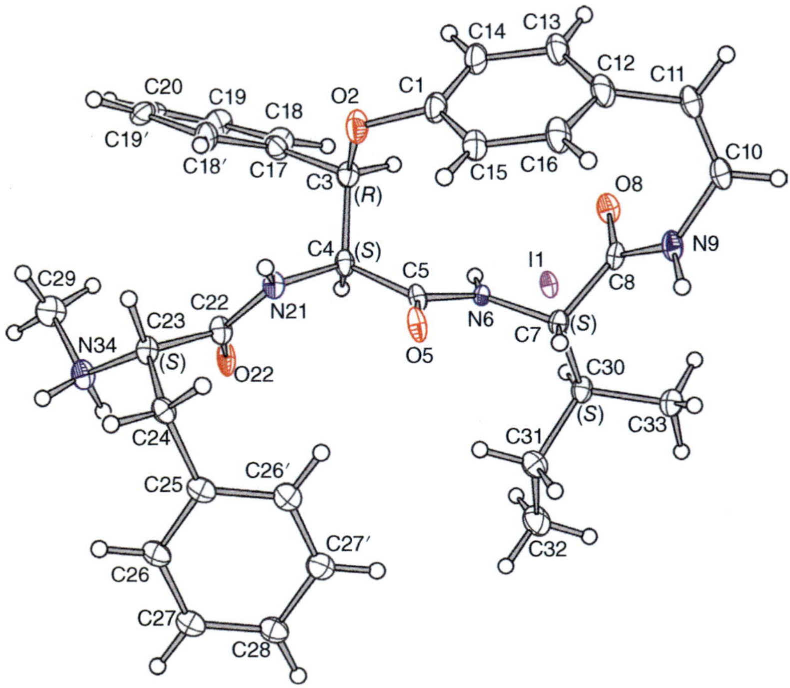

Figure 12.6 Absolute configurations of the methylated condaline A iodide salt, $(C_{33}H_{39}N_4O_4)I \cdot 3H_2O$, determined by the resonant scattering effects of the iodide anions.

12.10 CSD Deposit Numbers

All the crystallographic data have been deposited in the Cambridge Structural Database and are available for free from the Cambridge Crystallographic Data Centre [63] using the deposit number, listed in Table 12.8.

Table 12.7 Data for the X-ray crystal structure of the methylated condaline A iodide salt.

Parameter	Methylated condaline A iodide salt
Formula	$(C_{33}H_{39}N_4O_4)I{\cdot}3H_2O$
Crystal system	Monoclinic
Space group	$P2_1$
T	100 K
Radiation	Mo K_α
Reflections	80 659
Unique reflections	18 197
R_{int}	0.1271
R_1 $[F^2 > 2\sigma(F^2)]$	0.0885
wR_2 (all data)	0.2211
Resonant scattering	0.0377
Friedif	473
Bijvoet pairs	8 722 (99%)
Flack x	0.076(12)
Hooft y	0.043(11)
Parsons' z	0.093(12)
Probability P2(true)	1.000
Probability P3(true), P3(rac-twin)	1.000, 0.000

Table 12.8 Deposit numbers of the crystallographic data available from the CCDC.

Compound	Deposit number
(+)-Lanatine	798749
Diacetylated acrenol chloroform solvate	206790
Acetylated humirianthol	206789
(*R*)-MPA ester of clemateol	1551963
Waltherione A	626424
N-methylated vanessine	626425
O-methylated waltherione A	626426
Condaline A	1451898

12.11 Conclusions and Future Directions

The determination of the absolute stereochemistry of natural products has advanced enormously in recent years. The equipment is much more capable; the mathematical basis of the elucidation of the absolute structure is much better defined; and the software are much more user-friendly. The first leads to a considerably improved determination of the Bijvoet differences, even for small crystals, and sometimes with a radiation that is not optimal, such as Mo K_{α}. The second permits the confident establishment of the absolute structure even with tiny Bijvoet differences. Finally, the last aids in the knowledge if the absolute structure was correctly determined within a reasonable confidence interval.

A further aspect is the deposition of raw data under the fairness, accuracy, confidentiality, and transparency (FACT), and the findable, accessible, interoperable, and re-usable (FAIR) principles allow the future redetermination of the crystal structure and its absolute structure, using more advanced tools [43]. The improvement in structure analysis beyond the independent-atom model, such as using invarioms [64] or refinement of the resonant scattering correction parameters [65], might become mainstream.

References

1 le Pevelen, D.D. (2017). X-ray crystallography of small molecules: theory and workflow. *Encycl. Spectrosc. Spectrom.* 624–639. https://doi.org/10.1016/B978-0-12-409547-2.05260-4.

2 Yamano, A. (2019). Single crystal structure analysis. *Encycl. Anal. Chem.* 1–24. https://doi.org/10.1002/9780470027318.A9665.

3 Matsumoto, T., Yamano, A., Sato, T. et al. (2021). "What is this?" a structure analysis tool for rapid and automated solution of small molecule structures. *J. Chem. Crystallogr.* 51: 438–450. https://doi.org/10.1007/S10870-020-00867-W/FIGURES/9.

4 CheckCIF (n.d.). http://checkcif.iucr.org (accessed 21 October 2022).

5 Spek, A.L. (2020). checkCIF validation ALERTS: what they mean and how to respond. *Acta Crystallogr. Sect. E* 76: 1–11. https://doi.org/10.1107/S2056989019016244.

6 Baias, M., Widdifield, C.M., Dumez, J.-N. et al. (2013). Powder crystallography of pharmaceutical materials by combined crystal structure prediction and solid-state 1 H NMR spectroscopy powder crystallography of pharmaceutical materials by combined crystal structure prediction and solid-state H NMR spectroscopy. *Phys. Chem. Chem. Phys.* 15: 7899–8442.

7 Jones, C.G., Martynowycz, M.W., Hattne, J. et al. (2018). The CryoEM method MicroED as a powerful tool for small molecule structure determination. *ACS Cent. Sci.* 4: 1587–1592. https://doi.org/10.1021/acscentsci.8b00760.

8 Samperisi, L., Zou, X., and Huang, Z. (2022). Three-dimensional electron diffraction: a powerful structural characterization technique for crystal engineering. *CrystEngComm* 24: 2719–2728. https://doi.org/10.1039/D2CE00051B.

9 Gruene, T., Holstein, J.J., Clever, G.H., and Keppler, B. (2021). Establishing electron diffraction in chemical crystallography. *Nat. Rev. Chem.* 5 (9): 1–9. https://doi.org/10.1038/s41570-021-00302-4.

10 Flack, H.D. and Bernardinelli, G. (1999). Absolute structure and absolute configuration. *Acta Crystallogr. A* 55: 908–915. https://doi.org/10.1107/S0108767399004262.

11 Girard, É., Stelter, M., Vicat, J., and Kahn, R. (2003). A new class of lanthanide complexes to obtain high-phasing-power heavy-atom derivatives for macromolecular crystallography. *Acta Crystallogr. D Biol. Crystallogr.* 59: 1914–1922. https://doi.org/10.1107/S0907444903020511.

12 Flack, H.D. and Shmueli, U. (2007). The mean-square Friedel intensity difference in P1 with a centrosymmetric substructure. *Acta Crystallogr. A* 63: 257–265. https://doi.org/10.1107/S0108767307002802.

13 Determining chirality of a carbohydrate with Mo radiation | Rigaku Global Website (n.d.). https://exam.rigaku.com/applications/bytes/small-molecule/xtalab-mini-ii/1515532026 (accessed 22 October 2022).

14 Bijvoet, J.M., Peerdeman, A.F., and van Bommel, A.J. (1951). Determination of the absolute configuration of optically active compounds by means of X-rays. *Nature* 168: 271–272. https://doi.org/10.1038/168271a0.

15 Flack, H.D. (1983). On enantiomorph-polarity estimation. *Acta Crystallogr. A* 39: 876–881. https://doi.org/10.1107/S0108767383001762.

16 Bernardinelli, G. and Flack, H.D. (1985). Least-squares absolute-structure refinement. Practical experience and ancillary calculations. *Acta Crystallogr., Sect. A: Found. Adv.* 41: 500–511. https://doi.org/10.1107/S0108767385001064.

17 Flack, H.D. and Bernardinelli, G. (2000). Reporting and evaluating absolute-structure and absolute-configuration determinations. *J. Appl. Crystallogr.* 33: 1143–1148. https://doi.org/10.1107/S0021889800007184/KS0021SUP1.TXT.

18 Parsons, S. (2017). Determination of absolute configuration using X-ray diffraction. *Tetrahedron Asymmetry* 28: 1304–1313. https://doi.org/10.1016/J.TETASY.2017.08.018.

19 Watkin, D.J. and Cooper, R.I. (2016). Why direct and post-refinement determinations of absolute structure may give different results. *Acta Crystallogr. Sect. B: Struct. Sci. Cryst. Eng. Mater.* 72: 661–683. https://doi.org/10.1107/S2052520616012890.

20 Hooft, R.W.W., Straver, L.H., and Spek, A.L. (2008). Determination of absolute structure using Bayesian statistics on Bijvoet differences. *J. Appl. Crystallogr.* 41: 96–103. https://doi.org/10.1107/S0021889807059870.

21 Hooft, R.W.W., Straver, L.H., and Spek, A.L. (2010). Using the *t*-distribution to improve the absolute structure assignment with likelihood calculations. *J. Appl. Crystallogr.* 43: 665–668. https://doi.org/10.1107/S0021889810018601.

22 Hooft, R.W.W., Straver, L.H., and Spek, A.L. (2009). Probability plots based on Student's t-distribution. *Acta Crystallogr. A* 65: 319–321. https://doi.org/10.1107/S0108767309009908/ZM5057SUP1.TXT.

23 Spek, A.L. (2003). Single-crystal structure validation with the program PLATON. *J. Appl. Crystallogr.* 36: 7–13. https://doi.org/10.1107/S0021889802022112.

24 Parsons, S., Flack, H.D., and Wagner, T. (2013). Use of intensity quotients and differences in absolute structure refinement. *Acta Crystallogr. Sect. B: Struct. Sci. Cryst. Eng. Mater.* 69: 249–259. https://doi.org/10.1107/S2052519213010014.

25 Reibenspies, J. and Bhuvanesh, N. (2013). Absolute structure of R-(−)-2-methylpiperazine and S-(+)-2-methylpiperazine. *Acta Crystallogr. Sect. B: Struct. Sci. Cryst. Eng. Mater.* 69: 288–293. https://doi.org/10.1107/S2052519213008713.

26 Parsons, S. and Flack, H. (2004). Precise absolute-structure determination in light-atom crystals. *Acta Crystallogr. A* 60: s61.

27 Eccles, K.S., Deasy, R.E., Fábián, L. et al. (2011). The use of co-crystals for the determination of absolute stereochemistry: an alternative to salt formation. *J. Organomet. Chem.* 76: 1159–1162. https://doi.org/10.1021/JO102148P.

28 Chantrapromma, S., Fun, H.-K., Laphookhieo, S. et al. (2006). Determination of absolute configuration of natural products by X-ray diffraction: a novel approach of incorporating heavy-atom-containing solvent molecules into the single crystals and refinement of Flack parameter. *Front. Nat. Prod. Chem.* 1: 99–106. https://doi.org/10.2174/1574089054583759.

29 Linden, A. (2017). Best practice and pitfalls in absolute structure determination. *Tetrahedron Asymmetry* 28: 1314–1320. https://doi.org/10.1016/J.TETASY.2017.07.010.

30 Williams, A.E., Thompson, A.L., and Watkin, D.J. (2019). The role of multiple observations in small-molecule single-crystal service X-ray structure determination. *Acta Crystallogr. Sect. B: Struct. Sci. Cryst. Eng. Mater.* 75: 657–673. https://doi.org/10.1107/S2052520619006681/PS5073ISUP2.HKL.

31 Muller, P. (2009). Practical suggestions for better crystal structures. *Crystallogr. Rev.* 15: 57–83. https://doi.org/10.1080/08893110802547240.

32 Sanjuan-Szklarz, W.F., Hoser, A.A., Gutmann, M. et al. (2016). Yes, one can obtain better quality structures from routine X-ray data collection. *IUCrJ* 3: 61–70. https://doi.org/10.1107/S2052252515020941/FC5011SUP3.ZIP.

33 Thompson, A.L. (2019). Chemical crystallography: when are 'bad data' 'good data'? *Crystallogr. Rev.* 25: 3–53. https://doi.org/10.1080/0889311X.2019.1569643.

34 Powell, H.R. (2019). From then till now: changing data collection methods in single crystal X-ray crystallography since 1912. *Crystallogr. Rev.* 25: 264–294. https://doi.org/10.1080/0889311X.2019.1615483.

35 Thompson, A.L., Jenkinson, S.F., and Fleet, G.W.J. (2017). Some experimental aspects of absolute configuration determination using single crystal X-ray diffraction. *Tetrahedron Asymmetry* 28: 1330–1336. https://doi.org/10.1016/J.TETASY.2017.08.016.

36 Cooper, R.I. (2020). Recent developments in the refinement and analysis of crystal structures. *Struct. Bond.* 185: 43–67. https://doi.org/10.1007/430_2020_76/COVER.

37 Spek, A.L. (2016). Absolute structure determination: pushing the limits. *Acta Crystallogr. Sect. B: Struct. Sci. Cryst. Eng. Mater.* 72: 659–660. https://doi.org/10.1107/S2052520616014773.

38 Sheldrick, G.M. (2015). SHELXT – integrated space-group and crystal-structure determination. *Acta Crystallogr. A* 71: 3–8. https://doi.org/10.1107/S2053273314026370.

39 Müller, P., Herbst-Irmer, R., Spek, A.L. et al. (2006). *Crystal Structure Refinement: A Crystallographer's Guide to SHELXL*. Oxford: Oxford University Press.

40 Hübschle, C.B., Sheldrick, G.M., and Dittrich, B. (2011). ShelXle : a Qt graphical user interface for SHELXL. *J. Appl. Crystallogr.* 44: 1281–1284. https://doi.org/10.1107/S0021889811043202.

41 Macrae, C.F., Bruno, I.J., Chisholm, J.A. et al. (2008). *Mercury CSD 2.0* – new features for the visualization and investigation of crystal structures. *J. Appl. Crystallogr.* 41: 466–470. https://doi.org/10.1107/S0021889807067908.

42 Hall, S.R., Allen, F.H., and David, B.I. (1991). International Union of Crystallography Commission on crystallographic data commission on journals working party on crystallographic information the crystallographic information file (CIF): a new standard archive file for crystallography*. *Acta Cryst* 47: 655–685.

43 Helliwell, J.R. (2019). FACT and FAIR with big data allows objectivity in science: the view of crystallography. *Struct. Dyn.* 6: 054306. https://doi.org/10.1063/1.5124439.

44 Helliwell, J.R., McMahon, B., Guss, J.M., and Kroon-Batenburg, L.M.J. (2017). The science is in the data. *IUCrJ* 4: 714–722. https://doi.org/10.1107/S2052252517013690.

45 Westrip, S.P. (2010). publCIF: software for editing, validating and formatting crystallographic information files. *J. Appl. Crystallogr.* 43: 920–925. https://doi.org/10.1107/S0021889810022120.

46 (IUCr) (IUCr) publCIF (n.d.). Free software to edit and preview a CIF for publication. https://journals.iucr.org/services/cif/publcif (accessed 21 October 2022).

47 The Cambridge Crystallographic Data Centre (CCDC) (n.d.). EnCIFer. https://www.ccdc.cam.ac.uk/Community/csd-community/encifer (accessed 21 October 2022).

48 Groom, C.R., Bruno, I.J., Lightfoot, M.P., and Ward, S.C. (2016). The Cambridge structural database. *Acta Cryst* B72: 171–179.

49 The Cambridge Crystallographic Data Centre (CCDC) (n.d.). Deposit a structure. https://www.ccdc.cam.ac.uk/Community/depositastructure (accessed 21 October 2022).

50 Vaitkus, A., Merkys, A., and Grazulis, S. (2021). Validation of the crystallography open database using the crystallographic information framework. *J. Appl. Crystallogr.* 54: 661–672. https://doi.org/10.1107/S1600576720016532/YR5065SUP1.PDF.

51 Crystallography open database: validation and deposition interface (n.d.). https://www.crystallography.net/cod/initiate_deposition.php (accessed 21 October 2022).

52 Spek, A.L. (2018). What makes a crystal structure report valid? *Inorg. Chim. Acta* 470: 232–237. https://doi.org/10.1016/J.ICA.2017.04.036.

53 Neto, A.T., Oliveira, C.Q., Ilha, V. et al. (2011). Quinolizidine alkaloids from Lupinus lanatus. *J. Mol. Struct.* 1004: 174–177. https://doi.org/10.1016/j.molstruc.2011.07.061.

54 Graebner, I.B., Mostardeiro, M.A., Ethur, E.M. et al. (2000). Diterpenoids from *Humirianthera ampla*. *Phytochemistry* 53: 955–959. https://doi.org/10.1016/S0031-9422(99)00585-3.

55 König, W.A., Gehrcke, B., and Weseloh, G. (1994). Determination of the absolute configuration of secondary alcohols with Horeau's method including enantioselective gas chromatography. *Chirality* 6: 141–147. https://doi.org/10.1002/chir.530060215.

56 Burrow, R.A., Morel, A.F., Graebner, I.B. et al. (2003). The acetyl derivative of humirianthol. *Acta Crystallogr., Sect. E: Struct. Rep. Online* 59: o347–o349. https://doi.org/10.1107/S1600536803003374.

57 Pedroso, M., Gehn, A.Z., Stivanin, M.L. et al. (2018). Absolute configuration of clemateol. *J. Braz. Chem. Soc.* 29. https://doi.org/10.21577/0103-5053.20170189.

58 Flach, A., Gregel, B., Simionatto, E. et al. (2002). Chemical analysis and antifungal activity of the essential oil of *Calea clematidea*. *Planta Med.* 68: 836–838. https://doi.org/10.1055/s-2002-34414.

59 Seco, J.M., Quiñoá, E., and Riguera, R. (2001). A practical guide for the assignment of the absolute configuration of alcohols, amines and carboxylic acids by NMR. *Tetrahedron Asymmetry* 12: 2915–2925.

60 Gressler, V., Stüker, C.Z., de O C Dias, G. et al. (2008). Quinolone alkaloids from Waltheria douradinha. *Phytochemistry* 69: 994–999. https://doi.org/10.1016/j.phytochem.2007.10.018.

61 Hoelzel, S.C.S.M., Vieira, E.R., Giacomelli, S.R. et al. (2005). An unusual quinolinone alkaloid from Waltheria douradinha. *Phytochemistry* 66: 1163–1167. https://doi.org/10.1016/J.PHYTOCHEM.2005.03.019.

62 Gehm, A.Z., Cunha, S.B., da Silva, B.W. et al. (2022). New cyclopeptide alkaloid of Condalia buxifolia and the absolute stereochemistry of Condaline A. *Fitoterapia* 159: 105194. https://doi.org/10.1016/J.FITOTE.2022.105194.

63 Search – Access Structures (n.d.). https://www.ccdc.cam.ac.uk/structures (accessed 25 October 2022).

64 Dittrich, B., Strumpel, M., Schäfer, M. et al. (2006). Invarioms for improved absolute structure determination of light-atom crystal structures. *Acta Crystallogr. A* 62: 217–223. https://doi.org/10.1107/S0108767306010336.

65 Meurer, F., v. Dolomanov, O., Hennig, C. et al. (2022). Refinement of anomalous dispersion correction parameters in single-crystal structure determinations. *IUCrJ* 9: 604–609. https://doi.org/10.1107/S2052252522006844.

13

NMR for Stereochemical Elucidation

Xiaolu Li[1], *Xiaoliang Yang*[2], *and Han Sun*[1,3]

[1] *Group of Structural Chemistry and Computational Biophysics, Leibniz-Forschungsinsitut für Molekulare Pharmakologie, Berlin, Germany*
[2] *State Key Laboratory of Coordination Chemistry and Jiangsu Key Laboratory of Advanced Organic Materials, School of Chemistry and Chemical Engineering, Nanjing University, Nanjing, Jiangsu, China*
[3] *Institute of Chemistry, Technical University of Berlin, Berlin, Germany*

13.1 Conventional NMR Methods for Stereochemical Elucidation

Nuclear magnetic resonance (NMR) belongs to one of the mostly employed techniques for determining the constitution, conformation, and relative configuration of organic molecules in solution. For common organic molecules, ^{1}H and ^{13}C are the most commonly observed nuclei in NMR spectroscopy. Nonetheless, if nitrogen, phosphorous, and fluor are present, ^{15}N, ^{31}P, and ^{19}F can be also detected and often provide additional important structural information.

Relative configuration and conformation of chiral organic molecules are mostly elucidated using chemical shifts, nuclear Overhauser effects (NOEs) that are related to the distances between two hydrogens, and scalar couplings that provide valuable information about dihedral angles. Determination of the absolute configuration using NMR poses one of the major challenges in the structural elucidation of chiral organic molecules, as both enantiomers do not show difference in their NMR spectra. In achiral solvent system and without the presence of alignment media, two enantiomers can only be distinguished and their absolute configuration is able to be determined,

Chiral Separations and Stereochemical Elucidation: Fundamentals, Methods, and Applications, First Edition. Edited by Quezia Bezerra Cass, Maria Elizabeth Tiritan, João Marcos Batista Junior, and Juliana Cristina Barreiro.

when enantiomers are derivatized into diastereomers, as demonstrated using the Mosher ester analysis.

In Section 13.1, we will first briefly introduce conventional NMR approaches for establishing the planar structure of organic molecules. We will then describe NOE and *J*-coupling-based methods for establishing the relative configuration of chiral organic molecules. Aforementioned methods will be illustrated using a number of examples.

13.1.1 Determination of the Planar Structure Using 1D ^{1}H, ^{13}C NMR (DEPT), 2D HSQC, COSY, TOCSY, HMBC

In the past 50 years, with the development of modern NMR instruments and techniques, a variety of high-resolution NMR methods have emerged for establishing the planar structure of organic molecules. A routine structural assignment of organic compounds begins with acquiring the 1D ^{1}H and ^{13}C spectra. Many resonances can be directly assigned based on their characteristic chemical shifts and coupling constants. DEPT (**D**istortionless **E**nhancement by **P**olarization **T**ransfer) is used to determine the multiplicity of carbon atoms and thus often is helpful in completing the ^{13}C assignments [1].

2D NMR spectra are necessary if 1D spectra are not sufficient for deducing the planar structure of the organic molecules. With this respect, HSQC (**H**eteronuclear **S**ingle **Q**uantum **C**oherence) [2–5] experiment reveals correlations between the directly bounded proton and carbon atoms. COSY (**CO**rrelated **S**pectroscop**Y**) experiment [6, 7] lays the groundwork for building structure through ^{1}H–^{1}H connectivity, where two protons are separated by two (geminal protons) or three bonds (vicinal protons). In contrast to COSY, TOCSY (**TO**tal **C**orrelation **S**pectroscop**Y**) [8] creates correlations between all protons within a given spin system. COSY and TOCSY experiments are particularly important for establishing the constitution of proton-rich molecules. For proton-deficient molecules with many quaternary carbon atoms, quite often, correlation information between carbons and protons/carbons that are separated by two, three, and, sometimes even four bonds needs to be derived from HMBC (**H**eteronuclear **M**ultiple **B**ond **C**orrelation) [9] or INADEQUATE (**I**ncredible **N**atural **A**bundance **D**oubl**E** **QUA**ntum **T**ransfer **E**xperiment) [10].

There are a number of excellent book chapters and reviews illustrating the determination of planar structures using 1D and 2D NMR spectroscopy [11–16]. We encourage the readers to refer these works for detailed information.

13.1.2 Determination of Relative Configuration Using *J*-Couplings and NOEs/ROEs

13.1.2.1 Scalar Coupling

In high-resolution solution-state NMR spectroscopy, indirect or scalar coupling (J coupling) of nuclear spins through covalent bonds results in splitting of NMR signals into multiplicity. In contrast, direct or dipolar coupling between nuclear spins through space, which is a dominant effect in the solid-state NMR, is canceled out due to the isotropic molecular tumbling. In general, the scalar couplings between atoms that are coupled to each other are equal. Furthermore, J is independent of B_0 (the strength of the magnetic field), because the magnitude of the interaction depends only on nuclear properties. The coupling mechanism is based on the bonding electron transfer through chemical bonds. Notably, the coupling constant between two nuclei separated by an even number of chemical bonds is positive, while the coupling constant between two nuclei separated by an odd number of chemical bonds is negative.

13.1.2.1.1 *Proton-Proton Vicinal Couplings ($^3J_{HH}$)*

The vicinal coupling constant between two protons ($^3J_{\mathrm{HH}}$ coupling) is related to the corresponding dihedral angle. Therefore, this information is particularly useful for deducing the relative configuration of two adjacent stereocenters. In 1959, Karplus derived a mathematical relationship between the $^3J_{\mathrm{HH}}$ and H—C—C—H dihedral angle θ (Eq. 13.1), which is later named as Karplus equation [17]:

$$3J_{\mathrm{HH}} = A\cos^2\theta + B\cos\theta + C \tag{13.1}$$

where A, B, and C are empirically derived parameters, whose values depend on the atoms and substituents involved.

The Karplus equation provides useful quantitative interpretation of $^3J_{\mathrm{HH}}$ for different chemical systems. For example, in chair cyclohexanes, the coupling constant between adjacent axial protons ($^3J_{\mathrm{aa}}$) is large (8–13 Hz), because ϕ_{aa} is close to 180°, whereas the coupling between two equatorial protons ($^3J_{\mathrm{ee}} = 0$–5 Hz) or the coupling between an axial and an equatorial proton ($^3J_{\mathrm{ae}} = 1$–6 Hz) is usually small, as ϕ_{ee} and ϕ_{ae} are close to 60° [17]. Similarly, in acyclic carbon chains, $^3J_{\mathrm{HH}}$ is often used to distinguish *anti* (180°) and *gauche* (60° or − 60°) conformations of both coupled protons (Figure 13.1).

Experimentally, $^2J_{\mathrm{HH}}$ and $^3J_{\mathrm{HH}}$ can be directly measured from 1D-^{1}H spectra. In case of signal overlapping, 2D NMR experiments such as J-resolved,

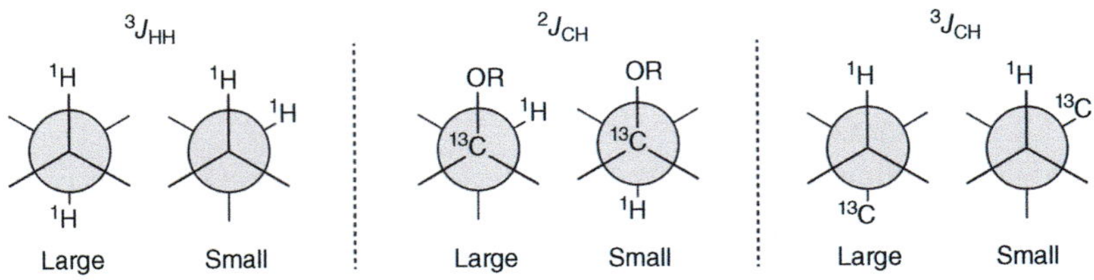

Figure 13.1 Dihedral dependency of $^{3}J_{HH}$, $^{2}J_{CH}$, and $^{3}J_{CH}$ couplings in acyclic systems.

Double **Q**uantum **C**oherence COSY (DQF-COSY) [18], **E**xclusive COSY (E.COSY) [19] experiments become helpful in obtaining $^{2}J_{HH}$ and $^{3}J_{HH}$. When the chemical shifts between the excited and observed nuclei are far away, J coupling can be easily extracted from homonuclear decoupling or multipoint irradiation homonuclear decoupling experiments. E.COSY [19] and **P**ure **E**xclusive COSY (P.E.COSY) [20] are the most accurate methods to extract the absolute value and the sign of $^{2}J_{HH}$ and $^{3}J_{HH}$ simultaneously if three spins are coupled with each other.

13.1.2.1.2 Proton-Carbon Vicinal Couplings ($^{3}J_{CH}$)

Similar to vicinal H—H coupling constants ($^{3}J_{HH}$), a vicinal CH coupling constant $^{3}J_{CH}$ depends on the C—C—C—H dihedral angle [21, 22], following the Karplus relationship. For example, $^{3}J_{CH}$ couplings may be used to derive the relative orientation of C and H as coupled nuclei in cyclohexane, pyranose rings, and alkenes. Substituted cyclohexanes and pyranoses have $^{3}J_{CH} \approx 2$–4 Hz for *cis* and 8–9 Hz for *trans* configuration, respectively. The information of $^{3}J_{CH}$(*trans*)> $^{3}J_{CH}$(*gauche*) (Figure 13.1) is highly useful for deducing the relative configuration of multi-substituted alkenes as well as acyclic alkanes [15, 22].

Experimentally, $^{2}J_{CH}$ and $^{3}J_{CH}$ couplings can be measured from a variety of experiments. **HE**teronuclear **C**ouplings from **A**rbitrarily scaled shift and coupling information **D**omain **E**xperiments (HSQC-HECADE experiment) [23] allow to measure not only very small couplings but also their sign, which is particularly useful for distinguishing $^{2}J_{CH}$ and $^{3}J_{CH}$ as they have opposite signs. However, HSQC-HECADE shows limitations in extracting the coupling constants when non-protonated carbons are involved. In cases where the values of coupling constants to quaternary carbon atoms are required, **SEL**ective ***J*****-RES**olved (SELJRES) [24] or 1D and 2D **H**eteronuclear **S**ingle **Q**uantum **M**ultiple **B**ond **C**orrelation (HSQMBC) [25] experiments can be used. It should be noted that a number of significant improvements in HSQMBC have been proposed. For example, LR-HSQMBC [26, 27],

HSQMBC-COSY [28], and HSQMBC-TOCSY [29] experiments allow to extract 4-, 5-, and even 6-bond long-range $^{n}J_{CH}$ heteronuclear couplings, while IPAP (**In**-**Phase** and **Anti**-**Phase**)-HSQMBC [30] enables the extraction of coupling constants by measurement of the relative displacement between α- and β-cross-peaks. In contrast to HSQC-HECADE, both SELJRES and HSQMBC experiments do not reveal the sign of the couplings.

13.1.2.1.3 Geminal Proton-Carbon Couplings ($^{2}J_{CH}$)

Geminal CH coupling constants also provide valuable structural information. As shown in Figure 13.1, when the proton is in *gauche* (−60° or 60°) conformation with an alkoxy group (OR), the coupling constant between the proton and the carbon where the alkoxy or alcohol is attached becomes large. In contrast, when two groups are in *anti* (180°) conformation, the respective coupling turns to be rather small. Combination of the $^{3}J_{HH}$ and $^{2,3}J_{CH}$ couplings is highly useful for establishing the relative configuration of two neighboring stereocenters in cyclic systems or in acyclic alkanes, where preferred conformations in solution exist.

13.1.2.1.4 Long-Range Couplings (^{4}J, ^{5}J)

Couplings between protons that are separated by more than three bonds are usually very small. They are less than 1 Hz and often cannot be measured with sufficient accuracy. Only under the following circumstances $^{4,5}J$ couplings could become significant.

1) σ-π overlap: For example, in the five-bonded doubly allylic (also called homoallylic) systems, the corresponding coupling constant depends on the orientation of two C—H bonds with respect to the π orbitals. For acyclic systems such as the 2-butenes, $^{5}J_{HH}$ is typically 2 Hz, with a range of 0–3 Hz. It is not unusual for the doubly allylic coupling to be larger than the allylic ones [$^{4}J(CH_3 - Ha) = 1.1$ Hz, $^{5}J(CH_3 - Hb) = 1.8$ Hz] (Figure 13.2a).

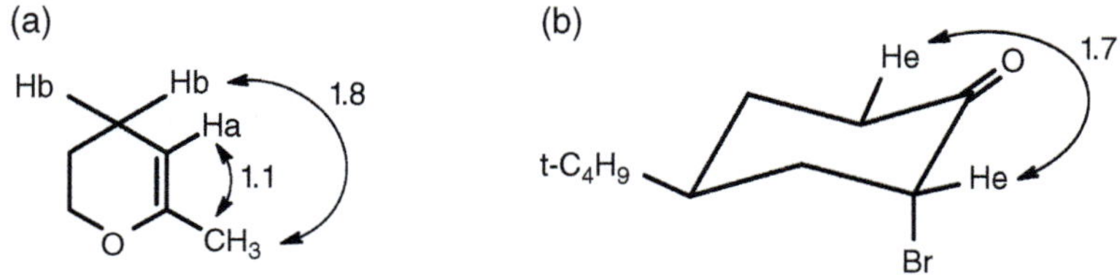

Figure 13.2 Long-range couplings (in Hz) in two cyclic systems: (a) 6-methyl-3,4-dihydro-2*H*-pyran and (b) (2*R*,4*S*)-2-bromo-(*tert*-butyl)cyclohexan-1-one. *Source:* Adapted from Gunther [12].

2) zigzag pathways: Enhanced long-range couplings are often observed between protons that are related to a planar W or zigzag pathway. This geometry is seen, for example, in the 1,3-diequatorial arrangement between protons in chair six-membered rings (Figure 13.2b, $^4J = 1.7$ Hz). With naphthalene and other condensed (hetero-) aromatics, the long-range coupling ($^5J_{HH} = 0.8$ Hz) is helpful in deducing substitution patterns.

13.1.2.2 NOE/ROE

NOE (**N**uclear **O**verhauser **E**ffect) was developed based on the theoretical work of Albert Overhauser [31, 32] and later experimentally confirmed by Anet and Bourn [33]. NOE is an intensity enhancement effect for spatially close protons, when one of these protons is saturated or inverted. A typical 2D NOESY (**N**uclear **O**verhauser **E**ffect **S**pectroscop**Y**) [34, 35] spectrum looks similar to a COSY spectrum, except that the cross-peaks correspond to pairs of protons that are close in space (<5 Å) and not necessarily close in the bonding network. The NOE intensity is proportional to the inverse sixth power of distance between two protons ($1/r^6$), where r is the space distance between two protons correlated by the cross-peak. NOEs can be only observed when two nuclei are in proximity. Therefore, they are very useful information for elucidation of three-dimensional conformations and for distinguishing relative configurations.

It is important to note that NOE also depends on the correlation time [16]. Small molecules show positive NOEs, while the NOEs of large biomolecules (proteins and nucleic acids) are normally negative. For medium-sized molecules such as large natural products, peptides, and oligosaccharides (1000–4000 Da depending on the molecular shape, rigidity, and solvent viscosity), NOEs may become zero. For these molecules an alternative experiment called ROESY (**R**otating-frame **O**verhauser **E**ffect **S**pectroscop**Y**) [36, 37] is a good alternative, as ROE has always a positive value.

13.1.2.3 Examples of Stereochemical Elucidation Using *J*-Couplings and NOEs/ROEs

In the following, we use three examples to demonstrate the assignment of the relative configuration of chiral organic molecules using NOE and *J*-coupling-based analysis.

13.1.2.3.1 Example 1 – J Coupling [38]

Example **1** is a six-membered cyclic olefin building block containing two nitrogen atoms. Two stereoisomers **1a** and **1b**, both as viscous liquids at room temperature, were separated by column chromatography (Figure 13.3).

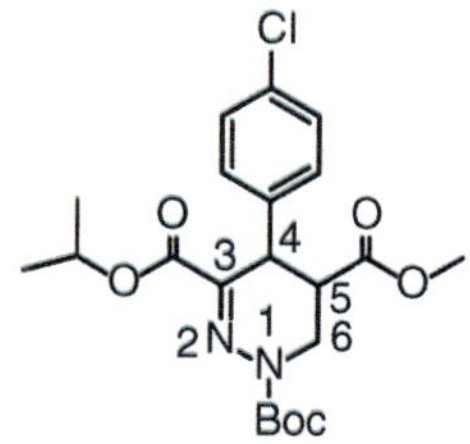

N-tert-Butoxycarboyl-3-isoproxycarboyl-4-(4-chlorophenyl)-5-methoxycarboyl-5,6-dihydropyridazine (**1**)

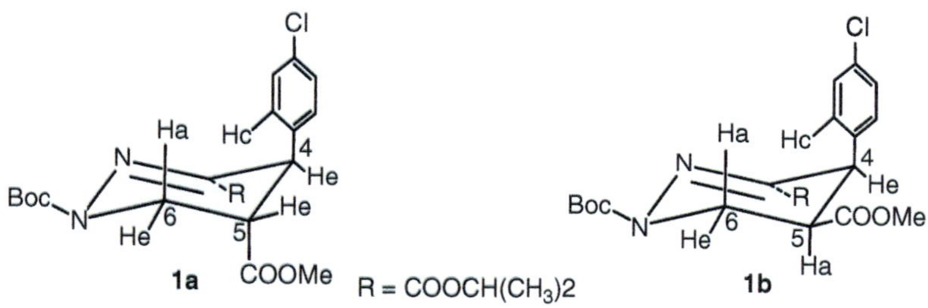

Figure 13.3 Planar structure of example **1** and its two possible stereoisomers **1a** and **1b**.

Their constitution was deduced by high-resolution mass spectrometry (HRMS), infrared (IR) spectroscopy, as well as ^{1}H-NMR and ^{13}C-NMR, revealing that **1a** and **1b** are two diastereomers that only differ in their relative configuration of C4 and C5 [38].

A selected portion of the ^{1}H NMR spectrum of the two stereoisomers is shown in Figure 13.4 for **1a** and Figure 13.5 for **1b**, respectively. It is interesting to note that the chemical shift of H4 moved to downfield (δ large) because of the vicinal groups of C=C and C=O, respectively. The chemical shifts of H6a/e were shifted to the lower field due to the induced effect of the vicinal nitrogen atom. In a cyclic six-membered ring [39], the chemical shift of the equatorial proton is in most cases larger than that of the axial one. The chemical shift order of H4, H5, and H6a/H6e, can be assigned as follows: δ (H4) > δ (H6e) > δ (H6a) > δ (H5). The chemical shifts of H4, H5, and H6 and coupling constants of compound **1a** and **1b** are given in Table 13.1.

In stereoisomer **1a**, H5 (δ, 2.88 ppm) signal splits into a quartet with intensity 1 : 3 : 3 : 1 with a coupling constant of 4.2 Hz. The small coupling constants (4.2 Hz) observed between H5 and H6a, H5 and H4 suggested an equatorial orientation of H5. According to the ^{3}J in the cyclohexane system, only when H5 is in an equatorial orientation, its vicinal couplings with H6a,

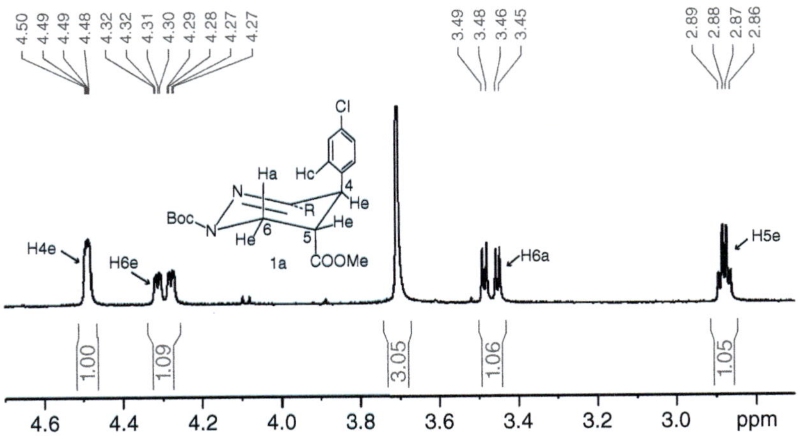

Figure 13.4 Selected portion of ^{1}H NMR spectrum of compound 1**a** (400 MHz, $CDCl_3$).

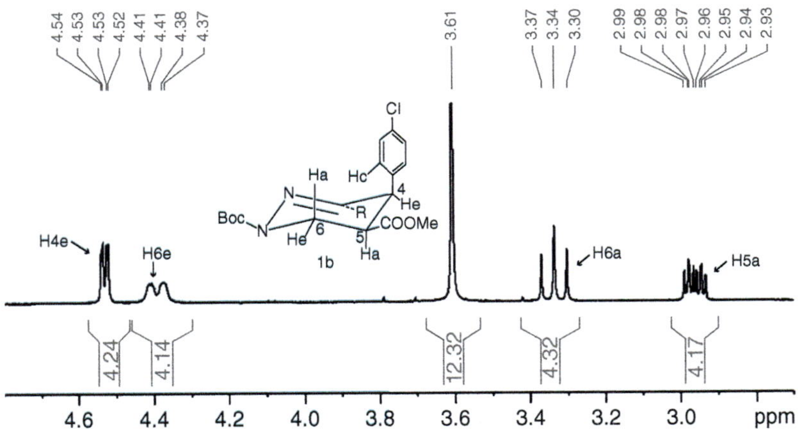

Figure 13.5 Selected portion of ^{1}H NMR spectrum of compound 1**b** (400 MHz, $CDCl_3$).

Table 13.1 Selected ^{1}H chemical shifts and $^nJ_{HH}$ couplings of **1a** and **1b**.

1a		**1b**	
Atom	δ(ppm)/J (Hz)	Atom	δ(ppm)/J (Hz)
H4	4.49 (dd, 4.2, 1.8)	H4	4.53 (dd, 5.6, 1.7)
H6e	4.30 (ddd, 13.5, 4.5, 1.8)	H6e	4.39 (ddd, 13.3, 4.5, 1.7)
H6a	3.47 (dd, 13.5, 4.2)	H6a	3.34 (t, 13.3)
H5	2.88 (q, 4.2)	H5	2.96 (ddd, 13.2, 5.5, 4.5)

H4 are both small and approximately equal. H6e (δ, 4.30 ppm) shows a ddd multiplicity with coupling constants of 13.5 Hz, 4.5 Hz, and 1.8 Hz, respectively. 13.5 Hz was assigned as the geminal proton-proton coupling of the methylene group (H6a/H6e), while 1.8 Hz corresponds to the small W-type long-range coupling constant between H6e and H4. In cyclohexane system, $^4J_{HH}$ (W type) coupling occurs between two equatorial-orientated meta-hydrogens. Furthermore, small coupling constant of 4.2 Hz between H5 and H4 suggested two protons in *gauche* conformation. Altogether, $^3J_{HH}$, $^4J_{HH}$ couplings clearly revealed the relative configuration between H4 and H5 in **1a** as *trans* configuration.

In stereoisomer **1b**, H4 (δ 4.53 ppm) splits into dd multiplet with coupling constants of 5.6 Hz and 1.7 Hz, respectively, which are similar to the ones in the stereoisomer **1a**. However, different from **1a**, the signal of H6a (δ 3.34 ppm) in **1b** implies a triplet splitting with peak intensity of 1 : 2 : 1 and a coupling constant of 13.3 Hz, suggesting a large vicinal coupling constant between H6a and H5(J = 13.3 Hz) induced by the axial orientation of H5. Furthermore, according to the small coupling constant between H4 and H5, H4 was assigned as an equatorial orientation. Altogether, an equatorial orientation of H4 and an axial orientation of H5 suggested the relative configuration of C4 and C5 in stereoisomer **1b** as *cis* configuration.

13.1.2.3.2 Examples 2 and 3 – J Coupling and NOE ([40])

The benzo[c]phenanthridine analogue (**2**), a synthetic product isolated by column chromatography, is employed here as a second example. Its constitution (Figure 13.6) was determined using a combination of HRMS, IR, 1D, and 2D NMR experiments. The assignments of ^{1}H, ^{13}C signals, and $^nJ_{HH}$ couplings are shown in Table 13.2. There is a benzocyclohexane fragment in the structure with C7, C8, C9, and C10 all being tertiary carbons. The relative configuration of the four tertiary carbons needed to be determined.

In the ^{1}H NMR spectrum (Figure 13.6), H9 splits into a triplet with an intensity of 1 : 2 : 1 and a coupling constant of 11.9 Hz. According to the $^3J_{HH}$ coupling characteristics between protons of axial orientation on cyclohexane, the equal large coupling constants observed between H9 and H8, H9 and H10 suggested an axial orientation for all three protons. Therefore, a *trans* configuration between H9 and H8, H9 and H10 (Figure 13.6) was determined, which was further confirmed by the absence of correlations between H9 and H8, H9 and H10 in 2D NOESY spectrum. H8 splits into a dd peak with coupling constants of 12.1 and 4.9 Hz, respectively. H8 is in an axial orientation and coupled to both H9 and H7. Large coupling constant of 12.1 Hz corresponds to vicinal proton-proton coupling between H8 and H9. Small coupling constant observed between H8 and H7 suggested an

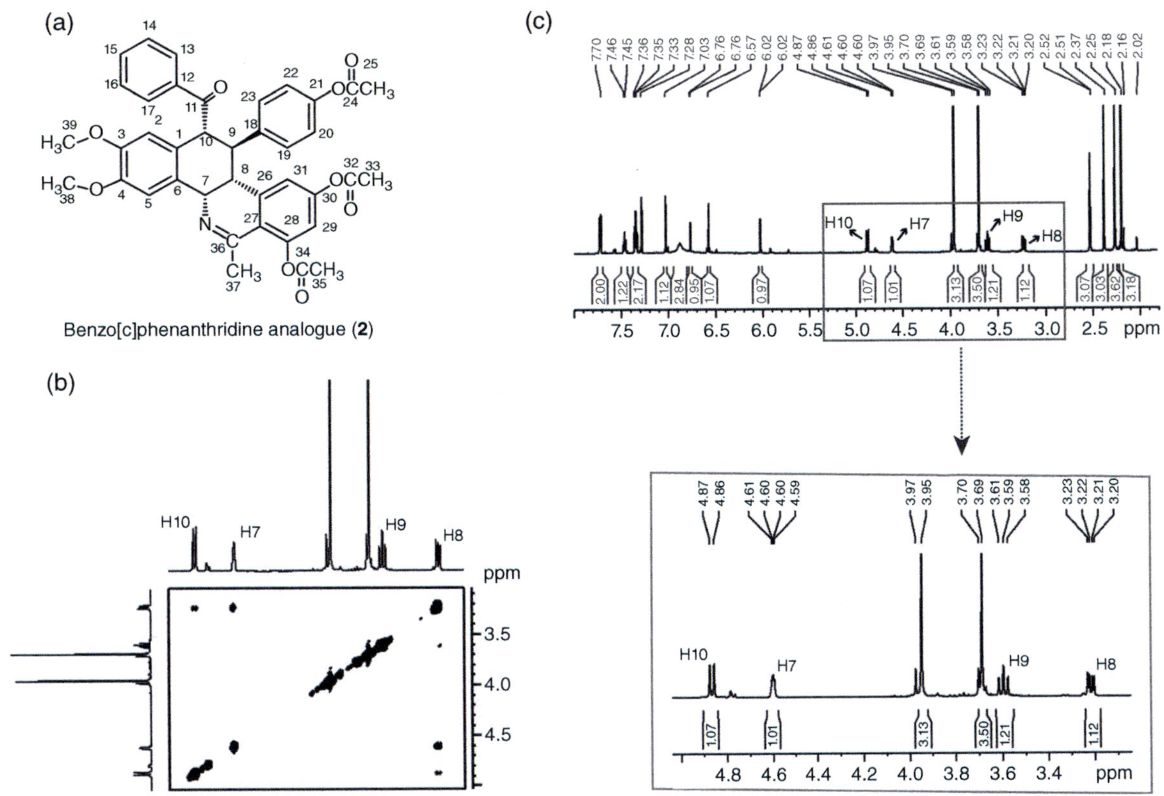

Figure 13.6 (a) Chemical structure of benzo[c]phenanthridine analogue; (b) selected portion of NOESY spectrum; (c) ^{1}H NMR spectrum (up) and a selected portion showing signals H7, H8, H9, and H10(down) (600 MHz, $CDCl_3$).

Table 13.2 ^{1}H and ^{13}C data (600 and 150 MHz) of compound **2** in $CDCl_3$.

Atom	δ_C (ppm)	δ_H (ppm)/J (Hz)	Atom	δ_C (ppm)	δ_H (ppm)/J (Hz)
1	125.36	—	21	149.72	—
2	112.68	7.03 (1H, s)	24	169.03	—
3	148.99	—	25	21.18	2.25 (3H, s)
4	148.88	—	26	141.62	—
5	109.71	6.57 (1H, s)	27	119.01	-
6	128.90	—	28	151.32	—
7	58.74	4.60 (1H, dd, 4.6, 2.3)	29	120.18	6.02 (1H, d, 2.2)
8	42.21	3.22 (1H, dd, 12.1, 4.9)	30	147.85	—
9	42.20	3.59 (1H, t, 11.9)	31	116.22	6.76 (1H, d, 2.2)
10	57.45	4.87 (1H, d, 11.1)	32	168.71	—
11	201.04	—	33	21.45	2.37 (3H, s)
12	136.24	—	34	168.43	—
13, 17	128.38	7.71 (1H, d, 7.4)	35	21.04	2.19 (3H, s)
14, 16	129.36	7.35 (1H, d, 7.4)	36	160.30	—
15	132.69	7.46 (1H, t, 7.4)	37	26.42	2.52 (3H, s)
18	137.59	—	38	55.87	3.69 (3H, s)
19, 20, 22, 23	121.68	6.88 (4H, br)	39	55.90	3.95 (3H, s)

equatorial orientation of H7, indicating a *cis* configuration between H8 and H7 (Figure 13.6). This relative configuration is supported by a NOESY cross-peak observed between H8 and H7. In conclusion, based on the analysis of $^{3}J_{HH}$ couplings and NOEs, the relative configuration of C7, C8, C9, and C10 was determined as 7*S**, 8*S**, 9*R**, and 10*R**, which was later confirmed by X-ray data [40].

Canescensterone (**3**) (Figure 13.7), a rare natural phytoecdysteroid isolated from the bark of *Vitex canescens*, is briefly introduced here as a third example [41]. This compound possesses an unusual ecdysteroid structure, including a pyrrole-2-carboxylate group attached to C-24. The structural assignment of this molecule was achieved by a combination of 1D ^{1}H and 2D NMR experiments.

The stereochemistry at C-24 in the side chain was previously unknown. Using a combination of $^{2,3}J_{CH}$, $^{3}J_{HH}$ couplings and NOE data, the stereochemical information was successfully transferred from the stereocenter C-20, where the stereochemistry is known, to C-24. As shown in Figure 13.7,

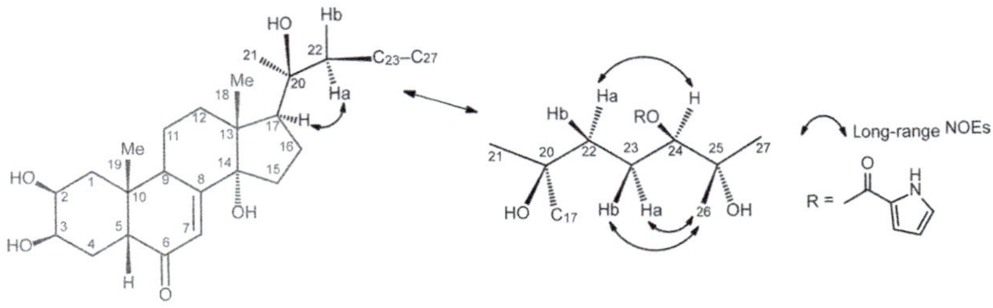

Figure 13.7 The main conformation of the side chain of canescensterone (**3**), as determined by extensive $^{2,3}J$ and short-range NOE analysis, which is further corroborated by the presence of several long-range NOEs.

long-range NOEs cross-validated the determined major conformer of canescensterone with the correct configuration *R* at the stereocenter C-24 [41].

13.2 Determination of the Relative Configuration Using Anisotropic NMR-Based Methods

Although the above-discussed NOE and *J*-coupling-based analyses become a well-established method for assigning the relative configuration of organic molecules, in many cases, the method reaches its limitation as NOE and *J*-coupling only provide information of two atoms in a 5 Å spatial distance or within three chemical bonds. In the past 20 years, a number of methods relying on anisotropic NMR parameters such as residual dipolar coupling (RDC), residual quadrupolar coupling (RQC), and residual chemical shift anisotropy (RCSA) have been developed [42–48]. Compared to NOE and *J*-coupling, anisotropic NMR parameters afford relative orientation of stereocenters without distance restrictions. Therefore, they are becoming essential for solving relative configuration when the unknown stereocenters are far apart from each other or separated by inert nuclei [49–51]. Also, the introduction of anisotropic NMR restraints may help to identify conformational states in flexible systems without the need for parametrization as is the case for the quantitative analysis of *J*-couplings [52–54].

In s13.2, we will briefly describe the principles of different anisotropic NMR parameters, which is followed by introduction of possible approaches to create weak alignment for small molecules and measurement of anisotropic NMR parameters. Finally, we will introduce various computational approaches to include these anisotropic NMR parameters in the structural elucidation process and discuss them using several published examples including combination of chiroptical methods and NMR for determination of absolute configuration.

13.2.1 Basic Principles of Anisotropic NMR Parameters

Orientation-dependent interactions such as dipolar coupling (DC), quadrupolar coupling (QC), and chemical shift anisotropy (CSA) are dominant effects in solid-state NMR spectroscopy. They result in broad line width in the NMR spectra that makes the spectra difficult to be interpretated. In solution-state NMR, DC, QC, and CSA all vanish because of isotropic tumbling of the molecules. Nevertheless, if a weakly aligning condition is introduced to the system, these interactions can be scaled at a very low degree, so that high-resolution NMR spectra that are similar to conventional solution NMR spectra are still obtained, but with additional structural information of internuclear or chemical shielding tensor orientations [13].

RDCs provide information about the relative orientation of internuclear vectors such as one-bond $^{13}C—^{1}H$ within one molecule [13, 47, 48]. The secular approximation of heteronuclear DC can be defined as:

$$D_{IS} = -\frac{3}{8\pi^2 R_{IS}^3}\mu_0 \gamma_I \gamma_S \hbar \left(\cos^2\theta_{IS} - \frac{1}{3} \right) \tag{13.2}$$

where R_{IS} is the spatial distance between nuclei I and S; θ_{IS} the angle between external magnetic field and the internuclear vector of I and S, as shown in Figure 13.8; μ_0 is the vacuum permeability; $\hbar$ is the Planck constant; and γ_I and γ_S are the gyromagnetic ratios of spins I and S. As the distance of a covalent bond is known, DC solely depends on the angle of the internuclear vector and the magnetic field.

RQC, an effect derived from quadrupolar interaction, can be detected only for nuclei with spin quantum number $I > 1/2$, such as deuterium (^2H). For ^2H, its RQC provides the same extent of information as one-bond carbon-proton RDC ($^1D_{CH}$) for a given hydrogen site. As the ^2H-RQC are more than five times larger than the corresponding $^1D_{CH}$, they are more sensitive to small orientational changes of the molecules [55].

Residual chemical shift anisotropy (RCSA) affords information about relative orientation of chemical shift tensor and hence gives orientational information on all NMR detectable atoms in any structure. So far, ^{13}C RCSA is among the mostly employed RCSAs in elucidation of relative configuration of organic molecules, which becomes especially useful for proton-deficient compounds [46, 56, 57]. Recently, the application of ^1H RCSA has been demonstrated for assigning the relative configuration of molecules with microgram availability [58].

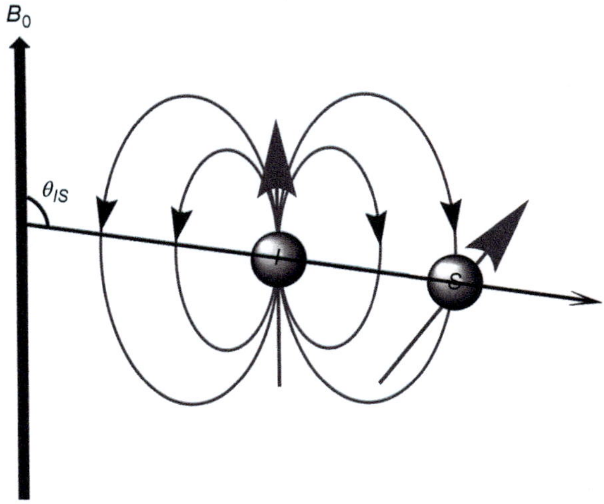

Figure 13.8 The induced magnetic moment by the spin I under B_0 also interacts with the one of the spin S, which is denoted as direct dipolar–dipolar interaction. θ_{IS} is defined as the angle between external magnetic field B_0 and the internuclear vector of nuclei I and S.

Interpretation of the RDC, RQC, and RCSA all require a term called as alignment tensor Â (Eq. 13.3), which is able to describe the orientational preference of the analyte in the magnetic field [59].

$$\hat{A} = \begin{pmatrix} A_{xx} & A_{xy} & A_{xz} \\ A_{yx} & A_{yy} & A_{yz} \\ A_{zx} & A_{zy} & A_{zz} \end{pmatrix} \tag{13.3}$$

It should be noted that the alignment tensor is symmetric ($A_{xy} = A_{yx}$, $A_{xz} = A_{zx}$, $A_{yz} = A_{zy}$) and traceless ($A_{xx}+A_{yy}+A_{zz} = 0$), indicating that the alignment tensor is determined by five independent parameters. If five independent RDC, RQC, or RCSA can be measured for a rigid molecule without internal flexibility, the alignment tensor of the analyte can be established unequivocally.

13.2.2 Alignment Media

A critical component of anisotropic NMR spectroscopy is the alignment medium, which is necessary for achieving a low molecular ordering for analytes. There are two main types of alignment media: (i) strain-induced alignment gel (SAG), that is able to be stretched or compressed

Table 13.3 Cross-linked polymer gels as alignment media together with their solvent compatibility and enantiodiscriminating property.

Year	SAG	Compatible solvents	Enantiodiscriminating property
2004	PS [60]	$CDCl_3$	N
2004	PDMS [61]	$CDCl_3$	N
2005	PVAC [62]	$CDCl_3$, DMSO-d_6, CD_3OD, CD_3CN,	N
2005	PH [63]	DMSO-d_6	Y (chiral monomers)
2005	Gelatin [64, 65]	D_2O	Y
2006	Collagen fibers [65, 66]	D_2O	Y
2007	PAN [67]	DMSO-d_6	N
2008	PMMA [68]	$CDCl_3$, CD_2Cl_2	N
2009	e^--gelatin [65, 69]	D_2O, DMSO-d_6	Y
2013	PEO [70]	$CDCl_3$, DMSO-d_6, CD_3OD, CD_3CN, THF, D_2O, benzene, toluene, dioxane	N
2016	*p*-HEMA [71]	DMSO-d_6	N
2017	*p*-DEGMEMA [72]	CD_3OD	N

mechanically in the NMR tube. Different types of SAGs are summarized in Table 13.3 with their compatible solvent systems. Different gel-based media are for example nicely discussed in a recent *Nature Protocols* paper [73]. (ii) Lyotropic liquid crystalline (LLC), forming an ordered phase with or without the presence of magnetic field, is able to partially transfer its orientation to the analytes (Table 13.4). The alignment degree of the gel-based media can be tuned by adjusting the degree of stretching or compressing, while changing the concentration of LLC phase allows to tune the alignment degree of the phases easily.

Some of the alignment media have been introduced as chiral alignment media, which is able to differentiate enantiomers based on the differences in the obtained anisotropic NMR data. The first introduced chiral alignment media are homopolypeptides, e.g. poly-γ-benzyl-L-glutamate (PBLG) or its enantiomer PDBG and poly-γ-ethyl-L-glutamate (PELG) that form LLC in apolar solvents such as $CDCl_3$ [74–76]. Strong enantiodiscriminating properties were so far observed with poly(acetylenes) for molecules with H-bond donors like alcohols and 2° amines [78]. Some of the gel-based

Table 13.4 LLC-based alignment media together with their solvent compatibility and enantiodiscriminating property.

Year	LLC	Compatible solvents	Enantiodiscriminating property
1981	PBLG, PDLG [74–76]	$CDCl_3$, CD_2Cl_2, THF	Y
2001	PCBLL/PCBDL [76]	$CDCl_3$, CD_2Cl_2, THF	Y
2004	PELG/PEDG [76]	$CDCl_3$, CD_2Cl_2, THF	Y
2007	*R*-PPEMG [77]	$CDCl_3$, CD_2Cl_2, THF	Y
2012	PLA [78]	$CDCl_3$	Y
2012	L-ala-based PPA [79]	$CDCl_3$	Y
2014	GO [80]	DMSO-d_6, CH_3COCH_3, CH_3CN	N
2016	GO-g-TFEMA [81]	DMSO-d_6	N
2017	AAKLVFF [82]	CD_3OD	N
2017	PPLA, PPDA [83]	$CDCl_3$, TCE-d_2	Y
2019	L-valine-based-PLA [84]	$CDCl_3$	Y
2020	OPA [85]	CD_3OD, DMSO-d_6, D_2O	Y (weak)
2021	DSCG [86]	D_2O	N

media (e.g. chiral polyacrylamide gels) have been introduced as chiral alignment media, but their enantiodifferentiating property is substantially weaker compared to many chiral LLCs [63].

We use here poly(methyl methacrylate) (PMMA) [68] and oligopeptide AAKLVFF [82] as two representative examples to illustrate the preparation of anisotropic samples using SAG and LLC phases, respectively.

13.2.2.1 Preparation of Anisotropic Sample with PMMA Gel

Gil and colleagues have pioneered the usage of cross-linked PMMA gel as alignment medium for aligning small organic molecules. We briefly discuss here the preparation of the PMMA gels for RDC and RCSA measurements in chloroform. For more detailed information, please refer to the original publications [68] and the recent *Nature Protocols* paper [73]. In general, preparation of the anisotropic phase using PMMA gel contains the following steps: (i) polymerization of the PMMA gel with non-deuterated solvent in an NMR tube; (ii) washing and drying the gel, so that a dried gel rod is obtained; (iii) reswelling the gel with deuterated solvent together with a stretching or a compressing device, so that a proper alignment is obtained for the analytes. Quantification of the alignment degree is generally monitored by the measurement of quadrupolar splitting of the deuterium signal of the solvent [87, 88]. Analysis of the deuterium splitting pattern provides additional information about the enantiodiscriminating property of the chiral alignment media [89].

For the RDC measurement, the analyte should be measured in the isotropic and aligned PMMA phases, respectively. For the RCSA measurement, in order to eliminate the contribution of the isotropic chemical shifts introduced by changing the chemical environment, usually ^{13}C spectra of two different alignment conditions are acquired [88, 90]. This can be achieved by (i) an NMR tube with two different diameters, so that two different alignment strengths can be obtained in a single tube; (ii) compressing device that enables to adjust the alignment degree via different compression levels. The RCSA measurement using compression devices requires a compensation of the compression-related isotropic chemical shift changes, which can be achieved by different approaches [56, 91].

13.2.2.2 Preparation of Anisotropic Sample with AAKLVFF

AAKLVFF was introduced by Lei and colleagues as the first LLC-based alignment media compatible with methanol [82]. This medium is stable at a wide range of concentrations and temperatures and show very weak background signals. AAKLVFF is compatible with most of the organic compounds. The oligopeptide can be synthesized either via established solid-state peptide synthesis methods or directly purchased commercially. Preparation of the anisotropic phase using AAKLVFF is simple and straightforward, requiring only addition of AAKLVFF peptide powder in solvent together with the analytes. Anisotropic NMR phase can be easily prepared for 3 or 5 mm NMR tubes. AAKLVFF self-assembles into nanotubes that are able to form anisotropic phases without the presence of magnetic field. However, equilibration of the anisotropic LLC phases of the oligopeptide usually takes 10–14 days to complete [92]. More recent studies based on the molecular self-assembly of oligopeptide amphiphile (OPA) enable fast acquisition of RDC without additional waiting times [85].

Based on the slow equilibration property of AAKLVFF, Sun and colleagues proposed a straightforward and accurate strategy for acquiring ^{13}C RCSA of organic molecules in AAKLVFF (Figure 13.9). A number of ^{13}C spectra are acquired at different time points during the equilibration process, so that RCSA at different alignment strength are obtained [46]. This method allows accurate RCSA measurement without any special instrument, tube, or correction method during postacquisition data analysis and thus can be easily applied in any chemistry laboratory.

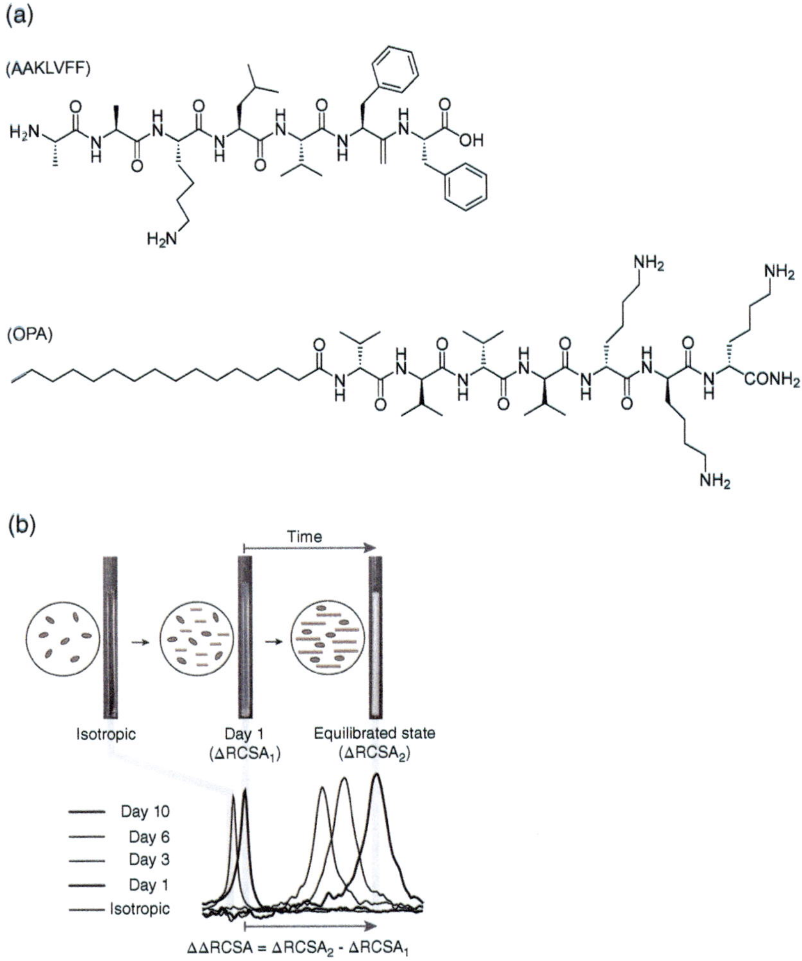

Figure 13.9 (a) Structure formula of AAKLVFF and oligopeptide amphiphile (OPA). (b) Schematic illustration showing the strategy of acquiring ^{13}C RCSA using AAKLVFF.

13.2.3 Acquisition of the Anisotropic NMR Data

Acquisition of RDCs requires measurements of the scalar coupling (J) under the isotropic condition and a sum of scalar and dipolar coupling ($J + D$) under the anisotropic condition. DC can be obtained by extracting the difference of the couplings measured under the anisotropic and isotropic conditions [13].

For organic molecules, mostly used RDCs for structural elucidation are one-bond carbon-proton ($^1D_{CH}$). One-bond scalar coupling under isotropic condition as well as the sum of scalar and dipolar coupling under anisotropic condition can be measured using different HSQC-type experiments. So far, F2-coupled CLIP-HSQC (**CL**ean **I**n**P**hase HSQC) [93] and F1-coupled BIRD-HSQC (**BI**linear **R**otation **D**ecoupling) [94] have been mostly employed for this purpose. CLIP-HSQC (Figure 13.10a) was proposed by Luy and colleagues to solve the phase distortion problem induced by residual dispersive antiphase coherences. BIRD-HSQC experiment (Figure 13.10b) was proposed to overcome the line broadening and asymmetry in the F2 dimension caused by strong coupling effects and extract the coupling in the F1 dimension instead. A major difference between the CLIP-HSQC and BIRD-HSQC experiments is the measurement of methylene proton-carbon couplings. While by using the CLIP-HSQC experiment individual C—H couplings of the CH_2 groups can be extracted, the BIRD-HSQC experiment measures their sum. This can be sometimes beneficial for the follow-up analysis, especially when the diastereotopic protons cannot be easily assigned using isotropic NMR data.

Although long-range ^1H—^{13}C RDCs ($^2D_{CH}$, $^3D_{CH}$) and long-range ^1H—^1H RDCs ($^2D_{HH}$, $^3D_{HH}$) are usually very small under weak alignment condition, they could become highly useful for solving stereochemical problems in cases only sparse $^1D_{CH}$ data are obtained. Selective *J*-scaled HSQC experiment [95] was proposed to extract $^{2,3}D_{CH}$ couplings at very high precision. An E.COSY type of pattern from selective *J*-scaled HSQC experiment enables the determination of the absolute value and the sign of the long-range coupling simultaneously in the F1 dimension. A drawback of this HSQC-based experiment is that the long-range couplings with quaternary carbon atoms cannot be observed. Instead, Griesinger and coworker proposed a simple proton-selective ^{13}C-detected NMR experiment that facilitates the extraction of long-range RDCs including non-protonated carbons [96]. This experiment however could not extract the sign of the coupling. Therefore, computational analysis tools that operate with absolute values of the couplings are necessary.

The absolute value and the sign of long-range ^1H—^1H RDCs can be extracted using E.COSY type of experiments [97, 98]. Nonetheless, due to the line broadening in the anisotropic samples, interpretation of the E.COSY multiplets becomes often a difficult task. An experiment combining an improved PSYCHEDELIC (**P**ure **S**hift **Y**ielded by **CH**irp **E**xcitation to **DEL**iver **I**ndividual **C**ouplings) method and a new selective constant-time β-COSY experiment could potentially overcome this problem [99].

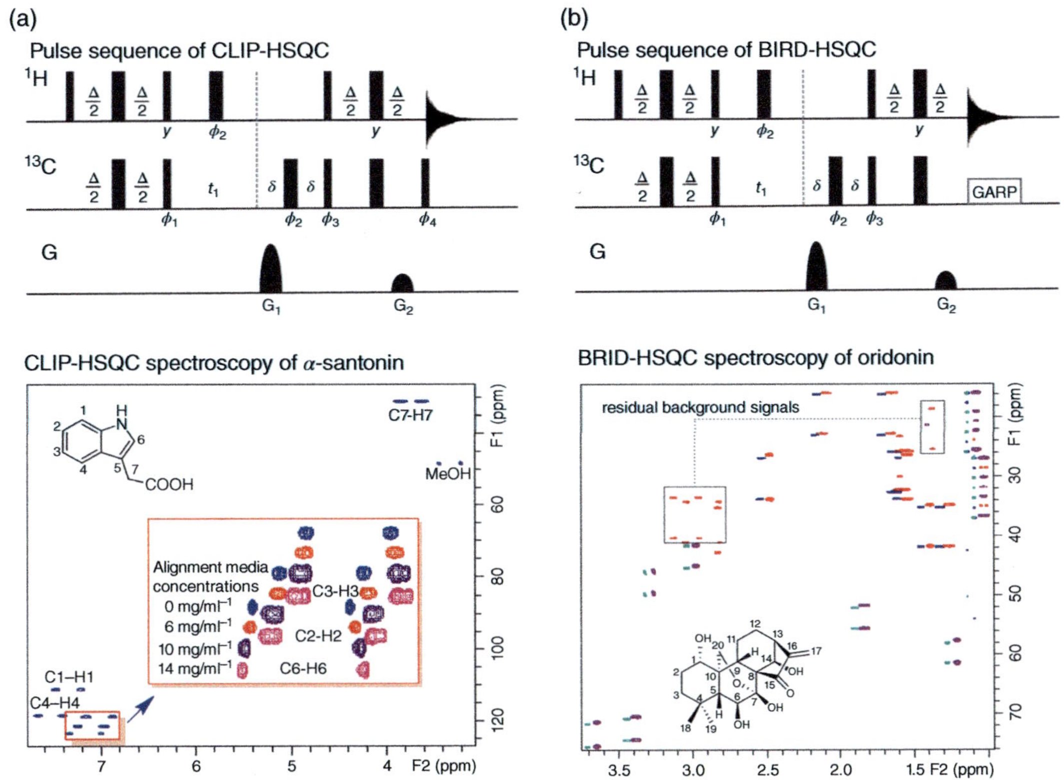

Figure 13.10 (a) Pulse sequences of F2-coupled CLIP-HSQC (left) and F1-coupled BRID-HSQC (right). (b) Experimental spectra of indole-3-acetic acid (left) (*Source:* Lei et al. [82]/John Wiley & Sons) and oridonin (right) (*Source:* Liu et al. [92]/John Wiley & Sons) acquired by CLIP-HSQC and BIRD-HSQC experiments, respectively, in AAKLVFF anisotropic phase.

Due to the low natural abundance of deuterium in conventional organic molecules and poor gyromagnetic ratio of deuterium ($\gamma(^1H)/\gamma(^2H) = 6.515$), acquiring the deuterium RQC data represents a general challenge. This can be partially overcome by introduction of a cryogenically cooled ^2H probe in a modern NMR spectrometer operating at high magnetic field. For small molecules with few ^2H, quite often 1D experiments are sufficient to extract the RQC. For molecules with substantial signal overlapping in 1D spectra, a number of **QU**adrupolar **O**rdered **S**pectroscop**Y** (QUOSY)-type 2D experiments have been proposed [100].

For the RCSA measurements, standard 1D ^{13}C and ^1H experiments are used to extract the chemical shifts under anisotropic conditions. As already discussed above, instead of recording the spectra under isotropic and anisotropic NMR conditions, respectively, the same 1D experiment is repeated twice under two alignment conditions with different alignment strengths. $\Delta\Delta$RCSA is determined by extracting the chemical shift difference in these two alignment states. In order to minimize the contribution of the isotropic chemical shift, it is desirable to choose a proper reference atom with very small or neglectable CSA [44, 101]. This atom is considered to have no RCSA in the measurement. Three different types of reference atoms have been used in previous investigations: (i) the carbon or proton atom with the smallest theoretically calculated CSA obtained from density functional theory (DFT) calculations; (ii) the methyl carbon of the solvent MeOH-d_4 or external referencing agents such as tetramethylsilane (TMS); (iii) the methyl carbon of the analyte showing the smallest chemical shift change under two anisotropic alignment states. Recent applications mostly chose the carbon atom with the smallest theoretically calculated CSA as the Refs. [46, 73].

13.2.4 Computational Approaches for Analyzing Anisotropic NMR Data

For organic molecules, the most common stereogenic center is a carbon atom with four different substituents in a tetrahedral geometry. If a molecule contains N stereocenters, 2^{N-1} diastereomeric possibilities may exist. So far, the most commonly used approach to employ anisotropic NMR data in the stereochemical determination requires generation of a single or an ensemble of energy-lowest structures for each possible configuration. This can be achieved by force-field-based calculations [102] or quantum-mechanical optimizations [103, 104] or a combination of both methods [105, 106]. After the structures of each configuration are obtained, RDC, RCSA, and RQC can be fitted individually or in combination to the structures of each possible relative configuration. During this procedure, the alignment tensor of each possible relative configuration is determined, which is subsequently used to back-calculate the anisotropic NMR data. Experimental and back-calculated anisotropic NMR data are then compared via the Q factor (Eq. 13.4): [107]

$$Q = \sqrt{\frac{\Sigma\left(D_i^{\text{exp}} - D_j^{\text{calc}}\right)^2}{\Sigma D_i^{\text{exp}}}} \tag{13.4}$$

where a small Q factor indicates a good agreement between experimental (D^{exp}) and back-calculated NMR data (D^{calc}). The configuration with substantially lowest Q factor (<0.1) can be determined as the correct relative configuration. If multiple independent anisotropic NMR data could be extracted and employed independently in the fitting, the inter-tensor angle between these calculated alignment tensors should be determined. As all types of anisotropic NMR data in principle have the same alignment tensor, a small inter-tensor angle should be obtained for the correct relative configuration [108, 109].

Calculation of the alignment tensor and comparison of experimental and back-calculated RDC data for small organic molecules can be realized by different programs, such as MSpin-RDC [42] and PALES [110]. MSpin-RDC, which is now also integrated in the program Stereofitter (Maestro), is able to include all types of anisotropic NMR data in the calculations. It should be noted that for RCSA analysis, it is necessary to calculate the chemical shielding tensors of each conformer beforehand using high-level quantum-mechanical calculations, usually based on DFT [44, 46].

For small to medium-sized structurally rigid molecules, quantum-mechanical calculations alone are sufficient for determining the energy-lowest conformers. For semi-flexible molecules, a combination of force-field-based conformational search and quantum-mechanical calculations are usually recommended to establish the conformational ensembles that are finally used for fitting of the anisotropic NMR data [73]. Conformational search can be performed using various programs, e.g. MacroModel [56, 111, 112] or GMMX [113]. A typical workflow of stereochemical elucidation process using anisotropic NMR data is illustrated in Figure 13.11. Previous examples also revealed molecular dynamics (MD) simulations as attractive methods for determining the conformational states of highly flexible molecules [114]. For structurally rigid or semi-flexible molecules, one alignment tensor is sufficient for interpretating the anisotropic NMR data. For highly flexible molecules, depending on the shape and amplitude of the conformational changes, single alignment tensor and multiple alignment tensor approaches have been proposed and compared in previous investigations [114, 115].

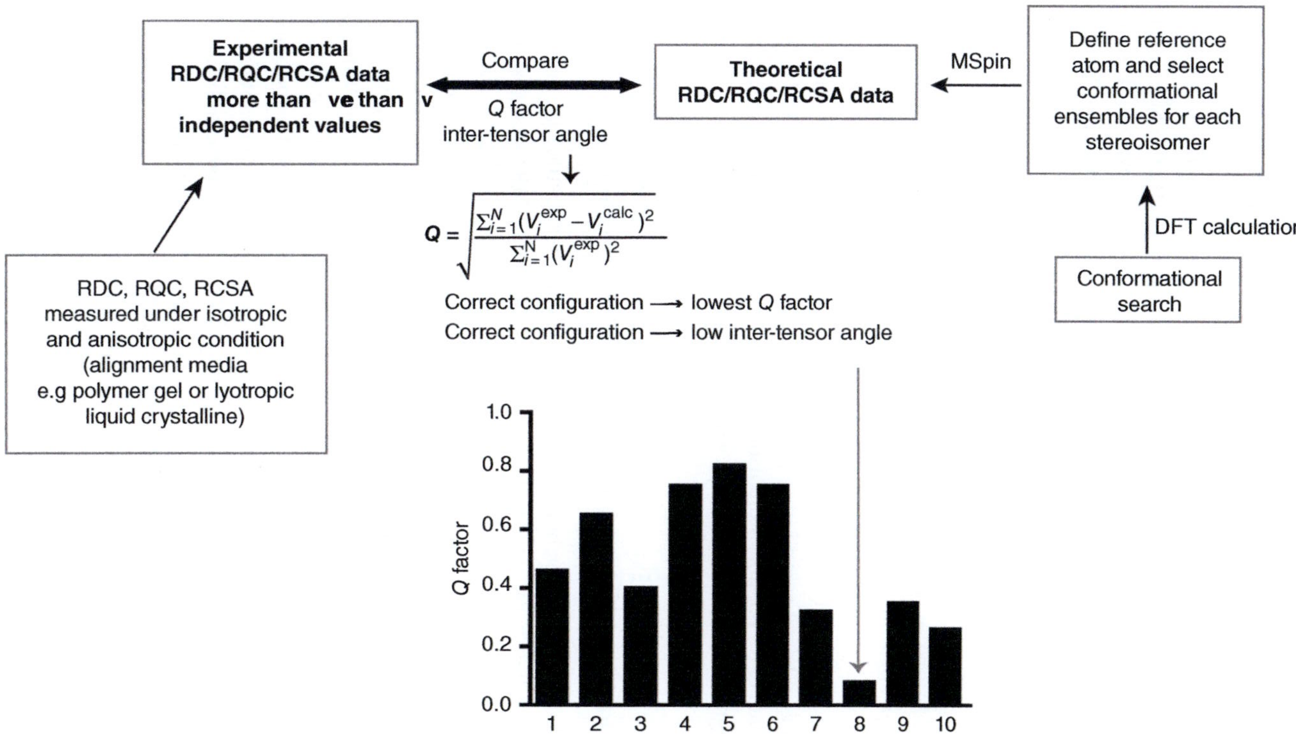

Figure 13.11 A typical workflow for the determination of the relative configuration of structurally rigid or semi-flexible molecules by means of anisotropic NMR.

In recent years, a number of more sophisticated computational approaches have been proposed to determine the relative configuration of small organic molecules in a semi-automatic manner. Instead of generating the configuration for each possible relative configuration that are then tested against the experimental data, correct configuration and conformation of the molecule can be directly determined simultaneously during the structural calculation process. These methods include progressive stereo locking (PSL) [116], molecular dynamics simulations with tensorial orientational constraints [117–120], and floating chirality distance geometry calculations (ConArch+) [121–124]. The first two methods intrinsically rely on force fields, which may introduce an artificial bias toward low-energy structures, while the last approach determines the configuration and conformation of common organic molecules solely based on isotropic and anisotropic NMR data. Additionally, a program called TITANIA is able to perform *de novo* structural calculations of organic molecules using a large number of independent RDC data [125]. PALES-3D (P3D) approach directly predicts the alignment tensor of the analytes in the presence of the alignment medium using a grid-based sampling approach [126, 127]. Comparison of the predicted data of all possible stereoisomers with the experimental ones enables the assignment of correct relative configuration.

13.2.5 Successful Examples of Determination of Relative Configuration of Challenging Molecules Using Anisotropic NMR

Until 2022, the stereochemistry of a large number (>100) of organic molecules has been determined using anisotropic NMR spectroscopy. This includes a number of highly challenging natural products, such as homodimericin A – a hexacyclic polyketide with a carbon backbone containing eight contiguous stereogenic carbons in a C20 hexacyclic core [128], sagittamide A – a long-chain acyclic α,ω-dicarboxylic acid with eight stereocenters [129], and fibrosterol sulfate A – a polysulfated bis-steroid with a highly flexible linker containing three unknown stereocenters [114]. Furthermore, with the aid of anisotropic NMR spectroscopy, the relative configuration of proton-deficient compounds or molecules containing multiple remote stereocenters that pose significant challenges for *J*-coupling and NOE analysis can be determined more rapidly and accurately than ever before [46, 56, 57].

We chose here two of our own examples to demonstrate the usage of anisotropic NMR spectroscopy for the elucidation of the relative configuration of chiral organic molecules. The first regards sarcomililate A [50] and the second bilobalide [46].

Sarcomililate A (**4,** Figure 13.12a) is a new diterpenoid with a previously undescribed tricyclo-[11.3.0.0^{2,16}]hexadecane scaffold. Sarcomililate A contains five unknown stereocenters in the macrocyclic ring, resulting in 16 possible relative configurations. It was particularly difficult to establish the relative configuration of two stereochemical domains (I: C1, C2, C16; II: C9, C10), as they were not able to be correlated with NOEs and *J*-couplings. As considerable flexibility exists in the macrocyclic ring, a conformational search was employed to sample possible conformations. Extensive *J*-coupling analysis was then used to limit the conformational space. The selected conformational ensemble of each possible configuration was fitted against the 14 $^{1}D_{CH}$ data acquired in AAKLVFF phase. Comparison of the *Q* factor of all possible stereoisomers revealed the relative configuration of sarcomililate A as *S***S***R***S***R** (Figure 13.12b). Long-range NOEs extracted from zero-quantum NOESY spectrum could additionally confirm the RDC-determined conformation of *S***S***R***S***R** as the major conformation in solution (Figure 13.12d).

Bilobalide (**5,** Figure 13.13a) is a proton-deficient sesquiterpenoid with neuroprotective effects. The absolute configuration of (−)-bilobalide was previously assigned based on X-ray diffraction. We use this example here to demonstrate the power of ^{13}C RCSA in the stereochemical assignment of molecules with a high number of non-protonated carbons [46]. A total of 7 RDCs and 11 ΔΔRCSAs could be acquired from the AAKLVFF phase for bilobalide. These data were fitted individually or in combination against the energy-lowest conformer of each possible configuration (Figure 13.13b). While RDC alone determined the relative configuration incorrectly, ΔΔRCSA selected the correct stereoisomer with high confidence. A combination of RDC and ΔΔRCSA is also able to differentiate the correct relative configuration from the wrong ones. Therefore, it is desirable to employ independent anisotropic NMR data in the stereochemical assignment of new challenging molecules.

13.3 Determination of the Relative Configuration Using DP4 Probability and CASE-3D

Computational methods based on ab initio calculations of NMR chemical shifts become increasingly popular in establishing the constitution and relative configuration of new organic compounds (Figure 13.14). This idea was first proposed by Bifulco [130, 131]. Based on this concept, DP4 was developed in 2010 by Goodman and coworkers [132]. To improve the performance of DP4, an upgraded version referred to as DP4+ was published in

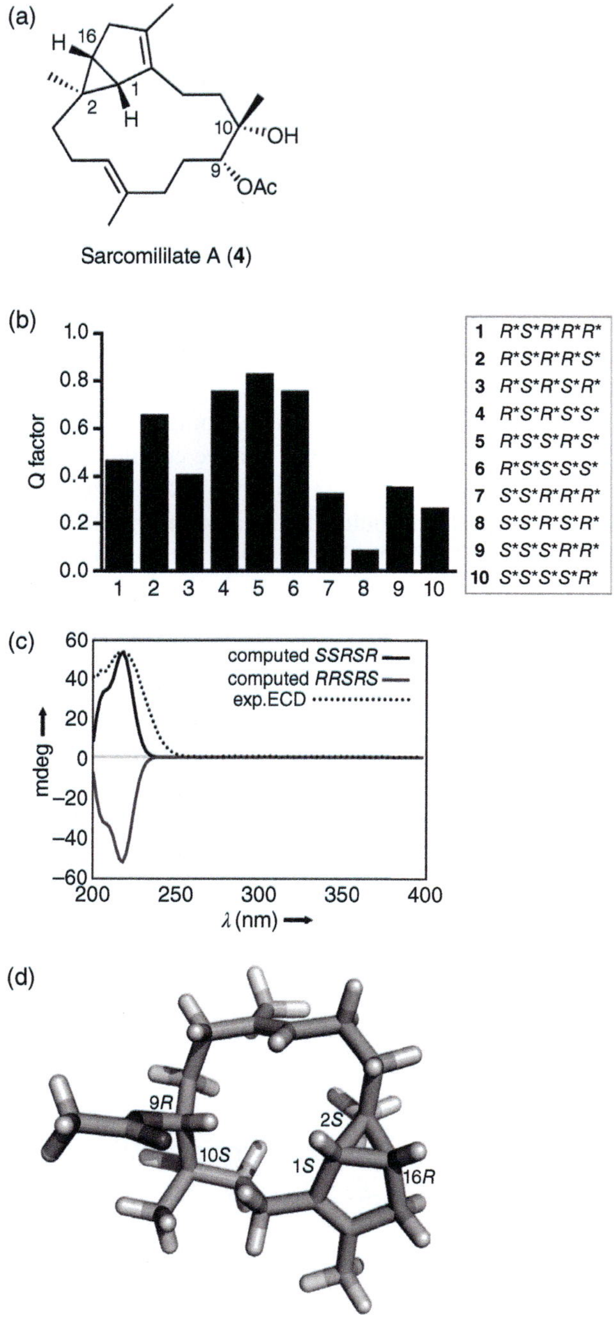

Figure 13.12 Determination of the absolute configuration of sarcomililate A using a combination of RDC and chiroptical spectroscopy: (a) structural formula of sarcomililate A; (b) Q factor of RDC analysis for 10 possible relative configurations; (c) Comparison of experimental and computed ECD spectra for deducing the absolute configuration; (d) Absolute configuration and the main conformation of sarcomililate A.

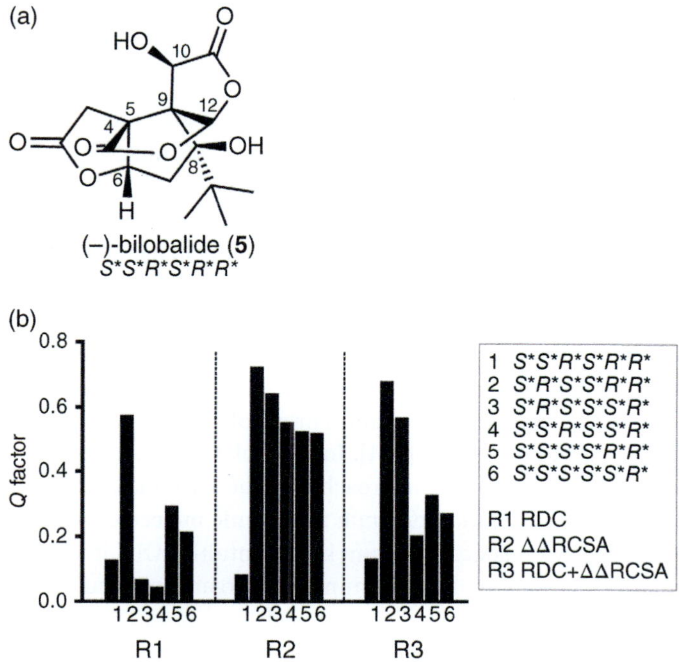

Figure 13.13 Determination of the relative configuration of (−)-bilobalide using ΔΔRCSA, RDC, and a combination of RDC and ΔΔRCSA. (a) Structural formula and the absolute configuration of (−)-bilobalide; (b) Q factor of RDC and RCSA analyses for six possible relative configurations.

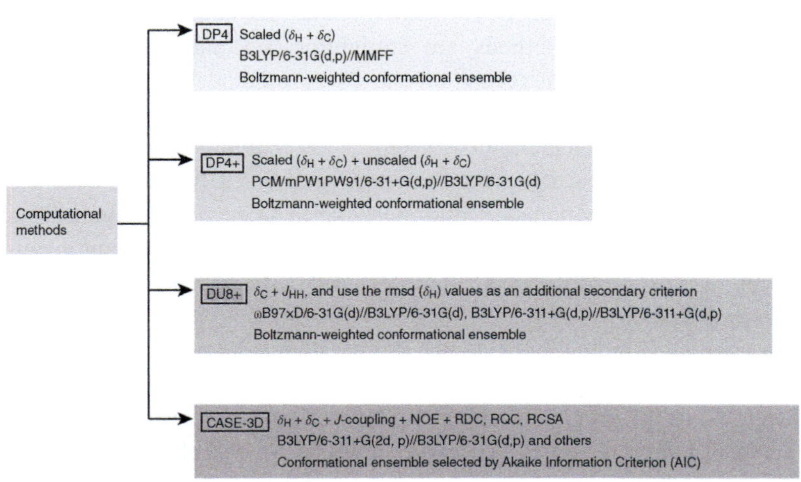

Figure 13.14 Comparison of NMR parameters, functional basis sets, and conformational selection methods applied in computational methods DP4, DP4+, DU8+, and CASE-3D.

2015 by Sarotti, which is based on prediction of the chemical shifts of $^{13}C/^{1}H$ nuclei from DFT Gauge-Independent Atomic Orbital (GIAO) calculations [133]. Both DP4 and DP4+ uses Bayes theorem, giving a probability of each possible relative configuration as the correct one. Among them, DP4+ becomes particularly popular as it has several unique advantages: (i) the estimation accuracy is improved by the introduction of both scaled and unscaled chemical shifts, (ii) each DP4+ term can be calculated using only ^{1}H data, ^{13}C data, or a combination of both and possibility to include results obtained from DFT calculations using different basis sets, (iii) DP4+ provides a user-friendly Excel sheet that is readily accessible to nonexperts. According to the report by Sarotti and coworkers, from 2015 to 2020 the structures of more than 200 organic molecules have been elucidated with the aid of DP4+. Advantages, problems, and a protocol of DP4+ were nicely summarized in a recent publication [134]. In parallel, Goodman and coworkers integrated machine learning approaches in the DP4 and developed an automatic structure elucidation program for organic molecules, named as DP4-AI [135]. This open-source program has an intuitive GUI interface, enabling structural elucidation of a large number of organic molecules in a high-throughput manner. In parallel, an approach named as DU8+ integrates predicted chemical shifts from DFT and relativistic force field (rff)-computed spin–spin couplings in the structure elucidation process. As a benchmark test for DU8+, structures including relative configuration of more than 100 natural products including a number of halogenated compounds were investigated, where the structures of 16 compounds could be revised [136].

Compared to anisotropic NMR-based methods, DP4 family gains its popularity because the methods are straightforward and time efficient. Nonetheless, in a recent review on the DP4+ method [134], the author stated that an "inherent problem associated with most computational methods for structural elucidation is related to the inability to unequivocally assess the correctness of a given structure," and that DP4/DP4+ becomes particularly problematic when intramolecular hydrogen bonds camouflage other conformers that are relevant in solution. To overcome this problem, an approach to avoid the dependency of DP4/DP4+ on DFT energies has been proposed.

In parallel, computer-assisted 3D structural elucidation (CASE-3D) has been developed by Navarro-Vázquez and colleagues [137, 138]. CASE-3D is integrated in the program Stereofitter (Maestro) and can also make use of chemical shift predictions by DFT computations. Nevertheless, a significant difference between these two methodologies exists in the conformational selection procedure. While original DP4/DP4+ method relies on DFT estimation of conformational amplitudes, CASE-3D combines a least-squares optimization of populations with an Akaike Information Criterion (AIC) selection of conformational ensembles. In addition to the chemical shift information, CASE-3D allows to integrate many different types of isotropic and anisotropic NMR data in the structural elucidation process, such as scalar coupling, NOE/ROE, RDC, RQC, and RCSA. This is especially helpful for solving stereochemistry of challenging natural products, in cases where the conformational spaces of the molecules are complex or the stereochemical domains are remote from each other.

13.4 Determination of the Absolute Configuration Using a Combination of NMR Spectroscopy and Chiroptical Spectroscopy

While anomalous X-ray diffraction is an established method for determining the absolute configuration of crystallizable organic molecules [139], determination of the absolute configuration using NMR spectroscopy [140, 141] poses a major challenge, as both enantiomers do not show difference in their NMR spectra. Chiroptical spectroscopic methods, such as vibrational (VCD) or electronic (ECD) circular dichroism as well as optical rotatory dispersion (ORD) and Raman optical activity (ROA) have become increasingly popular for studying the absolute configuration and conformations of chiral molecules in solution, but still with considerable difficulty for complex and flexible molecules (see Chapter 14) [142]. To determine the absolute configuration using chiroptical spectroscopy, in most cases, the experimental spectra need to be compared with the predicted ones obtained from cost-effective DFT or time-dependent-DFT computations [143–146].

Combination of NMR and chiroptical spectroscopy for determining the absolute configuration is a very attractive strategy, as NMR is able to simultaneously establish the relative configuration and preferred conformation of chiral organic molecules in solution. For the prediction of the chiroptical spectra, computed spectra of all dominant conformers need to be averaged according to their respective Boltzmann populations. Instead of solely relying on DFT estimation of conformational amplitudes, determination of major conformation in solution could be benefited from additional NMR restraints. Especially, anisotropic NMR data enables the determination of the population of preferred conformers, which turns to be rather difficult for NOE and *J*-coupling analysis. A number of published studies employed this strategy and established the relative and absolute configuration of semi-flexible and highly flexible compounds, such as mefloquine – an important antimalaria drug on the market [147], fusariumin A – a highly flexible alkylpyrrole derivative [148], vatiparol – a resveratrol oligomer [149], and sarcomililate A – an unusual diterpenoid [50]. As shown in Figure 13.12, a combination of RDC, *J*-coupling, and NOEs were employed to assign the relative configuration of sarcomililate A and its preferred conformation in solution. The ECD spectra of both enantiomers were predicted via time-dependent density-functional theory (TD-DFT) with the B3LYP/6-311G(d) basis set, where the NMR-determined conformers were used as input in the TD-DFT calculations. Comparison of experimental and computed ECD spectra enabled the determination of the absolute configuration of sarcomililate A unambiguously Figure 13.12c.

13.5 Determination of the Absolute Configuration Using NMR Alone

There are two general strategies for establishing the absolute configuration by means of NMR alone [140]: (i) without derivatization, e.g. using chiral solvating agents [150, 151]; (ii) with chiral derivatizing agents [141, 152–154], e.g. Mosher ester or amide analysis. In principle, the first approach is more attractive, as it does not require chemical modification of the molecules, but only rely on the chiral environment provided by a chiral solvent. However, in most of the cases since the interactions between chiral solvents and analytes are rather weak, only very small chemical shift differences could be detected for both enantiomers, so that the assignment of the absolute configuration became highly difficult and inaccurate. In contrast, the second method is substantially more popular, as derivatization of enantiomers into diastereomers enables higher differentiation of the molecules in the NMR spectra. Traditionally, assignment of the absolute configuration using chiral derivatization agents are empirical approaches, relying on comparison of different ^{1}H or ^{19}F chemical shifts of the derivatized diastereomers. Advances in this research field integrated several new strategies, such as structural and chemical shift calculations benefited from DFT calculations and acquiring NMR spectra at different temperatures, allowing this NMR-based method for determining the absolute configuration of chiral organic molecules with functional groups (e.g. hydroxyl, amino, and carboxyl) with sufficiently high accuracy and robustness [141].

13.5.1 Mosher Ester Analysis

The most commonly used approach relying on chiral derivation agents is the Mosher ester method for assigning the absolute configuration of secondary alcohols [155, 156]. It involves the coupling of the hydroxyl group of the alcohols with Mosher acid (*R*)- and (*S*)-α-methoxytrifluoromethylphenylacetic acid (MTPA), separately, resulting in an MTPA ester for each enantiomer. The chemical shifts of the obtained esters show an impact of the diamagnetic ring current from the phenyl substituent. According to the sign of the chemical shift difference of the relevant proton ($\Delta\delta^{SR} = \delta_S - \delta_R$) in the chiral ester, the absolute configuration of the chiral carbon attached to the hydroxyl group can be assigned [152]. This method can be also applied to the assignment of absolute configuration of α-chiral amines, which is called as Mosher amide analysis [157].

We demonstrate here the application of the Mosher ester method on the determination of the absolute configuration of (−)-menthol. Other examples and detailed procedures for performing Mosher ester analysis can be found in an excellent *Nature Protocol* paper [154]. For assigning the absolute configuration of (−)-menthol, the first step involved the coupling of the hydroxyl group attached to C1 of (−)-menthol with *S*- and *R*-MTPA, respectively. After the assignment of the proton resonances, differences in chemical shifts between *S*- and *R*-MTPA ester ($\Delta\delta^{SR}$) were calculated and sorted according to their sign. As illustrated in Figure 13.15, the resulting *S*- and *R*-MTPA esters show clear differences in proton chemical shifts, with positive $\Delta\delta^{SR}$ values for protons in front of the plane of the MTPA moiety, while those with negative values are all on the back side of that plane. Based on this result, the absolute configuration of C1 in (−)-menthol was successfully assigned as *R*.

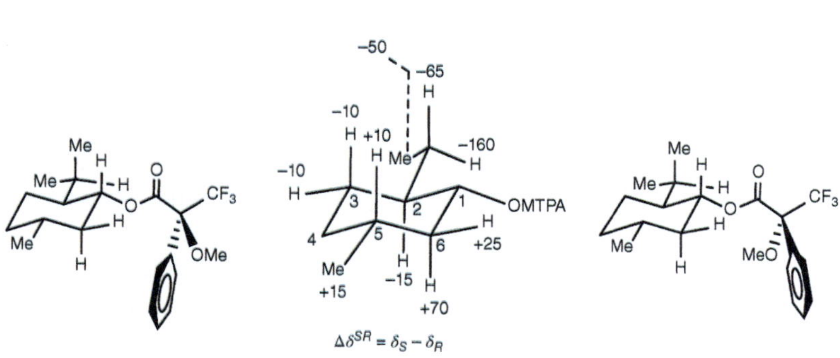

Figure 13.15 Conformations used for the analysis of the *S*- (left) and *R*-Mosher ester (right) of menthol. Comparison of $\Delta\delta^{SR}$ values (in Hz) for each proton on the front and back of the MTPA moiety (middle) allowed the determination of the absolute configuration of (−)-menthol.

13.5.2 Other Chiral Derivatizing Agents

Similar to the Mosher ester analysis, other chiral derivatizing agents have been proposed for determination of the absolute configuration of mono- and polyfunctional compounds. These include α-methoxyphenylacetic acid (MPA), 9-anthrylmethoxyacetic acid (9-AMA), and Boc-phenylglycine (BPG) [140]. Using MPA as derivatizing agent, instead of preparation of two derivatives from the two enantiomers of the chiral derivatizing agents and the substrate (double derivatization method), single derivatization method requiring only one enantiomer of the chiral derivatizing agent would also allow the assignment of the absolute configuration in some circumstances [158–160].

Other functional groups rather than alcohols and amides, e.g. sulfoxide, γ-lactones, imines, and lactams can be derivatized using Pirkle's alcohol (2,2,2-trifluoro-1-(9-anthryl) ethanol) [161]. Taking sulfoxide as an example, OH and $CH(OHCF_3)$ groups in Pirkle's alcohol can form hydrogen bonds with the sulfoxide group and largely influence the proton chemical shifts that can be used to assign the absolute configuration [162].

13.6 Future Perspective

NMR spectroscopy will continue to play a fundamental role in the determination of relative and absolute configuration of chiral organic molecules without a doubt. Further development is appreciated where different anisotropic and computational NMR approaches are combined, so that the relative configuration of novel complex molecules can be established with highest efficiency and accuracy. In this regard, we envision substantial progress will be made in the development of new computational approaches for structural calculation using isotopic and anisotropic NMR data, where correct configuration and conformation can be determined from single calculation run instead of extensive verification of each configurational possibility. We believe these advances will allow non-expert NMR users to apply the anisotropic NMR for determination of relative configuration on a routine basis.

Assignment of the absolute configuration without chiral derivatization is still highly challenging with only limited examples relying on the usage of chiral solvents or chiral nonracemic reagent. A number of studies revealed possibility of enantiodiscrimination based on anisotropic NMR data acquired in chiral alignment media [63–66, 69, 74–79, 83–85, 163]. Nevertheless, assignment of the absolute configuration is still not possible, as orientational order of the solute needs to be predicted reliably from first principles, while no computational approach so far produces satisfactory result. Active research will be continued in this field to finally enable the assignment of absolute configuration using anisotropic NMR data, which was experimentally demonstrated already in 1967 [163].

References

1 Doddrell, D.M., Pegg, D.T., and Bendall, M.R. (1982). Distortionless enhancement of NMR signals by polarization transfer. *J. Magn. Reson.* 48: 323–327.

2 Cavanagh, J., Palmer, A.G., Wright, P.E. et al. (1991). Sensitivity improvement in proton-detected two-dimensional heteronuclear correlation NMR spectroscopy. *J. Magn. Reson.* 93: 151–170.

3 Cavanagh, J., Palmer, A.G., Wright, P.E. et al. (1991). Sensitivity improvement in proton-detected two-dimensional heteronuclear relay spectroscopy. *J. Magn. Reson.* 91: 429–436.

4 Kay, L., Keifer, P., and Saarinen, T. (1992). Pure absorption gradient enhanced heteronuclear single quantum correlation spectroscopy with improved sensitivity. *J. Am. Chem. Soc.* 114: 10663–10665.

5 Schleucher, J., Schwendinger, M., Sattler, M. et al. (1994). A general enhancement scheme in heteronuclear multidimensional NMR employing pulsed field gradients. *J. Biomol. NMR* 4: 301–306.

6 Aue, W.P., Bartholdi, E., and Ernst, R.R. (1976). Two-dimensional spectroscopy. Application to nuclear magnetic resonce. *J. Chem. Phys.* 64: 2229–2246.

7 Nagayama, K., Kumar, A., Wüthrich, K. et al. (1980). Experimental techniques of two-dimensional correlated spectroscopy. *J. Magn. Reson.* 40: 321–334.

8 Bax, A. and Davis, D.G. (1985). MLEV-17-based two-dimensional homonuclear magnetization transfer spectroscopy. *J. Magn. Reson.* 65: 355–360.

9 Cicero, D.O., Barbato, G., and Bazzo, R. (2001). Sensitivity enhancement of a two-dimensional experiment for the measurement of heteronuclear long-range coupling constants, by a new scheme of coherence selection by gradients. *J. Magn. Reson.* 148: 209–213.

10 Bax, A., Freeman, R., and Kempsell, S.P. (1980). Natural abundance ^{13}C-^{13}C coupling observed via double-quantum coherence. *J. Am. Chem. Soc.* 102: 4849–4851.

11 Hesse, M., Meier, H., and Zeeh, B. (2007). *Spectroscopic Methods in Organic Chemistry*, 2e, 468. Stuttgart: Thieme.

12 Gunther, H. (2013). *NMR Spectroscopy: Basic Principles, Concepts and Applications in Chemistry*, 3e, 734. Weinheim: Wiley-VCH.

13 Levitt, M.H. (2008). *Spin Dynamics: Basic Principles of NMR Spectroscopy*, 2e, 740. Weinheim: Wiley-VCH.

14 Friebolin, H. (2010). *Basic One- and Two-Dimensional NMR Spectroscopy*, 5te, 418. Weinheim: Wiley-VCH.

15 Breitmaier, E. (2002). *Structure Elucidation by NMR in Organic Chemistry: A Practical Guide*, 3e, 272. Weinheim: Wiley-VCH.

16 Jacobsen, N.E. (2007). *NMR Spectroscopy Explained: Simplified Theory, Applications and Examples for Organic Chemistry and Structural Biology*, 688. Weinheim: Wiley-VCH.

17 Karplus, M. (1959). Contact electron-spin coupling of nuclear magnetic moments. *J. Chem. Phys.* 30: 11–15.

18 Piantini, U., Sorensen, O.W., and Ernst, R.R. (1982). Multiple quantum filters for elucidating NMR coupling networks. *J. Am. Chem. Soc.* 104: 6800–6801.

19 Griesinger, C., Sorensen, O.W., and Ernst, R.R. (1985). Two-dimensional correlation of connected NMR transitions. *J. Am. Chem. Soc.* 107: 6394–6396.

20 Mueller, L.P.E. (1987). COSY, a simple alternative to E.COSY. *J. Magn. Reson.* 7: 191–196.

21 Aydin, R. and Gunther, H. (1990). ^{13}C, ^{1}H spin-spin coupling. X†-norbornane: a reinvestigation of the karplus curve for $^{3}J(^{13}C, {}^{1}H)$. 28: 448–457.

22 Matsumori, N., Kaneno, D., Murata, M. et al. (1999). Stereochemical determination of acyclic structures based on carbon-proton spin-coupling constants. A method of configuration analysis for natural products. *J. Org. Chem.* 64: 866–876.

23 Koźmiński, W. and Nanz, D. (1997). HECADE: HMQC- and HSQC-based 2D NMR experiments for accurate and sensitive determination of heteronuclear coupling constants from E.COSY-type cross peaks. *J. Magn. Reson.* 124: 383–392.

24 Findeisen, M. and Berger, S. (2003). A selective pulse sequence for the determination of long-range C, H spin coupling constants. *Magn. Reson. Chem.* 41: 431–434.

25 Williamson, R.T., Marquez, B.L., Gerwick, W.H. et al. (2000). One- and two-dimensional gradient-selected HSQMBC NMR experiments for the efficient analysis of long-range heteronuclear coupling constants. *Magn. Reson. Chem.* 38: 265–273.

26 Williamson, R.T., Buevich, A.V., Martin, G.E. et al. (2014). LR-HSQMBC: a sensitive NMR technique to probe very long-range heteronuclear coupling pathways. *J. Org. Chem.* 79: 3887–3894.

27 Bigler, P. and Furrer, J. (2019). D-HMBC versus LR-HSQMBC: which experiment provides theoretically and experimentally the best results? *Magn. Reson. Chem.* 57: 129–143.

28 Saurí, J. and Parella, T. (2012). Efficient measurement of the sign and the magnitude of long-range proton-carbon coupling constants from a spin-state-selective HSQMBC-COSY experiment. *Magn. Reson. Chem.* 50: 717–721.

29 Saurí, J., Marcó, N., Williamson, R.T. et al. (2015). Extending long-range heteronuclear NMR connectivities by HSQMBC-COSY and HSQMBC-TOCSY experiments. *J. Magn. Reson.* 258: 25–32.

30 Sergi, G., Espinosa, J.F., and Parella, T. (2011). Accurate measurement of small heteronuclear coupling constants from pure-phase α/β HSQMBC cross-peaks. *J. Magn. Reson.* 213: 145–150.

31 Overhauser, A.W. (1953). Polarization of nuclei in metals. *Phys. Rev.* 89: 411415.

32 Neuhaus, D. and Williamson, M.P. (2000). *The Overhauser Effect in Structural and Conformational Analysis*, 2e, 656. Weinheim: Wiley-VCH.

33 Anet, F.A.L. and Bourn, A.J.R. (1965). Nuclear magnetic resonance spectral assignments from nuclear overhauser effects. *J. Am. Chem. Soc.* 87: 5250–5251.

34 Meier, B.H., Bachmann, P., and Ernst, R.R. (1979). Investigation of exchange processes by two-dimensional NMR spectroscopy. *J. Chem. Phys.* 71: 4546–4553.

35 Wagner, R. and Berger, S. (1996). Gradient-selected NOESY-A fourfold reduction of the measurement time for the NOESY experiment. *J. Magn. Reson. Ser. A* 123: 119–121.

36 Bax, A. and Davis, D.G. (1985). Practical aspects of two-dimensional transverse NOE spectroscop. *J. Magn. Reson.* 63: 207–213.

37 Hwang, T.L. and Shaka, A.J. (1992). Cross relaxation without TOCSY: transverse rotating-frame overhauser effect spectroscopy. *J. Am. Chem. Soc.* 114: 3157–3159.

38 Mao, H.B., Lin, A.J., Shi, Y. et al. (2013). Construction of enantiomerically enriched diazo compounds using diazo esters as nucleophiles: chiral Lewis base catalysis. *Angew. Chem. Int. Ed.* 52: 6288–6292.

39 Lambert, J.B., Gronert, S., Shurvell, H.F. et al. (1998). *Organic Structure Spectroscopy*, 2e. Prentice Hall. Inc.

40 Chen, G.G., Yang, X.L., and Yao, Z.J. (2016). Convenient one-step synthesis of benzo[c]phenanthridines by three-component reactions of isochromenylium tetrafluoroborates and stilbenes in acetonitrile. *Org. Lett.* 18: 1502–1505.

41 Sun, H., Dinan, L., Lafont, R. et al. (2010). Absolute configuration and docking study of canescensterone, a potent phytoecdysteroid, with non-lepidopteran ecdysteroid receptor selectivity. *Eur. J. Org. Chem.* 30: 5791–5709.

42 Thiele, C.M. and Berger, S. (2003). Probing the diastereotopicity of methylene protons in strychnine using residual dipolar couplings. *Org. Lett.* 5: 705–708.

43 Liu, Y.Z., Saurí, J., Mevers, E. et al. (2017). Unequivocal determination of complex molecular structurs using anisotropic NMR measurements. *Science* 356: eaam5349.

44 Hallwass, F., Schmidt, M., Sun, H. et al. (2011). Residual chemical shift anisotropy (RCSA): a tool for the analysis of the configuration of small molecules. *Angew. Chem. Int. Ed.* 50: 9487–9490.

45 Lesot, P., Gil, R.R., Berdagué, P. et al. (2020). Deuterium residual quadrupolar couplings: crossing the current frontiers in the relative configuration analysis of natural products. *J. Nat. Prod.* 83: 3141–3148.

46 Li, X.L., Chi, L.P., Navarro-Vázquez, A. et al. (2020). Stereochemical elucidation of natural products from residual chemical shift anisotropies in a liquid crystalline phase. *J. Am. Chem. Soc.* 142: 2301–2309.

47 Gil, R.R., Griesinger, C., Navarro-Vázquez, A. et al. (2014). *Structural Elucidation of Small Organic Molecules Assisted by NMR in Aligned Media. Structure Elucidation in Organic Chemistry*, 279–321. Weinheim: Wiley-VCH.

48 Gil, R.R. and Navarro-Vázquez, A. (2017). Application of residual dipolar couplings to the structural analysis of natural product. In: *Modern NMR Approaches to the Structure Elucidation of Natural Products: Volume 2: Data Acquisition and Applications to Compound Classes*, 117–176. Royal Society of Chemistry.

49 Schuetz, A., Murakami, T., Takada, N. et al. (2008). RDC-enhanced NMR spectroscopy in structure elucidation of sucro-neolambertellin. *Angew. Chem. Int. Ed.* 47: 2032–2034.

50 Yang, M., Li, X.L., Wang, J.R. et al. (2019). Sarcomililate A, an unusual diterpenoid with tricyclo[11.3.0.02,16] hexadecane carbon skeleton, and its potential biogenetic precursors from the Hainan soft coral *sarcophyton mililatensis*. *J. Org. Chem.* 84: 2568–2576.

51 Liu, C.P., Xie, C.Y., Zhao, J.X. et al. (2019). Dysoxylactam A: a macrocyclolipopeptide reverses P-glycoprotein-mediated multidrug resistance in cancer cells. *J. Am. Chem. Soc.* 141: 6812–6816.

52 Kummerlöwe, G. and Luy, B. (2009). Residual dipolar couplings for the configurational and conformational analysis of organic molecules. *Annu. Rep. NMR Spectrosc.* 68: 193–232.

53 Gil, R.R. (2011). Constitutional, configurational, and conformational analysis of small organic molecules on the basis of NMR residual dipolar couplings. *Angew. Chem. Int. Ed.* 50: 7222–7224.

54 Trigo-Mouriño, P., de la Fuente, M.C., Gil, R.R. et al. (2013). Conformational analysis of the anti-obesity drug lorcaserin in water: how to take advantage of long-range residual dipolar couplings. *Chem. Eur. J.* 19: 14989–14997.

55 Navarro-Vázquez, A., Berdagué, P., and Lesot, P. (2017). Integrated computational protocol for the analysis of quadrupolar splittings from natural-abundance deuterium NMR spectra in (chiral) oriented media. *ChemPhysChem* 18: 1252–1266.

56 Nath, N., Schmidt, M., Gil, R.R. et al. (2016). Determination of relative configuration from residual chemical shift anisotropy. *J. Am. Chem. Soc.* 138: 9548–9556.

57 Ndukwe, I.E., Brunskill, A., Gauthier, D.R. et al. (2019). ^{13}C NMR-based approaches for solving challenging stereochemical problems. *Org. Lett.* 21: 4072–4076.

58 Nath, N., Fuentes-Monteverde, J.C., Pech-Puch, D. et al. (2020). Relative configuration of micrograms of natural compounds using proton residual chemical shift anisotropy. *Nat. Commun.* 11: 4372.

59 Navarro-Vázquez, A. and MSpin-RDC. (2012). A program for the use of residual dipolar couplings for structure elucidation of small molecules. *Magn. Reson. Chem.* 50: S73–S79.

60 Luy, B., Kobzar, K., and Kessler, H. (2004). An easy and scalable method for the partial alignment of organic molecules for measuring residual dipolar couplings. *Angew. Chem. Int. Ed.* 43: 1092–1094.

61 Freudenberger, J.C., Spiteller, P., Bauer, R. et al. (2004). Stretched poly(dimethylsiloxane) gels as NMR alignment media for apolar and weakly polar organic solvents: an ideal tool for measuring RDCs at low molecular concentrations. *J. Am. Chem. Soc.* 126: 14690–14691.

62 Freudenberger, J.C., Knör, S., Kobzar, K. et al. (2005). Stretched poly(vinyl acetate) gels as NMR alignment media for the measurement of residual dipolar couplings in polar organic solvents. *Angew. Chem. Int. Ed.* 44: 423–426.

63 Haberz, P., Farjon, J., and Griesinger, C. (2005). A DMSO-compatible orienting medium: towards the investigation of the stereochemistry of natural products. *Angew. Chem. Int. Ed.* 44: 427–429.

64 Kobzar, K., Kessler, H., and Luy, B. (2005). Stretched gelatin gels as chiral alignment media for the discrimination of enantiomers by NMR spectroscopy. *Angew. Chem. Int. Ed.* 44: 3145–3147.

65 Luy, B. (2010). ChemInform abstract: distinction of enantiomers by NMR spectroscopy using chiral orienting media. *ChemInform* 41: 119–132.

66 Eliav, U. and Navon, G. (2006). Collagen fibers as a chiral agent: a demonstration of stereochemistry effects. *J. Am. Chem. Soc.* 128: 15956–15957.

67 Kummerlöwe, G., Auernheimer, J., Lendlein, A. et al. (2007). Stretched poly(acrylonitrile) as a scalable alignment medium for DMSO. *J. Am. Chem. Soc.* 129: 6080–6081.

68 Gil, R.R., Gayathri, C., Tsarevsky, N.V. et al. (2008). Stretched poly(methyl methacrylate) gel aligns small organic molecules in chloroform. Stereochemical analysis and diastereotopic proton NMR assignment in ludartin using residual dipolar couplings and ^3J coupling constant analysis. *J. Org. Chem.* 73: 840–848.

69 Kummerlçwe, G., Kiran, M.U., and Luy, B. (2009). Covalently cross-linked gelatin allows chiral distinction at elevated temperatures and in DMSO. *Chem. Eur. J.* 15: 12192–12195.

70 Merle, C., Kummerlöwe, G., Freudenberger, J.C. et al. (2013). Crosslinked poly(ethylene oxide) as a versatile alignment medium for the measurement of residual anisotropic NMR parameters. *Angew. Chem. Int. Ed.* 52: 10309–10312.

71 Gil-Silva, L.F., SantamaÍa-Fernández, R., Navarro-Vzquez, A. et al. (2016). Collection of NMR scalar and residual dipolar couplings using a single experiment. *Chem. Eur. J.* 22: 472–476.

72 García, M.E., Woodruff, S.R., Hellemann, E. et al. (2017). Di(ethylene glycol) methyl ether methacrylate (DEGMEMA)-derived gels align small organic molecules in methanol. *Magn. Reson. Chem.* 55: 206–209.

73 Liu, Y., Navarro-Vázquez, A., Gil, R.R. et al. (2019). Application of anisotropic NMR parameters to the confirmation of molecular structure. *Nat. Protoc.* 14: 217–247.

74 Czarniecka, K. and Samulski, E.T. (1981). Polypeptide liquid crystals: a deuterium NMR study. *Mol. Cryst. Liq. Cryst.* 63: 205–214.

75 Meddour, A., Berdague, P., Hedli, A. et al. (1997). Proton-decoupled carbon-13 NMR spectroscopy in a lyotropic chiral nematic solvent as an analytical tool for the measurement of the enantiomeric excess. *J. Am. Chem. Soc.* 119: 4502–4508.

76 Aroulanda, C., Sarfati, M., Courtieu, J. et al. (2001). Investigation of the enantioselectivity of three polypeptide liquid-crystalline solvents using NMR spectroscopy. *Enantiomer* 6: 281–287.

77 Arnold, L., Marx, A., Thiele, C.M. et al. (2010). Polyguanidines as chiral orienting media for organic compounds. *Chem. Eur. J.* 16: 10342–10346.

78 Meyer, N., Krupp, A., Schmidts, V. et al. (2012). Polyacetylenes as enantiodifferentiating alignment media. *Angew. Chem. Int. Ed.* 51: 8334–8338.

79 Krupp, A. and Reggelin, M. (2012). Phenylalanine-based polyarylacetylenes as enantiomer-differentiating alignment media. *Magn. Reson. Chem.* 50: S45–S52.

80 Lei, X., Xu, Z., Sun, H. et al. (2014). Graphene oxide liquid crystals as a versatile and tunable alignment medium for the measurement of residual dipolar couplings in organic solvents. *J. Am. Chem. Soc.* 136: 11280–11283.

81 Zong, W., Li, G.W., Cao, J.M. et al. (2016). An alignment medium for measuring residual dipolar couplings in pure DMSO: liquid crystals from graphene oxide grafted with polymer brushes. *Angew. Chem. Int. Ed.* 55: 3690–3693.

82 Lei, X., Qiu, F., Sun, H. et al. (2017). A self-assembled oligopeptide as a versatile NMR alignment medium for the measurement of residual dipolar couplings in methanol. *Angew. Chem. Int. Ed.* 56: 12857–12861.

83 Schwab, M., Herold, D., and Thiele, C.M. (2017). Polyaspartates as thermoresponsive e nantiodifferentiating helically chiral alignment media for anisotropic NMR spectroscopy. *Chem. Eur. J.* 23: 14576–14584.

84 Lesot, P., Berdagué, P., Meddour, A. et al. (2019). ^{2}H and ^{13}C NMR-based enantiodetection using polyacetylene versus polypeptide aligning media: versatile and complementary tools for chemists. *ChemPlusChem* 84: 144–153.

85 Qin, S.Y., Jiang, Y., Sun, H. et al. (2020). Measurement of residual dipolar couplings of organic molecules in multiple solvent systems using a liquid-crystalline-based medium. *Angew. Chem. Int. Ed.* 59: 17097–17103.

86 Berride, F., Troche-Pesqueira, E., Weiss, R.G. et al. (2021). *Effect of the Degree of Protonation of Amino Acid Dopants in Chirality Amplification in DSCG Lyotropic Nematic Phases*. ChemRxiv. Cambridge: Cambridge Open Engage This content is a preprint and has not been peer-reviewed.

87 Gayathri, C., Tsarevsky, N.V., and Gil, R.R. (2010). Residual dipolar couplings (RDCs) analysis of small molecules made easy: fast and tuneable alignment by reversible compression/relaxation of reusable PMMA gels. *Chem. Eur. J.* 16: 3622–3626.

88 Liu, Y.Z. and Prestegard, J.H. (2010). A device for the measurement of residual chemical shift anisotropy and residual dipolar coupling in soluble and membrane-associated proteins. *J. Biomol. NMR* 47: 249–258.

89 Lafon, O., Lesot, P., Zimmermann, H. et al. (2007). Chiral discrimination in the ^{13}C and ^{2}H NMR of the crown and saddle isomers of nonamethoxy-cyclotriveratrylene in chiral liquid-crystalline solutions. *J. Phys. Chem. B* 111: 9453–9467.

90 Hellemann, E. and Gil, R.R. (2018). New stretching method for aligning gels: its application to the measurement residual chemical shift anisotropies (RCSAs) without the need for isotropic shift correction. *Chem. Eur. J.* 24: 3689–3693.

91 Hallwass, F., Teles, R.R., Hellemann, E. et al. (2018). Measurement of residual chemical shift anisotropies in compressed polymethylmethacrylate gels. Automatic compensation of gel isotropic shift contribution. *Magn. Reson. Chem.* 56: 321–328.

92 Liu, H., Chen, P., Li, X.-L. et al. (2020). Practical aspects of oligopeptide AAKLVFF as an alignment medium for the measurements of residual dipolar coupling of organic molecules. *Magn. Reson. Chem.* 58: 404–410.

93 Enthart, A., Freudenberger, J.C., Furrer, J. et al. (2008). The CLIP/CLAP-HSQC: pure absorptive spectra for the measurement of one-bond couplings. *J. Magn. Reson.* 192: 314–322.

94 Snider, J.D., Troche-Pesqueira, E., Woodruff, S.R. et al. (2012). New strategy for RDC-assisted diastereotopic proton assignment using a combination of *J*-scaled BIRD HSQC and *J*-scaled BIRD HMQC/HSQC. *Magn. Reson. Chem.* 50: S86–S91.

95 Trigo-Mouriño, P., Navarro-Vázquez, A., Ying, J. et al. (2011). Structural discrimination in small molecules by accurate measurement of long-range proton-carbon NMR residual dipolar couplings. *Angew. Chem. Int. Ed. Eng.* 50: 7576–7580.

96 Nath, N., d'Auvergne, E.J., and Griesinger, C. (2015). Long-range residual dipolar couplings: a tool for determining the configuration of small molecules. *Angew. Chem. Int. Ed.* 54: 12706–12710.

97 Tzvetkova, P., Simova, S., and Luy, B.P.E. (2007). HSQC: a simple experiment for simultaneous and sign-sensitive measurement of (^{1}JCH+DCH) and (^{2}JHH+DHH) couplings. *J. Magn. Reson.* 186: 193–200.

98 Tzvetkova, P. and Luy, B. (2016). Q.E.COSY: determining sign and size of small deuterium residual quadrupolar couplings using an extended E.COSY principle. *Magn. Reson. Chem.* 54: 351–357.

99 Sinnaeve, D., Ilgen, J., Di Pietro, M.E. et al. (2020). Probing long-range anisotropic interactions: a general and sign-sensitive strategy to measure 1H–1H residual dipolar couplings as a key advance for organic structure determination. *Angew. Chem. Int. Ed.* 59: 5316–5320.

100 Gouilleux, B., Meddour, A., and Lesot, P. (2020). ^{2}H QUOSY 2D-NMR experiments in weakly aligning systems: from the conventional to the ultrafast approach. *ChemPhysChem* 21: 1548–1563.

101 Liu, Y.Z., Cohen, R.D., Gustafson, K.R. et al. (2018). Enhanced measurement of residual chemical shift anisotropy for small molecule structure elucidation. *Chem. Commun.* 54: 4254–4257.

102 Lewis-Atwell, T., Townsend, P.A., and Grayson, M.N. (2021). Comparisons of different force fields in conformational analysis and searching of organic molecules: a review. *Tetrahedron* 79: 131865.

103 Habgood, M., James, T., and Heifetz, A. (2020). Conformational searching with quantum mechanics. In: *Quantum Mechanics in Drug Discovery. Methods in Molecular Biology*, vol. 2114 (ed. A. Heifetz). New York: Humana https://doi.org/10.1007/978-1-0716-0282-9_14.

104 Kamachi, T. and Yoshizawa, K. (2016). Low-mode conformational search method with semiempirical quantum mechanical calculations: application to enantioselective organocatalysis. *J. Chem. Inf. Model.* 56: 347–353.

105 Abraham, R.J., Cooper, M.A., and Reid, M. (2021). The use of MM/QM calculations of 13C chemical shifts in the conformational analysis of some monosaccharides and sucrose. *New J. Chem.* 45: 2001–2009.

106 Vasilyev, V.V., Bliznyuk, A.A., and Voityuk, A.A. (1992). A combined quantum chemical/molecular mechanical study of hydrogen-bonded systems. *Int. J. Quantum Chem.* 44: 897–930.

107 Cornilescu, G., Marquardt, J., Ottiger, M. et al. (1998). Validation of protein structure from anisotropic carbonyl chemical shifts in a dilute liquid crystalline phase. *J. Am. Chem. Soc.* 120: 6836–6837.

108 Zweckstetter, M. and Bax, A. (2000). Prediction of sterically induced alignment in a dilute liquid crystalline phase: aid to protein structure determination by NMR. *J. Am. Chem. Soc.* 122: 3791–3792.

109 Zweckstetter, M., Hummer, G., and Bax, A. (2004). Prediction of charge-induced molecular alignment of biomolecules dissolved in dilute liquid-crystalline phases. *Biophys. J.* 86: 3444–3460.

110 Zweckstetter, M. (2008). NMR: prediction of molecular alignment from structure using the PALES software. *Nat. Protoc.* 3: 679–690.

111 *MacroModel Schrodinger Release 2017-3*. New York: Schrodinger LLC.

112 de Melo Sousa, C.M., Giordani, R.B., de Almeida, W.A.M. et al. (2021). Effect of the solvent on the conformation of monocrotaline as determined by isotropic and anisotropic NMR parameters. *Magn. Reson. Chem.* 59: 561–568.

113 Navarro-Vázqueza, A. (2021). Computational structural revision of a 4-hydroxy-3-(1′-angeloyloxy-2′,3′-epoxy-3′-methyl) butylacetophenone compound from *ageratina grandifolia*. *J. Nat. Prod.* 84: 2043–2047.

114 Sun, H., Reinscheid, U.M., Whitson, E.L. et al. (2011). Challenge of large-scale motion for residual dipolar coupling based analysis of configuration: the case of fibrosterol sulfate a. *J. Am. Chem. Soc.* 133: 14629–14636.

115 Thiele, C.M., Schmidts, V., Böttcher, B. et al. (2009). On the treatment of conformational flexibility when using residual dipolar couplings for structure determination. *Angew. Chem. Int. Ed.* 48: 6708–6712.

116 Cornilescu, G., Ramos Alvarenga, R.F., Wyche, T.P. et al. (2017). Progressive stereo locking (PSL): a residual dipolar coupling based force field method for determining the relative configuration of natural products and other small molecules. *ACS Chem. Biol.* 12: 2157–2163.

117 Di Pietro, M.E., Sternberg, U., and Luy, B. (2019). Molecular dynamics with orientational tensorial constraints: a new approach to probe the torsional angle distributions of small rotationally flexible molecules. *J. Phys. Chem. B* 123: 8480–8491.

118 Tzvetkova, P., Sternberg, U., Gloge, T. et al. (2019). Configuration determination by residual dipolar couplings: accessing the full conformational space by molecular dynamics with tensorial constraints. *Chem. Sci.* 10: 8774–8791.

119 Di Pietro, M.E., Tzvetkova, P., Gloge, T. et al. (2020). Fundamental and practical aspects of molecular dynamics using tensorial orientational constraints. *Liq. Cryst.* 47: 2043–2057.

120 Sager, E., Tzvetkova, P., Gossert, A.D. et al. (2020). Determination of configuration and conformation of a reserpine serivative with seven stereogenic centers using molecular dynamics with RDC-derived tensorial constraints. *Chem. Eur. J.* 26: 14435–14444.

121 Immel, S., Köck, M., and Reggelin, M. (2018). Configurational analysis by residual dipolar coupling driven floating chirality distance geometry calculations. *Chem. Eur. J.* 24: 13918–13930.

122 Immel, S., Köck, M., and Reggelin, M. (2019). Configurational analysis by residual dipolar couplings: a critical assessment of diastereomeric differentiabilities. *Chirality* 31: 384–400.

123 Köck, M., Reggelin, M., and Immel, S. (2020). The advanced floating chirality distance geometry approach – how anisotropic NMR parameters can support the determination of the relative configuration of natural products. *Mar. Drugs* 18: 330.

124 Immel, S., Köck, M., and Reggelin, M. (2022). NMR-based configurational assignments of natural products: gibbs sampling and bayesian inference using floating chirality distance geometry calculations. *Mar. Drugs* 20: 14.

125 Roth, F.A., Schmidts, V., and Thiele, C.M. (2021). TITANIA: model-free interpretation of residual dipolar couplings in the context of organic compounds. *J. Org. Chem.* 86: 15387–15402.

126 de Opakua, A.L., Klama, F., Ndukwe, I.E. et al. (2020). Determination of complex small-molecule structures using molecular alignment simulation. *Angew. Chem. Int. Ed.* 59: 6172–6176.

127 de Opakua, A.L. and Zweckstetter, M. (2021). Extending the applicability of P3D for structure determination of small molecules. *Magn. Reson. Chem.* 2: 105–116.

128 Mevers, E., Saurí, J., Liu, Y.Z. et al. (2016). Homodimericin a: a complex hexacyclic fungal metabolite. 138: 12324–12327.

129 Seike, H., Ghosh, I., and Kishi, Y. (2006). Stereochemistry of sagittamide a: prediction and confirmation. *Org. Lett.* 8: 3865–3868.

130 Barone, G., Gomez-Paloma, L., Duca, D. et al. (2002). Structure validation of natural products by quantum-mechanical GIAO calculations of 13C NMR chemical shifts. *Chem. Eur. J.* 15: 3233–3239.

131 Barone, G., Duca, D., Silvestri, A. et al. (2002). Determination of the relative stereochemistry of flexible organic compounds by ab initio methods: conformational analysis and boltzmann-averaged GIAO ^{13}C NMR chemical shifts. *Chem. Eur. J.* 8: 3240–3245.

132 Smith, S.G. and Goodman, J.M. (2010). Assigning stereochemistry to single diastereoisomers by GIAO NMR calculation: the DP4 probability. *J. Am. Chem. Soc.* 132: 12946–12959.

133 Grimblat, N., Zanardi, M.M., and Sarotti, A.M. (2015). Beyond DP4: an improved probability for the stereochemical assignment of isomeric compounds using quantum chemical calculations of NMR shifts. *J. Org. Chem.* 80: 12526–12534.

134 Marcarino, M.O., Cicetti, S., Zanardi, M.M. et al. (2022). A critical review on the use of DP4+ in the structural elucidation of natural products: the good, the bad and the ugly. A practical guide. *Nat. Prod. Rep.* 39: 58–76.

135 Howarth, A., Ermanis, K., and Goodman, J.M. (2020). DP4-AI automated NMR data analysis: straight from spectrometer to structure. *Chem. Sci.* 11: 4351–4359.

136 Kutateladze, A.G. and Reddy, D.S. (2017). High-throughput in silico structure validation and revision of halogenated natural products is enabled by parametric corrections to DFT-computed 13C NMR chemical shifts and spin–spin coupling constants. *J. Org. Chem.* 82: 3368–3381.

137 Troche-Pesqueira, E., Anklin, C., Gil, R.R. et al. (2017). Computer-assisted 3D structure elucidation of natural products using residual dipolar couplings. *Angew. Chem. Int. Ed.* 56: 3660–3664.

138 Navarro-Vázquez, A., Gil, R.R., and Blinov, K. (2018). Computer-assisted 3D structure elucidation (CASE-3D) of natural products combining isotropic and anisotropic NMR parameters. *J. Nat. Prod.* 81: 203–210.

139 Bijvoet, J.M., Peerdeman, A.F., and van Bommel, A.J. (1951). Determination of the absolute configuration of optically active compounds by means of X-rays. *Nature* 168: 271–272.

140 Seco, J.M., Quiñoá, E., and Riguera, R. (2004). The assignment of absolute configuration by NMR. *Chem. Rev.* 104: 17–118.

141 Seco, J.M., Quiñoá, E., and Riguera, R. (2012). Assignment of the absolute configuration of polyfunctional compounds by NMR using chiral derivatizing agents. *Chem. Rev.* 112: 4603–4641.

142 Pescitelli, G., Di Bari, L., and Berova, N. (2011). Conformational aspects in the studies of organic compounds by electronic circular dichroism. *Chem. Soc. Rev.* 40: 4603–4625.

143 Diedrich, C. and Grimme, S. (2003). Systematic investigation of modern quantum chemical methods to predict electronic circular dichroism spectra. *J. Phys. Chem. A* 107: 2524–2539.

144 Stephens, P.J., McCann, D.M., Butkus, E. et al. (2004). Determination of absolute configuration using concerted ab initio DFT calculations of electronic circular dichroism and optical rotation: bicyclo [3.3.1]nonane diones. *J. Org. Chem.* 69: 1948–1958.

145 Stephens, P.J., McCann, D.M., Devlin, F.J. et al. (2004). Determination of the absolute configuration of [3_2](1,4)barrelenophanedicarbonitrile using concerted time-dependent density functional theory calculations of optical rotation and electronic circular dichroism. *J. Am. Chem. Soc.* 126: 7514–7521.

146 Nugroho, A.E. and Morita, H. (2014). Circular dichroism calculation for natural products. *J. Nat. Med.* 68: 1–10.

147 Schmidt, M., Sun, H., Rogne, P. et al. (2012). Determining the absolute configuration of (+)-mefloquine HCl, the side-effect-reducing enantiomer of the antimalaria drug lariam. *J. Am. Chem. Soc.* 134: 3080–3083.

148 Liu, L.Y., Sun, H., Griesinger, C. et al. (2016). The use of a combination of RDC and chiroptical spectroscopy for determination of the absolute configuration of fusariumin A from the fungus fusarium sp. *Nat. Prod. Bioprospect.* 6: 41–48.

149 Ge, H.M., Sun, H., Jiang, N. et al. (2012). Relative and absolute configuration of vatiparol (1 mg): a novel anti-inflammatory polyphenol. *Chem. Eur. J.* 18: 5213–5221.

150 Pirkle, W.H. and Sikkenga, D.L. (1977). The use of chiral solvating agent for nuclear magnetic resonance determination of enantiomeric purity and absolute configuration of lactones. Consequences of three-point interactions. *J. Org. Chem.* 42: 1370–1374.

151 Kobayashi, Y., Hayashi, N., and Kishi, Y. (2002). Toward the creation of NMR databases in chiral solvents: bidentate chiral NMR solvents for assignment of the absolute configuration of acyclic secondary alcohols. *Org. Lett.* 4: 411–414.

152 Ohtani, I., Kusumi, T., Kashman, Y. et al. (1991). High-field FT NMR application of Mosher's method. The absolute configurations of marine terpenoids. *J. Am. Chem. Soc.* 113: 4092–4096.

153 Rieser, M.J., Hui, Y.H., Rupprecht, J.K. et al. (1992). Determination of absolute configuration of stereogenic carbinol centers in annonaceous acetogenins by proton and fluorine 19-NMR analysis of Mosher ester derivatives. *J. Am. Chem. Soc.* 114: 10203–10213.

154 Hoye, T.R., Jeffrey, C.S., and Shao, F. (2007). Mosher ester analysis for the determination of absolute configuration of stereogenic (chiral) carbinol carbons. *Nat. Protoc.* 2: 2451–2458.

155 Dale, J.A. and Mosher, H.S. (1973). Nuclear magnetic resonance enantiomer regents. Configurational correlations via nuclear magnetic resonance chemical shifts of diastereomeric mandelate, O-methylmandelate, and. Alpha.-methoxy-.Alpha.-trifluoromethylphenylacetate (MTPA) esters. *J. Am. Chem. Soc.* 95: 512–519.

156 Sullivan, G.R., Dale, J.A., and Mosher, H.S. (1973). Correlation of configuration and fluorine-19 chemical shifts of .Alpha.-methoxy-.Alpha. trifluoromethylphenyl acetate derivatives. *J. Org. Chem.* 38: 2143–2147.

157 Allen, D.A., Tomaso, A.E., Priest, O.P. et al. (2008). Mosher amides: determining the absolute stereochemistry of optically-active amines. *J. Chem. Educ.* 85: 698–700.

158 Latypov, S.K., Seco, J.M., Quiñoá, E. et al. (1998). Are both the (*R*)- and the (*S*)-MPA esters really needed for the assignment of the absolute configuration of secondary alcohols by NMR? The use of a single derivative. *J. Am. Chem. Soc.* 120: 877–882.

159 García, R., Seco, J.M., Vázquez, S.A. et al. (2006). Role of barium(II) in the determination of the absolute configuration of chiral amines by 1H NMR spectroscopy. *J. Org. Chem.* 71: 1119–1130.

160 Seco, J.M., Quiñoá, E., and Riguera, R. (1999). 9-Anthrylmethoxyacetic acid esterification shifts – correlation with the absolute stereochemistry of secondary alcohols. *Tetrahedron* 55: 569–584.

161 Pirkle, W.H., Sikkenga, D.L., and Pavlin, M.S. (1977). Nuclear magnetic resonance determination of enantiomeric composition and absolute configuration of. Gamma.-lactones using chiral 2,2,2-trifluoro-1-(9-anthryl)ethanol. *J. Org. Chem.* 42: 384–387.

162 Wenzel, T.J. (2017). Strategies for using NMR spectroscopy to determine absolute configuration. *Tetrahedron Asymmetry* 28: 1212–1219.

163 Sackmann, E., Meiboom, S., and Snyder, L.C. (1968). Nuclear magnetic resonance spectra of enantiomers in optically active liquid crystals. *J. Am. Chem. Soc.* 90: 2183–2184.

14

Absolute Configuration from Chiroptical Spectroscopy

Fernando Martins dos Santos Junior[1] *and João Marcos Batista Junior*[2]

[1] *Instituto de Química, Universidade Federal Fluminense, Niterói, RJ, Brazil*
[2] *Instituto de Ciência e Tecnologia, Universidade Federal de São Paulo, São José dos Campos, SP, Brazil*

14.1 Introduction

Chirality, meaning left- and right-handedness, is a universal concept in natural sciences, inherent to life and directly related to our daily activities [1]. It is present in all sorts of material objects ranging from fundamental particles to atoms, molecules, living cells, and even galaxies [1]. Particularly important to chemical sciences is molecular chirality, the manifestation of chirality in molecules. It is a key concept pervasive in different branches of chemistry and correlated areas [2]. Since the three-dimensional structure of molecules can lead to dissymmetry that impacts both physical–chemical properties and bioactivity, the exact orientation of atoms in space, known as absolute configuration (AC), has been an essential problem to be solved [3].

Enantiomers are defined as a pair of molecular entities that are mirror images of each other and non-superposable, which share, in an achiral environment, most physical and chemical properties. This fact makes them indistinguishable in many aspects. Their properties, however, may differ markedly when interacting with a chiral environment such as the biological systems, leading to different biological profiles [2–4]. As a consequence, there are a number of examples of commercially available drugs in which distinct pharmacological activity has been reported for enantiomers, as for example, the selective serotonin reuptake inhibitor (−)-(*S*)-citalopram (Lexapro; H. Lundbeck, Copenhagen, Denmark), which is 100-fold more

Chiral Separations and Stereochemical Elucidation: Fundamentals, Methods, and Applications, First Edition. Edited by Quezia Bezerra Cass, Maria Elizabeth Tiritan, João Marcos Batista Junior, and Juliana Cristina Barreiro.

potent than its enantiomer, and the antiparkinsonian agent L-DOPA (Sinemet; Organon, Jersey City, NJ, USA), where the (*S*)-enantiomer is active and safe while the (*R*)-enantiomer can cause agranulocytosis [5].

Currently, more than half of the commercially available drugs in use are chiral [6] and, based on the potential differences in pharmacokinetics, pharmacodynamics, and toxicological parameters for enantiomers, regulatory agencies have established strict requirements for the development of new chiral pharmaceuticals [7]. As a result, all new small molecule drugs registered by the US FDA (Food and Drug Administration) in 2015 were in the form of pure enantiomers with defined AC, with the exception of lesinurad (a racemate formed by two atropoisomers) [8]. This clearly demonstrates the current tendency in the pharmaceutical industry to switch from racemates to single enantiomers, a process known as "chiral switch." Hence, the determination of the AC of chiral molecules is of paramount importance in the discovery pipeline, being indispensable before the investigation of biological activities [9, 10].

The AC determination of chiral molecules, however, may be considered a major bottleneck for many small molecule-based research areas, such as natural product chemistry, organic synthesis, and drug discovery [11]. Thereby, this topic has historically played a prominent role in the field of molecular stereochemistry [12, 13]. The increased attention paid to chiral molecules over the years has guided the development of new methods for the resolution of racemates, for asymmetric synthesis, as well as for the structural characterization, including the determination of AC and conformations in solution [12, 13].

Considering AC determination, one of the first methodologies successfully applied was the stereocontrolled organic synthesis, where the molecule in question is transformed through a series of well-understood steps, into a chiral molecule with known optical rotation (OR) and AC. However, this methodology can often become complex and expensive [14]. The golden standard for stereochemical studies is single crystal X-ray crystallography [4]. However, although it is considered the most reliable methodology and presents unquestionable importance, it has limitations that can often either slow the analysis of some molecules or, in some cases, preclude it entirely [15, 16]. Its application is restricted to substances in the solid state (with single crystals of high quality). The presence of at least one strong anomalous scatterer (a heavy atom) is another important aspect of the methodology being necessary to directly attribute the AC by X-ray diffraction (Bijvoet method), otherwise an internal chiral reference of known AC has to be introduced in the crystal structure. Recent progress, especially the use of Cu $k\alpha$ radiation, has enabled confident assignment of absolute structure even for hydrocarbons [17].

Another important technique is nuclear magnetic resonance (NMR) spectroscopy. In the recent decades, NMR methods have grown in both sensitivity and signal dispersion by the introduction of spectrometers with stronger magnetic fields, leading to relevant instrumental progress. This advance has allowed faster acquisition of spectra with smaller sample requirements. Another important development was two-dimensional (2D) NMR spectroscopy. These methods provide important information based on the correlation of nuclei via scalar or dipolar couplings, making it possible to obtain a complete structural assignment relying entirely on NMR methods [18]. With these advancements, in conjunction with the development of quantum chemical calculation methods to predict NMR parameters [19], NMR has become the main analytical tool for determining the structure of organic compounds of either synthetic or natural origin, in solution, solid state, or even in the gas phase [4]. However, it is intrinsically achiral in isotropic media resulting in no sensitivity to the differentiation of enantiomers. To be used for this purpose, a chiral auxiliary must be introduced during the experiment such as chiral derivatizing agents, chiral solvation agents, chiral hosting compounds, metal complexes, among others [20]. More information on the use of X-ray crystallography and NMR to determine AC can be found in Chapters 12 and 13 of Part III – Methods, respectively.

As an alternative, chiroptical spectroscopic methods have faced a renaissance in the past decades, following important developments in both instrument and quantum chemical software to predict spectra [21]. They are non-destructive spectroscopic methods that investigate different aspects of optical activity [22]. The chiroptical methods originally included OR, optical rotatory dispersion (ORD), and electronic circular dichroism (ECD), all of them based on electronic transitions and referred to as electronic optical activity (EOA). In the early 1970s, two other methods evolved, namely, vibrational circular dichroism (VCD) and vibration Raman optical activity (ROA), which are, on the other hand, based on vibrational transitions and consequently referred to as vibrational optical activity (VOA) methods [22]. The chiroptical methods arise from the differential interaction of a chiral molecule in a non-racemic enantiomeric mixture with left- and right-circularly polarized light beams, which are chiral entities themselves, with one being the mirror image of the other [21–24].

The use of chiroptical spectroscopy is less laborious, time-consuming, and expensive than methods such as X-ray crystallography, NMR, and stereocontrolled organic synthesis. Additionally, it can be evaluated directly in solution without the need for crystallization and/or chemical derivatizations [21–24].

Of the various chiroptical spectroscopic methods developed for AC assignment, the most widely used are OR and ECD. Despite their widespread application, OR and ECD were first utilized as empirical methods, through a number of useful rules and procedures based on correlations with similar compounds reported in the literature [22]. However, all the empirical correlations suffer from exceptions that in several cases led to incorrect assignments [25]. The advent of accurate ab initio quantum mechanical calculations for predicting theoretical spectra, available in commercial software, resulted in significantly enhanced reliability for the interpretation of experimental data [26]. VOA methods have come to age after nearly 50 years of evolution and also greatly benefited from the aforementioned software developments. VCD and ROA now represent powerful and reliable tools for both conformational and configurational analysis of chiral molecules, even for those lacking UV–vis chromophores, which are indispensable for ECD analysis [14, 21].

The best approach for unambiguous assignments of AC in solution-state using chiroptical spectroscopy involve the combination of chiroptical methods aided by their corresponding quantum chemical predictions [27]. As the different chiroptical spectroscopic methods available probe distinct physical chemical phenomena resulting from the interaction of radiation with molecular chromophores, they are considered complementary in nature [28]. Therefore, the simultaneous application of more than one method provides additional levels of confidence for molecular stereochemistry assessments [21–26, 28].

In this chapter, we aim to provide the reader with an overview of the main chiroptical methods available for stereochemical assignments of organic compounds as well as the best practices for their application in different classes of molecules. It will cover the current state-of-the-art of chiroptical spectroscopy in combination with quantum chemistry calculations for determining both the AC and conformations of chiral molecules. Finally, recent examples of application of chiroptical methods, either alone or in combination, for the analysis of pharmaceuticals, as well as synthetic and natural product compounds will be discussed.

14.2 Chiroptical Methods

14.2.1 Optical Rotation and Optical Rotatory Dispersion

When linearly polarized light (LPL), also known as plane polarized light, passes through a sample containing chiral non-racemic molecules (optically active), its plane of polarization rotates. Such a rotation is referred to as OR. At a given wavelength, concentration, path length, and temperature, the sign and magnitude of the rotation are related to the molecular stereostructure of that substance [21]. Nevertheless, there is no simple and accurate procedure to deduce 3D structures from OR values, except when working with enantiomers of molecules of known AC.

The phenomenon of OR was discovered more than 200 years ago by Arago and Biot, and, after Fresnel's work in 1824, an explanation was possible. The incident LPL can be described as the superposition of two circularly polarized components of identical amplitude, which oscillates in phase. Thus, OR can be described as the difference in the refraction indices for left (n_L) and right (n_R) circular polarized light (CPL) in a chiral medium (optical active), known as circular birefringence. Therefore, when LPL goes through a medium presenting circular birefringence, the two circularly polarized components slow to a different extent and, as a result, the plane of polarization is rotated with respect to the original polarization plane (dextrorotatory for clockwise and levorotatory for anticlockwise rotations when viewed toward the light source) [25]. The angle of rotation in radians at a given wavelength (λ) can be written as:

$$\alpha = \frac{\pi}{\lambda}\left(n_L - n_R\right) \tag{14.1}$$

Important experimental quantities include the specific rotation $[\alpha]$. For liquids at wavelength λ (in nm) and temperature t (in °C) it is given by:

$$[\alpha]_\lambda^t = 100\alpha/l\rho \tag{14.2}$$

where α is the measured OR in degrees, l is the path length of light traversing through the liquid sample in decimeters, and ρ is the density of the liquid. In the case of solutions, specific rotations are given as:

$$[\alpha]_\lambda^t = 100\alpha/lc \tag{14.3}$$

where c is grams of optically active substance in 100 cm^3 of solution. These relations yield $[\alpha]$ in units of deg cm^3 dm^{-1} g^{-1}. Specific rotations $[\alpha]$ depend directly on concentration and temperature of the sample solution, in addition to the path length and wavelength of measurement. Therefore, it is recommended that the wavelength (λ) and temperature (t) of the measurements be indicated by subscript and superscript, as seen in Eqs. (14.2) and (14.3).

Historically, specific rotation has been used to determine the enantiomeric purity (in fact, enantiomeric excess, %ee) of chiral substances mainly measured at the wavelength of sodium D-line (589.3 nm, symbolized by $[\alpha]_D$) [21]. As the magnitude of the OR increases, the wavelength of the incident light decreases (Law of Inverse Squares); when the OR is evaluated at different wavelengths, a plot of specific rotation as a function of wavelength can be generated, at a given temperature. This plot is referred to as an ORD curve [21].

For many years, specific rotation was used as an empirical methodology to determine the AC of chiral compounds. However, the lack of a reliable method to correlate the observed specific rotation with the molecular structure limited its use as a structural tool. Hence, it was restricted to teaching laboratories for a long period [21]. This scenario has changed dramatically in the recent years with the development of quantum mechanics. Nowadays, standard quantum mechanical software, such as Gaussian [29] and Dalton [30], can be used to simulate the specific rotation at the wavelengths of interest for a given input structure. This development allowed the resurgence of the methodology as an important auxiliary tool in assessing molecular stereochemistry [21, 22].

14.2.1.1 Instrumentation

The instrument used to measure OR, known as polarimeter, is conceptually simple. First, a light beam of appropriate wavelength must be generated. This can be achieved by passing polychromatic light either through an appropriate monochromator or through a narrowband wavelength filter. Alternatively, a monochromatic laser beam may be used. The most commonly used light sources are Na and Hg lamps. Next, light has to be linearly polarized. This can be done by using an appropriate linear polarizer, referred to as input polarizer. Subsequently, the LPL passes through the sample contained in an appropriated sample cell, with known path length, and then through a second polarizer, referred to as output analyzer that is set at 90° with respect to the input polarizer, to finally reach an appropriate light detector, generally a photomultiplier [21, 25]. Figure 14.1 illustrates a block diagram of a typical OR instrument.

If the sample cell is empty or if the sample is optically inactive, as the output analyzer axis is set perpendicular to that of the input polarizer, no light reaches the detector, which is referred to as the "null point." However,

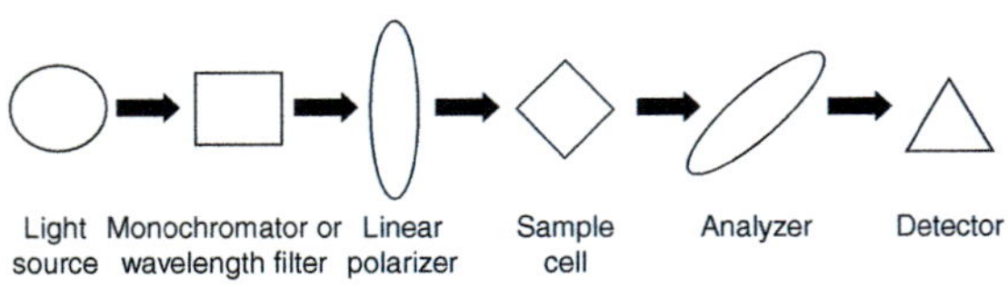

Figure 14.1 Block diagram of a simple polarimeter using polychromatic light. *Source:* Fernando M. dos Santos Jr.

in the case of optically active samples, the original incident LPL is rotated by the sample. Thus, the original null point is no longer observed since now light reaches the detector. Therefore, the analyzer axis must be rotated either clockwise or counterclockwise to reestablish the null point. The angle by which the analyzer axis is rotated to find this new null point is the measured OR value (in deg), which will be positive for clockwise rotation and negative for counterclockwise rotation. The specific rotation is then deduced using Eq. (14.2) or (14.3) [21, 25].

Currently, digital polarimeters are commercially available from a variety of manufacturers. To obtain ORD curves, dedicated ORD spectropolarimeters can be used, although they are not easily found. ORD plots can also be constructed from specific rotation values recorded at discrete wavelengths by using conventional digital polarimeters equipped with lamps and proper filters that yield wavelengths ranging from 880 to 254 nm, such as tungsten-halogen (WI) and Hg lamps. Another alternative is to use ORD accessories available for some ECD spectrometers [21, 25].

14.2.1.2 Measurements

The measurements of both OR and ORD are simple and straightforward. However, some points must be observed. Both methods are extremely dependent on factors such as the solvent used in the measurements, the temperature, the path length of the sample cell, and the sample concentration. This way, comparisons of experimental OR/ORD properties with literature reported values should be performed only for those recorded under the same conditions. The influence of the solvent in the experiment is so relevant that it is even possible to obtain inverted signals of OR for the same sample in different solvents [31]. The main reasons for the strong solvent dependence are based on electrostatic and nonelectrostatic effects, different conformations of the sample molecules, and solute–solvent interactions in different solvents [32, 33]. Regarding the temperature dependence, it directly affects the conformer population in solution, which can be particularly critical for flexible molecules. As a result, the observed rotation changes with temperature. Regarding the concentration dependence, higher concentrations commonly yield larger ORs than diluted samples. This dependency, however, is not necessarily linear; it can be a quadratic function or even a polynomial function of the concentration. Therefore, temperature and concentration at which OR and ORD are measured should be specified.

Finally, it should be highlighted that OR values should only be compared for molecules that share a direct enantiomeric relationship (enantiomers). No stereochemical information should be inferred from OR obtained for structurally related molecules, including diastereomers. There are cases reported in the literature of molecules that share the same skeleton, except for small structural modifications that surprisingly display opposite OR signs regardless of their identical ACs [11, 34, 35]. This clearly emphasizes the risks of assigning AC based on comparison of OR values of similar structures.

14.2.2 Electronic Circular Dichroism

The phenomenon of circular dichroism (CD) was discovered by the French physicist Aimé Cotton (1859–1961) in 1896, and the positive and negative bands observed in the CD spectrum became known as Cotton effects. It is important to mention that the term *Cotton effect* was first used to describe anomalous ORD. The CD phenomenon obtained in the ultraviolet–visible (UV–vis) spectral region (170–800 nm), where electronic transitions take place, is now known as ECD [23].

ECD, as the other chiroptical spectroscopy methods, also originates from the differential interaction of a chiral non-racemic molecule with left and right CPL. However, unlike OR and ORD, discussed in the previous section, which results from different refractive indices for each circular component, ECD results from the difference between the absorption coefficients for left and right CPL, defined as:

$$\Delta A = A_L - A_R \tag{14.4}$$

where A_L and A_R are the absorbances for left and right CPL, respectively. It may also be defined in terms of molar absorptivity ($\Delta\varepsilon$) in analogy to the Lambert–Beer law as:

$$\Delta\varepsilon = \varepsilon_L - \varepsilon_R = \frac{\Delta A}{c \cdot b \cdot \%ee} \tag{14.5}$$

which is dependent on the concentration c, expressed in mol l^{-1}, and on the pathlength b, expressed in cm. The differential absorbance is also directly proportional to the enantiomeric excess (%ee) of the sample.

ECD information may be expressed in another unit, ellipticity θ, which can be related to ΔA by Eq. (14.6) (θ in mdeg). Although modern CD instruments usually present their output in ellipticity units, they actually measure differential absorption.

$$\theta = 32{,}980 \cdot \Delta A \tag{14.6}$$

Another important factor to be highlighted in ECD is the anisotropy factor (g factor), also called dissymmetric factor, which is independent of sample concentration and pathlength, as long as the CD and absorbance measurements are performed on the same sample [23]; g is defined by:

$$g = \frac{\Delta A}{A} = \frac{\Delta \varepsilon}{\varepsilon} \tag{14.7}$$

Based on these equations, it becomes clear that ECD can be measured only in correspondence to absorption bands. A CD band (or Cotton effect) can be either positive or negative since ε_L may be either smaller or larger than ε_R (and consequently A_L and A_R). Thus, a positive Cotton effect represents that, for a given electronic transition, left CPL is absorbed to a higher extent. Accordingly, negative Cotton effects will result from right CPL being absorbed to a higher extent. Another important feature is that for each absorption band, the ECD sign and intensity for two enantiomers will always be exactly opposite (mirror image) [23].

As mentioned before, ECD probes electronic transitions, which only occur in the presence of specific structural motifs called chromophores. A chromophore can be defined as a portion of the molecule responsible for one or more absorption bands in the UV–vis spectrum. In the context of organic chemistry, a chromophore is usually a functional group, or the combination of functional groups, which presents extended π electron systems [36]. This may be considered the main limitation of ECD, as UV–vis transparent molecules cannot be analyzed. However, at the same time, it may be seen as an advantage, since in some cases it is possible to focus on specific chromophore moieties, without much influence from the rest of the molecular structure [36]. When the molecule contains chromophores, its ECD will reflect the electronic structure, chemical environment, orientation, and interactions of such chromophores. All this information is directly related to the configuration and conformations of the analyzed molecule. Hence, ECD is considered a powerful technique for stereochemical analysis given its sensitiveness to conformational features and AC, the former not available from an ordinary UV–vis absorption spectrum [23, 25].

Despite the enhanced stereochemical sensitivity, it is noteworthy that, as discussed for OR and ORD, ECD investigations are ideally suitable for enantiomers, whose ECD curves will be fully opposite. In the case of diastereomers for example, the relative configuration (RC) has to be first determined by other methods, such as NMR or X-ray diffraction, since there are numerous examples in the literature in which different diastereomers display indistinguishable ECD spectra [37].

Another asset of ECD (and also OR) is the possibility of hyphenation with high-performance liquid chromatography (HPLC) systems, equipped with chiral columns [38]. By this combination, it is possible to simultaneously measure the absorption and optical activities of eluting compounds either at a single wavelength, in the on-flow mode, or to record a full ECD spectrum in the stopped-flow mode. This interesting approach allows the determination of enantiomeric (or diastereomeric) ratio, enantiomeric elution order, and even the AC of a given molecule without the necessity to isolate the compound beforehand. Therefore, it is very useful for unstable analytes that are present in small amounts in complex mixtures such as natural product extracts or reaction mixtures [25].

14.2.2.1 Instrumentation

The basic instrumentation for ECD measurements consists of a conventional spectrophotometer equipped with a xenon lamp and a photomultiplier detector, in which a linear polarizer and photoelastic modulator (PEM), an optical element capable of generating left and right CPL, are introduced between the monochromator and the sample cell [23].

To create CPL, the PEM introduces a quarter-wave phase retardation between the two linearly polarized components of equal amplitude that compose the linearly polarized incident beam. The PEM consists of a piezoelectric quartz crystal typically oscillating with a frequency in the 50 kHz range, that is, during each cycle the light polarization changes between the left and right circular components 50,000 times per second, while the intensity remains constant. Then, after passing through a chiral non-racemic sample, the two CPL components become absorbed to different extents. The light reaching the detector is then time modulated in the same frequency as that of the PEM, and the signal is then lock-in amplified, processed, and displayed by a computer. The phase-locked amplification of the signal provides absorbance and CD spectra simultaneously. Aqueous (+)-(1*S*)-10-camphorsulfonate placed in a 10 mm cell at a concentration of 0.06% is a typical standard for CD intensity calibration and should give a signal of +190.4 mdeg for the band at around 290 nm. Figure 14.2 depicts a block diagram of a typical ECD instrument [23, 25].

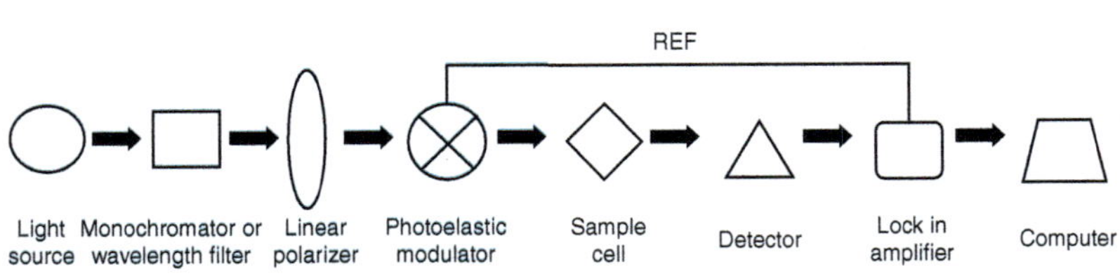

Figure 14.2 Block diagram of a conventional ECD instrument. *Source:* Fernando M. dos Santos Jr.

ECD spectroscopy is one of the most popular and widespread techniques for characterization of biomolecules and chiral compounds in both pharmaceutical and academic laboratories worldwide. This success is related to the variety of instruments commercially available from different manufacturers, ranging from typical ECD spectrometers to different accessories allowing magnetic, fluorescent, HPLC, and temperature-dependent CD detection [25].

14.2.2.2 Measurements

Essentially, ECD spectral acquisition is a spectrophotometric measurement. Consequently, the same usual precautions applied for the collection of an ordinary UV–vis absorption spectrum should be taken for CD as well. The first point regards the concentration of the sample. The total sample absorption must not exceed 1.5 absorbance units (a.u.), with 0.8 being the recommended value. This means that it may be necessary to use different pathlength (0.1–10 mm) cuvettes or even to adequately dilute the sample, taking into account the extinction coefficient of the relevant absorption bands. Another point worth mentioning is the choice of solvent. Although achiral solvents do not explicitly contribute to the ECD spectrum, a highly absorbent solvent may prevent correct measurements as it can overlap UV bands from the molecule of interest in certain regions of the spectrum and saturate the detector. Hence, the UV solvent cut-off must be considered when choosing the solvent for the experiments. Usually, spectrograde solvents are used, such as water, methanol, ethanol, acetonitrile, and hexane. Another important measurement detail is the need to purge the ECD instrument with high purity nitrogen gas to remove oxygen. This is a necessary step for two main reasons. First, oxygen can absorb light at shorter wavelengths that would interfere with the acquisition of the ECD spectrum; and second, to prevent the production of ozone, which in addition to being toxic, may cause degradation of the optical parts of the equipment, considerably decreasing its lifetime. Finally, the ECD spectrum needs to be baseline corrected. This can be done by subtracting the sample ECD spectrum from a measurement of the racemate of the pertinent compound, obtained under the same conditions or, as a common alternative that also provides acceptable results, subtracting from a blank measurement of the same solvent used.

14.2.3 Vibrational Circular Dichroism and Raman Optical Activity

As mentioned before, VCD and ROA are two spectroscopic techniques in the broader field of VOA [39]. As for the other chiroptical methods already discussed, VOA also probes the differential response of a chiral molecule to left versus right CPL, however, this time, arising from molecular vibrational transitions [14]. Both VCD and ROA were discovered in the early 1970s [40–42] and have evolved considerably over the last 45 years. They are now considered mature fields of research and application, focused mainly on the field of molecular stereochemistry. Both methods can be used to determine AC and the predominant conformations of chiral molecules in solution, bringing many advantages over other chiroptical methods. The application of ROA, however, still lags behind that of VCD [25, 43].

VCD can be considered as an extension of ECD (Eq. 14.5) into the infrared (IR) and near-IR regions of the electromagnetic spectrum, where vibrational transitions occur within the ground electronic state of the molecule. It is obtained by recording alternating left and right CPL absorbance spectra and taking their difference (Eq. 14.8), while the conventional vibrational spectrum arises from the absorbance of unpolarized IR light (defined as the average of left and right CPL absorptions).

$$\Delta A(\bar{\nu}) = A_{\mathrm{L}}(\bar{\nu}) - A_{\mathrm{R}}(\bar{\nu}) \tag{14.8}$$

where A_{L} and A_{R} represent absorbance for left and right CPL, respectively, and $(\bar{\nu})$ is the wavenumber frequency of the radiation. It can also be defined in terms of molar absorptivity as:

$$\Delta \varepsilon(\bar{\nu}) = \varepsilon_{\mathrm{L}}(\bar{\nu}) - \varepsilon_{\mathrm{R}}(\bar{\nu}) \tag{14.9}$$

ROA, in turn, can be visualized as the chiral version of ordinary vibrational Raman spectroscopy. An ordinary Raman spectrum is usually obtained using unpolarized, or linearly polarized, visible laser light and measuring the scattered light intensities as a function of the wavenumber shift associated with the scattered light. ROA, on the other hand, can be obtained by using alternating left and right CP incident laser light and measuring the difference in scattered intensities for a given linear scattered polarization [39]. This form is called incident circular polarization (ICP) ROA. Equivalently, ROA can be measured as the difference in intensity of small circularly polarized components in the scattered light when using incident light of fixed nonelliptical polarization, known as scattered circular polarization (SCP) ROA (Eq. 14.10 and Eq. 14.11, respectively) [25].

$$\Delta I_{\alpha}(\bar{\nu}) = I_{\alpha}^{\mathrm{R}}(\bar{\nu}) - I_{\alpha}^{\mathrm{L}}(\bar{\nu}) \quad \text{ICP} - \text{ROA} \tag{14.10}$$

$$\Delta I_{\alpha}(\bar{\nu}) = I_{\mathrm{R}}^{\alpha}(\bar{\nu}) - I_{\mathrm{L}}^{\alpha}(\bar{\nu}) \quad \text{SCP} - \text{ROA} \tag{14.11}$$

where α represents the fixed nonelliptical polarization state for scattered, in Eq. (14.10), or incident, in Eq. (14.11), radiation. I^{R} and I^{L} are the intensities of scattered light in right and left CP incident light, whereas I_R and I_L are the intensities of scattered right and left CPL [25]. The Raman spectrum of the sample is obtained by summation of the detected components. ROA intensities are proportional to the ROA cross-section (σ_{ROA}), incident laser intensity (I_0), number of molecules (N_A), and the enantiomeric excess (%ee) (Eq. 14.12):

$$I_{ROA} = \sigma_{ROA} \cdot I_0 \cdot N_A \cdot \%ee \tag{14.12}$$

VOA methods present advantages and also some disadvantages over other chiroptical methods. The accessible stereochemical information from VOA is far greater than that from the corresponding methods based on electronic transitions. The substantial number of accessible vibrational molecular transitions allows VOA methods, at least in theory, to investigate all degrees of freedom ($3N-6$, where N is the number of atoms) of the molecule of interest. In addition, vibrational transitions involving different functional groups within a molecule may behave independently (uncoupled) resulting in higher selectivity. The opposite happens to ECD, which is restricted to the UV–vis chromophoric portions of the molecules, where electronic transitions take place within a much narrower spectral range and with high degrees of overlap. VOA methods can be measured virtually for any molecule, in solid, liquid (more common), and gas phases without the requirement of either UV–vis chromophores as for ECD, derivatizations, or simulations of excited-state wavefunctions. On the other hand, extracting stereochemical information from VOA spectra becomes more difficult given the wealth of vibrational transitions and smaller intensities, which increases the dependence on the accuracy of the theoretical simulations [25, 39]. The high stereochemical discriminatory power of VOA methods culminated in the recognition of VCD as a new "standard method" in the US Pharmacopeia. Two chapters dedicated to VCD were published in 2016, USP 39-NF34, PF 41(5), Chapters 782 and 1782. The first chapter discusses in detail the aspects of VCD usage including qualification of VCD spectrometers, sample measurements, validation, and verification of measured spectra. The second provides specific examples on instrumentation, qualitative and quantitative analysis, comparisons between measured and calculated spectra, determination of enantiomeric excess (%ee), and concomitant use for determination of AC and %ee [40].

14.2.3.1 Instrumentation

For detailed information regarding VCD instrumentation, please refer to references [41, 42, 44, 45].

VCD measurements are usually carried out in the mid-infrared (mid-IR) region (~2000–900 cm^{-1}). The observed VCD bands in this region can be satisfactorily interpreted using theoretical predictions within harmonic approximations. Those in the higher energy region (~3500–2500 cm^{-1}) are usually complicated due to the presence of anharmonic effects, while the measurements in the lower energy region (~900–100 cm^{-1}) have not been established yet [39].

The basic VCD instrumentation consists of a Fourier transform infrared spectrometer (FT-IR) in which a ZnSe PEM is used to modulate the IR beam in the mid-IR region between left and right CPL at a typical frequency between 35 and 45 kHz.

There are two interferograms present at the liquid nitrogen-cooled mercury-cadmium-telluride (MCT) detector, one corresponding to the usual pathway for the measurement of the ordinary IR spectrum of the sample (I_{DC}), while the other carries spectral information at that PEM frequency (I_{AC}), which is then demodulated by a lock-in amplifier (or digital time-sampling) to give the VCD spectrum. Fourier transformation of these two interferograms leads to the simultaneous acquisition of IR and VCD intensities. If their calibrated ratio is taken, the final VCD spectrum is obtained; see Figure 14.3 for more details.

The first dedicated commercial FT-VCD spectrometer was the Chiral*IR* from Bomen/BioTools, introduced in 1997. Currently, other manufacturers, such as Bruker, Jasco, and BioTools, offer either VCD accessories or standalone dedicated instrumentation. The second generation of the Chiral*IR* instrument (Chiral*IR*-2X) from BioTools allows for a dual-PEM setup in which a second PEM is introduced right after the sample cell leading to real-time artifact subtraction and improved baseline correction [25, 39]. Typical standards for VCD intensity calibration include neat (−)-(1*S*)-α-pinene measured with a path length of ~75 μm and a 0.9 M solution of (−)-(1*S*)-camphor in CCl_4.

ROA instrumentation is more complex and only one spectrometer is commercially available, the Chiral*RAMAN* from BioTools. This instrument uses the SCP-ROA setup based on Werner Hug's design [46]. In this setup, laser light at 532 nm passes through two half-wave plates (HWPs) to create

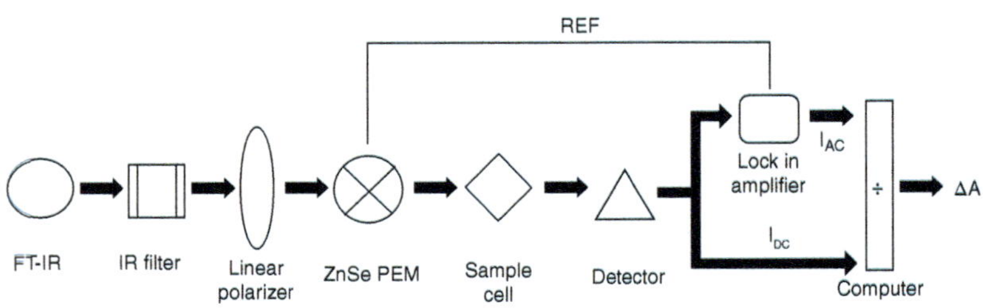

Figure 14.3 Block diagram of a conventional VCD instrument (see text for the definitions of the abbreviations used). *Source:* Fernando M. dos Santos Jr.

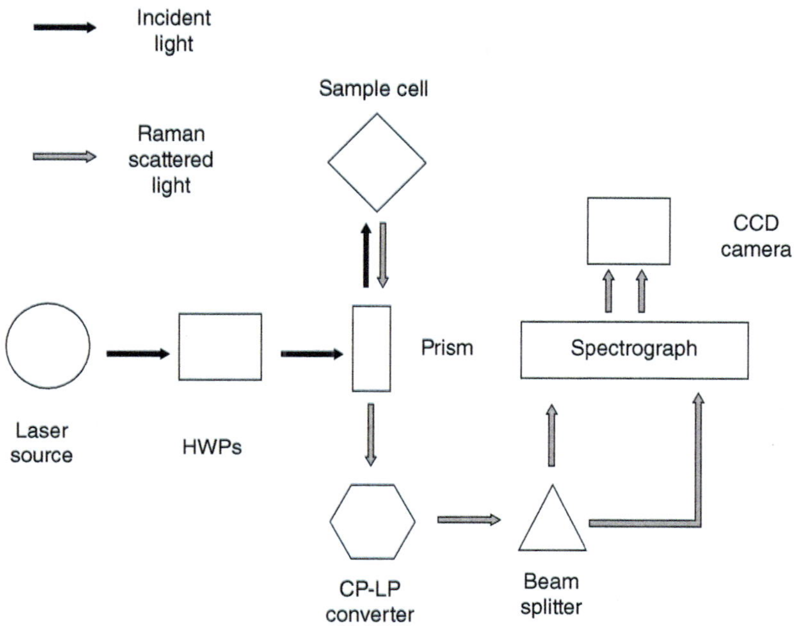

Figure 14.4 Block diagram of an instrument for the measurement of backscattered SCP-ROA spectra (see text for the definitions of the abbreviations used). *Source:* Fernando M. dos Santos Jr.

unpolarized visible laser light, which is then guided by prisms to the sample for 180° backscattering geometry. The two scattered circularly polarized components are converted into orthogonally linear polarization states (CP-LP converter) and measured separately, but simultaneously, through two different optical paths focused on the upper and lower halves of a multi-channel charge-coupled device (CCD) detector. The ROA spectrum is obtained by subtracting the intensities of the two scattered circularly polarized components, whereas the summation of these components yields the SCP-Raman spectrum. For detailed information regarding ROA instrumentation, including latest developments, please refer to references 45–49. See Figure 14.4 for more details.

14.2.3.2 Measurements

The correct measurement of IR and VCD spectra starts with the appropriate selection of solvent, sample concentration, and path length so that the IR bands in the mid-IR region remain within 0.1–0.9 absorption units. The optimum absorbance value is approximately 0.5. It may be then necessary to measure different regions of the spectrum separately to achieve appropriate absorbance levels and thereby a reliable VCD spectrum. The reliability of a VCD measurement is often assessed by checking the VCD noise level, which is obtained by dividing the VCD collection into two halves and subtracting them [25].

Most VCD measurements are performed in solution, making the choice solvent an important step. The desired characteristics of a suitable solvent include complete solubilization of the sample, large mid-IR window, and relatively small interaction with the target molecule (e.g. hydrogen bonding), avoiding interference. Solvents presenting multiple absorption bands in the IR region of interest should be avoided. Examples of appropriate inert solvents include CCl_4 and CS_2. More polar solvents can also be used, however, to avoid overlaps with absorption bands from the sample, deuterated analogues are preferred. Deuterium substitution typically enlarges the spectral window in the mid-IR by eliminating solvent absorption in the CH-bending region from 1500 to 1200 cm^{-1}. Some solvents typically used for VCD measurements include: CCl_4, CS_2, $CDCl_3$, DMSO-d_6, CD_3OD, CD_3CN, and D_2O [50, 51].

As the phenomenon of CD in the infrared region is ca. 10–100 times weaker than that in the UV–vis, the measurement of a quality VCD spectrum requires larger amounts of sample (2–5 mg) and longer collection times when compared to its electronic counterpart (ECD). Generally, concentrations of 0.01–0.5 M are required. Sample cells of different volumes (~50–200 μl), different path lengths (~5–15 μm for aqueous samples and ~50–200 μm for nonaqueous samples), and with either CaF_2 or BaF_2 windows are the most commonly used.

IR, VCD, and Noise spectra are generally collected with a spectral resolution in the range of 4–8 cm^{-1} using a collection of blocks obtained with variable time duration, which are then averaged to yield the final spectra. The total collection time for a high-quality VCD spectrum varies from 1 to 12 hours. The signal-to-noise ratio is proportional to the square root of the number of blocks used. As already mentioned, once a VCD spectrum is recorded, its baseline should be corrected. This can be achieved either by subtracting the VCD spectra of both enantiomers followed by division by two or by subtracting the VCD spectrum of the corresponding racemate. In case neither are available, subtraction of the VCD spectrum of the solvent should be carried out, with the best results being obtained in this case using a dual-PEM system [25, 45].

ROA spectra are usually recorded in a wider spectral range compared to that of VCD (100–2000 cm^{-1}). The preferred solvents for measuring ROA spectra are those that do not have strong vibrational Raman bands. H_2O is an excellent choice due to its poor Raman scattering. When organic solvents are used, they do not necessarily need to be deuterated. Neat liquid samples can also be used. Even though the ROA signal is intrinsically small requiring concentrated solutions (30–50 mg ml^{-1}), the smaller volumes needed in ROA cells (40–50 μl) result in similar amounts of sample to those used for VCD measurements (i.e. 2–5 mg).

Additional challenges need to be overcome for the measurement of quality ROA spectra, such as the reduction of background fluorescence, the sample stability under laser radiation, and the appropriate sample concentration to provide approximately 10^{10} Raman counts for the major bands in the Raman spectrum. Once these points are dealt with, the ROA spectrum of neat liquids may be acquired in less than one hour; however, for aqueous solutions of biological molecules, it can take up to several hours or even days. Finally, the excitation wavelength of the incident laser radiation is typically 532 nm, and its power can be adjusted from approximately 200 to 1000 mW [25, 45].

14.2.4 Simulation of Chiroptical Properties

Computational chemistry based on first-principles theory has undergone an impressive revolution in the past three decades, moving from being a highly specialized field into a mainstream [52]. Such a development has changed the way chemists face the important step of assigning the AC of chiral molecules [27]. It is now relatively easy to correlate the quantum mechanical theoretical prediction with observed chiroptical properties, which allows the assignment of AC without the need for any reference system or chemical derivatization, and often without the necessity of establishing the molecular physicochemical mechanisms responsible for the observed property [27].

The simulation of chiroptical spectra is often the last step in a series of computations, which usually involves the following common steps: (1) construction of an input containing information about the three-dimensional structure, (2) conformational sampling at a low-level method, (3) refined geometry optimization at the density functional theory (DFT) level, (4) computation of the Boltzmann populations, (5) chiroptical property simulations for each relevant conformer, (6) creation of Boltzmann weighted composite spectra, and (7) comparison of simulated and experimental data. Following each of these steps correctly leads to an unambiguous AC assignment (8) (Figure 14.5).

Herein, we are going to discuss in detail the common steps (1–4) that should be taken before the actual simulation of the different chiroptical properties. The following section is aimed at guiding non-specialists through

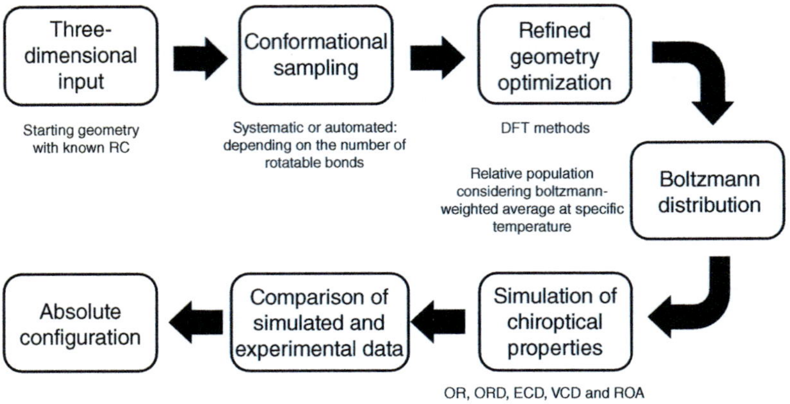

Figure 14.5 Flowchart for the typical steps in chiroptical properties calculations. *Source:* Fernando M. dos Santos Jr.

the standard computational workflow applied to molecular stereochemistry studies. Then, we will go through the theoretical background for the simulation of each chiroptical method separately.

14.2.4.1 Common Theoretical Steps

The initial step (1) for each chiroptical property simulation is to construct an input containing the three-dimensional structural information of a given molecule, with AC arbitrarily chosen. A molecular editor software with a graphical interface may be used to draw and manipulate the chemical structures, such as ArgusLab [53], Avogadro [54], ChemSketch [55], Chem3D [56], HyperChem [57], GaussView [58], among others [19]. Even at this initial stage, an important point should be highlighted. All structural features of the molecule of interest must be known in advance, including molecular constitution and, principally, the RC of the various chirality elements present in the molecule. Although, in principle, the RC may also be determined by chiroptical spectroscopies through the combination of different methods [37, 59], the use of independent tools, such as X-ray crystallography (whenever available) and NMR spectroscopy is highly recommended to secure the RC [27].

Subsequently, a thorough conformational search should be performed (2) at a low computational level using different methods, which will depend on the degree of freedom and the number of rotatable bonds for each molecule. For structures with few rotatable bonds, a systematic conformational search is indicated. However, for molecules with higher conformational mobility, an automated conformational search is recommended. The most commonly used method is the Monte Carlo algorithm combined with molecular mechanics force fields, like MMFF and MM+, among others, incorporated into several conformational search software, such as Spartan [60], HyperChem [57], MacroModel [61], Conflex [62], among others [19]. The calculated conformers are filtered according to an energy criterion. A threshold around 10–15 kcal mol^{-1} above the lowest-energy conformer is usually enough to account for all candidates that might be relevant to the chiroptical property simulation [27].

Then, the selected conformers are subjected to a geometry optimization step (3) using quantum mechanical methods. This is typically done using the B3LYP/6-31G(d) model chemistry; however, the use of more efficient functionals for structure and energetics is recommended. In some cases, a higher level of theory may be necessary for the calculations. This may include larger basis sets, such as a triple-ζ and/or the inclusion of additional polarization or even diffuse functions. The inclusion of a solvent model such as the polarizable continuum model (PCM), the concductor-like screening model (COSMO), or the solvation model based on density (SMD) is also highly recommended. The use of dispersion-corrected functionals may also improve the accuracy of the simulations in specific cases. Next, vibrational normal modes are computed with the same theoretical method used in step 3 to ensure that the optimized structures are true minimum energy points on the potential energy surface [27].

After obtaining refined geometries and energy values with a higher level of theory, the Boltzmann distribution (4) is calculated for the set of low-energy conformers obtained in step (3). At this stage, Gibbs free energies or zero-point corrected electronic energies are preferred. A new cut-off threshold based on the Boltzmann population is then applied to select those conformers accounting for more than 90% (the higher the better) of the total distribution, as these are expected to contribute to the experimental spectrum at the working temperature, generally 298 K. In practice, this usually refers to all the conformers that coexist within a 2.0 kcal mol^{-1} free energy window relative to the most stable conformer. These selected optimized conformers need to be considered in the following steps for the simulation of the chiroptical properties [27].

The chiroptical property of interest (5) is then calculated for all populated conformers found in steps 3 and 4. The final Boltzmann-averaged simulated spectrum is created by weighting each individual conformer spectrum with their appropriate Boltzmann factor, and then adding all weighted spectra to each other (6). A dedicated subtopic will be presented for the simulation of each chiroptical method in more detail (5). Currently, commercial or open-source packages are available for the DFT calculations, such as Gaussian [29], TURBOMOLE [63], Tinker [64], and GAMESS [65, 66].

Finally, the comparison of the experimental spectrum with the final calculated spectrum obtained in step (6) over the whole available spectral range is performed (7). If a good match is obtained, the AC assumed in the calculations is the correct one; if a mirror-image relationship is observed, the opposite AC is correct (8). Since, by definition, the spectra of two enantiomers are mirror images of each other, it is not necessary to simulate the spectral properties of both enantiomers. Although the comparison between calculation and experiment is usually done visually, when available, the use of comparison tools that allow a quantitative assessment is recommended, such as SpecDis [67], SpecTech [68], and CompareVOA [69]. Additionally, the use of a single chiroptical spectroscopic method may provide ambiguous or incomplete results. In such cases, the simultaneous use of more than one chiroptical method may provide complementary information. Hence, it is strongly recommended that, whenever possible, different chiroptical methods be used in combination to determine the stereostructure of a given molecule [70].

14.2.4.2 OR and ORD Simulations

The complete description of the quantum mechanical methods used for calculating OR and ORD is beyond the scope of this chapter; however, some guidelines will be provided for an accurate prediction of this molecular property. For a more complete and comprehensive review on this subject, including quantum mechanical methods, empirical, and semiclassical models, see [23, 71].

The calculations of OR and ORD are performed using time-dependent density functional theory (TD-DFT), gauge-invariant (including) atomic orbitals (GIAOs), to yield gauge origin-independent rotations, and the dynamic (frequency-dependent) method [25]. This type of calculation can be done in different software packages, and should include the desired incident frequencies (e.g. 589.3 nm for OR and 633, 589.3, 578, 546, 436, 365, and 355 nm for ORD).

OR and ORD are sensitive to any approximations made in the electronic structure calculation, thus, it is advisable to use the most accurate methods possible. This includes the use of a large basis set containing diffuse functions, such as Pople's 6-311G++(2d,2p) or Dunning's aug-cc-pVDZ and aug-cc-pVTZ, combined with the hybrid functional B3LYP. These theoretical levels are considered a good compromise between computational cost and accuracy for OR and ORD calculations [23, 71].

The theoretical calculation of OR is related to the frequency-dependent, electric-dipole–magnetic-dipole polarizability tensor, $G'_{\alpha\beta}$, given by:

$$G'_{\alpha\beta} = \frac{-4\pi}{h}\sum_{n\neq 0}\frac{\omega}{\omega_{n0}^{2}-\omega^{2}}\mathrm{Im}\left[\left\langle 0\middle|\mu_{\alpha}\middle|n\right\rangle\left\langle n\middle|m_{\beta}\middle|0\right\rangle\right] \tag{14.13}$$

where ω_{n0} is the angular transition frequency; 0 and n represent the ground and excited electronic state wave functions, respectively, and μ_α and m_β are the electric and magnetic dipole moment operators, respectively. Equation (14.13) is related to the OR parameter β as:

$$\beta = -\frac{1}{3\omega}\left(G'_{xx} + G'_{yy} + G'_{zz}\right) \tag{14.14}$$

where ω is the frequency of the incident radiation, which in turn is correlated with the OR in units of deg $cm^3\ dm^{-1}\ g^{-1}$ through:

$$\left[\alpha\right] = 0.1343 \times 10^{-3}\frac{\beta\bar{\upsilon}^2}{M} \tag{14.15}$$

with β in atomic units (Bohr), M in g mol^{-1}, and $\bar{\nu}$ is the wavenumber where the OR is measured, in cm^{-1}.

Polavarapu and coworkers, in 2003, suggested that intrinsic rotation, the specific rotation at infinite dilution, should be compared with the calculated data since calculations are generally carried out for isolated molecules. Intrinsic rotations can be determined by measuring the specific rotation as a function of finite concentration and extrapolating the data to zero concentration. By doing so, it is possible to avoid experimental errors generally caused by chemical impurities, incorrect %ee values, and solute–solute interactions, when high concentrations are employed [31].

Another point that should be discussed is related to DFT functional and basis set errors. These errors are expected in any quantum mechanical calculation since it is based on approximations of atomic wave functions. The accuracy of such predictions is sensitive to the choice of the functional and directly proportional to the size of the basis set used. Therefore, based on that, Stephens and coworkers, in 2005, proposed that for rigid compounds the comparisons between experimental and calculated specific rotation values should be reliable only for those molecules yielding experimental values much greater than 30. In the case of conformationally flexible molecules, the uncertainty of specific rotation further increases with the possibility of errors in the conformational population average at the initial step of calculation [72].

Finally, it is important to highlight that the use of comparisons between experimental and calculated OR data obtained at a single wavelength (e.g. specific rotation at 589.3 nm) in order to determine the AC of chiral molecules is strongly discouraged, especially for systems with multiple electronic transitions [73]. Instead, the method of choice should be ORD, since it is possible to assess the trend of the dispersion, which is opposite for two enantiomers. Moreover, at shorter wavelengths, the magnitude of experimental OR becomes larger than that at 589 nm, making theoretical predictions more reliable [27].

14.2.4.3 ECD Simulations

Although it is relatively easy to obtain a good quality ECD spectrum, its interpretation to extract stereochemical information is not a trivial task. Over the years, different methods were developed to help solving this problem. A comprehensive discussion about empirical, semiempirical, and/or exciton chirality methods are beyond the scope of this chapter and may be found elsewhere [23, 74, 75]. The present chapter will be focused on ab initio quantum mechanical calculations of ECD spectra, which allows an accurate assignment of AC of chiral molecules by comparisons with experimental data.

UV and ECD intensities are obtained theoretically by calculating the dipole (D) and rotational strengths (R), respectively. The quantity D is defined as:

$$D_{0n} = \left|\langle 0|\mu|n\rangle\right|^2 \tag{14.16}$$

where μ represents the electric dipole transition moment vector, where 0 and n are the wavefunctions of ground and excited states, respectively.

The rotational strength, on the other hand, is given by the scalar product:

$$R_{0n} = \mathrm{Im}\{\langle 0|\mu|n\rangle\langle \cdot n|m|0\rangle\} = |\mu|\cdot|m|\cdot\cos\xi \tag{14.17}$$

where m is the corresponding magnetic dipole transition moment. It is interesting to highlight that dipole strength is always positive, whereas the rotational strength can be either positive or negative, depending on the absolute value of both vectors as well as their relative orientation (pointing to the same or opposite directions). R will be zero if either $\mu = 0$ or $m = 0$ or if the orientation of these two vectors is orthogonal ($\xi = 90°$). In achiral molecules, one of these two conditions is met for each transition [25].

The method of choice for calculating ECD excited-state wavefunctions and transition energies is TD-DFT. Generally, this is the most computationally demanding step in the whole process, thus, the choice of the level of theory is an important step where both functional and basis set should be chosen carefully. As a general guide, range-separated functionals such as CAM-B3LYP and ωB97X are the best choice, since they perform better than hybrid ones in ECD calculations. Regarding the basis sets, it is recommended the use of those of double or triple-ζ quality with a sufficiently wide set of polarization functions. Diffuse functions may be necessary whenever Rydberg states contribute substantially to the ECD spectrum; good choices include Ahlrichs' TZVP and def2-TZVP or Dunning's and

Pople's aug-cc-pVDZ and 6-311++G(2d,2p), respectively. However, one of the most important points to consider is the use of more than one functional/basis set combination for UV/ECD simulations [27]. Solvation is commonly taken into account in UV/ECD simulations by means of PCM.

The calculation accuracy is also affected by the number of excited states to be simulated. Usually, best results are obtained with the use of larger basis set and a larger number of excited states; however, the higher the accuracy the higher is the computational cost [25].

Another important point when comparing experimental and calculated absorption/ECD spectra is the application of wavelength correction factors to the theoretical spectra to take into account systematic errors in the transition energies. This shift is determined considering a comparison of experimental and calculated UV absorption spectra [27]. The theoretical UV and ECD spectra are generated by plotting the calculated dipole and rotational strengths, respectively, against discrete transition energies. These plots are then to be converted into a true spectrum through approximation of UV and ECD bands by a Gaussian distribution. UV and ECD are better presented in stack mode, with ECD over UV so that a visual correlation between absorbance and CD bands can be established.

14.2.4.4 VCD and ROA Simulations

IR and VCD intensities are also proportional to the quantities dipole (D) and rotational (R) strengths, respectively, which are calculated using Eqs. (14.16) and (14.17), as described for UV and ECD. For IR and VCD, however, D and R should be calculated at the same level of theory employed in the geometry optimization step. Additionally, in the case of VCD, the electric (μ) and magnetic (m) dipole transition moment vectors include the wavefunctions of the ground (0) and the first excited vibrational state (1), within the ground electronic state of the molecule. Moreover, in the harmonic approximation, the dipole strength for the normal mode displacement Q_a is proportional to:

$$D_{01}^{a} \sim \left| \left(\frac{\partial \mu}{\partial Q_a} \right)_0 \right|^2 \tag{14.18}$$

While the rotational strength for a fundamental transition of mode a is proportional to:

$$R_{01}^{a} \sim \left| \left(\frac{\partial \mu}{\partial Q_a} \right)_0 \right| \cdot \left| \left(\frac{\partial m}{\partial P_a} \right)_0 \right| \tag{14.19}$$

where P_a represents the conjugated momentum for the normal mode a. It is interesting to note that the second term of Eq. (14.19) requires going beyond the Born–Oppenheimer approximation, in order to include the correlation between nuclear velocities and electronic current densities. The methodology for obtaining this derivative was developed by Stephens in 1985. The Stephens equation for vibrational rotational strengths uses the magnetic field perturbation method with atomic polar tensors (APTs), atomic axial tensors (AATs), and GIAOs [76].

As VCD is related to the electronic ground state of the molecule, VCD calculations are usually carried out at the DFT level. The use of B3LYP or B3PW91 functionals combined with 6-31G(d) basis set, which is considered the minimum basis set for VCD calculations, represents a good compromise between accuracy and computational cost. In some cases, however, it might be necessary to increase the size of the basis set. At least for small molecules, a larger basis set such as cc-pVDZ, cc-pVTZ, TZ2P, and TZVP provide better results compared with 6-31G(d). For larger molecules, on the other hand, the triple-ζ basis set TZVP does not provide practical advantages over 6-31G(d) concerning the accuracy-to-time ratio (23), while for heptapeptides 6-31G(d) performs similarly to cc-pVDZ [77]. Specific basis sets may be necessary depending on the functional group present in the target molecule. For instance, in the case of sulfones, 6-311G(3df,2dp) basis (or equivalent Dunning and Aldrich's variants) should be used to correctly reproduce the asymmetric and symmetric S=O stretching frequencies [78]. Another interesting point regarding basis set for VCD calculation is that polarization functions are necessary, while diffuse functions do not seem to improve significantly the results. Furthermore, solvent effects on vibrational rotational strengths are typically small. However, it clearly affects the optimized geometries and the conformer population in solution. Therefore, when experiments are performed in highly coordinating solvents, such as DMSO-d_6 or methanol-d_4, inclusion of explicit solvation may be necessary to improve the correlation between observed and calculated data [79, 80]. The calculated dipole and rotational strengths for each mode are then converted to a full IR and VCD spectra to be compared with experiment assuming Lorentzian band shapes. A bandwidth of 6 cm^{-1} (half width at half maximum) is generally considered for comparison with experimental spectra measured at 4 cm^{-1} resolution. Spectra should be plotted also in stack mode with VCD above the IR on the same wavenumber frequency scale. The calculated vibrational frequencies are based on harmonic force fields, whereas the observed experimental frequencies originate from anharmonic oscillations, thus, the calculations usually yield larger frequencies when compared to their respective experimental values. For this reason, calculated vibrational frequencies are multiplied by a global scaling factor that depends on the model chemistry used in the calculation. For B3LYP/6-31G(d), for example, a scale factor of 0.97 is typically used. Higher levels of theory, such as B3PW91/cc-pVTZ, may require a larger scale factor of 0.98 [21].

ROA originates from a distinct physical chemical phenomenon compared to VCD. While VCD arises from changes in dipole moments (absorption), ROA involves changes in polarizabilities (inelastic scattering) resulting in vibrational transitions within the electronic ground state [52]. It makes ROA calculations more computationally demanding than those for VCD. Backscattered (180°) ROA intensities are determined by the magnetic $\langle 0|G'|1\rangle$ and quadrupole $\langle 0|A|1\rangle$ transition polarizabilities and require the evaluation of analogue anisotropic invariants [43, 45].

$$I_{\text{ROA}} \sim 12\beta\left(G'\right)^2 + 4\beta\left(A\right)^2 \tag{14.20}$$

Regarding basis set, larger basis sets with diffuse functions (i.e. 6-31++G(d,p), aug-cc-pVDZ or aug-cc-pVTZ) are required to describe the more tenuous behavior of the polarizability of the molecule in response to the incident radiation [51]. A low-cost alternative often used is the so-called rDPS basis set [81]. Although the measurement of ROA in water is considered as an advantage, it generally complicates the simulations. The water environment is difficult to reproduce theoretically as a result of its polarity and high affinity for making hydrogen bonds [25].

ROA calculations can be performed using either a one-step or a two-step process, meaning that the Raman and ROA tensors are calculated using the same or different levels of theory, respectively, than that employed in the optimization step. This procedure reduces the computational cost of the whole process. An analytical time-dependent protocol for the calculation of the ROA property-tensor derivatives is currently implemented in some commercial software packages. The full description of the methods available for calculating ROA spectra are beyond the scope of this chapter. For comprehensive descriptions of ROA simulations, please see references [45, 82, 83].

14.2.5 Examples of Application

In this section, we present a list of carefully selected articles to demonstrate the use of different chiroptical methods to determine the AC of different compounds.

14.2.5.1 OR

The first example of application to be discussed is the case of the iridoid natural product oruwacin, reported in 2008 by Stephens and coworkers [84]. In this paper, the AC of was determined by TD-DFT calculations of the specific rotation, [α], at the sodium D line. This compound was first isolated in 1979 from the leaves of the plant *Morinda lucida* and, based on chemical and spectroscopic data, the structure of oruwacin was determined as closely related to that of the iridoid plumericin (Figure 14.6). In its first report, the AC of oruwacin was determined as (1*R*,5*S*,8*S*,9*S*,10*S*), the same as that of plumericin, since both compounds presented similar $[\alpha]_D$ values of +198 ($CHCl_3$) for plumericin and + 193 ($CHCl_3$) for oruwacin. This conclusion was reached assuming that the replacement of the C-14 methyl group of plumericin by the *O*-methylcatechol group in oruwacin would have little impact on the OR.

After a conformational search based on the energy variation of fragment molecules as a function of different dihedral angles at the B3LYP/6-31G(d) level, 32 conformers were identified, where only 8 presented free energies within 3 kcal mol^{-1} of the lowest energy conformation. Therefore, only these eight conformers, with the same proposed AC (1*R*,5*S*,8*S*,9*S*,10*S*), were selected for TD-DFT calculations of the $[\alpha]_D$ values at the B3LYP/aug-cc-pVDZ level. The obtained results showed that different conformers of oruwacin yielded $[\alpha]_D$ values directly sensitive to the conformation of the tetracyclic core and the orientation of the *O*-methylcatechol group. These results clearly demonstrate that in fact the *O*-methylcatechol group had a considerable impact on the OR properties. Moreover, the conformationally averaged $[\alpha]_D$ value obtained for (1*R*,5*S*,8*S*,9*S*,10*S*)-oruwacin was −193, the opposite of that obtained for the natural product (+193.3 in $CHCl_3$). Therefore, this result led to an unambiguous conclusion that (+)-oruwacin

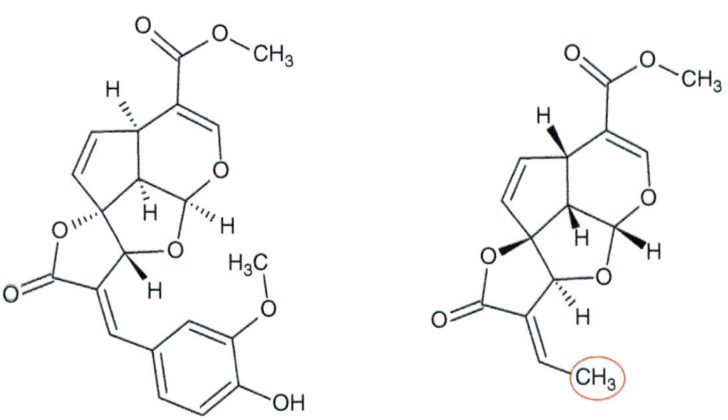

Figure 14.6 Chemical structures of the iridoids natural products (+)-oruwacin (left) and (+)-plumericin (right). Methyl group C-14 is marked in red. *Source:* Adapted from Stephens et al. [84]. (DOI 10.1021/np070502r).

in fact had the 1*S*,5*R*,8*R*,9*R*,10*R* AC, opposite to the one reported previously based on the assumption that structurally related compounds would present similar OR values.

This study demonstrated the power and applicability of TD-DFT methodology in simulating the OR of chiral organic molecules, including complex natural products. Besides, this paper highlights the risk of AC assignments based on comparison of $[\alpha]_D$ value with related molecules that, as previously mentioned, can be unreliable.

14.2.5.2 ORD

The next example to be discussed is the case of the natural product centratherin, reported in 2015 by Junior and coworkers [85]. In this paper, the AC was determined by a combination of ECD, ORD, and VCD, supported by quantum chemical calculations. However, only the ORD results will be presented.

Centratherin is a furanoheliangolide sesquiterpenoid lactone extracted from *Eremanthus crotonoides*. This compound displayed numerous biological activities, such as antimicrobial, anti-inflammatory, and trypanocidal. Due to its complex structure (Figure 14.7), different structural assignments for centratherin were found in the earlier literature. Only in 1982 its RC was determined based on X-ray diffraction. However, its AC remained unassigned, being only proposed based on its germacran precursor [85]. Therefore, in order to verify the proposed configuration, the stereostructure of centratherin was analyzed by means of chiroptical spectroscopy in correlation with its corresponding quantum chemical predictions.

After a random conformational analysis using the MMFF94s force field carried out on previously established RC (the 6*R*,7*R*,8*S*,10*R*,2′*Z* AC was arbitrarily chosen), all the geometries within a 25 kcal mol^{-1} energy range

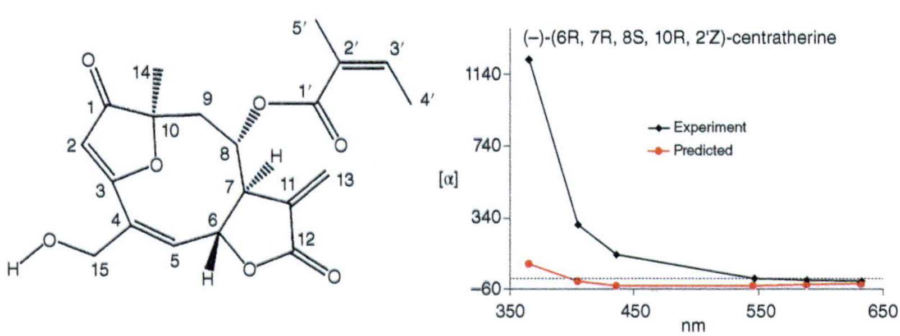

Figure 14.7 Structure of centratherin (left) and comparison of experimental ORD of (–)-centratherin with that predicted for the (6*R*,7*R*,8*S*,10*R*,2′*Z*) stereoisomer (right). *Source:* Reproduced from Junior et al. [85]/American Chemical Society. Scheme 1 and Figure 3 in original publication (DOI 10.1021/acs.jnatprod.5b00546).

were reoptimized using the semiempirical PM6 method. The resulting geometries within a 5 kcal mol^{-1} energy window were then reoptimized using DFT at the B3LYP/6-31G(d) level. The resulting 16 lowest-energy conformers within a 3 kcal mol^{-1} energy range had their geometries once more reoptimized at the B3LYP/aug-cc-pVDZ level of theory, and were used for ORD calculations using the long-range corrected CAM-B3LYP functional combined with the Dunning's aug-cc-pVDZ basis set. The PCM was used to take the influence of solvent (CH_3CN) into account. Experimentally, the ORD data were measured in CH_3CN, at a concentration of 2.76 mg ml^{-1}, using an Autopol IV polarimeter, at six different wavelengths (633, 589, 546, 436, 405, and 365 nm). The experimental ORD curve of (−)-centratherin (Figure 14.7) displayed negative values at longer wavelengths with a change of sign at 546 nm and increasing positive values at shorter wavelengths. This same pattern was reproduced correctly by the predicted ORD curve. Thereby, it was possible to assign the (6*R*,7*R*,8*S*,10*R*,2′*Z*) configuration to (−)-centratherin.

14.2.5.3 ECD

The next example to be discussed was published in 2019 by Jimenez and coworkers [86] and presents the AC determination of a synthetic compound using the hyphenation of HPLC and ECD assisted by TD-DFT calculations, The UV/ECD spectra were recorded using a stopped-flow mode where the compound eluting from the chiral column was trapped into the ECD detector.

In this paper, the authors reported the green organic chemistry synthesis of *E*-2-cyano-3(furan-2-yl) acrylamide under microwave radiation, followed by bioreduction (ene-reduction) using marine-derived and terrestrial fungi to obtain for the first time the (*R*)-2-cyano-3-(furan-2-yl) propanamide (Figure 14.8). In this study, the chiral separation of 2-cyano-3-(furan-2-yl) propanamide racemate was performed using a CHIRALPAK AD-H column and hexane and ethanol (60 : 40 v/v) as mobile phase. The ECD spectrum of each enantiomer eluting from the chiral LC system was recorded on a Jasco CD-2095 detector by trapping in a quartz cell through a switching valve. Then, the experimental UV/ECD spectra of both enantiomers were compared with those calculated for the Boltzmann average of the lowest-energy conformers identified for the (*R*) configuration at the CAM-B3LYP/TZVP and wB97XD/aug-cc-pVDZ levels. The very good agreement between the observed and theoretical data allowed the unambiguous determination of the AC of the first eluting peak as (*R*)-2-cyano-3-(furan-2-yl) propanamide and, consequently, that of the last eluted peak as the *S*-enantiomer.

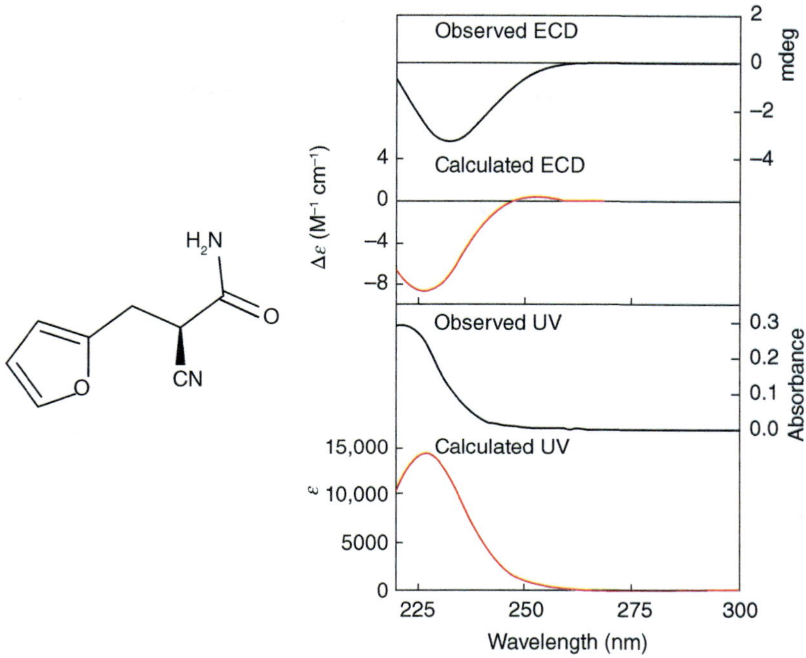

Figure 14.8 Structure of (*R*)-2-cyano-3-(furan-2-yl) propanamide (left) and comparison of the observed UV and ECD spectra for the first eluting peak of (±)-2-cyano-3-(furan-2-yl) propanamide with the calculated [CAM-B3LYP/TZVP//B3LYP/6-31G(d)] UV and ECD spectra for the (*R*)-enantiomer. *Source:* Reproduced from Jimenez et al. [86]/John Wiley & Sons. Figure 3 in original publication (DOI 10.1002/chir.23078).

14.2.5.4 VCD

The next paper to be discussed was published in 2019 by Gimenes and coworkers [87] and presents the determination of both the relative and absolute configurations of two undescribed natural products, picraviane A and B. The AC of both compounds was unambiguously determined by a combination of ECD and VCD spectroscopies supported by TD-DFT calculations; however, only the VCD results will be discussed.

The two undescribed highly oxygenated nortriterpenes were isolated from *Picramnia glazioviana* Engl. (Picramniaceae). The RC determinations were carried out by a combination of 2D NMR spectroscopy and single-crystal X-ray crystallography. The combined use of COSY and HMBC were employed to identify the cyclic backbone. The RC at C-1, C-5, C-7, C-8, C-9, C-10, and C-14 was established using the NOESY spectrum. However, due to the flexibility of the E-ring, it was not possible to unambiguously assign the configuration at C-19 based on the observed NOE signals. Thus, DFT simulations of IR and VCD spectra were performed for the lowest-energy conformations of (1*S*,5*R*,7*R*,8*R*,9*S*,10*R*,14*R*,19*S*)-picraviane A and its C-19-epimer (Figure 14.9). After a conformational search carried out with the Monte Carlo algorithm employing the MM+ force field for both epimers, three conformers with relative energy within 6 kcal mol^{-1} of the lowest-energy conformer were selected and further geometry optimized at the B3LYP/6-31G(d) level. In both cases, a single conformer was identified with relative energy <2.2 kcal mol^{-1}, corresponding to more than 97% of the total Boltzmann distribution. These conformers were also further geometry optimized and had their IR and VCD spectra calculated at the wB97XD/PCM($CHCl_3$)/6-311G(d,p) level (Figure 14.10). The opposite configuration for the chiral center at C-19 resulted in some VCD bands with reversed signs, especially fundamentals 132 and 171 (Figure 14.10), which were safely used to discriminate the two diastereomers. These bands arise mainly from the C—O stretching vibration of the acetyl group at C-19 coupled with CH_3 deformation modes of ring E, (fundamental 132), as well as from CH_2 twisting of ring A′ and the in-plane C—H bending of H-19 coupled with the CH_2 wagging of ring E (fundamental 171). The excellent correlation between experimental and simulated VCD data for picraviane A unambiguously established its AC as 1*S*,5*R*,7*R*,8*R*,9*S*,10*R*,14*R*,19*S*. For picraviane B, on the other hand, a high-quality crystal suitable for X-ray diffraction studies was obtained, which confirmed the molecular structure of picraviane B and allowed the authors to establish the same RC as that observed for picraviane A. Then, the AC of picraviane B was subsequently determined by comparing experimental and calculated UV and ECD spectra (not shown).

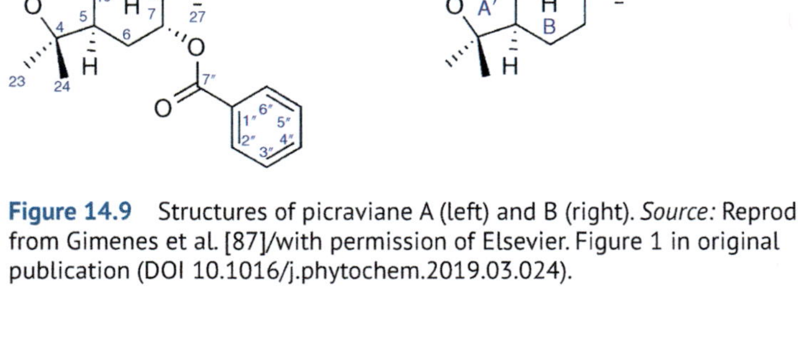

Figure 14.9 Structures of picraviane A (left) and B (right). *Source:* Reproduced from Gimenes et al. [87]/with permission of Elsevier. Figure 1 in original publication (DOI 10.1016/j.phytochem.2019.03.024).

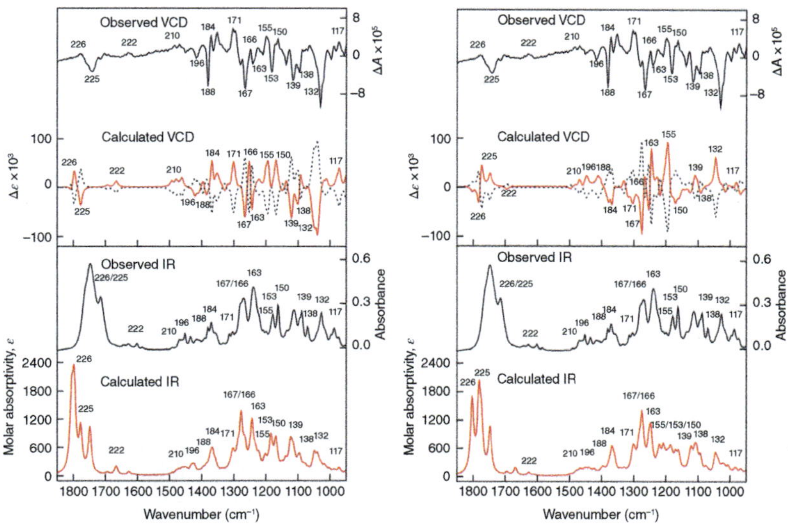

Figure 14.10 Comparison of the observed IR and VCD spectra of (+)-picraviane A with the calculated IR and VCD spectra. Left: C-19*S* epimer (solid line) and the corresponding enantiomer (dotted line). Right: C-19 *R* epimer (solid lines) and the corresponding enantiomer (dotted lines). Numbers represent selected fundamental vibrational modes. *Source:* Reproduced from Gimenes et al. [87]/with permission of Elsevier. Figure 3 in original publication (DOI 10.1016/j.phytochem.2019.03.024).

14.2.5.5 ROA

The next example deals with galantamine, which is a competitive inhibitor of acetylcholinesterase and an allosteric modulator of the nicotinic acetylcholine receptor. This compound is clinically used to treat mild-to-moderate Alzheimer's disease [88].

In this paper, the AC of the alkaloid galantamine was studied using a range of solution-state techniques, such as NMR, VCD, and ROA. While the use of NMR and VCD provided an unambiguous determination of the AC, both techniques were needed in combination to achieve this goal. ROA, on the other hand, proved to be sensitive enough to assign the AC without relying on other techniques. Therefore, only the ROA results will be presented.

Following a conformational analysis using the MMFF94s force field on the four possible diastereomeric pairs of enantiomers of galantamine, all conformers found were subsequently optimized at the B3LYP/6-31G(d) level of theory. Herein, only one enantiomer of each diastereomeric pair was considered in the calculations. Then, the conformers with a relative energy within 5 $\mathrm{kcal\,mol^{-1}}$ of the most stable one were further optimized at the B3LYP/

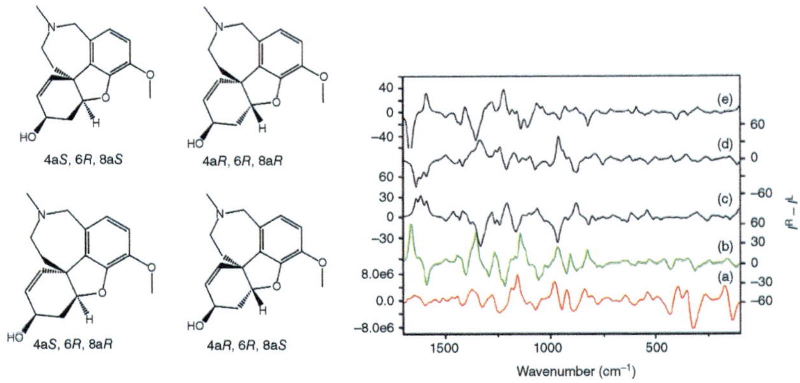

Figure 14.11 Four possible diastereomers of galantamine (left) and the comparison of Boltzmann weighted calculated ROA spectra for the possible candidates with experimental ROA spectrum (right). *Source:* Reproduced from De Gussem et al. [88]/American Chemical Society. Figures 2 and 7 in original publication (DOI 10.1021/acsomega.9b02108).

aug-cc-pVDZ level and used for the Raman and ROA spectral simulations at the same level of theory. Raman and ROA intensities were calculated with an incident light wavelength of 532 nm, the same excitation wavelength used for the backscattered SCP-ROA measurements. Figure 14.11 shows the comparison of calculated and experimental ROA spectra. Simply by visual comparison it is possible to conclude that galantamine (trace a, red) has the 4*aS*,6*R*,8*aS* configuration (trace b, green) as the predicted spectrum for this configuration presents the best correlation to experiment. Thus, with a qualitative visual analysis, ROA was able to positively identify the correct AC, even without prior knowledge on the RC obtained from NMR spectroscopy.

14.2.5.6 Association of Different Chiroptical Methods

The next example to be discussed illustrates the difficulties that can be faced in the AC determination of certain chiral compounds for which a combination of chiroptical methods becomes necessary.

The paper published by Junior and coworkers [37] deals with the AC of a rare sesquiterpene, 3-ishwarone, which presents a complex structure with numerous fused rings (Figure 14.12). This rare natural compound was isolated from the aerials parts of *Peperomia scandens* Ruiz & Pavon (Piperaceae). The proposed structure was based on 1D and 2D NMR studies, and the RCs at carbons 1, 2, 9, and 11 were unambiguously determined from NMR data. However, the RCs at carbons 4 and 5, and therefore their

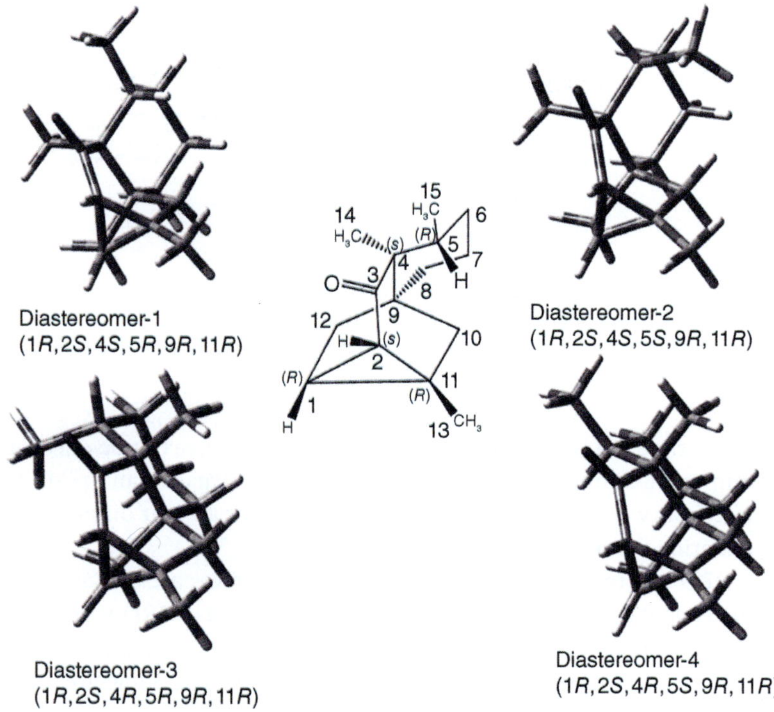

Figure 14.12 The four possible diastereomers of 3-iswarone. *Source:* Reproduced from Junior et al. [37]/American Chemical Society. Figure 1 in original publication (DOI 10.1021/np500363e).

AC, remained unspecified. Thus, a combination of chiroptical methods, supported by calculations, was used to distinguish the four possible diastereomeric pairs of enantiomers of this rare compound and to unambiguously determine its AC.

In this study, comparisons of the calculated ECD, ORD, and VCD spectra of the four diastereomers (1*R*,2*S*,4*S*,5*R*,9*R*,11*R*; 1*R*,2*S*,4*S*,5*S*,9*R*,11*R*; 1*R*,2*S*,4*R*,5*R*,9*R*,11*R*; and 1*R*,2*S*,4*R*,5*S*,9*R*,11*R*; see Figure 14.12) of 3-ishwarone with the experimental data were carried out. Due to the fused ring structure, only one stable conformer was identified for each diastereomer at the conformational search; the geometries were then further optimized at the B3LYP/6-31G(d) level. These reoptimized structures were used for ORD, ECD, and VCD calculations at the B3LYP/aug-cc-pVDZ with PCM to take the influence of solvent into consideration ($CHCl_3$ for VCD and ORD and MeCN for ECD).

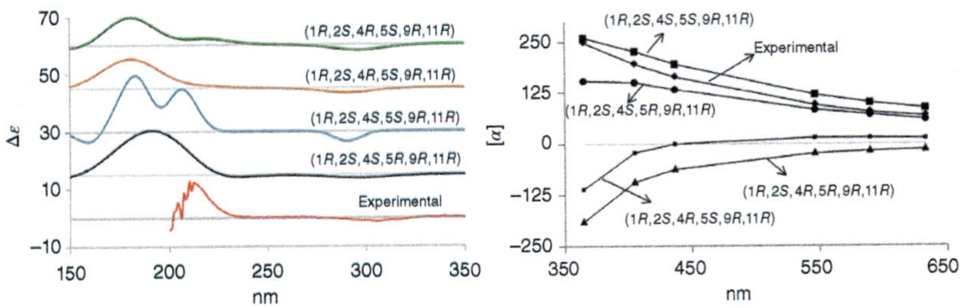

Figure 14.13 Comparison calculated ECD (left) and ORD (right) spectra of the four possible diastereomers of 3-ishwarone with experimental data. *Source:* Reproduced from Junior et al. [37]/American Chemical Society. Figures 3 and 5 in original publication (DOI 10.1021/np500363e).

The experimental ECD spectrum showed a weak negative Cotton effect at ~300 nm and a strong positive one at ~210 nm. The negative and positive Cotton effects were reproduced in the calculations for all four diastereomers, rendering them indistinguishable (Figure 14.13).

The ORD spectra calculated for the four diastereomers are compared to the experimental data in Figure 14.13. The experimental ORD spectra of (+)-3-ishwarone in $CHCl_3$ were positive at all six wavelengths measured, with increasing magnitudes at shorter wavelengths. The same trend is observed only for the (1*R*,2*S*,4*S*,5*R*,9*R*,11*R*) and (1*R*,2*S*,4*S*,5*S*,9*R*,11*R*) diastereomers. Based on that, the other two diastereomeric structures could be ruled out for (+)-3-ishwarone; however, these data alone did not provide enough discrimination between the two diastereomers with positive ORD. This required the analysis of the VCD spectra.

The experimental VCD spectrum is compared to those calculated for the four diastereomers in Figure 14.14. Since there are numerous vibrational bands in the 1600–1000 cm^{-1} region, the visual assessment of the level of agreement between the experimental spectrum and the calculated VCD spectra for each diastereomer is difficult. This problem is avoided by using the similarity analysis that numerically evaluates the overlap of experimental and predicted spectra. The similarity of the experimental VCD spectrum to the corresponding calculated spectrum is ~0.6 for the (1*R*,2*S*,4*S*,5*R*,9*R*,11*R*) diastereomer, 0.36 for the (1*R*,2*S*,4*S*,5*S*,9*R*,11*R*) diastereomer, and less than 0.1 for the diastereomers that were discarded following the ORD analysis. Therefore, it can be concluded that the predicted VCD spectra for the (1*R*,2*S*,4*S*,5*R*,9*R*,11*R*) diastereomer have the greatest similarities to the corresponding experimental observations. Therefore, based on analyses of the three different chiroptical methods, the AC of (+)-3-ishwarone was unequivocally assigned as (1*R*,2*S*,4*S*,5*R*,9*R*,11*R*).

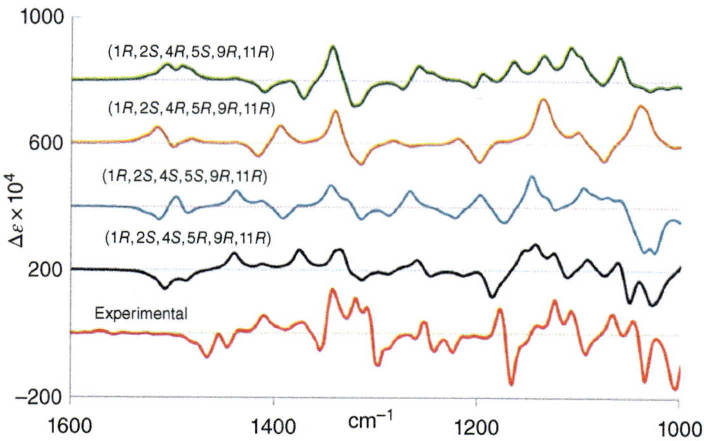

Figure 14.14 Comparison calculated VCD spectra of the four possible diastereomers of 3-ishwarone with experimental VCD spectrum. *Source:* Reproduced from Junior et al. [37]/American Chemical Society. Figure 6 in original publication (DOI 10.1021/np500363e).

14.3 Concluding Remarks

In this chapter, we aimed to provide the reader with an overview of the potential of chiroptical spectroscopy associated with quantum chemical calculations to unambiguously determine the AC of chiral organic molecules. The advantages and limitations of each chiroptical method, as well as their main theoretical, instrumentation, and measurement aspects were provided. Good practices for the correct use of OR, ORD, ECD, VCD, and ROA for stereochemical analysis of chiral molecules were also presented. Special attention was paid to the complementarity of each method and the benefits of their combined use. Although chiroptical methods are powerful tools to probe molecular stereochemistry (including conformations) directly in solution, and without the need of either single crystals or derivatization, their use is still limited especially among natural product and synthetic chemists. Such limitation is even more evident in the case of the vibrational chiroptical methods VCD and ROA. One of the main drawbacks of chiroptical spectroscopy is the amount of sample required in the case of VCD and ROA as well as the need of DFT calculations to interpret experimental data. Further developments in microsampling, streamlined computational procedures, as well as pattern recognition tools will certainly facilitate the use of chiroptical spectroscopy, even by non-experts. Finally, with this chapter we hope more researchers will feel encouraged to join this expanding research field.

References

1 Davankov, V. (2006). Chirality as an inherent general property of matter. *Chirality* 1: 459–461.

2 Batista, J.M. and Bolzani, V.S. (2014). Determination of the absolute configuration of natural product molecules using vibrational circular dichroism. *Stud. Nat. Prod. Chem.* 41: 383–417.

3 Allenmark, S. and Gawronski, J. (2008). Determination of absolute configuration – an overview related to this special issue. *Chirality* 20: 606–608.

4 Eliel, E.L. and Wilen, S.H. (1994). *Stereochemistry of Organic Compounds.* Wiley-Interscience.

5 Nunez, M.C., Garcia-Rubino, M.E., Conejo-Garcia, A. et al. (2009). Homochiral drugs: a demanding tendency of the pharmaceutical industry. *Curr. Med. Chem.* 16: 2064–2074.

6 Coelho, M.M., Fernandes, C., Remião, F., and Tiritan, M.E. (2021). Enantioselectivity in drug pharmacokinetics and toxicity: pharmacological relevance and analytical methods. *Molecules* 26: 3113–3136.

7 U.S. Food and Drug Administration. Development of new stereoisomeric drugs. https://www.fda.gov/regulatory-information/search-fda-guidance-documents/development-new-stereoisomeric-drugs (accessed 14 August 2021).

8 Kasprzyk-Hordern, B. (2010). Pharmacologically active compounds in the environment and their chirality. *Chem. Soc. Rev.* 39: 4466–4503.

9 Calcaterra, A. and D'Acquarica, I. (2018). The market of chiral drugs: chiral switches versus de novo enantiomerically pure compounds. *J. Pharm. Biomed. Anal.* 147: 323–340.

10 Pescitelli, G., Kurtán, T., Flörke, U., and Krohn, K. (2009). Absolute structural elucidation of natural products – a focus on quantum-mechanical calculations of solid-state CD spectra. *Chirality* 21: E181–E201.

11 Batista, A.N.L., Angrisani, B.R.P., Lima, M.E. et al. (2021). Absolute configuration reassignment of natural products: an overview of the last decade. *J. Braz. Chem. Soc.* 1–20.

12 Batista, A.N.L., dos Santos, F.M., Batista, J.M., and Cass, Q.B. (2018). Enantiomeric mixtures in natural product chemistry: separation and absolute configuration assignment. *Molecules* 23: 1–18.

13 Polavarapu, P.L. (2012). Molecular structure determination using chiroptical spectroscopy: where we may go wrong? *Chirallity* 24: 909–920.

14 Freedman, T.B., Cao, X., Dukor, R., and Nafie, L.A. (2003). Absolute configuration determination of chiral molecules in the solution state using vibrational circular dichroism. *Chirality* 15: 743–758.

15 Harada, N. (2008). Determination of absolute configurations by X-ray crystallography and ^{1}H NMR anisotropy. *Chirality* 20: 691–723.

16 Flack, H.D. and Bernardinelli, G. (2008). The use of X-ray crystallography to determine absolute configuration. *Chirality* 20: 681–690.

17 Parsons, S. (2017). Determination of absolute configuration using X-ray diffraction. *Tetrahedron Asymmetry* 28: 1304–1313.

18 Bross-Walch, N., Kühn, T., Moskau, D., and Zerbe, O. (2005). Strategies and tools for structure determination of natural products using modern methods of NMR spectroscopy. *Chem. Biodivers.* 2: 147–177.

19 Costa, F.L.P., De Albuquerque, A.C.F., Fiorot, R.G. et al. (2021). Structural characterisation of natural products by means of quantum chemical calculations of NMR parameters: new insights. *Org. Chem. Front.* 8: 2019–2058.

20 Wenzel, T.J. and Chisholm, C.D. (2011). Using NMR spectroscopic methods to determine enantiomeric purity and assign absolute stereochemistry. *Prog. Nucl. Magn. Reson. Spectrosc.* 59: 1–63.

21 Polavarapu, P.L. (2006). Determination of molecular stereochemistry using optical rotatory dispersion, vibrational circular dichroism and vibrational Raman optical activity. In: *Chiral Analysis* (ed. K.W. Busch and M.A. Busch), 461–504. Elsevier.

22 Polavarapu, P.L. (2007). Renaissance in chiroptical spectroscopic methods for molecular structure determination. *Chem. Rec.* 7: 125–136.

23 Berova, N., Polavarapu, P.L., Nakanishi, K., and Woody, R.W. (2011). *Comprehensive Chiroptical Spectroscopy*. Wiley.

24 Batista, J.M., Blanch, E.W., and Bolzani, V.S. (2015). Recent advances in the use of vibrational chiroptical spectroscopic methods for stereochemical characterization of natural products. *Nat. Prod. Rep.* 32: 1280–1302.

25 Batista, J.M. (2013). Determination of absolute configuration using chiroptical methods. In: *Stereoselective Synthesis of Drugs and Natural Products* (ed. V. Andrushko and A. Andrushko), 1571–1599. Wiley.

26 Dos Santos, F.M., Bicalho, K.U., Calisto, I.H. et al. (2018). Scope of the 2(5H)-furanone helicity rule: a combined ECD, VCD, and DFT investigation. *Org. Biomol. Chem.* 16: 4509–4516.

27 Pescitelli, G. and Bruhn, T. (2016). Good computational practice in the assignment of absolute configurations by TDDFT calculations of ECD spectra. *Chirallity* 28: 466–474.

28 Nicu, V.P., Mándi, A., Kurtán, T., and Polavarapu, P.L. (2014). On the complementarity of ECD and VCD techniques. *Chirality* 26: 525–531.

29 Frisch, G., Trucks, G.W., Schlegel, H.B. et al. (2009). *Gaussian 09, Revision E. 01; Gaussian.* Wallingford, CT: Gaussian, Inc.

30 Aidas, K., Angeli, C., Bak, K.L. et al. (2014). The Dalton quantum chemistry program system. *Wiley Interdiscip. Rev.: Comput. Mol. Sci.* 4: 269–284.

31 Polavarapu, P.L., Petrovic, A., and Wang, F. (2003). Intrinsic rotation and molecular structure. *Chirality* 15: 143–149.

32 Mennucci, B., Cappelli, C., Cammi, R., and Tomasi, J. (2011). Modeling solvent effects on chiroptical properties. *Chirality* 23: 717–729.

33 D'Cunha, R. and Crawford, D. (2021). Modeling complex solvent effects on the optical rotation of chiral molecules: a combined molecular dynamics and density functional theory study. *Phys. Chem. A* 125: 3095–3108.

34 Zhou, H., Li, L., Wu, C. et al. (2016). Penipyridones A–F, pyridone alkaloids from *Penicillium funiculosum*. *J. Nat. Prod.* 79: 1783–1790.

35 Nakahashi, A., Yaguchi, Y., Miura, N. et al. (2011). A vibrational circular dichroism approach to the determination of the absolute configurations of flavorous 5-substituted-2(5 H)-furanones. *J. Nat. Prod.* 74: 707–711.

36 Cimino, P., Neese, F., and Barone, V. (2010). Calculation of magnetic tensors and EPR spectra for free radicals in different environments. In: *Computational Spectroscopy: Methods, Experiments and Applications* (ed. J. Grunenberg), 63–104. Wiley-VCH Verlag GmbH & Co. KGaA.

37 Junior, F.M.S., Covington, C.L., De Amorim, M.B. et al. (2014). Absolute configuration of a rare sesquiterpene: (+)-3-ishwarone. *J. Nat. Prod.* 77: 1881–1886.

38 Bringmann, G., Gulder, T.A.M., Reichert, M., and Gulder, T. (2008). The online assignment of the absolute configuration of natural products: HPLC-CD in combination with quantum chemical CD calculations. *Chirality* 20: 628–642.

39 Polavarapu, P.L. and Santoro, E. (2020). Vibrational optical activity for structural characterization of natural products. *Nat. Prod. Rep.* 37: 1661–1699.

40 United State Pharmacopeia:39-NF34. 2016. Chapters 782 and 1782.

41 Lu, X., Li, H., Nafie, J.W. et al. (2017). A vibrational circular dichroism microsampling accessory: mapping enhanced vibrational circular dichroism in amyloid fibril films. *Appl. Spectrosc.* 71: 1117–1126.

42 Keiderling, T.A. (2018). Instrumentation for vibrational circular dichroism spectroscopy: method comparison and newer developments. *Molecules* 23: 2404–2424.

43 Bogaerts, J., Aerts, R., Vermeyen, T. et al. (2021). Tackling stereochemistry in drug molecules with vibrational optical activity. *Pharmaceuticals* 14: 877–825.

44 Keiderling, T.A. and Lakhani, A. (2018). Mini review: instrumentation for vibrational circular dichroism spectroscopy, still a role for dispersive instruments. *Chirality* 30: 238–253.

45 Nafie, L.A. (2011). *Vibrational Optical Activity: Principles and Applications*. Wiley.

46 Hug, W. and Hangartner, G. (1999). A novel high-throughput Raman spectrometer for polarization difference measurements. *J. Raman Spectrosc.* 30: 841–852.

47 Barron, L.D. (2015). The development of biomolecular Raman optical activity spectroscopy. *Biomed. Spectrosc. Imaging* 4: 223–253.

48 Kapitán, J., Barron, L.D., and Hecht, L. (2015). A novel Raman optical activity instrument operating in the deep ultraviolet spectral region. *J. Raman Spectrosc.* 46: 392–399.

49 Zhang, Y., Wang, P., Jia, G. et al. (2017). A short-wavelength Raman optical activity spectrometer with laser source at 457 nm for the characterization of chiral molecules. *Appl. Spectrosc.* 71: 2211–2217.

50 Urbanova, M. and Malon, P. (2007). Circular dichroism spectroscopy. In: *Analytical Methods in Supramolecular Chemistry* (ed. C. Schalley), 265–304. Wiley-VCH Verlag GmbH & Co. KGaA.

51 Stephens, P.J., Devlin, F.J., and Cheeseman, J.R. (2012). *VCD Spectroscopy for Organic Chemists*. CRC Press.

52 Autschbach, J. (2009). Computing chiroptical properties with first-principles theoretical methods: background and illustrative examples. *Chirallity* 21: E116–E152.

53 Thompson, M.A. (2004). Molecular docking using ArgusLab, an efficient shape-based search algorithm and the AScore scoring function. ACS Meeting, Philadelphia.

54 Hanwell, M.D., Curtis, D.E., Lonie, D.C. et al. (2012). Avogadro: an advanced semantic chemical editor, visualization, and analysis platform. *J Cheminform.* 4: 1–s.

55 Advanced Chemistry Development I (2020). ACD/ChemSketch. Toronto, Canada.www.acdlabs.com (accessed 2021).

56 Cousins, K.R. (2005). ChemDraw Ultra 9.0. Chemistry Software Package ChemOffice Ultra 2005 CambridgeSoft, 100 CambridgePark Drive, Cambridge, MA 02140. *J. Chem. Inf. Model.* 45: 1474–1477.

57 Froimowitz, M. (1993). HyperChem(TM): a software package for computational chemistry and molecular modeling. *Biotechniques* 14: 1010–1013.

58 Dennington, R., Keith, T.A., and Millam, J.M. (2016). *GaussView*. Shawnee Mission, KS: Semichem Inc.

59 Derewacz, D.K., McNees, C.R., Scalmani, G. et al. (2014). Structure and stereochemical determination of hypogeamicins from a cave-derived actinomycete. *J. Nat. Prod.* 77: 1759–1763.

60 Wavefunction Inc. (2011). Spartan'10. Irvine.

61 Schrödinger, L. (2020). *MacroModel*. New York, NY: Schrödinger.

62 CONFLEX Corporation. Conflex Program. Tokyo. www.conflex.net (accessed October 2021).

63 Furche, F., Ahlrichs, R., Hättig, C. et al. (2014). Turbomole. *Wiley Interdiscip. Rev.: Comput. Mol. Sci.* 4: 91–100.

64 Rackers, J.A., Wang, Z., Lu, C. et al. (2018). Tinker 8: software tools for molecular design. *J. Chem. Theory Comput.* 14: 5273–5289.

65 Schmidt, M.W., Baldridge, K.K., Boatz, J.A. et al. (1993). General atomic and molecular electronic structure system. *J. Comput. Chem.* 14: 1347–1363.

66 Gordon, M.S. and Schmidt, M.W. (2005). Chapter 41 – Advances in electronic structure theory: GAMESS a decade later. In: *Theory and Applications of Computational Chemistry* (ed. C. Dykstra, G. Frenking, K. Kim and G. Scuseria), 1167–1189. Elsevier Science.

67 Bruhn, T., Schaumlöffel, A., Hemberger, Y., and Bringmann, G. (2013). SpecDis: quantifying the comparison of calculated and experimental electronic circular dichroism spectra. *Chirality* 25: 243–249.

68 Polavarapu, P.L. and Covington, C.L. (2014). Comparison of experimental and calculated chiroptical spectra for chiral molecular structure determination. *Chirallity* 26: 539–552.

69 Debie, E., De Gussem, E., Dukor, R.K. et al. (2011). A confidence level algorithm for the determination of absolute configuration using vibrational circular dichroism or raman optical activity. *ChemPhysChem* 12: 1542–1549.

70 Polavarapu, P.L. (2008). Why is it important to simultaneously use more than one chiroptical spectroscopic method for determining the structures of chiral molecules? *Chirallity* 20: 664–672.

71 Polavarapu, P.L. (2002). Optical rotation: recent advances in determining the absolute configuration. *Chirality* 14: 768–781.

72 Stephens, P.J., McCann, D.M., Cheeseman, J.R., and Frisch, M.J. (2005). Determination of absolute configurations of chiral molecules using ab initio time-dependent density functional theory calculations of optical rotation: how reliable are absolute configurations obtained for molecules with small rotations? *Chirality* 17: 52–64.

73 Polavarapu, P.L. (2006). Protocols for the analysis of theoretical optical rotations. *Chirality* 18: 348–356.

74 Berova, N., Di Bari, L., and Pescitelli, G. (2007). Application of electronic circular dichroism in configurational and conformational analysis of organic compounds. *Chem. Soc. Rev.* 36: 914–931.

75 Pescitelli, G., Di Bari, L., and Berova, N. (2011). Conformational aspects in the studies of organic compounds by electronic circular dichroism. *Chem. Soc. Rev.* 40 (9): 4603–4625.

76 Stephens, P.J. (1985). Theory of vibrational circular. *J. Phys. Chem.* 89: 748–752.

77 Yokomichi, M.A.S., Silva, H.R.L., Brandao, L.E.V.N. et al. (2022). Conformational preferences induced by cyclization in orbitides: a vibrational CD study. *Org. Biomol. Chem.* 20: 1306–1314.

78 Scholten, K., Engelage, E., and Merten, C. (2020). Basis set dependence of S [double bond, length as m-dash] O stretching frequencies and its consequences for IR and VCD spectra predictions. *Phys. Chem. Chem. Phys.* 22: 27979–27986.

79 Bünnemann, C., Pollok, C.H., and Merten, C. (2018). Explicit solvation of carboxylic acids for vibrational circular dichroism studies: limiting the computational efforts without losing accuracy. *J. Phys. Chem. B* 122: 8056–8064.

80 Weirich, L., Blanke, K., and Merten, C. (2020). More complex, less complicated? Explicit solvation of hydroxyl groups for the analysis of VCD spectra. *Phys. Chem. Chem. Phys.* 22: 12515–12523.

81 Zuber, G. and Hug, W. (2004). Rarefied basis sets for the calculation of optical tensors. The importance of gradients on hydrogen atoms for the Raman scattering tensor. *J. Phys. Chem. A* 108: 2108–2118.

82 Pecul, M. (2009). New applications and challenges for computational ROA spectroscopy. *Chirality* 21: E98–E104.

83 Ruud, K. and Thorvaldsen, A.J. (2009). Theoretical approaches to the calculation of Raman optical activity spectra. *Chirality* 21: E54–E67.

84 Stephens, P.J., Pan, J.J., Devlin, F.J., and Cheeseman, J.R. (2008). Determination of the absolute configurations of natural products using TDDFT optical rotation calculations: the iridoid oruwacin. *J. Nat. Prod.* 71: 285–288.

85 Junior, F.M.S., Covington, C.L., De Albuquerque, A.C.F. et al. (2015). Absolute configuration of (−)-centratherin, a sesquiterpenoid lactone, defined by means of chiroptical spectroscopy. *J. Nat. Prod.* 78: 2617–2623.

86 Jimenez, D.E.Q., Barreiro, J.C., dos Santos, F.M. et al. (2019). Enantioselective ene-reduction of E-2-cyano-3-(furan-2-yl) acrylamide by marine and terrestrial fungi and absolute configuration of (R)-2-cyano-3-(furan-2-yl) propanamide determined by calculations of electronic circular dichroism (ECD) spectra. *Chirality* 31: 534–542.

87 Gimenes, L., Batista, J.M., Junior, F.M.S. et al. (2019). Picraviane A and B: nortriterpenes with limonoid-like skeletons containing a heptanolide E-ring system from Picramnia glazioviana. *Phytochemistry* 163: 38–45.

88 De Gussem, E., Tehrani, K.A., Herrebout, W.A. et al. (2019). Comparative study of the vibrational optical activity techniques in structure elucidation: the case of galantamine. *ACS Omega* 4: 14133–14139.

Index

Chiral Separations and Stereochemical Elucidation: Fundamentals, Methods, and Applications, First Edition. Edited by Quezia Bezerra Cass, Maria Elizabeth Tiritan, João Marcos Batista Junior, and Juliana Cristina Barreiro.

d

e

f

l

m

X

Z

Printed and bound by CPI Group (UK) Ltd, Croydon, CR0 4YY

17/09/2026

14964253-0008